T0332644

Theory of Random Determinants

Mathematics and Its Applications (*Soviet Series*)

Volume 45

Theory of
Random Determinants

by

V. L. Girko

Department of Cybernetics,
Kiev State University, U.S.S.R.

KLUWER ACADEMIC PUBLISHERS
DORDRECHT / BOSTON / LONDON

Library of Congress Cataloging-in-Publication Data

Girko, V. L. (Vīacheslav Leonidovich).
    [Teorīīa sluchaĭnykh determinantov. English]
    Theory of random determinants / by V.L. Girko.
        p.   cm. -- (Mathematics and its applications (Soviet series) ;
    45)
    Translation of: Teorīīa sluchaĭnykh determinantov.
    Includes index.
    ISBN 0-7923-0233-8 (alk. paper)
    1. Stochastic matrices. 2. Determinants.   I. Title. II. Series:
Mathematics and its applications (Kluwer Academic Publishers).
Soviet series ; 45.
QA188.G5813 1990
512.9'434--dc20   90-45262

ISBN 0-7923-0233-8

Published by Kluwer Academic Publishers,
P.O. Box 17, 3300 AA Dordrecht, The Netherlands.

Kluwer Academic Publishers incorporates
the publishing programmes of
D. Reidel, Martinus Nijhoff, Dr W. Junk and MTP Press.

Sold and distributed in the U.S.A. and Canada
by Kluwer Academic Publishers,
101 Philip Drive, Norwell, MA 02061, U.S.A.

In all other countries, sold and distributed
by Kluwer Academic Publishers Group,
P.O. Box 322, 3300 AH Dordrecht, The Netherlands.

Printed on acid-free paper

Printed in the Netherlands

SERIES EDITOR'S PREFACE

'Et moi, ..., si j'avait su comment en revenir,
je n'y serais point allé.'

     Jules Verne

The series is divergent; therefore we may be
able to do something with it.

     O. Heaviside

One service mathematics has rendered the
human race. It has put common sense back
where it belongs, on the topmost shelf next
to the dusty canister labelled 'discarded non-
sense'.

     Eric T. Bell

Mathematics is a tool for thought. A highly necessary tool in a world where both feedback and non-linearities abound. Similarly, all kinds of parts of mathematics serve as tools for other parts and for other sciences.

Applying a simple rewriting rule to the quote on the right above one finds such statements as: 'One service topology has rendered mathematical physics ...'; 'One service logic has rendered computer science ...'; 'One service category theory has rendered mathematics ...'. All arguably true. And all statements obtainable this way form part of the raison d'être of this series.

This series, *Mathematics and Its Applications*, started in 1977. Now that over one hundred volumes have appeared it seems opportune to reexamine its scope. At the time I wrote

> "Growing specialization and diversification have brought a host of monographs and textbooks on increasingly specialized topics. However, the 'tree' of knowledge of mathematics and related fields does not grow only by putting forth new branches. It also happens, quite often in fact, that branches which were thought to be completely disparate are suddenly seen to be related. Further, the kind and level of sophistication of mathematics applied in various sciences has changed drastically in recent years: measure theory is used (non-trivially) in regional and theoretical economics; algebraic geometry interacts with physics; the Minkowsky lemma, coding theory and the structure of water meet one another in packing and covering theory; quantum fields, crystal defects and mathematical programming profit from homotopy theory; Lie algebras are relevant to filtering; and prediction and electrical engineering can use Stein spaces. And in addition to this there are such new emerging subdisciplines as 'experimental mathematics', 'CFD', 'completely integrable systems', 'chaos, synergetics and large-scale order', which are almost impossible to fit into the existing classification schemes. They draw upon widely different sections of mathematics."

By and large, all this still applies today. It is still true that at first sight mathematics seems rather fragmented and that to find, see, and exploit the deeper underlying interrelations more effort is needed and so are books that can help mathematicians and scientists do so. Accordingly MIA will continue to try to make such books available.

If anything, the description I gave in 1977 is now an understatement. To the examples of interaction areas one should add string theory where Riemann surfaces, algebraic geometry, modular functions, knots, quantum field theory, Kac-Moody algebras, monstrous moonshine (and more) all come together. And to the examples of things which can be usefully applied let me add the topic 'finite geometry'; a combination of words which sounds like it might not even exist, let alone be applicable. And yet it is being applied: to statistics via designs, to radar/sonar detection arrays (via finite projective planes), and to bus connections of VLSI chips (via difference sets). There seems to be no part of (so-called pure) mathematics that is not in immediate danger of being applied. And, accordingly, the applied mathematician needs to be aware of much more. Besides analysis and numerics, the traditional workhorses, he may need all kinds of combinatorics, algebra, probability, and so on.

In addition, the applied scientist needs to cope increasingly with the nonlinear world and the extra mathematical sophistication that this requires. For that is where the rewards are. Linear models are honest and a bit sad and depressing: proportional efforts and results. It is in the non-linear world that infinitesimal inputs may result in macroscopic outputs (or vice versa). To appreciate what I am hinting at: if electronics were linear we would have no fun with transistors and computers; we would have no TV; in fact you would not be reading these lines.

There is also no safety in ignoring such outlandish things as nonstandard analysis, superspace and anticommuting integration, p-adic and ultrametric space. All three have applications in both electrical engineering and physics. Once, complex numbers were equally outlandish, but they frequently proved the shortest path between 'real' results. Similarly, the first two topics named have already provided a number of 'wormhole' paths. There is no telling where all this is leading - fortunately.

Thus the original scope of the series, which for various (sound) reasons now comprises five sub-series: white (Japan), yellow (China), red (USSR), blue (Eastern Europe), and green (everything else), still applies. It has been enlarged a bit to include books treating of the tools from one subdiscipline which are used in others. Thus the series still aims at books dealing with:

- a central concept which plays an important role in several different mathematical and/or scientific specialization areas;
- new applications of the results and ideas from one area of scientific endeavour into another;
- influences which the results, problems and concepts of one field of enquiry have, and have had, on the development of another.

Matrices turn up just about everywhere in science and technology. And as soon as matrices play a role one usually has to consider determinants. Thus it should be no surprise that random determinants (determinants of random matrices) have a vast array of applications. These range from multivariate statistics (of course) to such things as nuclear physics (the names of Dysen and Wigner are associated to some important results there), stability questions, and stochastic differential equations, control of dynamical systems under stochastic perturbations, linear stochastic programming, molecular chemistry and quite generally any theories dealing with sudden events, ripples and resonances. As the determinant is a complicated function of a matrix the theoretical depth of the subject is considerable making this a doubly interesting book.

The original Russian version dates from 1980. Taking the opportunity offered by the publication of an English edition the author expanded the book by over 50% to make it also an encyclopaedic source of knowledge on the topic. As such it is a unique book which I am most happy to be able to publish in this series.

The shortest path between two truths in the real domain passes through the complex domain.

J. Hadamard

Never lend books, for no one ever returns them; the only books I have in my library are books that other folk have lent me.

Anatole France

La physique ne nous donne pas seulement l'occasion de résoudre des problèmes ... elle nous fait pressentir la solution.

H. Poincaré

The function of an expert is not to be more right than other people, but to be wrong for more sophisticated reasons.

David Butler

Bussum, August 1990

Michiel Hazewinkel

CONTENTS

List of Basic Notations and Assumptions

By the expressions $A_n = (\xi_{ij})$, we shall understand respectively a square matrix of order n,

I — is the identity matrix,
A' — the transposed matrix,
A^{-1} — the inverse matrix,
$\det A$ — the determinant of a matrix,
$A_{k\ell}$ — the cofactor of the element $\xi_{k\ell}$ of the matrix A_n,
$\operatorname{Tr} A$ — the trace of a matrix,

the random entries $\xi_{ij}^{(n)}$ of the matrix Ξ_n be given on some probability space;
all vectors considered in this book are vector-columns;
a matrix A of the dimension $m \times n$ has m rows and n columns;

R^n — Euclidean space,
$L_2(\alpha, \beta)$ — the Hilbert space of functions $\varphi(x), x \in (\alpha, \beta)$,
  for which $\int_0^1 \varphi^2(x)dx < \infty$,
$W_s^{(p,k)}$ — Sobolev space,
$\mathbf{E}$ — mathematical expectation,
$\mathbf{V}$ — variance,
$\approx$ — the symbol denoting the equality of distributions of
  random variables, vectors, and matrices;

by *convergence of the distributions functions* we shall mean weak convergence.

We say that the vectors $\vec{\xi}_p^{(n)} = (\xi_{p_1}^{(n)}, \ldots, \xi_{p_n}^{(n)}), p = \overline{1,n}$ are asymptotically constant if there exist constant vectors $\vec{a}_p^{(n)} = (a_{p_1}^{(n)}, \ldots, a_{p_n}^{(n)})$ such that for every $\varepsilon > 0$,

$$\lim_{n \to \infty} \sup_{p=\overline{1,n}} \mathbf{P}\{(\vec{\xi}_p^{(n)} - \vec{a}_p^{(n)}, \vec{\xi}_p^{(n)} - \vec{a}_p^{(n)}) \geq \varepsilon\} = 0,$$

the random vectors $\vec{\xi}_{k_n}$, $k = \overline{1,n}$ are said to be infinitesimal if for every $\varepsilon > 0$,

$$\lim_{n\to\infty} \sup_{k=\overline{1,n}} \mathbf{P}\{(\vec{\xi}_{kn}, \vec{\xi}_{kn}) \geq \varepsilon\} = 0;$$

by the notation $\xi_n \sim \eta_n$, where ξ_n and η_n are sequences of random variables, we mean that, for almost all x,

$$\lim_{n\to\infty} [\mathbf{P}\{\xi_n < x\} - \mathbf{P}\{\eta_n < x\}] = 0;$$

by definition, the notation := is understood to be equality; in order to simplify formulas, we sometimes do not write the index (n); when the domain of integration is not indicated, the integration is taken over the whole domain in which the variables vary.

INTRODUCTION TO THE ENGLISH EDITION

The book contains main assertions of the theory of random determinants which appear on the boundary between probability theory and related sciences. Distributions of random determinants in the multidimensional statistical analysis were first investigated in which the following elegant assertion was obtained: let $\Xi = (\xi_{ij})_{i,j=1}^n$ be a real matrix whose elements are independent and distributed according to a normal law $N(0,1)$, then $\det \Xi_n^2 \approx \prod_{i=1}^n \sum_{j=1}^i \xi_{ij}^2$. At present, the distributions of random determinants also find applications in numerical analysis, in the theory of pattern recognition, in the theory of control of linear stochastic systems, in linear stochastic programming, in the theory of unordered crystalline structures, in statistical and nuclear physics, in the theory of experiment planning, and in the theory of signal filtration.

Especially fruitful is the use of determinants in the general analysis of observations of random vectors which made it possible to considerably reduce the volume of optional values when solving practical problems.

The material of the book is arranged in the following way.

Chapters 1–4 contain proofs of different assertions concerning the distributions of random determinants. Because the distributions of random determinants are complicated and since determinants of high order occur in different applied sciences, the study of the properties of random determinants when their orders tend to infinity is of interest. Chapters 5–16 deal with proofs of limit theorems for random determinants. The theory developed in Chapters 1–16 is applied in Chapters 17-27 to different problems of an applied nature in which random determinants are used.

Here we must mention that the matrix determinant is one of the most widely used matrix functions. It must be examined in almost all the problems where matrices are used. For example, in many tasks the inverse matrix and resolvent matrix are used, which are expressed by matrix determinants, with eigenvalues being solutions of the characteristic equation, etc.

The first chapter is an introductory one. It deals with the main properties of the Haar measure, which must be used when computing distributions of random determinants, and generalized Wishart density, which is necessary when evaluating moments of random determinants.

We must note that the random determinant is a rather complicated function of the matrix; therefore, in order to study it, different integral representations allowing the reduction of the study of the determinant to the study of some random quadratic forms are described in this chapter. However, these integral representations are not good in some cases where inverse matrices should be used. It may be avoided if we use some formulas of integration on Grassmann and Clifford algebras. Berezin was the first to apply these formulas to the study of random determinants.

Chapter 2 deals with methods of computing random determinants as well as some formulas of disturbances. Since the distribution functions of random determinants are complicated, it is reasonable to find moments for them. However, this problem proved to be rather difficult, and in some cases it cannot be solved to this day. Section 1 shows conditions found with the help of the Wishart density, which, after being executed, permit the evaluation of moments of random determinants in the explicit form. Moments of some random Vandermonde determinants are found in Section 2. Here, one theorem of Mehta (the proof of which he personally communicated to the author) is given, as well as the proof of the Dyson hypothesis. Section 3 contains a collection of methods of computing the moments of random determinants. Section 4 of the same chapter gives some moments of random permanents. A matrix permanent possesses some properties of determinants, and, therefore, it is interesting to study its functions of distribution. The last Section of the second chapter describes formulas of disturbances of random determinants which play an important part when proving limit theorems.

Chapter 3 is, if we may say so, the triumph of the theory of random determinants. Here formulas for densities of roots of the characteristic equation are found. First, the problem of evaluating the distribution function of the distribution of eigenvalues of random matrices appeared in the multidimensional statistical analysis where the following assertion was obtained: let Ξ_n be a random symmetric real matrix with the distribution density $c \exp(-\operatorname{Tr} X^2/2)$, where X is a real symmetric matrix and $\lambda_1 \geq \cdots \geq \lambda_n$ are its eigenvalues; then the distribution density of eigenvalues $\lambda_i, i = \overline{1, n}$ will be equal to

$$c_1 \prod_{i>j}(x_i - x_j)\exp\{-\frac{1}{2}\sum_{i=1}^{n}x_i^2\},$$

where c_1 and c are normalization constants.

We must note that the problem of computation of roots of the characteristic (age–old) equation $\det(Iz - A) = 0$, where A is a square matrix and z is a complex parameter, has attracted the attention of mathematicians for many centuries. This problem is difficult; and it seemed that if matrix A is random, the problem of computing distributions of roots of the characteristic equation becomes even more difficult. However, if we assume that elements of matrix A have a common density of distribution $p(X)$, then the density of the distribution of eigenvalues of the roots of the characteristic equation of such a matrix has a simple form. We can illustrate this result with the help of a simple example. Let equation $f(X) = \Xi$ be given, where f is a mutually unique differential transformation over a set L of matrices X, Ξ is a random matrix with the density of distribution $p(z), z \in L$. It is obvious that then the density of distribution of matrix X equals $p(f(z))J(z)$, where $J(z)$ is a Jacobian of the transformation $Y = f(z), Y, z \in L$. It is clear from this example that if the Jacobian $J(z)$ has a simple form, the solution of the distribution density of the equation $f(X) = \Xi$ also has a simple form, although it may not be expressed in the explicit form by elements of the matrix Ξ.

With the help of the same simple idea in the third chapter, formulas for common distributions of eigenvalues and eigenvectors of symmetric, Hermitian, nonsymmetric, complex, Gaussian, orthogonal, and unitary random matrices will be found.

Chapter 4 contains proofs of inequalities for random determinants as well as application of random determinants when studying the Frechet hypothesis. The English translation does not include the Section on the Van der Waerden hypothesis of the original Russian text, because it has meanwhile been proved by Falikman and Yegoritchev, and random matrices are not used in the proof. At the end of the chapter there are some inequalities for random quadratic forms which will be often used when proving limit theorems for random determinants.

The study of the distribution of functions of random matrices, such as the permanent, the determinant, and others, is an interesting problem in probability theory. However, as a rule, the exact distribution of these functions is complicated. Therefore, various limit theorems are of interest as the order of the matrix increases.

Chapter 6 contains proofs of limit theorems similar to the law of large numbers and of the central limit theorem for random determinants.

By a central limit theorem for random determinants we mean any assertion saying that

$$\lim_{n\to\infty} \mathbf{P}\{b_n^1[\ln|\det \Xi_n| - a_n] < x\} = (2\pi)^{-1/2} \int_{-\infty}^{x} e^{-y^2/2}dy,$$

for some choice of constants a_n and b_n under certain conditions imposed on the elements of Ξ_n. The choice of a logarithmic normalizing function is justified on the grounds that $\ln|\det \Xi_n|$ is equal to the sum of the logarithms of the moduli of the eigenvalues of the random matrix Ξ_n; and this fact suggests that after a suitable normalization of these sums, a central limit theorem can be obtained.

At present, three methods have been designed for proving the central limit theorem for random determinants: the perturbation method (Chapter 6), the orthogonalization method (Chapter 6), and the integral representation method (Chapter 8). The perturbation method is based on the formula

$$\ln|\det A| - \ln|\det B| = \ln|\det[I + B^{-1}(A - B)]|,$$

where A and B are square matrices with $\det A \neq 0, \det B \neq 0$. With this method, $\ln |\det \Xi_n|$ can be represented in certain instances as a sum of weakly dependent random variables to which the central limit theorem may be applied.

The orthogonalization method is based on the following well–known result: if the elements $\xi_{ij}, i, j = \overline{1, n}$ of a random matrix $\Xi_n = (\xi_{ij})$ are independent and each has a standard normal distribution $N(0, 1)$, then $\det \Xi_n^2$ is distributed in the same manner as the product of independent random variables $\chi_i^2, i = \overline{1, n}$, each having a χ^2- distribution with i degrees of freedom. If Ξ_n is an arbitrary random matrix, then by means of such an orthogonal transformation, $\ln \det \Xi_n^2$ can be represented as a sum of n random variables to which the central limit theorem is applicable.

The integral representation method is based on the formula

$$\det A^{-1/2} = \pi^{-n/2} \int \cdots \int \exp(-A\boldsymbol{x}, \boldsymbol{x}) \prod_{i=1}^{n} d\boldsymbol{x}_i,$$

where A is a positive definite matrix and $\vec{x} = (x_1, \ldots, x_n)$. By means of this formula, the study of $\ln |\det A|$ may be reduced to that of sums of weakly dependent random variables to which the central limit theorem may also be applied after certain transformations.

Each of these methods has its advantages and disadvantages. The perturbation method is convenient when $\mathbf{E}\Xi_n^{-2}$ exists, $n = 1, 2, \ldots$. The orthogonalization method yields good results when the elements of the matrix Ξ_n are independent and belong to the domain of attraction of the normal law with parameters $(0, 1)$. The integral representation method is used chiefly for determinants of matrices $(I + \Xi_n)$ under the condition that

$$\lim_{h \to \infty} \overline{\lim_{n \to \infty}} \mathbf{P}\{|\det(I + \Xi_n)| \geq h\} = 0$$

The central limit theorem for random determinants was first proved by using these methods in [65].

The main result of Chapter 6 is the so-called logarithmic law: if for every value n random elements $\xi_{ij}^{(n)}, i, j, = \overline{1, n}$ of the matrix Ξ_n are independent, $\mathbf{E}\xi_{ij}^{(n)} = 0, \mathbf{V}\xi_{ij}^{(n)} = 1, \mathbf{E}[\xi_{ij}^{(n)}]^4 = 3$, for some $\delta > 0$

$$\sup_n \sup_{i,j=\overline{1,n}} \mathbf{E}|\xi_{ij}^{(n)}|^{4+\delta} < \infty,$$

then

$$\lim_{n \to \infty} \mathbf{P}\{[\ln \det \Xi_n^2 - \ln(n-1)!](2\ln n)^{-1/2} < x\} = (2\pi)^{-1/2} \int_{-\infty}^{x} e^{-y^2/2} dy.$$

Chapter 7 shows conditions after the execution of which distributions of random determinants and permanents are weakly convergent to the infinitely divisible law. Here the limit theorems proved in Chapter 6 are used.

Chapter 8 contains the most advanced part of the theory of limit theorems for random determinants. The proofs of limit theorems of this chapter are based on the use of integral representations for random determinants, limit theorems for random analytic functions, and also analytic continuation of functions.

Chapter 9 deals with the interrelations between the convergence of distributions of random determinants and functionals of random functions. In Section 1 of this chapter, the proof of limit theorems for random determinants is reduced to the proof of limit theorems for some functionals of random functions by using the integral representation method. In Section 2 the use of the method of spectral functions for proving limit theorems for random determinants is described. It is the most powerful method for studying distributions of random determinants with the help of which random determinants may be expressed by means of spectral functions of random matrices. A productive and advanced theory is developed for them in this chapter. In Section 3 a canonical spectral equation for limit spectral functions under rather general conditions has been obtained. In particular, the Wigner semicircle law which was described in Section 4 follows from this equation. Section 5 contains proofs of limit theorems of the general form for spectral functions of random matrices.

Chapter 10 is dedicated to the study of limit theorems for Gram random determinants. When proving the theorems, the method of spectral functions is used. In this chapter, as well as in the previous one, a canonical spectral equation for limit spectral functions has been obtained.

Chapter 11 deals with the study of determinants of Toeplitz and Hankel random matrices. Here limit theorems of the type of the law of large numbers have been proved, methods of integral representations and disturbances for determinants of Toeplitz and Hankel random determinants have been examined, as well as the theorem which is a stochastic analogue of the Szegö theorem is proved.

Chapter 12 covers a very important class of random determinants—the random determinants of Jacobi. A significant number of problems of numeric analysis, physics, and theory of control require the study of such determinants. Limit theorems of the type of large numbers for the determinants of Jacobi are proved in Section 1 of this chapter. The Dyson equation is obtained in Section 2. Section 3 describes the stochastic problem of Sturm–Liouville, and Section 4 is dedicated to the Sturm oscillation theorem. Sections 5 and 6 are devoted to the proof of the central limit theorem for determinants and normalized spectral functions of Jacobi random matrices, respectively. These theorems are used in applied problems for finding confidence intervals for estimations of Jacobi determinants.

Chapter 13 deals with the distribution of Fredholm random determinants. With the help of these determinants, limit theorems for eigenvalues of symmetric and nonsymmetric random matrices are proved in this chapter. Section

4 covers Fredholm determinants of random linear operators in Hilbert space.

Chapter 14 is dedicated to the study of the solutions distributions of the systems of linear algebraic equations with random coefficients. Here are found the distribution densities of such systems solutions; the stochastic method of least squares is considered, and the spectral method of evaluating moments of inverse random matrices is discussed.

Since the functions of solutions distributions of the systems of linear algebraic equations have an awkward form, the limit theorem for solutions distributions of such systems will be of interest. Such limit theorems are discussed in Chapter 15. Section 1 contains the proof of the so-called arctangent law, stating that under some rather general conditions the limit distribution for components of the vector solution of the system of random algebraic equations is equal to $\frac{1}{2} + \pi^{-1} \operatorname{arctg} z$. In the following sections of this chapter, the integral representation method and the resolvent method of solving systems of linear random algebraic equations heve been discussed.

Chapter 16 deals with integral equations with random confluent kernels. These are the simplest integral equations; however, the study of their solutions distributions is connected with great analytical difficulties, and therefore various limit theorems which are analyzed in this chapter are of interest.

Some questions of the limit theorem of the spectral theory of random matrices are proved in Chapter 17. The above theorem is a logical conclusion of the following statements: in 1958 E. Wigner, in connection with the consideration of mathematical models for energetic levels of heavy atomic nuclei, proved a theorem which was called the semicircular law. Later investigations, in particular, were directed to finding a spectral limit of random non-self-adjoint matrices. Then Ginibr and Mehta proved a theorem which is called the circle law. In order to prove this theorem, the author offered to use a V-transformation with the help of which the proof of limit theorems for spectral functions of random non-self-adjoint matrices was reduced to the proof of limit theorems for random determinants.

In chapter 17, the density of the limit spectral function which is equal to some constant value on the domain whose boundary is an ellipse has been found by using the V-transformation for non-self-adjoint matrices. This is the main assertion of the chapter.

In Chapter 18, it is shown how stochastic differential equations for eigenvalues and eigenvectors of matrix random processes with independent increments can be found with the help of perturbation formulas for random determinants and eigenvalues of random matrices. These equations are used in the following chapters when solving applied problems in which random determinants are used.

Chapter 19 describes applications of methods for studying random determinants to the stochastic Ljapunov problem which consists of finding a probability of the fact that the system of linear differential equations with random coefficients is asymptotically stable.

Chapter 20 discusses the uses of the theory of random determinants in problems of evaluating parameters of linear and nonlinear recurrent systems of equations.

Chapter 21 deals with the solutions of some main problems concerning the theory of control obtained with the help of methods of the theory of random determinants. Here the stochastic condition of Kalman, the problem of control of the spectrum of stochastic systems of control, and some models for manipulator robots have been discussed.

The following chapter, Chapter 22, gives solutions for some problems of linear stochastic programming, which are obtained with the help of the results of the theory of random determinants.

In this chapter, based on the integral representation for determinants, the limit theorem is proved under general assumptions for the solution x_n^* of the equation

$$\mathbf{E}f((C_n(\omega), x_n) = \mathbf{E}f((C_n(\omega), x_n^*), \min x_n : A_n(\omega)x_n \leq b_m(\omega), x_n \geq 0$$

where $A_n(\omega)$ is a random matrix of dimension $n \times m$ and $\vec{C}_n(\omega), \vec{b}_m(\omega)$ are random vectors, $(\vec{C}_n, \vec{x}_n)$ is the scalar product of the vectors $\vec{C}_n$ and $\vec{x}_n$, and f is a measurable function. The notation $\vec{a}_n \leq \vec{b}_n$, where $\vec{a}_n$ and $\vec{b}_n$ are vectors of the same dimensions, will be understood as the correspondent inequality for their components.

The main result presented in the chapter is that the matrix $A_n(\omega)$ under certain conditions may be replaced by an approximate matrix, with diagonal elements equal to the sums of the elements of the matrix $A_n(\omega)$. Provided that the law of large numbers holds for such sums, these diagonal elements may be replaced by determined variables. The result obtained makes it possible to considerably simplify the calculation of the solution x_n; the original stochastic problem may be reduced to a determinate one.

The theory of random determinants found its most powerful and fruitful application in general statistical analysis (Chapter 23). The consistent and asymptotically normal estimation method of some covariance matrices functions $\varphi(R_{m_n})$ under large dimensions m_n of observed vectors is offered in this chapter. Earlier, in most cases the estimates $\varphi(\hat{R}_{m_n})$, where $\hat{R}_{m_n}$ is the empirical covariance matrix, required so many observations n, that there has been serious doubt about the usefulness of multivariate statistical analysis for the solution of practical problems at large m_n. These doubts were partly dispersed after the works [6,97,98] were published, in which corrections to the estimates $\varphi(\hat{R}_{m_n})$ were found under the Kolmogorov condition: $\lim_{n \to \infty} m_n n^{-1} = c, 0 < c < \infty$. After many years of research, it seemed that under the Kolmogorov conditions, consistent and asymptotically normal estimates of functions $\varphi(R_{m_n})$ do not exist. However, due to the developed matrix spectral theory, one succeeded to establish that under the G-condition:

$$\overline{\lim_{n \to \infty}} \, m_n n^{-1} < \infty, \operatorname{plim}_{n \to \infty}[\varphi(\hat{R}_{m_n}) - \psi(R_{m_n})] = 0,$$

where ψ is some known measurable function of the matrix R_{m_n} elements. This equation will be called the G-equation. This is the principal statement making up the basis of the G-analysis of observations of large dimensions. For the proof of the G-equation, one uses as a rule the limit theorems for sums of Martingale differences and perturbation formulas for random matrix resolvents, which form the theoretical foundation of G-analysis.

Using the G-equation, we can find such a measurable function $G(\hat{R}_{m_n})$ (G-estimate), that

$$[G(\hat{R}_{m_n}) - \varphi(R_{m_n})]c_n^{1/2} \Rightarrow N(0,1),$$

where c_n is a sequence of numbers. It is shown in this treatise that the error of G-estimates of some functions $\varphi(R_{m_n})$ has the order $(m_n n)^{-1/2}$, while the estimates $\varphi(\hat{R}_{m_n})$ have an error equivalent to $m_n n^{-1/2}$.

The following three chapters are a continuation of Chapter 23. Chapter 24 gives G-estimates of solutions of the Kolmogorov-Wiener equations; and Chapter 25 gives G-estimates of the Mahalanobis distance and of Anderson–Fisher statistics. The results obtained in this chapter are extremely important in the theory of pattern recognition. Chapter 26 describes applications of the theory of random determinants in scheduling experiments.

The 27th chapter deals with applications of the theory of random determinants in nuclear physics, in the theory of unordered crystalline structures, in the theory of scattering, and in statistical physics. In brief, the applications of random determinants in physics may be described in the following way. In accordance with quantum mechanics, energy levels of the atomic nucleus must be described by means of eigenvalues of the Hermitian operator, which is called a Hamiltonian. The Hamiltonian is valid in infinite-dimensional Hilbert space. By selecting the basis of this space, we may in some cases replace the Hamiltonian in an approximation by a finite dimensional matrix of high order.

Wigner formulated the hypothesis that the local statistical behaviour of energy levels of heavy atomic nuclei can be described by eigenvalues of random matrices of high order. The distribution of eigenvalues of random matrices of high order has a complicated form; therefore, it seemed reasonable to apply limit theorems for spectral functions of random matrices. Among the first limit theorems, the most widely known is the Wigner semicircular law, stating that the limit density of spectral functions in certain cases has the form of a semicircle. Wigner described his ideas in his paper "Random Matrices in Physics" [166]. At the same time, Lifshits [122] began to study mathematical models of crystals with impurities by using perturbation formulas for eigenvalues of random matrices. In [30] Dyson proposed a method for studying limit spectral functions of Jacobi symmetric matrices with arbitrary elements. Mehta in his work [132] proposed some original methods of studying the densities of the distributions of eigenvalues. Some of his results are described in Chapter 27. The last chapter, Chapter 28, deals with applications of the theory of random determinants in the numerical analysis.

Apparently, after many main problems of the theory of random determinants are solved, we may expect that they will be used in nuclear and statistical physics. At present, some works on the applications of spectral theory of random matrices in physics have appeared; thus, for example, the review "Matrix Random Physics" [19] was published.

Kiev, 1986

CHAPTER 1

GENERALIZED WISHART DENSITY AND INTEGRAL REPRESENTATION FOR DETERMINANTS

In this chapter, basic information concerning the Haar measure and integration with respect to the Haar measure is given. Generalized Wishart density and some new integral representations for determinants are found. All of this material will be necessary to find random determinants distributions.

§1 The Haar Measure

The majority of formulas for random determinants distribution contains some integrals on invariant measure; therefore, in order to study these formulas, one needs to know invariant measure properties.

Let T be a separated topological locally compact group, and let E be a space in which a group of transformations G acts. The measure $\mu(A)$ defined on a Borel σ-algebra B of the space $\mathbf{E}$ is called invariant with respect to G if for any $A \in B$ and $s \in G$, such that the sA set is measurable, $\mu(sA) = \mu(A)$. Here, as sA, we consider the set $\{sg : g \in A\}$.

If the function $f(p), P \in \mathbf{E}$ is measurable with respect to the σ-algebra B and nonnegative, then, under the assumption that at least one of these integrals exists, $\int f(p)\mu(dp) = \int f(sp)\mu(dp)$. We call the measure μ the left- invariant Haar measure (left Haar measure) if the equalities $\mu(sA) = \mu(A), \int f(x)\mu(dx) = \int f(sx)\mu(dx)$ hold. If μ is the left Haar measure, then the function ν defined by the equality $\nu(K) = \mu(K^{-1})$ on a σ-algebra of measurable subsets of K of G elements is the right Haar measure. By the set K^{-1} we mean $\{K^{-1} : k \in \mathbf{E}\}$. Obviously, with μ being the right Haar measure, ν is the left Haar measure.

We state now a basic theorem for the Haar measure. Its proof can be found in [165], for example.

Theorem 1.1.1. *The left Haar measure exists on any separated topological locally compact group T. If μ and μ' are two left Haar measures on T, then $\mu' = c\mu$, where c is a positive number.*

1

Note that in the general case $\mu(T)$ may be equal to infinity; therefore, we are not always able to choose such a normalization of the measure μ that $c = 1$. If the group T is compact, then the measure μ can be normalized by the condition $\mu(T) = 1$.

Let us formulate the principal properties of the Haar measure.

Theorem 1.1.2. *1) There exists a continuous positive function Δ on T, such that $\Delta(t) > 0, \forall t \in T, \Delta(e) = 1$, where e is the unit element of the group T, and $\Delta(tt') = \Delta(t)\Delta(t'), t, t' \in T$.*

2) If μ is the left Haar measure on T, then $\mu(Pg) = \Delta(g) \times \mu(P)$, $\int f(ta)\mu dt = \Delta(a) \int f(t)\mu(dt)$, where $a, g \in T, P \in \mathfrak{S}, \mathfrak{S}$ is the σ-algebra of the Borel subsets of the group T, f is the measurable nonnegative function on T, with respect to $\mathfrak{S}$, such that at least one of the given integrals exists.

3) If ν is the right Haar measure on T, defined by the condition $\nu(p) = \mu(p^{-1})$, then $\nu(g\mathbf{P}) = \Delta(g^{-1})\nu(\mathbf{P})$.

The function Δ is called the modular function on T.

Suppose that μ and ν are left and right Haar measures on T, respectively. Then a number $c > 0$ exists, such that $\int f(t)\nu(dt) = c \int f(t)\Delta(t)\mu(dt)$, $\int f(at)\nu(dt) = [\Delta(a)]^{-1} \times \int f(t)\mu(dt)$. Obviously, if the group T is compact, then $\Delta(t) \equiv 1, \forall t \in T$.

We shall now give some examples of Haar measures on the group of matrices.

Let G be a locally compact group of real matrices of the order n, and B the σ-algebra of Borel sets on it. There are left and right Haar measures on the group G. Consider the following integral $\int_G f(x)p(x)dx$, where $p(x)$ is a measurable nonnegative scalar function defined on G, $f(x)$ is a measurable scalar function defined on G, $dx = \Pi_{i,j=1}^n dx_{ij}$ in such a way that the integral exists. We make the change of variables $x = Ay, A, y \in G$. Then $\int f(x)p(X)dx = \int f(Ay)p(Ay)J(A)dy$, where $J(A)$ is the Jacobian of the transformation $x = Ay$. We find the properties of the Jacobian $J(A)$. Obviously, $J(A, B) = J(A)J(B), A, B \in G$, for any diagonal matrix $\Lambda \in GJ(\Lambda) = |\det \Lambda|^n, J(A)$, is a continuous function of G. Now assume that the matrix A is of the form $A = T\Lambda T^{-1}, \Lambda, T \in G$. Then by properties of the Jacobian, $J(A) = |\det A|^n$. One can extend this formula to arbitrary matrices $a \in G$ by continuity, i.e., A is replaced by the matrix $A + \varepsilon B$, where $\varepsilon > 0$, and B is such that $A + \varepsilon B$ and $A + \varepsilon B = T_\varepsilon \Lambda_\varepsilon T_\varepsilon^{-1}, \Lambda_\varepsilon, T_\varepsilon \in G, \Lambda_\varepsilon$ is a diagonal matrix. Then $J(A + \varepsilon B) = |\det(A + \varepsilon B)|^n$. Hence, tending to a limit as $\varepsilon \to 0$, we get

$$J(A) = |\det A|^n \forall A \in G. \tag{1.1.1}$$

Using this formula for the Jacobian of transformation, we shall write $\int f(x) p(x)dx = \int f(Ay)p(Ay)|\det A|^n dy$. For $p(x)$ being a density of the Haar measure with respect to the Lebesgue measure dx, it is necessary that $P(Ay) |\det A|^n = p(y)$. By setting $A = y^{-1}$, we find that $p(y) = c|\det y|^{-n}$, where $c > 0$ is a certain constant. Thus, the density of the Haar measure on G

with respect to the Lebesgue measure on G is equal to $c|\det x|^{-n}, x \in G$.
Obviously, the left and right Haar measures coincide on such a group with
accuracy up to an arbitrary constant positive coefficient.

Similarly, we define the Haar measure on the group K of complex matrices
of order n. Here the densities of left and right Haar measures, with respect
to the Lebesgue measure on K, are equal to $|\det x|^{-2n}, x \in K$ with accuracy
up to a constant positive coefficient.

Proving this fact, we used the following lemma,

Lemma 1.1.1.   *Let $w_s = u_s + iv_s, s = \overline{1,n}$ be the analytical functions of the
complex variables $z_p = x_p + iy_p, p = \overline{1,n}$. Then the Jacobian of the functions
$u_s, v_s, s = \overline{1,n}$ by variables $x_s, y_s, s = \overline{1,n}$ is equal to the modulus square of
the Jacobian of the functions $w_s, s = \overline{1,n}$ by variables z_s; that is,*

$$\left|\det\begin{bmatrix} \frac{\partial u_s}{\partial x_p} & \frac{\partial v_s}{\partial x_p} \\ \frac{\partial u_s}{\partial y_p} & \frac{\partial v_s}{\partial y_p} \end{bmatrix}\right|_{s,p=\overline{1,n}} = \left|\det\begin{bmatrix} \frac{\partial \omega_s}{\partial z_p} \end{bmatrix}\right|^2_{s,p=\overline{1,n}}.$$

Proof.   Since w_s are analytical functions of the complex variables z_p, then

$$\frac{\partial u_i}{\partial x_k} = \frac{\partial v_i}{\partial y_k}, \quad \frac{\partial v_i}{\partial x_k} = -\frac{\partial u_i}{\partial y_k}, \quad i, k\overline{1,n}.$$

On the basis of these ratios and after some simple transformations, we get

$$\det\begin{bmatrix} \frac{\partial u_s}{\partial x_p} & \frac{\partial v_s}{\partial x_p} \\ \frac{\partial u_s}{\partial y_p} & \frac{\partial v_s}{\partial y_p} \end{bmatrix}_{s,p=\overline{1,n}} = \det\begin{bmatrix} A & B \\ -B & A \end{bmatrix},$$

where $A = (a_{sp}, B = (b_{sp}), a_{sp} = \partial u_s/\partial x_p, b_{sp} = \partial v_s/\partial x_p$. Then, owing to
$|\det\begin{bmatrix} A & B \\ -A & A \end{bmatrix}| = |\det(A+iB)|^2, \partial w_s/\partial z_p = a_{sp}+ib_{sp}$, we obtain the statement
of Lemma 1.1.1.

Now, we find the Haar measure on the group of real triangular matrices with
positive entries on the main diagonal. Let T be a group of lower real triangular
matrices of order n with positive entries on the main diagonal, and let B be
the σ- algebra of Borel sets on it. Consider the integral $\int f(x)p(x)d(x)$ on the
group T, where $p(x)$ is a nonnegative, B-measurable function, $f(x)$ is a B-
measurable function chosen so that this integral is finite, and $dx = \prod_{i \geqslant j} dx_{ij}$.
Let us change the variables in this integral $x = Ay, A, y \in T$. The Jacobian of
such a transformation satisfies the condition $J(AB) = J(A)J(B), A, B \in T$.
For any diagonal matrix $\Lambda(\lambda_i d_{ii}) \in T, J(\Lambda) = \prod_{i=1}^{n} \lambda_i^{n+1-i}$. As in the proof
of (1.1.1), we get

$$J(A) = \prod_{i=1}^{n} a_{ii}^{n+1-i}. \tag{1.1.2}$$

It follows from this formula that, with respect to the Lebesgue measure on T, the density of the left Haar measure is equal to $c \prod_{i=1}^{n} x_{ii}^{-(n+1-i)}, x_{ii} > 0$, where $c > 0$ is an arbitrary constant. Analogous arguments also hold for the right Haar measure given on T. In this case,

$$J(\Lambda) = \prod_{i=1}^{n} \lambda_i^i. \tag{1.1.3}$$

Consequently, the density of the right Haar measure with respect to the Lebesgue measure on T, is equal to $c \prod_{i=1}^{n} x_{ii}^{-i}, x_{ii} > 0$, where $c > 0$ is an arbitrary constant.

The modular function given on T has the form

$$\Delta(x) = \prod_{i=1}^{n} x_{ii}^{2i-n-1}, \quad x \in T.$$

Similarly, we find the Haar measure on the group of upper triangular matrices with positive entries on the main diagonal as well as the Haar measure on the group of complex lower (upper) triangular matrices.

§2 The Haar Measure On the Group of Orthogonal Matrices

Let G be a group of real orthogonal matrices of order n and let μ be the invariant normalized Haar measure on it. The entries of a matrix $H \in G$ satisfy the equations $n(n-1)/2$. Solving these equations, we obtain $n(n-1)/2$ independent parameters of the matrix H. So-called Euler angles are rather convenient parameters of the group G. First, the functions by which the entries of the matrix H are expressed in terms of the Euler angles, are almost everywhere differentiable by these angles. Second, the Haar measure expressed in terms of the Euler angles has a simple form.

Let us show how one can express the matrix H in terms of the Euler angles. We consider the orthogonal matrices $R_{ks}(\Theta_{ks})$ defining a rotation in the coordinate plane: $x'_k = \cos \Theta_{ks} x_k + \sin \Theta_{ks} x_s$, $x'_s = -\sin \Theta_{ks} x_k + \cos \Theta_{ks} x_k$, $x'_l = x_l$, $l \neq k$, $l \neq s$, $k < s$, $k, s = \overline{1, n}$.

Let hij be the entries of the matrix H. On the intersection of the first line and the second column of the matrix $HR_{12}(\Theta_{12})$, there is the entry $h_{11} \sin \Theta_{12} + h_{12} \cos \Theta_{12}$. If $\Theta_{12} = \arctan(-h_{12}/h_{11})$ then this entry is zero. In the matrix $HR_{12}(\Theta_{12})R_{13}(\Theta_{13})$, we choose Θ_{13} in such a manner that the element on the intersection of the first line and the third column is equal to zero. In this case, the element standing on the intersection of the first line and the second column will be equal to zero, irrespective of the value of angle Θ_{13}. By continuing this process, one can select the angles $\Theta_{1k}, k = \overline{2, n}$ in such a manner that all elements of the first line of the matrix

$H R_{12}(\Theta_{12}) R_{13}(\Theta_{13}) \ldots R_{1n}(\Theta_{1n})$, expect for the first one are equal to zero. But this matrix is orthogonal, and consequently

$$H R_{12}(\theta_{12}) R_{13}(\theta_{13}) \ldots R_{1n}(\theta_{1n}) = \begin{bmatrix} \pm 1 & 0 \\ 0 & H_{n-1} \end{bmatrix},$$

where H_{n-1} is a certain orthogonal matrix of the order $n-1$.

Let us make the same transformations for the matrix H_{n-1} as for the matrix H_n. Using induction by n, we obtain

$$H_n \prod_{k=1}^{n} \prod_{s=k+1}^{n} R_{ks}(\theta_{ks}) = (\pm 1 \delta_{ij}). \tag{1.2.1}$$

By substitution of the angles Θ_{kn} for $\Theta + \pi_{kn}$ one can always achieve that a unit matrix is on the right side of this equality. It follows from the equality (1.2.1), that

$$H_n = \prod_{k=n}^{1} \prod_{s=n}^{k+1} R_{ks}(\varphi_{ks}), \quad \varphi_{ks} = -\theta_{ks}. \tag{1.2.2}$$

For $n = 3$, the equality (1.2.2) turns into a well- known representation of an orthogonal third-order matrix with the Euler angles

$$H_3 = \begin{bmatrix} \cos\psi & \sin\psi & 0 \\ -\sin\psi & \cos\psi & 0 \\ 0 & 0 & 1 \end{bmatrix} \begin{bmatrix} \cos\theta & 0 & -\sin\theta \\ 0 & 1 & 0 \\ \sin\theta & 0 & \cos\theta \end{bmatrix} \begin{bmatrix} \cos\varphi & \sin\varphi & 0 \\ -\sin\varphi & \cos\varphi & 0 \\ 0 & 0 & 1 \end{bmatrix}.$$

The representation of a matrix of the form (1.2.2) is unique for almost all values of the angles Θ_{ks}. The exception is the cases in which one of the angles Θ_{ks} is equal to 0 or π. Obviously, the Euler angles vary within the following bounds:

$$0 \le \theta_{ki} < 2\pi, \quad 0 \le \theta_{ks} < \pi, \quad s = \overline{k+1,n}, \quad k = \overline{1,n}.$$

Theorem 1.2.1. [135]. *The Haar measure μ of the group G of the matrices H, defined by means of the Euler angles Θ_{ks} by Eq. (1.2.2), is absolutely continuous with respect to the Lebesgue measure given on a set of variations of Euler angles Θ_{ks} with the density*

$$c_n \prod_{k=1}^{n-1} \prod_{j=1}^{k} \sin^{j-1}(\theta_{jk}), \tag{1.2.3}$$

where $c_n = \prod_{k=1}^{n} \Gamma(k/2)(2\pi)^{-k/2}$.

Proof. We introduce in the Euclidean real space E_n the spherical coordinates $r, \Theta_1, \ldots, \Theta_{n-1}$ connected with Cartesian coordinates by formulas

$$x_1 = r \sin \theta_{n-1} \ldots \sin \theta_2 \sin \theta_1,$$
$$x_2 = r \sin \theta_{n-1} \ldots \sin \theta_2 \cos \theta_1,$$
$$\cdots \quad \cdots \quad \cdots \tag{1.2.4}$$
$$x_{n-1} = r \sin \theta_{n-1} \cos \theta_{n-2}, \quad x_n = r \cos \theta_{n-1},$$
$$0 \le r < \infty, \quad 0 \le \theta_1 < 2\pi, \quad 0 \le \theta_k < \pi, \quad k = \overline{2, n}.$$

Consider a random n-dimensional vector $\vec{\eta}_n := (\sin \Theta_{n-1} \ldots \sin \Theta_2 \times \sin \Theta_1, \ldots, \cos \Theta_{n-1})$, where the random values $\Theta_i, i = \overline{1, n}$ have the distribution density $(2\pi)^{-n/2} \Gamma(n/2) \times \sin^{n-2} \Theta_{n-1} \ldots \sin \Theta_2, 0 \le \Theta_1 < 2\pi, 0 \le \Theta_k < \pi$, $k = \overline{2, n}$. We show that the distribution of this vector is invariant with respect to the orthogonal transformation T_n.

In fact,

$$\mathbf{E} \exp\{i(\vec{q}, T_n \vec{\eta}_n)\} = \int_0^\infty \mathbf{E} \exp\{i(\vec{q}, T_n r \vec{\eta}_n)\} \delta(r^2 - 1) dr,$$

where $\delta(r^2 - 1)$ is the generalized function, and $\vec{q}$ is the vector of the parameters.

By using a reverse transformation of (1.2.4), we obtain (see the calculation of the Jacobian of the transformation (1.2.3) in [35])

$$\mathbf{E} \exp\{i(\vec{q}, T_n \vec{\eta}_n)\} = \int_0^\infty \exp\{i(\vec{q}, T_n \vec{x}_n)\} \delta(\sum_{i=1}^n x_i^2 - 1) \prod_{i=1}^n dx_i$$

$$= \int_0^\infty \exp\{i(\vec{q}, \vec{x}_n)\} \delta(\sum_{i=1}^n x_i^2 - 1) \prod_{i=1}^n dx_i = \mathbf{E} \exp\{i(\vec{q}, \vec{\eta}_n)\}.$$

Obviously, the vector $\vec{\eta}_n$ can be represented in the following form

$$\vec{\eta} = \prod_{i=1}^n T_{1i}(\theta_i) \vec{l}_1, \quad \vec{l}_1 = (1, 0, \ldots, 0). \tag{1.2.5}$$

Column vectors of the matrix H, whose entries are expressed by the Euler angles θ_{ij}, are equal to

$$\vec{\eta}_1 = \prod_{i=1}^n T_{1i}(-\theta_{1i}) \vec{l}_1,$$

$$\vec{\eta}_2 = \prod_{i=1}^n T_{1i}(-\theta_{1i}) [\prod_{i=2}^n T_{2i}(-\theta_{2i})] \vec{l}_2, \tag{1.2.6}$$

$$\cdots \quad \cdots \quad \cdots$$

$$\vec{\eta}_n = \prod_{i=1}^n T_{1i}(-\theta_{1i}) [\prod_{i=2}^n T_{2i}(-\theta_{2i})] \ldots [\prod_{i=n-1}^n T_{nn-1}(-\theta_{nn-1})] \vec{l}_n.$$

Suppose that $\Theta_{ij}, j = \overline{1,n}$ are random variables, independent for different i, and joint densities of vector components $(\theta_{ii}, \theta_{ii+1}, \ldots, \theta_{in})$ are equal to $(2\pi)^{(n-i+1)/2}\Gamma((n-i+1)/2) \times \sin^{n-i-1}\theta_{ii} \ldots \sin\theta_{in-1}$. The matrix H_n will be random, its entries will be expressed through the random Euler angles θ_{ij} by (1.2.2), whose distribution density equals (1.2.3). The distribution of such a matrix will be invariant in the group G. To make sure of that, we multiply the matrix H on the left-hand side by the arbitrary nonrandom matrix $T \in G$. From the invariance of the distribution of the vectors $\vec{\eta}_n$ (1.2.5) and (1.2.6), the validity of Theorem 1.2.1 follows.

§3 The Generalized Wishart Density

Let us consider the rectangular matrix $\Xi = (\xi_{ij})$ whose random row vectors $\vec{\xi}_i = (\xi_{i1}, \ldots, \xi_{im}), i = \overline{1,n}, m \geq n$ are independent and distributed according to the multivariate normal law with zero vector of mean and nondegenerate matrix of covariances R_n. A matrix $\Xi\Xi'$ is called the Wishart matrix. By the density of a symmetric random matrix we mean the distribution density of its entries located on and above the diagonal. The density of the matrix $\Xi\Xi'$ is called the Wishart density. We designate it by $\omega(X_n, R_n, m)$, where X_n is a symmetric positive-definite matrix. The density of the matrix HH', where H is a random rectangular matrix with density $p(X)$, will be called a generalized Wishart density.

Theorem 1.3.1. *Let G be a group of real orthogonal $m \times m$-matrices and μ the normalized Haar measure on it, H a random rectangular $n \times m$-matrix, $m \geq n$ with distribution density $p(X)$. Then the generalized Wishart density is equal to*

$$c_{n,m} \int_G p(\sqrt{Z_n}\Theta^{(n)})\mu(d\Theta) \det Z_n^{(m-n-1)/2}, \qquad (1.3.1)$$

where Z_n is a positive definite $n \times n$-matrix, $\Theta = (\theta_{ij}) \in G, \Theta^{(n)} = (\theta_{ij}), i = \overline{1,n}, j = \overline{1,m}$,

$$c_{n,m} = [\pi^{(n(n-1)/4)-nm/2}\prod_{i=1}^{n}\Gamma((m+1-i)/2)]^{-1}.$$

Let K_1 be a set of real $m \times n$-matrices, K_2 a set of positive definite matrices of $n \times n$-dimension, K_3 a set of real orthogonal matrices of $m \times n$-dimension. To prove the theorem we need the following lemma [10,38].

Lemma 1.3.1. *The real $m \times n$-matrix $Z(m \geq n)$, under the condition that $\det ZZ' \neq 0$, can be represented by $Z = XU$, where $X \in K_2, U \in K_3$, and this representation is unique.*

Proof. Represent the matrix Z by $Z = XU$, where $U = X^{-1}Z, X = (ZZ')^{1/2}$. Obviously, $X \in K_2, U \in K_3$. We claim that a representation of the matrix Z by $Z = XU$ is unique. Assume that one more representation $Z = X_1 U_1, X_1 \in K_2, U \in K_3$ exists. Then $XUU'X = X_1 U_1 U_1' X_1$. Hence $X_1 = X, U_1 = X^{-1}Z$. This proves Lemma 1.3.1.

Proof of Theorem 1.3.1. Consider the integral $\int f(ZZ') \times p(Z)d(Z)$ for any bounded continuous function $f(X), X \in K_2$. In this integral, we change the variables $Z = Y\Theta, Y \in K_1, \Theta \in K_3$; the matrix Θ is fixed. The Jacobian of such a change of variables does not depend on the matrix Y and is equal to 1, it is clear from the following correlation:

$$\int \exp(-\operatorname{Tr} ZZ')dZ = J \int \exp(-\operatorname{Tr}(Y\Theta)(Y\Theta)')dY. \qquad (1.3.2)$$

Then

$$\int f(ZZ')p(Z)dZ = \int f(YY')p(Y\Theta)dY. \qquad (1.3.3)$$

The expression (1.3.3) does not depend on the matrix Θ; therefore it will not be changed if we integrate it with respect to the measure μ defined on the group G of matrices Θ. Formula (1.3.3) takes the following form:

$$\int f(ZZ')p(Z)dZ = \int f(YY')p(Y\Theta^{(n)})dY\mu(d\Theta). \qquad (1.3.4)$$

Introduce the change of variables $Y = X^{1/2}U, X \in K_2, J \in K_3, \det YY' \neq 0$. As parameters of the matrix U, one can choose $mn - n(n+1)/2$ of the Euler angles (see the previous section), which can be changed independently in certain intervals. A number of independent parameters of the matrix X is equal to $n(n+1)/2$. Thus, a number of independent parameters on the left-hand and right-hand sides of $Y = X^{1/2}U$ are the same. This transformation is differentiable with respect to the chosen parameters and is one-to-one on the measurable set of $\{Y : \det YY' \neq 0\}$. Let us denote a Jacobian of the transformation $Y = X^{1/2}U$ by $J(X, U)$. By taking into account that $\int_{\det ZZ'=0} f(ZZ') \times p(Z)dZ = 0$, we find that

$$\int f(ZZ')p(Z)dZ = \int f(X)p(\sqrt{X}U\Theta^{(n)})J(X, U)dX\mu(d\Theta)dU, \qquad (1.3.5)$$

where $dU = \prod_i d\alpha_i, \alpha_i$ are the Euler angles of the matrix U, and dX is an entry of Lebesgue measure on K_2.

Nothing will be changed in this equality if we substitute $p(\tilde{X}^{1/2}\tilde{U}\Theta)$ for $p(X^{1/2}U\Theta)$, where

$$\sqrt{X} = \begin{bmatrix} \sqrt{X} & 0 \\ 0 & 0 \end{bmatrix}, \quad \tilde{U} = \begin{cases} u_{ij}, & i = \overline{1, n}, \quad j = \overline{1, m}, \\ q_{ij}, & i = \overline{n+1, m}, \quad j = \overline{1, m}. \end{cases}$$

We supplement the matrix $X^{1/2}$ by zero entries, so that $\tilde{X}^{1/2}$ is $m \times n$, and take the entries q_{IJ} so that $U \in G$. Then, by using the invariance of the measure μ in (1.3.5), we get

$$\int f(ZZ')p(Z)dZ = \int f(X)p(\sqrt{X}\Theta^{(n)})\varphi(X)\mu(d\Theta)dX, \qquad (1.3.6)$$

where $\varphi(X) = \int T(X,U)dU$.

Now, make the change of variables $X = CYC', X, Y, C \in K_2$, where the matrix C is fixed. The Jacobian of this change of the variables $\alpha(C)$ is equal to $|\det C|^{n+1}$. This is clear from the following considerations. Obviously, $\alpha(C)$ does not depend on Y and is $|\det C|^{n+1}$ if C is a diagonal matrix. Moreover $\alpha(C_1 C_2) = \alpha(C_1)\alpha(C_2)$, where $C_1 C_2 \in K_2$. Hence, taking into account that the matrix C can be reduced to the diagonal form by orthogonal transformation, we obtain $\alpha(C) = |\det C|^{n+1}$. Similarly, the Jacobian of the change of the variables $Z = CX, C \in K_2, Z, X \in K_1$ is equal to $|\det C|^m$. By making use of these two changes of variables and of Eq. (1.3.6), we find

$$\int f(CZZ'C)p(CZ)dZ|\det C|^m = \int f(CXC)p(C\sqrt{X}\Theta^{(n)})$$
$$\times \varphi(X)\mu(d\Theta)dX|\det C|^m,$$
$$\int f(ZZ')p(Z)dZ = \int f(CXC)p(\sqrt{CXC}\Theta^{(n)})\varphi(CXC) \qquad (1.3.7)$$
$$\times \mu(d\Theta)dX|\det C|^{n+1}.$$

Let the density $p(Z), Z \in K_1$ be of the form

$$p(Z) = (2\pi)^{-nm/2}\exp(-(\operatorname{tr} ZZ')/2).$$

Then, bearing in mind that the left sides of Eqs. (1.3.7) are equal and that function f can be chosen arbitrarily, we obtain from (1.3.7),

$$\varphi(X)|\det C|^m = \varphi(CXC)|\det C|^{n+1}.$$

As $C = X^{-1/2}$,
$$\varphi(X) = c\det X^{(m-n-1)/2}, \qquad (1.3.8)$$

where $c > 0$ is some constant.

To find the constant c, we prove the following Lemma [129].

Lemma 1.3.2. *Any nonnegative definite $(n \times n)$- matrix A can be represented uniquely in the form $A = SS'$, where S is the upper (lower) triangular matrix with positive entries on the diagonal. The Jacobian of the transformation $A = SS'$ equals $J(S) = 2^n \prod_{i=1}^{n} s_{ii}^{n+1-i}$.*

Proof. Any square real $n \times n$-matrix A whose main minors are not equal to zero can be represented in the form $A = SQ$, where S is the lower and Q the upper triangular matrix. In fact, consider the system of equations $SA = Q$, from which we obtain the following systems of equations,

$$s_{i1}a_{11} + s_{i2}a_{21} + \cdots + s_{ii}a_{i1} = 0,$$

$$\cdots \quad \cdots \quad \cdots$$

$$s_{i1}a_{1i-1} + s_{i2}a_{2i-1} + \cdots + s_{ii}a_{ii-1} = 0, \quad i = \overline{2, n}.$$

Obviously, the solutions of these systems of equations exist. Consequently, $A = S^{-1}Q$. One can choose diagonal entries of the matrices S and Q at will; the only requirement is that they not be equal to zero. Since the matrix A is a symmetric one, $S^{-1}Q = Q'(S^{-1})'$. Hence, $(Q')^{-1}S^{-1} = (S-1)'Q^{-1}$. A lower triangular matrix is on the left-hand side and an upper one is on the right-hand side, respectively. Therefore, $Q = \Lambda(S^{-1})'$, where Λ is a diagonal matrix. By introducing the notation $S^{-1} = P$, we obtain $A = P\Lambda P'$. The matrix A is a positive-definite matrix; therefore, the diagonal entries of matrix Λ will be positive. Thus, $A = \widetilde{P}\widetilde{P}'$, where $\widetilde{P} = P\Lambda^{1/2}$. Let us prove that this representation is unique. Assume that there is a lower triangular matrix P_1 distinct from $\widetilde{P}, A = P_1 P_1'$. Then $P_1^{-1}\widetilde{P} = P_1'(\widetilde{P}')^{-1}$. When the lower triangular matrix is on the left and the upper one is on the right-hand side, $P_1 = P$.

Let us find the Jacobian of the transformation $A = SS'$. Denote it by $J(S)$. Let T be a group of lower real nondegenerate triangular $n \times n$-matrices with nonnegative entries on the diagonal, and let $f(A)$ be a bounded continuous function defined on the set of nonnegative definite $n \times n$-matrices. This function is such that the integral $\int f(A)dA$ exists, where dA is an element of the Lebesgue measure of the set K_1. We make the following change of variables in this integral: first, $A = UBU', B = PP', U, P \in T, B \in K_1$, and then $A = SS', S = UP, S, U, P \in T$. Using the Jacobian of the change of variables $X = CYC$ (see the proof of Eq. (1.3.7) as well as (1.1.3)), we obtain

$$\int f(A)dA = \int f(UBU')dB \det U^{n+1} = \int f(UPP'U')J(P) \det U^{n+1}dP,$$

$$\int f(A)dA = \int f(SS')J(S)dS = \int f(UPP'U')J(UP)\prod_{i=1}^{n} u_{ii}^i dP.$$

Hence, by virtue of the arbitrariness of function f, we obtain

$$J(P) \det U^{n+1} = J(UP)\prod_{i=1}^{n} u_{ii}^i. \tag{1.3.9}$$

Let us determine the value $J(I)$. A simple calculation shows that the matrix $[\delta a_{ij}/\delta s_{pl}]_{S=I}, i \geq j$ is a triangular one, and there will be n of twos on its diagonal, and all other diagonal entries are 1. It follows that $J(I) = 2^n$. Making use of this and setting $P = I$ in (1.3.9), we find $J(U) = 2^n \prod_{i=1}^{n} u_{ii}^{n+1-i}$. Thus, Lemma 1.3.2 is proved.

Set $f(X) \equiv 1$ in (1.3.6) and $p(Z) = (2\pi)^{-mn/2} \exp(-(\operatorname{Tr} ZZ')/2)$. Then for the constant c, we obtain the following formula,

$$c^{-1} = (2\pi)^{-nm/2} \int_{K_2} \det X^{(m-n-1)/2} \exp(-0.5 \operatorname{Tr} X) dX.$$

By virtue of Lemma 1.3.1,

$$c^{-1} = 2^n (2\pi)^{-nm/2} \int_{s_{ii}>0} \exp\left(-0.5 \sum_{i>j}^{n} s_{ij}^2\right) \prod_{i=1}^{n} s_{ii}^{m-i} \prod_{i \geq j} ds_{ij}$$

$$= 2^n (2\pi)^{-nm/2} \prod_{i=1}^{n} \left[\int_0^\infty e^{-y^2/2} y^{-m-i} dy \right] (2\pi)^{(n^2-n)/4}$$

$$= (2\pi)^{-nm/2} \prod_{i=1}^{n} \Gamma((m+1-i)/2) \pi^{n(n-1)/4} 2^{mn/2}.$$

Theorem 3.3.1 is proved.

Corollary 1.3.1. *The Wishart density is equal to*

$$\omega(Z_n, R_n, m) = c_{n,m} (2\pi)^{-nm/2} \det R_n^{-m/2} \det Z_n^{(m-n-1)/2}$$
$$\times \exp(-0.5 \operatorname{Tr} R_n^{-1} Z_n),$$

where Z_N is a nonnegative-definite matrix.

We obtain the proof from (1.3.1) by substituting the density of Ξ for density p,

$$p(X) = (2\pi)^{-nm/2} \exp(-0.5 \operatorname{Tr} R_n^{-1} XX') \det R_n^{-m/2}, \qquad (1.3.10)$$

where X is a real rectangular $(m \times n)$-matrix.

Similarly, we set up the following result [112].

Corollary 1.3.2. *Let a random matrix Ξ have the density*

$$(2\pi)^{-mn/2} \det R_n^{m/2} \exp\{-\operatorname{Tr}(X-M) R_n (X-M)'\},$$

where R_n is a positive definite matrix, X and M are real $(m \times n)$-matrices, with $m \geq n$.

Then the density of the matrix $\Xi\Xi'$ is

$$(2\pi)^{-mn/2} c_{n,m} \int \exp\{-\operatorname{Tr}(\sqrt{Z_n}\Theta^{(n)} - M) R_n (\sqrt{Z_n}\Theta^{(n)} - M)'\}$$
$$\times \mu(d\Theta) \det R^{m/2} \det Z^{(m-n-1)/2}. \qquad (1.3.11)$$

The expression (1.3.11) is called a noncentral Wishart density.

Using the proof of Theorem 1.3.1, one can obtain a more profound result. Let us denote a set of nonnegative definite $(n \times n)$-matrices by K_2; K_3 is a set of real orthogonal $(m \times n)$-matrices, B_2, B_3 are σ-algebras of Borel sets in K_2 and K_3, respectively.

Theorem 1.3.2. *Let the conditions of Theorem 1.3.1 hold. Then a joint distribution of the matrices HH' and $(HH')^{-1/2}H$ is equal to*

$$\mathbf{P}\{HH' \in M_2, (HH')^{-1/2}H \in M_3\} = c_{n,m} \int_{Z_n \in M_1, \Theta^{(n)} \in M_2}$$
$$\times\, p(\sqrt{Z_n}\Theta^{(n)}) \det Z^{(m-n-1)/2} \mu(d\Theta)dZ_n,$$

where $M_2 \in B_2, M_3 \in B_3$.

If $p(Z_n^{1/2}\Theta^{(n)}) \equiv q(Z_n)$, then the matrices HH' and $(HH')^{-1/2}H$ are independent and have the distributions

$$\mathbf{P}\{HH' \in M_2\} = c_{n,m} \int_{Z \in M_2} q(Z_n) \det Z_n^{(m-n-1)/2} dZ_n,$$

$$\mathbf{P}\{(HH')^{-1/2}H \in M_3\} = \int_{\Theta^{(n)} \in M_3} \mu(d\Theta).$$

respectively.

The proof is the same as that of Theorem 1.3.1 but the integral $\int \tilde{f}(ZZ', (ZZ')^{-1/2}Z)p(Z)dZ$, where f is an arbitrary measurable function on $K_2 \times K_3$ such that the integral exists, has to be considered instead of that of (1.3.3) Similarly, other functions of random matrices can be found.

Theorem 1.3.3. *Let $\Xi_n = US$ be a real random $(m \times n)$-matrix with density $p(Xn)$, let U be an orthogonal, and S an upper triangular matrix with positive diagonal entries, and (G_1, B_1) a measurable group of matrices U. Then for all $E_1 \in B_1, E_2 \in B_2$,*

$$\mathbf{P}\{U \in E_1, S \in E_2\} = 2^n c_{n,n} \int_L p(HZ) \prod_j z_{jj}^{n-j} \mu(dH) \prod_{i \geq j} dZ_{ij},$$
$$L = \{Z \in E_2, H \in E_1\}, \tag{1.3.12}$$

where μ is the normalized Haar measure on (G_1, B_1).

If $p(HZ) \equiv p(Z)$, then the matrices U and S are independent, the matrix U is distributed with respect to the Haar measure, and the density of the matrix S is equal to $2^n c_{n,n} p(Z) \prod_{j=1}^n Z_{jj}^{n-j}, Z \in T$.

Proof. We show that any real matrix $A_n(\det A_n \neq 0)$ can be represented uniquely in the form of $A_n = HZ, H \in G_1, Z \in T$. By virtue of Lemma 1.3.1, $A = KB$, provided that $\det A \neq 0$, where $K \in G, B$ is a positive-definite matrix; and according to Lemma 1.3.2, $B^2 = SS', S' \in T$. Again, by using Lemma 1.3.1, we find $S = CH_1$, where C is a positive definite matrix, $H_1 \in G$. Hence, $B = C$, and $B = SH_1' = H_1S'$. Therefore, $A = KH_1S'$. The UH_1 is the orthogonal, and S' is the upper triangular matrix. We prove that this representation is unique. Assume on the contrary that there are two representations $A = U_1S_1, A = U_2S_2$.

Then $U_1'U_2 = S_1S_2^{-1}$. On the left side, there is an orthogonal matrix, and therefore $S_2S_2' = S_1S_1'$. Hence $S_1^{-1}S_2 = S_1'S_2^{-1}$. On the left side there is an upper triangular matrix, and on the right, there is a lower one. Therefore, $S_1^{-1}S_2 = \Lambda$, where Λ is a real diagonal matrix. Since the diagonal entries of the matrices S_1 and S_2 are positive, the diagonal entries of the matrix Λ will be positive, too. But it follows from $U_1 = U_2\Lambda$ that $\Lambda \equiv I$. Let us determine the Jacobian of the transformation $A = HZ$. Let f be a measurable function on $(G \times T)$ such that $\int f(H, Z)dX_n$ exists, where H and Z are found from $X = HZ$. In this integral, we start with the change of variables $X = UY$, where U is fixed. Then

$$\int f(H, Z)dX_n = \int f(\tilde{H}, \tilde{Z})dY_n, \qquad (1.3.13)$$

where we find the matrices $\tilde{H}$ and $\tilde{Z}$ from the equation $\tilde{H}\tilde{Z} = UY$. After the change of variables $Y = HZ$, (1.3.13) takes the form

$$\int f(H, Z)dX_n = \int f(UH, Z)J(H, Z)dZ\,dH,$$

where $J(H, Z)$ is the Jacobian of the transformation $Y = HZ$. By integrating this expression with respect to a normalized Haar measure, we find

$$\int f(H, Z)dX_n = \int f(\Theta, Z)\varphi(Z)dZ\,\mu(d\Theta),$$

where $\varphi(Z) = \int J(\Theta, Z)d\Theta$.

By now using the formula of the Wishart density as well as the Jacobian of the transformation $Z = SS'$, $S \in T$ (see Lemma 1.3.2), where Z is a positive-definite matrix and $S \in T$, we find

$$\int f(U, ZA)\varphi(ZA)dZ\,\mu(dU)\prod_{i=1}^{n} a_{ii}^i = c\int f(U, ZA)\varphi(Z)\mu(dU)dZ\prod_{i=1}^{n} a_{ii}^n,$$

$$c_{n,n}\int e^{-\operatorname{Tr}Z}\det Z^{-1/2}dZ = c\int e^{-\operatorname{Tr}SS'}\prod_{i=1}^{n} s_{ii}^{n-i}ds_{ii}, \qquad A \in T.$$

Hence $\varphi(S) = c_{n,n}2^n\prod_{j=1}^{n} S_{jj}^{n-j}$. This proves Theorem 1.3.3.

Similarly, we obtain the following assertion.

Theorem 1.3.4. *If Ξ_n is a positive-definite matrix with density $p(X)$ and $\Xi_n = SS'$, where S is an upper triangular matrix with nonnegative diagonal entries, the density of the matrix S is equal to $2^n p(ZZ')\prod_{j=1}^{n} Z_{jj}^{n+1-j}$, where Z is an upper triangular matrix with nonnegative diagonal entries.*

When proving Lemma 1.3.2, we make use of the fact that any real square matrix A, whose main minors are not equal to zero, can be represented in the form $A = SQ$, where S is the lower and Q is the upper triangular matrix. This representation will be unique if diagonal entries of any of the matrices S and Q are equal to fixed numbers. Of course, there are many other ways of choosing the unique representation $A = SQ$.

We introduce the following notations: T_1 is a group of upper triangular $n \times n$-matrices, T_2 is a group of lower triangular $n \times n$-matrices, T_3 is a group of triangular lower matrices with diagonal entries equal to 1, and M is a set of real $n \times n$-matrices.

Theorem 1.3.5. *Let $\Xi = SQ$ be a real random $n \times n$-matrix with the density $p(X), S = (s_{ij})$ and $Q = (q_{ij})$ be random matrices with the values in T_1 and T_2, respectively, and $q_{ii} = 1, i = \overline{1,n}$. Then a joint density of matrices S and Q entries is equal to $p(TK) \prod_{i=1}^{n} t_{ii}^{i-1}, T \in T_1, K \in T_3$.*

Proof. Let f be a measurable function on $T_1 \times T_3$, chosen in such a way that $\int f(T, K) p(X) dX$ exits; the matrices T and K are found from the equation $X = TK$. We denote by $J(T, K)$ the Jacobian of the transformation $X = TK$. By first changing the variables $X = TK$ and then $T = P_1 U, U, P_1 \in T, K = V P_2, V, P_2 \in T_3$, we obtain

$$\int f(P_1 U, V P_2) J(P_1 U, V P_2) \prod_{i=1}^{n} (P_1)_{ii}^{n+1-i} J(P_2) dU\, dV, \qquad (1.3.14)$$

where $J(P_2)$ is the Jacobian of the transformation $K = V P_2, dU = \prod_{i<j} du_{ij}$, $dV = \prod_{i<j} dv_{ij}$. Let us find the Jacobian of the transformation $J(P_2)$. The matrices K, V, and P_2 can be represented in the following form, $K = I + \tilde{K}, V = I + \tilde{V}, P_2 = I + \tilde{P}_2$, where the matrices $\tilde{K}, \tilde{V}$, and $\tilde{P}_2$ are obtained by replacing the diagonal entries with nulls. Then $\tilde{K} = \tilde{V} + \tilde{P}_2 + \tilde{V}\tilde{P}_2 = \tilde{P}_2 + \tilde{V}(I + \tilde{P}_2)$. This transformation can be made in two steps: first, by a shift on $\tilde{P}_2$, then $\tilde{K} = \tilde{V}(I + \tilde{P}_2)$. Obviously, the Jacobian of this transformation is 1. Thus, $J(P_2) = 1$. We make the change of the variables $X = P_1 Y P_2$ in the integrand expression $\int f(TK) p(X) dX$, then $Y = UV, Y \in M_1$, then (1.3.14) will be equal to

$$\int f(P_1 U, V P_2) J(U, V) dU\, dV |\det P_1|^n |\det P_2|^n.$$

By virtue of arbitrariness of the function f, we obtain

$$J(P_1 U, V P_2) \prod_{i=1}^{n} (P_1)_{ii}^{n+1-i} = J(U, V) \prod_{i=1}^{n} (P_1)_{ii}^{n}.$$

By setting $U = V = I$ in this equality, we find $J(P_1 P_2) = J(I, I) \prod_{i=1}^{n} (P_1)_{ii}^{i-1}$. By a simple calculation, we make sure that $J(I, I) = 1$. Thus, Theorem 1.3.5 is proved.

We shall now study the polar decomposition of complex random rectangular $m \times n$-matrices, $m \geq n$. Assume that the matrix Ξ has a joint distribution density of random entries ξ_{ij} equal to $p(X)$, where X is a complex matrix. Let L_1 be a set of complex $m \times n$-matrices, L_2 a set of Hermitian nonnegative definite $m \times n$-matrices, L_3 a set of unitary $m \times n$-matrices, and D_1 and D_2 the σ- algebras of Borel sets in L_2 and L_3, respectively. The polar decomposition of Ξ is a representation of Ξ in the form $\Xi = SU$, where $S = (\Xi\Xi^*)^{1/2}, U = (\Xi\Xi^*)^{-1/2}\Xi$.

Theorem 1.3.6. *Let Γ be the group of unitary $m \times m$-matrices and ν the normalized Haar measure on it, Ξ a random complex rectangular $m \times n$-matrix, $m \geq n$, with distribution density $p(X)$. Then the joint distribution of the matrices $\Xi\Xi^*$, and $(\Xi\Xi^*)^{1/2}\Xi$ is equal to*

$$\mathbf{P}\{\Xi\Xi^* \in M_1, (\Xi\Xi^*)^{-1/2}\Xi \in M_2\} = \tilde{c}_{n,m} \int_{Z_n \in M_1, H^{(n)} \in M_2}$$

$$\times \, p(\sqrt{Z_n} H^{(n)}) \det Z_n^{m-n-1} \nu(dH) dZ_n,$$

where

$$M_1 \in D_1, M_2 \in D_2, H = (h_{ij}) \in \Gamma, H^{(}n) = (h_{ij}), i = \overline{1,n},$$

$$j = \overline{1,m}, \tilde{c}_{n,m} = [\prod_{i=1}^{n} \Gamma(m - i + 1)\pi^{n(n-1)-nm} 2^{n^2 - n(n+1)/2}]^{-1}.$$

Proof. For any bounded and continuous function f of entries of the matrices $\Xi\Xi^*$ and $(\Xi\Xi^*)^{-1/2}\Xi$, we consider the integral $\int f(ZZ^*, (ZZ^*)^{-1/2}Z)p(Z)dZ$. In this integral, we introduce the change of the variables $Z = YH, Y \in L_1, H \in L_3$; the matrix H is fixed. The Jacobian J of such a change of variables is independent of the matrix Y and is equal to 1.

Then $\int f(ZZ^*, (ZZ^*)^{-1/2}Z)p(Z)dZ = \int f(YY^*, YY^*)^{-1/2}YH)p(YH)dY$. Using the proof of Theorems 1.3.1 and 1.3.2, we obtain

$$\int f(ZZ^*, (ZZ^*)^{-1/2}Z)p(Z)dZ = c \int_{L_2 \times L_3} f(X, H^{(n)})$$

$$\times \, p(\sqrt{X} H^{(n)}) \det X^{m-n-1} \nu(dH) dX.$$

Let us find the constant c. Let $f = 1, p(Z) = (2\pi)^{-nm} \exp(-(\mathrm{Tr}\, ZZ^*)/2)$. Then $c^{-1} = (2\pi)^{-nm} \int_{L_2} \det X^{m-n} \exp(-(\mathrm{Tr}\, X)/2)dX$. We make the change of the variables $X = SS^*$, where S is the complex upper triangular matrix with real nonnegative entries on the diagonal. The Jacobian of such a change of variables is $2^n \prod_{i=1}^{n} s_{ii}^{2n-2i+1}, s_{ii}$ are the entries of the matrix S (see the Proof of Theorem 1.3.4). As a result, we obtain

$$c^{-1} = 2^n (2\pi)^{-nm} \int \exp(-0.5 \sum_{i>j} |s_{ij}|^2) \prod_{i>j} d\,\mathrm{Re}\, s_{ij} d\,\mathrm{Im}\, s_{ij} \prod_{i=1}^{n} \int_{s_{ii}>0}$$

$$\times \, \exp(-0.5 s_{ii}^2) s_{ii}^{2m-2i+1} ds_{ii} = \prod_{i=1}^{n} \Gamma(m - i + 1)\pi^{n(n-1)-nm} 2^{n^2 - n(n+1)/2}.$$

Thus, Theorem 1.3.6 is proved.

Corollary 1.3.3. *If in addition to the conditions of Theorem 1.3.6*

$$p(Z_n^{1/2} H^{(n)}) \equiv q(Z_n), Z^n \in L_2, H^{(n)} \in L_3,$$

then the matrices $\Xi\Xi^*$ and $(\Xi\Xi^*)^{-1/2}\Xi$ are independent and have the distributions

$$\mathbf{P}\{\Xi\Xi^* \in M_1\} = \tilde{c}_{n,m} \int_{Z_n \in M_1} q(Z_n) \det Z_n^{m-n-1} dZ_n,$$

$$\mathbf{P}\{(\Xi\Xi^*)^{-1/2}\Xi \in M_2\} = \int_{H^{(n)} \in M_2} \nu(dH).$$

respectively.

§4 Integral Representations for Determinants

The characteristic function of the Wishart density $\omega(X_n, R_n, m)$ is equal to

$$\mathbf{M} \exp(i \operatorname{Tr} \Xi\Xi' T_n) = \det R_n^{-m/2} \det(R_n^{-1} - 2iT_n)^{-m/2},$$

where T is a matrix of parameters.

In this formula, under the fractional degree of complex numbers, we assume the principal value of complex numbers. By making use of the reversal formula for the characteristic function, we obtain the following integral representation for determinants,

$$(2\pi)^{-n(n+1)/2} \int \exp(-i \operatorname{Tr} B_n C_n) \det(A_n - iC_n)^{-k} dC_n = \exp(-\operatorname{Tr} B_n A_n)$$

$$\times \det B_n^{k-(n+1)/2} (2\sqrt{\pi})^{-n(n-1)/2} \prod_{m=1}^{n} \{\Gamma(k - (m-1)/2)\}^{-1}, \tag{1.4.1}$$

where $k > (n+1)/2$ is the real number, A_n and B_n are the positive-definite matrices, and C_n is a symmetric matrix.

The Wishart density can be defined in a manner different from that of the previous section: first, Eq. (1.4.1) has to be proved; then $\omega(X_n, R_n, m)$ has to be found by the reversal formula for the characteristic functions.

Since the integral from the density of the probability distribution is 1, by using the Wishart density, we obtain [10]

$$\int \exp(-\operatorname{Tr} A_n B_n) \det B_n^{q-(n+1)/2} dB_n = \pi^{n(n-1)/2} \prod_{i=0}^{n-1} \Gamma(q - i/2) \det A_n^{-q}, \tag{1.4.2}$$

where $q > (n-1)/2$, the matrices A_n and B_n are the same as in (1.4.1), and dB_n is the entry of the Lebesgue measure of a set of positive definite $n \times n$-matrices.

By virtue of orthogonal change of variables, the integral representations for determinants are also defined [10],

$$\pi^{n/2} \det A_n^{-1/2} = \int \exp[-(A_n \vec{x}, \vec{x})] \prod_{i=1}^{n} dx_i; \tag{1.4.3}$$

$$\int \exp[-((A_n - iC_n)\vec{x}, \vec{x})] \prod_{i=1}^{n} dx_i = \pi^{n/2} \det(A_n + iC_n)^{-1/2}; \tag{1.4.4}$$

$$\exp[0.5(A_n \vec{y}, \vec{y})] = (2\pi)^{-n/2} \det A_n^{-1/2} \int \exp\{\sum_{i=1}^{n} y_i x_i$$
$$- 0.5(A_n^{-1} \vec{x}, \vec{x})\} \prod_{i=1}^{n} dx_i. \tag{1.4.5}$$

Equations (1.4.1)–(1.4.5) are true for symmetrical matrices, but in most cases one needs to find the distributions of the determinants of nonsymmetric random matrices.

Theorem 1.4.1. *Let A be a real square $n \times n$- matrix. Then for any real $0 \leq t < 1$*

$$\det(I + \alpha_t A)^{-1} = \mathbf{E} \exp\{i\alpha_t((A - A')\vec{\xi}, \vec{\eta}) - \alpha_t(A\vec{\xi}, \vec{\xi}) - a_t(A\vec{\eta}, \vec{\eta})\}, \tag{1.4.6}$$

where $\alpha_t = t[1 + |\operatorname{Tr}(A + A')|/2 + \operatorname{Tr}(A + A')^2/4]^{-1}, \vec{\xi}$, and $\vec{\nu}$ are independent random vectors distributed according to normal laws $N(0, 0.5I)$.

Proof. The matrix $(I + \alpha_t(A + A')/2)$ is symmetric and positive definite. Therefore, there exists $(I + \alpha_t(A + A')/2)^{-1/2}$. We introduce the change of variables in (1.4.6), $\vec{x}_1 = (I + \alpha_t(A + A')/2)\vec{y}_1$, $\vec{x}_2 = (I + \alpha_t A + A'/2)\vec{y}_2$, $\vec{x}_1$, $\vec{y}_1, \vec{x}_2, \vec{y}_2 \in R_n$. Then

$$(2\pi)^{-n} \int \cdots \int \exp\{i\alpha_t((A - A')\vec{x}_1, \vec{x}_2) - 0.5\alpha_t((A + A')\vec{x}_1, \vec{x}_1)$$
$$- 0.5\alpha_t((A + A')\vec{x}_2, \vec{x}_2) - (\vec{x}_1, \vec{x}_1) - (\vec{x}_2, \vec{x}_2)\} \prod_{i=1}^{n} dx_{1i} dx_{2i}$$

$$= (2\pi)^{-n} \int \cdots \int \exp\{i\alpha_t(R(A - A')R\vec{y}_1, \vec{y}_2) - (\vec{y}_1, \vec{y}_1)$$
$$- (\vec{y}_2, \vec{y}_2)\} \prod_{i=1}^{n} dy_{1i} dy_{2i} \det(I + 0.5\alpha_t(A + A'))^{-1}, \tag{1.4.7}$$

where $R = (I + 0.5\alpha_t(A + A'))^{-1/2}$.

By integrating (1.4.7) over the variables y_{2i}, $i = \overline{1,n}$, we obtain

$$(2\pi)^{-n/2} \int \cdots \int \exp\{-\alpha_t^2[(R(A - A')R)(R(A - A')R)'\vec{y}_1, \vec{y}_1]/4 - (\vec{y}_1, \vec{y}_1)\}$$

$$\times \prod_{i=1}^{n} dy_{1i} \det(I + 0.5\alpha_t(A + A'))^{-1} = \det(I + 0.5\alpha_t(A + A'))^{-1}$$

$$\times \det[I + 0.25\alpha_t^2(R(A - A')R)(R(A - A')R)']^{-1/2}.$$

Denote $K = R(A - A')R/2$. The matrix K is antisymmetric. Therefore, $\det(I + KK') = \det(I + K)^2$. Making use of this, we obtain

$$\det(I + 0.5\alpha_t(A + A'))^{-1} \det(I + 0.5\alpha_t R(A - A')R)^{-1} = \det[I + 0.5\alpha_t$$
$$\times (A + A') + 0.5\alpha_t(A - A')]^{-1} = \det(I + \alpha_t A)^{-1}.$$

Theorem 1.4.1 is proved.

Another factor can be chosen as α_t, for example, $\alpha_t = \min_{i=\overline{1,n}} t(1 + |\lambda_i|)^{-1}$, where λ_i are the eigenvalues of the matrix $(A + A')/2$. However, in order to prove the limit theorems for random determinants, we need the distribution properties of the eigenvalues of random matrices, the study of which is a difficult task. It is rather simple to prove limit theorems for the traces of degrees of random matrices, because the integral representations also hold for them,

$$\mathrm{Tr}(A + A')^k = \frac{\partial^k}{\partial t^k} \ln \det(I + it(A + A'))i^{-k}(-1)^{k-1}(k!)^{-1}|_{t=0},$$

where $\det(I + it(A + A'))$ is defined by Eq. (1.4.4).

In the following chapters, in order to prove the limit theorems for random determinants, we shall need the following auxiliary assertion.

Lemma 1.4.1. *For any real square $n \times n$-matrix A and all $0 \leq t < 1 \det(I + \alpha_t A) \geq \exp(-t/(1 - t))$.*

Proof. Since

$$\det(I + \alpha_t A) = \det(I + 0.5\alpha_t(A + A')) \det(I + KK')^{1/2}$$
$$\geq \det(I + 0.5\alpha_t(A + A')),$$

then

$$\det(I + \alpha_t A) \geq \prod_{k=1}^{n}[1 + \lambda_k t(1 + |\sum_{p=1}^{n}\lambda_p| + \sum_{p=1}^{n}\lambda_p^2)^{-1}],$$

where λ_k, $k = \overline{1,n}$ are the eigenvalues of the matrix $(A + A')/2$.

Let us introduce the notation $\theta_k = \lambda_k t[I + |\sum_{k=1}^{n}\lambda_k| + \sum_{k=1}^{n}\lambda_k^2]^{-1}$. Obviously, for any integer $s > 0|\sum_{k=1}^{n}\theta_k^s| \leq t^s$. Therefore, $\ln \det(I + \alpha_t A) \geq -\sum_{s=1}^{\infty} s^{-1}|\sum_{k=1}^{n}\theta_k^s| \geq t(1 - t)$. From this follows the assertion of Lemma 1.4.1.

§5 Integration on Grassmann and Clifford Algebras

An algebra whose generatrices $x_i, x_i^*, i = \overline{1, n}$ satisfy the following conditions,

$$x_i x_j + x_j x_i = 0, \quad x_i^* x_j^* + x_j^* x_i^* = 0, \quad x_i x_j^* + x_j^* x_i = 0. \tag{1.5.1}$$

is called the Grassmann algebra with $2n$ generatrices. In particular, from the relations (1.5.1), it follows that $x_i^2 = 0, (x_i^*)^2 = 0, i = \overline{1, n}$. We denote by Γ_{2n} the Grassmann algebra with $2n$ generatrices. The basis monomials

$$1, x_1, \ldots, x_n, x_1^*, \ldots, x_n^*, x_1 x_2, \ldots, x_{n-1} x_n, x_1^* x_2, \ldots, x_{n-1}^* x_n^*, \ldots, x_1^*, \ldots, x_n^*$$

can be singled out in it. Any entry of the algebra Γ_{2n} is a polynomial of the form,

$$\sum_{c_{k_1 \ldots k_n p_1 \ldots p_n}} x_1^{k_1} \ldots x_n^{k_n} (x_1^*)^{p_1} \ldots (x_n^*)^{p_n}, \tag{1.5.2}$$

where $c_{k_1 \ldots k_n p_1 \ldots p_n}$ are complex values.

From the correlations (1.5.1) it follows that the generatrices in (1.5.2) do not have degrees beyond the first one. Any entry of the algebra can be reduced to the form (1.5.2) by means of the correlations (1.5.1).

In general, the representation of any entry of the algebra by way of generatrices is ambiguous. The uniqueness can be achieved in different ways, for example, by requiring that the entry coefficient is skew-symmetric or that the entry is always of the form (1.5.2).

We denote the algebra whose generatrices $k_1, \ldots, k_n$ satisfy the correlations $k_i k_j + k_j k_i = 0, i \neq j, k_i^2 = 1, i, j = \overline{1, n}$ by the Clifford algebra K_n with generatrices n.

Let us introduce dx_i, dx_i^* which satisfy the conditions $dx_i dx_j + dx_j dx_i = 0$, $dx_i^* dx_j^* + dx_j^* dx_i^* = 0, dx_i^* dx_j + dx_j dx_i^* = 0, x_i dx_j + dx_j x_i = 0, x_i^* dx_j^* + dx_j^* d_i^* = 0, x_i dx_j^* + dx_j^* x_i = 0, x_i^* dx_j^* + dx_j^* x_i = 0$.

We define the single integrals $\int dx_i = 0, \int dx_i^* = 0, \int x_i^* dx_i^* = 1, \int x_i dx_i = 1$. By iterated integrals we mean multiple ones. The integral introduced on Γ_{2n} satisfies the linear property

$$\int (c_1 f_1 + c_2 f_2) dx_i^* = c_1 \int f_1 dx_i^* + c_2 \int f_2 dx_i^*,$$

$$\int (c_1 f_1 + c_2 f_2) dx_i = c_1 \int f_1 dx_i + c_2 \int f_2 dx_i, \quad i = \overline{1, n},$$

where c_1, c_2 are complex values, and $f_1, f_2 \in \Gamma_{2n}$.

Obviously, the integral of the entry (1.5.2) over generatrices $dx_i, dx_i^*, i = \overline{1, n}$ is equal to $c_1 \ldots_1$. Using an integral conception on the Grassmann algebra, we shall prove the following integral representations for the determinants [14].

Let A be a square complex $n \times n$–matrix, then

$$\int \exp(A\vec{x}, \vec{x}^*) d\vec{x} d\vec{x}^* = \det A; \qquad (1.5.3)$$

$$\int \exp[(A\vec{x}, \vec{x}^*) + (\vec{x}, \vec{\eta}^*) + (\vec{x}^*, \vec{\eta})] d\vec{x}^* d\vec{x}/$$

$$\times \int \exp[(A\vec{x}, \vec{x}^*)] d\vec{x}^* d\vec{x} = \exp(A^{-1}\vec{\eta}^*, \vec{\eta}), \qquad (1.5.4)$$

where $\vec{x} = (x_1, \ldots, x_n)$, $\vec{x}^* = (x_1^*, \ldots, x_n^*)$, $d\vec{x} = \prod_{i=1}^n dx_i$, $x_i, x_i^*, \nu_i, \nu_i^*$ are the generatrices of the Grassmann algebra Γ_{4n}, and the matrix A^{-1} exists (by exponential function we mean its expansion in series).

Obviously,

$$\int \exp(A\vec{x}, \vec{x}^*) d\vec{x}^* d\vec{x} = (n!)^{-1} \int (A\vec{x}, \vec{x}^*)^n d\vec{x} d\vec{x}^*$$

$$= (n!)^{-1} \int [\sum a_{i_1 j_1} a_{i_2 j_2} \ldots a_{i_n j_n} x_{i_1} \ldots x_{i_n} x_{j_1}^* \ldots x_{j_n}^*] d_x^* d_x,$$

where a sum is taken for all possible permutations $\langle i_1, \ldots, i_n \rangle, \langle j_1, \ldots, j_n \rangle$ of the numbers $1, 2, \ldots n$. Now, taking into account that $\int x_{i_1} \ldots x_{i_n} d\vec{x} = (-1)^m, \int x_{i_1}^* \ldots x_{i_n}^* d\vec{x}^* = (-1)^m$, where m is the number of transpositions in the substitution

$$\begin{pmatrix} 1 & 2 & \ldots & n \\ i_1 & i_2 & \ldots & i_n \end{pmatrix},$$

we arrive at Eq. (1.5.3). By using the correlations

$$\int \exp(A(\vec{x}+\vec{y}), (\vec{x}^* + \vec{y}^*)) d\vec{x}^* d\vec{x} = \int \exp(A\vec{x}, \vec{x}^*) d\vec{x}^* d\vec{x},$$

$$\int \exp(A(\vec{x}+\vec{y}), (\vec{x}^* + \vec{y}^*)) d\vec{x}^* d\vec{x}$$

$$= \exp(A\vec{y}, \vec{y}^*) \int \exp[(A\vec{x}, \vec{x}^*) + (A\vec{y}, \vec{x}^*) + (A\vec{x}, \vec{y}^*)] d\vec{x}^* d\vec{x},$$

where the elements of the vector y^* are the polynomials of the elements of the vectors $\vec{\nu}$ and $\vec{\nu}^*$, and by setting $\vec{y} = A^{-1}\vec{\nu}, y^* = (A')^{-1}\vec{\nu}^*$, we obtain (1.5.4). Note that for some matrices A, the formula (1.5.3) can be written in the simpler form. Let B be antisymmetric complex $n \times n$–matrices. We prove that

$$\int \exp(B\vec{x}, \vec{x}) d\vec{x} = \sqrt{\det 2B}, \qquad (1.5.5)$$

where by the square root is meant its primary value.

Let B be real antisymmetric matrices. They can be represented in the form $B = T\Lambda T'$, where T is a real orthogonal matrix, and

$$\Lambda = \mathrm{diag}\left\{ \begin{pmatrix} 0 & \lambda_i \\ -\lambda_i & 0 \end{pmatrix}, \quad i = \overline{1, k}, \quad 0, \ldots, 0 \right\},$$

where k is an integer smaller than $n + 1$.

It is easy to check that $\prod_{i=1}^{n} y_i = \det T \prod_{i=1}^{n} x_i$, where $\vec{y} = T\vec{x}$. By virtue of this,

$$\int \exp(B\vec{x}, \vec{x})d\vec{x} = \int \exp[2(\lambda_1 y_1 y_2 + \lambda_2 y_3 y_4 + \cdots + \lambda_k y_{2k-1} y_{2k}) \prod_{i=1}^{n} d_{x_i}$$

$$= \begin{cases} 2^{n/2}\lambda_1 \ldots \lambda_n, & \text{if } n \text{ is even and } k = n/2, \\ 0, & \text{if } n \text{ is odd} \end{cases} = \sqrt{\det 2B}.$$

The expression $(\det B)^{1/2}$ is called the Pfaffian of matrix B. For the complex matrix B, the Pfaffian is a certain polynomial of its entries. The integral $\int \exp(B\vec{x}, \vec{x})d\vec{x}$ is also a certain polynomial of complex matrix B entries. Since these two polynomials coincide under real values of arguments, it follows that they will coincide under the complex values of arguments as well. Equation (1.5.5) is proved.

Note that we can consider another algebra, whose generatrices $x_i, x_i^*, i = \overline{1, n}$ satisfy the correlations $x_i x_j = x_j x_i, x_i^* x_j^* = x_j^*, x_i^*, x_i x_j = x_j x_i, i \neq j, i, j = \overline{1, n}$. Then for any square complex matrix A, we have

$$\int \exp(A\vec{x}, \vec{x}^*)d\vec{x}^* d\vec{x} = \text{per } A, \tag{1.5.6}$$

where the permanent of the matrix A is defined as follows: $\text{per } A = \sum a_{1i_1} a_{2i_2} \ldots a_{ni_n}$, where the sum is taken by all permutations $\langle i_1, \ldots, i_n \rangle$ of the numbers $1, 2, \ldots n$.

From Eq. (1.5.5), an expression for the Pfaffian of the antisymmetric matrix B can be obtained. Obviously, for the even n,

$$\int \exp(B\vec{x}, \vec{x})d\vec{x} = ((n/2)!)^{-1} \int \left(\sum_{i,j=1}^{n} b_{ij} x_i x_j \right)^{n/2} d\vec{x}$$

$$= ((n/2)!)^{-1} \int \sum b_{i_1 i_{n/2+1}} \ldots b_{i_{n/2} i_n} x_{i_1} \ldots x_{i_{n/2}} x_{i_{1+n/2}} \ldots x_{i_n} d\vec{x},$$

where the sum is taken over all permutations $\langle i_1, \ldots, i_n \rangle$ of the numbers $1, 2, \ldots n$. Hence, we obtain

$$\text{pfaff } B = 2^{-n}((n/2)!)^{-1} \sum (-1)^m b_{i_1 i_{1+n/2}} \ldots b_{i_{n/2} i_n} = ((n/2)!)^{-1}$$

$$\times \sum (-1)^m b_{i_1 i_2} b_{i_3 i_4} \ldots b_{i_{n-1} i_n}, \tag{1.5.7}$$

where m is the number of permutations in the substitution

$$\begin{pmatrix} 1 & 2 & \ldots & n \\ i_1 & i_2 & \ldots & i_n \end{pmatrix}$$

and the sum is taken over all permutations $\langle i_1, \ldots, i_n \rangle, i_1 > i_2, \ldots, i_{n-1} < i_n$.

Similarly, a concept of integration by the Clifford algebra can be introduced.

CHAPTER 2

MOMENTS OF RANDOM MATRIX DETERMINANTS

Random determinant distributions have a cumbersome form; therefore, it is of interest to find their moments.

§1 Moments of Random Gram Matrix Determinants

Let G be the group of m-dimensional real orthogonal matrices, and μ the normalized Haar measure on it.

Theorem 2.1.1. *If a random real $m \times n(m \geq n)$ matrix has the probability density $p(X)$ and $\mathbf{E}(det \Xi\Xi')^k$ exists, where k is a nonnegative integer, then*

$$\mathbf{E}(\det \Xi\Xi')^k = (c_{n,m}/c_{n,m+2k}) \int p(\sqrt{ZZ'}H^{(n)})\mu(dH)dZ, \qquad (2.1.1)$$

where Z is a real $(m+2k) \times n$ matrix, $H^{(n)} = (h_{ij})$, $i = \overline{1,n}$, $j = \overline{1,m}$, and the quantities $c_{n,m}$ are defined in Theorem 1.9.1.

Proof. By using a generalized Wishart density, we find that

$$\mathbf{E}(\det \Xi\Xi')^k = c_{n,m} \int p(\sqrt{Z_n}h^{(n)}) \det Z_n^{(m+2k-n-1)/2}\mu(dH)dZ_n,$$

where Z_n is a nonnegative-definite matrix of order n.

By using the integral

$$J_\varepsilon = c_{n,m} \int q(\sqrt{Z_n}) \det Z_n^{(m+2k-n-1)/2} \exp(-0.5\varepsilon \operatorname{Tr} Z_n)dZ_n,$$

where $\varepsilon > 0$ is a constant, we obtain $q(\sqrt{Z_n}) = \int p(\sqrt{Z_n}H^{(n)})\mu(dH)$. Obviously,

$$\lim_{\varepsilon\downarrow 0} J_\varepsilon = \mathbf{E}(\det \Xi\Xi')^k. \qquad (2.1.2)$$

22

Let Q^ε be a random $(m + 2k) \times n$ matrix with the probability density $\varepsilon^{(m+2k)n/2}(2\pi)^{-(m+2k)n/2}\exp(-0.5\varepsilon \operatorname{Tr} ZZ')$. Using Theorem 1.3.1 and Eq. (2.1.2),

$$
\int p(\sqrt{ZZ'}H^{(n)})\mu(dH)dZ = \lim_{\varepsilon\downarrow 0}(\varepsilon/2\pi)^{-(m+2k)n/2}\mathbf{E}\int p(\sqrt{Q^{(\varepsilon)}Q'^{(\varepsilon)}}H^{(n)})
$$

$$
\times\ \mu(dH) = \lim_{\varepsilon\downarrow 0}(\varepsilon/2\pi)^{-(m+2k)n/2}\int p(\sqrt{Z_n}H^{(n)})\exp(-0.5\varepsilon \operatorname{Tr} Z_n)
$$

$$
\times\ \det Z_n^{(m+2k-n-1)/2}dZ_n\mu(dH)c_{n,m+2k}(\varepsilon/2\pi)^{(m+2k)n/2}
$$

$$
= \lim_{\varepsilon\downarrow 0}(c_{n,m+2k}/c_{n,m})J_\varepsilon = (c_{n,m+2k}/c_{n,m})\mathbf{E}(\det \Xi\Xi')^k.
$$

The theorem is proved.

Corollary 2.1.1. *If, in addition to the conditions of Theorem 2.1.1, $p(X) \equiv \tilde{p}(XX')$ for all $m \times n$ matrices X, then*

$$
\mathbf{E}(\det \Xi\Xi')^k = (c_{n,m}/c_{n,m+2k})\int \tilde{p}(ZZ')dZ.
$$

We have the proof from Eq. (2.1.1).

An example of a random matrix satisfying the condition $p(X) \equiv \tilde{p}(XX')$ is a matrix with density $p(X) = (2\pi)^{-mn/2} \times \exp(-0.5 \operatorname{Tr} R_n^{-1}XX')\det R_n^{-m/2}$, where R_n is a positive-definite matrix.

In this case,

$$
\int p(Z)dZ = (2\pi)^{-nm/2}\int \det R_n^{-m/2}\exp(-0.5 \operatorname{Tr} R_n^{-1}ZZ')dZ
$$

$$
= (2\pi)^{kn}\det R_n^k.
$$

Corollary 2.1.2. *[3] If the matrix W has the Wishart density $\omega(Z_n, R_n, m)$, then*

$$
\mathbf{E}(\det W)^k = 2^{nk}\prod_{i=1}^{n}\Gamma((m + 2k + 1 - i)/2)\left[\prod_{i=1}^{n}\Gamma((m + 1 - i)/2)\right]^{-1}\det R_n^k.
$$

$$
(2.1.3)
$$

Another example of the matrix Ξ density satisfying Corollary 2.1.1 is: $p(X) = \Gamma((nm)/2 + 1)\pi^{-(nm)/2}, \operatorname{Tr} XX' \leq 1$. In this case, the kth order moments for the random variables $\det \Xi\Xi'$ are equal to

$$
c_{n,m}\Gamma((nm/2 + 1)\pi^{nk}[c_{n,m+2k}\Gamma(1 + m(n + 2k)/2)].
$$

Therefore, all moments of Gram determinants of certain classes of random matrices exist in obvious form.

It is easy to obtain that in Corollary 2.1.2 the random variable $\det W$ has such moments as the random variable $\det R_n \prod_{i=1}^n X_{m+1-i}^2$, where $X_i^2, i = 1, 2, \ldots$ are independent random variable and have X^2 distribution with i degrees of freedom.

Let μ_k be the moments of a random variable $\det W$.

It is easy to obtain that the Carleman sum $\sum_{k=1}^\infty \mu_k^{-1/k}$ for moments μ_k is not convergent.

Consequently the distribution function which has moments μ_k is single. We can assert this fact with the help of another approach called the method of random orthogonal transformations.

Theorem 2.1.2. *Let Ξ be a random real $m \times n(m \geq n)$ matrix whose row vectors $\vec{\xi} = (\xi_{i_1}, \ldots, \xi_{i_m}), i = \overline{1, n}$ are independent and whose distribution functions $F_i(\vec{x})$ are satisfying the condition that for every real orthogonal matrix H_k of kth order,*

$$F_i(x_1, \ldots, x_k, H_{n-k}\vec{\tilde{x}}_{n-k}) \equiv F_i(x_1, \ldots, x_k, \vec{\tilde{x}}_{n-k}), \qquad (2.1.4)$$
$$x_i \in R_1, \quad i = \overline{1, n}, \quad k = \overline{0, n-1},$$

where $\vec{\tilde{x}}_{n-k} = (x_{k+1}, \ldots, x_n)$.
    Then

$$\det \Xi\Xi' \approx \prod_{i=1}^n \left(\sum_{j=i}^n \xi_{ij}^2 \right). \qquad (2.1.5)$$

Proof. Let H be a random real orthogonal matrix of order m, the first column vector of which is collinear to the first row vector of the matrix Ξ, and the other column vectors of the matrix H are measurable with respect to the minimal σ- algebra generated by the vector $\vec{\xi}_1$. By using (2.1.4), we have

$$\det \Xi\Xi' = \det \Xi H H' \Xi' \approx \det \begin{bmatrix} \vec{\nu}_m \\ \Xi_{m \times (n-1)} \end{bmatrix} \begin{bmatrix} \vec{\nu}_m \\ \Xi_{m \times (n-1)} \end{bmatrix}, \qquad (2.1.6)$$

where the matrix $\Xi_{m \times (n-1)}$ is obtained from Ξ by deleting the first row vector, and $\vec{\nu}_m = ([\sum_{i=1}^m \xi_{1i}^2]^{1/2}, 0, \ldots, 0)$ is an m-dimensional row vector.

Complete the matrix

$$\begin{bmatrix} \vec{\nu}_m \\ \Xi_{m \times (n-1)} \end{bmatrix}$$

by a random matrix C, so that the new matrix D_m has $m \times m$ order; and the elements of a random matrix C are satisfying the conditions; the elements of the first column vector are only zeros, the row vectors are orthogonal to the row vectors of the matrix $\Xi_{m \times (n-1)}$, and $\det CC' \neq 0$.

Obviously, such a matrix C always exists. Using the properties of matrix C, we have

$$\det D_m^2 = \det \begin{bmatrix} \vec{\nu}_m \\ \Xi_{m \times (n-1)} \end{bmatrix} \begin{bmatrix} \vec{\nu}_m \\ \Xi_{m \times (n-1)} \end{bmatrix} \det CC'$$

$$= \det \Xi_{m \times (n-1)} \Xi'_{m \times (n-1)} \sum_{i=1}^{m} \xi_{1i}^2 \det CC'.$$

Using this and (2.1.6), we find

$$\det \Xi_{m \times n} \Xi'_{m \times n} \approx \det \Xi_{(m-1) \times (n-1)} \Xi'_{(m-1) \times (n-1)} \sum_{i=1}^{m} \xi_{1i}^2.$$

Repeating this operation n times, we have (2.1.5), and Theorem 2.1.2 is proved.

The following examples of the functions satisfy the condition (2.1.4):

$$F(x_1, \ldots, x_n) = c \int_{(\vec{y},\vec{y}) \le 1, y_i < x_i} d\vec{y}, \quad c = \pi^{-n/2} \Gamma(1 + n/2),$$

$$F(x_1, \ldots, x_n) = \pi^{-n/2} \int_{y_i < x_i, i = \overline{1,n}} \exp(-\vec{y}, \vec{y}) d\vec{y}.$$

Corollary 2.1.3. *[3] If under the conditions of Theorem 2.1.2 $m = n$, then*

$$\det \Xi \approx \prod_{i=1}^{n-1} \left(\sum_{j=i}^{n} \xi_{ij}^2 \right)^{1/2} \xi_{nn}.$$

To prove Corollary 2.1.3, we use at every step the orthogonal real matrix T, whose first column vector is orthogonal to the first row vector $\vec{\xi}_1$ of the matrix Ξ; and the other volume vectors are measurable with respect to the minimal σ-algebra of events generated by the vector $\vec{\xi}_1$.

Theorem 2.1.3. *Let $\Xi_{m,n} = (\xi_{ij})$ and $\theta_{k,n} = (\theta_{ij})$ be independent random $m \times n$ and $k \times n (m \ge n, k \ge n)$ matrices, satisfying the conditions of Theorem 2.1.2 and the additional condition that the distribution of the random vectors $(\xi_{ji}, \theta_{jl}, l = \overline{i, n+i-1}), j = \overline{1,n}$ with fixed $\xi_{jk}, \theta_{js}, k \ne i, s = \overline{i, n+i-1}$ is invariant according to orthogonal transformation. Then*

$$(\det \Xi_{m \times n} \Xi'_{m,n}, \det[\Xi_{m,n} \Xi'_{m,n} + \theta_{k,n} \theta'_{k,n}])$$

$$\approx \left(\prod_{i=1}^{n} \left(\sum_{j=i}^{m} \xi_{ij}^2 \right), \prod_{i=1}^{n} \left(\sum_{j=i}^{m} \xi_{ij}^2 + \sum_{j=1}^{k} \theta_{ij}^2 \right) \right).$$

Proof. By using the proof of Theorem 2.1.2, we have

$$(\det \Xi_{m,n}\Xi'_{m,n}, \det[\Xi_{m,n}\Xi'_{m,n} + \theta_{k,n}\theta'_{k,n}])$$

$$\approx \left(\det \begin{bmatrix} \vec{\xi}_m \\ \Xi_{m\times(n-1)} \end{bmatrix}\begin{bmatrix} \vec{\xi}_m \\ \Xi_{m\times(n-1)} \end{bmatrix}', \det\left\{\begin{bmatrix} \vec{\xi}_m \\ \Xi_{m\times(n-1)} \end{bmatrix}\right.\right.$$

$$\times \left.\left.\begin{bmatrix} \vec{\xi}_m \\ \Xi_{m\times(n-1)} \end{bmatrix}' + \begin{bmatrix} \vec{\theta}_k \\ \theta_{k\times(n-1)} \end{bmatrix}\begin{bmatrix} \vec{\theta}_k \\ \theta_{k\times(n-1)} \end{bmatrix}'\right\}\right), \qquad (2.1.7)$$

where $\vec{\xi}_m = ([\sum_{i=1}^m \xi_{1i}^2]^{1/2}, 0, \ldots, 0)$, $\vec{\theta}_k = ([\sum_{i=1}^k \theta_{1i}^2]^{1/2}, 0, \ldots, 0)$ are m and k-dimensional vectors; and we represent the matrix

$$\begin{bmatrix} \vec{\xi}_m \\ \Xi_{m\times(n-1)} \end{bmatrix}\begin{bmatrix} \vec{\xi}_m \\ \Xi_{m\times(n-1)} \end{bmatrix}' + \begin{bmatrix} \vec{\theta}_k \\ \theta_{k\times(n-1)} \end{bmatrix}\begin{bmatrix} \vec{\theta}_k \\ \theta_{k\times(n-1)} \end{bmatrix}'$$

in the following form

$$\begin{bmatrix} \vec{a}_1 \\ \vec{a}_2 \\ \cdot \\ \cdot \\ \cdot \\ \vec{a}_n \end{bmatrix}\begin{bmatrix} \vec{a}_1 \\ \vec{a}_2 \\ \cdot \\ \cdot \\ \cdot \\ \vec{a}_n \end{bmatrix} + \begin{bmatrix} \vec{0}_{m-1} \\ \Xi_{(m-1)\times(n-1)} \end{bmatrix}\begin{bmatrix} \vec{0}_{m-1} \\ \Xi_{(m-1)\times(n-1)} \end{bmatrix}'$$

$$+ CC', \quad C = \begin{bmatrix} 0_{k-n+1} \\ \theta_{(k-n+1)\times(n-1)} \end{bmatrix},$$

where $\vec{a}_1 = ([\sum_{i=1}^m \xi_{1i}^2]^{1/2}, [\sum_{i=1}^n \theta_{1i}^2]^{1/2}, 0, \ldots, 0)$; $a_k = (\xi_{k1}, \theta_{k1}, \ldots, \theta_{kn-1})$, $k = 2, n$ are n-dimensional row vectors; $\vec{0}_{m-1}, \vec{0}_{k-n+1}$ are $m-1$ and $k-n+1$-dimensional vectors which have only zero elements; matrix $\Xi_{(m-1)\times(n-1)}$ is obtained from the matrix $\Xi_{m\times n}$ by deleting the first column vector; and the matrix $\theta_{(k-n+1)\times(n-1)}$ is obtained from the matrix $\theta_{k\times(n-1)}$ by deleting the first $n-1$ column vectors.

Let H_n be a real orthogonal matrix whose first column vector equals $\vec{a}_1$, and all the rest are measurable with respect to the minimal σ-algebra of events generated by the vector $\vec{a}_1$. Then (see the theorem conditions),

$$\begin{bmatrix} \vec{a}_1 \\ \vec{a}_2 \\ \cdot \\ \cdot \\ \cdot \\ \vec{a}_n \end{bmatrix} H_n H'_n \begin{bmatrix} \vec{a}_1 \\ \vec{a}_2 \\ \cdot \\ \cdot \\ \cdot \\ \vec{a}_n \end{bmatrix} \approx \begin{bmatrix} \vec{b}_n \\ \vec{a}_2 \\ \cdot \\ \cdot \\ \cdot \\ \vec{a}_n \end{bmatrix}, \qquad (2.1.8)$$

where $\vec{b}_n = ([\sum_{i=1}^m \xi_{1i}^2 + \sum_{i=1}^n \theta_{1i}^2]^{1/2}, 0, \ldots, 0)$ is an n-dimensional vector.

From Theorem 2.1.2, we find that

$$\det \begin{bmatrix} \vec{\xi}_m \\ \Xi_{m \times (n-1)} \end{bmatrix} \begin{bmatrix} \vec{\xi}_m \\ \Xi_{m \times (n-1)} \end{bmatrix}'$$

does not depend on first column vector elements of the matrix $\Xi_{m \times (n-1)}$. Using this and (2.1.8), we have from (2.1.7) and with the help of Theorem 2.1.2,

$$\{\det \Xi_{m,n} \Xi'_{m,n}, \det[\Xi_{m,n} \Xi'_{m,n} + \theta_{k,n} \theta'_{k,n}]\}$$

$$\approx \left\{ \sum_{i=1}^{m} \xi_{1i}^2 \det \Xi_{(m-1) \times (n-1)} \Xi'_{(m-1) \times (n-1)}, \right.$$

$$\det \left(\begin{bmatrix} \vec{b}_m \\ \Xi_{m \times (n-1)} \end{bmatrix} \begin{bmatrix} \vec{b}_m \\ \Xi_{m \times (n-1)} \end{bmatrix}' + \begin{bmatrix} \vec{0}_k \\ \theta_{k \times (n-1)} \end{bmatrix} \begin{bmatrix} \vec{0}_k \\ \theta_{k \times (n-1)} \end{bmatrix}' \right) \right\}$$

$$\approx \left\{ \sum_{i=1}^{m} \xi_{1i}^2 \det \Xi_{(m-1) \times (n-1)} \Xi'_{(m-1) \times (n-1)}, \right.$$

$$\left. \left(\sum_{i=1}^{m} \xi_{1i}^2 + \sum_{i=1}^{k} \theta_{1i}^2 \right) \det[\Xi_{(m-1) \times (n-1)} \Xi'_{(m-1) \times (n-1)} + \theta_{k \times (n-1)} \theta'_{k \times (n-1)}] \right\}.$$

By repeating this process n times, we obtain the statement of Theorem 2.1.3.

Corollary 2.1.4. *[102] If the elements of the matrices $\Xi_{m,n}$ and $\theta_{k,n}$ are independent and distributed according to the standard normal law $N(0,1)$, then*

$$\det \Xi_{m,n} \Xi'_{m,n} \det[\Xi_{m,n} \Xi'_{m,n} + \theta_{k,n} \theta'_{k,n}]^{-1} \approx \prod_{i=1}^{n} \eta_i,$$

where

$$\eta_i = \sum_{j=1}^{m} \xi_{ij}^2 [\sum_{j=1}^{m} \xi_{ij}^2 + \sum_{j=1}^{k} \theta_{ij}^2]^{-1}, i = \overline{1,n}.$$

The random variables $\eta_i, i = \overline{1,n}$ are independent and have Beta densities:

$$B\left(x, \frac{(m+1-i)}{2}, \frac{k}{2}\right) = x^{(m+1-i)/2} \times (1-x)^{(k/2)-1} \left[B\left(\frac{m+1-i}{2}, \frac{k}{2}\right) \right]^{-1}$$

where $B(a,b)$ is the Beta function.
The moments of the random variables η_i are equal to

$$\mathbf{E}\eta_i^s = \prod_{i=1}^{n} \{ \Gamma[(m+1-i)/2 + s] \Gamma[(m+k+1-i)/2] (\Gamma[(m+1-i)/2]$$

$$\times \Gamma[(m+k+1-i)/2 + s])^{-1} \}, \quad i = \overline{1,n}, \quad s = 0,1,2,\ldots$$

Let us find the moments of some random complex matrices determinants. Let $\vec{\xi} = (\vec{\nu} + i\vec{\mu})$ be a complex Haussian vector with the mean $\vec{m} = \vec{a} + i\vec{b}$ and the Hermitian covariance matrix R, if for any real vectors $\vec{s}$ and $\vec{q}$,

$$\mathbf{E} \exp\{i(\vec{\nu}, \vec{s}) + i(\vec{\mu}, \vec{q})\} = \exp\{-0.5(R(\vec{s} + i\vec{q}), (\vec{s} - i\vec{q})) + i(\vec{a}, \vec{s}) + i(\vec{b}, \vec{q})\}.$$

If we use unitary transformations as in Theorems 2.1.2 and 2.1.3 instead of orthogonal transformations we obtain the following results.

Theorem 2.1.4. *If the row vectors of the random complex $m \times n$ matrix $\Xi = (\xi_{ij})$ are independent and distributed according to the Haussian laws with parameters $\vec{0}, I$, then $\det \Xi\Xi^* \approx \prod_{i=1}^{n}(\vec{\xi}_i, \vec{\xi}_i)$, where $\vec{\xi}_i = (\xi_{ij}, j = \overline{2(i-1), m})$.*

Theorem 2.1.5. *[102] If the complex random $m \times n(m \geq n)$ and $k \times n(k \geq n)$ matrices $\Xi = (\xi_{ij}), \theta = (\theta_{ij})$ are independent and are satisfying the conditions of Theorem 2.1.4, then*

$$\det \Xi\Xi^* \det[\Xi\Xi^* + \theta\theta^*]^{-1} \approx \prod_{i=1}^{n} \eta_i,$$

where η_i are independent random variables $[(\vec{\xi}_i, \vec{\xi}_i) + (\vec{\theta}_i, \vec{\theta}_i)]^{-1}(\vec{\xi}_i, \vec{\xi}_i)$, $\vec{\theta}_i$ are row vectors of matrix θ.

We get the proof for these results from the fact that for any unitary matrix U of order i, $U\vec{\xi}_i \approx \vec{\xi}_i$. The remainder of the proof is the same as for Theorem 2.1.2 and 2.1.3. The first row vector of the matrices Ξ and θ in Theorems 2.1.2–2.1.5 can be distributed according to any arbitrary law, and all the other row vectors are satisfying the conditions of the Theorem.

Let $\Xi = (\xi_{ij})$ be a random nonnegative-definite $n \times n$ matrix with probability density $p(Z)$. Now we find the density of the elements of a random matrix $H = (\eta_{ij}), \eta_{ij} = \xi_{ij}(\xi_{ii}\xi_{jj})^{-1/2}, i \neq j$. By using the transformation $z_{ij} = r_{ij}(z_{ii}z_{jj})^{1/2}, i \neq j$, we obtain the density of matrix H,

$$q(r_{ij}, i \neq j) = \int p(t_{ij}) \prod_{i=1}^{n} z_{ii}^{(n-1)/2} dz_{ii}, t_{ij} = r_{ij}(z_{ii}z_{jj})^{1/2}, i \neq j, \ t_{ii} = z_{ii}.$$

The moments of the determinants of the matrices H are

$$\mathbf{E} \det H^k = \int \det Z^k p(Z) \prod_{i=1}^{n} z_{ii}^{-k} dZ,$$

where dZ is the Lebesgue measure of the positive-definite matrices Z.

If $p(Z)$ is the Wishart density $\omega(Z_n, I, m)$, then we have

$$q(r_{ij}, i \neq j) = \Gamma^n(m/2) \det(r_{ij})^{(m-n-1)/2} \left[\prod_{i=1}^{n} \Gamma((m+1-i)/2) \pi^{n(n-1)/4} \right]^{-1},$$

$$\mathbf{E} \det H^k = c_{n,m} (2\pi)^{-nm/2} \int \det Z^{k+(m-n-1)/2} \exp(-0.5 \operatorname{Tr} Z) dZ$$

$$\times \left[\int_0^\infty (1+2y)^{-k-m} y^{k-1} dy \right]^n = \mathbf{E} \prod_{i=1}^{n} \eta_i^k,$$

where $\eta_i, i = \overline{1,n}$ are independent random variables which have beta density,

$$B(x, (n-1)/2, (i-1)/2), 0 < x < 1.$$

§2 Moments of Random Vandermond Determinants. Hypothesis by Mehta and Dyson

The determinant of the matrix $(\eta_i^j), i = \overline{1,n}, j = \overline{0, n-1}$, where $\eta_i, i = 1, 2, \ldots$ are some random variables called the random Vandermond determinant.

Theorem 2.2.1. *(Selberg[151]) If the random variables $\vec{\xi}_i, i = \overline{1,n}$ are independent, identically distributed, and have B-distribution with density $[B(\alpha, \beta)]^{-1} x^{\alpha-1}(1-x)^{\beta-1}, 0 < x < 1, \alpha > 0, \beta > 0$, then for $n = 2, 3, \ldots$,*

$$J = \mathbf{E}[\prod_{1 \leq i < j \leq n} (\xi_i - \xi_j)]^{2k} \tag{2.2.1}$$

$$= \prod_{j=1}^{n} \{ \Gamma(1 + jk) \Gamma(\alpha + (j-1)k) \Gamma(\beta + (j-1)k)$$

$$\times [\Gamma(1+k) \Gamma(\alpha + \beta + (n+j-2)k)]^{-1} \} \quad B^{-n}(\alpha, \beta),$$

where

$$\operatorname{Re} \alpha > 0, \quad \operatorname{Re} \beta > 0, \quad \operatorname{Re} k > -\min\{n^{-1}, \operatorname{Re} \alpha(n-1)^{-1}, \operatorname{Re} \beta(n-1)^{-1}\}.$$

Proof. Let $k \geq 1$ be an integer. It is evident that

$$\Delta^{2k}(x) = \sum_{(j)} c_{j1\ldots jn} x_1^{j1} x_2^{j2} \ldots x_n^{jn}, \tag{2.2.2}$$

where

$$\Delta(x) := \Delta(x_1, \ldots, x_n) := \prod_{1 \leq i < j \leq n} (x_i - x_j),$$

$$c_{j1\ldots jn}, j_1, \ldots, j_n \quad \text{are integers},$$

$$j_s \geq 0, s = \overline{1,n} \quad \text{and} \quad j_1 + \cdots + j_n = kn(n-1).$$

Without loss of generality, we may assume $j_1 \leq j_2 \leq \cdots \leq j_n$. Obviously, $j_n \geq k(n-1)$, but $\Delta(x_1,\ldots,x_l)$ divides $\Delta(x_1,\ldots,x_n)$ for $l = \overline{1,n}$. Hence $j_l \geq k(l-1)$ for $l = \overline{1,n}$. Obviously,

$$\Delta^{2k}(x_1^{-1},\ldots,x_n^{-1}) = \prod_{i=1}^{n} x_i^{-2k(n-1)} \Delta^{2k}(x).$$

Therefore,

$$\Delta^{2k}(x^{-1}) = \sum_{(s)} c_{s_1\ldots s_n} x_1^{-s_1} \ldots x_n^{-s_n}$$

$$= \sum_{(j)} c_{j_1\ldots j_n} \prod_{l=1}^{n} x_l^{-2k(n-1)+jl}$$

$$(s) = (s_1 \leq \cdots \leq s_n), \quad (j) = (0 \leq j_1 \leq \cdots \leq j_n).$$

Hence,

$$-S_{n-l+1} = -2k(n-1) + jl, \qquad s_l \geq k(l-1),$$
$$j_l = 2k(n-1) - S_{n-l+1} \leq 2k(n-1) - k(n-l) = k(n+l-2).$$

But $j_l \geq k(l-1)$. Therefore,

$$k(n+l-2) \geq j_l \geq k(l-1) \qquad \text{for} \quad l = \overline{1,n}. \tag{2.2.3}$$

We set

$$\Gamma(\alpha+j_l)[\Gamma(\alpha+\beta+j_l)]^{-1}$$
$$= \Gamma(\alpha+(l-1)k)[\Gamma(\alpha+\beta+(n+l-2)k]^{-1} q_{jl}(\alpha,\beta), \tag{2.2.4}$$

where

$$q_{jl}(\alpha,\beta) = \Gamma(\alpha+j_l)[\Gamma(\alpha+(l-1)k)]^{-1}$$
$$\times \Gamma(\alpha+\beta+(n+l-2)k)[\Gamma(\alpha+\beta+j_l)]^{-1}.$$

By using the formula

$$B(n,\alpha) = B(\alpha,n) = (n-1)![\alpha(\alpha+1)(\alpha+2)\ldots(\alpha+n-1)]^{-1},$$

we get from (2.2.4) that $q_{jl}(\alpha,\beta)$ is a polynomial in α and β of degree $k(n+l-2) - j_l$ in β.

Therefore,

$$\prod_{l=1}^{n} \frac{\Gamma(\alpha+j_l)\Gamma(\beta)}{\Gamma(\alpha+\beta+j-l)} = \prod_{l=1}^{n} \frac{\Gamma(\alpha+(l-1)k)\Gamma(\beta)}{\Gamma(\alpha+\beta+(n+l-2)k)} Q_{(j)}(\alpha,\beta),$$

and $Q_{(j)}(\alpha, \beta)$ is a polynomial in α and β, whose degree in β is

$$\sum_{l=1}^{n}\{(n+l-2)k - j_l\} = k\{n(n-2) + n(n+1)/2 - n(n-1)\}$$

$$= kn(n-1)/2.$$

Then,

$$\dot{J} := B^n(\alpha, \beta)\mathbf{E}\sum_{0\le j_1\le j_2\le\cdots\le j_n} c_{j_1\ldots j_n}$$

$$\times \xi_1^{j_1+\alpha-1}\xi_2^{j_2+\alpha-1}\ldots\xi_n^{j_n+\alpha-1}(1-\xi_1)^{\beta-1}\ldots(1-\xi_n)^{\beta-1}$$

$$= \sum_{0\le j_1\le\cdots\le j_n} c_{j_1\ldots j_n}\prod_{l=1}^{n} B(\alpha+j_l,\beta)$$

$$= \prod_{l=1}^{n}\frac{\Gamma(\alpha+(l-1)k)\Gamma(\beta+(l-1)k)}{\Gamma(\alpha+\beta+(n+l-2)k)}\frac{Q(\alpha,\beta)}{R(\beta)},$$

where $Q(\alpha,\beta) = \sum_{(j)} c_{j_1\ldots j_n}Q_{(j)}(\alpha,\beta)$, $Q(\alpha,\beta)$ is a polynomial, and the degree of $Q(\alpha,\beta)$ in β is at most equal to $kn(n-1)/2$; $R(\beta) = \prod_{l=1}^{n}[\Gamma(\beta)]^{-1}\Gamma(\beta+(l-1)k)$ is a polynomial in β of degree $\sum_{l=1}^{n}(l-1)k = (n-1)nk/2$. Now $\Delta x = \pm\Delta(1-x)$. Hence $\dot{J}$ is symmetric in α and β, and

$$Q(\alpha,\beta)[R(\beta)]^{-1} = Q(\beta,\alpha)[R(\alpha)]^{-1},$$

where $Q(\beta,\alpha)[R(\alpha)]^{-1}$ is a polynomial in β; $Q(\alpha,\beta)$ is divisible by $R(\beta)$; and the degree of $Q(\alpha,\beta)$ is less than or equal to that of degree $R(\beta)$. Therefore,

$$Q(\alpha,\beta)(R(\beta))^{-1} \equiv c(n,k),$$

and the constant c is independent of α and β.

Let us find the constant $c(n,k)$; we set $\alpha = \beta = 1$.

Then,

$$J = c(k,n)\prod_{l=1}^{n}\frac{\Gamma(1+(l-1)k)}{\Gamma(2+(n+l-2)k)}.$$

Let y be the largest of the $x_1, x_2, \ldots, x_n$, and replace the other x_j by $x_j = yz_j$, $0 \le z_j \le 1$. Then,

$$J = n\int_0^1 y^{n-1}y^{kn(n-1)}dy\int_0^1\cdots\int_0^1[\prod_{j=1}^{n-1}(1-z_j)\Delta(Z)]^{2k}dz_1\ldots dz_n$$

$$= [k(n-1)+1]^{-1}\int_0^1\cdots\int_0^1|\Delta(Z)|^{2k}\prod_{j=1}^{n-1}(1-z_j)^{2k}dz$$

$$= [k(n-1)+1]^{-1}c(k,n-1)\prod_{l=1}^{n-1}\frac{\Gamma(1+(l-1)k)\Gamma(2k+1+(l-1)k)}{\Gamma(1+2k+1+(n+l-1)k)}.$$

Or, on simplification,

$$c(k,n)c^{-1}(k,n-1) = \Gamma(1+nk)(\Gamma(1+k))^{-1}.$$

Hence,

$$c(k,n) = c(k,1) \prod_{l=1}^{n} \{\Gamma(1+lk)[\Gamma(1-k)]^{-1}\}.$$

Obviously $c(k,1) = 1$.

Theorem 2.2.1 is proved, when k is an integer. Let k be complex.

A function $\varphi(x)$ on $(0,\infty)$ is called completely monotone if it has derivatives $\varphi^k(x)$ of all orders, and $(-1)^k\varphi^k(x) \geq 0$, for all $x \geq 0$.

Lemma 2.2.1. *For some random variable $\xi \geq 0$, assume that $\mathbf{E}\xi^k = \mu_k$ exists, and that for some function $m(t)$, we have $m(k) = \mu_k$, $k = 1, 2, \ldots$.*

In order that $\mathbf{E}\xi^t = m(t), t > 0$ and $\xi \leq 1$ with probability 1, it is necessary and sufficient that $m(t)$ be completely monotone and that $m(0) = 1$.

Proof. By Bernstein's theorem, ([34], Chapter 12, §4, Theorem 12.4.1), the function $m(t), t \geq 0$ is completely monotone, and $m(0) = 1$ if and only if it is the Laplace transform of some random variable $\eta \geq 0$. The necessity of the conditions of Lemma 2.2.1 follows from this.

Let us prove sufficiency. Since $\mathbf{E}\exp(-t\eta) = m(t)$, $t > 0$, the random variables $e^{-\eta}$ and ξ have the same moments. From the divergence of the Carleman series $\sum_{k=1}^{\infty} m(2k)^{-1/2k}$ [Chapter 7, Eq. (3.10)], it follows that the random variables $e^{-\eta}$ and ξ have the same distribution function. Lemma 2.2.1 is proved.

Obviously, by using (2.2.1), we obtain

$$\mathbf{E}[\prod_{1 \leq i < j \leq n} (\xi_i - \xi_j)]^{2k} \mathbf{E}\nu_1^k = c\mathbf{E}\nu_2^k,$$

where ν_1 and ν_2 are random variables and k is an integer.

$$\mathbf{E}\nu_1^k = c_1 \prod_{j=1}^{n} \{\Gamma(1+k)\Gamma(\alpha+\beta+(n+j-2)k\} := \psi_1(k),$$

$$\mathbf{E}\nu_2^k = c_2 \prod_{j=1}^{n} \{\Gamma(1+jk)\Gamma(\alpha+(j-1)k)\Gamma(\beta+(j-1)k)\} := \psi_2(k),$$

where $\psi_1(k), \psi_2(k)$ are completely monotone and analytical functions.

Hence, making use of Lemma 2.2.1, we obtain (2.2.1). Theorem 2.2.1 is proved.

Theorem 2.2.2.   *(Mehta) If the $\eta_i, i = \overline{1, n}$ are independent normally distributed random variables, with law $N(0, 1)$, then for $n = 2, 3, \ldots$,*

$$\mathbf{E}[\prod_{1 \le i < j \le n} (\eta_i - \eta_j)]^{2k} = \prod_{j=1}^{n} \frac{\Gamma(1 + jk)}{\Gamma(1 + k)}, \qquad (2.2.5)$$

where $Re k > -n^{-1}$.

Proof.   Let $x = \frac{1}{2}(1 + \frac{x'}{\sqrt{l}})$, $\alpha = \beta = l + 1$, and let k be an integer. Using Theorem 2.2.1, we find that

$$\int_0^1 \int_0^1 \cdots \int_0^1 \Delta^{2k}(x) \prod_{i=1}^{n} x_i^{\alpha-1}(1 - x_i)^{\beta-1} dx_i$$

$$= \int_{-\sqrt{l}}^{\sqrt{l}} \cdots \int_{-\sqrt{l}}^{\sqrt{l}} \Delta^{2k}(x') \prod_{i=1}^{n}(1 - x_i'^2/l)^l dx_i' 2^{-2nl}(2\sqrt{l})^{-n-kn(n-1)}$$

$$= \prod_{j=1}^{n} \left[\frac{(jk)!}{k!} \Gamma(\alpha + (j-1)k) \frac{\{\Gamma(l + (j-1)k + 1)\}^2}{\Gamma(2l + 1 + (n + j - 2)k + 1)} \right].$$

Hence

$$\gamma := \int_{-\infty}^{\infty} \cdots \int_{-\infty}^{\infty} |\Delta(x)|^{2k} \exp(-\sum_{i=1}^{n} x_i^2) \prod_{j=1}^{n} dx_j \{ \prod_{j=1}^{n} \frac{(jk)!}{k!} \}^{-1}$$

$$= \lim_{l \to \infty} 2^{2nl+n+kn(n-1)} l^{(n+kn(n-1))/2} \prod_{j=1}^{n} \frac{\{\Gamma(l + 1 + (j-1)k)\}^2}{\Gamma(2l + (n + j - 2)k + 2)}.$$

By using the Stirling formula, we find that

$$\gamma = \lim_{l \to \infty} 2^{2nl+n+kn(n-1)} l^{(n+kn(n-1))/2}$$

$$\times \prod_{j=1}^{n} \left\{ \frac{[(l + (j-1)k)^{l+(j-1)k} e^{-(l+(j-1)k)} \sqrt{2\pi[l + 1(j-1)k]}]^2}{(2l + (n + j - 2)k + 1)^{2l+(n+j-2)k+1} e^{-[2l+(n+j-2)k+1]}} \right.$$

$$\times \left. \frac{1}{\sqrt{2\pi[2l + (n + j - 2)k + 1]}} \right\} = (2\pi)^{n/2} 2^{-n(1+k(n-1))/2}.$$

Hence, making use of Lemma 2.2.1, we obtain (2.2.5).
    Theorem 2.2.2 is proved.

    Let $u(x)$ and $v(x)$ be real bounded continuous functions.

Theorem 2.2.3. *[132]*

$$
\mathbf{E} \prod_{i=1}^{n} u(\eta_{2i-1})v(\eta_{2i}) \prod_{i>j} |\eta_i - \eta_j|
$$

$$
= n! [2^{n^2/2} \prod_{j=0}^{n-1} (j!)2^{n/2}]^{1/2} \Big\{ [\det(b_{ij})_{i,j=0}^{n-1}]^{1/2}, \; n = 2m;
$$

$$
\sum_{s=1}^{n} (-1)^{s+n} [\det(b_{ij})_{i,j=0}^{n-1}, \quad j \neq n-1, \quad i \neq s]^{1/2}
$$

$$
\times \int \varphi_{s-1}(y)dy, \quad n = 2m - 1 \Big\},
$$

where

$$
b_{ij} = \int\!\!\int_{y<x} dy\,dx\, u(y)v(x)[\varphi_i(y)\varphi_i(x) - \varphi_i(x)\varphi_i(y)], \quad i,j = \overline{0, k-1}
$$

$$
\varphi_j(x) = (2^j j! \sqrt{\pi})^{-1/2} \exp(-x^2/2) H_j(x),
$$

$H_j(x)$ *are Hermitian polynomials.*

Proof.   Consider $p(u, v) = \mathbf{E} \prod_i u(\eta_{2i-1})v(\eta_{2i}) \prod_{i>j} |\eta_i - \eta_j|$. Obviously,

$$
p(u, v) = (2\pi)^{-n/2} n! \int \cdots \int_{L} \Big[\prod_i u(x_{2i-1})v(x_{2i}) \Big]
$$

$$
\times \exp\left(-0.5 \sum_{i=1}^{n} x_i^2\right) \prod_{i>j}(x_i - x_j) \prod_{i=1}^{n} dx_i, \tag{2.2.6}
$$

$$
L = \{x_1 < x_2 < \cdots < x_n\},
$$

where $\prod_{i>j}(x_i - x_j)$ is a determinant of the Vandermond matrix, the jth row of which equals $x_1^{j-1}, x_2^{j-1}, \ldots, x_n^{j-1}, j = \overline{1, n}$.

By multiplying the jth row by 2^{j-1} and by adding to it the linear combination of the other rows we have $H_{j-1}(x_1), H_{j-1}(x_2), \ldots, H_{j-1}(x_n)$, where $H_j(x_i)$ is a Hermitian multinomial of order j, $H(x) = e^{x^2}(-1)^i d^j/dx^j e^{-x^2}$. Because $\int e^{-x^2} H_n(x)H_m(x)dx = \sqrt{\pi} 2^n n\delta_{n,m}$, and by multiplying the jth row to the constant $(2^{j-1}(j-1)!\sqrt{\pi})^{-1/2}$, we obtain

$$
p(u, v) = 2^{-n/2} n! \prod_{j=1}^{n} (2^{1-j}(j-1)!)^{1/2} \int \cdots \int_{L} \prod_i u(x_{2i-1})
$$

$$
\times v(x_{2i}) \det[\varphi_{j-1}(x_i)]_{i,j=1}^{n} \prod_{i=1}^{n} dx_i,
$$

where

$$\varphi_j(x) = (2^j j! \sqrt{\pi})^{-1/2} e^{-x^2/2} H_j(x).$$

The functions $\varphi_j(x)$ have the property

$$\int \varphi_j(x)\varphi_k(x)dx = \delta_{jk}.$$

Suppose that $n = 2m$. By integrating by x_1, we obtain from the first column of the matrix $(\varphi_{j-1}(x_i))$ the column

$$[F_0(x_2), F_1(x_2), \ldots, F_{n-1}(x_2)],$$

where

$$F_j(x) = \int_{-\infty}^{x} u(y)\varphi_i(y)dy.$$

By integrating by x_3, we obtain from the third column the column, $F_0(x_4) - F_0(x_2), \ldots, F_{n-1}(x_4) - F_{n-1}(x_2)$, to which we can add the first column.

When integrating by $x_3, x_5, \ldots$, we obtain the columns which have F_j elements.

As a result,

$$p(u,v) = 2^{-n/2} n! \prod_{j=0}^{n-1} (2^{-j} j!)^{1/2} \int \cdots \int_{L_1} \prod_i v(x_{2i})$$

$$\times \det[F_{i-1}(x_{2j}), \varphi_{i-1}(x_{2j})] \prod_{l=1}^{m} dx_{2l}, \quad L_1 = \{x_2 < x_4 < \cdots < x_{2m}\}.$$

$$(2.2.7)$$

It is easy to check that $\det[F_{i-1}(x_{2j}), \varphi_{i-1}(x_{2j})]$ does not change its value when the two arguments x_{2k}, x_{2l} are permutated. Therefore, in Eq. (2.2.7), we can interchange the integration order:

$$p(uv) = 2^{-n/2} n! \prod_{j=1}^{n-1} (2^{-j} j!)^{1/2} \int \cdots \int \prod_i v(x_{2i})$$

$$\times \det[F_{i-1}(x_{2j}), \varphi_{i-1}(x_{2j})] \prod_{l=1}^{m} dx_{2l}.$$

$$(2.2.8)$$

Lemma 2.2.2. *[132] Let $\Xi_n = (\xi_{ij}), n = 2m$ be a random matrix, the elements of which are satisfying the condition that the vectors $(\xi_{kj}, \xi_{k+1j}, j = \overline{1,n}), k = 1, 3, \ldots$ are independent, and $\mathbf{E}\xi_{ki}\xi_{k+1j} = a_{ij}, a_{ij}$ are not depending on k.*
Then

$$\mathbf{E} \det \Xi_{2m} \equiv 2^n m! [\det(b_{ij})_{i,j=1}^n]^{1/2}$$

$$= 2^n m! \operatorname{pfaff}(b_{ij})_{i,j=1}^n, \quad b_{ij} = a_{ij} - a_{ji}.$$

Proof. By using the expression for the determinant of a matrix, we find

$$\mathbf{E}\det\Xi_{2m} = \sum(-1)^{\alpha}\mathbf{E}\xi_{1i_1}\xi_{2i_2}\ldots\xi_{ni_n}$$

$$= \sum_{i_1>i_2>\cdots>i_n}(-1)^{\alpha}b_{i_1i_2}b_{i_3i_4}\ldots b_{i_{n-1}i_n},$$

where α is the number of permutations in the substitution

$$\begin{pmatrix} 1 & 2 & \ldots & n \\ i_1 & i_2 & \ldots & i_n \end{pmatrix}.$$

Using (1.5.7), we obtain Lemma 2.2.2.
Now we can transform Eq. (2.2.8),

$$p(u,v) = 2^{-n/2}n!\prod_{j=0}^{n-1}(2^{-j}j!)^{1/2}[\det(b_{ij})_{i,j=0}^{n-1}]^{1/2},$$

$$b_{ij} = \iint_{y<x}dydxu(y)v(x)[\varphi_i(y)\varphi_j(x) - \varphi_i(x)\varphi_j(y)],$$

$$i,j = \overline{0,n-1}. \tag{2.2.9}$$

If $n = 2m + 1$, then we obtain (2.2.7) in the following form:

$$p(u,v) = 2^{-n/2}n!\prod_{j=0}^{n-1}(2^{-j}j!)^{1/2}\int\cdots\int_{L_1}\det[F_{i-1}(x_{2j}),\varphi_{i-1}(x_{2j}),$$

$$j = \overline{1,m},\int\varphi_{i-1}(y)dy - F_{i-1}(x_{2m})]\prod_{j=1}^{m}dx_{2j}.$$

From this formula, we obtain [see (2.2.9)],

$$p(u,v) = 2^{-n/2}n!\prod_{j=0}^{n-1}(2^{-j}j!)^{1/2}\sum_{s=1}^{n}(-1)^{s\times n}$$

$$\times [\det(b_{ij})_{i,j=0}^{n-1}, j \neq n-1, i \neq s]\int\varphi_{s-1}(y)dy. \tag{2.2.10}$$

Theorem 2.2.3 is proved.

Transform the formula (2.2.9). We represent it as following (for $n = 2m$),

$$p(u,v) = 2^{-n/2}n!\prod_{j=0}^{n-1}(2^{-j}j!)^{1/2}\prod_{j=0}^{m-1}(8/(2_j + 1))^{1/2}$$

$$\times \{\det[f_{ij}]_{i,j=0}^{n-1}\}^{1/2},$$

where

$$f_{2i,2j} = \int\int_{y<x} dy\,dx\,u(y)v(x)[\varphi_{2i}(y)\varphi_{2j}(x) - \varphi_{2i}(x)\varphi_{2j}(y)],$$

$$f_{2i+1,2j+1} = ((2i+1)(2j+1)/64)^{1/2} \int\int_{y<x} dy\,dx\,u(y)v(x)$$

$$\times\,[\varphi_{2i+1}(y)\varphi_{2j+1}(x) - \varphi_{2j+1}(x)\varphi_{2j+1}(y)].$$

$$(2.2.11)$$

Using $\sqrt{2}\varphi'_n(x) = \sqrt{n}\varphi_{n-1}(x) - \sqrt{n+1}\varphi_{n+1}(x)$ and (2.2.11), we obtain

$$g_{ij} := f_{2i,2j+1} - ((2j-1)/2j)^{1/2}f_{2i,2j-1} = -0,5\int\int_{y<x} dy\,dx$$

$$\times\,u(y)v(x)[\varphi_{2i}(y)\varphi'_{2j}(x) - \varphi_{2i}(x)\varphi'_{2j}(y)];$$

$$(2.2.12)$$

$$\mu_{ij} := f_{2i+1,2j-1} - ((2i-1)/2i)^{-1/2}f_{2i-1,2j+1}(-(2j/(2j-1))^{1/2}$$

$$\times\,f_{2i+1,2j-1} + (4ij/(2i-1)(2j-1))^{1/2}f_{2i-1,2j-1}$$

$$= 0,25\int\int_{y<x} dy\,dx\,u(y)v(x)[\varphi'_{2i}(y)\varphi'_{2j}(x) - \varphi'_{2i}(x)\varphi'_{2j}(y)].$$

$$(2.2.13)$$

After some calculations, we have

$$\pi^{n/4}\prod_{j=0}^{n-1}(j!)^{1/2} = \prod_{j=0}^{m-1}[\pi(2j)!(2j+1)]^{1/2}$$

$$= \prod_{j=0}^{m-1}[2^{2j}\Gamma(j+1/2)\Gamma(j+1)(2j+1)^{1/2}]$$

$$= 2^{m(m-1)}\prod_{j=0}^{m-1}(2j+1)^{1/2}\prod_{k=1}^{2m}[2\Gamma(1+k/2)/k].$$

$$(2.2.14)$$

To reduce the order of the determinant in Eq. (2.2.5), we subtract from $2j$ columns multiplied by $(2j/(2j-1)/2$ from $2j$ columns. Doing the same for the rows and using (2.2.14), we obtain

$$p^2(u,v) = \det\begin{bmatrix} \lambda_{ij} & g_{ij} \\ -g_{ij} & \mu_{ij} \end{bmatrix}_{i,j=\overline{0,n-1}},$$

where $\lambda_{ij} = f_{2i,2j}$. If $u(-x)v(-y) = u(x)v(y)$, then $\lambda_{ij} = \mu_{ij} = 0$, and

$$p(u,v) = \prod_{j=1}^{n}\Gamma(1+j/2)\det(g_{ij}), \quad i,j = \overline{0,m-1},$$

$$(2.2.15)$$

where

$$g_{ij} = \int\int_{y<x} dy dx u(y) v(x) \varphi_{2i}(x) \varphi'_{2j}(y). \qquad (2.2.16)$$

Examine Eq. (2.2.1) when $k = 1$. Let $u(x) = v(x) = 1$. Then $g_{ij} = \delta_{ij}$. Using (2.2.11) and (2.2.12) from (2.2.5), we have (2.2.1), when $k = 1$ and n is even. Analogously taking into account (2.2.6), we find that (2.2.1) is valid when $k = 1$ and n is odd.

Examine formula (2.2.1) when $k = 2$. In this case, as in the proof of Theorem 2.2.3, we obtain

$$\mathbf{E} \prod_{i>j} |\eta_i - \eta_j|^2 = (2\pi)^{-n/2} \int \cdots \int \prod_{i>j} (x_i - x_j)^2$$

$$\times \exp\{-0.5 \sum_{i=1}^{n} x_i^2\} \prod_{i=1}^{n} dx_i = (2\pi)^{-n/2}$$

$$\times \prod_{j=0}^{n-1} [2^{-j/2}(j!\sqrt{\pi})^{1/2}]^2 \int \cdots \int [\varphi_{k-1}(x_j)]^2_{k,j=\overline{1,n}}$$

$$\times \prod_{i=1}^{n} dx_i = n! 2^{-n^2/2} \prod_{j=0}^{n-1} j!;$$

$$\int \cdots \int_{x_i \notin (-\theta,\theta)} \prod_{i>j} (x_i - x_j)^2 \exp\{-0.5 \sum_{i=1}^{n} x_i^2\} \prod_{i=1}^{n} dx_i$$

$$= \det[\delta_{ij} - \int_{-\theta}^{\theta} \varphi_i(x) \varphi_j(x) dx]. \qquad (2.2.17)$$

Let $k = 4$ in (2.2.1). We shall prove the formula

$$\prod_{i<j} (x_i - x_j)^4 = \det[x_i^j, jx_i^{j-1}], \qquad i = \overline{1,n}, \qquad (2.2.18)$$

$$j = \overline{0, 2n-1}.$$

As $\det[x_i^j, jx_i^{j-1}]$ is a polynom of order $2n(n-1)$, it is equal to zero when $x_i = x_j$, and the matrix $[x_i^j, jx_i^{j-1}]$, $i = \overline{1,n}$, $j = \overline{0, 2n-1}$ has two pairs of rows which are equal when $x_i = x_p$, and the row vector $\frac{\partial}{\partial x_i}(x_i^j, j = \overline{0, 2n-1})$ equals

$$\lim_{x_p \to x_i} (x_p - x_i)^{-1}(x_p^j - x_i^j, j = \overline{0, 2n-1}).$$

This uniform polynomial must be divided by $(x_i - x_p)^4$ without a remainder.

Therefore, Eq. (2.2.18) is proved. Using Hermitian polynomials with the variables $y_i = x_i\sqrt{2}$, we obtain

$$\det[x_i^j, jx_i^{j-1}] = 2^{-n(n-1)}(\prod 2^j) \det[H_j(y_i), H'_j(y_j)]$$

$$= 2^{-n(3n-2)} \det[H_j(y_i), H'_j(y_i)]. \qquad (2.2.19)$$

As in the proof of Theorem 2.2.1, we obtain from (2.2.15),

$$\exp\{-2\sum_{i=1}^{n} x_i^2\} \prod_{i<l}(x_i - x_l)^4 = a\,\det[\varphi_j(y_i), \sqrt{2}\varphi_{j-1}(y_i)], \qquad (2.2.20)$$

where

$$a = 2^{-n(2n-3/2)}\pi^{n/2}\prod_{l=0}^{n-1}[(2l)!(2l+1)^{1/2}].$$

Using (2.2.20),

$$\int \exp(-2\sum_{i=1}^{n} x_i^2)\prod_{i>j}(x_i - x_j)^4$$

$$= a\int \det[\varphi_j(y_i), \sqrt{2}\varphi_{j-1}(y_i)]\prod_{i=1}^{n} dy_i. \qquad (2.2.21)$$

Applying Lemma 2.2.2 to (2.2.21), we find

$$\int \exp(-2\sum_{i=1}^{n} x_i^2)\prod_{i>j}(x_i - x_j)^4\prod_{i=1}^{n} dx_i$$

$$= a2^{2n}n!(\det(g_{ij}))^{1/2}, g_{ij} = \int dy[\sqrt{2}\varphi_j(y)\varphi_{j-1}(y) \qquad (2.2.22)$$

$$- \sqrt{2}\varphi_{j-1}(y)\varphi_i(y)] = \sqrt{2}(\delta_{ij-1} - \delta_{ji-1}).$$

Since $\det(\delta_{ij-1} - \delta_{ji-1}) = 1$, then we obtain (2.2.1) from (2.2.22) when $k = 4$ and with arbitrary $n \geq 1$.

Now we shall find the moments of the complex random Vandermond matrices $(\xi_i^{j-1})i, j = \overline{1,n}$, where $\xi_p = \nu_p + i\mu_p$, $\nu_p, \mu_p, p = \overline{1,n}$ are independent random variables distributed according to the standard normal law.

Theorem 2.2.4. *[132]*

$$\mathbf{E}|\det(\xi_i^{j-1})|^2_{i,j=\overline{1,n}} = 2^{n(n-1)/2}n!\prod_{j=1}^{n-1} j!$$

Proof. Using the Vandermond determinant, we have

$$\prod_{i>j}|\xi_i - \xi_j|^2 = \det[\sum_p \overline{\xi_p^i}\xi_p^j]_{i,j=0}^{n-1}.$$

Then

$$\mathbf{E}|\det(\xi_i^{j-1})/^2 = (2\pi)^{-n} \int \cdots \int \det[\sum_p z_p^i \bar{z}_p^j]_{i,j=0}^{n-1}$$

$$\times \exp\{-0,5\sum_{i=1}^n |z_i|^2\} \prod_{i=1}^n dx_i dy_i.$$

Taking into consideration that the integral function is symmetric on all z_i, and by transforming the determinant by the first row, we obtain:

$$\mathbf{E}|\det(\xi_i^{j-1})|^2$$

$$= (2\pi)^{-n} \int \cdots \int \begin{bmatrix} 1 & z_1 & \cdots & z_1^{n-1} \\ \vdots & \vdots & & \vdots \\ \sum \bar{z}_i^{(n-1)} & \cdots & \cdots & \sum_i \bar{z}_i^{(n-1)} z_i^{(n-1)} \end{bmatrix}$$

$$\times \exp\{-0,5\sum_{i=1}^n |z_i|^2\} \prod_{i=1}^n dx_i dy_i.$$

By multiplying the first row by z_1 and by subtracting it from the second row, we transform the determinant. And if we decompose every row in the same manner, then

$$\mathbf{E}|\det \xi_i^{j-1}|^2 = (2\pi)^{-n} n! \int \cdots \int \det[\bar{z}_{i+1}^i z_{i+1}^j]_{i,j=0}^{n-1}$$

$$\times \exp\{-0,5\sum_{i=1}^n |z_i|^2\} \prod_{i,j=1}^n dx_i dy_j.$$

Taking into account that

$$\int \exp\{-0,5|z|^2\}\bar{z}^j z^k dx dy = \pi \delta_{jk} j! 2^{j+l},$$

we find

$$\mathbf{E}|\det(\xi_i^{j-1})|^2 = 2^{n(n-1)/2} n! \prod_{j=1}^{n-1} j!$$

Theorem 2.2.4 is proved.

The Dyson Hypothesis

If the random variables $\theta_i, i = \overline{1,n}$ are independent and have uniform distribution on the interval $(0, 2\pi)$, then for any integer $k \geq 0$,

$$\mathbf{E}\prod_{p\neq l} |e^{i\theta_p} - e^{i\theta_l}|^k = \Gamma(1 + \frac{kn}{2})[\Gamma(1 + \frac{k}{2})]^{-n}. \tag{2.2.23}$$

We shall present another form of this hypothesis. Let $k = 2m$, where m is an integer, then

$$\prod_{p>l} |e^{i\theta_p} - e^{i\theta_l}|^{2m} = \prod_{p>l}(e^{i\theta_p} - e^{i\theta_l})^m(e^{-i\theta_p} - e^{-i\theta_l})^m.$$

Now we consider the integral

$$(2\pi)^{-n}\int_0^{2\pi}\cdots\int_0^{2\pi}\prod_{p>l}(e^{i\theta_p} - e^{i\theta_l})^m(e^{-i\theta_p} - e^{-\theta_l})^m\prod_{p=1}^n d\theta_p.$$

It is equal to zero for all the items of the integral sum except one which does not depend on $\exp(\pm i\theta_p)$.

Consequently, this integral equals the spare item at the decomposition $\prod_{p\neq l}(1 - z_p/z_l)^m$ to the Loran series.

Dyson proposed the following statement:

If $a_i, i = \overline{1,n}$ are the nonnegative integers, then the constant item in the Loran series of the quantity $\prod_{i\neq j}(1 - z_j/z_i)^{a_j}$ equals $(a_1 + \cdots + a_n)![a_1! \ldots a_n!]^{-1}$. this statement was proved by Wilson and Gunson. Let us consider Wilson's proof.[108],[168].

The unknown expression equals the coefficient $t_1^{a_1}\ldots t_n^{a_n}$ in the Taylor series of the function

$$G(t_1,\ldots,t_n) := (2\pi i)^{-n}\int_{\Gamma(\rho)}\prod_{j=1}^n(1 - t_j\prod_{i\neq j}(1 - z_i/z_j))^{-1}\prod_{i=1}^n z_i^{-1}dz_i, \quad (2.2.24)$$

where $\Gamma(\rho) = \{z_j : |z_j| = \rho_j, \rho_i \neq \rho_j, i \neq j\}$, and for all $z_i \in \Gamma(\rho), |t| \leq \varepsilon$,

$$|t_j\prod_{i\neq j}(1 - z_i/z_j)| < 1, j = \overline{1,n}. \quad (2.2.25)$$

Transforming (2.2.24), we obtain

$$G(t_1,\ldots,t_n) = (2\pi i)^{-n}\int_{\Gamma(\rho)}\prod_{j=1}^n(z_j^{n-1} - t_j\prod_{i\neq j}(z_j - z_i))^{-1}$$

$$\times\prod_{i=1}^n z_i^{n-2}dz_i = (-1)^{n(n-1)/2}(2\pi i)^{-n}\int_{\Gamma(\rho)}\prod_{j=1}^n(z_j^{n-1}\Delta_{n-1}^{(j)}$$

$$- (-1)^{j-1}t_j\Delta_n)^{-1}(z\Delta_n)^{n-1}\prod_{i=1}^n dz_i, \quad (2.2.26)$$

where $\Delta_n = \prod_{i>j}(z_i - z_j), \Delta_{n-1}^{(j)} = \prod_{i<k}(z_i - z_k)$.

The function Δ_n satisfies the equations:

$$\sum_{j=1}^{n}(-1)^{j-1}z_j^{n-1}\Delta_{n-1}^{(j)} = \Delta_n,$$

$$\prod_{j=1}^{n}(-1)^{j-1}z_j^{n-1}\Delta_{n-1}^{(j)} = (-1)^{n(n+1)/2}\prod_{i=1}^{n}z_i^{n-1}\Delta_n^{n-2}.$$

$$\tag{2.2.27}$$

Using inequality (2.2.21) and the Rouché theorem, when $|t| < \epsilon$, we find that the equation system

$$z_j^{n-1} - t_j\prod_{i\neq j}(z_j - z_k) = 0, \quad j = \overline{1,n}$$

has $n(n-1)$ different solutions: $z_i(t)$ are lying in the vicinity of the origin of the coordinates, and $z_i(0) = 0$, for all $i = \overline{1, n(n-1)}$.

We shall use the change of variables in the integral (2.2.22),

$$\omega_j = (-1)^{j-1}z_j^{n-1}\Delta_{n-1}^{(j)}, \quad j = \overline{1,n}. \tag{2.2.28}$$

The variables ω_j satisfy the equations

$$F_1(\vec{\omega}, \vec{z}) = \omega_1 + \cdots + \omega_n - \Delta_n = 0,$$

$$F_2(\vec{\omega}, \vec{z}) = \frac{\omega_1}{z_1} + \cdots + \frac{\omega_n}{z_n} = 0, \tag{2.2.29}$$

$$F_n(\vec{\omega}, \vec{z}) = \frac{\omega_1}{z_1^{n-1}} + \cdots + \frac{\omega_n}{z_n^{n-1}} = 0,$$

which we obtain by using the properties of the Vandermond determinant.

The Jacobian of the transformation equals

$$\partial(\omega_1,\ldots,\omega_n)/\partial(z_1,\ldots,z_n) = \partial(F_1,\ldots,F_n)/\partial(z_1,\ldots,z_n)$$
$$\times[\partial(F_1,\ldots,F_n)/\partial(\omega_1,\ldots,\omega_n)]^{-1} = (2n-3)(n-1)!(z\Delta_n)^{n-2},$$

and $\partial(\omega_1,\ldots,\omega_n)/\partial(z_1,\ldots,z_n) \neq 0$, when $z_i \in \Gamma(\rho)$.

After the change of variables, the integral (2.2.26) takes the following form:

$$\times \int_{\gamma(r)}\prod_{j=1}^{n}(\omega_j - t_j(\omega_1 + \cdots + \omega_n) -)^{-1}\prod_{i=1}^{n}d\omega_i, \tag{2.2.30}$$

where $\gamma(r) = \{\omega_i : |\omega_i| = r_i, i = \overline{1,n}\}$, and c_n is a constant.

As $G(0,\ldots,0) = 1$ and assuming that $t = 0$ in (2.2.30), we obtain $c_n = (2n-3)(n-1)!$. Consequently,

$$G(t_1,\ldots,t_n) = (2\pi i)^{-n}\int_{\gamma(r)}\prod_{j=1}^{n}(\omega_j - t_j(\omega_1 + \cdots + \omega_n))^{-1}\prod d\omega_i. \tag{2.2.31}$$

After the change of variables,

$$v_j = \omega_j - t_j(\omega_1 + \cdots + \omega_n), \quad j = \overline{1,n}.$$

This integral can be calculated easily. As a result, $G = (1 - t_1 - \cdots - t_n)^{-1}$.

By decompositing $G(t_1,\ldots,t_n)$ in the Taylor series, we obtain the necessary result.

§3 Methods of Calculating the Moments of Random Determinants

In this section, the main methods of calculating the moments of random determinants are introduced. We shall consider that all the moments of random elements of matrices do exist.

1) *The change of variables method.* Let Ξ be a random real matrix of order n with the probability density $p(x)$. Obviously, $\mathbf{E} \det \Xi_n^k = \int p(X) \det X^k dX$. We use the change of variables $X = \varphi(Y)$, where φ is a one-to-one differential transformation. Then $\mathbf{E} \det \Xi_n^k = \int p(\varphi(Y)) \det(\varphi(Y))^k j(Y) dY$, where $j(Y)$ is a Jacobian of the transformation. As we see in §1 and 2, in some cases there exists transformation φ such that the moments $\mathbf{E} \det \Xi_n^k$ can be calculated in obvious form.

2) *The orthogonal transformations method.* It is applied successfully if the elements of the matrix Ξ_n are independent and distributed according to the standard normal law $N(0,1)$. In this case, the distribution of the matrix $\Xi_n \times H_n$ does not change, where H_n is some nonrandom orthogonal matrix. If the elements of the random matrix H_n are independent and distributed according to the normal $N(0,1)$ law, and if B_n is a square nonrandom matrix of order n, then the distribution of a random variable $\det[B_n + \Xi_n]$ depends only on singular eigenvalues of the matrix B_n and does not depend on the other functions of the elements of the matrix B_n. The proof is obtained from the fact that $B_n = U_n \Lambda V_n$, where U_n and V_n are orthogonal matrices, and Λ is a diagonal matrix of eigenvalues of the matrix $B_n B_n'$.

Many results which were obtained by this method can be seen in §1 of this chapter.

3) *Differentiation by the parameter method [11].* Let $D = \det(\partial/\partial s_{ij})_{i,j=1}^n$ be a differential operator. Obviously,

$$\det(\xi_{pl}) = D \exp \left\{ i \sum_{p,l=1}^{n} \xi_{pl} s_{pl} \right\}_{s_{pl}=0}, \quad p,l = \overline{1,n}.$$

From this formula, we obtain

$$\mathbf{E} \det{}^k(\xi_{pl}) = D^k \mathbf{E} \exp \left\{ i \sum_{p,l=1}^{n} \xi_{pl} s_{pl} \right\}_{s_{pl}=0}, \quad p,l = \overline{1,n},$$

where $k > 0$ is an integer.

If the random variables ξ_{pl} are independent, then we have

$$\mathbf{E} \det{}^k(\xi_{pl}) = D^k \prod_{p,l=1}^{n} f_{pl}(s_{pl})|_{s_{pl}=0}, \quad p,l = \overline{1,n},$$

where f_{pl} are the characteristic functions of the random variables ξ_{pl}. If ξ_{pl}, $p,l = \overline{1,n}$ are independent and distributed according to the normal $N(0,\sigma_{pl}^2)$ laws, then

$$\mathbf{E}\det{}^k(\xi_{pl}) = D^k \exp\left\{-0.5\sum_{p,l=1}^{n} s_{pl}^2\sigma_{pl}^2\right\}\Bigg|_{s_{pl}=0}.$$

4) *The method of integration on Grassmann and Clifford algebras.* For any positive integer k, we shall take k Grassmann algebras with $2n$ generators $(x_{is}, x_{is}^*, i = \overline{1,n})$, $s = \overline{1,k}$ (see §5, Chapter 1); herewith the generators of the different Grassmann algebras commute. Then, using (1.5.3), we obtain

$$\mathbf{E}\det{}^k(\xi_{pl}) = \mathbf{E}\int \exp\left\{i\sum_{p,l=1}^{n}\xi_{pl}\left(\sum_{s=1}^{n}x_{ps}x_{ls}^*\right)\right\}\prod_{p=\overline{1,n},s=\overline{1,k}}dx_{ps}dx_{ps}^* i^{-kn}.$$

(2.3.1)

If the random variables ξ_{pl} are independent and their characteristic functions are analytical, then we obtain formula (2.3.1) in the form

$$\mathbf{E}\det{}^k(\xi_{pl}) = \int \prod_{p,l=1}^{n} f_{pl}\left(\sum_{s=1}^{k}x_{ps}x_{ls}^*\right)\prod dx_{ps}dx_{ps}^* i^{-kn}.$$

If $\xi_{pl}, p, l = \overline{1,n}$ are independent and distributed according to the normal $N(0,\sigma_{pl}^2)$ laws, then

$$\mathbf{E}\det{}^k(\xi_{pl}) = \int \exp\left\{-0.5\sum_{p,l=1}^{n}\sigma_{pl}^2\left(\sum_{s=1}^{k}x_{ps}x_{ls}^*\right)^2\right\}\prod dx_{ps}dx_{ps}^* i^{-kn}.$$

(2.3.2)

5) *The integral representation method.* This method is applied to the calculation of the moments of random variables $\det \Xi^{-1}$. Using formula (1.4.3), we obtain

$$\mathbf{E}[\det(I + \Xi_n\Xi_n')]^{-k} = \mathbf{E}\exp\left[i\sum_{p,l=1}^{n}\xi_{pl}\left(\sum_{s=1}^{2k}\eta_{ps}\zeta_{ls}\right)\right],$$

(2.3.3)

where ξ_{pl} are the elements of the square matrix Ξ_n, $\eta_{ps}, \zeta_{ls}, p, l, s = 1, 2, \ldots$ are independent random variables (not depending on ξ_{pl}) and distributed according to the normal $N(0,1)$ law. If $\xi_{pl}, p, l = \overline{1,n}$ are independent and distributed according to the normal $N(0, \sigma^2 pl)$ law, then

$$\mathbf{E}[\det(I + \Xi_n\Xi_n')]^{-k} = \mathbf{E}\exp\left[-0.5\sum_{p,l=1}^{n}\sigma_{pl}^2\left(\sum_{s=1}^{2k}\eta_{ps}\zeta_{ls}\right)^2\right].$$

By expanding the exponent into series, we can obtain the formulas for the moments of random determinants $\det(I + \Xi_n\Xi_n')^{-1}$.

With the help of the integral representation (2.3.3), in some cases we can obtain the asymptotic formulas for $\mathbf{E}\det(I + \Xi_n\Xi_n')$ for sufficiently large n (see Chapter 8).

6) *The spectral method.* If the elements of a random square matrix Ξ_n have a probability density, then the eigenvalues λ_i, $i = \overline{1,n}$ of such a matrix have a probability density $p(x_i, i = \overline{1,n})$, and in some cases it has a common form. Using this fact, we can calculate the moments of random determinants $\det\Xi_n$ because $\det\Xi_n = \prod_{i=1}^{n}\lambda_i$.

7) *The recurrent equations method* [173, 36, 138]. By decomposing the determinant of a random matrix by the row vector (column vector), we can obtain some recurrent equations for the moments of random determinants. But they have a cumbersome form; therefore, we can find a solution for special cases only. Let us give an example of calculating the moments of random determinants with this method.

If the elements ξ_{ij} of a random matrix Ξ_n are independent and distributed identically, then

$$\mathbf{E}\det\Xi_n^{2k+1} = 0, \quad n > 1, \quad k = 0,1,2,\ldots, \tag{2.3.4}$$

$$\mathbf{E}\det\Xi_n^2 = n!(m_2 - m_1^2)^{n-1}(m_2 - m_1^2 + nm_1^2),$$

where $m_k = \mathbf{E}\xi_{11}^k$.

Consider the matrix $\Xi_{j_1j_2\ldots j_k}^{i_1i_2\ldots i_k}$, where the $i_1, i_2, \ldots, i_k$ rows and $j_1, j_2, \ldots, j_k$ columns are deleted. If we change any two rows of the matrix Ξ_n, then the distribution of a new matrix determinant and the distribution of $\det\Xi_n$ are equal. Therefore the distribution of $\det\Xi_n$ is symmetric when $n > 1$, and consequently, $\mathbf{E}\det\Xi_n^{2k+1} = 0, n > 1, k = 0,1,2,\ldots$.

Let us decompose $\det\Xi$ by the first column. Then

$$\mathbf{E}\det\Xi_n^2 = nm_2\mathbf{E}\det\Xi_{n-1}^2 + \sum_{k\neq l}m_1^2\mathbf{E}\det\Xi_1^k\det\Xi_1^l(-1)^{k+1}. \tag{2.3.5}$$

The distribution of $\mathbf{E}\det\Xi_1^k\det\Xi_1^l(-1)^{k+l}$ does not depend on the choice of numbers k and $l, k \neq l$. It is easily obtained by changing the places of corresponding rows of matrices Ξ_1^k, Ξ_1^l. We obtain

$$\mathbf{E}\det\Xi_1^1\det\Xi_1^2 = (n-1)m_2\mathbf{E}\det\Xi_{n-2}^2 - (n-1)(n-2)m_1^2\mathbf{E}(\det\Xi_{12}^{12}\det\Xi_{13}^{21}). \tag{2.3.6}$$

From Eq. (2.3.5), it follows that

$$\mathbf{E}\det\Xi_{n-1}^2 = (n-1)m_2\mathbf{E}\det\Xi_{n-2}^2 - (n-1)(n-2)m_1^2\mathbf{E}(\det\Xi_{12}^{12}\det\Xi_{13}^{21}). \tag{2.3.7}$$

By excluding $\mathbf{E}(\det\Xi_{12}^{12}\det\Xi_{13}^{21})$ from (2.3.6) and (2.3.7), we find

$$\mathbf{E}\det\Xi_n^2 = nm_2\mathbf{E}\det\Xi_{n-1}^2 - n(n-1)m_1^2\mathbf{E}\det\Xi_{n-1}^2 - n(n-1)^2m_1^4$$
$$\times \mathbf{E}\det\Xi_{n-2}^2 + n(n-1)^2m_2m_1^2\mathbf{E}\det\Xi_{n-2}^2. \tag{2.3.8}$$

Let $\mathbf{E}\det\Xi_n^2 = m_1^{2n}n!y_n, m_1 \neq 0$. Consequently, from (2.3.8), it follows that

$$y_n = y_{n-1}(\gamma + 2 + n) + (n-1)\gamma y_{n-2}, \quad y_1 = \gamma + 1, \quad y_2 = \gamma(\gamma + 2), \quad (2.3.9)$$

where $\gamma = m_2/m_1^2 - 1$.

Suppose that $y_0 = 1$, and we find the generator function $f(t) = \sum_{n=0}^{\infty} y_n t^n / n!$. By using (2.3.9), we obtain $f'(t)(1+t) = f(t)(1 + \gamma + \gamma t)$. The solution of this equation is $f(t) = c(1+t)e^{\gamma t}$, where c is an arbitrary constant. Hence, by using (2.3.9), we obtain (2.3.4).

If $m_1 = 0$, then (2.3.4) follows from Eq. (2.3.5). We obtain

$$\mathbf{E}\det\Xi_n^4 = m_2^{2n}(n!)^2 n^{-1} \sum_{k=0}^{n}(n-k+1)(n-k+2)(k!)^{-1}((m_4/m_2^2) - 3)^k.$$

$$(2.3.10)$$

where $m_1 = 0$.

8) *The "straight" method of calculating the moments of random determinants.* By decomposing the determinant by some rows and columns and by calculating the expectation of the obtained sum, we can obtain some simple results.

Let the elements ξ_{ij} of a random matrix Ξ_n be independent. Obviously, $\mathbf{E}\det\Xi_n = \det \mathbf{E}\Xi_n$. If $\mathbf{E}\xi_{ij} = 0$, $\mathbf{E}\xi_{ii}^2 = m_{ij}^2$, then $\mathbf{E}\det\Xi_n^2 = \mathrm{per}\, K_n$, where $K_n = (m_{ij}^2)$ is a square matrix of order n.

9) *The perturbations method.* Using the formula matrices perturbations (see §5 of this chapter), we can obtain different asymptotic formulas for sufficiently large n for the moments of the random determinants $\mathbf{E}|\det\Xi_n|^{k/c_n}$, where c_n is a sequence of normalized numbers, and k are positive integers which do not depend upon n.

§4 Moments of Random Permanents

The permanent of the random matrix has some properties of the matrix determinant. Therefore, we apply the methods of studying moments of random permanents to the study of moments of random determinants. In this paragraph we consider an example of random matrices for moments of permanents for which the approximate formulas can be found.

Theorem 2.4.1. *[152].* *If the elements ξ_{ij} of the random matrices $\Xi_n = (\xi_{ij})_{i,j=1}^n$ are independent and if there exist constants $c > 0, d > 0$ such that $c \leq \xi_{ij} \leq d$, for all $i,j = 1,2,\ldots$ with probability one, $\mathbf{E}\xi_{ij} = a$, $\mathrm{Var}\,\xi_{ij} = \sigma^2$, then for any integer $k > 1$,*

$$\lim_{n\to\infty} \mathbf{E}[\mathrm{per}\,\Xi_n/n!a^n]^k = \exp\{k(k-1)\sigma^2/2a^2\}. \quad (2.4.1)$$

Proof. The matrices $T_s, s = \overline{1,n}$ are obtained from the matrix Ξ by replacing the elements of the first s rows by the variables a. Let $T_0 = \Xi_n$. Then,

obviously,

$$\mathbf{E}[\text{per}\,\Xi_n/n!a^n]^k = \mathbf{E}\left(\prod_{s=0}^{n-1}[\text{per}\,T_s/\text{per}\,T_{s+1}]\right)^k$$

$$= \mathbf{E}\left\{\prod_{s=1}^{n}\left(1 + \sum_{i=1}^{n}(\xi_{si} - a)T_{si}/a\sum_{i=1}^{n}T_{si}\right)\right\}, \quad (2.4.2)$$

where T_{si} is a permanent of the matrix, which is obtained from the matrix Ξ_n by striking off the sth row and column.

Consider

$$\eta_s = \sum_{i=1}^{n}(\xi_{si} - a)T_{si}\sigma^{-1}[\sum_{i=1}^{n}T_{si}]^{1/2}, b_s = \sum_{i=1}^{n}T_{si}^2(\sum_{i=1}^{n}T_{si})^{-2}.$$

Obviously, $\mathbf{E}(\eta_s/\sigma_s) = 0$, $\mathbf{E}(\eta_s^2/\sigma_s) = l$, where σ_s is a minimal σ- algebra with respect to which the vector rows of the matrix Ξ_n are measurable, except the sth row. We transform

$$b_s = \sum_{j=1}^{n}(\sum_{i=1}^{n}(T_{si}/T_{sj}))^{-2}.$$

Decomposing the permanent T_{si} by the jth column and T_{sj} by the ith column, we obtain $c \le \frac{T_{si}}{T_{sj}} \le d$, $n^{-1}d^{-2} \le b_s \le n^{-1}c^{-2}$. Using this inequality, we obtain $p\lim_{n\to\infty}(T_{si}/T_{sj}) = 1$, and for any $\varepsilon > 0$,

$$\mathbf{P}\{|T_{si}/T_{sj} - 1| > \varepsilon\} \le \varepsilon^{-m}(n-1)^{-m}c, \quad (2.4.3)$$

where $c > 0$ is a constant, and $m > 0$ is an integer.

For any integer m,

$$\sup_s \mathbf{E}[|\eta_s|^m/\sigma_s] < \infty.$$

Then by using (2.4.2), it follows that

$$\mathbf{E}[\text{per}\,\Xi_n/n!a^n]^k = \mathbf{E}[\mathbf{E}[1 + a^{-1}\sigma\eta_1\sqrt{b_1}/\sigma_1]^k \prod_{s=2}^{n}(1 + a^{-1}\sigma\eta_s\sqrt{b_s})^k]$$

$$= \mathbf{E}[1 + 0.5a^{-2}k(k-1)\sigma^2\mathbf{E}(b_1^2/\sigma_1) + \varepsilon_{kn}n^{-3/2}]\prod_{s=2}^{n}(1 + \sigma a^{-1}\eta_s\sqrt{b_s})^k$$

$$= [1 + (k(k-1)\sigma^2a^{-2}/2 + o(1))n^{-1}\mathbf{E}\prod_{s=2}^{n}(1 + \sigma\eta_s\sqrt{b_s}/a)^k$$

$$+ \mathbf{P}\{|b_1^2 - n^{-1}| > \varepsilon\} + \mathbf{E}\{\varepsilon_{kn}n^{-3/2}\prod_{s=2}^{n}(1 + \sigma\eta_s\sqrt{b_s}/a)^k$$

$$\times \prod_{s=2}^{n}(1 + \sigma\eta_s\sqrt{b_s}/a)^k\}, \quad (2.4.4)$$

where ε_{kn} is a random variable with probability 1,

$$\sup_{n,k} |\varepsilon_{kn}| < \infty.$$

Using (2.4.3), we have $\mathbf{P}\{|b_1^2 - n^{-1}| > \varepsilon/\sigma_1\} \le cn^{-3/2}$. By using this inequality and by continuing the process of calculating the variables $\prod_{s=2}^n (1 + \sigma\eta_s)\sqrt{b_s}/a)^k$ which are included in (2.4.4), we obtain the theorem.

Theorem 2.4.2.   *Let the random elements* ξ_{ij}, $i,j = 1,2,\ldots$ *of the random matrices* $\Xi_n = (\xi_{ij})$ *be independent and there exist the constants* $c > 0, d > 0$, *such that with probability 1,*

$$c \le \xi_{ij} \le d, \qquad i,j = 1,2,\ldots, \qquad \mathbf{E}\xi_{ij} = a_{ij}, \quad \mathrm{Var}\,\xi_{ij} = \sigma_{ij}^2,$$

Then for any $\varepsilon > 0$,

$$\mathrm{plim}_{n\to\infty} [\mathrm{per}\,\Xi_n / \mathrm{per}\,\mathbf{E}\Xi_n]^{1/c_n} = 1(c_n \to \infty),$$

and if $a_{ij} \equiv a, i,j = 1,2,\ldots$, *then*

$$\mathbf{E}[\mathrm{per}\,\Xi_n / n! a^n]^k = \exp\left\{ 0.5k(k-1)n^{-2}a^{-2} \sum_{i,j=1}^n \sigma_{ij}^2 \right\} + o(1).$$

To prove Theorem 2.4.2, we use the equality

$$\mathrm{per}\,\Xi_n / \mathbf{E}\,\mathrm{per}\,\Xi_n = \prod_{s=0}^{n-1} (\mathrm{per}\,\widetilde{T}_s / \mathrm{per}\,\widetilde{T}_{s+1}),$$

where

$$\widetilde{T}_s = (a_{ij}\delta_{\min(s,i),i} + \xi_{ij}\delta_{\max(s+1,i)i}).$$

§5 Formulas of Random Determinant Perturbation

To calculate the moments of random determinants is rather difficult; therefore, it is of interest to find the limit theorems for the moments of random determinants. To prove them, we need a method for the perturbation of random determinants which is based on the formula:

$$\det A / \det B = \det(I + B^{-1}(A - B)),$$

where $\det B \neq 0$.

By choosing special built matrices as the matrix $A - B$, we can formulate limit theorems for some random determinants. Consider the main assertions and formulas for the proof of these limit theorems.

Lemma 2.5.1. *Let $H_n, \tilde{H}_n, c_n$ be the nonsingular matrices of order n, and $H_n - \tilde{H}_n = (\vec{\xi}\vec{\eta}\,')$ be the square matrix of order n.*
Then,

$$\operatorname{Tr} C_n H_n^{-1} - \operatorname{Tr} C_n \tilde{H}_n^{-1} = -(\tilde{H}_n^{-1} C_n \tilde{H}_n^{-1}\vec{\xi}, \vec{\eta})[1 + (\tilde{H}_n^{-1}\vec{\xi}, \vec{\eta})]^{-1}. \quad (2.5.1)$$

Proof. Obviously,

$$\operatorname{Tr} C_n H_n^{-1} - \operatorname{Tr} C_n \tilde{H}_n^{-1}$$
$$= \frac{\partial}{\partial z} \ln \det[Iz + C_n^{-1} H_n]_{z=0} - \frac{\partial}{\partial z} \ln \det[Iz + C_n^{-1} \tilde{H}_n]_{z=0}$$
$$= \frac{\partial}{\partial z} \ln \det I + (Iz + C_n^{-1}\tilde{H}_n)^{-1} C_n^{-1}(H_n - \tilde{H}_n)_{z=0}. \quad (2.5.2)$$

Using the formula $\det[I + R\vec{\xi}\vec{\eta}\,'] = 1 + (R\vec{\xi}\vec{\eta})$, where R is the matrix of order n, we have (2.5.1). Lemma 2.5.1 is proved.

Using (2.5.1), we have the formulas

$$H_n^{-1} - \tilde{H}_n^{-1} = \left\{ -\sum_{i=1}^{n} \tilde{h}_{ip}^{(-1)}\xi_j \sum_{j=1}^{n} \tilde{h}_{lj}^{(-1)}\eta_j [1 + (\tilde{H}_n^{-1}\vec{\xi}, \vec{\eta})]^{-1} \right\}_{p,l=\overline{1,n}}, \quad (2.5.3)$$

where $\tilde{h}_{ip}^{(-1)}$ are the elements of the matrix $\tilde{H}_n^{-1}$.
To obtain Eq. (2.5.3), we differentiate Eq. (2.5.1) by the variable c_{pl}. If the matrices H_n and $\tilde{H}_n$ are symmetric, $\vec{\xi} = \vec{\eta}$, then

$$h_{pp}^{(-1)} - \tilde{h}_{pp}^{(-1)} = -\left(\sum_{i=1}^{n} \tilde{h}_{ip}^{(-1)}\xi_i \right)^2 [1 + (\tilde{H}_n^{-1}\vec{\xi}, \vec{\xi})]^{-1},$$
$$\operatorname{Tr} H_n^{-1} - \operatorname{Tr} \tilde{H}_n^{-1} = -[(\tilde{H}_n^{-1})^2\vec{\xi}, \vec{\xi}][1 + (\tilde{H}_n^{-1}\vec{\xi}, \vec{\xi})]^{-1}, \quad (2.5.4)$$

where $h_{pp}^{(-1)}$ are the elements of the matrix H_n^{-1}.
Let A be a square matrix of order n. We represent it in the form

$$A = \begin{pmatrix} A_{11} & A_{12} \\ A_{21} & A_{22} \end{pmatrix},$$

where

$$A_{11} = (a_{ij}), \quad i = \overline{1,p}, \qquad j = \overline{1,q};$$
$$A_{12} = (a_{ij}), \quad i = \overline{1,p}, \qquad j = \overline{q+1,n};$$
$$A_{21} = (a_{ij}), \quad i = \overline{p+1,n}, \quad j = \overline{1,q};$$
$$A_{22} = (a_{ij}), \quad i = \overline{p+1,n}, \quad j = \overline{q+1,n}.$$

If $p = q$, then [3],

$$BAB' = \begin{bmatrix} A_{11} - A_{12}A_{22}^{-1}A_{21} & 0 \\ 0 & A_{22} \end{bmatrix}, \tag{2.5.5}$$

where

$$B = \begin{bmatrix} I & -A_{12}A_{22}^{-1} \\ 0 & I \end{bmatrix}, \quad \det A_{22} \neq 0.$$

If $\det A_{22} \neq 0$ and $p = q$, then from (2.5.5), we obtain

$$\det A = \det[A_{11} - A_{12}A_{22}^{-1}A_{21}]\det A_{22}. \tag{2.5.6}$$

If $p = q = 1$, then

$$\det A = (a_{11} - (A_{22}^{-1}\vec{a}, \vec{b}))\det A_{22}, \tag{2.5.7}$$

where $\vec{a} = (a_{i1}, i = \overline{2,n}), \vec{b} = (a_{1i}, i = \overline{2,n}.$

Let Ξ_n be the symmetric matrices; and the matrix Ξ_n^k is obtained from Ξ_n by replacing the elements of the kth row and the kth column by zeros. Let $A_n = (a_{ij})$ be the matrix which satisfies the condition $\det A_n \neq 0$,

$$R_t = (I + it\Xi_n)^{-1}, \qquad R_t^k = (I + it\Xi_n^k) = (r_{ij}^k),$$

where t is a real parameter.

Lemma 2.5.2.

$$\operatorname{Tr} R_t A_n - \operatorname{Tr} R_t^k A_n = - \left\{ it \sum_{p=1}^{n} \left(\sum_{\nu,\mu=1}^{n} r_{p\nu}^k a_{\nu\mu} r_{\mu k}^k \right) \xi_{pk} \left(1 + it \sum_{j=1}^{n} r_{kj}^k \xi_{kj} \right) \right.$$

$$+ \left(1 + it \sum_{p=1}^{n} r_{pk}^k \xi_{pk} \right) \left(it \sum_{j=1}^{n} \left(\sum_{\nu,\mu=1}^{n} r_{k\nu}^k a_{\nu\mu} r_{\mu j}^k \right) \xi_{kj} + it \right.$$

$$\times \left(\sum_{\nu,\mu=1}^{n} r_{k\nu}^k a_{\nu\mu} r_{\mu k}^k \right) \xi_{kk} + t^2 \left(\sum_{\nu,\mu=1}^{n} r_{k\nu}^k a_{\nu\mu} r_{\mu k}^k \right) \sum_{j,l=1}^{n} r_{lj}^k \xi_{lk} \xi_{kj}$$

$$+ t^2 r_{kk}^k \sum_{j,l=1}^{n} \left(\sum_{\nu,\mu=1}^{n} r_{l\nu}^k a_{\nu\mu} r_{\mu j}^k \right) \xi_{lk} \xi_{kj} \right\} \left\{ \left(1 + it \sum_{p=1}^{n} r_{pk}^k \xi_{pk} \right) \right.$$

$$\times \left. \left(1 + it \sum_{j=1}^{n} r_{kj}^k \xi_{kj} \right) - it r_{kk}^k \xi_{kk} + t^2 r_{kk}^k \sum_{j,l=1}^{n} r_{lj}^k \xi_{lk} \xi_{kj} \right\}^{-1}.$$

Proof. Obviously,

$$\operatorname{Tr} R_t A_n - \operatorname{Tr} R_t^k A_n$$

$$= \frac{\partial}{\partial z} \ln \det[I + A_n z + it\Xi_n]_{z=0} - \frac{\partial}{\partial z} \ln \det[I + A_n z + it\Xi_n^k]_{z=0}$$

$$= \frac{\partial}{\partial z} \ln \det[I + itR_{tz}^k(\Xi_n - \Xi_n^k)]_{z=0},$$

where $R_{tz}^k = [I + A_n z + it\Xi_n^k]^{-1} = (r_{pl}^k(z))$.

The matrix $\Xi_n - \Xi_n^k$ has the form $H_1 + H_2$, where $H_1 = [\xi_{ij}\delta_{jk}]$, $H_2 = [\xi_{ij}\delta_{ik}(1 - \delta_{jk})]$.

Therefore, we have

$$\det[I + itR_{tz}^k H_1 + itR_{tz}^k H_2] = \det Q_{tz} \det[I + itQ_{tz}^{-1} R_{tz}^k H_2],$$

where $Q_{tz} = [I + itR_{tz}^k H_1]$.

Using formula (2.5.2), we obtain

$$\det[I + itR_{tz} H_1 + itR_{tz} H_2]$$

$$= \left[1 + it \sum_{p=1}^n r_{pk}^k(z)\xi_{pk}\right]\left[1 + it \sum_{j \neq k, l=\overline{1,n}} q_{kl} r_{lj}^k \xi_{kj}\right]. \tag{2.5.8}$$

For every matrix C_n, $\det C_n \neq 0$,

$$\operatorname{Tr} C_n Q_{tz}^{-1} = \frac{\partial}{\partial \theta} \ln \det[I + \theta C_n + itR_{tz}^k H_1]_{\theta=0}$$

$$= \frac{\partial}{\partial \theta} \ln \det[I + \theta C_n]_{\theta=0} + \frac{\partial}{\partial \theta} \ln \det[I + it(I + \theta C_n)^{-1} R_{tz}^k H_1]_{\theta=0}$$

$$= \operatorname{Tr} C_n + \frac{\partial}{\partial \theta} \ln \left[1 + it \sum_{p,l=1}^n (I + \theta C_n)_{pl}^{-1} r_{lk}^k(z)\xi_{pk}\right]_{\theta=0}$$

$$= \operatorname{Tr} C_n - it \sum_{p,l=1}^n c_{pl} r_{lk}^k(z)\xi_{pk} \left[1 + it \sum_{p=1}^n r_{pk}^k(z)\xi_{pk}\right]^{-1}.$$

We obtain

$$q_{kl} = \delta_{kl} - itr_{kk}^k(z)\xi_{lk}[1 + it \sum_{p=1}^n r_{pk}^k(z)\xi_{pk}].$$

Substituting q_{kl} in (2.5.8), we have

$$\det[I + itR_{tz}H_1 + itR_{tz}H_2] = \left[1 + it\sum_{p=1}^{n} r_{pk}^k(z)\xi_{pk}\right]$$

$$\times \left[1 + it\sum_{j\neq k}\left\{\delta_{kl} - itr_{kk}^k(z)\xi_{lk}\left[1 + it\sum_{p=1}^{n} r_{pk}^k(z)\xi_{pk}\right]^{-1}\right\}r_{lj}^k(z)\xi_{kj}\right]$$

$$= \left(1 + it\sum_{p=1}^{n} r_{pk}^k(z)\xi_{pk}\right)\left(1 + it\sum_{j=1}^{n} r_{kj}^k(z)\xi_{kj}\right)$$

$$- itr_{kk}^k(z)\xi_{kk} + t^2 r_{kk}^k(z)\sum_{j,l=1}^{n} r_{lj}^k(z)\xi_{lk}\xi_{kj}. \tag{2.5.9}$$

Obviously,

$$\frac{\partial}{\partial z}R_{tz}^k|_{z=0} = -R_t^k A_n R_t^k = (b_{ij}),$$

where $b_{ij} = -\sum_{p,l=1}^{n} r_{ip}^k a_{pl}r_{lj}^k$. Therefore,

$$\frac{\partial}{\partial z}r_{ij}^k(z)|_{z=0} = -\sum_{p,l=1}^{n} r_{ip}^k a_{pl}r_{lj}^k, \quad r_{ij}^k(z)|_{z=0} = r_{ij}^k.$$

From (2.5.9), we have

$$\frac{\partial}{\partial z}\ln\det[I + itR_{tz}^k(\Xi_n - \Xi_n^k)]_{z=0}$$

$$= \left\{-it\sum_{p=1}^{n}\left(\sum_{\nu,\mu} r_{p\nu}^k a_{\nu\mu}r_{\mu k}^k\right)\xi_{pk}\left(1 + it\sum_{j=1}^{n} r_{kj}^k\xi_{kj}\right)\right.$$

$$- \left(1 + it\sum_{p=1}^{n} r_{pk}^k\xi_{pk}\right)\left(it\sum_{j=1}^{n}\left(\sum_{\nu,\mu=1}^{n} r_{k\nu}^k a_{\nu\mu}r_{\mu j}^k\right)\xi_{kj}\right)$$

$$+ it\left(\sum_{\nu,\mu=1}^{n} r_{k\nu}^k a_{\nu\mu}r_{\mu k}^k\right)\xi_{kk} - t^2\left(\sum_{\nu,\mu=1}^{n} r_{k\nu}^k a_{\nu\mu}r_{\mu k}^k\right)$$

$$\times \sum_{j,l=1}^{n} r_{lj}^k\xi_{lk}\xi_{kj} - t^2 r_{kk}^k\sum_{j,l=1}^{n}\left(\sum_{\nu,\mu=1}^{n} r_{l\nu}^k a_{\nu\mu}r_{\mu j}^k\right)\xi_{lk}\xi_{kj}\right\}$$

$$\times \left\{\left(1 + it\sum_{p=1}^{n} r_{pk}^k\xi_{pk}\right)\left(1 + it\sum_{j=1}^{n} r_{kj}^k\xi_{kj}\right)\right.$$

$$\left.- itr_{kk}^k\xi_{kk} + t^2 r_{kk}^k\sum_{j,l=1}^{n} r_{lj}^k\xi_{lk}\xi_{kj}\right\}^{-1}.$$

Lemma 2.5.2 is proved.

Using (2.5.9), we obtain

$$r_{pp} = \left[1 + it\xi_{pp} + t^2 \sum_{i,j \neq k} r_{ij}^k \xi_{ik} \xi_{jk} \right]^{-1}, \tag{2.5.10}$$

where $r_{pk}^k = 0, r_{kp}^k = 0, p \neq k, r_{kk}^k = 1$.

If Ξ_n is a symmetric matrix, then from Lemma 2.5.2 we have

$$r_{pp} - r_{pp}^k = -t^2 \left(\sum_{i \neq k} r_{ip}^k \xi_{ik} \right)^2 \left[1 + it\xi_{kk} + t^2 \sum_{i,j \neq k} r_{ij}^k \xi_{ik} \xi_{kj} \right]^{-1},$$

$$p \neq k; \tag{2.5.11}$$

$$r_{kk} - r_{kk}^k = \left[1 + it\xi_{pp} + t^2 \sum_{i,j \neq k} r_{ij}^k \xi_{ik} \xi_{kj} \right]^{-1} - 1; \tag{2.5.12}$$

$$\mathrm{Tr}\, R_t - \mathrm{TR}\, R_t^k = \frac{\partial}{\partial t} \ln \left[1 + it\xi_{kk} + t^2 \sum_{i,j=1}^n r_{ij}^k \xi_{ik} \xi_{kj} \right]^{-1}. \tag{2.5.13}$$

By using these formulas, we can find the asymptotic expressions for the determinants of random matrices with independent elements for a sufficiently large order of them.

CHAPTER 3

DISTRIBUTION OF EIGENVALUES AND
EIGENVECTORS OF RANDOM MATRICES

In this chapter, a connection between the distributions of eigenvalues and eigenvectors of random matrices and the distributions of random determinants is studied. It will be proved that with matrix entries being random and having a joint distribution density, there is a joint density of roots of simple characteristic equations. The moments of some random determinants are found. Formulas for densities of the eigenvalues and eigenvectors of random matrices will be used since the determinants of many functions of random matrices are expressed in terms of its eigenvalues and eigenvectors.

§1 Distribution of Eigenvalues and Eigenvectors of Hermitian Random Matrices

We begin with finding the distribution of eigenvalues and eigenvectors of symmetric random matrices and the Hermitian random matrices.

We consider a real symmetric random matrix Ξ_n of the order n whose entries $\xi_{ij}, i \geq j, j = \overline{1, n}$ have the joint distribution density $p(Z_n)$, where $Z_n = (z_{ij})$ is a real symmetric $n \times n$ matrix. The eigenvalues λ_k of the matrix Ξ_n are real and are some functions of its entries. It is known that these functions can be Borel functions. Therefore, the eigenvalues of Ξ_n can be chosen so that they will be random variables. In order to show that eigenvalues can be chosen so that they will not be random variables, we set

$$\tilde{\lambda}_1(\omega) = \left\{ \begin{array}{ll} \lambda_1(\omega), & \omega \in C, \\ \lambda_2(\omega), & \omega \bar{\in} C, \end{array} \right. \quad \tilde{\lambda}_2(\omega) = \left\{ \begin{array}{ll} \lambda_1(\omega), & \omega \bar{\in} C, \\ \lambda_2(\omega), & \omega \in C, \end{array} \right.$$

$$\tilde{\lambda}_k(\omega) = \lambda_k(\omega), \quad k = \overline{3, n},$$

where C is a certain unmeasurable set.

Obviously, $\lambda_1(\omega), \lambda_2(\omega)$ and the set C now can be chosen so that $\tilde{\lambda}_1(\omega)$ will not be a random variable.

54

In general, there are no exact formulas for eigenvalues as $n \geq 5$, and therefore it is difficult to find such a change of variables in the expression for the joint distribution of the eigenvalues that one-to-one correspondence between old and new variables is established. It is easy to select a proper change of variables if the eigenvalues $\lambda_k(\omega)$ are arranged in nondecreasing order. One can be convinced of the measurability of such eigenvalues by considering the following relations:

$$\nu_1(\omega) = \lim_{s \to \infty} [\mathrm{Tr}\, \Xi^{2s}]^{1/2s}, \quad \mu_1(\omega) = \lim_{s \to \infty} [\mathrm{Sp}(\Xi + \nu_1(\omega))0^{2s}]^{1/2s},$$

$$\mu_2(\omega) = \lim_{s \to \infty} [\mathrm{Sp}[\Xi + \nu_1(\omega)]^{2s} - \mu_1^{2s}(\omega)]^{1/2s}, \ldots,$$

$$\lambda_i(\omega) = \mu_i(\omega) - \nu_1(\omega), \quad i = \overline{1,n}. \tag{3.1.1}$$

In what follows, the eigenvalues $\lambda_i(\omega)$ are assumed to be arranged in nondecreasing order. We need the following statement.

Lemma 3.1.1. *The eigenvalues $\lambda_i, i = \overline{1,n}$ of Ξ are distinct with probability 1.*

Proof. If at least two roots λ_i coincide, then the expression for the Vandermond determinant which we obtain is $\Delta^2 := \det(s_{i+j})$, $i,j = \overline{0,n-1}$, $s_0 = n$, $s_k = \mathrm{Tr}\, \Xi_n^k$, $k = \overline{1,2n-2}$; Δ^2 is some polynomial of the entries of Ξ_n which is not identically equal to zero. Since the entries of Ξ_n have the joint distribution density, $\mathbf{P}\{\Delta^2 = 0\} = 0$. Lemma 3.1.1 is proved.

Let us choose n measurable eigenvectors $\overrightarrow{\theta_i}$ of the matrix Ξ_n, corresponding to the eigenvalues $\lambda_i, i = \overline{1,n}$. Note that the eigenvectors $\overrightarrow{\theta}_i$ are defined uniquely to within a coefficient 1 with probability 1 from the equations $(\Xi_n - \lambda_i I)\overrightarrow{\theta}_i = 0$, $(\overrightarrow{\theta}_i, \overrightarrow{\theta}_i) = 1$. One can rather easily choose n distinct eigenvectors by fixing a sign of some nonzero element of each vector. If some eigenvalues $\lambda_i, i = \overline{1,n}$ coincide, then the eigenvector is chosen by fixing some of its components and a sign of some nonzero element of each vector. Thus, the eigenvalues λ_i, arranged in nondecreasing order and corresponding eigenvectors $\overrightarrow{\theta}_i$ of the matrix Ξ_n thus chosen are random values and vectors.

Let θ_n be a random matrix whose column vectors are equal to $\overrightarrow{\theta}_i, i = \overline{1,n}$; let G be a group of real $n \times n$ matrices; B the σ-algebra of Borel sets of orthogonal $n \times n$ matrices on it, and μ the normalized Haar measure on G.

Theorem 3.1.1. *If a real random symmetric matrix Ξ_n has the density $p(Z_n)$, then for any subset $E \in B$ and the real numbers $\alpha_i, \beta_i, i = \overline{1,n}$,*

$$\mathbf{P}\{\theta_n \in \mathbf{E}, \alpha_i < \lambda_i < \beta_i, i = \overline{1,n}\}$$

$$= c_{1n} \int p \times (X_n Y_n X_n') \prod_{i>j}(y_i - y_j)\mu(dX_n)dY_n, \tag{3.1.2}$$

where the integration is over the domain $y_1 > y_2 > \cdots > y_n$, $\alpha_i < y_i <$ β_i, $X_n \in E$, $x_{1i} \geq 0$, $i = \overline{1,n}$, *where* X_n *is an orthogonal matrix*, $c_{1n} =$ $2^n \pi^{n(n+1)/4} \prod_{i=1}^{n} \{\Gamma[(n-i+1)/2]\}^{-1}$, $Y_n = (\delta_{ij} y_i)$, $dY_n = \prod dy_i$.

In this and the following analogous theorems, we assume that the density $p(Z_n)$ is such that the integral on the right side of Eq. (3.1.2) exists.

Proof. By Lemma 3.1.1, we have for any bounded continuous functions $f(\theta_n, \Lambda_n)$ of the entries of the matrices θ_n and $\Lambda_n = (\delta_{ij} \lambda_i)$,

$$\mathbf{E}[f(\theta, \Lambda)/\theta_{1i} \geq \varepsilon, i = \overline{1,n}]\mathbf{P}\{\theta_{1i} \geq \varepsilon, i = \overline{1,n}\}$$

$$= \int f(X_n, Y_n)p(Z_n)dZ_n,$$

where the integration is over the domain $\{x_{1i} \geq \varepsilon, i = \overline{1,n}, y_1 > \cdots > y_n\}$, $1 > \varepsilon > 0$, and the matrices Y_n and X_n satisfy $Z_n = X_n Y_n X_n'$. The Jacobian of the transformation $Z_n = H_n T_n H_n'$, where T_n is a real symmetric matrix, $H_n \in G$ is 1 (which is easy to verify by considering the transformation in the integral $\int \exp(-\operatorname{Tr} Z_n^2)dZ_n = \int \exp(-\operatorname{Tr} T_n^2)dT_n$), therefore we find that

$$\mathbf{E}[f(\theta_n, \Lambda_n)/\theta_{1i} \geq \varepsilon, i = \overline{1,n}]\mathbf{P}\{\theta_{1i} \geq \varepsilon, i = \overline{1,n}\}$$

$$= \int f(H_n X_n, Y_n)p(H_n Z_n H_n')dZ_n,$$

$$(3.1.3)$$

where the domain of integration is $\{\sum_{k=1}^{n} h_{1k} x_{ki} \geq \varepsilon, i = \overline{1,n}, y_1 > y_2 > \cdots > y_n\}$.

Let us consider a change of variables $Z_n = U_n Y_n U_n'$, where $U_n \in G$, and the entries of U_n are defined uniquely by the Euler angles (see §2, Chapter 1) which take values from the set K so that the first nonzero component of each column vector of U_n is positive. Obviously, the number of variables on the left-hand and right-hand sides of the equality $Z_n = U_n Y_n U_n'$ are the same; on the set $\{Z_n : y_1 > y_2 > \cdots > y_n\}$, this transformation is one-to-one, and the entries of $U_n Y_n U_n'$ are continuously differentiable with respect to the parameters u_i, y_i. After the change of the variables $Z_n = U_n Y_n U_n'$, the integral (3.1.3) takes the form

$$\int f(H_n U_n, Y_n)p(H_n U_n Y_n U_n' H_n')J(U_n, Y_N)dY_n \prod_{i=1}^{l} du_i,$$

$$(3.1.4)$$

where $J(U_n, Y_n)$ is the Jacobian of the transformation $Z_n = U_n Y_n U_n'$, $l = n(n-1)/2$, and the domain of integration equals $\{\sum_{k=1}^{n} h_{1k} u_{ki} \geq \varepsilon, i = \overline{1,n}, y_1 > y_2 > \cdots > y_n, u_j \in K, j = \overline{1,l}\}$.

The expression (3.1.4) does not depend on H_n. Therefore, if we integrate it with respect to the Haar measure μ defined on the group G of matrix H_n, we obtain

$$\int f(H_n, Y_n)p(H_n Y_n H_n')\varphi(Y_n)\mu(dH_n)dY_n,$$

$$(3.1.5)$$

where $\varphi(Y_n) = \int_{u_i \in K} J(Y_n, U_n) \prod_i du_i$, and the domain of integration equals $\{h_{1i} \geq \varepsilon, i = \overline{1,n}, y_1 > \cdots > y_n\}$.

We find the function $\varphi(Y_n)$ in the following way. Let us introduce the change of variables $y_i = az_i + b$, $i = \overline{1,n}$ in the integral (3.1.5), where $a \neq 0$ and b are arbitrary constants. Then we change the variables $z_{ij} = at_{ij} + b\delta_{ij}$, $i \geq j$, $j, i = \overline{1,n}$ in the integral (3.1.3). We obtain two equal integrals,

$$\int f(H_n, aY_n + Ib)p(aH_nY_nH'_n + Ib)\psi(aY_n + Ib)\mu(dH_n)dY_n a^n$$

$$= \int f(H_n, aY_n + Ib)p(aH_nY_nH'_n + Ib)\psi(Y_n)\mu(dH_n)dY_n a^{n(n+1)/2}$$

(the domain of integration is equal to $\{h_{1i} > \varepsilon, i = \overline{1,n}, y_1 > \cdots > y_n\}$). Hence by the arbitrariness of the function f,

$$\psi(aY_n + Ib) = \psi(Y_n)a^{n(n-1)/2}. \tag{3.1.6}$$

Since

$$J(U_n, Y_N) = \left|\det\left[\begin{array}{c}\partial z_{ij}/\partial y_p \\ \partial z_{ij}/\partial u_l\end{array}\right]\right|,$$

then $J(U_n, Y_n)$ is a polynomial of values y_p. It is easy to show that it is zero if at least two values y_i coincide; therefore, it must be divided without remainder into the values $y_i - y_j, i \neq j$. But (3.1.6) implies that this polynomial is homogeneous of the degree $n(n-1)/2$. Thus, $\varphi(Y_n) = c\prod_{i>j}(y_i - y_j)$, where c is a constant. Substituting $\varphi(Y_n)$ in (3.1.5) and bearing in mind the ε is arbitrary, we obtain $\mathbf{P}\{\theta_{1i} = 0, i = \overline{1,n}\} = 0$. Consequently, we can regard θ_n as a matrix whose column vectors are $\vec{\theta}_i$ with nonnegative first components.

Let us calculate c. To do this, we set $f \equiv 1$ in (3.1.5),

$$p(Z_n) = 2^{-n/2}\pi^{-n(n+1)/4}\exp\{-\operatorname{Tr} Z_n^2/2\}.$$

Then

$$c_{1n}\int_{h_{1i}>0}\mu(dH)2^{-n/2}\pi^{-n(n+1)/4}\int\exp\left(-0.5\sum_{i=1}^n y_i^2\right)$$

$$\prod_{i>j}(y_i - y_j)dY_n = 1(y_1 > \cdots > y_n). \tag{3.1.7}$$

Let us divide the group G into measurable nonintersecting subsets G_i, $i = \overline{1, 2^n}$, so that in each subset of matrices H either $h_{1i} > 0$ or $h_{1i} < 0$. Making use of the change of variables $H_n = U_nT$, $U_n \in G$, $T = (\delta_{ij}\operatorname{sign} h_{1i})$, we obtain $\int_{G_p}\mu(dH) = \int_{u_{1i}>0}\mu(dU_n)$. Hence

$$\int_{h_{1i}>0}\mu(dH) = 2^{-n}. \tag{3.1.8}$$

In accordance with §2 of Chapter 2, we find

$$(2\pi)^{-n/2} \int_L \exp(-0.5 \sum_{i=1}^n y_i^2) \prod_{i>j}(y_i - y_j)dY_n = (n!)^{-1}\mathbf{E}|\det V_n|$$

$$= (n!)^{-1}[\Gamma(1 + 1/2)]^{-n} \prod_{j=1}^n \Gamma(1 + j/2, \quad L = \{y_1 > \cdots > y_n\},$$

$$(3.1.9)$$

where $V_n = (\eta_i^{j-1}), i, j = \overline{1, n}$ is a Vandermonde $n \times n$ matrix whose entries $\eta_i, i = \overline{1, n}$ are independent and are distributed according to the normal law $N(0, 1)$.

Substituting (3.1.8) and (3.1.9) in (3.1.7), we find

$$c_{1n} = 2^n \pi^{n(n+1)/4} \{\prod_{j=1}^n \Gamma(j/2)\}^{-1}.$$

Theorem 3.1.1 is proved.

Let G_1 be a subset of matrices of the group G with $h_{1i} > 0, i = \overline{1, n}$.

Corollary 3.1.1.. *If $p(H_n Z_n H_n') \equiv p(Z_n)$ for all matrices $H_n \in G$ and Z_n, then θ_n is stochastically independent of the eigenvalues of Ξ_n and has the following distribution:*

$$\mathbf{P}\{\theta_n \in E\} = 2^n \int_E \mu(dH), E \subset G_1.$$

The distribution density of the eigenvalues of Ξ is

$$2^{-n}c_{1n}p(Y_n) \prod_{i>j}(y_i - y_j), \quad y_1 > \cdots > y_n. \quad (3.1.10)$$

We call the measure $2^n \mu$ defined on the set G_1, where μ is a Haar measure on the group G, a conditional Haar measure on the set G.

Corollary 3.1.2.. *[3]. If the entries $\xi_{ij}, i \geq j, i, j = \overline{1, n}$ of Ξ_n are independent and distributed according to the normal laws $N(0, (1 + \delta_{ij})/2$, then the matrix θ_n is stochastically independent of the eigenvalues of the matrix Ξ_n and is distributed according to the conditional Haar measure. The distribution density of the eigenvalues of Ξ_n is*

$$2^{-3n/2}\pi^{-n(n+1)/4}c_{1n} \exp(-\sum_{i=1}^n y_i^2/2) \prod_{i>j}(y_i - y_j), \quad y_1 > \cdots > y_n. \quad (3.1.11)$$

We proceed with the study of the eigenvalues and eigenvectors of Hermitian random matrices. Let $H_n = (\eta_{ij})$ be a Hermitian $n \times n$ matrix whose entries are complex random variables, and X_n a nonrandom Hermitian $n \times n$ matrix. We assume that the real and imaginary parts of entries of the matrix H_n arranged on the diagonal and above have the joint distribution density $p(X_n)$ (the function $p(X_n)$ depends on imaginary and real parts of X_n entries). It is easy to check, as in the proof of Lemma 3.1.1 above, that the eigenvalues $\lambda_1 \geq \cdots \geq \lambda_n$ of the matrix are real, do not coincide with probability 1, and are random variables. The matrix $\theta = (\theta_{ij})$ whose columns are the eigenvectors of H_n is unitary. The eigenvectors $\vec{\theta}_i$ with probability 1 are defined from the system of equations

$$(H_n - \lambda_i I)\vec{\theta}_i = 0, \quad (\vec{\theta}_i, \bar{\vec{\theta}}_i) = 1. \tag{3.1.12}$$

The vectors $\vec{\theta}_i$ can be chosen uniquely by fixing an argument of some nonzero element of each vector $\vec{\theta}_i$. If some numbers $\lambda_i, i = \overline{1,n}$ coincide, then the eigenvectors $\vec{\theta}_i$ are chosen as in Theorem 3.1.1. The proof of the following theorem is similar to that of Theorem 3.1.1. It implies, in particular, that the random variables $\arg \theta_{1i}$ have a continuous distribution. Therefore, we consider the matrix θ_n being chosen so that $\arg \theta_{1i} = c_i, i = \overline{1,n}$, where c_i are some nonrandom values, $0 \leq c_i \leq 2\pi, i = \overline{1,n}$.

Let Γ be a group of unitary $n \times n$ matrices and ν the normalized Haar measure on it, and B the σ-algebra of the Borel sets of the group Γ.

Theorem 3.1.2.. *If a Hermitian random matrix H_n has the distribution density $p(X_n)$, then for any subset E of B and any real numbers $\alpha_i, \beta_i, i = \overline{1,n}$,*

$$\mathbf{P}\{\theta_n \in \mathbf{E}, \alpha_i < \lambda_i < \beta_i, i = \overline{1,n}\}$$
$$= c_{2n} \int p(U_n Y_n U_n^*) \prod_{i>j}(y_i - y_j)^2 \nu(dU_n / \arg u_{1i}$$
$$= c_i, i = \overline{1,n})dY_n, \tag{3.1.13}$$

where the integration is over the domain $y_1 > \cdots > y_n$, $\alpha_i < y_i < \beta_i$, $i = \overline{1,n}$, $U_n \in E$; $\nu(U_n / \arg u_{1i} = c_i, i = \overline{1,n})$ is the regular conditional Haar measure, $c_{2n} = [\pi^{(-n^2+n)/2} \prod_{j=0}^{n-1} j!]^{-1}$.

Proof.. When $y_1 > \cdots > y_n$ and $\arg u_{1i} = c_i$, the transformation $X_n = U_n Y_n U_n^*$ is one-to-one, since the number of independent parameters from the left and from the right side of $X_n = U_n Y_n U_n^*$ is the same. Let $x_{ij}^{(0)}$ and $x_{ij}^{(1)}$ be real and imaginary parts of the element x_{ij}, respectively. Note that the number of real parameters P_k, upon which the matrix U_n depends, is $n(n-1)$, and the entries of the matrix U_n are almost everywhere continuously differentiable according to these parameters (see Chapter 1). We calculate the Jacobian of the transformation $X_n = U_n Y_n Y_n^*$ [132]. If we differentiate the identity $U_n U_n^* = U^* U = 1$, we find $(\partial U^* / \partial p_k)U + U^*(\partial U / \partial p_k) = 0$. Hence, we can see that the matrix $S^{(k)} := U^*(\partial U/\partial p_k) = -(\partial U^*/\partial p_k)U$ is an antisymmetric Hermitian. Further, $\partial X/\partial p_k = (\partial U/\partial p_k)YU^* + UY(\partial U^*/\partial p_k)$. This implies $U^*(\partial X/\partial p_k)U = S^k Y - YS^k$.

$$\sum_{i,j=1}^{n} (\partial x_{ij}/\partial p_k)\bar{U}_{ip}U_{jl} = s_{pl}^{(k)}(y_l - y_p),\qquad (3.1.14)$$

$$p,l = \overline{1,n}, \quad k = \overline{1,n(n-1)}.$$

Similarly,

$$\sum_{i,j=1}^{n} (\partial x_{ij}/\partial y_k)\bar{u}_{ip}u_{jl} = (\partial Y_{(pl)}/\partial y_k) = \delta_{pl}\delta_{pk},$$

$$k = \overline{1,n}, \quad p,l = \overline{1,n},\qquad (3.1.15)$$

where $Y_{(pl)}$ is the entry of Y on the intersection of the pth row and the lth column.

The Jacobian is represented in the following form:

$$\text{mod det}\begin{bmatrix} A_1 & A_2 & A_3 \\ B_1 & B_2 & B_3 \end{bmatrix},\qquad (3.1.16)$$

where

$$A_1 = [\partial x_{jj}^{(0)}/\partial y_l]_{j,l=1}^{n}, \quad A_2 = [\partial x_{jk}^{(0)}/\partial y_l],$$

$$A_3 = [\partial x_{jk}^{(l)}/\partial y_l]_{j>k,j,k,l=\overline{1,n}}, \quad B_1 = [\partial x_{jj}^{(0)}/\partial p_m]_{j=\overline{1,n}}^{m=\overline{1,n(n-1)}},$$

$$B_2 = [\partial x_{jk}^{(0)}/\partial p_m], \quad B_3 = [\partial x_{jk}^{(l)}/\partial y_l]_{j>k,m\overline{1,n(n-1)}}.$$

Multiply (3.1.16) by the matrix determinant

$$\begin{bmatrix} C_1 & D_1 \\ C_2 & D_2 \\ C_3 & D_3 \end{bmatrix},$$

where

$$C_1 = (\bar{u}_{jp}u_{jp})_{j,p=\overline{1,n}}, \quad C_2 = (2\bar{u}_{ip}u_{jp})_{i>j,p=\overline{1,n}}, \quad C_3 = 0_{n\times n(n-1)/2},$$

$$D_1 = (\bar{u}_{jp}u_{jl})_{p\neq l}, \quad D_2 = (2u_{ip}u_{jl})_{i>j,p\neq l}, \quad D_3 = i(\bar{u}_{sp}u_{lj} - \bar{u}_{jp}u_{sl})_{s>j,p\neq l},$$

where $0_{n\times n(n-1)/2}$ is a matrix of the dimension $n \times n(n-1)/2$ whose entries are equal to zero.

By multiplying the matrices and taking into account (3.1.14) and (3.1.15), we find that the Jacobian equals

$$\text{mod det} \begin{bmatrix} I_n & 0_{n(n-1)\times n} \\ 0_{n\times n(n-1)} & [S_{pl}^{(k)}(y_l - y_p)]_{\substack{k=\overline{1,n(n-1)} \\ p\neq l, p, l=\overline{1,n}}} \end{bmatrix}.$$

Thus

$$J = \prod_{p>l}(y_p - y_l)^2 \varphi(U_n), \tag{3.1.17}$$

where $\varphi(U_n)$ is some Borel function of the parameters p_k of the matrix U_n.

For any bounded and continuous function $f(\theta_n, \Lambda_n)$ of the entries of the matrices θ_n and Λ_n $Ef(\theta, \Lambda) = \int f(U_n, Y_n)p(X_n)dX_n$, where the integration is over the domain $\{X_n; \arg u_{1i} = c_i, i = \overline{1,n}, y_1 > \cdots > y_n\}$, the matrices Y_n and U_n satisfy the equation $X_n = U_n Y_n U_n^*$. Since the Jacobian of the transformation $X_n = H_n Z_n H_n^*$ (where $H_n \in \Gamma$, Z_n is an Hermitian matrix) is equal to 1,

$$\mathbf{E}f(\theta_n, \Lambda_n) = \int f(H_n U_n, Y_n)p(H_n X_n H_n^*)dX_n, \tag{3.1.18}$$

where the domain of integration is $L := \{X_n : \arg \sum_{k=1}^n h_{1k}u_{ki} = c_i, i = \overline{1,n}, y_1 > \cdots > y_n\}$. After the change of variables $X_n = \tilde{U}_n Y_n \tilde{U}_n^*$, where the parameters $\tilde{j}_{1k}$, $k = \overline{1.n}$ are chosen so that this transformation is one-to-one on the set L, taking into account (3.1.17), we obtain from (3.1.18),

$$\mathbf{E}f(\theta_n, \Lambda_n) = \int f(H_n \tilde{U}_n, Y_n)p(H_n(\tilde{U}_n Y_n \tilde{U}_n^*)H_n^*)$$
$$\times \prod_{p>l}(y_p - y_l)^2 \varphi(\tilde{U}_n)dY_n \prod_{i=1}^l dp_i, \tag{3.1.19}$$

where the domain of integration is equal to

$$\{y_1 > y_2 > \cdots > y_n, \arg \sum_{k=1}^n h_{1k}\tilde{u}_{ki} = c_i, i = \overline{1,n}\},$$

the parameters p_i are changing in some domains. Without loss of generality, we assume that the functions p and φ in (3.1.19) are continuous.

The expression on the right-hand side of (3.1.19) is independent of the matrix H_n. Therefore, we can integrate both parts of the equality (3.1.19) by the Haar measure ν defined on the group Γ of matrices H. Then

$$\mathbf{E}f(\theta_n, \Lambda_n) = c_{2n} \int f(H_n, Y_n)p(H_n Y_n H_n^*)$$
$$\times \prod_{i=1}^n \delta(\arg h_{1i} - c_i)\prod_{p>l}(y_p - y_l)^2 dY_n \nu(dH),$$

where c_{2n} is a constant. Hence (3.1.13) follows.

Define the constant c_{2n}. To do this, we set $f = 1$,

$$p(X_n) = 2^{-n/2}\pi^{-n^2/2}\exp(-0,5\operatorname{Tr}X_nX_n^*).$$

Taking into account (2.2.1), we find

$$c_{2n}2^{-n/2}\pi^{-n^2/2}\int_L \exp\left(-\sum_{i=1}^n y_i^2\right)\prod_{i>j}(y_i - y_j)^2 dY_n$$

$$= c_{2n}\pi^{-n^2+n/2}(n!)^{-1}[\Gamma(2)]^{-n}\prod_{j=1}^n \Gamma(1+j) = 1.$$

From this equation, we obtain the value of the constant c_{2n}. Theorem 3.1.2 is proved.

Corollary 3.1.3.. *If $p(U_nY_nU_n^*) \equiv p(Y_n)$ for all unitary matrices $U_n \in \Gamma$, then θ_n is independent of the eigenvalues of H_n and has the distribution*

$$\mathbf{P}\{\Theta_n \in E\} = \int_E \nu(dU/\arg U_{1i} = c_i, i = \overline{1,n}).$$

The density of the eigenvalues is $c_{2n}p(Y_n)\prod_{p>l}(y_p - y_l)^2$, $y_1 > \cdots > y_n$.

An important special case of Hermitian random matrices are the matrices H_n with entries whose real and imaginary parts are independent and distributed according to the standard normal laws. In this case, the density of H_n is $p(X_n) = 2^{-n/2}\pi^{-n^2/2}\exp(\operatorname{Tr}X_nX_n^*/2)$, and that of the eigenvalues has the form [132]:

$$c_{2n}'\exp\left\{-\sum_{i=1}^n y_i^2/2\right\}\prod_{i>j}(y_i - y_j)^2, c_{2n}'$$

$$= (2\pi)^{-n/2}\left(\prod_{j=1}^{n-1} j!\right)^{-1}, \quad y_1 > y_2 > \cdots > y_n.$$

§2 Distribution of the Eigenvalues and Eigenvectors of Antisymmetric Random Matrices

We call a matrix $\Xi_n = (\xi_{ij})$, an antisymmetric $n \times n$ matrix, if $\Xi_n + \Xi_n' = 0$ and its entries are random variables. Suppose that the random variables ξ_{ij}, $i > j$, $i,j = \overline{1,n}$ have a joint distribution density which we denote by $p(X_n)$. As in §1 of the chapter, we verify that the eigenvalues of such a matrix will be the imaginary conjugate for even n and with probability

1 and not equal to one another. We arrange the eigenvalues as follows: $\{i\lambda_1, -i\lambda_1, i\lambda_2, -i\lambda_2, \ldots, i\lambda_{n/2}, -i\lambda_{n/2}\}$ for even n and $\{i\lambda_1, -i\lambda_1, i\lambda_2, -i\lambda_2, \ldots, i\lambda_{(n-1)/2}, -i\lambda_{(n-1)/2} -, 0\}$ for odd n, $\lambda_1 \geq \lambda_2 \geq \cdots \geq \lambda_{[n/2]} \geq 0$.

First, suppose n is even. The eigenvectors corresponding to the eigenvalues $i\lambda_k, -i\lambda_k$, are conjugate. We denote them by $\vec{u}_k + i\vec{v}_k$, $\vec{u}_k - i\vec{v}_k$. Obviously, the eigenvectors $\vec{u}_k$, $\vec{v}_k$ satisfy the following system of equations

$$\Xi \vec{u}_k = -\lambda_k \vec{v}_k, \quad \Xi \vec{v}_k = \lambda_k \vec{v}_k. \qquad (3.2.1)$$

It is easy to show that the vectors $\vec{v}_k$ and $\vec{u}_k$ are orthogonal with probability 1. In the basis $\{\vec{u}_1, \vec{v}_1, \vec{u}_2, \vec{v}_2, \ldots, \vec{u}_{n/2}, \vec{v}_{n/2}\}$ with probability 1, Ξ_n can be represented in the form

$$\Xi_n = T\left\{\begin{pmatrix} 0 & \lambda_1 \\ -\lambda_1 & 0 \end{pmatrix}, \ldots, \begin{pmatrix} 0 & \lambda_{n/2} \\ -\lambda_{n/2} & 0 \end{pmatrix}\right\} T', \qquad (3.2.2)$$

where T is an orthogonal matrix whose columns are the vectors $\vec{u}_k$, $\vec{v}_k$, and $\{\ldots\}$ is a diagonal matrix with second-order matrices on its diagonal. The representation (3.2.2) cannot yet be used for finding the joint density of eigenvalues, since the matrix T is not chosen uniquely. One can see from the system (3.2.1) that the vectors $\vec{v}_k$ and $\vec{u}_k$ belong to the subspace of dimension 2 of the eigenvalue λ_k^2 of multiplicity 2 of $\Xi_n \Xi_n'$. In order to define these vectors uniquely with probability 1, we require that $(\vec{v}_k, \vec{v}_k) = 1, (\vec{u}_k, \vec{u}_k) = 1$ and that some nonzero coordinate of the vector $\vec{v}_k$ be equal to some constant $c_k.|c_k| < 1$. In what follows, we assume for simplicity, that the first coordinate of the vector $\vec{v}_k$ satisfies this condition. Of course, there is an infinite number of other ways of fixing the coordinates of the vectors $\vec{v}_k$. For example, we can require the sum of the vector $\vec{v}_k$ elements to be equal to a constant.

For an odd n,

$$\Xi_n = T\left\{\begin{pmatrix} 0 & \lambda_1 \\ -\lambda_1 & 0 \end{pmatrix}, \ldots, \begin{pmatrix} 0 & \lambda_{[n/2]} \\ -\lambda_{[n/2]} & 0 \end{pmatrix}, 0\right\} T', \qquad (3.2.3)$$

where T is an orthogonal matrix, $t_{1k} = c_k$, $k = 1, 3, \ldots$ are some constants, and $|c_k| < 1$.

Let G be the group of orthogonal $n \times n$ matrices and μ the normalized Haar measure on it, B the σ-algebra of Borel sets of elements of the group G, and X_n be the real antisymmetric matrix.

Theorem 3.2.1. *If a random antisymmetric matrix X_{in} has the distribution density $p(X_n)$, then for any subsets $E \in B$ and the real numbers $\alpha_i, \beta_i, i = \overline{1,n}$, $n \geq 3$,*

$$\mathbf{P}\{T_n \in E, \alpha_i < \lambda_i < \beta_i, i = \overline{1, n/2}\} = c'_{3n} \int p(H_n Y_n H'_n)$$

$$\times \prod_{i>j}(y_i^2 - y_j^2)\mu(dH/h_{1,2i} = c_i)dY_n, \quad n = 2m, \qquad (3.2.4)$$

$$\mathbf{P}\{T_n \in E, \alpha_i < \lambda_i < \beta_i, i = \overline{1, [n/2]}\} = c''_{3n} \int p(H_n Y_n H'_n)$$

$$\times \prod_{i=1}^{[n/2]} y_i^2 \prod_{i>j}(y_i^2 - y_j^2)\mu(dH/h_{1,2i} = c_i)dY_n, \quad n = 2m+1,$$

where the integration is over the domain $y_1 > y_2 > \cdots > y_{[n/2]} > 0$, $H_n \in E, \alpha_i < y_i < \beta_i, i = \overline{1, [n/2]}$,

$$Y_n = \begin{cases} \text{diag}\left\{\begin{pmatrix} 0 & y_1 \\ -y_1 & 0 \end{pmatrix}, \ldots, \begin{pmatrix} 0 & y_{n/2} \\ -y_{n/2} & 0 \end{pmatrix}\right\}, & n = 2m, \\ \text{diag}\left\{\begin{pmatrix} 0 & y_1 \\ -y_1 & 0 \end{pmatrix}, \ldots, \begin{pmatrix} 0 & y_{[n/2]} \\ -y_{[n/2]} & 0 \end{pmatrix}, 0\right\}, & n = 2m+1, \end{cases}$$

$$c'_{3n} = [\prod_{i=1}^{n/2} \Gamma(\frac{n}{2} + 2^{-1} - i) \prod_{i=0}^{(n/2)-1} j!]^{-1}\pi^{n(n-1)/2}2^{n/2},$$

$$c''_{3n} = [\prod_{i=1}^{(n-1)/2} \Gamma(\frac{n}{2} - i + 1) \prod_{i=0}^{((n-1)/2)-1} j!]^{-1}\pi^{n(n-1)/2}2^{(n-1)/2}.$$

Proof. Let $n \geq 3$ be even. Let us calculate the Jacobian of the transformation $X_n = H_n Y_n H'_n$. Obviously, it will coincide with the Jacobian of the transformation $X = UVU^{-1}$, where $V = \{i\tilde{y}_l\delta_{lj}\}$, U is a complex matrix formed by the vectors $\overrightarrow{u}_k$, $\overrightarrow{v}_k$, and $\tilde{y}_l$, $l = \overline{1, n}$ are the eigenvalues $iy_1, -iy_1, \ldots, iy_{n/2} - iy_{n/2}$. If $y_1 > y_2 > \cdots > y_{n/2}$ and $h_{1,2i} = c_i, i = \overline{1, n/2}$, then the transformation $X = UVU^{-1}$ is one-to-one. Let $p_m, m = \overline{1, n^2/2 - n}$ be the coordinates of U. If we differentiate the identity $UU^* = U^*U = I$, we find $(\partial U^*/\partial p_k)U + U^*(\partial U/\partial p_k) = 0$. Hence,

$$S^{(k)} := \left(\frac{\partial U^*}{\partial p_k}\right)U = -U^*\left(\frac{\partial U}{\partial p_k}\right). \tag{3.2.5}$$

By the method of the proof of Theorem 3.1.2 and by using (3.2.5), we obtain $U^*(\partial X/\partial p_k)U = S^{(k)}V - VS^{(k)}$. For the entries of this matrix,

$$\sum_{i,j=1}^{n} \left(\frac{\partial x_{ij}}{\partial p_k}\right)\tilde{U}_{pi}u_{jl} = s_{pl}^{(k)}(\tilde{y}_l - \tilde{y}_p), \quad p,l = \overline{1,n},$$

$$k = \overline{1, n^2/2 - n}. \tag{3.2.6}$$

Similarly,

$$\sum_{i,j=1}^{n} \left(\frac{\partial x_{ij}}{\partial y_k}\right)\tilde{u}_{pi}u_{jl} = \left(\frac{\partial V_{(pl)}}{\partial y_k}\right) = i\delta_{pl}(\delta_{2k-1,p} - \delta_{2k,p}),$$

$$p,l = \overline{1,n}, \quad k = \overline{1, n/2}, \tag{3.2.7}$$

where $\tilde{u}_{pi}$ are the entries of U^*.

Let us represent the Jacobian in the form mod $\det \left\| \begin{matrix} A_1 \\ A_2 \end{matrix} \right\|$, where

$$A_1 = (\partial x_{ij}/\partial y_l), \quad l = \overline{1, n/2}, \quad i > j, \quad i,j = \overline{1,n};$$
$$A_2 = (\partial x_{ij}/\partial p_m), \quad i > j, i, j = \overline{1,n}, \quad m = \overline{1, n^2/2 - n}.$$

We multiply mod $\det \left\| \begin{matrix} A_1 \\ A_2 \end{matrix} \right\|$ by the determinant of the matrix $[C_1, C_2]$, where

$$C_1 = (\tilde{u}_{pi}u_{jp} - \tilde{u}_{pj}u_{ip}), \quad p = 2, 4, \ldots, n, \quad i > j, \quad i,j = \overline{1,n};$$
$$C_2 = (\tilde{u}_{pi}u_{jl} - \tilde{u}_{pj}u_{il}), \quad p = 2, 4, \ldots, n, \quad l \neq \overline{p, p+1},$$
$$l = \overline{1,n}, \quad i > j, \quad i,j = \overline{1,n}.$$

Using (3.2.6) and (3.2.7), we obtain

$$\text{mod } \det \begin{bmatrix} A_1 \\ A_2 \end{bmatrix} \det[C_1, C_2]$$

$$= \text{ mod } \det \begin{bmatrix} I_{n/2} & 0_{(n^2/2-n)\times(n/2)} \\ 0_{(n/2)\times(n^2/2-n)} & [S_{pl}^{(k)}(\tilde{y}_l - \tilde{y}_p))]_L \end{bmatrix}$$

$$= \prod_{p \neq l}(y_p^2 - y_l^2)\varphi(p_m), \quad L = \{k = \overline{1, n^2/2 - n},$$

$$p = 2, 4, \ldots, n, l \neq \overline{p, p+1}, l = \overline{1,n}\},$$

where φ is a Borel function. All the performed transformations are valid for odd n. In this case, the Jacobian is equal to n,

$$\prod_{p \neq l}(y_p^2 - y_l^2) \prod_{l=1}^{(n-1)/2} y_l^2 \varphi(p_m, m = \overline{1, n^2/2 - n + 1/2}),$$

where φ is a Borel function.

One can easily check that $\tilde{y}_n = 0$. The rest of the proof is similar to that of Theorem 3.1.2. Let us find the constants c'_{3n} and c''_{3n}. The density of Ξ_n is chosen in the form $p(X_n) = \pi^{-n(n-1)/2} \exp\{- \text{Tr } X_n X'_n\}$. Then for an even n,

$$c'_{3n} \int_{L_{[n/2]}} \exp\left\{-\sum_{i=1}^{n/2} y_i^2\right\} \prod_{i \neq j}(y_i^2 - y_j^2) dY_n = 1,$$

$$L_{[n/2]} = \{y_1 > \cdots > y_{[n/2]} > 0\}$$

and for an odd n,

$$c''_{3n} \int_{L_{[n/2]}} \exp\left\{-\sum_{i=1}^{(n-1)/2} y_i^2\right\} \prod_{i\neq j}(y_i^2 - y_j^2) \prod_{i=1}^{(n-1)/2} y_i^2 dY_n = 1.$$

Consider a complex random $m \times p$ matrix Ξ whose entries are independent and distributed according to the normal law $N(0, 1/2)$. It follows from Corollary 1.3.3 that the density of $\Xi\Xi^*$ is equal to $\tilde{c}_{m,p}\pi^{-mp}\exp(-\operatorname{Tr} Z_p)\det Z_p^{m-p}$, where Z_p is a nonnegative definite Hermitian matrix. Making use of Theorem 2.1.4., we find

$$\int \exp(-\operatorname{Tr} Z_p)\det Z_p^{m-p} dZ_p = \pi^{p(p-1)/2} \prod_{i=1}^{p} \Gamma(m - i + 1), \qquad (3.2.8)$$

where the integration is over the set of all nonnegative definite Hermitian matrices Z_p. By corollary 3.3.1, (3.2.8) implies

$$\int_{L_{[n/2]}} \exp\left\{-\sum_{i=1}^{n/2} x_i\right\} \prod_{i=1}^{n/2} x_i^{m-p} \prod_{i>j}(x_i - x_j)^2 \prod_{i=1}^{n} dx_i$$

$$= \prod_{i=1}^{p} \Gamma(m - i + 1) \prod_{i=0}^{p-1} i!$$

We introduce the change of the variables $x_i = y_i^2, y_i > 0$. Then

$$\int_{L_{[n/2]}} \exp\left\{-\sum_{i=1}^{n/2} y_i^2\right\} \prod_{i=1}^{n/2} y_i^{2m-2p+1} \prod_{i>j}(y_i^2 - y_j^2)^2$$

$$\times \prod_{i=1}^{n} dy_i = 2^{-p} \prod_{i=1}^{p} \Gamma(m - i + 1) \prod_{i=0}^{p-1} i!$$

Assuming in this equality an even n, $m = (n - 1)/2$, $p = n/2$, and an odd n, $m = [n/2] + 1/2$, $p = (n - 1)/2$, we obtain the value of the constants c'_{3n} and c''_{3n}.

When $n = 2$,

$$\mathbf{P}\{T_2 \in E, \alpha_1 < \lambda_1 < \beta_1\}$$

$$= c'_{32} \int_E \int_{y>0} p\left(H_2 \begin{pmatrix} 0 & y \\ -y & 0 \end{pmatrix} H_2'\right) dy\mu(dH_2/h_{11} = c_1).$$

Theorem 3.2.1 is proved.

Corollary 3.2.1.   *If the entries ξ_{ij}, $i \geq j$, $i,j = \overline{1,n}$ of the random matrix Ξ_n are independent and distributed according to the normal law with the parameters 0 and 1, then the eigenvalues of Ξ_n are stochastically independent of the eigenvectors of Ξ_n and have the following distribution density [132],*

$$(2\pi)^{-n(n-1)/2} c'_{3n} \exp\{-\sum_{i=1}^{n/2} y_i^2/2\} \prod_{i>j} (y_i^2 - y_j^2)^2 \quad n = 2m \geq 3,$$

as

$$(2\pi)^{-n(n-1)/2} c''_{3n} \exp\{-\sum_{i=1}^{(n-1)/2} y_i^2/2\} \prod_{i>j} (y_i^2 - y_j^2)^2 \prod_{i=1}^{(n-1)/2} y_i^2$$

$$n = 2m + 1 \geq 3, \qquad y_1 \geq y_2 \geq \cdots \geq y_{[n/2]} \geq 0.$$

The distribution of the matrix T_n is

$$\mathbf{P}\{T_n \in E\} = \int_E \mu(dH/h_{1,2i} = c_i, i = \overline{1,[n/2]}).$$

§3 Distribution of Eigenvalues and Eigenvectors of Nonsymmetric Random Matrices

Let $\Xi_n = (\xi_{ij})$ be a real square random matrix with the distribution density $p(X_n)$, where $X_n = (x_{ij})$ is a real square $n \times n$ matrix. Let us introduce the notations $\lambda_k + i\mu_k$, $\lambda_k - i\mu_k$, $k = \overline{1,s}$, λ_l, $l = \overline{s+1, n-2s}$ are the eigenvalues of Ξ_n, and $\overrightarrow{z}_k = \overrightarrow{x}_k + i\overrightarrow{y}_k$, $\overrightarrow{z}_k = \overrightarrow{x}_k - i\overrightarrow{y}_k$, $k = \overline{1,s}$, $\overrightarrow{z}_l$, $l = \overline{s+1, n-2s}$ are the eigenvectors.

Before studying the distribution of the eigenvalues and eigenvectors of Ξ_n, we need to choose the eigenvalues and eigenvectors so that they are random variables. It is known that the eigenvalues are continuous functions of the entries of the matrix Ξ_n. If a matrix has the distribution density, then the eigenvalues of Ξ_n are distinct with probability 1 (see Lemma 3.1.1). We arrange the complex eigenvalues of Ξ_n in increasing order of their moduli. If some of these complex numbers (among which there are no conjugate pairs) have equal moduli, then we arrange them in increasing order of their arguments. Among pairs of conjugate eigenvalues, the first one is the number with a positive imaginary part. The real eigenvalues are arranged in increasing order. The eigenvalues thus chosen are random variables. There are many other ways of ordering the eigenvalues, but we adhere to this principle as the most natural. We require that the vectors $\overrightarrow{x}_k$, $\overrightarrow{y}_k$, $k = \overline{1,s}$, $\overrightarrow{x}_l$, $l = \overline{s+1, n-2s}$ are of unit length and that the first nonzero component of each vector is positive. If the eigenvalues coincide, they are chosen arbitrarily only if they are random (see §1 of this chapter). As in the proof of Theorem 3.1.1, the moduli of the eigenvalues (among which there are no conjugate ones) are distinct with probability 1 and the probability that the first component of the eigenvectors are zero is equal to zero. Therefore, we consider that the first components of the eigenvectors $\overrightarrow{x}_k$, $\overrightarrow{y}_k$, $k = \overline{1,s}$, $\overrightarrow{x}_l$, $\overline{s+1, n-2s}$ are nonnegative and that the moduli of the eigenvalues (among which there are no conjugate ones) are distinct with probability 1.

The vectors $\overrightarrow{x}_k$, $\overrightarrow{y}_k$, $k = \overline{1,s}$, $\overrightarrow{x}_l$, $l = \overline{s+1, n-2s}$ with probability 1 form a basis in the Euclidian space R_n. Herewith, $\Xi \overrightarrow{x}_k = \lambda_k \overrightarrow{x}_k - \mu_k \overrightarrow{y}_k$, $\Xi \overrightarrow{y}_k = \lambda_k \overrightarrow{x}_k + \mu_k \overrightarrow{y}_k$, $k = \overline{1,s}$, $\Xi \overrightarrow{x}_l = \lambda_l \overrightarrow{x}_l$, $l = \overline{s+1, n-2s}$.

With probability 1, the matrix Ξ_n can be represented in the following form,

$$\Xi_n = T \operatorname{diag}\left\{ \begin{pmatrix} \lambda_1 & \mu_1 \\ -\mu_1 & \lambda_1 \end{pmatrix}, \ldots, \begin{pmatrix} \lambda_s & \mu_s \\ -\mu_s & \lambda_s \end{pmatrix}, \lambda_{s+1}, \ldots, \lambda_{n-2s} \right\} T^{-1},$$

where the diagonal matrix is the one with the diagonal comprising first the second-order matrices and then the entries λ_i, with the rest of the elements being zero, and T being a real nondegenerate matrix with probability 1 whose column vectors are $\overrightarrow{x}_k$, $\overrightarrow{y}_k$, $k = \overline{1,s}$, $\overrightarrow{x}_l$, $l = \overline{s+1, n-2s}$.

Let K be a group of real nondegenerate $n \times n$ matrices, B the σ-algebra of the Borel subsets of the group K, and θ_i, $i = \overline{1,n}$ the eigenvalues of Ξ_n, chosen as described above.

Theorem 3.3.1. *If a random matrix Ξ_n has the distribution density $p(X_n)$, then for any subset $E \in B$ and any real numbers α_i, β_i, $i = \overline{1,n}$,*

$$\mathbf{P}\{T_n \in E, \operatorname{Re}\theta_i < \alpha_i, \operatorname{Im}\theta_i < \beta_i, i = \overline{1,n}\}$$

$$= \sum_{s=0}^{[n/2]} c_s \int_{K_s} p(X_n Y_s X_n^{-1}) J_s(Y_s) \varphi(Y_s) |\det X_n|^{-n}$$

$$\times \prod_{i=2}^{n} \left\{ \left(1 - \sum_{j=2}^{n} x_{ji}^2 \right)^{-1/2} \right\} dX_n dY_s, \qquad (3.3.1)$$

where

$$Y_s = \operatorname{diag}\left\{ \begin{pmatrix} x_1 & y_1 \\ -y_1 & x_1 \end{pmatrix}, \ldots, \begin{pmatrix} x_s & y_s \\ -y_s & x_s \end{pmatrix}, x_{s+1}, \ldots, x_{n-2s} \right\},$$

$$J_s(Y_s) = |\prod_{p \neq l}(q_p - q_l)|,$$

$q_p, p = \overline{1,n}$ *are the eigenvalues of Y_s, and the domain of integration K_s is equal to $X_n \in E$, $x_{1i} = [1 - \sum_{j=2}^{n} x_{ji}^2]^{1/2}$, $\sum_{j=2}^{n} x_{ji}^2 \leq 1$, $i = \overline{1,n}$, $x_1 < \alpha_1$,*

$y_1 < \beta_1, \ldots x_s < \alpha_{2s-1}, -y_s < \alpha_{2s}, x_{s+1} < \alpha_{2s+1}, 0 < \beta_{2s+1}, \ldots, x_{n-2s} <$
$\alpha_s, 0 < \beta_s, dX_n = \prod_{i=\overline{2,n}, j=\overline{1,n}} dx_{ij}, dY_s = \prod dx_i dy_j,$

$$\varphi(Y_s) = \begin{cases} 1 \text{ if the eigenvalues } x_k + iy_k, x_k - iy_k, k = \overline{1,s}, x_l, \\ \quad l = \overline{s+1, n-2s} \text{ are arranged in increasing order of} \\ \quad \text{their moduli, and if among conjugate eigenvalues the} \\ \quad \text{first is the one with } y_k \geq 0, \text{ and if the values } x_l, l = \\ \quad \overline{s+1, n-2s} \text{ are arranged in increasing order;} \\ 0 \text{ otherwise.} \end{cases}$$

The constants c_s satisfy the system of equations

$$\mathbf{E}\psi(\mathrm{Tr}\,\Xi^k) = \sum_{s=1}^{[n/2]} c_s a_{sk}, \quad k = \overline{0, [n/2]-1}, \tag{3.3.2}$$

where φ is a measurable function chosen so that the integrals $\mathbf{E}(\mathrm{Tr}\,\Xi^k)$, $k = \overline{0, [n/2]-1}$ exist (in particular, if $\mathbf{E}\,\mathrm{Tr}\,\Xi^k$ exist, then we can put $\varphi(x) = x$),

$$a_{sk} = \int_Q \psi_k(\mathrm{Tr}\, X_n Y_s^k X_n^{-1}) J_s(Y_s)\varphi(Y_s)|\det X_n|^{-n}$$

$$\times \prod_{i=2}^n \left\{ 1 - \sum_{j=2}^n x_{ji}^2 \right\}^{-1/2} dX_n dY_s,$$

$$Q = \{x_{1i} = \left[1 - \sum_{j=2}^n x_{ji}^2 \right]^{1/2}, \quad \sum_{j=2}^n x_{ji}^2 \leq 1, \quad i = \overline{1,n}\}.$$

Proof. Let H_s be the event that the matrix Ξ_n has exactly s pairs of the complex conjugate eigenvalues and that the other eigenvalues are real. This is a random event, since the eigenvalues are the continuous functions of the entries Ξ_n. Then

$$\mathbf{E}f(T_n, \Theta_n) = \sum_{s=0}^{[n/2]} \mathbf{E}[f(T_n, \Theta_n)/H_s]p(H_s),$$

where f is an arbitrary bounded continuous function of the entries of T_n and $\Theta_n = (\delta_{pl}\theta_p)$.

As in the proof of Theorem 3.1.2, we have

$$\mathbf{E}f(T_n, \Theta_n) = \sum_{s=0}^{[n/2]} \int_{x_{1i}>0} f(X_n, Y_s) J_s(X_n, Y_s)\varphi(Y_s)$$

$$\times p(X_n Y_s X_n^{-1}) \prod_{i=1}^n \delta(\sum_{j=1}^n x_{ij}^2 - 1) dX_n dY_s, \tag{3.3.3}$$

where $J_s(X_n, Y_s)$ is the Jacobian of the transformation $Z_n = X_n Y_s X_n^{-1}$.

Let us calculate it. To do this instead of the equality $Z_n = X_n Y_s X_n^{-1}$, we use $Z_n = P_{(s)} Q_{(s)} P_{(s)}^{-1}$, where $Q_{(s)} = (\delta_{pl} q_l)$, q_l are the eigenvalues of the matrix Y_s, and where $P_{(s)}$ is the complex matrix of the eigenvectors corresponding to the eigenvalues q_l. It is easy to show that the number of independent variables on the right and on the left of this equality is the same. Let p_m, $m = \overline{1, n^2 - n}$ be the independent coordinates of the matrix $P_{(s)}$. By differentiating the equalities $P_{(s)}^{-1} P_{(s)} = I$, $Z_n = P_{(s)} Q_{(s)} P_{(s)}^{-1}$ by the variables p_m, we find

$$P_{(s)}^{-1}\left(\frac{\partial Z}{\partial p_m}\right) P_{(s)} = S^{(m)} Q_{(s)} - Q_{(s)} S^{(m)},$$

193

where $S^{(m)} = P_{(s)}^{-1}(\partial P_{(s)}/\partial p_m)$.

Similarly,

$$P_{(s)}^{-1}\left(\frac{\partial Z}{\partial \operatorname{Re} q_k}\right) P_{(s)} = \frac{\partial Q_{(s)}}{\partial \operatorname{Re} q_k},$$

$$P_{(s)}^{-1}\left(\frac{\partial Z}{\partial \operatorname{Im} q_k}\right) P_{(s)} = \frac{\partial Q_{(s)}}{\partial \operatorname{Im} q_k}(\operatorname{Im} q_k \neq 0).$$

For the elements of these two matrix equations we obtain,

$$\sum_{i,j=1}^{n} \frac{\partial z_{ij}}{\partial p_m} \tilde{P}_{\alpha i} P_{j\beta} = s_{\alpha\beta}^{(m)}(q_\beta - q_\alpha), \qquad \sum_{i,j=1}^{n} \frac{\partial z_{ij}}{\partial \operatorname{Re} q_k} \tilde{P}_{\alpha i} P_{j\beta} = \delta_{\alpha\beta}\delta_{\alpha k},$$

$$\alpha, \beta = \overline{1, n}, \quad k = \overline{1, n}, \quad m = \overline{1, n^2 - n}, \tag{3.3.4}$$

where $\tilde{P}_{\alpha i}$ are the entries of the matrix $P_{(s)}^{-1}$.

Making use of (3.3.4), we find

$$J_s(X_n, Y_s) = |\prod_{l \neq p}(q_l - q_p)|\kappa(p_m, m = \overline{1, n^2 - 2}), \tag{3.3.5}$$

where $\kappa(p_m)$ is some measurable function of the variables p_m.

Since the function $\kappa(p_m)$ can be expressed in terms of some measurable functions of the entries of the matrix X_n, we can assume that $kappa(p_m) = \tilde{\kappa}(X_n)$. By introducing the change of variables $Z_n = A_n \tilde{Z}_n A_n^{-1}$ in formula (3.3.3), then $\tilde{Z}_n = X_n Y_s X_n^{-1}$, and by making the change of variables $X_n = A_n \tilde{X}_n$ in formula (3.3.2), where A_n is a real nondegenerate matrix, bearing in mind that the function f is arbitrary, we obtain the following equation for the Jacobian,

$$J_s(A_n X_n, Y_s)|\det A_n|^n = J_s(X_n, Y_s).$$

This, together with (3.3.5), yields

$$J_s(X_n, Y_s) = |\prod_{p \neq l}(q_p - q_l)| \det X_n|^{-n} c_s,$$

where c_s is a constant. For the constants c_s, the system of equations (3.3.2) is easily derived.

Theorem 3.3.1 is proved.

Corollary 3.3.1. *If a matrix Ξ_n has the distribution density of the matrix $p(X_n)$, then*

$$p(H_s) = c_s \int_{L_s} p(X_n Y_s X_n^{-1}) J_s(Y_s) \varphi(Y_s) |\det X_n|^{-n}$$

$$\times \prod_{i=2}^{n}\left(1 - \sum_{j=2}^{n} x_{ji}^2\right)^{-1/2} dX_n dY_s,$$

where the domain of integration L_s is

$$\{X_n, Y_s : x_{1i} = [1 - \sum_{j=2}^{n} x_{ji}^2]^{1/2}, \sum_{j=2}^{n} x_{ji}^2 \leq 1, i = \overline{1,n}, Y_s \in K\}.$$

If the event H_0 is given, then the distribution density of the eigenvalues is

$$c_0 \int_{L_0} p(X_n Y_0 X_n^{-1}) \prod_{i>j}(x_i - x_j)^2 \prod_{i=1}^{n}\left(1 - \sum_{j=2}^{n} x_{ji}^2\right)^{-1/2}$$

$$\times |\det X_n|^{-n} dX_n, \quad x_1 > x_2 > \cdots > x_n,$$

the distribution density of the eigenvalues, provided the event $H_{[n/2]}$, is fulfilled and n is even, is equal to

$$c_{[n/2]} \int_{L_{[n/2]}} p(X_n Y_{n/2} X_n^{-1}) 4^{n/2} \prod_{k=1}^{n/2} y_k^2 \prod_{p \neq l}^{n/2}\prod_{l=1}^{n/2} |(x_l + iy_l - x_k)^2$$

$$+ y_k^2|^2 \prod_{i=1}^{n}\left(1 - \sum_{j=2}^{n} x_{ji}^2\right)^{-1/2} |\det X_n|^{-n} dX_n,$$

$$x_1^2 + y_1^2 > x_2^2 + y_2^2 > \cdots > x_{n/2}^2 + y_{n/2}^2.$$

If the entries of the matrix Ξ_n are independent and distributed according to the standard normal law, then the density of Ξ_n is

$$p(X_n) = (2\pi)^{-n^2/2} \times \exp\{-0.5 \operatorname{Tr} X_n X_n'\}.$$

However, in this case, we fail to obtain the simple formulas for the distribution of eigenvalues and eigenvectors of random matrices, since the density $p(X_n)$ is not invariant with respect to the transformation $X_n = T_n Y_n T_n^{-1}$.

§4 **Distribution of the Eigenvalues and Eigenvectors of Complex Random Matrices**

Let $\Xi_n = (\xi_{pl} + i\eta_{pl})$ be a complex random $n \times n$ matrix. Suppose that the random variables ξ_{pl} and η_{pl}, $p, l = \overline{1, n}$ have a joint distribution density which in the following is denoted by $p(Z_n)$, where Z_n is a complex random $n \times n$ matrix. As in Lemma 3.1.1, it is easy to check that the eigenvalues $\lambda_1, \lambda_2, \ldots, \lambda_n$ of the matrix Ξ_n are distinct with probability 1. The eigenvalues are assumed to be arranged in increasing order of their arguments. Let $\vec{\theta}_1, \vec{\theta}_2, \ldots, \vec{\theta}_n$ be the eigenvectors corresponding to the eigenvalues $\lambda_1, \ldots, \lambda_n$, and Θ_n is a matrix whose columns are $\vec{\theta}_i$, $i = \overline{1, n}$. With probability 1, Ξ_n can be represented in the form $\Xi_n = \Theta_n \Lambda_n \Theta_n^{-1}$, where $\Lambda_n = (\lambda_i \delta_{ij})$. For Θ_n to be uniquely determined with probability 1, we require that $(\vec{\theta}_p, \vec{\theta}_p) = 1$, and that $\arg \theta_{1p} = c_p$, $p = \overline{1, n}$, where $c_p (0 \le c_p \le 2\pi)$ are arbitrary real numbers.

We denote the group of nondegenerate complex $n \times n$ matrices by L and the Borel σ-algebra of L by B.

Theorem 3.4.1. *If Ξ_n has the distribution density $p(Z_n)$, then for any subset $E \in B$ and the complex numbers α_i, β_i, $i = \overline{1, n}$,*

$$P\{\Theta_n \in E, \operatorname{Re} \alpha_k < \operatorname{Re} \lambda_k < \operatorname{Re} \beta_k, \operatorname{Im} \alpha_k < \operatorname{Im} \lambda_k$$
$$< \operatorname{Im} \beta_k, k = \overline{1, n}\} = c \int p(X_n Y_n X_n^{-1}) \prod_{i \ne j} (|y_i - y_j|^2)$$
$$|\det X_n|^{-2n} \prod_{i = \overline{2,n}, j = \overline{1,n}} r_{ij} dr_{ij} d\varphi_{ij} dY_n, \qquad (3.4.1)$$

where $X_n = (r_{pl} e^{i\varphi_{pl}})$, $\varphi_{1l} = c_l$, $l = \overline{1, n}$, $r_{1i} = [1 - \sum_{j=2}^n r_{ji}^2]^{1/2}$, and the integration is over the domain $\arg y_1 > \arg y_2 > \cdots > \arg y_n$, $\operatorname{Re} \alpha_k < \operatorname{Re} y_k < \operatorname{Re} \beta_k$, $\operatorname{Im} \alpha_k < \operatorname{Im} y_k < \operatorname{Im} \beta_k$, $k = \overline{1, n}$, $X_n \in E$, $\sum_{j=2}^n r_{ji}^2 \le 1$, $i = \overline{1, n}$, $dY_n = \prod_{k=1}^n d \operatorname{Re} y_k d \operatorname{Im} y_k$, $Y_n = (\delta_{pl} y_l)$. The constant c is determined by the condition that the integral over the whole domain of the variables is 1.

Proof. Let us calculate the Jacobian of the transformation $Z_n = X_n Y_n X_n^{-1}$. Let p_m, $m = \overline{1, 2n^2 - 2n}$ be the parameters of the matrix $X_n, q_m, m = \overline{1, 2n}$ real and imaginary parts of the roots y_i. We represent the Jacobian in the form $\operatorname{mod} \det \begin{bmatrix} A_1 \\ A_2 \end{bmatrix}$, where

$$A_1 = \left[\frac{\partial \operatorname{Re} z_{ij}}{\partial q_m}, \frac{\partial \operatorname{Im} z_{ij}}{\partial q_m} \right]_{i,j = \overline{1,n}, m = \overline{1,2n}},$$

$$A_2 = \left[\frac{\partial \operatorname{Re} z_{ij}}{\partial p_m}, \frac{\partial \operatorname{Im} z_{ij}}{\partial p_m} \right]_{i,j = \overline{1,n}, m = \overline{1,2n^2-n}}.$$

By using the equalities

$$X^{-1}(\partial Z/\partial p_m)X = S^{(m)}Y - YS^{(m)},$$
$$\overline{X}^{-1}(\partial \overline{Z}/\partial p_m)\overline{X} = \overline{S}^{(m)}\overline{Y} - \overline{Y}\overline{S}^{(m)},$$
$$S^{(m)} := X^{-1}(\partial X/\partial p_m),$$
$$X^{-1}(\partial Z/\partial q_m)X = \partial Y/\partial q_m, \qquad \overline{X}^{-1}\partial \overline{Z}/\partial q_m \overline{X} = \partial \overline{Y}/\partial q_m,$$

we obtain the following relations,

$$\sum_{i,j=1}^{n}\left(\frac{\partial z_{ij}}{\partial p_m}\right)\tilde{x}_{pi}x_{lj} = s_{pl}^{(m)}(y_p - y_l),$$

$$\sum_{i,j=1}^{n}\left(\frac{\partial z_{ij}}{\partial q_m}\right)\tilde{x}_{pi}x_{lj} = \delta_{pl}\delta_{pm}, \qquad (3.4.2)$$

where $\tilde{x}_{pi}$ are the entries of the matrix X^{-1}.

We multiply $\mathrm{mod}\,\det\begin{bmatrix}A_1\\A_2\end{bmatrix}$ by the absolute value of the determinant of matrix (B_1, B_2),

$$B_1 = \left[\begin{pmatrix} \tilde{x}_{pi}x_{pj} & \tilde{x}_{pi}x_{pj} \\ ix_{pi}x_{pj} & -i\tilde{x}_{pi}\tilde{x}_{pj} \end{pmatrix}\right]_{p=\overline{1,n},i,j=\overline{1,n}},$$

$$B_2 = \left[\begin{pmatrix} \tilde{x}_{pi}x_{lj} & \tilde{\overline{x}}_{pi}\tilde{x}_{lj} \\ ix_{pi}x_{lj} & -i\tilde{x}_{pi}\tilde{x}_{lj} \end{pmatrix}\right]_{i,j,p,l=\overline{1,n},p\neq l}.$$

After multiplication, and following (3.4.2), we obtain

$$J = \mathrm{mod}\,\det\begin{bmatrix} I_{2n} & 0_{(2n^2-2n)\times n} \\ 0_{(2n^2-2n)\times n} & (s_{pl}^{(m)}(y_p - y_l), \overline{s}_{pl}^{(m)}(\overline{y}_p - \overline{y}_l)) \end{bmatrix}_L$$
$$\times \kappa'(p_m, m = \overline{1, 2n^2 - 2n}) = \prod_{p\neq l}|y_p - y_l|^2 \kappa''(p_m,$$

$$m = \overline{1, 2n^2 - 2n}), L = \{p \neq l, p, l = \overline{1,n}, m = \overline{1, 2n^2 - 2n}\},$$

where κ', κ'' are some Borel functions of the entries p_m. The rest of the proof is similar to that of Theorem 3.3.1. Theorem 3.4.1 is proved.

We note that Eq. (3.4.1) is not simplified if the entries ξ_{pl} and η_{pl} are independent and distributed according to the normal law $N(0,1)$. The integrand expression in (3.4.1) in this case takes the form,

$$c(2\pi)^{-n^2}\exp\{-\mathrm{Tr}(X_n Y_n X_n^{-1})(X_n Y_n X_n^{-1})^*/2\}$$
$$\times \prod_{i\neq j}|y_i - y_j|^2 \prod_{i=\overline{2,n},j=\overline{1,n}} r_{ij}|\det X_n|^{-2}. \quad (3.4.3)$$

Similarly, we can determine the distribution of the eigenvalues and eigen-vectors of complex symmetric and antisymmetric random matrices.

Although in the Gaussian case, the (3.4.3) and (3.3.1) have cumbersome form, we can obtain simpler formulas for the distribution density of the eigen-values. We need the following theorem of Schur [129].

Theorem 3.4.2.   *If A is a complex $n \times n$ matrix, then there is a unitary $n \times n$ matrix U_n, such that $T = U^*AU$ is the upper triangular matrix, and the entries on its main diagonal are the eigenvalues of A.*

If A is real and has real eigenvalues, the U can be chosen to be real orthog-onal.

The matrix A is normal if and only if T is diagonal.

Proof.   Since any square matrix A can be represented by the Jordan form $A = B \Lambda B^{-1}$, and the matrix B can be represented by the form $B = US$ (see the Proof of Theorem 1.3.3), where U is a unitary and S is an upper trian-gular matrix with real nonnegative diagonal entries, then $A = U(S \Lambda S^{-1})U^*$. Hence, all the statements of Theorem 3.4.2 follow.

Let us show that if the eigenvalues of the matrix A are distinct and if some nonzero component of each column of the U has a fixed argument, then the representation $A = UTU^*$ is unique. First, we note, that if the diagonal entries of the triangular upper matrix S are distinct, then it can be represented in the form $S + Q \Lambda Q^{-1}$, where Q is some upper triangular matrix, and Λ is the diagonal matrix of the eigenvalues of the matrix S.

Assume that the eigenvalues of the matrix A are distinct and that the matrix U is chosen as above, then the system of equations $A = UTU^*$ defines the U and S ambiguously. Suppose there are two representations $A = U_1 T_1 U_1^*$, $A = U_2 T_2 U_2^*$. For T_1 and T_2 we have $T_1 = Q_1 \Lambda Q_1^{-1}$, $T_2 = Q_2 \Lambda Q_2^{-1}$, where Q_1 and Q_2 are upper triangular matrices with positive entries on the diagonals, Λ the diagonal matrix of the eigenvalues of A. Then $U_1 Q_1$ and $U_2 Q_2$ are the matrices of the eigenvectors of A. Consequently, $U_1 Q_1 = U_2 Q_2 C_1$, where C_1 is some diagonal matrix. Since $U_2^* U_2 = Q_2 C_1 Q_1^{-1}$, then $Q_2 C_1 Q^{-1} = (Q_1 C_1^{-1} Q_1^{-1})^*$. On the left there is an upper triangular matrix and on the right a lower one. Therefore, $Q_2 C_1 Q_1^{-1} = C_2$, where C_2 is a diagonal matrix. But then $U_1 U_2^* = C_2$. By virtue of the choice of U_1 and U_2, we obtain $C_2 \equiv I$, $U_1 = U_2$, and so $T_1 = T_2$.

Assume that Ξ_n is a complex $n \times n$ random matrix whose entries have a joint distribution density $p(X)$. Let $\Xi_n = USU^*$ be the Schur representation of Ξ_n, and let the diagonal entries s_{ii}, $i = \overline{1,n}$ of S be chosen so that their arguments are arranged in nonincreasing, $\arg u_{1i} = c_i$, $i = \overline{1,n}$, where $c_i (0 \le c_i \le 2\pi)$ are arbitrary real numbers. As in the previous sections, we easily find that U and S are random, and that the eigenvalues of the matrix Ξ_n are distinct with probability 1.

Let Γ be the group of unitary $n \times n$ matrices, B the σ-algebra of the Borel subsets of Γ, ν the normalized Haar measure defined on Γ, and T the group of upper triangular complex $n \times n$ matrices.

Theorem 3.4.3. *For any subset $E \in B$ and any measurable set $C \in T$,*

$$P\{U \in E, S \in C\} = c \int p(HYH^*) \prod_{p \neq l} |y_{pp} - y_{ll}|$$

$$\times \nu(dH / \arg h_{1p} = c_p, p = \overline{1, n}) dY,$$

(3.4.4)

where the integration is over the domain

$$\arg y_{11} > \arg y_{22} > \cdots > \arg y_{nn}, Y \in C, H \in E,$$

$$dY = \prod_{i \geq j} d \operatorname{Re} y_{ij} d \operatorname{Im} y_{ij},$$

$$c = [(2\pi)^{-n(n-1)/2} 2^{n(n-1)/2} \prod_{j=1}^{n-1} j!]^{-1}.$$

Proof. For any measurable and finite function $f(U, S)$ of the entries of U and S, we have

$$\mathbf{E} f(U, S) = \int_D f(H < S) p(X) dX,$$

(3.4.5)

where the matrices H and S are determined by the equation $X = HSH^*$, and the integration is over the domain $D : S \in T$, $H \in T$, $\arg h_{1p} = c_p$, $p = \overline{1, n}$, $\arg s_{11} > \cdots > \arg s_{nn}$.

As in the proof of Theorems 3.4.1 and 3.4.2, we see that (3.4.5) is equal to

$$\mathbf{E} f(U, S) = \int_D f(H, S) p(HSH^*) J(S) \prod_{p=1}^{n} \delta(\arg h_{1p} - c_p) \nu(dH) dS, \quad (3.4.6)$$

where $J(S)$ is some measurable function of the entries of the matrix S (without loss of generality, we consider the functions p and J to continuous).

We need the following statement.

Lemma 3.4.1. *Let $S = Q\Lambda Q^{-1}$ be an upper complex $n \times n$ triangular matrix, where $Q \in T$, $\Lambda = (\lambda_p \delta_{pl})$, $(\overrightarrow{q}_i, \overrightarrow{q}_i) = 1$, $\arg q_{p1} = c_p$, $\overrightarrow{q}_i$ are column vectors of Q, and $0 \leq c_p \leq 2\pi$, $p = \overline{1, n}$ are some real numbers. The entries of the matrix S have the distribution density $P(X), X \in T$. Then for any bounded and continuous function f of the entries of Q and Λ, we have*

$$\mathbf{E} f(Q, \Lambda) = c \int f(R, Y) p(RYR^{-1}) \prod_{p \neq t} |y_p - y_t|$$

$$\times \prod_{s=1}^{n} r_{ss}^{-2s} \prod_{i>j} r_{ij} dr_{ij} d\varphi_{ij} dY,$$

(3.4.7)

where the integration is over the domain $r_{pl} \geq 0$, $0 \leq \varphi_{ij} \leq 2\pi$, $\varphi_{p1} = c_p$, $p = \overline{1,n}$, $r_{pp} = [1 - \sum_{l>p} r_{pl}^2]^{1/2}$, $\sum_{l\geq p} r_{pl}^2 \leq 1$, $R = (r_{pj}e^{i\varphi_{pj}})_{i\geq j}$, $0 \leq \arg y_1 \leq \cdots \leq \arg y_n$, $c > 0$ is some constant.

Proof. Making use of Lemma 1.1.1, we find from Eqs. (1.1.2) and (1.1.3) that where $A, S, \widetilde{S} \in T$, A is fixed, the Jacobian of the transformation $S = A\widetilde{S}A^{-1}$ is equal to

$$\prod_{p=1}^{n} |a_{pp}|^{2(n+1-2p)}. \tag{3.4.8}$$

Using the proof of Theorem 3.4.1, we obtain the following expression for the Jacobian of the transformation $S = Q\Lambda Q^{-1}$:

$$\prod_{p\neq l} |\lambda_p - \lambda_l| f(Q), \tag{3.4.9}$$

where $f(Q)$ is some Borel function of the parameters of the matrix Q.

Now, by making the transformations $S = AYA^{-1}$, $Y = Q\Lambda Q^{-1}$ and then $S = Y\Lambda Y^{-1}$, $Y = AQ$, and by using (3.4.8) and (3.4.9), we obtain two identical Jacobians,

$$\prod_{p=1}^{n} |a_{pp}^{n+1-2p}|^2 \prod_{p\neq l} |\lambda_p - \lambda_l| f(Q),$$

$$\prod_{p=1}^{n} |a_{pp}^{n+1-p}|^2 \prod_{p\neq l} |\lambda_p - \lambda_l| f(AQ),$$

and by equating them, we arrive at (3.4.7). Lemma 3.4.1 is proved.

Using (3.4.6) and Lemma 3.4.1, we find that

$$\mathbf{E}f(U,S) = c \int_D f(H, Q\Lambda Q^{-1}) p(HQ\Lambda Q^{-1}H') J(Q\Lambda Q^{-1})$$

$$\times \prod_{p\neq l} |\lambda_p - \lambda_l| \prod_{p=1}^{n} |q_{pp}|^{-2p} \prod_{p=1}^{n} \delta(\arg h_{1p} - c_p)$$

$$\times \prod_{p=1}^{n} [\delta((\vec{q}_i, \bar{\vec{q}}_i) - 1)\delta(\arg q_{pp} - c_p)]\nu(dH)dQd\Lambda. \tag{3.4.10}$$

Theorem 3.4.1 yields

$$\mathbf{E}f(U,S) = c' \int_D f(H,S)p(X\Lambda X^{-1}) \prod_{p\neq l} |\lambda_p - \lambda_l|^2$$

$$\times \prod_{p=1}^{n} [\delta(\arg x_{1p} - c_p)\delta(\vec{x}_p, \bar{\vec{x}}_p) - 1]|\det X|^{-2n}dX_n dY_n,$$

where the matrices H and S are determined from equation $HSH^* = X\Lambda X^{-1}$. We make the change of the variables $X = HQ$ in this integral (see Theorem 1.3.3). Then,

$$
\mathbf{E}f(U, S) = c' \int_D f(H, Q\Lambda Q^{-1}) p(HQ\Lambda Q^{-1}H') \prod_{p \neq l} |\lambda_p - \lambda_l|^2
$$
$$
\times \prod_{p=1}^n [\delta(\arg \sum_k h_{1k}q_{kp} - c_p'')\delta((\bar{q}_p, \bar{q}_p) - 1)\delta(\arg q_{pp} - c_p')]
$$
$$
|\det HQ|^{-2n} \prod_{p=1}^n |q_{pp}|^{2n-2p}\nu(dH)dQd\Lambda, \tag{3.4.11}
$$

where c_p', c_p'', c' are some constants.

Bearing in mind that the function f is arbitrary from (3.4.10) and (3.4.11), it follows that $J(Q\Lambda Q^{-1}) = \prod_{p \neq l} |\lambda_p - \lambda_l| c''$. Hence (3.4.4) follows.

Let us find the normalizing constant c. To do this, we choose the density of Ξ_n in the form $p(X) = (2\pi)^{-n^2} \times \exp\{-\operatorname{Tr} XX^*/2\}$. Then

$$
c(2\pi)^{-n^2} \int \exp\{-\sum_{k=1}^n |y_{pp}|^2/2\} \times \prod_{p \neq l} |y_{pp} - y_{ll}| \prod_{p=1}^n d\operatorname{Re} y_{pp} d\operatorname{Im} y_{pp}
$$
$$
\times \int \exp\{-\sum_{p>l} |y_{pl}|^2/2\} \prod_{p>l} d\operatorname{Re} y_{pl} d\operatorname{Im} y_{pl} = 1.
$$

Hence, by using Theorem 2.2.4, we find c. Theorem 3.4.3 is proved.

Corollary 3.4.1. *If the distribution density of the entries of the matrix Ξ_n is invariant with respect to the unitary transformation $X = UYU^*$, then the distribution density of the entries of the matrix S is*

$$
c'p(Y) \prod_{p \neq l} |y_{pp} - y_{ll}|, \arg y_{11} > \cdots > \arg y_{nn}, c' = \pi^{n(n-1)/2}[\prod_{j=1}^{n-1} j!]^{-1},
$$

and the matrix U does not depend on the matrix S and has the distribution

$$
\mathbf{P}\{U \in E\} = \int_E \nu(dH/ \arg h_{1p} = c_p).
$$

If the real and imaginary parts of the entries of the matrix Ξ_n are independent and distributed according to the normal law $N(0, 1)$, then the distribution density of the eigenvalues of the matrix Ξ_n is [132],

$$
c'' \exp\left\{-0.5 \sum_{k=1}^n |y_k|^2\right\} \prod_{p \neq l} |y_{pp} - y_{ll}|, \quad \arg y_{11} > \cdots > \arg y_{nn},
$$
$$
c'' = \left[2^{n(n+1)/2} n! \prod_{j=1}^n j!\right]^{-1},
$$

the real and imaginary parts of the entries of s_{ij}, $i > j$ of the matrix S are independent, do not depend on s_{ii} and the matrix U, and are distributed according to the normal law $N(0,1)$.

§5 Distribution of Eigenvalues of Gaussian Real Random Matrices

In this paragraph, we find the distribution of the eigenvalues of a random nonsymmetric real matrix Ξ_n whose entries are independent and distributed according to the standard normal law. The density of the distribution of the entries of such a matrix is equal to

$$p(X) = (2\pi)^{-n^2/2} \exp\{\frac{-\operatorname{Tr} XX'}{2}\}.$$

Let

$$\Lambda_n = \operatorname{diag}\left\{\begin{pmatrix} \lambda_1 & \mu_1 \\ -\mu_1 & \lambda_1 \end{pmatrix}, \ldots, \begin{pmatrix} \lambda_s & \mu_s \\ -\mu_s & \lambda_s \end{pmatrix}, \lambda_{s+1}, \ldots \lambda_{n-2s}\right\},$$

where $(\lambda_k \pm i\mu_k)$, and λr are the eigenvalues of Ξ_n; and let K be the set of real random Jacobi matrices of order n. The eigenvalues $\lambda_k + i\mu_k$, $\lambda_k - i\mu_k$, $k = \overline{1,s}$ are arranged in increasing order of their moduli; and among the conjugate eigenvalues, the one with $\mu_k \geq 0$ comes first, and the values λ_l, $l = \overline{s+1, n-2s}$ are arranged in increasing order. We introduce the notation

$$Y_s = \operatorname{diag}\left\{\begin{pmatrix} x_1 & y_1 \\ -y_1 & x_1 \end{pmatrix}, \ldots, \begin{pmatrix} x_s & y_s \\ -y_s & x_s \end{pmatrix}, x_{s+1}, \ldots, x_{n-2s}\right\}.$$

Let $\varphi(Y_s) = 1$, if the values $x_k \pm y_k$, $k = \overline{1,s}$, x_l, $l = \overline{k+1, n-2k}$ are ordered as described above, and $\varphi(Y_s) = 0$ otherwise.

Theorem 3.5.1. *If the entries of the matrix Ξ_n are independent and distributed according to the standard normal law, then for $n \geq 2$,*

$$P\{\Lambda_n \in B\} = \sum_{s=0}^{[n/2]} c_s \int_{Y_s \in B} \exp\{\frac{-\operatorname{Tr} Y_s Y_s'}{2}\} \times$$

$$|J_s(Y_s)|^{1/2} \varphi(Y_s) \psi(Y_s) dY_s, \tag{3.5.1}$$

where B is any measurable subset of the matrices of K_n, and the constants c_s

satisfy the system of equations

$$\sum_{s=0}^{[n/2]} c_s \int \det Y_s^{2k} \exp\{-\operatorname{Tr} Y_s Y_s'/2\}[J_s(Y_s)]^{1/2} \varphi(Y_s)\psi(Y_s)dY_s$$

$$= 2^k \prod_{i=1}^{k} \Gamma((n+2k+1-i)/2)[\prod_{i=1}^{n} \Gamma((n+1-i)/2)]^{-1}, \quad k = \overline{0,[n/2]};$$

$$\psi(Y_s) = \prod_{p=1}^{s} \int_{-1}^{1} \exp\{-2y_p^2((1-x^2)^{-1}-1)\}(1-x^2)^{1/2}dx|y_p|, \quad \psi(Y_0) = 1,$$

$$\psi(Y_1)_{n=2} = \int_{-1}^{1} \exp\{-2y_1^2((1-x^2)^{-1}-1)\}(1-x^2)^{-1}dx|y_1|,$$

$$J_s(Y_s) = |\prod_{p\neq l}(q_p - q_l)|, \tag{3.5.2}$$

where $q_p, p = \overline{1,n}$ are the eigenvalues of the matrix Y_s.

Proof. In accordance with the proof of Theorem 3.3.1, we make the change of the variables $Z = XY_kX^{-1}$. Here the entries of the matrix satisfy the following conditions: X may be represented in the form $X = H(X)S(X)$, where $H(X)$ is an orthogonal real matrix, and $S(X)$ is an upper triangular matrix whose diagonal elements are equal to 1,

$$s_{11} = (1-s_{21}^2)^{1/2}, \quad s_{22} = 1, \quad s_{33} = (1-s_{43}^2)^{1/2}, \ldots, s_{kk} = 1, \tag{3.5.3}$$

$$s_{k+1k+1} = 1, \ldots, s_{nn} = 1, |s_{21}| < 1, |s_{43}| < 1, \ldots, |s_{kk-1}| < 1.$$

$$k = 2, 4, \ldots.$$

We show that such a choice of the matrix X is possible. Instead of the transformation $Z = HSY_kS^{-1}H$, we consider the following:

$$Z = H\tilde{S}\tilde{Y}_k\tilde{S}^{-1}H', \tag{3.5.4}$$

where

$$\tilde{Y}_k = \operatorname{diag}(\lambda_1 + i\mu_1, \lambda_1 - i\mu_1, \ldots, \lambda_k + i\mu_k, \lambda_k - i\mu_k, \lambda_{k+1}, \ldots, \lambda_{n-2k}),$$

$$\tilde{S} = \{(\vec{s}_1 + i\vec{s}_2), (\vec{s}_1 - i\vec{s}_2), \ldots, (\vec{s}_{2k-1} - i\vec{s}_{2k}), \vec{s}_{2k+1}, \ldots, \vec{s}_n\},$$

and the $\vec{s}_i$ are the column vectors of the matrix S.

In (3.5.4), we can make the following transformation on the matrix $\tilde{S}$: $Z = H(\tilde{S}\Lambda)\tilde{Y}(\tilde{S}\Lambda^{-1})H'$, where Λ is an arbitrary diagonal complex matrix. Suppose that the first two diagonal elements of the matrix Λ are, respectively, equal to $d_1 + id_2$, $d_1 - id_2$, where d_1 and d_2 are arbitrary constants. After such a transformation, the first two components of the vectors $\vec{s}_1$ and $\vec{s}_2$ of the matrix S take the form

$$\begin{bmatrix} s_{11}d_1 & s_{11}d_2 \\ s_{21}d_1 - s_{22}d_2 & s_{22}d_1 + s_{21}d_2 \end{bmatrix} := \begin{bmatrix} \tilde{\tilde{s}}_{11} & \tilde{\tilde{s}}_{12} \\ \tilde{\tilde{s}}_{21} & \tilde{\tilde{s}}_{22} \end{bmatrix}.$$

We choose d_1 and d_2 so that the column vectors of this matrix have unit length. After a few simple calculations, and using (3.5.3), we find that this is possible when

$$d_1 = d_2 \{2s_{21}s_{22} \pm [4s_{21}^2 s_{22}^2 + (s_{11}^2 - s_{22}^2 + s_{21}^2)]^{1/2}\}(s_{11}^2 - s_{22}^2 + s_{21}^2)^{-1},$$
$$d_2 = \pm[2^{-1}(s_{22}^2 + s_{11}^2 + s_{21}^2)(1 + \{2s_{21}s_{22} \pm [4s_{21}^2 s_{22}^2 + (s_{11}^2 - s_{22}^2 + s_{21}^2)^2]^{1/2}\}$$
$$\times (s_{11}^2 - s_{22}^2 + s_{21}^2)^{-1})^2]^{-1/2}.$$

Consequently, for almost all

$$\begin{bmatrix} \tilde{\tilde{s}}_{11} & \tilde{\tilde{s}}_{12} \\ \tilde{\tilde{s}}_{21} & \tilde{\tilde{s}}_{22} \end{bmatrix} = \tilde{H}\begin{bmatrix} (1 - u^2)^{1/2} & 0 \\ u & 1 \end{bmatrix},$$

where $\tilde{H}$ is an orthogonal matrix of the second order and $|u| < 1$.

Thus, the matrix S can be chosen with probability 1 as described above.

We show that with such a choice of X, the system of equations $Z = XY_kX^{-1}$, $q_i \neq q_j$, $i \neq j$ uniquely determines the matrices X and Y_k.

Suppose that this is not so and that the matrix XC as well as X also satisfies this system of equations, where C is a diagonal nondegenerate real matrix. Then $XC = HSC$. The matrix SC is uniquely determined by the matrix XC, but the $c_i = 1$, since the diagonal entries of C must be equal to 1.

From Theorem 3.3.1, for any continuous bounded function $f(\Lambda_n)$, with the help of the transformation $Z = XY_kX^{-1}$, we obtain

$$\mathbf{E}f(\Lambda_n)$$
$$= \sum_{k=0}^{[n/2]} C_k \int f(Y_k) \exp\{-\operatorname{Tr}(X_n Y_k X_n^{-1})(X_n Y_k X_n^{-1})'/2\}$$
$$\times J_k(Y_k)\varphi(Y_k)dY_k|\det X_n|^{-n} \prod_{i=1}^{k} \tag{3.5.5}$$
$$\times \delta(s_{2i,2i}(x) - 1)\delta(s_{2i-1,2i-1}(x) - [1 - s_{2i,2i-1}^2(x)]^{1/2}$$
$$\times \prod_{i=k+1}^{n} \delta(s_{ii}(x) - 1)dX_n,$$

where the $s_{ii}(x)$ are determined by the system of the equations $X = H(X)S$ (X); $\delta(\cdot)$ is the delta function; the C_k are arbitrary constants; and the domain of integration is $|s_{ll-1}| < 1$, $l = \overline{2, k}$.

In the integral (3.5.5), we change the variables $X = HS_k$, where H is an orthogonal matrix, and S_k is an upper triangular matrix with positive entries on the diagonal. The Jacobian of such a change of variables is $\kappa(\theta_i)\prod_{i=1}^{n} s_{ii}^{i-1}$ (Theorem 1.3.5), where θ_i are the Euler angles of the matrix H, and κ is a Borel function. Then (3.5.5) takes the form

$$\mathbf{E}f(\Lambda_n) = \sum_{k=0}^{[n/2]} \tilde{c}_k \int f(Y_k)\exp\{-\operatorname{Tr}(\tilde{S}_k Y_k \tilde{S}_k^{-1})(\tilde{S}_k Y_k \tilde{S}_k^{-1})'/2\}J_k(Y_k)\varphi(Y_k)$$

$$\prod_{i=1}^{k}(\sqrt{1 - s_{2i,2i-1}^2})^{-(n+2i-2)}dY_k d\tilde{S}_k, |s_{2i,2i-1}| < 1, \qquad (3.5.6)$$

where the diagonal elements $s_{2i,2i}$, $i = \overline{1,k}$ of the matrix $\tilde{S}_k$ equal to 1.

In (3.5.6), we consider the change of the variables $\tilde{S}_k Y_k \tilde{S}_k^{-1} = Q_k$, where the matrix Q_k is constructed in the following way:

$$Q_k = \begin{bmatrix} L_1 & & & & & & 0 \\ & L_2 & & & & & \\ & & \ddots & & & & \\ & & & x_{2k+1} & & & \\ & & & & \ddots & & \\ q_{ij} & & & & & x_{n-2k} \end{bmatrix};$$

where

$$L_p = \begin{bmatrix} x_p - y_p s_{2p,2p-1} & (1 - s_{2p,2p-1}^2)^{1/2} y_p \\ (-y_p - s_{2p,2p-1}^2 y_p) & (1 - s_{2p,2p-1}^2)^{1/2} s_{2p,2p-1} y_p + x_p \end{bmatrix},$$

and the entries q_{ij} do not depend on the variables $s_{2p,2p-1}$, $p = \overline{1,k}$, x_p, y_p, $p = 1, 2, \ldots$.

It is obvious that

$$L_p = \begin{bmatrix} (1 - s_{2p,2p-1}^2)^{1/2} & 0 \\ s_{2p,2p-1} & 1 \end{bmatrix}\begin{bmatrix} x_p & y_p \\ -y_p & x_p \end{bmatrix}\begin{bmatrix} (1 - s_{2p,2p-1}^2)^{-1/2} & 0 \\ -s_{2p,2p-1}(1 - s_{2p,2p-1}^2)^{-1/2} & 1 \end{bmatrix}.$$

By using this equality, it is easy to verify that the transformation $\tilde{S}_k Y_k \tilde{S}_k^1 = Q_k$ is one-to-one for almost all matrices Q_k.

Let us calculate the Jacobian of the transformation $Q_k = \tilde{S}_k Y_k \tilde{S}_k^{-1}$. For this, we need to calculate the Jacobian of the transformation

$$Q_k = P_k \tilde{Y}_k P_k^{-1}, \qquad (3.5.7)$$

where

$$P_k = \tilde{S}_k H_k, H_k = \text{diag}\left\{\begin{pmatrix} 1 & 1 \\ i & -i \end{pmatrix}, \ldots, \begin{pmatrix} 1 & 1 \\ i & -i \end{pmatrix}, 1, \ldots, 1\right\},$$

$$\tilde{Y}_k = \text{diag}\{x_1 + iy_1, x_1 - iy_1, \ldots, x_k - iy_k, x_{2k+1}, \ldots, x_{n-2k}\};$$

the number of matrices $\begin{pmatrix} 1 & 1 \\ i & -i \end{pmatrix}$ in H_k is equal to k. The transformation (3.5.7) is equivalent to the change of variables

$$q_{lj} = \sum_{m=1}^{n} P_{lm} Z_m P_{mj}^{(-1)}, \quad l > j, \quad l \neq j+1 \quad j = 1,3,\ldots,2k-1, x_p = x'_p,$$

$$p = \overline{1, n-2k}, \quad y_p = y'_p, \quad p = \overline{1, k}, \quad s_{2p,2p-1} = s'_{2p,2p-1}, \quad p = \overline{1, k},$$

where Z_m is the mth diagonal element of the matrix $\tilde{Y}_k$, and $p_{mj}^{(-1)}$ are elements of the matrix P_k^{-1}.

As in Theorem 3.3.1, we check that the Jacobian of (3.5.7) has the form

$$\text{mod det} \left[\frac{\partial q_{ij}}{\partial S_{pm}}\right]_{p>m,p\neq m+1,m=1,3,\ldots,2k-1}^{l>j,l\neq j+1,j=1,3,\ldots,2k-1,}.$$

It is obvious that

$$\sum_{t<l,t\neq l+1,l=1,3,\ldots,2k-1} P_{it}^{(-1)} (\frac{\partial q_{tl}}{\partial s_{pm}}) P_{lj} = \theta_{ij}^{(p,m)}(z_i - z_j),$$

$$i < j, \quad i \neq j+1, \quad j = 1,3,\ldots,2k-1,$$

where $\theta_{ij}^{(p,m)}$ are the entries of the matrix $P_k^{-1}(\partial p_k / \partial s_{pm})$ and $P_{it}^{(-1)}$ is an element of the matrix P_k^{-1}.

Furthermore, $\text{mod det}[\partial q_{tl} / \partial s_{pm}] \det[P_{it}^{-1} P_{lj}] = \text{mod det}[\theta_{ij}^{(p,m)}(z_i - z_j)] = \prod_{p>l} |z_p - z_l| \{\prod_{p=1,3,\ldots,2k-1} |z_p - z_{p+1}|^{-1}\}\gamma(\tilde{s}_k)$, where $\gamma(\tilde{s}_k)$ is a Borel function of the entries of the matrix.

Let us find γ. We introduce the matrix $\tilde{Q}_k = (\tilde{q}_{pl})$, where $\tilde{q}_{pl} = q_{pl}$ if $p < l$, $p \neq l+1$, $l = 1,3,\ldots,2k-1$, $p = \overline{1,2k}$, and $\tilde{q}_{pl} = q_{pl}$ if $p \leq l$; $\tilde{q}_{pl} = 0$ otherwise; q_{pl} are the entries of the matrix Q_k. For any bounded integrable function $f(\tilde{Q}_k)$, we consider the integral $J := \int f(\tilde{Q}_k) d\tilde{Q}_k$. In this integral, we make the change of the variables $Q_k = AQ'_k A^{-1}$, where $A = (a_{pl})$ is the real square matrix of order n in which $a_{pp} > 0$ and $a_{pl} = 0$, $p > l$. The Jacobian of such a change of variables has the form

$$\Theta_1(A) := \prod_{i=1}^{n} a_{ii}^{i} \prod_{i=1}^{n} a_{ii}^{-(n-i+1)} \prod_{i=1}^{k} \alpha_{2i,2i}^{-1} \alpha_{2i-1,2i-1}. \tag{3.5.8}$$

The first and second factors are, respectively, the left and right Haar measures on the group of matrices A. After this change of variables, we introduce the change $Q_k = \widetilde{S}_k Y_k \widetilde{S}_k^{-1}$, where the matrices $\widetilde{S}_k$ and Y_k are those described above. Then, using (3.5.7) and (3.5.8), we obtain

$$J = \int f(A S_k Y_k S_k^{-1} A^{-1}) \prod_{p>l} |z_p - z_l| \{ \prod_{p=1,3,\dots,2k-1} |z_p - z_{p+1}|^{-1} \}$$

$$\times \gamma(\widetilde{S}_k) \prod_{i=1}^{k} [\delta(s_{2i,2i-1}) \delta(s_{2i-1,2i-1} - (1 - s_{2i,2i-1}^2)^{1/2})]$$

$$\times \prod_{i=k+1}^{n} \delta(s_{ii} - 1) dS_k \theta_1(A), \qquad (3.5.9)$$

where $S_k = (s_{ij})$ is a lower triangular matrix, and $dS_k = \prod_{i \leq j} ds_{ij}$. Using (3.5.7) and the change of variables $S_k = A S_k'$, we find that

$$J = \int f(A S_k Y_k S_k^{-1} A^{-1}) \prod_{p>l} |z_p - z_l| \{ \prod_{p=1,3,\dots,2k-1} |z_p - z_{p+1}|^{-1} \gamma(A\widetilde{S}_k)$$

$$\times \prod_{i=1}^{k} [\delta(\tilde{s}_{2i,2i-1}) \delta(s_{2i-1,2i-1} - (1 - s_{2i,2i-1}^2)^{1/2})]$$

$$\times \prod_{i=k+1}^{n} \delta(\tilde{s}_{ii} - 1) dS_k \prod_{i=1}^{n} a_{ii}^{i} \prod_{i=1}^{k} \alpha_{2i,2i}^{-1} \}. \qquad (3.5.10)$$

From (3.5.9) and (3.5.10) and by virtue of the fact that the function of f may be chosen arbitrarily, we obtain

$$\gamma(\widetilde{S}_k) = c \prod_{i=1}^{n} s_{ii}^{-(n-i+1)} \prod_{i=1}^{k} S_{2i-1,2i-1},$$

where $c > 0$ is a constant, and s_{ii} are the entries of the matrix $\widetilde{S}_k$.

It is obvious that $\operatorname{Tr} Q_k Q_k' = 2 \sum_{p=1}^{k} (x_p^2 + y_p^2) + 4 \sum_{p=1}^{k} y_p^2 ((1 - s_{2p,2p-1}^2)^{-1} - 1) + \sum_{p<l, p \neq l+1, l=1,3,\dots,2k-1} q_{pl}^2$. Using this equality and the expression for $\gamma(\widetilde{S}_k)$ and introducing the change of variables $\widetilde{S}_k Y_k \widetilde{S}_k^{-1} = Q_k$, we write the integral (3.5.6) in the form

$$\mathbf{E} f(\Lambda_n) = \sum_{s=0}^{[n/2]} c_s' \int f(Y_s) \exp\{-\operatorname{Tr} Y_s Y_s'/2\} [J_s(Y_s)]^{1/2} \times$$

$$\varphi(Y_s) \psi(Y_s) dY_s.$$

From this equality, we obtain (3.5.1). We find the system of equations (3.5.2) with the help of the result of Corollary 2.1.2.

$$\mathbf{E} \det \Xi^{2k} = 2^k \prod_{i=1}^{n} \Gamma((n + 2k + 1 - i)/2) [\prod_{i=1}^{n} \Gamma(n + 1 - i/2)]^{-1}.$$

This proves the theorem.

Corollary 3.5.1. *On condition that among the eigenvalues exact s pairs are conjugate and the rest are real, the conditional density of the distribution of the eigenvalues of the matrix* Ξ *is equal to*

$$c_s \exp\{\operatorname{Tr} Y_s Y_s'/2\} [J_s(Y_s)]^{1/2} \varphi(Y_s) \psi(Y_s).$$

§6 Distribution of Eigenvalues and Eigenvectors of Unitary Random Matrices

Suppose that we are given the Euler angles φ_i of the unitary random matrix u_n, which are random variables, along with their joint distribution function. Assume that the random variables φ_i have a joint distribution density, and denote it by $p(x_1, \ldots, x_n^2)$. This density can be represented in the form $p = \tilde{p}(T_n(x_1, \ldots, x_n^2))$, where T_n is a unitary matrix defined with the help of the angles x_i.

The distribution of the matrix U_n is equal to $P\{U_n \in B\} = \int_{H_n \in B} \tilde{p}(H_n) dH_n$, where B is a measurable set of elements of the group $\Gamma_n, dH_n = \prod_{i=1}^{n^2} dx_i$, and the x_i are the Euler angles of the matrix H_n. The group Γ_n is compact, and thus there is a normalized Haar measure ν on it. This measure can be represented in the form $\nu(B) = \int_{H_n \in B} q(H_n) dH_n$, where $q(H_n)$ is a certain function of the angles x_i, called the density of ν.

A random matrix U_n is said to have a distribution that is absolutely continuous with respect to the Haar measure ν if

$$\mathbf{P}\{U_n \in B\} = \int_{H_n \in B} p(H_n) \nu(dH_n),$$

where $p(H_n)$ is some Borel function of the angles φ_i. We determine the random eigenvalues and eigenvectors of the unitary random matrix U_n, which can be represented in the form $U_n = H_n \Theta_n H_n^*$, where H_n is a unitary matrix, $\Theta_n = (\exp(i\theta_p) \delta_{pl})$, and the $\exp(i\theta_p)$ are the eigenvalues of U_n. Let us arrange the arguments of the eigenvalues in nondecreasing order: $0 \le \theta_1 \le \theta_2 \le \cdots \le \theta_n \le 2\pi$. Such eigenvalues are random variables. The eigenvectors are uniquely determined if $\arg h_{1p} = c_p, p = \overline{1, n}$, where $0 \le c_p \le 2\pi$ are certain nonrandom numbers. It is assumed that the first components of the eigenvectors have fixed arguments c_i. We can now proceed to the determination of the distribution for the eigenvalues and eigenvectors of U_n.

Let Γ be the group of n-dimensional unitary matrices, and B the σ-algebra of Borel sets of Γ.

Theorem 3.6.1. *If the Euler angles of the random matrix U_n have a distribution density $p(\cdot)$, then for any set $E \in B$ and any real numbers α_i and β_i, $i = \overline{1,n}$,*

$$\mathbf{P}\{H_n \in E, \alpha_k < \theta_k < \beta_k, k = \overline{1,n}\}$$

$$= c \int \tilde{p}(X_n Y_n X_n^*) \tilde{q}^{-1}(X_n Y_n X_n^*) \prod_{k,l} |e^{iy_k} - e^{iy_l}|^2$$

$$\times \nu(dX_n | \arg x_{1p} = c_p, p = \overline{1,n}) dY_n, \qquad (3.6.1)$$

where $dY_n = \prod_{i=1}^n dy_i$, $Y_n = (e^{iy_p} \delta_{pl})$; the integration is over the domain $0 < y_1 < y_2 < \cdots < y_n < 2\pi$, $\alpha_k < y_k < \beta_k$, $k = \overline{1,n}$, $X_n \in E$; and $c = ((2\pi)^n)^{-1}$.

Proof. For any continuous and bounded function f of the elements of the matrices H_n and of $\theta = (\exp(i\theta_p)\delta_{pl})$, we have

$$\mathbf{E}f(H_n, \Theta) = \int f(X_n, Y_n) \tilde{p}(T_n) dT_n$$

$$= \int f(X_n, Y_n) \bar{p}(T_n) [\bar{q}(T_n)]^{-1} \nu(dT_n)$$

$$= e^{n/2} \int_{\Gamma \times L} f(X_n, Y_n) \bar{p}(T_n \sqrt{s}) [\bar{q}(T_n \sqrt{s})]^{-1}$$

$$\times \exp(-\operatorname{Tr} S/2) \delta(I - S) \det S dS \nu(dT_n),$$

$$\delta(I - S) = \prod_{i=1}^n \delta(1 - s_{ii}) \prod_{i>j} \delta(s_{ij}),$$

where $\delta(X)$ is the delta function; X_n, Y_n form a solution of the equation $X_n Y_n X_n^* = H_n$, $\arg x_{ip} = c_p$; S_n is a nonnegative-definite Hermitian matrix; L is the set of nonnegative-definite Hermitian matrices; and the functions $\bar{p}$ and $\bar{q}$ are defined as follows:

$$\bar{q}(A), \bar{p}(A) = \begin{cases} p(A), q(A), & \text{if } A \text{ is a unitary matrix,} \\ 0, & \text{otherwise.} \end{cases}$$

(Without loss of generality, we assume that $\bar{q}(A)$ and $\bar{p}(A)$ are replaced by continuous functions.)

In this integral, we make the change of variables $T_n S^{1/2} = A$. Using the proof of Theorem 3.1.1, we get

$$\mathbf{E}f(H_n, \Theta) = \int f(X_n(A), Y_n(A)) \bar{p}(A) [\bar{q}(A)]^{-1}$$

$$\times \exp(-\operatorname{Tr} AA^*/2) \delta(I - (AA^*)^{1/2}) dA e^{n/2} c_1, \qquad (3.6.2)$$

$$c_1 = \pi^{n(n-1)-n^2} 2^{n^2 - n(n+1)/2} \prod_{i=1}^n \Gamma(n + 1 - i),$$

where $X_n(A), Y_n(A)$ form a solution of the equation $X_n Y_n X_n^* = (AA^*)^{-1/2}A$, $\arg x_{1p} = c_p$, $p = \overline{1,n}$, $0 < y_1 < \cdots < y_n < 2\pi$, and $dA = \prod_{i,j=1}^n d\operatorname{Re} a_{ij} d\operatorname{Im} a_{ij}$.

In (3.6.2) we make the change of variables $A = UQU^*$, where U is a unitary matrix; $\arg u_{1p} = c_p$, $p = \overline{1,n}$ is an upper triangular matrix; and $0 < \arg q_{11} < \arg q_{22} < \cdots < \arg q_{nn} < 2\pi$. As a result of §4, Chapter 3,

$$
\begin{aligned}
\mathbf{E}f(H_n, \Theta) = \int & f(X_n(UQU^*), Y_n(UQU^*))\overline{p}(UQU^*) \\
& \times [\overline{q}(UQU^*)]^{-1} \exp(-\operatorname{Tr} QQ^*/2)\delta(I - QQ^*) \\
& \times \prod_{p \neq l} |q_{pp} - q_{ll}|\nu(dU|\arg u_{1p} = c_p, p = \overline{1,n})dQ e^{n/2}c_1 c_2,
\end{aligned}
$$

$$
c_2 = [(2\pi)^{-n(n-1)/2} 2^{n(n+1)/2} \prod_{j=1}^{n-1} j!]^{-1}.
$$

In this integral, we first make the change of variables $q_{ll} = r_l e^{i\varphi_l}$, $0 < r_l < \infty$, $0 < \varphi_l < 2\pi$, and then $q_{ij} = p_{ij}\exp(i\varphi_j)$, $i > j$. The change of variables $PP^* = S$ (see §4, Chapter 3), where S is a Hermitian nonnegative-definite matrix, and $P = ((1 - \delta_{ij})p_{ij} + \delta_{ij}r_i)$, $i \geq j$ gives (3.6.1). After simple calculations we get $c_1 c_2 = c$. Theorem 3.6.1 is proved.

Corollary 3.6.1. *If the distribution density of the Euler angles of U_n is equal to the density of the Haar measure ν, then the eigenvectors of U_n are stochastically independent of its eigenvalues. The distribution density of the arguments of the eigenvalues of U_n is equal to*

$$
((2\pi)^n)^{-1} \prod_{k<l} |e^{iy_k} - e^{iy_l}|^2, 0 < y_1 < \cdots < y_n < 2\pi.
$$

The distribution of the matrix H_n is

$$
\mathbf{P}\{H_n \in E\} = \int_{X_n \in E} \nu(dX_n|\arg x_{1p} = c_p, p = \overline{1,n}). \tag{3.6.3}
$$

Corollary 3.6.2. *If the distribution of U_n is absolutely continuous with respect to the Haar measure ν with density p, and the density p satisfies the relation $\tilde{p}(Y_n) \equiv \tilde{p}(X_n Y_n X_n^*)$, where $X_n \in \Gamma$ and $Y_n = (\exp(iy_p)\delta_{pl})$, then the eigenvalues of U_n are independent of its eigenvectors. The distribution density of the arguments of the eigenvalues of the matrix is equal to*

$$
\prod_{k<l} |e^{iy_k} - e^{iy_l}|p(Y_n)((2\pi)^n)^{-1}, 0 < y_1 < \cdots < y_n < 2\pi.
$$

The distribution of H_n is given by (3.6.3).

§7 Distribution of Eigenvalues and Eigenvectors of Orthogonal Random Matrices

Let H_n be a real orthogonal $n \times n$ random matrix. Suppose that there is a joint distribution density $p(x_1, \ldots, x_{n(n-1)/2})$ of its Euler angles φ_i. For almost all values of x_i, we can write $p = \tilde{p}(T_n(x_i, i = 1, n(n-1)/2))$, since the Euler angles φ_i can be represented in terms of the entries of the orthogonal matrix $T_n(x_i, i = 1, n(n-1)/2)$ (see §2, Chapter 1).

It is easy to check that when the distribution density of the Euler angles of H_n exists, the arguments of the eigenvalues of H_n are distinct with probability 1. The eigenvalues of H_n are $\{e^{\pm i\lambda_k}, k = 1, n/2\}$ if n is even, and $\{e^{\pm i\lambda_k}, k = 1, (n-1)/2\}$ if n is odd, where the λ_k are real numbers with $0 \leq \lambda_k \leq 2\pi$. Let $\overrightarrow{\theta}_k$ be the eigenvectors that correspond to the eigenvalues $e^{\pm i\lambda_k}$. The vectors $\overrightarrow{\theta}_k$ corresponding to nonconjugate eigenvalues are orthogonal.

We order the eigenvalues as follows:

$$\{e^{i\lambda_1}, e^{-i\lambda_1}, \ldots, e^{i\lambda_{n/2}}, e^{-i\lambda_{n/2}}, 2\pi \geq \lambda_1 \geq \cdots \geq \lambda_{n/2} \geq 0\}$$

if n is even; and it can happen that some eigenvalues are ± 1. Since the eigenvalues λ_k are distinct with probability 1, the case of interest to us is when two of the eigenvalues λ_k are $+1$ and -1. In this case, we order the eigenvalues as follows:

$$\{e^{i\lambda_1}, e^{-i\lambda_1}, \ldots, e^{i\lambda_{(n-2)/2}}, e^{-i\lambda_{(n-2)/2}}, +1, -1, 2\pi \geq \lambda_1 \geq \cdots \geq \lambda_{(n-2)/2} \geq 0\}.$$

For an odd n, we order them as follows:

$$\{e^{i\lambda_1}, e^{-i\lambda_1}, \ldots, e^{i\lambda_{(n-1)/2}}, e^{-i\lambda_{(n-1)/2}}, \xi, 2\pi \geq \lambda_1 \geq \cdots \geq \lambda_{(n-1)/2} \geq 0\},$$

where the last eigenvalue ξ is a random variable which takes the values $+1$ or -1.

The matrix H_n can be represented almost surely in the following form:

$$H_n = \Theta_n \operatorname{diag}\left\{\begin{pmatrix} \cos\lambda_1 & \sin\lambda_1 \\ -\sin\lambda_1 & \cos\lambda_1 \end{pmatrix}, \ldots, \begin{pmatrix} \cos\lambda_q & \sin\lambda_q \\ -\sin\lambda_q & \cos\lambda_q \end{pmatrix}, +1, -1\right\} \Theta_n'$$

for even n, and

$$H_n = \Theta_n \operatorname{diag}\left\{\begin{pmatrix} \cos\lambda_1 & \sin\lambda_1 \\ -\sin\lambda_1 & \cos\lambda_1 \end{pmatrix}, \ldots, \begin{pmatrix} \cos\lambda_p & \sin\lambda_p \\ -\sin\lambda_p & \cos\lambda_p \end{pmatrix}, \xi\right\} \Theta_n',$$

$$p = \frac{n-1}{2},$$

for odd n, where Θ_n is an orthogonal matrix whose column vectors are Re $\vec{\theta}_k$, Im $\vec{\theta}_k$. In the first of these equalities, there may be no eigenvalues $+1$ or -1, for example, when $n = 3$.

However, such a representation is not unique. To make it unique we must fix some entries of Θ_n. Let $\vec{\theta}_p = \vec{x}_p + i\vec{y}_p$. Then

$$H_n \vec{x}_p = \cos \lambda_p \vec{x}_p - \sin \lambda_p \vec{y}_p, H_n \vec{y}_p = \sin \lambda_p \vec{x}_p + \cos \lambda_p \vec{y}_p.$$

From these equalities, we find that $\vec{x}_p$ and $\vec{y}_p$ are orthogonal and the $[(H_n - \cos \lambda_p I)^2 + I \sin^2 \lambda_p] \vec{y}_p = 0$. The matrix $(H_n - \cos \lambda_p I)^2$ has real eigenvalues $-\sin^2 \lambda_p$ of multiplicity 2. Therefore, we can require that $(\vec{x}_p, \vec{x}_p) = 1$, $(\vec{y}_p, \vec{y}_p) = 1$, and $x_{1p} = c_p$, where c_p is a fixed number with $|c_p| \le 1$.

If n is even and H_n has no eigenvalues $+1$ or -1, then we set $x_{1p} = c_p (p = 2, 4, \ldots, n)$ if n is even; and if H_n has the eigenvalues $+1$ and -1, we set

$$x_{1p} = c_p(p = 2, 4, \ldots, n - 2), x_{1n-1} \ge 0, x_{1n} \ge 0;$$

if n is odd, we set $x_{1p} = c_p(p = 2, 4, \ldots, n - 1), x_{1n} \ge 0$.

Let G be the group of real orthogonal $n \times n$ matrices, μ the normalized Haar measure on G, B the σ-algebra of Borel subsets of G, and n an odd integer.

Theorem 3.7.1. *If the Euler angles of a random matrix H_n have the distribution density p, then for any $E \in B$ and the real numbers α_i, $\beta_i (i = \overline{1, (n-1)/2}$, where $0 \le \alpha_i, \beta_i = 2\pi$,*

$$P\{\Theta_n \in E, \alpha_k < \lambda_k < \beta_k, k = \overline{1, (n-1)/2}, \xi = \pm 1\}$$
$$= c_n^\pm \int_{L_1} \tilde{p}(T_n Y_n^\pm T_n') \tilde{q}^{-1}(T_n Y_n^\pm T_n') \prod_{s=1}^{(n-1)/2}$$
$$\times \{[\sin^2(x_s/2)(1 \pm 1) + \cos^2(x_s/2)(1 \mp 1)]|\sin x_s|\}$$
$$\times \prod_{s>m} \sin^2((x_s - x_m)/2) \sin^2((x_s + x_m)/2)$$
$$\times \prod_s dx_s \mu(dT_n/t_{1p} = c_p, p = 2, 4, \ldots, (n-1), t_{1n} \ge 0);$$

and the integration is over the domain

$$L_1 = \{0 < x_1 < \cdots < x_{(n-1)/2} < 2\pi, \alpha_k < x_k < \beta_k,$$

$$k = \overline{1, (n-1)/2}, T_n \in E\},$$

$$Y_n^{\pm} = \text{diag}\left\{\begin{pmatrix} \cos y_1 & \sin y_1 \\ -\sin y_1 & \cos y_1 \end{pmatrix}, \ldots, \begin{pmatrix} \cos y_{(n-1)/2} & \sin y_{(n-1)/2} \\ -\sin y_{(n-1)/2} & \cos y_{(n-1)/2} \end{pmatrix}, \pm 1\right\},$$

$$c_n^+ = (2a^+)^{-1}, c_n^- = (2a^-)^{-1},$$

$$a^{\pm} = \int_{0 < x_1 < \cdots < x_{(n-1)/2} < 2\pi} \prod_{s=1}^{(n-1)/2} [\sin^2(\frac{x_s}{2})(1 \mp 1)$$

$$+ \cos^2(\frac{x_s}{2})(1 \mp 1)]|\sin x_s| \prod_{s>m} \sin^2(\frac{x_s - x_m}{2})$$

$$\times \sin^2(\frac{x_s - x_m}{2}) \prod_s dx_s.$$

Proof.   For any bounded continuous function f of the entries of Θ_n and

$$\Lambda_n = \text{diag}\left\{\begin{pmatrix} \cos \lambda_1 & \sin \lambda_1 \\ -\sin \lambda_1 & \cos \lambda_1 \end{pmatrix}, \ldots, \begin{pmatrix} \cos \lambda_{(n-1)/2} & \sin \lambda_{(n-1)/2} \\ -\sin \lambda_{(n-1)/2} & \cos \lambda_{(n-1)/2} \end{pmatrix}, \xi\right\},$$

under the condition that $\xi = 1$, we have

$$\mathbf{E}[f(\Theta_n, \Lambda_n)/\xi = 1]\mathbf{P}\{\xi = 1\} = \int f(T_n, Y_n^+)p(H_n)dH_n$$

$$= c^{-1} \int_{G \times K} f(T_n, Y_n^+)\tilde{p}(H_n S_n^{1/2})\tilde{q}^{-1}(H_n S_n^{1/2})$$

$$\times \delta(I - S_n)dS\mu(dH_n), \qquad (3.7.1)$$

where T_n and Y_n^+ are the solutions of the equation $T_n Y_n^+ T_n' = H_n, t_{1p} = c_p, p = 2, 4, \ldots, (n-1), t_{1n} \geq 0, 0 < y_1 < \cdots < y_{(n-1)/2} < 2\pi$, the S_n are nonnegative-definite real matrices; K is the set of nonnegative definite $n \times n$ matrices; $\tilde{p}(A) = p(A), \tilde{q}(A) = q(A)$ if A is orthogonal; $\tilde{p}(A) = 0, \tilde{q}(A) = 0,$ $\tilde{p}(A)\tilde{q}^{-1}(A) = 0$

if A is nonorthogonal; and the δ-function is defined by the following relation:

$$\int_k \varphi(S)\delta(I - S)dS = \lim_{\varepsilon \downarrow 0} \varepsilon^{-n(n+1)/4} \int_k \varphi(S)\exp[-\varepsilon^{-1}$$

$$\times \text{Tr}(I - S)^2]dS = c\varphi(I),$$

where φ is a continuous function on K, and c is a positive constant. Without loss of generality, we can make the change of variables in the integral (3.7:1): $H_n S_n^{1/2} = A_n$, with A_n being an $n \times n$ matrix. It follows from Theorem 3.1.1

that the Jacobian of the transformation $A_n = H_n \sqrt{S_n}$ is $c_n \det S_n^{-1/2} q(H_n)$, where c_n is a constant. Therefore,

$$\mathbf{E}[f(\Theta_n, \Lambda_n)/\xi = 1] \mathbf{P}\{\xi = 1\} = c_1 \int f(T_n(A), Y_n^+(A))$$
$$\times \tilde{p}(A) \tilde{q}^{-1}(A) \delta(I - AA') \det(AA')^{1/2} dA,$$
$$(3.7.2)$$

where $T_n(A)$ and $Y_n^+(A)$ are the solutions of the system of the equations

$$T_n(A) Y_n^+(A) T_n'(A) = (AA')^{-1/2} A, \quad t_{1p}(A) = c_p,$$
$$p = 2, 4, \ldots, n - 1, \quad t_{1n}(A) \geq 0.$$

In (3.7.2) we make the change of variables $A = H\widetilde{A}H'$, where H is an orthogonal matrix, and we integrate the resulting expression with respect to the Haar measure on H. Again we write $\widetilde{A}$ instead of A. Then

$$\mathbf{E}[f(\Theta_n, \Lambda_n)/\xi = 1] \mathbf{P}\{\xi = 1\} = c_1 \int f(T_n(HAH'), Y_n^+$$
$$\times (HAH')) \tilde{p}(HAH') \tilde{q}^{-1}(HAH') \delta(I - AA')$$
$$\times \prod_p \delta(t_{1p}(HAH') - c_p) \det(AA')^{1/2} dA\mu(dH).$$
$$(3.7.3)$$

We consider the change of variables $A = XYX^{-1}$ (see Theorem 3.5.1), where X can be represented as $X = H(X)S(H)$; where $H(X)$ is real orthogonal; and $S(X)$ is lower triangular.

$$s_{2p-1,2p-1} = (1 - s_{2p,2p-1}^2)^{1/2}, |s_{2p,2p-1}| \leq 1,$$
$$p = 1, 2, \ldots, \frac{n-1}{2}, s_{2p} = 1, p = 1, 2, \ldots, \frac{n-1}{2}, s_n = 1,$$
$$Y = \text{diag} \left\{ \begin{pmatrix} x_1 & y_1 \\ -y_1 & x_1 \end{pmatrix}, \ldots, \begin{pmatrix} x_{(n-1)/2} & y_{(n-1)/2} \\ -y_{(n-1)/2} & x_{(n-1)/2} \end{pmatrix}, x \right\}.$$

The Jacobian of the transformation $A = XYX^{-1}$ is calculated in §5, Chapter 3. It is

$$J = \prod_{p \neq l} |z_p - z_l| |\det X_n|^{-n} c,$$

where the $z_l (l = 1, \ldots, n)$ are the eigenvalues of $Y, c > 0$.

By making use of this transformation, we obtain from (3.7.3)

$$\mathbf{E}[f(\Theta_n, \Lambda_n)/\xi = 1]\mathbf{P}\{\xi = 1\} = c_2 \int_{t_{1n}(R) \geq 0} \tilde{p}(R)$$

$$\times f(T_n(R), Y_n^+(R))\tilde{q}^{-1}(R)\delta(I - (XYX^{-1}) \tag{3.7.4}$$

$$\times (XYX^{-1})') \prod_p \delta(t_{1p}(R) - c_p)| \det Y|| \prod_{l \neq p}(z_l - z_p)|$$

$$\times |\det X_n|^{-n} \prod_{p=\overline{1,(n-1)/2}} \delta(S_{2p-1,2p-1}(X) - (1 - S_{2p,2p-1}^2(X))^{1/2})$$

$$\times \prod_{p=\overline{1,(n-1)/2}} \delta(S_{2p}(X) - 1)\delta(S_n(X) - 1)dX\,dY\,\mu(dH),$$

where the $S_{ii}(X)$ are determined by the system of equations

$$X = H(X)S(X), \qquad R = HXYX^{-1}H'.$$

In the integral (3.7.4), we make the change of variables $X = US$, where U is orthogonal and S is lower triangular with positive entries on the diagonal. The Jacobian of this change of variables is $q(\Theta_i)\prod_{i=1}^n S_{ii}^{i-1}$, where q is the density of the Euler angles Θ_i of H. Then (3.7.4) takes the form

$$\mathbf{E}[f(\Theta_n, \Lambda_n)/\xi = 1]\mathbf{P}\{\xi = 1\}$$

$$= c_3 \int f(T_n(L), Y_n^+(L))\tilde{p}(L)\tilde{q}^{-1}(L)\delta(I - (SYS^{-1})(SYS^{-1})') \tag{3.7.5}$$

$$\times \prod_{p=1}^{(n-1)/2} \delta(s_{2p} - 1)\delta(s_n - -1) \prod_p (t_{1p}(L) - c_p)$$

$$\times \prod_{p=1}^{(n-1)/2} \delta(s_{2p-1,2p-1} - \sqrt{1 - s_{2p,2p-1}^2})| \det Y| \prod_{l \neq p}|z_l - z_p|$$

$$\times \prod_i s_{ii}^{-n+i-1}dY\,dS\,\mu(dH),$$

where $L = HSYS^{-1}H'$ and the s_{ii} are the entries of S.

We now consider the change of variables $\tilde{S}Y\tilde{S}^{-1} = Q$ in the integral (3.7.5), where Q has the following form:

$$Q = \begin{pmatrix} L_1 & & & & 0 \\ & L_2 & & & \\ & & \ddots & & \\ & & & L_{(n-1)/2} & \\ q_{ij} & & & & x_n \end{pmatrix};$$

$(n-1)/2$ square (2×2) matrices of the following stand on the diagonal of Q:

$$
L_p = \begin{bmatrix} \sqrt{1 - s^2_{2p,2p-1}} & 0 \\ s_{2p,2p-1} & 1 \end{bmatrix} \begin{bmatrix} x_p & y_p \\ -y_p & x_p \end{bmatrix} \begin{bmatrix} (1 - s^2_{2p,2p-1})^{-1/2} & 0 \\ -s_{2p,2p-1}(1 - s^2_{2p,2p-1})^{-1/2} & 1 \end{bmatrix},
$$

the diagonal entries of the lower triangular matrix S are

$$
\sqrt{1 - s^2_{21}}, 1, \sqrt{1 - s^2_{32}}, \ldots, \sqrt{1 - s^2_{nn-1}}, 1,
$$

and the other entries are the corresponding entries of S. The Jacobian of the change of variables $Q = \widetilde{S} Y \widetilde{S}^{-1}$ is calculated in §5, Chapter 3. It is

$$
c \prod_{p>l} |z_p - z_l| \{ \prod_{p=1,3,\ldots,2k-2} |z_p - z_{p+1}|^{-1} \} \prod_{i=1}^{k} s_{2i-1,2i-1},
$$

where the s_{ii} are the entries of $\widetilde{S}$.

After the change of variables $\widetilde{S} Y \widetilde{S}^{-1}$, the integral (3.7.5) takes the form

$$
\mathbf{E}[f(\Theta_n, \Lambda_n)/\xi = 1] \mathbf{P}\{\xi = 1\}
$$

$$
= c_4 \lim_{\varepsilon \downarrow 0} \int_{t_{1n} \geq 0} f(T_n(R), Y^+_n(R)) \tilde{p}(R) \tilde{q}^{-1}(R) \tag{3.7.6}
$$

$$
\times \exp\{-\varepsilon^{-1} \operatorname{Tr}(I - QQ')^2\} \varepsilon^{-n(n+1)/4} \prod_{l>p} |z_l - z_p|
$$

$$
\times \prod_{p=1,3,\ldots,2k-2} |z_p - z_{p+1}| \prod_p \delta(c_p - t_{1p}(R)) |\det Y|
$$

$$
\times \prod_{p=1}^{(n-1)/2} \sqrt{1 - s^2_{2p,2p-1}} \, ds_{2p,2p-1} \prod dx_i dy_i dq_{pl} \mu(dH),
$$

where $R = HQH'$.

In (3.7.6) we make the change of variables

$$
x_p = r_p \cos \varphi_p, \qquad y_p = r_p \sin \varphi_p (p = 1, \ldots, (n-1)/2),
$$

$$
0 < r_p < \infty, 0 < \varphi_p < 2\pi, \qquad x_n = 1 + \varepsilon r_n, \qquad r_p = 1 + \varepsilon r'_p,
$$

$$
s_{2p,2p-1} = \varepsilon s'_{2p,2p-1}, \qquad q_{ij} = \varepsilon q'_{ij}, \quad i < j,
$$

$$
i \neq j + 1.
$$

It is easy to check that for sufficiently small ε we then have

$$
\exp\{-\varepsilon^{-1} \operatorname{Tr}(I - QQ')^2\} = \exp\{-\sum_{i=1}^{m} \varepsilon^i f_i(Y^+, Q)\},
$$

where the $f_i(Y^+, Q)$ are certain bounded Borel functions of the

$$\prod_{l>p}(z_l - z_p) \prod_{p=1,3,\ldots 2k-2} |z_p - z_{p+1}| = \prod_{l>p}(\gamma_l - \gamma_p) \prod_{p=1,2,\ldots,2k-2} |\gamma_p - \gamma_{p=1}|$$

$$+ \sum_{i=1}^{k} \varepsilon^i \tilde{\varphi}_i(Y^+, r_p, p = 1, \ldots, (n-1-)/2, r_n),$$

where the $\tilde{\varphi}_i$ are bounded Borel functions. It is easy to see that

$$\prod_{l>p} |\gamma_l - \gamma_p| \prod_{p=1,3,\ldots,2k-2} |\gamma_p - \gamma_{p+1}|$$

$$= \prod_{s=1}^{(n-1)/2} \{\sin^2(\varphi_s/2)|\sin \varphi_s|\} \prod_{s>m} \sin^2[(\varphi_s - \varphi_m)/2] \sin^2[(\varphi_s + \varphi_m)/2],$$

where the γ_l are the eigenvalue of Y^+.

We obtain a similar expression when $\xi = -1$. By making use of these relations and by passing to the limit as $\varepsilon \downarrow 0$, we obtain the result stated in the Theorem (3.7.1), in which the constants $c_n^{\pm}$ have not yet been found. Let us suppose that $\tilde{p}(A) = \tilde{q}(A)$. Then $E \det H = 0$ and $E \det H^2 = 1$. By using these equalities, we obtain

$$c_n^+ a^+ - c_n^- a^- = 0, \quad c_n^+ a^+ + c_n^- a^- = 1.$$

This proves Theorem 3.7.1. For even n we obtain similar results.

§8 Distribution of Roots of Algebraic Equations with Random Coefficients

In general, the entries of the random matrices can have no distribution density, but in some cases the coefficients of the characteristic equation can have the distribution density. Therefore, it is of interest to find the distribution of roots of random polynomials.

Let $f(t) : t^n + \xi_1 t^{n+1} + \cdots + \xi_n = 0$ be the algebraic equation whose coefficients ξ_i, $i = \overline{1, n}$ are random variables. Consider the solution of such an equation in the field of complex numbers. It is known from algebra that the equation $f(t) = 0$ has n roots ν_i, $i = \overline{1, n}$; and the roots ν_i, $i = \overline{1, n}$ are continuous functions of the coefficients ξ_i, $i = \overline{1, n}$. Therefore, the roots ν_i, $i = \overline{1, n}$ can be selected in such a way that they will be random variables.

The roots of the equation $f(t) = 0$ have the following form:

$$\nu_1 = \lambda_1 + i\mu_1, \qquad \nu_2 = \lambda_1 - i\mu_1, \ldots, \qquad \nu_{2k-1} = \lambda_k + i\mu_k$$

$$\nu_{2k} = \lambda_k - i\mu_k, \qquad \nu_{2k+1} = \tau_1, \ldots, \qquad \nu_n = \tau_n - 2k,$$

where $\lambda_i, \mu_i, i = \overline{1,k}, \tau_j, j = \overline{1, n - 2k}$ are real variables, and index R is the random variable taking the values from 0 to $[n/2]$.

Arrange the complex roots in increasing order of their moduli. If the moduli of the complex roots coincide, then we arrange them in increasing order of their arguments; among the conjugate pairs of roots, the one with negative imaginary part comes first. Real roots are arranged in increasing order. Roots selected in such a way are random variables. Of course, there are many other ways of ordering eigenvalues, but we adhere to this one as the most natural procedure. We need the following statement.

Lemma 3.8.1.   *If there is a joint distribution density $p(x_1, \ldots, x_n)$ of the co-efficients ξ_i, $i = \overline{1,n}$, then the roots of the equation $f(t) = 0$ are distinct with probability 1.*

Proof.   If at least two roots ν_i, ν_j coincide, then the expression $\prod_{k>p}(\nu_k - \nu_p)$ must be equal to zero. Using the expression for the Vandermonde determinant, we get

$$|\prod_{k>p}(\nu_k - \nu_p)| = |\det(\operatorname{Tr} H^{i+j})_{i,j=0}^{n-1}|^{1/2},$$

where H is the Frobenius random matrix whose characteristic equation is equal to $f(t)$, $\det(\operatorname{Tr} H^{i+j})^{n-1}$ and is some polynomial function of coefficients ξ_i which is not identically $i, j = 0$ equal to zero. Since the distribution density of the coefficients ξ_i exists, then $\mathbf{P}\{|\prod_{i>j}(\nu_i - \nu_j)| = 0\} = 0$.

Lemma 3.8.1. is proved.

Theorem 3.8.1.   *If the random coefficients ξ_i, $i = \overline{1,n}$ of the equation $f(t) = 0$ have the joint distribution density $p(x_1, \ldots, x_n)$, then for any real numbers α_i, β_i, $i = \overline{1,n}$,*

$$\mathbf{P}\{\operatorname{Re} \nu_i < \alpha_i, \operatorname{Im} \nu_i < \beta_i, i = \overline{1,n}\} = \sum_{s=0}^{[n/2]} 2^s$$

$$\times \int_{L_s} p(\Delta_1, \Delta_2, \ldots \Delta_n) \cdot \varphi(z_1, z_2, \ldots, z_n)|\prod_{i>j}(z_i - z_j)|$$

$$\times \prod_{i=1}^{s} dx_i dy_i \prod_{i=2s+1}^{n} dz_i, \tag{3.8.1}$$

where $z_{2p-1} = x_p + iy_p, z_{2p} = x_p - iy_p, p = \overline{1,s}; z_l, l = \overline{2s+1, n}$ are the real variables, the domain of integration L_s is equal to

$$\{x_i, y_i, z_i : x_1 < \alpha_1, y_1 < \beta_1, \ldots, x_s < \alpha_{2s}, -y_s < \beta_{2s},$$

$$z_{2s+1} < \alpha_{2s+1}, 0 < \beta_{2s+1}, \ldots, z_n < \alpha_n, 0 < \beta_n\},$$

and

$$\Delta_k = (-1)^k \sum_{i_1 < i_2 < \cdots < i_k} z_{i_1}, z_{i_2}, \ldots z_{i_k}$$

are the symmetric functions of the variables $z_i, i = \overline{1,n}; \varphi(z_1, \ldots z_n) = 1$, *if the values* $z_i, i = \overline{1,n}$ *are ordered as described above, and* $\varphi(z_i, \ldots z_n) = 0$ *otherwise.*

The statement of the theorem is valid for all $n \geq 2$. (The case $n = 1$ is trivial). For $s = 0$ in the formula (3.8.1), we get a summand of the following form, $\int_{L_0} p(\Delta_1, \ldots, \Delta_n) \varphi(Z_1, \ldots Z_n) \prod_{i>j} (z_i - z_j) \prod_{i=1}^n dz_i$, and the domain of integration L_0 has the form $\{Z_i < \alpha_i, 0 < \beta_i, i = \overline{1,n}\}$.

Proof. It is easy to see that

$$\mathbf{P}\{\operatorname{Re} \nu_i < \alpha_i, \operatorname{Im} \nu_i < \beta_i, i = \overline{1,n}\}$$

$$\sum_{s=0}^{[n/2]} \int_{k_s} p(u_1, \ldots, u_n) \varphi(\Theta_1, \ldots, \Theta_n) \prod_{i=1}^n du_i, \qquad (3.8.2)$$

where the domain of integration K_s is equal to

$$\{(u_1, \ldots, u_n) : \operatorname{Re} \theta_i < \alpha_i, \operatorname{Im} \theta_i < \beta_i, i = \overline{1,n}\} \cap B_s,$$

where B_s is a measurable set for R_n such that for all $(u_1, \ldots, u_n) \in B_s$, the equation $F(t) := t^n + t^{n-1} u_1 + \cdots + u_n = 0$ has exactly S pairs of conjugate complex roots, and $\theta_i(u_1, \ldots, u_n)$ are the roots of the equation $F(t) = 0$. The sets B_s will be measurable since the roots θ_i are continuous functions of the equation coefficients $F(t) = 0$.

In the formula (3.8.2) we make the following change of variables $u_k = \Delta_k(z_i, \ldots, z_n)$ in the Sth integral, where

$$z_{2p-1} = x_p + iy_p, z_{2p} = x_p - iy_p, p = \overline{1,s}, z_1, l = \overline{2s+1, n}$$

are real variables, the values $z_i, i = \overline{1,n}$ are ordered as described above. Obviously, the function Δ_k is a real, and the number of the variables on the left side and on the right side of the equalities $u_k = \Delta_k(z_1, \ldots, z_n), k = \overline{1,n}$ is the same. By Lemma 3.8.1, we consider that $z_i \neq z_j, i \neq j$, and that therefore this is a one-to-one transformation on the set K_s. Let us calculate the Jacobian J of this transformation. Multiply and divide the Jacobian by the Vandermonde determinant W:

$$W = \det(Z_i^{j-1})_{i=\overline{1,n}}^{j=\overline{1,n}} = \prod_{i>j}(z_i - z_j).$$

By multiplying the matrices, we get

$$J = |\det(\frac{\partial \Delta_i}{\partial q_j})^{i,j=\overline{1,n}} \|W\| |W|^{-1}$$

$$= |\det(\sum_{k=1}^n \frac{\partial \Delta_k}{\partial q_j} z_i^{k-1})_{i,j=\overline{1,n}} \| \prod_{i>j}(z_i - z_j)|^{-1},$$

where $q_1 = x_1, q_2 = y_1, \ldots, q_{2s-1} = x_s, q_{2s} = y_s, q_{2s+1} = z_{2s+1}, \ldots, q_n = z_n$.
Write $g(z) = \prod_{i=1}^{n}(z - z_i)$. It is easy to check that

$$\sum_{k=1}^{n} \frac{\partial \Delta_k}{\partial x_1} z_m^{k-1} = \frac{\partial}{\partial x_1}(z^n + z^{n-1}\Delta_1 + \cdots + \Delta_n)_{z=z_m}$$

$$= \begin{cases} \dfrac{\partial}{\partial x_1} g(z_1) & \text{if } m = 1, \\[2mm] \dfrac{\partial}{\partial x_1} g(z_2) & \text{if } m = 2, \\[2mm] 0 & \text{if } m > 2. \end{cases}$$

If

$$g(z) = \prod_{l=1}^{s}[(z - x_l)^2 + y_l^2] \prod_{l=2s+1}^{n} (z - z_l),$$

then

$$\frac{\partial}{\partial y_k} g(x_k + iy_k) = 2y_k \prod_{l=\overline{1,n},\, l \neq 2k,2k-1} (z_{2k-1} - z_l),$$

$$\frac{\partial}{\partial x_k} g(x_k + iy_k) = -2iy_k \prod_{l=\overline{1,n},\, l \neq 2k,2k-1} (z_{2k-1} - z_l),$$

$$\frac{\partial}{\partial y_k} g(x_k + iy_k) = 2y_k \prod_{l=\overline{1,n},\, l \neq 2k,2k-1} (z_{2k} - z_l),$$

$$\frac{\partial}{\partial x_k} g(x_k + iy_k) = 2iy_k \prod_{l=\overline{1,n},\, l \neq 2k,2k-1} (z_{2k} - z_l), k = \overline{1,s}.$$

By using these correlations, we get

$$J = | \prod_{k=1}^{s} [\frac{\partial}{\partial x_k} g(z_{2k-1}) \frac{\partial}{\partial y_k} g(z_{2k}) - \frac{\partial}{\partial x_k} g(z_{2k}) \frac{\partial}{\partial y_k} g(z_{2k-1})]$$

$$\times \prod_{k=2s+1}^{n} \frac{\partial}{\partial z_k} g(z_k) \|W\|^{-1} = \prod_{k=2s+1}^{s} [\prod_{l=\overline{1,n},\, l \neq 2k} , (z_k - z_l)$$

$$\times (z_{2k-1} - z_l)] \prod_{k=1}^{s} 8y_k^2 \prod_{k=1}^{n} \prod_{l \neq k}(z_k - z_l) \|W\|^{-1}.$$

Taking into account that $\prod_{k=1}^{n} 4y_k^2 = |\prod_{k=1}^{s}(z_{2k} - z_{2k-1})^2|$, we transform
the expression of the Jacobian

$$J = |2^s \prod_{i \neq j}(z_i - z_j)| \|W\|^{-1} = 2^s |\prod_{i>j}(z_i - z_j)|.$$

Obviously, after such a change of variables, the domain of integration K_s will be changed into L_s. When $s = 0$, evidently $J = |\prod_{i>j}(z_i - z_j)|$, the domain of integration K_0 will be changed into L_0.

Theorem 3.8.1 is proved.

Corollary 3.8.1. *If the conditions of Theorem 3.8.1 hold, then the probability that the equation $f(t) = 0$ has exactly s pairs of conjugate complex roots is*

$$2^s \int_{R^n} p(\Delta_1, \ldots, \Delta_n) \varphi(z_1, \ldots, z_n) | \prod_{i>j}(z_i - z_j)| \prod_{i=1}^{3} dx_i dy_i \prod_{i=2s+1}^{n} dz_i.$$

Corollary 3.8.2. *Let C be some measurable set of the complex plane whose Lebesque measure given on this plane is equal to zero and the linear measure of the Lebesque set which is equal to the intersection of the set C with the real direct is also zero. Then if the conditions of Theorem 3.8.1 hold, the roots of the equation $f(t) = 0$ get into C with zero probability. The probability that the roots of the equation $f(t) = 0$ are on the real direct is*

$$\int_{z_1 > \cdots > z_n} p(\Delta_1, \ldots, \Delta_n) \prod_{i>j}(z_i - z_j) \prod_{i=1}^{n} dz_i.$$

Suppose that the coefficients of the equation $f(t) = 0$ are complex random variables. The roots of such an equation ν_i, $i = \overline{1, n}$ will be complex. Order the roots ν_i, $i = \overline{1, n}$ in increasing order of their arguments. The density of the real and imaginary parts of coefficients ξ_i, $i = \overline{1, n}$ will be denoted by $p(\operatorname{Re} z_i, \operatorname{Jm} z_i, i = \overline{1, n})$, where z_i, $i = \overline{1, n}$ are complex variables.

Theorem 3.8.2. *[109] If the real and imaginary parts of the random coefficients ξ_i, $i = \overline{1, n}$ have the joint density $p(\operatorname{Re} z_i, \operatorname{Jm} z_i, i = \overline{1, n})$ then the density of roots ν_i, $i = \overline{1, n}$ is*

$$p(\operatorname{Re} \Delta_i, \operatorname{Jm} \Delta_i) \prod_{i>j} |z_i - z_j|^2, \arg z_1 > \cdots > \arg z_r,$$

where $\Delta_i, i = \overline{1, n}$ are symmetric functions of the complex variables $z_i, i = \overline{1, n}$.

Proof. The proof is similar to that for Theorem 3.8.1, except for the calculation of the Jacobian of the transformation $x_i = \Delta_i$, $i = \overline{1, n}$. Let $z_k = x_k + iy_k$, $W = (z_i^j)_{i=\overline{1,n}, j=\overline{0,n-1}}$.

Using Lemma 1.1.1, we obtain

$$J = |\det(\frac{\partial \Delta_s}{\partial z_p})|^2| \det W|^{-2} = |\prod_{i=1}^{n} \prod_{j \neq i}(z_i - z_j)|| \det W|^{-2} = |\det W|^2.$$

Theorem 3.8.2 is proved.

For the study of the distributions of real roots of algebraic equations with random coefficients, we can use the formula first established by Kac.

Lemma 3.8.2. *Let the function $f(t)$ be continuous on the segment $a \le t \le b$, let the continuous derivatives be on the interval $a < t < b$, and have a finite number of points in which the derivative $f(t)$ vanishes. Then the number of zeroes of the function $f(t)$ on the interval $(a.b)$ is equal to*

$$n(a, b) = (2\pi)^{-1} \int dy \int_a^b \cos[yf(t)]|f'(t)dt. \qquad (3.8.3)$$

Moreover, the multiple zero is counted once but the zero coinciding with a or b gives the contribution in $n(a, b)$, equal to $1/2$.

Proof. Denote by $\alpha_1 < \alpha_2 < \cdots < \alpha_n$ the points in which $f'(t)$ is equal to zero. Let us introduce also the notations $a = \alpha_0$, $b = \alpha_{n+1}$. Then

$$\frac{1}{2\pi} \int dy \int_a^b \cos[yf(t)]|f'(t)|dt = \frac{1}{2\pi} \int dy \{ \sum_{j=0, \alpha_j < t < \alpha_{j+1}}^n \text{sign} f'(t)$$

$$\times \int_{\alpha_j}^{\alpha_{j+1}} \cos[yf(t)]f'(t)dt \} = \frac{1}{2\pi} \int dy \{ \sum_{j=0, \alpha_j < t < \alpha_{j+1}}^n \text{sign} f'(t) \frac{1}{y}$$

$$\times [\text{sign}(yf(\alpha_{j+1})) - \text{sign}(yf(\alpha_i)))] \}$$

$$= \sum_{j=0, \alpha_j < t < \alpha_{j+1}}^n \text{sign} f'(t)[\text{sign} f(\alpha_{j+1}) - \text{sign} f(\alpha_j)] = n(a, b).$$

Lemma 3.8.2 is proved.

Using formula 3.8.3, in some cases we can calculate $\mathbf{E}_n(a, b)$. Suppose that $f(t) = \sum_{k=1}^n \xi_k t^k = 0$ is the algebraic equation of the nth degree, whose coefficients are independent and distributed according to the standard normal law. Then

$$\mathbf{E}_n(a, b) = (2\pi)^{-1} \mathbf{E} \int \int_a^b \cos[yf(t)]|f'(t)dtdy$$

$$= (2\pi)^{-1} \frac{\partial}{\partial c} \mathbf{E} \int \int_a^b \exp\{i\xi yf(t) + ic\eta f'(t)dtdy|^{c=0},$$

where ξ and η are independent random variables which do not depend on $f(t)$; moreover, η is distributed according to the Cauchy law, and ξ has the following distributions: $\mathbf{P}\{\xi = +1\} = \mathbf{P}\{\xi = -1\} = 1/2$.

It is not difficult to show that the order of integration can be changed. By changing the order of integration, we have

$$\int \mathbf{E} \exp\{i\xi yf(t) + ic\eta f'(t)\}dy = \int \mathbf{E} \exp\{i\xi y \sum_k t^k \xi_k + i\eta c \sum_k kt^{k-1} \xi_k\}dy$$

$$= \int dy \mathbf{E} \exp\{-0, 5 \sum_k [\xi yt^k - c\eta kt^{k-1}]^2\}$$

$$= \int dy \mathbf{E} \exp\{-0, 5y^2 A - y\xi\eta cB - 0, 5c^2\eta^2 \mathbf{V}\},$$

where $A = \sum_k t^{2k}, B = \sum_k k t^{2k-1}, V = \sum_k k^2 t^{2k-1}$.

Hence, after the simple calculations, we get

$$E_n(a,b) = (2\pi)^{-1}\frac{\sqrt{2}}{\sqrt{A}}\frac{\partial}{\partial c}\sqrt{2\pi}\int_a^b E\exp\{-0,5c^2\eta^2[V - \frac{B^2}{A}]\}_{c=0}dt$$

$$= (2\pi)^{-1/2}\sqrt{2}\int_a^b\sqrt{VA - B^2}A^{-1/2}dt(2\pi)^{-1/2}\int|y|e^{-y^2/2}dy$$

$$= \pi^{-1}\int_a^b\sqrt{VA - B^2}A^{-1}dt = \pi^{-1}\int_a^b[1 - h_n^2(t)]^{1/2}(1 - t^2)^{-1}dt,$$

where $h_n(t) = nt^{n-1}(1 - t^2)(1 - t^{2n})^{-1}$.

Formula 3.8.3 can be used also for finding the mean number of roots which lie on some curve of the complex plane. Let G be the field in the complex plane whose boundary Γ is given by means of functions $x = \varphi(t), y = \psi(t), t \in [a, b]$. Then the number of roots of the equation $f(t) = \sum_{k=1}^n z^k\xi_k = 0$ lying on the boundary Γ is equal the number of real roots of the equation $|f(\varphi(t) + i\psi(t))|^2 = 0$. Let the functions $\varphi(t), \psi(t)$ which are continuous on the segment $a \le t \le b$ have the continuous derivatives on the interval $a < t < b$ and have the finite number of points in which the derivatives $\varphi'(t)$ and $\psi'(t)$ vanish, and let $n(\Gamma)$ be the number of roots of function $g(t) := |f(\varphi(t) + i\psi(t))|^2 = 0$. Then according to the formula 3.8.3,

$$n(\Gamma) = (2\pi)^{-1}\int dy\int_a^b\cos[yg(t)]|g'(t)|dt;$$

moreover, the multiple zero is considered once, but the zero coinciding with a or b gives the contribution in $n(\Gamma)$ equal to $1/2$.

Let $\mu_n(B) = \sum_{k=1}^n X_B(\lambda_k)$, where $X_B(\lambda_k) = 1$ if $\lambda \in B$ and $X_B(\lambda_k) = 0$ if $\lambda_k \overline{\in} B$; B is the arbitrary Borel set of the complex plane.

It is easy to see that

$$\sum_{k=1}^n(Z - \lambda_k)^{-1} = \frac{\partial}{\partial Z}\ln\prod_{k=1}^n(Z - \lambda_k) = (\sum_{k=0}^k kZ^{k-1}\xi_k)(\sum_{k=0}^n Z^k\xi_k)^{-1}.$$

In order to find $E\mu_n(B)$, we use the integral formula by Cauchy,

$$(2\pi i)^{-1}\int_{\partial\Gamma}(u - Z)^{-1}du = X_D(Z),$$

where $X_D(Z) = 1$ if $Z \in D$, $X_D(Z) = 0$; if $Z \in C\backslash\overline{D}$, G is compactly belonging to the D field, which is bounded by the finite number of the continuous curves, and ∂G is the oriented boundary.

For μG, we have

$$\mu(G) = (2\pi i)^{-1} \int_{\partial G} \sum_{k=0}^{n} kZ^{k-1}\xi_k \left(\sum_{k=0}^{n} Z^k \xi_k\right)^{-1} dZ.$$

If there exists $\mathbf{E}(\sum_{k=0}^{n} kZ^{k-1}\xi_k)(\sum_{k=0}^{n} Z^k \xi_k)^{-1}$, then the signs of mathematical expectations and the integral in the formula for $\mathbf{E}\mu(G)$ can be changed.

Let us deal with a particular but very important case when the variables ξ_k, $k = \overline{1,n}$ are independent and distributed according to the normal law $N(0,0,5)$. Then

$$\begin{aligned}
\varphi(Z) : &= \mathbf{E} \sum_{k=0}^{n} kZ^{k-1}\xi_k \left(\sum_{k=0}^{n} Z^k \xi_k\right)^{-1} \\
&= \mathbf{E}(\eta_1\eta_3 - \eta_2\eta_4 + i(\eta_2\eta_3 + \eta_1\eta_4))(\eta_3^2 + \eta_4^2)^{-1},
\end{aligned}$$

$(3.8.4)$

where

$$\eta_1 = \sum_k kr^{k-1}\cos((k-1)\varphi)\xi_k, \eta_3 = \sum_k r^k \cos(k\varphi)\xi_k,$$

$$\eta_2 = \sum_k kr^{k-1}\sin((k-1)\varphi)\xi_k, \eta_4 = \sum_k r^k \sin(k\varphi)\xi_k.$$

In these equations, r is the module, and φ is the statement of the complex number Z.

Let us note that as $\operatorname{Jm} Z \neq 0$, the integral 3.8.4. exists.

Transform $(3.8.4)$ into the following form,

$$\begin{aligned}
\mathbf{E}\frac{\eta_3\eta_1 + \eta_3\eta_2}{\eta_3^2 + i\eta_3\eta_4} \\
&= \frac{\partial}{\partial u} \int_0^{\infty} \exp\{a\eta_3\eta_1 + ia\eta_3\eta_2 - x\eta_3^2 - ix\eta_3\eta_4\}dx|_{a=0} \\
&= \frac{\partial}{\partial a} \int_0^{\infty} \det R^{-1/2}[\det(B + R^{-1})]^{-1/2}dx|_{a=0} \\
&= -\frac{1}{2} \int_0^{\infty} \frac{\det R \frac{\partial}{\partial a} \det(B + R^{-1})}{[\det(BR + I)]^{3/2}} dx|_{a=0},
\end{aligned}$$

$(3.8.5)$

where

$$B = \begin{pmatrix} 0 & 0 & -a & 0 \\ 0 & 0 & ia & 0 \\ -a & -ia & 2x & ix \\ 0 & 0 & ix & 0 \end{pmatrix}, \qquad R = (r_{ij}) = (M\eta_i\eta_j)_{i,j=1}^{4}.$$

Using some simple transformations, we have

$$\det(I + BR) = 1 + x^2(r_{33}r_{44} - r_{34}^2) + 2x(r_{33} + ir_{43}),$$

$$\det R \frac{\partial}{\partial a} \det(B - R^{-1})|_{a=0}$$
$$= 2r_{13} - 2ir_{32} + 2x[r_{32}r_{43} - r_{33}r_{42} - ir_{31}r_{43} + ir_{41}r_{33}.$$

Using this expression and (3.8.5), we obtain

$$\varphi(Z) = -\frac{1}{2} \int_0^\infty \frac{+2r_{13} - 2r_{32}i + 2x[r_{32}r_{43} - r_{33}r_{42} - ir_{31}r_{43} + ir_{41}r_{33}]}{[1 + x^2(r_{33}r_{44} - r_{34}^2) + 2x(r_{33} - ir_{43})]^{3/2}} dx.$$

$$\tag{3.8.6}$$

Using this equality and the integral formula by Cauchy, we get

$$\mathbf{E}\mu(G) = \frac{1}{2\pi i} \int_{\partial G} \varphi(Z) dZ,$$

where the boundary ∂G does not intersect the real axis.

In order to calculate the mathematical expectation of the number of roots ν which got on the real axis, it is necessary to calculate the number of roots ν_ε lying above and below the real axis and at the distance $\varepsilon > 0$ from it.

Then $\mathbf{E}\nu = \lim_{\varepsilon \downarrow 0}[n - \mathbf{E}\nu_\varepsilon]$.

CHAPTER 4

INEQUALITIES FOR RANDOM DETERMINANTS

The study of moments as well as the proof of limit theorems for random determinants are rather complicated: many tasks of calculation of moments are not yet solved. Therefore, various inequalities for the distribution of random determinants are of interest. In this chapter, some of the inequalities are proved and their connection with the Frechet hypothesis is studied.

§1 The Stochastic Hadamard Inequality

In matrix theory, the Hadamard inequality is well known. Let $A_n = (a_{ij})$ be a square complex matrix, then

$$|\det A_n|^2 \le \prod_{i=1}^{n} \left(\sum_{j=1}^{n} |a_{ij}|^2 \right). \qquad (4.1.1)$$

This inequality may be verified in the following way. Let H_n be an orthogonal real matrix of order n whose first vector column is equal to

$$\{a_{1i}(\sum_{i=1}^{n} a_{1i}^2)^{-1/2}, i = \overline{1,n}\},$$

then

$$\det A_n^2 = \sum_{i=1}^{n} a_{1i}^2 \det((\overrightarrow{\xi}_i, \overrightarrow{h}_i)), \quad i,j = \overline{2,n},$$

where $\overrightarrow{\xi}_i$ are vector rows of the matrix A_n, $\overrightarrow{h}_i$ are vector columns of the matrix H_n. As a result of such choice of orthogonal matrices, we obtain the formula

$$\det A_n^2 = \prod_{k=1}^{n} \gamma_{n-k+1}, \qquad (4.1.2)$$

102

where

$$\gamma_n = \sum_{i=1}^{n} a_{1i}^2, \quad \gamma_{n-k+1} = \sum_{j=k}^{n} (\sum_{L_k} a_{kp_1} h_{p_1 p_2}^{(1)}$$
$$\times h_{p_2 p_3}^{(2)} \ldots h_{p_{k-1}j}^{(k-1)})^2,$$
$$L_k = \{p_1 = \overline{1,n}, p_2 = \overline{2,n}, \ldots, p_{k-1} = \overline{h-1,n}\}$$
$$k = \overline{2,n}; h_{ij}^{(k)}, i,j = \overline{k,n}$$

are elements of orthogonal matrix H_k, the first vector-column of which is equal to the vector

$$\{\sum_{L_k} a_{kp_1} h_{p_1 p_2}^{(1)} h_{p_2 p_3}^{(2)} \ldots h_{p_{k-1}j}^{(k-1)}[\gamma_{n-k+1}]^{-1/2}, \qquad j = \overline{k,n}\}.$$

For formula (4.1.2) we assume that $\det A_n \neq 0$.
It is easy to show that

$$\gamma_{n-k+1} \leq \sum_{p_1=1}^{n} a_{kp_1}^2,$$

therefore, the Hadamard inequality (4.1.1) follows from (4.1.2).

If the elements ξ_{ij} of the random matrix A_n are independent and distributed according to the normal law $N(0,1)$, then by choosing the elements of the H_k matrix so that they are measurable with respect to the smallest σ-algebra induced by the random variables

$$\xi_{pi}, \quad i = \overline{1,n}, \quad p = \overline{1,k}$$

from §1, Chapter 2, we obtain

$$\{\gamma_{n-k+1} \approx X_{n-k+1}^2, \quad k = \overline{1,n}\},$$

where $X_p^2, p = \overline{1,k}$ are independent random variables distributed according to the X^2 law with p degrees of freedom.

In §1, Chapter 2, more general results for random matrices are given. Vector rows of these matrices are independent and have distributions invariant with respect to orthogonal transforms.

Furthermore, by the expression $\zeta \leq \xi < \eta$, where ξ, η, ζ are some random variables, we mean the inequality of their distribution functions

$$\mathbf{P}\{\zeta < x\} \geq \mathbf{P}\{\xi < x\} \geq \mathbf{P}\{\eta < x\}.$$

Theorem 4.1.1. *Let the elements ξ_{ij} of the random matrix Ξ_n be independent and distributed according to the normal law $N(0, \sigma_{ij}^2)$. Then*

$$\prod_{i=1}^{n} \min_{j=\overline{1,n}} \sigma_{ij}^2 \prod_{k=1}^{n} \chi_k^2 \lesssim \det \Xi_n^2 \lesssim \prod_{k=1}^{n} \chi_k^2 \prod_{i=1}^{n} \max_{j=\overline{1,n}} \sigma_{ij}^2.$$

The proof follows from the inequality

$$\det[\Xi_n^{(k)}]^2 \min_{j=\overline{1,n}} \sigma_{kj}^2 \lesssim \det \Xi_n^2 \lesssim \max_{j=\overline{1,n}} \sigma_{kj}^2 \det[\Xi_n^{(k)}]^2.$$

Here $\Xi_n^{(k)}$ is a random matrix obtained from the matrix Ξ_n by substitution of the kth vector row by the vector $(\nu_{ki}, i = \overline{1,n})$, where $\nu_{ki}, i = \overline{1,n}$ are independent random variables which do not depend on $\xi_{ij}^{(n)}$ and are distributed according to the normal law $N(0,1)$.

Analogously, we obtain the following assertion. Let the elements of the matrix $\Xi_n = (\xi_{ij})$ be independent and distributed according to the normal law $N(a_{ij}, \sigma_{ij}^2)$. Then

$$\det \Xi_n^2 \lesssim 2^n \prod_{j=1}^{n} \left(\sum_{i=1}^{n} a_{ij}^2 + \sum_{i=1}^{j} \nu_{ij}^2 \max_{i=\overline{1,n}} \sigma_{ij}^2 \right),$$

where ν_{ij} are independent random variables distributed according to the normal law $N(0,1)$.

Indeed from (4.1.2) it follows that

$$\gamma_{n-k+1} \leq \sum_{j=k}^{n} 2 \left(\sum_{p_1=1}^{n} a_{kp_1} \theta_{p_1 j} \right)^2 + 2 \sum_{j=k}^{n} \left(\sum_{p_1=1}^{n} \xi_{kp_1} \theta_{p_1 j} \right)^2,$$

where $\theta_{p_1 j} = \sum h_{p_1 p_2}^{(1)} \ldots h_{p_{k-1} j}^{k-1}$.

Further it is necessary to use the proof of Theorem 4.1.1.

Let us call the distribution of the random variable $X_k^2(c)$ non-central, X^2 the distribution with k degrees of freedom and the parameter of noncentrality c if its Laplace transformation equals

$$\mathbf{E} \exp(-s\chi_k^2(c)) = (1+2s)^{-k/2} \exp(-sc + c(s+1/2)^{-1}), \quad s \geq 0.$$

Theorem 4.1.2. *Let the elements of the random matrix Ξ_n be independent and distributed according to the normal law $N(a_{ij}, 1)$. Then*

$$\det \Xi_n^2 \approx \prod_{k=1}^{n} \chi_k^2(\beta_k), \tag{4.1.3}$$

where $X_k^2(x)$ are independent noncentral X^2 distributions with the parameter of noncentrality x, β_k- *dependent* random variables which are independent of the variables $X_s^2(x)$, $s = \overline{k+1, n}$ and

$$\beta_{n-k+1} = \sum_{j=k}^n \left(\sum_{L_k} a_{kp_1} h_{p_1 p_2}^{(1)} h_{p_2 p_3}^{(2)} \dots h_{p_{k-1}j}^{(k-1)} \right)^2, \quad k = \overline{1, n}.$$

Proof. Using formula (4.1.2) we obtain

$$\gamma_{n-k+1} = \sum_{j=k}^n \left(\sum_{p_1=1}^n a_{kp_1} \theta_{p_1 j} + \sum_{p_1=1}^n \nu_{kp_1} \theta_{p_1 j} \right)^2. \tag{4.1.4}$$

The random variable γ_{n-k+1} is obtained by means of consequent orthogonal transformations of the vector $(a_{kp_1} + \nu_{kp_1}, p_1 = \overline{1, n})$; therefore, the vector $(\sum_{p_1=1}^n \nu_{kp_1} \theta_{p_1 j}, j = \overline{k, n})$ is distributed according to the normal law $N(0, 1)$ and does not depend on the variables $\sum_{p_1=1}^n a_{kp_1} \theta_{p_1 j}, j = \overline{k, n}$.
Theorem 4.1.2 is proved.

Note that the variables β_{n-k+1} satisfy the inequality $\beta_{n-k+1} \le \sum_{p_1=1}^n a_{kp_1}^2$.
The inequalities for moments of random determinants follow from (4.1.2).
From §1, Chapter 2, we obtain stochastic Hadamard inequalities for determinants of the random matrices $\Xi_{n,m} R \Xi'_{n,m}$, where $\Xi_{n,m}$ is the random rectangular matrix of size $n \times m$, $m \ge n$, and R is the random square matrix of order n; the elements of matrix $\Xi_{n,m}$ are independent and distributed according to the normal law.

§2 Inequalities for Random Determinants

Let $\Xi_n = (\xi_{kl}), C_n = (c_{kl})$ be square matrices of order n, Ξ_{kl} the cofactor of the element ξ_{lk} of the matrix Ξ_n, and

$$f(t, \alpha, c) = \exp(-c|t|^\alpha), \quad c \ge 0, \quad 0 < \alpha < 2$$

the characteristic function of the stable symmetric law.
Consider the matrices $c_n = (c_{ij}^{\alpha_i^{-1}}, \tilde{c}_n = (\pm c_{ij}^{\alpha_i^{-1}}), c_{ij} \ge 0$, where pluses and minuses under $c_{ij}^{\alpha_i^{-1}}$ are chosen to make $\det \tilde{c}_n$ maximal.

Theorem 4.2.1. *If the random elements $\xi_{ij}, i, j = \overline{1, n}$ of the matrix Ξ_n are independent and have the characteristic functions*

$$f(t, \alpha_i, c_{ij}) = \exp(-c_{ij}|t|^{\alpha_i}), c_{ij} \ge 0, 0 < \alpha_i \le 2,$$

then

$$|\det \Xi_n| \min_{k=\overline{1,n}} \left[\sum_{i=1}^{n} c_{ik}|\tilde{C}_{ik}|^{\alpha_k}\right]^{-\alpha_k} \prod_{p \neq k} n_p^{\beta_p} \gtrsim \prod_{i=1}^{n} \eta(\alpha_i),$$

where n_p is the number of nonzero elements in the pth row of matrix C_n, per $C_n \neq 0$,

$$\beta_p = \begin{cases} (\alpha_p - 1)/\alpha_p, & 1 < \alpha_p \leq 2, \\ 0, & 0 < \alpha_p \leq 1, \end{cases}$$

$\eta(\alpha_i), i = \overline{1,n}$ *are independent random variables with the characteristic functions $f(t, \alpha_i, 1)$.*

Proof. Let

$$T^{(k)} = (\pm c_{ij}^{\alpha_i^{-1}}(1 - \delta_i, \max(i, k)) + \xi_{ij}(1 - \delta_i, \min(i, k - 1))),$$
$$k = \overline{2,n}, \quad T^{(1)} = \Xi_n, \quad T^{(n+1)} = \tilde{C}_n.$$

The first $k - 1$ rows of the matrix $T^{(k)}$ are the first $k - 1$ rows of the matrix $\tilde{C}_n$.

Denote

$$\varphi_k(T^{(k+1)}) = \left\{\sum_{j=1}^{n} c_{kj}|T_{kj}^{(k+1)}|^{\alpha_k}\right\}^{\alpha_k^{-1}}.$$

Consider the random variables

$$\eta(\alpha_k) = \begin{cases} \det T^{(k)}/\varphi_k(T^{(k+1)}), & \text{if } \varphi_k(T^{(k+1)}) \neq 0 \\ 1, & \text{if } \varphi_k(T^{(k+1)}) = 0. \end{cases}$$

Obviously, if $0 < \alpha_k \leq 1$, then

$$\varphi_k(T^{(k+1)}) \geq \sum_{j=1}^{n} c_{kj}^{\alpha_k^{-1}}|T_{kj}^{(k+1)}| \geq |\det T^{(k+1)}|.$$

If $1 < \alpha_k \leq 2$, then from the inequality

$$\sum_{i=1}^{n} n^{-1}|a_i|^p \geq n^{-p} \left|\sum_{i=1}^{n} a_i\right|^p, \quad p > 1$$

if follows that $\varphi_k(T^{(k+1)}) \geq |\det T^{(k+1)}|n_k^{\alpha_k^{-1}-1}$.

Therefore, with probability 1,

$$\left|\prod_{k=1}^{n} \eta(\alpha_k)\right| \lesssim |\prod_{k=1}^{n}[\det T^{(k)} n_k^{-\beta_k}/\det T^{(k+1)}][\det T^{(n)}/\varphi_n(\tilde{C}_n)]$$

$$= |\det \Xi_n| \prod_{k=1}^{n-1} n_k^{-\beta_k}/\varphi_n(\tilde{C}_n). \tag{4.2.1}$$

Replacing the rows of the matrix Ξ_n, we obtain the following inequality,

$$\prod_{k=1}^{n} |\eta(\alpha_k)| \leq |\det \Xi_n| / \max_{k=\overline{1,n}} \varphi_k(\tilde{C}_n) \prod_{p \neq k} n_p^{\beta_p}.$$

Consider the characteristic function of the random variables $\eta(\alpha_k), k = \overline{1,n}$. It is clear that

$$\mathbf{P}\{\varphi_k(T^{(k+1)}) = 0\} = 0, \quad k = \overline{1,n}.$$

Therefore using the conditional mathematical expectation for every integer $m = \overline{1,n}$,

$$\mathbf{E}\exp\left\{i\sum_{k=m}^{n} t_k \eta(\alpha_k)\right\} = \mathbf{E}[\exp\left\{i\sum_{k=m+1}^{n} t_k \eta(\alpha_k)\right\}$$

$$\times \mathbf{E}(\exp\left\{it_m \sum_{j=1}^{n} \xi_{mj} T_{mj}^{(m+1)}\varphi_m^{-1}(T^{(m+1)})\right\}/\xi_{pl},$$

$$p = \overline{m+1,n}, \quad l = \overline{1,n})] = \prod_{k=m}^{n} \exp\{-|t_k|^{\alpha_k}\}.$$

Theorem 4.2.1 is proved.

Corollary 4.2.1. *If in addition to the conditions of Theorem 4.2.1, $\alpha_k = 1, k = \overline{1,n}$, then*

$$\mathbf{P}\{|\det \Xi_n| \det \tilde{C}_n^{-1} < x\} \leq \pi^{-n} \int \cdots \int_{|\prod_i y_i| < x} \prod_i (1 + y_i^2)^{-1} dy_i.$$

Proving limit theorems for random determinants, we often have to calculate integrals such as $\mathbf{E}\ln|\det \Xi_n|$. The exact formulas for such integrals are complicated. Therefore, some estimates may be useful. Let us prove the following theorems.

Theorem 4.2.2. *If in addition to the conditions of Theorem 4.2.1, $\alpha_k = \alpha, k = \overline{1,n}$, then*

$$n\mathbf{E}\ln|\eta(\alpha)| + \alpha^{-1}\ln \operatorname{per} C_n \leq \mathbf{E}\ln|\det \Xi_n|$$

$$\leq \sigma^{-1}\ln \operatorname{per} B_n + n\sigma^{-1}\ln \mathbf{E}|\eta(\sigma)|^{\sigma}, \qquad (4.2.2)$$

where $0 < \sigma < \alpha, B_n = (c_{kl}^{\sigma/\alpha})$ is a square matrix of order n.

Proof. Consider the following formulas,

$$f_{il}(\Xi_n) = \sum \operatorname{per} B_{k_1 \ldots k_i}^{l \ldots i} \psi(\Xi_{k_{i+1} \ldots k_n}^{i+1 \ldots n}), \quad i = \overline{1,n-1},$$

$$f_{0l}(\Xi_n) = \psi_l(\Xi_n), \quad f_{nl}(\Xi_n) = \operatorname{per} B_n, \quad l = 1,2,$$

$$\psi_1(\Xi_{k_{i+1} \ldots k_n}^{i+1 \ldots n}) = |\det \Xi_{k_{i+1} \ldots k_n}^{i+1 \ldots n}|^{\sigma},$$

$$\psi_2(\Xi_{k_{i+1} \ldots k_n}^{i+1 \ldots n}) = \left(\sum_{j=i+1}^{n} C_{i+1k_j} |\det(\Xi_{k_{i+1} \ldots k_n}^{i+1 \ldots n})_{i+1,k_j}|^{\alpha}\right)^{\sigma/\alpha},$$

where the summation is taken over all possible samples $< k_1, \ldots, k_i >$ from the numbers $1, 2, \ldots, n,$; the matrix $\Xi^{i+1\ldots n}_{k_{i+1}\ldots k_n}$ is obtained from Ξ_n by suppressing the first i rows and the $k_1, k_2, \ldots k_i$ columns. Let

$$\eta_i = \begin{cases} f_{1i}(\Xi_n)/f_{i2}(\Xi_n), & \text{if } f_{i2}(\Xi_n) \neq 0, \\ 1, & \text{if } f_{i2}(\Xi_n) = 0. \end{cases}$$

Since

$$f_{i2}(\Xi_n) \leq \sum \operatorname{per} B_{k_1\ldots k_n} \Big(\sum_{j=i+1}^{n} C_{i+1k_j}$$

$$\times |\det(\Xi^{i+1\ldots n}_{k_{i+1}\ldots k_n})_{i+1k_j}|^{\sigma} = f_{i+1l}(\Xi_n),$$

then, with probability 1,

$$|\det \Xi_n|^{\sigma}/\operatorname{per} B_n = \prod_{i=0}^{n-1} [f_{i1}(\Xi_n)/f_{i+1i}(\Xi_n)] \leq \prod_{i=0}^{n-1} \eta_i. \qquad (4.2.3)$$

Denote

$$\rho^{(i)}_{l_i} = \operatorname{per} B^{1\ldots i}_{k_1\ldots k_n} \psi_2(\Xi^{i+1\ldots n}_{k_{i+1}\ldots k_n}), \quad \rho^{(0)}_{l_0} \equiv 1,$$

$$\beta^{(i)}_{l_i} = \rho^{(i)}_{l_i}/\Xi_{l_i}\rho^{(i)}_{l_i}, \quad l_i = \overline{1, C_n^i},$$

$$\eta^{(i)}_{l_i} = \begin{cases} \det \Xi^{i+1\ldots n}_{k_{i+1}\ldots k_n} \psi_2^{-\sigma^{-1}}(\Xi^{i+1\ldots n}_{i_{k+1}\ldots k_n}), & \text{if } \psi_2 \neq 0, \\ 1, & \text{if } \psi_2 = 0, \end{cases}$$

where a one-to-one correspondence between the values l_i and the values of the vector $(k_1, \ldots k_i)$ is established. Then $\eta_i = \sum_{l_i} \beta^{(i)}_{l_i} |\eta^{(i)}_{l_i}|^{\sigma}$. Since $\operatorname{per} C_n \neq 0$, then the variables $\psi_2(\Xi^{i+1\ldots n}_{k_{i+2}\ldots k_n})$ can be found which are not identically equal to zero.

From Theorem 4.2.1 it follows that all $\eta^{(i)}_{l_i}$, for which $\psi_2(\Xi^{i+1\ldots n}_{l_i k_n}) \neq 0$, are distributed identically and have the characteristic functions $f(t, \alpha, 1)$, and $\eta^{(i)}_{l_i}$ do not depend on $\eta^{(k)}_{l_k}$ and $\beta^{(m)}_{l_m}$ if $i \neq k, m < i$. Therefore, using conditional mathematical expectations we obtain $\mathbf{E}\eta_i = \mathbf{E}|\eta(\alpha)|^{\sigma}$. Then the right side of inequality (4.2.2) follows from inequality (4.2.3),

$$\mathbf{E} \ln[|\det \Xi_n|^{\sigma}/\operatorname{per} B_n] \leq \sum_{i=0}^{n-1} \ln \mathbf{E}\eta_i.$$

We shall prove the left side of inequality (4.2.2). Let $\sigma = \alpha$. Then

$$|\det \Xi_n|^{\alpha}/\operatorname{per} C_n = \prod_{i=0}^{n-1} \eta_i.$$

Using the inequality

$$\ln \prod_{i=0}^{n-1} \eta_i \geq \alpha \sum_{i=0}^{n-1} \sum_{l_i} \beta_{l_i}^{(i)} \ln |\eta_{l_i}^{(i)}|,$$

we obtain the assertion of Theorem 4.2.2.

Corollary 4.2.2. *If the random variables $\xi_{kl}, k, l = \overline{1, n}$ are independent, normally distributed $N(0, C_{pl})$ and per $C_n \neq 0$, then $\mathbf{E} \ln \det \Xi_n^2 = \ln \operatorname{per} C_n + n\varepsilon_n$, where $-C - \ln 2 \leq C_n \leq 0$, and C is the Euler constant.*

The proof follows from Theorem (4.2.2) and from the fact that $\mathbf{E} \ln |\xi| = -C$, and ξ is the random variable, distributed normally $N(0, 2)$.

Theorem 4.2.3. *Let random vector rows $\overrightarrow{\xi}_p, p = \overline{1, n}$ of the matrix $\Xi_n = (\xi_{pl})$ be independent and distributed normally with zero mean vectors and nonsingular matrices of covariance $R_p = (r_{ij}^p), p = \overline{1, n}$. Then*

$$\mathbf{E} \ln \det \Xi_n^2 = \ln \sum_{j_k = \overline{1,n}, k = \overline{1,n}} \det(t_{lj_p}^p)_{p, l = \overline{1, n}}^2 + n\delta_n,$$

where $-C - \ln 2 \leq \delta_n \leq 0$, and t_{ij}^p are elements of the matrix $\sqrt{R_p}$.

Proof. Consider the matrices $R_{j_1 \ldots j_{k+1}} = (t_{lj_p}^p (1 - \delta_p, \max(p, k - 1)) + \xi_{pl}(1 - \delta_p, \min(p, k))), j_1, \ldots, j_k = \overline{1, n}, k = \overline{0, n - 1}, R_{j_0} = \Xi_n$. Denote

$$\varphi_k = \sum_{j_l = \overline{1,n}, l = \overline{1,n}} \det R_{j_1 \ldots j_k}^2,$$

$$\eta_{j_1 \ldots j_{n+1}} = \begin{cases} \det R_{j_1 \ldots j_k} / \psi_{k+1}, & \psi_{k+1} \neq 0, \\ 1, & \psi_{k+1} = 0, \end{cases}$$

where $\psi_{k+1} = (\sum_{j_{k+1}=1}^n \det R_{j_1 \ldots j_{k+1}}^2)^{1/2}$. Then, with probability 1,

$$\det \Xi_n^2 / \varphi_n = \prod_{k=1}^n (\varphi_{k+1} / \varphi_k).$$

Let us show that the random variables $\eta_{j_1}, \ldots, \eta_{j_{k+1}}$ are distributed normally. Indeed, using conditional mathematical expectations, we obtain

$$\mathbf{E} \exp\{it\eta_{j_1 \ldots j_{k+1}}\} = \mathbf{E}[\mathbf{E} \exp\{it\eta_{j_1 \ldots j_{k+1}}\} / \xi_{lp}, p = \overline{1, n}, l = \overline{k + 1, n}]$$

$$= \mathbf{E} \exp\{-0, 5t^2 (\sum_{j_{k+1}=1}^n \det R_{j_1 \ldots j_{k+1}}^2)^{-1}$$

$$\times \sum_{l, m=1}^n (R_{j_1 \ldots j_k})_{kl} (R_{j_1 \ldots j_k})_{km} r_{lm}^k\} = \exp\{-t^2/2\},$$

where $(R_{j_1,\ldots,j_k})_{kl}$ is the cofactor of the $R_{j_1\ldots j_k}$ element of the matrix which stands on the intersection of kth row and lth column.

It is easy to notice that $\eta_{j_1\ldots j_{k+1}}$, does not depend on ψ_{k+1}/ψ_{k+1}. Therefore, the assertion of the theorem follows from the inequality

$$\sum_{k=1}^{n} \sum_{j_l=\overline{1,n},l=\overline{1,k}} \mathbf{E}\beta_{j_1\ldots j_k} \ln \eta_{j_1\ldots j_k}^2 \leq \mathbf{E}\ln[\det \Xi_n^2/\varphi_n]$$

$$\leq \ln \sum_{k=1}^{n} \sum_{j_l=\overline{1,n},l=\overline{1,k}} \mathbf{E}\beta_{j_1\ldots j_k} \eta_{j_1\ldots j_k}^2, \beta_{j_1\ldots j_{k+1}} = \psi_{k+1}/\varphi_{k+1}.$$

§3 The Frechet Hypothesis

Let X_n be a set of square matrices of the order n whose elements take values of either $+1$ or -1. The Frechet hypothesis consists of the fact that under $n \geq 4, \max_{A_n \in X_n} \det A_n = n^{n/2}$, if and only if $n \equiv 0(\bmod 4)$. In other words, the matrix $A_n, n \geq 4$, which consists of the elements ± 1, may be chosen orthogonally if and only if $n = 0(\bmod 4)$.

The hypothesis was known long ago. There exist many examples when it is true [33], but there is no proof for the general case.

The necessity of the condition $n = 0(\bmod 4)$ can be proved quite easily. Choose the elements $a_{ij}, i, j = \overline{1,n}$ of the matrix A_n in such a way that $a_{i1}, a_{1i} = 1, i = \overline{1,n}$.

It is clear that such a choice of elements does not restrict the generality of reasonings as well as the assumption that in the second row on the first $n/2$ places there are $+1$ and on the other -1. Denote the number $+1$ on the first $n/2$ places in the third row by t_1, and the number -1 on the other $n/2$ places by t_2. When the matrix A is orthogonal, $t_1 - t_2 = 0, t_1 + t_2 = n/2$.

Therefore n is a multiple of four under $n \geq 4$.

The Frechet hypothesis may be formulated in terms of random matrices. Let $\Xi_n = (\xi_{ij})$ be a square random matrix of order $n, \xi_{ij}, i, j = \overline{1,n}$ independent random variables and

$$\mathbf{P}\{\xi_{ij} = +1\} = \mathbf{P}\{\xi_{ij} = -1\} = 1/2.$$

Then [115],

$$\max|\det \Xi_n| = \lim_{k \to \infty} \sqrt[2k]{\mathbf{E}\det \Xi_n^{2k}}. \tag{4.3.1}$$

By using (4.3.1), the following lower estimates may be found (Chapter 2),

$$\max|\det \Xi_n| \geq \sqrt{\mathbf{E}\det \Xi_n^2} = \sqrt{n!},$$

$$\max|\det \Xi_n| \geq [n^{-1}(n!)^2 \sum_{k=0}^{n} (k!)^{-1}(n - k + 1)$$

$$\times (n - k + 2)(-2)^k]^{1/4}.$$

In the same manner, we obtain the lower estimates for $\det \widetilde{C}_n$ (the matrix is defined in § 2 of this chapter):

$$\det \widetilde{C}_n \geq (\mathrm{per}(C_{ij}^{2/\alpha_j}))^{1/2}.$$

To prove this inequality, we have to use the formula

$$\max |\det H_n| = \lim[\mathbf{E} \det H_n^{2\cdot}]^{1/2\cdot},$$

where $H = (h_{ij})$ is the random matrix of order n, the random variables h_{ij} are independent and $\mathbf{P}\{h_{ij} = c_{ij}^{1/\alpha_i}\} = \mathbf{P}\{h_{ij} = -c_{ij}^{1/-a_i}\} = 1/2$.

Lemma 4.3.1. *Let $\eta_i, i = \overline{1,n}$ be random variables, distributed according to joint normal law, $\mathbf{E}\eta_i = 0, \mathbf{E}\eta_i^2 = 1, i = \overline{1,n}$. Then $\mathbf{E}\prod_{i=1}^n \eta_i^2$ is minimal only in the case, if $\eta_i, i = \overline{1,n}$ are independent.*

Proof. Any nonnegative-definite matrix R may be introduced as $R = SS'$ (Chapter 4). Where $S = (s_{ij})$ is lower triangular matrix, therefore random variables $\eta_i, i = \overline{1,n}$ distributed in the same way as random variables $\nu_i = \sum_{k=1}^i s_{ik}\xi_k$ where $\xi_k, k = 1, 2, \ldots$ are independent random variables distributed normally $N(0,1)$.

When the random vector $(\nu_1, \ldots, \nu_n, \xi_1, \ldots, \xi_n)$ has symmetric distribution, then

$$\mathbf{E}\prod_{i=1}^n \eta_i^2 = \mathbf{E}\prod_{i=1}^n \nu_i^2 (\sum_{k=1}^{n-1} s_{nk}\xi_k + s_{nn}\xi_n)^2$$

$$= \sum_{k=1}^{n-1} \mathbf{E}\xi_k^2 \prod_{i=1}^{n-1} \nu_i^2 s_{nk}^2 + s_{nn}^2 \mathbf{E}\prod_{i=1}^{n-1} \nu_i^2. \qquad (4.3.2)$$

It is easy to show the fact that $\mathbf{E}(\xi_k^2 - 1)\prod_{i=1}^{n-1} \nu_i^2 \geq 0$. Hence,

$$\mathbf{E}\prod_{i=1}^n \eta_i^2 = \sum_{k=1}^{n-1} \mathbf{E}(\xi_k^2 - 1)\prod_{i=1}^{n-1} \nu_i^2 s_{nk}^2 + \mathbf{E}\prod_{i=1}^{n-1} \nu_i^2.$$

The minimum of this expression takes place if and only if $s_{nk} = 0, k = \overline{1, n-1}$. By using the method of mathematical induction, we came to the assertion of Lemma 4.3.1.

Analogously, we may prove that $\mathbf{E}\prod_{i=1}^n \eta_i^2$ is maximal if and only if the random variables $\eta_i, i = \overline{1,n}$ are equal to one another. This is easy to show. Let us consider (4.3.2). As $\sum_{k=1}^n s_{nk}^2 = 1$, then (4.3.2) takes maximal value, if some $s_{ni} = 1$ and all the other $s_{nj}, j \neq i$ are equal to zero. Hence the needed assertion follows. If η_i are equal to one another, then

$$\mathbf{E}\prod_{i=1}^n \eta_i^2 = \mathbf{E}\eta_1^{2n} = (2n)!(n!)^{-1}. \qquad (4.3.3)$$

Consider the expression

$$n^{-n}\mathbf{E}[\prod_{j=1}^{n}(\sum_{k=1}^{n}\xi_k\varepsilon_{kj})^2/\varepsilon_{kj}], \qquad (4.3.4)$$

where $\varepsilon_{kj}, k, j = \overline{1, n}$ are independent random variables, not depending on ξ_k, and their distributions are $\mathbf{P}\{\varepsilon_{kj} = 1\} = \mathbf{P}\{\varepsilon_{kj} = -1\} = 1/2$. On the basis of Lemma 4.3.1, (4.3.4) will take minimal value equal to 1 if and only if the matrix (ε_{ij}) will be orthogonal.

Therefore, if under $n \equiv 0 \pmod 4$,

$$\lim_{s\to\infty}[(2n)!(n!)^{-1} - \{\mathbf{E}((2n)!(n!)^{-1} - n^{-n}\mathbf{E}[\prod_{j=1}^{n}\sum_{k=1}^{n}\xi_k$$
$$\times \varepsilon_{kj})^2/\varepsilon_{kj}])^{2s}\}^{1/2s}] = 1,$$

then the Frechet hypothesis is true, otherwise is not true.

§4 On Inequalities for Sums of the Martingale Difference and Random Quadratic Forms

In the course of the proof of the limit theorems for random determinants, we shall need the following inequalities.

A certain probability space $(\Omega, F, \mathbf{P})$ with a singled out family of σ-algebras $F_n n \geq 0$, such that $F_1 \leq \cdots \leq F_n$ is given and a sequence of martingale-differences $\eta_n : \mathbf{E}|\eta_n| < \infty, n = 1, 2, \ldots$ and $\mathbf{E}(\eta_{k+1}/F_k) = 0$ on $(\Omega, F, \mathbf{P})$ is given.

Then for all $p \geq 2$ and $n = 1, 2, \ldots$, the inequality

$$\mathbf{E}|\sum_{k=1}^{n}\eta_k|^p \leq c_p n^{p/2} \sum_{k=1}^{n}\mathbf{E}|\eta_k|^p n^{-1},$$
$$c_p = [8(p-1)\max(1, 2^{p-3})]^p$$

is correct. [27 p. 1719]

We shall need the following statement the proof of which was told to the author by Pinelis.

Lemma 4.4.1. *Let $\xi_1, \xi_2, \ldots$ be independent random variables,*

$$v_k = \sum_{1\leq i<j\leq k} a_{ij}\xi_i\xi_j, \quad x_k = v_k - v_{k-1}$$

$$= \xi_k \sum_{i=1}^{k-1} a_{ik}\xi_i, \quad \sigma_k^2 = \mathbf{V}\xi_k, \quad \alpha_{kp} = \mathbf{E}|\xi_k|^p,$$

$$F_k = \sigma(\xi_1, \ldots, \xi_k), \quad k \geq 1, \quad F_0 = \{\phi, \Omega\}, \quad \mathbf{E}\xi_k = 0;$$

a_{ij} are real numbers. Then,

$$E|v_n|^p \le c_p^2[(\sum_{k=1}^{n} \sigma_k^2 \mu_{kp}^{2/p})^{p/2} + \sum_{k=1}^{n} \alpha_{kp}\mu_{kp}], \tag{4.4.1}$$

where

$$\mu_{kp} = (\sum_{i=1}^{n-1} a_{ik}^2 \sigma_k^2)^{p/2} + \sum_{i=1}^{k-1} |a_{ik}|^p \alpha_{ip}.$$

If $\sigma_k^2 \le c, \alpha_{kp} \le c$, then it follows from (4.4.1) that

$$E|v_n|^p \le c_1(\sum_{1 \le i < j \le n} a_{ij}^2)^{p/2}. \tag{4.4.2}$$

Proof. When v_k is a martingale, then Burkholder's value is correct under $p \ge 2$,

$$E|v_n|^p \le c(p)[E(\sum_{k=1}^{n} E(x_k^2/F_{k-1}))^{p/2} + \sum_{k=1}^{n} E|x_k|^p], \tag{4.4.3}$$

where $c(p)$ depends only on p. According to the Minkovsky inequality,

$$E(\sum_{k=1}^{n} E(x_k^2/F_{k-1}))^{p/2} \le \{\sum_{k=1}^{n} [E(E(x_k^2/F_{k-1}))^{p/2}]^{2/p}\}^{p/2}$$

$$= \{\sum_{k=1}^{n} \sigma_k^2 (E|\sum_{i=1}^{k-1} a_{ik}\xi_i|^p)^{2/p}\}^{p/2}. \tag{4.4.4}$$

Again using Burkholder's inequality, we obtain [21]

$$E|\sum_{i=1}^{k-1} a_{ik}\xi_i|^p \le c(p)\mu_{kp}. \tag{4.4.5}$$

If

$$\sum_{k=1}^{n} E|x_k|^p = \sum_{k=1}^{n} \alpha_{kp} E|\sum_{i=1}^{k-1} a_{ik}\xi_i|^p,$$

then the estimates (4.4.1) and (4.4.2) arise from (4.4.3)– (4.4.5). Lemma 4.4.1 is proved.

CHAPTER 5

LIMIT THEOREMS FOR THE BOREL FUNCTIONS
OF INDEPENDENT RANDOM VARIABLES

Limit theorems for the Borel functions of independent random variables are proved in this chapter. We need them for studying the determinant distributions of matrices with independent random elements.

§1 Limit Theorems with the Lindeberg Condition

For every n let the random variables $\xi_i^{(n)}$, $i = \overline{1,n}$ be independent. Consider them measurable by the Borel functions $f_n(\xi_i^{(n)}, i = \overline{1,n})$, which we denote as $f_n(\overrightarrow{\xi}_n)$ for simplicity, where $\overrightarrow{\xi}_n = (\overrightarrow{\xi}_1^{(n)}, \ldots, \xi_n^{(n)})$. By $\nu_n \sim \mu_n$ we shall mean $\lim_{n \to \infty}[\mathbf{E}\exp(it\nu_n) - \mathbf{E}\exp(it\mu_n)] = 0$, for any finite t, where ν_n and μ_n are certain sequences of random variables. The superscript (n) on the random variables $\xi_i^{(n)}$ will be omitted if it does not give rise to any misunderstanding.

Let the independent random variables $\nu_i, \mu_i, i = \overline{1,n}$ and a certain Borel function $g_n(x_1, \ldots, x_n)$ with n variables be given, such that

$$\mathbf{E}g_n^2(\nu_1^{(i_1)}, \ldots, \nu_n^{(i_n)}) < \infty, \quad i_k = \pm 1, \quad \nu_p^{(-1)} = \mu_p, \quad \nu_p^{(+1)} = \nu_p.$$

A certain rule operating on the function $g_n(\nu_1^{(i_1)}, \ldots, \nu_n^{(i_n)})$ according to which the random variables $g_n(\nu_1^{(i_1)}, \ldots, \nu_n^{(i_n)})$ are replaced with the following expression:

$$\theta_s^l g_n(\nu_1^{(i_1)}, \ldots, \nu_{s-1}^{(i_{s-1})}, \nu_s, \ldots, \nu_n) = g_n(\nu_1^{(i_1)}, \ldots, \nu_{s-1}^{(i_{s-1})}, \mu_s,$$

$$\mu_{s+1}, \ldots, \mu_{l-1}, \nu_l, \nu_{l+1}, \ldots, \nu_n) - \mathbf{E}[g_n(\nu_1^{(i_1)}, \ldots, \nu_{s-1}^{(i_{s-1})},$$

$$\mu_s, \mu_{s+1}, \ldots, \mu_{l-1}, \nu_l, \nu_{l+1}, \ldots, \nu_n)/\sigma_{nl}], \tag{5.1.1}$$

will be called the operator $\theta_s^l, n \geq l \geq s \geq 1$, where σ_{nl} is the minimal σ-algebra, and the random variables $\nu_k, \mu_k, k \neq l, k = \overline{1,n}$ being measurable with respect to it.

114

The operator $\tilde{\theta}_s^l$ is defined analogously:

$$\tilde{\theta}_s^l g_n(\nu_1^{(i_1)}, \ldots, \nu_s^{(i_s-1)}, \nu_s, \ldots, \nu_n) = g_n(\nu_1^{(i_1)}, \ldots, \nu_{s-1}^{(i_s-1)},$$

$$\mu_s, \ldots, \mu_l, \nu_{l+1}, \ldots, \nu_n) - \mathbf{E}\{g_n(\nu_1^{(i_1)}, \ldots, \nu_{s-1}^{(i_s-1)},$$

$$\mu_s, \ldots, \mu_l, \nu_{l+1}, \ldots, \nu_n)/\sigma_{nl}\}.$$

Theorem 5.1.1. *Let the random variables* $\xi_i^{(n)}, \eta_i^{(n)}, i = \overline{1,n}$ *be independent for every* n *and the Borel functions* $f_n(x_1, \ldots, x_n)$ *be given such that*

$$\mathbf{E}f_n^2(\eta_1, \ldots, \eta_{k-1}, \xi_k, \ldots, \xi_n) < \infty, \quad k = \overline{1, n+1}. \qquad (5.1.2)$$

For the functions f_n *and the random variables* $\xi_i, \eta_i, i = \overline{1,n}$, *the operators* $\theta_s^l, \tilde{\theta}_s^l$ *are found from formulas (5.1.1), and*

$$\mathbf{E}f_n^i(x_1, \ldots, x_{k-1}, \xi_k, x_{k+1}, \ldots x_n)$$
$$= \mathbf{E}f_n^i(x_1, \ldots, x_{k-1}, \eta_k, x_{k+1}, \ldots, x_n), \quad i = \overline{1,2}, \quad k = \overline{1,n} \qquad (5.1.3)$$

holds for all $x_i, i = \overline{1,n}$, *with the exception of a Borel set with* $f_n(x_1, \ldots, x_n)$ *values of the Lebesgue measure 0 and*

$$\sup_n \sum_{k=1}^n [\mathbf{E}(\theta_1^k f_n(\overrightarrow{\xi}_n))^2 + \mathbf{E}(\tilde{\theta}_1^k f_n(\overrightarrow{\xi}_n))^2] < \infty, \qquad (5.1.4)$$

and the Lindeberg condition is fulfilled, for any $\tau > 0$,

$$\lim_{n \to \infty} \sum_{k=1}^n \int_{|x| > \tau} x^2 d[\mathbf{P}\{\theta_1^k f_n(\overrightarrow{\xi}_n) < x\}$$

$$+ \mathbf{P}\{\tilde{\theta}_1^k f_n(\overrightarrow{\xi}_n) < x\}] = 0. \qquad (5.1.5)$$

Then $f_n(\xi_1, \ldots, \xi_n) \sim f_n(\eta_1, \ldots, \eta_n)$.

Proof. Using condition (5.1.3), let us make the following obvious transforms:

$$|\mathbf{E}\exp[isf_n(\overrightarrow{\xi}_n)] - \mathbf{E}\exp[isf_n(\overrightarrow{\eta}_n)]|$$

$$\leq \sum_{k=1}^n \mathbf{E}|\mathbf{E}\{(\exp(is\theta_1^k f_n) - (is\theta_1^k f_n)^2/2 - 1)/\sigma_{nk}\}$$

$$+ \sum_{k=1}^n \mathbf{E}|\mathbf{E}\{(\exp(is\tilde{\theta}_1^k f_n) - (is\tilde{\theta}_1^k f_n)^2/2 - 1)/\sigma_{nk}\}|$$

$$\leq |s|^3 (\tau/3) \sum_{k=1}^n \{\mathbf{E}(\theta_1^k f_n)^2$$

$$+ 2s^2 \sum_{k=1}^n \int_{|x| \geq \tau} x^2 d[\mathbf{P}\{\theta_1^k f_n < x\} + \mathbf{P}\{\tilde{\theta}_1^k f_n < x\}].$$

Choosing τ sufficiently small, we obtain that the assertion of Theorem 5.1.1 is valid.

Condition (5.1.5) can be changed for Ljapunov's condition: for some $\delta > 0$,

$$\lim_{n\to\infty}\sum_{k=1}^{n}[\mathbf{E}|\theta_1^k f_n|^{2+\delta} + \mathbf{E}|\tilde{\theta}_1^k f_n|^{2+\delta}] = 0.$$

Condition (5.1.5) is difficult to verify in general. To simplify it for several cases, let us prove one assertion. We need the next definition.

Let the independent random variables $\nu_i, \mu_i, i = \overline{1, n}$ and a Borel function $g(x_1, \ldots, x_n)$ with n variables be given. A certain rule according to which the variables ν_s are replaced with μ_s for all $s = \overline{p+1, n}$ in the function $g_n(\nu_1^{(i_1)}, \ldots, \nu_n^{(i_n)})$ will be called the operator $\kappa_p, p = \overline{1, n-1}$.

Theorem 5.1.2. *For every n let the random variables $\xi_i^{(n)}, \eta_i^{(n)}, i = \overline{1, n}$ be independent, with the Borel functions $f_n(x_1, \ldots, x_n)$ given for them with the defined operators $\kappa_p, \theta_s^l, \tilde{\theta}_s^l$. For all $i_k = \pm 1, k = \overline{1, m}$ (m is a natural number independent of n), conditions (5.1.2) and (5.1.3) hold, and $n \ge l_k > l_{k-1} > \cdots > l_1 > 1$*

$$\mathbf{E}[\theta_{l_{k-1}+1}^{l_k}(i_k)\ldots\theta_{l_1+1}^{l_2}(i_2)\theta_1^{l_1}(i_1)f_n]^2$$
$$= \mathbf{E}[\kappa_{l_k}\theta_{l_{k-1}+1}^{l_k}(i_k)\ldots\theta_{l_1+1}^{l_2}(i_2)\theta_1^{l_1}(i_1)f_n]^2, \quad (5.1.6)$$

where $\theta_p^l(+1) = \theta_p^l, \theta_p^l(-1) = \tilde{\theta}_p^l$, for any $\tau > 0$,

$$\lim_{n\to\infty}\sum_{L_m}\int_{|x|>\tau}x^2 d\mathbf{P}\{\theta_{l_{m-1}+1}^{l_m}(i_m)\ldots\theta_1^{l_1}(i_1)f_n < x\} = 0, \tag{5.1.7}$$

$$\lim_{n\to\infty}\sum_{L_k}\int_{|x|>\tau}x^2 d\mathbf{P}\{\kappa_{l_{k-1}}\theta_{l_{k-2}}^{l_{k-1}}(i_k)\ldots\theta_1^{l_1}(i_1)f_n < x\} = 0, \tag{5.1.8}$$

$$\sup_n\sum_{L_k}\mathbf{E}(\theta_{l_{k-1}+1}^{l_k}(i_k)\ldots\theta_1^{l_1}(i_1)f_n)^2 < \infty, \tag{5.1.9}$$

$$L_k = \{n \ge l_{k-1} > \cdots > l_1 > 1\}, \quad k = \overline{2, m}.$$

Then

$$f_n(\xi_1, \ldots, \xi_n) \sim f_n(\eta_1, \ldots, \eta_n). \tag{5.1.10}$$

Proof. According to Theorem 5.1.1, condition (5.1.10) holds, provided conditions (5.1.3)–(5.1.5) hold. It follows from conditions (5.1.6) and (5.1.8) that the variables $\theta_1^k f_n, \tilde{\theta}_1^k f_n$ are infinitesimal. (The random variables $\xi_i^{(n)}, i = \overline{1, n}, n = 1, 2, \ldots$ are called infinitesimal if for any $\varepsilon > 0$,

$$\lim_{n\to\infty}\sup_{i=\overline{1,n}}\mathbf{P}\{|\xi_i| > \varepsilon\} = 0).$$

We show that condition (5.1.5) will be derived from the relation

$$\lim_{n\to\infty} \Big[\prod_{k=1}^{n} \mathbf{E}\exp(is\theta_1^k(i_1)f_n)$$

$$- \prod_{k=1}^{n} \exp\{-s^2\mathbf{E}(\theta_1^k(i_1)f_n)^2/2\}\Big] = 0, \quad i_1 = \pm 1. \tag{5.1.11}$$

Indeed, since the variables $\theta_1^k(i_1)f_n$ are infinitesimal and (5.1.9) holds,

$$\Big| \prod_{k=1}^{n} \mathbf{E}\exp(is\nu_k) - \exp\{\sum_{k=1}^{n}[\mathbf{E}\exp(is\nu_k) - 1]\}\Big|$$

$$\leq e^2 \sum_{k=1}^{n} |\mathbf{E}\exp(is\nu_k) - 1|^2 \leq [|s|\varepsilon$$

$$+ \sup_{k=\overline{1,n}} \mathbf{P}\{|\nu_k| > \varepsilon\}] \sum_{k=1}^{n} \mathbf{E}\nu_k^2 \to 0(n \to \infty),$$

where $\nu_k = \theta_1^k(i_1)f_n, \varepsilon > 0$. According to the above relation and (5.1.11),

$$\lim_{n\to\infty} \sum_{k=1}^{n} \{\mathbf{E}[\exp(is\nu_k) - 1] + s^2 M\nu_k^2/2\} = 0.$$

Therefore,

$$s^2 \sum_{k=1}^{n} \mathbf{E}\nu_k^2/2 - \sum_{k=1}^{n} \int_{|x|\leq\varepsilon} \mathrm{Re}(1 - e^{isx})d\mathbf{P}\{\nu_k < x\}$$

$$= \sum_{k=1}^{n} \int_{|x|>\varepsilon} \mathrm{Re}(1 - e^{isx})d\mathbf{P}\{\nu_k < x\} + 0(1).$$

The integrand on the right does not exceed x^2e^{-2}, and the one on the left does not exceed $x^2\varepsilon^2$. Consequently,

$$\sum_{k=1}^{n} \int_{|x|>\varepsilon} x^2 d\mathbf{P}\{\nu_k < x\} \leq e^{-2}s^{-2} \sum_{k=1}^{n} \mathbf{E}\nu_k^2 + 0(1).$$

Letting s tend to infinity, we obtain (5.1.5).

Using (5.1.6), we represent (5.1.11) as follows:

$$\lim_{n\to\infty}\{[\prod_{k=1}^{n}\mathbf{E}\exp(is\theta_1^k(i_1)f_n)$$

$$-\prod_{k=1}^{n}\mathbf{E}\exp(is\kappa_k\theta_1^k(i_1)f_n)]$$

$$+[\prod_{k=1}^{n}\mathbf{E}\exp(is\kappa_k\theta_1^k(i_1)f_n)$$

$$-\prod_{k=1}^{n}\exp\{-s^2\mathbf{E}(\kappa_k\theta_1^k(i_1)f_n)^2/2\}]\}=0.$$

According to the proof of Theorem 5.1.1, the first difference in this expression tends to zero, if for any $\tau>0$,

$$\lim_{n\to\infty}\sum_{n\geq l_2>l_1>1}\int_{|x|>\tau}x^2d\mathbf{P}\{\theta_{l_1+1}^{l_2}(i_2)\theta_1^{l_1}(i_1)f_n<x\}=0,\qquad(5.1.12)$$

the second difference tends to zero, if

$$\lim_{n\to\infty}\sum_{l_1=1}^{n}\int_{|x|>\tau}x^2d\mathbf{P}\{\kappa_{l_1}\theta_1^{l_1}(i_1)f_n<x\}=0,\quad i_1=\pm1.$$

Repeating these reasonings in order to sum (5.1.12) m times, and taking into account that for any $k=\overline{1,m}$ the variables $\theta_{l_k-1}^{l_k}(i_k)\ldots\theta_1^{l_1}f_n$ are infinitesimal thereby (5.1.6) and (5.1.8), we obtain the assertion of Theorem 5.1.2.

We can generalize Theorem 5.1.1 for the case when some ξ_k can considerably influence the behavior of the Borel function of independent random variables. This generalization looks like the central limit theorem for the sum of independent random variables under the Lindeberg–Zolotarev condition.

Theorem 5.1.3. *Let the Lindeberg–Zolotarev condition hold in Theorem 5.1.1 instead of (5.1.5): for any $\tau>0$,*

$$\lim_{n\to\infty}\sum_{k=1}^{n}\int_{|x|>\tau}x^2|d(\mathbf{E}|\mathbf{P}\{\theta_1^kf_n<x/\sigma_{nk}\}-\mathbf{P}\{\tilde\theta_1^kf_n<x/\sigma_{nk}\}|)|=0.$$

$$(5.1.13)$$

Then $f_n(\xi_1,\ldots,\xi_n)\sim f_n(\eta_1,\ldots,\eta_n)$.

The proof follows from that of Theorem 5.1.1 on the basis of the following simple inequality,

$$\mathbf{E}|\mathbf{E}\{\exp(is\theta_1^kf_n)-\exp(is\tilde\theta_1^kf_n)/\sigma_{nk}\}|$$

$$\leq\int|e^{isx}-1-isx-(isx)^2/2||d\mathbf{E}|\mathbf{P}\{\theta_1^kf_n<x/\sigma_{nk}$$

$$-\mathbf{P}\{\tilde\theta_1^kf_n<x/\sigma_{nk}\}||.$$

We cite one more generalization of Theorem 5.1.1. Let $\{\Omega, \mathfrak{S}\}$, $\{X, \mathfrak{B}\}$ be measurable spaces.

The mapping $g : \omega \to x (x \in X)$ is called the measurable mapping $\{\Omega, \mathfrak{S}\}$ in $\{X, \mathfrak{B}\}$, if $g^{-1}(\mathfrak{B}) = \{\omega : g(\omega) \in \mathfrak{B}\} \in \mathfrak{S}$ for arbitrary $B \in \mathfrak{B}$. The measurable mapping $\{\Omega, \mathfrak{S}, \mathbf{P}\}$ in $\{X, \mathfrak{B}\}$ is called a random element ξ with values on the measurable space.

If X is a vector space, then ξ is called a random vector.

Theorem 5.1.4. *For every n, let the random elements $\xi_i^{(n)}, \eta_i^{(n)}, i = \overline{1, n}$ with values on the measurable space $\{X, \mathfrak{B}\}$ be independent and the functionals $f_n(x_1, \ldots, x_n), x_i \in X$ be given such that $f_n(\xi_1, \ldots, \xi_k, \eta_{k+1}, \ldots, \eta_n), k = \overline{0, n}$ are the random variables, $\mathbf{E}[f_n^i(\eta_i, \ldots, \eta_{k-1}, \xi_k, \ldots, \xi_n)/\sigma_{nk}] = \mathbf{E}[f_n^i(\eta_1, \ldots, \eta_k, \xi_{k+1}, \ldots, \xi_n)/\sigma_{nk}]$ the minimal σ_{nk}- algebra with respect to which the random elements $\xi_p, \eta_p, p \neq k, p = \overline{1, n}$ are measurable, for the functionals f_n and the random elements ξ_p, η_p the operators $\theta_s^l, \tilde{\theta}_s^l$ found from formulas (5.1.1), $i = 1, 2,$*

$$\sup_n \sum_{k=1}^{n} [\mathbf{E}(\theta_1^k f_n(\xi_1, \ldots, \xi_n))^2 + \mathbf{E}(\tilde{\theta}_1^k f_n(\xi_1, \ldots, \xi_n))^2] < \infty,$$

and the Lindeberg condition hold: for any $\tau > 0$,

$$\lim_{n \to \infty} \sum_{k=1}^{n} \int_{|x| > \tau} x^2 d[\mathbf{P}\{\theta_1^k f_n(\xi_1, \ldots, \xi_n) < x\} + \mathbf{P}\{\tilde{\theta}_1^k f_n(\xi_1, \ldots, \xi_n) < x\}] = 0.$$

Then $f_n(\xi_1, \ldots, \xi_n) \sim f_n(\eta_1, \ldots, \eta_n)$.

Analogous assertions can be formulated for Theorems 5.1.2 and 5.1.3.

§2 Limit Theorems for Polynomial Functions of Independent Random Variables

We consider different special cases when conditions of Theorems 5.1.1–5.1.3 are satisfied. We obtain the following assertion on the basis of Theorem 5.1.3.

Theorem 5.2.1.[147]. *Let the random variables $\xi_i^{(n)}, \eta_i^{(n)}, i = \overline{1, n}$ be independent for every n, $f_n(x_1, \ldots, x_n) = \sum_{i=1}^{n} x_i$, $\mathbf{E}\xi_i = \mathbf{E}\eta_i$, $\mathrm{Var}\,\xi_i = \mathrm{Var}\,\eta_i$, $\sup \sum_{i=1}^{n} \mathrm{Var}\,\xi_i < \infty$ exist and for any $\tau > 0$*

$$\lim_{n \to \infty} \sum_{k=1}^{n} \int_{|x| > \tau} x^2 |d\mathbf{P}\{\xi_k^{(n)} - \mathbf{E}\xi_k^{(n)} < x\} - \mathbf{P}\{\eta_k^{(n)} - \mathbf{E}\eta_k^{(n)} < x\}| = 0. \quad (5.2.1)$$

Then

$$\sum_{i=1}^{n} (\xi_i^{(n)} - \mathbf{E}\xi_i^{(n)}) \sim \sum_{i=1}^{n} (\eta_i^{(n)} - \mathbf{E}\eta_i^{(n)}).$$

Proof. It is easy to show that conditions (5.1.2)–(5.1.4) hold and the operators $\theta_1^k, \tilde{\theta}_1^k$ have the following property: $\theta_1^k f_n = \xi_k - \mathbf{E}\xi_k, \tilde{\theta}_1^k f_n = \eta_k - \mathbf{E}\eta_k$. Consequently, (5.2.1) follows from (5.1.13). Theorem is proved.

Note that random variables distributed by the normal laws $N(\mathbf{E}\xi_i, \mathbf{V}\xi_i)$ can be chosen as $\eta_i^{(n)}$.

Theorem 5.2.2. *For every n, let the random variables $\xi_i^{(n)}, \eta_i^{(n)}, i = \overline{1, n}$ be independent $f_n(x_1, \ldots, x_n) = \sum_{i<j} a_{ij} x_i x_j, a_{ij}$ nonrandom real numbers, $\mathbf{E}\xi_i = \mathbf{E}\eta_i, \operatorname{Var}\xi_i = \operatorname{Var}\eta_i, i = \overline{1, n}$,*

$$\sup_n [\sum_{i>j} a_{ij}\sigma_i^2\sigma_j^2 + \sum_{i=1}^{n} \sigma_i^2 \mathbf{E}(\sum_{j=i+1}^{n} a_{ij}\xi_j)^2] < \infty, \qquad (5.2.2)$$

where $\sigma_i^2 = \operatorname{Var}\xi_i$, for any $\tau > 0, i_1, i_2 = \pm 1$,

$$\lim_{n\to\infty} \sum_{p>l} \int_{|x|>\tau} x^2 d\mathbf{P}\{a_{lp}(\xi_l^{(i_1)} - \mathbf{E}\xi_l^{(i_1)})(\xi_p^{(i_2)} - \mathbf{E}\xi_p^{(i_2)}) < x\} = 0; \qquad (5.2.3)$$

$$\lim_{n\to\infty} \sum_{i=1}^{n} \int_{|x|>\tau} x^2 d\mathbf{P}\{[(\xi_i^{(i_1)} - \mathbf{E}\xi_i^{(i_1)})[\sum_{j=i+1}^{n} a_{ij}\eta_j + \sum_{j=1}^{i-1} a_{ij}\eta_j] < x\} = 0, \qquad (5.2.4)$$

where $\xi_i^{(+1)} = \xi_i, \xi_i^{(-1)} = \eta_i$.
Then $\sum_{j>i} a_{ij}\xi_i\xi_j \sim \sum_{j>i} a_{ij}\eta_i\eta_j$.

Proof. Let us verify the conditions of Theorem 5.1.2. Obviously, the operators $\theta_1^l, \tilde{\theta}_1^l, \kappa_l$ have the following property:

$$\theta_1^l f_n = (\xi_l - \mathbf{E}\xi_l)(\sum_{j=l+1}^{n} a_{lj}\xi_j),$$

$$\tilde{\theta}_1^l f_n = (\eta_l - \mathbf{E}\eta_l) \sum_{j=l+1}^{n} a_{lj}\xi_j,$$

$$\theta_{l+1}^q \theta_1^l f_n = (\xi_l - \mathbf{E}\xi_l)(\xi_q - \mathbf{E}\xi_q)a_{lq}, \quad q > l.$$

Conditions (5.1.6) and (5.1.9) are fulfilled. We obtain (5.2.3) and (5.2.4) from conditions (5.1.7) and (5.1.8) when $m = 2$. Theorem 5.2.2 is proved.

If the random variables η_i are distributed according to the normal laws $N(\mathbf{E}\xi_i, \mathbf{V}\xi_i)$, then (5.2.4) follows from

$$\lim_{n\to\infty} \sum_{i=1}^{n} \int_{|x|>\tau} x^2 d[\mathbf{P}\{(\xi_i - \mathbf{E}\xi_i) \sum_{j=i+1}^{n} a_{ij} \mathbf{E}\xi_j < x\}$$

$$+ \mathbf{P}\{(\xi_i - \mathbf{E}\xi_j)[\sum_{j=i+1}^{n} a_{ij}^2 \mathbf{V}\xi_j]^{1/2} < x\}] = 0,$$

and condition (5.2.3) holds from the condition, for any $\tau > 0$,

$$\lim_{n\to\infty} \sum_{j>i} \int_{|x|>\tau} x^2 dP\{a_{ij}(\xi_i - E\xi_i)(\xi_j - E\xi_j) < x\} = 0.$$

Let us prove an analogous assertion for any sequence of polylinear functions.

Theorem 5.2.3. *For every n, let the random variables $\xi_i^{(n)}, \eta_i^{(n)}$ be independent,*

$$f_n(\vec{x}_n) = \sum_{m=1}^{k} \sum_{L_m} a_{i_1}^{(m)}, i_2 \ldots i_m x_{i_1} x_{i_2} \ldots x_{i_m},$$

$$L_m = \{i_s, s = \overline{1,m} : 1 \le i_1 \le \cdots < i_m \le n\},$$

where k is a natural number independent of n, $i_s = \overline{1,n}, s = \overline{1,k}, a_{i_1 i_2}^{(m)} \ldots i_m$ are nonrandom real numbers, $E\xi_i = E\eta_i = 0, \operatorname{Var}\xi_i = \operatorname{Var}\eta_i, i = \overline{1,2},$

$$\sup_n \sum_{m=1}^{k} \sum_{L_m} [a_{i_1 i_2}^{(m)} \ldots i_m]^2 \operatorname{Var}\xi_{i_1} \ldots \operatorname{Var}\xi_{i_m} < \infty, \qquad (5.2.5)$$

for any $\tau > 0, s = \overline{1,k}, l_1 = \pm 1, \ldots, l_s = \pm 1,$

$$\lim_{n\to\infty} \sum_{1 \le p_1 < \cdots < p_s \le n} \int_{|x|>\tau} x^2 dP\{\xi_{p_1}^{(l_1)} \ldots \xi_{p_s}^{(l_s)} \sum_{m=s}^{k}$$

$$\sum_{1 \le i_{s+1} < \cdots < i_m \le n} a_{q_1 q_2 \ldots q_m}^{(m)} \eta_{i_{s+1}} \ldots \eta_{i_m} < x\} = 0, \quad i_j \ne p_j,$$

$$(5.2.6)$$

where $q_1 < \cdots < q_m, q_i \in \{P_j, j = \overline{1,s}, i_l, l = \overline{s+1,m}\}; \xi_p^{(+1)} = \xi_p, \xi_p^{(-1)} = \eta_p$.
Then

$$\sum_{m=1}^{k} \sum_{L_m} a_{i_1 i_2 \ldots i_m}^{(m)} \xi_{i_1} \ldots \xi_{i_m} \sim \sum_{m=1}^{k} \sum_{L_m} a_{i_1 i_2 \ldots i_m}^{(m)} \eta_{i_1} \ldots \eta_{i_m}. \qquad (5.2.7)$$

Proof. Let us use Theorem 5.1.2 again. Obviously, conditions (5.1.6) and (5.1.9) hold, and thereby (5.2.5). Condition (5.1.8) can be represented as (5.2.6). Condition (5.1.7) follows from (5.2.6) when $s = k$. Therefore, all the conditions of Theorem 5.1.2 for the polylinear function f_n are satisfied. Consequently, condition (5.2.7) holds. Theorem 5.2.3 is proved.

Any polylinear function of n variables is equal to the sum of homogeneous polylinear functions. Let us denote such a polylinear homogeneous function of order m by $f_n^{(m)}(x_1, \ldots, x_n)$. If the expectations of ξ_i, η_i are nonzero, then substituting the variables $\xi_i - \mathbf{E}\xi_i + \mathbf{E}\xi_i$ by ξ_i in the function f_n, we arrive again at a certain polylinear function of n random variables $\xi_i - \mathbf{E}\xi_i$, which, as noted above, is equal to the sum of the homogeneous polylinear functions.

If the random variables $\eta_i^{(n)}$ have absolute moments of order $2 + \delta$, where $\delta > 0$, then condition (5.2.6) follows from the condition, for any $\tau > 0$ and $s = \overline{1, k}$,

$$\lim_{n \to \infty} \sum_{1 \le p_1 < \cdots < p_s \le n} \int_{|x| > \tau} x^2 d\mathbf{P}\{\xi_{p_1} \cdots \xi_{p_s}$$
$$\times \left[\sum_{m=s}^{k} \sum_{1 \le i_{s+1} < \cdots < i_m \le n} \operatorname{Var}\xi_{i_{s+1}} \cdots \operatorname{Var}\xi_{i_m} \{a_{q_1 q_2 \cdots q_m}^{(m)}\}^2 \right]^{1/2} < x\} = 0,$$
$$\sup_n \sup_{i=\overline{1,n}} \mathbf{E}|\eta_i^{(n)}(\operatorname{Var}\eta_i^{(n)})^{-1/2}|^{2+\delta} < \infty,$$

and condition (5.2.5) is fulfilled.

Really, using the simple inequality

$$\mathbf{E}\int_{|\alpha x| > \tau} x^2 \alpha^2 dF(x) \le \varepsilon^{-\delta} \mathbf{E}|\alpha|^{2+\delta} \int x^2 dF(x)$$
$$+ \mathbf{E}\alpha^2 \varepsilon^2 \int_{|x| > \tau/\varepsilon} x^2 dF(x), \varepsilon, \tau > 0, \tag{5.2.8}$$

where α is a random variable, and $F(x)$ is a distribution function, we find

$$\sum_{1 \le p_1 < \cdots < p_s \le n} \int_{|x| > \tau} x^2 d\mathbf{P}\{\xi_{p_1}\xi_{p_2} \cdots \xi_{p_j}\eta_{p_{j+1}} \cdots \eta_{p_s}$$
$$\times \sum_{m=s}^{k} \sum_{1 \le i_{s+1} < \cdots < i_m \le n} a_{q_1 \cdots q_m}^{(m)} \eta_{i_{s+1}} \cdots \eta_{i_m} < x\}$$
$$\le c_1 \varepsilon^{-\delta} \sum_{1 \le p_1 < \cdots < p_s \le n} \sigma_{p_1}^2 \cdots \sigma_{p_s}^2 \sum_{m=s}^{k} \sum_{1 \le i_{s+1} < \cdots < i_m \le n} [a_{q_1 \cdots q_m}^{(m)}]^2$$
$$\times \sigma_{i_{s+1}}^2 \cdots \sigma_{i_m}^2 + \varepsilon^2 \sum_{1 \le p_1 < \cdots < p_m \le n} x^2 d\mathbf{P}\{\xi_{p_1} \cdots \xi_{p_j}\sigma_{p_{j+1}}^2 \cdots \sigma_{p_s}^2$$
$$\times [\sum_{m=s}^{k} \sum_{1 \le i_{s+1} < \cdots < i_m \le n} [a_{q_1 \cdots q_m}^{(m)}]^2 \sigma_{i_{s+1}}^2 \cdots \sigma_{i_m}^2]^{1/2} < x\},$$

where $\sigma_i^2 = \mathbf{E}\xi_i^2, j = \overline{1, s}, c_1$ is a certain constant depending only on k. By taking ε sufficiently large, letting n tend to infinity, and using (5.2.5) and (5.2.8), we obtain (5.2.6) on the basis of the following simple inequality:

$$\sum_{i=1}^{n} \int_{|x|>\tau} x^2 d\mathbf{P}\{\xi a_i < x\} \le \int_{|x|>\tau} x^2 d\mathbf{P}\{\xi[\sum_{i=1}^{n} a_i^2]^{1/2} < x\}.$$

Let us generalize Theorem 5.2.3 for the case when the random variables ξ_k can influence the behavior of the polylinear function considerably.

Theorem 5.2.4. *Let condition (5.2.6) be changed for the Lindeberg–Zolotarev condition in Theorem 5.2.3, for any $\tau > 0$ and $s = \overline{1,k}$,*

$$\lim_{n\to\infty} \sum_{1\le p_1<\cdots<p_s\le n} \int_{|x|>\tau} x^2 d\mathbf{P}\{\gamma_{p_1}\ldots\gamma_{p_s} \sum_{m=s}^{k}$$

$$\times \sum_{1\le i_{s+1}<\cdots<i_m\le n} a_{q_1\ldots q_m}^{(m)} \eta_{i_{s+1}}\ldots\eta_{i_m} < x\} = 0,$$
$$(5.2.9)$$

where the generalized random variables $\gamma_i, i = \overline{1,n}$ with the distribution functions

$$\mathbf{P}\{\gamma_i < x\} = \int_{-\infty}^{x} |d[\mathbf{P}\{\xi_i < x\} - \mathbf{P}\{\eta_i < x\}]|$$

are independent of each other and independent of ξ_i, η_i.
Then (5.2.7) holds.

Proof. It follows from Theorem 5.1.3 that under condition (5.1.13) with the polylinear function f_n, condition (5.2.7) is satisfied. Obviously, for f_n condition (5.1.13) follows from the following relation; for any $\tau > 0$,

$$\lim_{n\to\infty} \sum_{k=1}^{n} \int_{|x|>\tau} x^2 d\mathbf{P}\{\Delta_1^k f_n < x\} = 0, \qquad (5.2.10)$$

where Δ_s^k is operating on the function f_n such that it changes all the variables $\xi_i, s \le i < k$ in f_n for the variables η_i and ξ_k for γ_k and all the members of the polynomial without ξ_k going to zero. It follows from condition (5.2.9) that the random variables $\Delta_1^k f_n$ are infinitesimal, consequently, thereby the proof of Theorem 5.1.2, we obtain the assertion of Theorem 5.2.4.

If the random variables η_i have absolute moments of order $2 + \delta$ and

$$\sup_n \sup_{i=\overline{1,n}} \mathbf{E}|\eta_i(\operatorname{Var}\eta_i)^{-1/2}|^{2+\delta} < \infty$$

(taking into account that $\eta_i(\operatorname{Var}\eta_i)^{-1/2} = 0$, when $\operatorname{Var}\eta_i = 0$), then condition (5.2.9) can be changed in the following manner:

$$\lim_{n\to\infty} \sum_{1\le p_1<\cdots<p_s\le n} \int_{|x|>\tau} x^2 d\mathbf{P}\{\gamma_{p_1}\ldots\gamma_{p_s}[\sum_{m=s}^{k}$$

$$\times \sum_{1\le i_{s+1}\cdots<i_m\le n} \{a_{q_1\ldots q_m}^{(m)}\}^2 \mathbf{V}\eta_{i_{s+1}}\ldots\mathbf{V}\eta_{i_m}]^{1/2} < x\} = 0, \quad i_j \ne p_j.$$

Now, let us examine the random polynomial functions g_n of n independent random variables $\xi_i^{(n)}, i = \overline{1,n}$. Denote $\overrightarrow{\xi}_i = (\xi_i, \xi_i^2, \ldots, \xi_i^k)$, where k is an order of the polynomial function g_n. It is more convenient to represent this function as follows: $g_n = g_n(\overrightarrow{\xi}_1, \overrightarrow{\xi}_2, \ldots, \overrightarrow{\xi}_n)$. Let us prove the following assertion by using Theorem 1.5.4.

Theorem 5.2.5. *For every n, let the random variables $\xi_i^{(n)}, \eta_i^{(n)}, i = \overline{1,n}$ be independent $\mathbf{E}\xi_i^{s+p} = \mathbf{E}\eta_{is}\eta_{ip}, \mathbf{E}\xi_i^s \mathbf{E}\eta_{is}, p, s = \overline{1,k}$ for the functions g_n and the random vectors $\overrightarrow{\xi}_i, \overrightarrow{\eta}_i$ the operators $\theta_s^l, \tilde{\theta}_s^l, \kappa_p$ found from formulas (5.1.1), where σ_{nl} is the minimal σ- algebra with respect to which the random vectors $\overrightarrow{\xi}_k, \overrightarrow{\eta}_k, k \neq l, k = \overline{1,n}$ are measurable, κ_p operating on $g_n(\overrightarrow{\xi}_1, \ldots, \overrightarrow{\xi}_n)$ such that it changes the vectors ξ_s for $\overrightarrow{\eta}_s, s = \overline{p+1,n}$. The Lindeberg condition holds for any $\tau > 0, i_k = \pm 1, k = \overline{1,m}$,*

$$\lim_{n \to \infty} \sum_{L_m} \int_{|x|>\tau} x^2 d\mathbf{P}\{\theta_{l_{m+1}}^{lm}(i_m)\ldots\theta_1^{l_1}(i_1)g_n < x\} = 0,$$

$$\lim_{n \to \infty} \sum_{L_{k-1}} \int_{|x|>\tau} x^2 d\mathbf{P}\{\kappa_{l_{k-1}}\theta_{l_{k-2}+1}^{l_{k-1}}(i_k)\ldots\theta_1^{l_1}(i_1)g_n < x\} = 0,$$

$$L_m = \{1 \leq l_1 \cdots < l_m \leq n\},$$

where

$$\theta_p^l(+1) = \theta_p^l, \qquad \theta_p^l(-1) = \tilde{\theta}_p^l,$$

$$\sup_n \sum_{L_{k-1}} \mathbf{E}(\theta_{l_{k-1}+1}^{l_k}(i_k)\ldots\theta_1^{l_1}(i_1)g_n)^2 < \infty.$$

Then $g_n(\overrightarrow{\xi}_1, \ldots, \overrightarrow{\xi}_n) \sim g_n(\overrightarrow{\eta}_1, \ldots, \overrightarrow{\eta}_n)$.

We need an important corollary from this theorem—the limit theorem for random quadratic forms $\sum_{i,j=1}^n a_{ij}\xi_i\xi_j$.

Corollary 5.2.1. *Let the random variables $\xi_i, i = \overline{1,n}$ be independent for every n, and $\mathbf{E}\xi_i = 0, i = \overline{1,n}; \mathbf{E}\xi_i^4$ exist,*

$$\sup_n \sum_{i=1}^n a_{ii}^2 \mathbf{E}(\xi_i^2 - \mathbf{E}\xi_i^2)^2 < \infty, \tag{5.2.11}$$

$$\sup_n \sum_{i \neq j} a_{ij}^2 \sigma_i^2 \sigma_j^2 < \infty, \sigma_i^2 = \text{Var } \xi_i. \tag{5.2.12}$$

and the Lindeberg condition holds: for any $\tau > 0$

$$\lim_{n \to \infty} \sum_{l=1}^{n} \int_{|x|>\tau} x^2 dP\{(\xi_p^2 - \sigma_l^2)a_{ll} < x\} = 0,$$

$$\lim_{n \to \infty} \sum_{i \neq j} \int_{|x|>\tau} x^2 dP\{a_{ij}\xi_i\xi_j < x\} = 0, \qquad (5.2.13)$$

$$\lim_{n \to \infty} \sum_{i=1}^{n} \int_{|x|>\tau} x^2 dP\{\xi_i(\sum_{i>j} a_{ij}^2 \sigma_j^2 + \sum_{j>i} a_{ij}^2 \sigma_i^2)^{1/2} < x\} = 0. \qquad (5.2.14)$$

Then

$$\sum_{i \geq j} a_{ij}\xi_i\xi_j \sim \sum_{i=1}^{n} a_{ii}\nu_i^{(1)} + \sum_{i>j} a_{ij}\nu_i^{(2)}\nu_j^{(2)} + \sum_{i=1}^{n} a_{ii}\sigma_i^2, \qquad (5.2.15)$$

where the random vectors $(\nu_i^{(1)}, \nu_i^{(2)})$ *are independent for every n and independent of the random variables* $\xi_i, i = \overline{1,n}$. *They are normally distributed with the vector of expectations with mean zero and the covariance matrix*

$$\begin{bmatrix} E(\xi_i^2 - E\xi_i^2)^2 & E\xi_i(\xi_i^2 - E\xi_i^2) \\ E\xi_i(\xi_i^2 - E\xi_i^2) & E\xi_i^2 \end{bmatrix}.$$

Proof. According to Theorem 5.2.5, when $m = 1$, condition (5.2.15) is satisfied, if for any $\tau > 0$,

$$\lim_{n \to \infty} \sum_{l=1}^{n} \int_{|x|>\tau} x^2 dP\{(\xi_{ll}^2 - \sigma_{ll}^2)a_{ll} + \xi_l(\sum_{j=l+1}^{n} a_{lj}\xi_j + \sum_{j=1}^{l-1} a_{jl}\nu_j^{(2)}) < x\} = 0, \qquad (5.2.16)$$

$$\lim_{n \to \infty} \sum_{l=1}^{n} \int_{|x|>\tau} x^2 dP\{\nu_l^{(1)}a_{ll} + \nu_l^{(2)}(\sum_{j=l+1}^{n} a_{lj}\xi_j + \sum_{j=1}^{l-1} a_{jl}\nu_j^{(2)}) < x\} = 0, \qquad (5.2.17)$$

and conditions (5.2.11) and (5.2.12) hold.

We find, using the inequality

$$\iint_{|x+y|>\tau} x^2 dF(x)dG(y) \leq \iint_{|x+y|>\tau, |y| \leq \frac{1}{2}\tau} x^2 dF(x)dG(y)$$

$$+ \iint_{|y|>\frac{1}{2}\tau} x^2 dF(x)dG(y) \leq \int_{|x|>\frac{1}{2}\tau} x^2 dF(x) + \int x^2 dF(x) \int_{|y|>\tau/2} dG(y),$$

where $F(x)$ and $G(x)$ are distribution functions, that conditions (5.2.16) and (5.2.17) are equivalent to

$$\lim_{n\to\infty} \sum_{l=1}^{n} \int_{|x|>\tau} x^2 dP\{(\xi_l^2 - \sigma_{ll}^2)a_{ll} < x\} = 0,$$

$$\lim_{n\to\infty} \sum_{l=1}^{n} \int_{|x|>\tau} x^2 dP\{\xi_l(\sum_{j=l+1}^{n} a_{lj}\xi_j + \sum_{j=1}^{l-1} a_{jl}\xi_j) < x\} = 0,$$

$$\lim_{n\to\infty} \sum_{l=1}^{n} \int_{|x|>\tau} x^2 dP\{\nu_l^{(1)} a_{ll} < x\} = 0,$$

$$\lim_{n\to\infty} \sum_{l=1}^{n} \int_{|x|>\tau} x^2 dP\{\nu_l^{(2)}(\sum_{j=l+1}^{n} a_{lj}\xi_j + \sum_{j=1}^{l-1} a_{jl}\xi_j) < x\} = 0. \tag{5.2.18}$$

Obviously, condition (5.2.18) follows from conditions (5.2.11), (5.2.13), and (5.2.14). Corollary 5.2.1 is proved.

Now let us find necessary and sufficient conditions of the limit theorems for the Borel functions of independent random variables. Let the random variables $\xi_i^{(n)}, i = \overline{1,n}$ be independent for every $n, f_n(\xi_1^{(n)}, \ldots, \xi_n^{(n)})$ measurable by Borel functions of these random variables.

Theorem 5.2.6. *Let the functions $f_n(\xi_1^{(n)}, \ldots, \xi_n^{(n)})$ be such that the independent random variables $\eta_i^{(n)}, i = \overline{1,n}$ exist independent of the random variables $\xi_i^{(n)}, i = \overline{1,n}$ and satisfying the following relations for almost all the values of $x_i, i = \overline{1,n}$,*

$$\mathbf{E} \exp\{is[f_n(x_1, \ldots, x_{l-1}, \eta_l, x_{l+1}, \ldots, x_n)$$
$$- \mathbf{E} f_n(x_1, \ldots, x_{l-1}, \xi_l, x_{l+1}, \ldots, x_n)]\}$$
$$= \exp\{-s^2 V f_n(x_1, \ldots, x_{l-1}, \xi_l, x_{l+1}, \ldots, x_n)/2\}; \tag{5.2.19}$$

for the functions f_n and the random variables $\xi_i, \eta_i, i = \overline{1,n}$ the operators $\theta_1^k, k = \overline{1,n}$ are found from formula (5.1.1), for some $\delta > 0$,

$$\sup_{n} \sum_{k=1}^{n} \mathbf{E}|\theta_1^k f_n|^{4+\delta} < \infty,$$

$$\lim_{h\to\infty} \overline{\lim}_{n\to\infty} \mathbf{P}\{|\mathbf{E}f_n(\eta_1, \ldots, \eta_{k-1}, \xi_k, \ldots, \xi_n)/\rho_{nk}| > h\} = 0, \quad k = \overline{1,n}; \tag{5.2.20}$$

for some $\delta_1 > 0$,

$$\lim_{n\to\infty} \sum_{k=1}^{n} \mathbf{E}[\mathbf{E}\{(\theta_1^k f_n)^2/\rho_{nk}\}]^{1+\delta_1} = 0, \tag{5.2.21}$$

$$\mathbf{E}[(\theta_1^k f_n)^3/\rho_{nk}] = 0, \quad k = \overline{1,n}, \quad \rho_{nk} = \sigma\{\eta_l, \xi_l, l \neq k, k = \overline{1,n}\}. \tag{5.2.22}$$

Then in order that $f_n(\overrightarrow{\xi}) \sim f_n(\overrightarrow{\eta})$, it is necessary and sufficient that the Lindeberg condition hold: for any $\tau > 0$,

$$\lim_{n \to \infty} \sum_{k=1}^{n} \int_{|x| > \tau} x^2 dP\{\theta_1^k f_n < x\} = 0. \qquad (5.2.23)$$

Proof. The sufficiency of condition (5.2.23) is proved in Theorem 5.1.1. Let us now prove the necessity of condition (5.2.23).

Obviously,

$$\mathbf{E} \exp\{is f_n(\xi_1, \ldots, \xi_n)\} - \mathbf{E} \exp\{is f_n(\eta_1, \ldots, \eta_n)\}$$

$$= \sum_{k=1}^{n} [\mathbf{E} \exp\{is f_n(\eta_1, \ldots, \eta_{k-1}, \xi_k, \ldots, \xi_n)\}$$

$$- \mathbf{E} \exp\{is f_n(\eta_1, \ldots, \eta_k, \xi_{k+1}, \ldots, \xi_n)\}]$$

$$= \sum_{k=1}^{n} \mathbf{E}\{[\mathbf{E} \exp(is\theta_1^k f_n) - \exp\{-\frac{1}{2}s^2 \mathbf{E}[(\theta_1^k f_n)^2 / \rho_{nk}]\} / \rho_{nk}]$$

$$\times \exp(is\Delta_k)\} = \sum_{k=1}^{n} \mathbf{E}\{[\mathbf{E}\{(\exp(is\theta_1^k f_n) - 1$$

$$- is\theta_1^k f_n) / \rho_{nk}\} + \frac{1}{2}s^2 \mathbf{E}\{(\theta_1^k f_n)^2 / \rho_{nk}\}] \exp(is\Delta_k)\}$$

$$+ \sum_{k=1}^{n} \mathbf{E}[\exp(-\frac{1}{2}s^2 \mathbf{E}\{(\theta_1^k f_n)^2 / \rho_{nk}\})$$

$$- 1 - \frac{1}{2}s^2 \mathbf{E}\{(\theta_1^k f_n)^2 / \rho_{nk}\}] \exp(is\Delta_k), \qquad (5.2.24)$$

where $\Delta_k = \mathbf{E} f_n(\eta_1, \ldots, \eta_{k-1}, \xi_k, \ldots, \xi_n) / \rho_{nk}\}$.

Denote

$$G_n(x, y) = \sum_{k=1}^{n} \int_{-\infty}^{x} z^2 dP\{\theta_1^k f_n < z, \Delta_k < y\}.$$

Using (5.2.20)–(5.2.22), we obtain from (5.2.24) for all $|s| \le S < \infty$,

$$\lim_{n \to \infty} \iint [(e^{isx} - isx - 1)x^{-2} + \frac{1}{2}s^2] e^{isy} dG_n(x, y) = 0.$$

From this, on the basis of (5.2.20), $\lim_{n \to \infty} \int x^2 dG_n(x, +\infty) = 0$. Condition (5.2.23) follows from this relation. Theorem 5.2.2 is proved.

§3 Accompanying Infinitely Divisible Distributions for Borel Functions of Independent Random Variables

Let $f_n(\xi_1^{(n)}, \ldots, \xi_n^{(n)})$ be a Borel function of independent random variables $\xi_1^{(n)}, \ldots, \xi_n^{(n)}$ for every n,

Theorem 5.3.1. *Let the random variables* $\xi_i^{(n)}, i = \overline{1,n}$ *be independent for every* n, *the random variables* $\eta_i^{(n)}$ *be independent of each other and of the random variables* $\xi_i, i = \overline{1,n}$ *exist. Suppose also that the Borel functions* $c_{kn}(x_1, \ldots, x_{k-1}, x_{k+1}, \ldots, x_n)$ *are such that for almost all the values* x_k,

$$\mathbf{E}\exp\{is[f_n(x_1, \ldots, x_{k-1}, \eta_k, x_{k+1}, \ldots, x_n) - c_{kn}(x_1, \ldots, x_{k_1}, x_{k+1}, \ldots, x_n)]\}$$

$$= \exp\{\int [\exp(is[f_n(x_1, \ldots, x_{k-1}, y, x_{k+1}, \ldots, x_n)$$

$$- c_{kn}(x_1, \ldots, x_{k+1}, \ldots, x_n)]) - 1]dP\{\xi_{kn} < y\}; \qquad (5.3.1)$$

$$\lim_{n \to \infty} \mathbf{E}\sum_{k=1}^{n} |\mathbf{E}[\exp(is[f_n(\eta_1, \ldots, \eta_{k-1}, \xi_k, \ldots, \xi_n)$$

$$- c_{kn}(\eta_1, \ldots, \eta_{k-1}, \xi_{k+1}, \ldots, \xi_n)] - 1/\sigma_{kn}]|^2 = 0, \qquad (5.3.2)$$

where σ_{kn} *is the minimal* σ-*algebra with respect to which the random variables* $\eta_i, i = \overline{1, k-1}, \xi_l, l = \overline{k+1, n}$ *are measurable.*

Then $f(\xi_1, \ldots, \xi_n) \sim f_n(\eta_1, \ldots, \eta_n)$.

Proof. Let us consider the obvious inequalities

$$|\mathbf{E}\exp\{isf(\xi_1, \ldots, \xi_n)\} - \mathbf{E}\exp\{isf(\eta_1, \ldots, \eta_n)\}|$$

$$= |\sum_{k=1}^{n}[\mathbf{E}\exp\{isf(\eta_1, \ldots, \eta_{k-1}, \xi_k, \ldots, \xi_n)\}$$

$$- \mathbf{E}\exp\{isf(\eta_1, \ldots, \eta_k, \xi_{k+1}, \ldots, \xi_n)\}]| \leq \sum_{k=1}^{n} \mathbf{E}|\varphi_k + 1 - e^{\varphi_k}|,$$

where

$$\varphi_k = \mathbf{E}\{\exp[is(f(\eta_1, \ldots, \eta_{k-1}, \xi_k, \ldots, \xi_n)$$

$$- c_{kn}(\eta_1, \ldots, \eta_{k-1}, \xi_{k+1}, \ldots, \xi_n))]/\sigma_{kn}\} - 1.$$

Hence by virtue of (5.3.2), the statement of Theorem 5.3.1 follows.

Conditions (5.3.1) and (5.3.2) can formulated for polylinear functions of independent random variables. Relation (5.3.1) shows that in this case η_i is distributed by the infinitely divisible law. Let us consider a special case of polylinear functions: linear and quadratic forms which admit effective conditions for (5.3.2).

First, we consider the sums of independent random variables and prove the following theorems.

Theorem 5.3.2[101]. *For each n let the random variables $\xi_{in}, i = \overline{1,n}$ be independent, infinitesimal and for some constants A_n appropriately chosen,*

$$\lim_{h \to \infty} \overline{\lim_{n \to \infty}} \, \mathbf{P}\{|\sum_{k=1}^{n} \xi_{kn} - A_n| \geq h\} = 0. \qquad (5.3.3)$$

Then there exists a constant $0 < c < \infty$ such that

$$\sup_{n} \sum_{k=1}^{n} \int x^2(1+x^2)^{-1} d\mathbf{P}\{\xi_{kn} - \alpha_{kn} < x\} \leq c, \qquad (5.3.4)$$

where $\alpha_{kn} = \int_{|x| < \tau} x d\mathbf{P}\{\xi_k < x\}$, and $\tau > 0$ is an arbitrary constant.

Proof. By virtue of (5.3.3) for any sequences of constants $\xi_n \to 0$,

$$p \lim_{n \to \infty} \varepsilon_n(\sum_{k=1}^{n} \xi_{kn} - A_n) = 0.$$

Hence

$$\lim_{n \to \infty} \prod_{k=1}^{n} |f_{kn}(t)| = 1, \qquad (5.3.5)$$

where $f_{kn}(t) = \mathbf{E} \exp(it\nu_{kn}), \nu_{kn} = \varepsilon_n(\xi_{kn} - \alpha_{kn})$. It follows from (5.3.5) that starting from some $n_0, \sup_{k=\overline{1,n}} |f_{kn}(t)| > 0$ for all $t \in [-T, T]$, where $T > 0$ is an arbitrary constant and

$$\lim_{n \to \infty} \sum_{k=1}^{n} \ln |f_{kn}(t)| = 0. \qquad (5.3.6)$$

Now use the following elementary inequality

$$2|\ln |f_{kn}(t)|| = -\ln |f_{kn}(t)|^2 \geq 1 - |f_{kn}(t)|^2$$

$$\geq 1 - (\int \cos(tx) dF_k(x))^2 - (\int \sin(tx) dF_k(x))^2$$

$$\geq \int (1 - \cos(tx)) dF_k(x) - (\int \sin(tx) dF_k(x))^2, \qquad (5.3.7)$$

where $F_k(x) = \mathbf{P}\{\nu_{kn} < x\}, t \in [-T, T], n \geq n_0$. It is easy to verify that for all $x \in R^1$ and for any constant b,

$$0 < c_1(b) \leq (1 - [\sin(bx)/bx])(1 + x^2)x^{-2} \leq c_2(b) < \infty.$$

By using this inequality, we obtain

$$\int_0^b [\int (1 - \cos(tx)) dF_k(x)] dt = b \int (1 - [\sin bx](bx)^{-1})$$

$$\times dF_k(x) \geq bc_1(b) \mathbf{E}\nu_{kn}^2 (1 + \nu_{kn}^2)^{-1}. \qquad (5.3.8)$$

We need the following assertion [142].

Lemma 5.3.1. *For any finite t and $\varepsilon_n > 0$,*

$$|f_{kn}(t) - 1| \le c_{kn}(t)\mathbf{E}\nu_{kn}^2(1 + \nu_{kn}^2)^{-1}, \qquad (5.3.9)$$

where

$$
\begin{aligned}
c_{kn}(t) =&(2 + |t^2\alpha_{kn}\varepsilon_n|)[1 + \varepsilon_n^2(\tau - \alpha_{kn})^2]\varepsilon_n^{-2} \\
&\times (\tau - \alpha_{kn})^{-2} + t^2\varepsilon_n^2[1 + \varepsilon_n^2(\tau + |\alpha_{kn}|)^2].
\end{aligned}
$$

Proof. Applying the inequality $|e^{itx} - itx - 1| \le t^2x^2/2$ and the expression α_{kn}, we obtain

$$
\begin{aligned}
\Big| \int (\exp\{it\varepsilon_n(x - \alpha_{kn})\} &- 1)dG_{kn}(x)\Big| \le 2\int_{|x| \ge \tau} dG_{kn}(x) \\
&+ \int_{|x| < \tau} (\exp\{it\varepsilon_n(x - \alpha_{kn})\} - it\varepsilon_n(x - \alpha_{kn}) - 1)dG_{kn}(x) \\
&+ it\varepsilon_n \int_{|x| < \tau} (x - \alpha_{kn})dG_{kn}(x)\Big| \le (2 + |\alpha_{kn}t\varepsilon_n|) \\
&\times \int_{|x| \ge \tau} dG_{kn}(x) + \int_{|x| < \tau} (x - \alpha_{kn})^2 dG_{kn}(x)t^2\varepsilon_n^2/2,
\end{aligned}
\qquad (5.3.10)
$$

where $G_{kn}(x) = \mathbf{P}\{\xi_{kn} < x\}$.

When $x \ge \tau$ by virtue of $|\alpha_{kn}| < \tau$,

$$\varepsilon_n^2(x - \alpha_{kn})^2[1 + \varepsilon_n^2(x - \alpha_{kn})^2]^{-1} \ge \varepsilon_n^2(\tau - \alpha_{kn})^2[1 + \varepsilon_n^2(\tau - \alpha_{kn})^2]^{-1};$$

when $x < \tau, [1 + \varepsilon_n^2(\tau + |\alpha_{kn}|)^2][1 + \varepsilon_n^2(x - \alpha_{kn})^2]^{-1} \ge 1$. From both the inequalities and (5.3.10), (5.3.9) follows. Lemma 5.3.1 is proved.

It is easy to show that for all $0 \le t \le b$ and any $\delta > 0$,

$$\Big| \int \sin(tx)dF_{kn}(x)\Big| \le \mathbf{P}\{|\xi_{kn} - \alpha_{kn}| > \delta\} + b\varepsilon_n\delta.$$

Hence $\delta_n := \sup_{k=\overline{1,n}} |\int \sin(tx)dF_{kn}(x)|$ tends to 0 as $n \to \infty$, since the random variables ξ_{kn} are infinitesimal and $\alpha_{kn} \to 0$ as $n \to \infty$.

According to Lemma 5.3.1,

$$\Big(\int \sin(tx)dF_k(x)\Big)^2 \le \delta_n|f_{kn}(t) - 1| \le \delta_n c_{kn}(t)\mathbf{E}\nu_{kn}^2(1 + \nu_{kn}^2)^{-1}. \quad (5.3.11)$$

By using (5.3.8) and (5.3.11), we obtain from (5.3.7)

$$2\int_0^b |\ln|f_{kn}(t)||dt \ge [bc_1(b) - \delta_n \int_0^b c_{kn}(t)dt]\mathbf{E}\nu_{kn}^2(1 + \nu_{kn}^2)^{-1}. \quad (5.3.12)$$

Obviously, $\sup_n \sup_{k=\overline{1,n}} \int_0^b c_{kn}(t)dt\varepsilon_n^2$ is bounded with some constant.
The sequence ε_n is arbitrary tending to 0 as slowly as necessary.
Therefore, ε_n can be chosen such that

$$\lim_{n\to\infty} \delta_n \int_0^b c_{kn}(t)dt = 0, \qquad \lim_{n\to\infty} \varepsilon_n = 0.$$

Denote the set of sequences ε_n' which satisfy this relation as $K_n = \{\varepsilon_n'\}$. It follows from (5.3.12) and (5.3.6) that for any sequence ε_n',

$$\lim_{n\to\infty} \sum_{k=1}^n \mathbf{E}\nu_{kn}^2[1 + \nu_{kn}^2]^{-1} = 0.$$

Taking into consideration that

$$\varepsilon^2 x^2(1 + \varepsilon^2 x^2)^{-1} \geq \varepsilon^2 x^2(1 + x^2)^{-1} \qquad \text{as} \quad 0 < \varepsilon < 1,$$

we obtain

$$\lim_{n\to\infty} (\varepsilon_n')^2 \sum_{k=1}^n \int x^2(1 + x^2)^{-1}d\mathbf{P}\{\xi_{kn} - \alpha_{kn} < x\} = 0.$$

Hence, since $\varepsilon_n' \in K_n$ is a sequence tending to 0 as slowly as necessary, we obtain the statement of Theorem 5.3.2.

Theorem 5.3.3[101]. *For each n let the random variables $\xi_{kn}, \eta_{kn}, k = \overline{1,n}$ be independent and infinitesimal, η_{kn} be distributed by the infinitely divisible laws with the characteristic functions $\exp\{\mathbf{E}\exp[is(\xi_{kn} - \alpha_{kn})] - 1\}$, where*

$$\alpha_{kn} = \int_{|x|<\tau} x d\mathbf{P}\{\xi_{kn} < x\},$$

$\tau > 0$ is an arbitrary constant, and condition (5.3.3) is fulfilled. Then $\sum_{k=1}^n \xi_{kn} \sim \sum_{k=1}^n \eta_{kn}$.

Proof. According to Theorem 5.3.1, to prove Theorem 5.3.3, we show that

$$\lim_{n\to\infty} \sum_{k=1}^n |g_{kn}(t) - 1|^2 = 0, \qquad (5.3.13)$$

where $g_{kn}(t) = \mathbf{E}\exp[it(\xi_{kn} - \alpha_{kn})]$.
Using Lemma 5.3.1 and assuming $\varepsilon_n = 1$, we obtain

$$\sum_{k=1}^n |g_{kn}(t) - 1|^2 \leq \sum_{k=1}^n c_{kn}^2(t)[1 - \mathbf{E}\{1 + (\xi_k - \alpha_k)^2\}^{-1}]^2.$$

With the random variables ξ_{kn} being infinitesimal and (5.3.4) fulfilled, the assertion of Theorem 5.3.3 follows.

We now consider the limit theorems for random quadratic forms. Any symmetric matrix can be represented as a difference of nonnegative-definite matrices, and thus the proof of the limit theorems for quadratic forms of random variables can be reduced to the proof of the limit theorems for nonnegative-definite quadratic forms.

Consider the nonnegative-definite quadratic forms

$$(A_n \overrightarrow{\xi}_n, \overrightarrow{\xi}_n) = \sum_{i,j=1}^{n} a_{ij}(\xi_{in} - \beta_{in})(\xi_{jn} - \beta_{jn}),$$

$$A_n = (a_{ij}),$$

$$\overrightarrow{\xi}_n = (\xi_{1n} - \beta_{1n}, \dots, \xi_{nn} - \beta_{nn}).$$

Theorem 5.3.4. *For each n let the random variables $a_{kk}^{1/2}\xi_{kn}, \eta_{kn}, k = \overline{1, n}$ be independent, the random variables $\xi_{kn}a_{kk}^{1/2}$ be infinitesimal, and η_{kn} distributed by the infinitely divisible laws with characteristic functions*

$$\exp\{\mathbf{E}\exp[is(\xi_{kn} - \beta_{kn})] - 1\},$$

$$\beta_{kn} = \int_{|xa_{kk}^{1/2}|<\tau} x^2 d\mathbf{P}\{\xi_{kn} < x\}a_{kk}^{1/2}, \quad k = \overline{1, n}, \quad a_{kk} \neq 0,$$

$$\lim_{h \to \infty} \overline{\lim_{n \to \infty}} \mathbf{P}\{\sum_{i,j=1}^{n} a_{ij}(\xi_{in} - \beta_{in})(\xi_{jn} - \beta_{jn}) \geq h\} = 0. \tag{5.3.14}$$

Then

$$\sum_{i,j=1}^{n} a_{ij}(\xi_{in} - \beta_{in})(\xi_{jn} - \beta_{jn}) \sim \sum_{i,j=1}^{n} a_{ij}\eta_{in}\eta_{jn}. \tag{5.3.15}$$

Proof. Obviously that

$$\mathbf{E}\exp\{-s(A_n \overrightarrow{\xi}_n, \overrightarrow{\xi}_n)\} = \mathbf{E}\exp\{i\sqrt{s}\sum_{k=1}^{n} a_{kk}^{1/2}(\xi_{kn} - \beta_{kn})\gamma_{kn}\}, \tag{5.3.16}$$

where $s \geq 0$, the random variables $\gamma_{kn}, k = \overline{1, n}$ do not depend on the random variables $\xi_{kn}, \eta_{kn}, k = \overline{1, n}$ for each n, and the random vector $\overrightarrow{\gamma}_n = (a_{kk}^{1/2}\gamma_{kn}, k = \overline{1, n})$ is distributed by the multidimensional normal law. In this case the vector of mean values equals 0, and the covariance matrix is $2A_n$.

It follows from (5.3.16) and (5.3.14) that for any sequence ε_n that tends to 0 as slowly as necessary,

$$\text{plim}_{n \to \infty} \varepsilon_n \sum_{k=1}^{n} a_{kk}^{1/2}(\xi_{kn} - \beta_{kn})\gamma_{kn} = 0, \quad \lim_{n \to \infty} \prod_{k=1}^{n} |f_{kn}(t\gamma_{kn})| = 1, \tag{5.3.17}$$

for all $\overrightarrow{\gamma}_n \in C_n$, where C_n is a sequence of Borel sets of the Euclidean space R^n such that

$$\lim_{n \to \infty} \mathbf{P}\{\overrightarrow{\gamma}_n \in C_n\} = 0, f_{kn}(t) = \mathbf{E}\exp\{ita_{kk}^{1/2}(\xi_{kn} - \beta_{kn})\}.$$

Using the proof of Theorem 5.3.2, we obtain

$$2\mathbf{E}|\sum_{k=1}^{n} \int_0^b |\ln|f_{kn}(t\gamma_{kn})||dt/\overrightarrow{\gamma}_n \overline{\in} C_n|$$

$$\geq \sum_{k=1}^{n} \mathbf{E}\{bc_1(b, \gamma_{kn}) - \delta_n(\gamma_{kn}) \int_0^b c_{kn}(t\gamma_{kn})$$

$$\times \, dt/\overrightarrow{\gamma}_n \overline{\in} C_n, |\gamma_{kn}| \leq 1\}\mathbf{E}\varepsilon_n^2\mu_{kn}^2(1 + \mu_{kn}^2\varepsilon_n^2)^{-1}\mathbf{P}\{|\gamma_{kn}| \leq 1\}, \tag{5.3.18}$$

where $c_1(b, \gamma_{kn})$ is a constant bounding $(1 - [\sin(b\gamma_{kn}x)](b\gamma_{kn}x)^{-1}(1 + x^2)$ $\times \, x^{-2} \geq c_1(b, \gamma_{kn}) > 0$ below;

$$\delta_n(\gamma_{kn}) = \int \sin(t\gamma_{kn}x)d\mathbf{P}\{\varepsilon_n\mu_{kn} < x\}, \mu_{kn} = a_{kk}^{1/2}\xi_{kn} - \beta_{kn},$$

the expression for $c_{kn}(t)$ is given in Lemma 5.3.1.

It is obvious that under the condition $|\gamma_{kn}| \leq 1$ the sequence ε_n can be chosen such (see proof of Theorem 5.3.2) that

$$\lim_{n \to \infty} \inf_{k=\overline{1,n}} \mathbf{E}\{bc_1(b, \gamma_{kn}) - \delta_n(\gamma_{kn}) \int_0^b c_{kn}(t\gamma_{kn})$$

$$\times \, dt/\overrightarrow{\gamma}_n \overline{\in} C_n, |\gamma_{kn}| \leq 1\}\mathbf{P}\{\gamma_{kn}| \leq 1\} \geq c(b) > 0,$$

where $c(b)$ is some constant.

Therefore, according to (5.3.17), it follows from (5.3.18) that

$$\lim_{n \to \infty} \sum_{k=1}^{n} \mathbf{E}[1 - (1 + \varepsilon_n^2\mu_{kn}^2)^{-1}] = 0.$$

Hence

$$\sup_n \sum_{k=1}^{n} \mathbf{E}(1 - (1 + \mu_{kn}^2)^{-1}) < \infty. \tag{5.3.19}$$

Using (5.3.16), we obtain for all $\varepsilon > 0$

$$|\mathbf{E}\exp\{-s(A_n \vec{\xi}_n, \vec{\xi}_n\} - \mathbf{E}\exp\{-s\sum_{i,j=1}^{n} a_{ij}\eta_{in}\eta_{jn}\}|$$

$$\leq \mathbf{E}[|\prod_{k=1}^{n} f_{kn}(\sqrt{s}\gamma_{kn}) - \exp\{\sum_{k=1}^{n}[f_{kn}(\sqrt{s}\gamma_{kn})$$

$$- 1]\}|/\sum_{k=1}^{n}|f_{kn}(\sqrt{s}\gamma_{kn}) - 1|^2 \leq \varepsilon]\mathbf{P}\{\sum_{k=1}^{n}|f_{kn}(\sqrt{s}\gamma_{kn})$$

$$- 1|^2 \leq \varepsilon\} + 2\mathbf{P}\{\sum_{k=1}^{n}|f_{kn}(\sqrt{s}\gamma_{kn}) - 1|^2 > \varepsilon\}$$

$$\leq e^2\varepsilon\mathbf{P}\{\sum_{k=1}^{n}|f_{kn}\sqrt{s}\gamma_{kn}) - 1|^2 \leq \varepsilon\}$$

$$+ 2\mathbf{P}\{\sum_{k=1}^{n}|f_{kn}\sqrt{s}\gamma_{kn}) - 1|^2 > \varepsilon\}. \tag{5.3.20}$$

From Lemma 5.3.1 we obtain for $\varepsilon_n \equiv 1$

$$\mathbf{E}\sum_{k=1}^{n}|f_{kn}(\sqrt{s}\gamma_{kn}) - 1|^2 \leq \sum_{k=1}^{n}\sup_{k=\overline{1,n}}\mathbf{E}[c_{kn}(\sqrt{s}\gamma_{kn})$$

$$\times \mathbf{E}\mu_{kn}^2(1+\mu_{kn}^2)^{-1}]^2 \leq c\sum_{k=1}^{n}[\mathbf{E}\mu_{kn}^2(1+\mu_{kn}^2)^{-1}]^2,$$

where $c > 0$ is some constant depending only on t and bounded for all $|s| \leq S$, and $S > 0$ is an arbitrary constant.

Hence

$$\lim_{n\to\infty}\mathbf{E}\sum_{k=1}^{n}|f_{kn}(\sqrt{s}\gamma_{kn}) - 1|^2 = 0, \tag{5.3.21}$$

because (5.3.19) is fulfilled, and the variables $a_{kk}^{1/2}\xi_{kn}$ are infinitesimal. From (5.3.21) and (5.3.20) the assertion of Theorem 5.3.4 follows.

§4 Limit Theorems for Sums of Martingale Differences

The limit theorems for sums of martingale differences are of great importance for studying the limiting distributions for measurable maps of independent random elements. A characteristic feature of these theorems is that verification of their conditions requires proving some limit theorem type of law of large numbers for dependent random variables.

It has been shown in the previous paragraphs that under some assumptions the distribution of a Borel function of random variables can be approximated by the distribution of the same function but with different arguments.

This approach, however, usually fails because, with trivial exceptions, it is difficult to match random variables such that the distribution of their Borel function can be found. It turns out that the representation of Borel functions as sums of martingale differences yields nontrivial theorems for their distributions. Let $f_n(\xi_1^{(n)}, \dots, \xi_n^{(n)})$ be a sequence of measurable maps of random elements $\xi_1^{(n)}, \dots, \xi_n^{(n)}$. Suppose that $El_n(f_n(\cdot))$ exists, where l_n is a functional. Then, obviously, with probability 1 (in the following, we shall omit the words "with probability 1"),

$$l_n(f_n(\cdot)) - El_n(f_n(\cdot)) = \sum_{k=1}^{n} \{ \mathbf{E}_{\sigma_{k-1}^{(n)}} l_n(f_n(\cdot)) - \mathbf{E}_{\sigma_k^{(n)}} l_n(f_n(\cdot)) \}, \qquad (5.4.1)$$

where $\mathbf{E}_{\sigma_k^{(n)}}$ is the conditional expectation with respect to the minimal σ-algebra $\sigma_k^{(n)}$ generated by the random elements $\xi_{k+1}^{(n)}, \dots, \xi_n^{(n)}$.

The sequence of the random variables $\gamma_k = \mathbf{E}_{\sigma_{k-1}^{(n)}} l_n(f_n(\cdot)) - \mathbf{E}_{\sigma_k^{(n)}} l_n(f_n(\cdot))$, $k = \overline{1, n}$ referred to as a martingale difference for the variables $\sum_{p=1}^{k} \mathbf{E}_{\sigma_p^{(n)}} l_n \times (f_n(\cdot))$ forms a martingale. Sometimes it is also referred to as an absolutely unbiased sequence.

It is obvious that for this sequence the equalities

$$E[\gamma_k / \sigma_k^{(n)}] = 0, E[\gamma_k \gamma_p / \sigma_{\min(k,p)}^{(n)}] = 0, \quad k \neq p$$

are satisfied. On the basis of these equalities, we get the limit theorem type of law of large numbers.

Theorem 5.4.1. *If for some sequence of constants c_n*

$$\lim_{n \to \infty} c_n^{-2} \sum_{k=1}^{n} \mathbf{E}\gamma_k^2 = 0,$$

then $\operatorname{plim}_{n \to \infty} c_n^{-1} [\sum_{k=1}^{n} \gamma_k] = 0.$

The condition of the existence of the moments $\mathbf{E}\gamma_k^2$ can be weakened. Consider the reduced random variables

$$\nu_k = \gamma_k X(|\gamma_k| < \delta_n), \qquad \mu_k = \gamma_k X(|\gamma_k| \geq \delta_n),$$

where δ_n are arbitrary constants.

Corollary 5.4.1. *If for some sequence of constants c_n*

$$\lim_{n \to \infty} c_n^{-1} \sum_{k=1}^{n} \mathbf{E}|\mu_k| = 0,$$

$$\lim_{n \to \infty} c_n^{-2} \delta_n \sum_{k=1}^{n} \mathbf{E}|\nu_k| = 0,$$

$$\lim_{n \to \infty} c_n^{-1} \sum_{k=1}^{n} \mathbf{E}|\mathbf{E}[\nu_k/\sigma_k^{(n)}]| = 0,$$

then $\operatorname{plim}_{n \to \infty} c_n^{-1} \sum_{k=1}^{n} \gamma_k = 0$.

The proof follows from the simple remarks: the random variables $\nu_k - \mathbf{E}[\nu_k/\sigma_k^{(n)}]$ form the sequence of martingale differences

$$c_n^{-1} \mathbf{E}|\sum_{k=1}^{n} \gamma_k| \leq [c_n^{-2} \mathbf{E}(\sum_{k=1}^{n} [\nu_k - \mathbf{E}(\nu_k/\sigma_k^{(n)})])^2)]^{1/2}$$

$$+ c_n^{-1} \sum_{k=1}^{n} \mathbf{E}|\mu_k| + \mathbf{E}|c_n^{-1}| \sum_{k=1}^{n} \mathbf{E}[\nu_k/\sigma_k^{(n)}]| \leq \{c_n^{-2} \delta_n$$

$$\times \sum_{k=1}^{n} \mathbf{E}|\nu_k|\}^{1/2} + c_n^{-1} \sum_{k=1}^{n} \mathbf{E}|\mu_k| + c_n^{-1} \mathbf{E}|\sum_{k=1}^{n} \mathbf{E}[\nu_k/\sigma_k^{(n)}]|.$$

In particular, if such $\varepsilon > 0$ exists, that

$$\sup_{n,k} \mathbf{E}|\gamma_k|^{1+\varepsilon} < \infty,$$

then $\mathbf{E}|\mu_k| \leq \delta_n^{-\varepsilon} \mathbf{E}|\gamma_k|^{1+\varepsilon}, |\mathbf{E}[\nu_k/\sigma_k^{(n)}]| \leq \delta_n^{-\varepsilon} \mathbf{E}|\gamma_k|^{1+\varepsilon}$, and therefore, if we choose such δ_n that $\lim c_n^{-2} \delta_n^n = 0, \lim_{n \to \infty} c_n^{-1} \delta_n^{-\varepsilon} = 0$ obtain $\operatorname{plim}_{n \to \infty} c_n^{-1} \times (f_n - \mathbf{E}f_n) = 0$.

In the following we shall assume that all random variables, defined in a set scheme, are given on a common probability space.

Theorem 5.4.2. *If $\mathbf{E}\gamma_k^4$ exists, for some sequence of constants c_n,*

$$\sum_{n=1}^{\infty} (\sum_{k=1}^{n} [\mathbf{E}c_n^{-4} \gamma_k^4]^{1/2})^2 < \infty, \tag{5.4.2}$$

then $\lim_{n \to \infty} c_n^{-1}(f_n - \mathbf{E}f_n) = 0$ *with probability 1.*

Proof. We estimate the fourth moment of the variable $f_n - \mathbf{E}f_n$:

$$\mathbf{E}(\sum_{i=1}^{n} \gamma_i)^4 = \mathbf{E}(\sum_{i=1}^{n} \gamma_i)^2(\sum_{i=1}^{n} \gamma_i)^2 = \mathbf{E}(\sum_{i=1}^{n} \gamma_i^2$$

$$+ \sum_{i \neq j} \gamma_i \gamma_j)(\sum_{p=1}^{n} \gamma_p^2 + \sum_{p \neq l} \gamma_p \gamma_l) = \mathbf{E} \sum_{i=1}^{n} \gamma_i^2 (\sum_{p,l=1}^{n} \gamma_p \gamma_l)$$

$$+ \mathbf{E} \sum_{p=1}^{n} \gamma_p^2 (\sum_{i,j=1}^{n} \gamma_i \gamma_j) + \mathbf{E} \sum_{i \neq j, p \neq l} \gamma_i \gamma_j \gamma_p \gamma_l$$

$$\leq 2 \sum_{i=1}^{n} \sqrt{\mathbf{E}\gamma_i^4} \sqrt{\mathbf{E}(\sum_p \gamma_p)^4} + \mathbf{E} \sum_{i \neq j, i \neq l, j \neq l} \gamma_i^2 \gamma_j \gamma_l$$

$$+ \mathbf{E} \sum_{j \neq i, p \neq i, j \neq p} \gamma_i^2 \gamma_j \gamma_p + \mathbf{E} \sum_{i \neq p, l \neq p, i \neq l} \gamma_p^2 \gamma_i \gamma_l + \mathbf{E} \sum_{i \neq l, p \neq l, i \neq p} \gamma_l^2 \gamma_i \gamma_p$$

$$\leq 2 \sum_{i=1}^{n} \sqrt{\mathbf{E}\gamma_i^4} \sqrt{\mathbf{E}(\sum_p \gamma_p)^4} + 4\mathbf{E} \sum_{i=1}^{n} \gamma_i^2$$

$$\times (\sum_j \gamma_j - \gamma_i)^2 - 4\mathbf{E} \sum_{l \neq i} \gamma_i^2 \gamma_l^2 \leq 2 \sum_{i=1}^{n} \sqrt{\mathbf{E}\gamma_i^4}$$

$$\times \sqrt{\mathbf{E}(\sum_p \gamma_p)^4} + 8\mathbf{E} \sum_{i=1}^{n} \gamma_i^2 (\sum_j \gamma_j)^2$$

$$+ 8\mathbf{E} \sum_{i=1}^{n} \gamma_i^4 \leq 10 \sum_{i=1}^{n} \sqrt{\mathbf{E}\gamma_i^4} \sqrt{\mathbf{E}(\sum_p \gamma_p)^4} + 8\mathbf{E} \sum_{i=1}^{n} \gamma_i^4.$$

Here we employed the fact, that $\mathbf{E}\gamma_i \gamma_j \gamma_p \gamma_l = 0$, if all indexes i, j, p, l do not coincide.

Denote $\mathbf{E}(\sum_i \gamma_i)^4 = x$, $10\sum_i \sqrt{\mathbf{E}\gamma_i^4} = a$, $8\sum_i \mathbf{E}\gamma_i^4 = b$. Then $x \leq a\sqrt{x} + b$. Solving this quadratic equation, we find $x \leq c\sum_i(\mathbf{E}\gamma_i^4)^{1/2}$, where $c > 0$ is a constant. To complete the proof we use the Borel-Cantelly lemma.

Let us prove the central limit theorem for the sums $\sum_k \gamma_k$.

Theorem 5.4.3[123]. *Let $\mathbf{E}\gamma_k^2$ exist,*

$$\lim_{n \to \infty} \sum_{k=1}^{n} \mathbf{E}\gamma_k^2 > 0; \tag{5.4.3}$$

$$\lim_{n \to \infty} c_n^{-1} \sum_{k=1}^{n} \mathbf{E}|\mathbf{E}\gamma_k^2 - \mathbf{E}(\gamma_k^2/\rho_k^{(n)})| = 0, \tag{5.4.4}$$

where $\rho_k^{(n)}$ is minimal σ-algebra generated by the random variables $\xi_l, l \neq k, k = \overline{1, n}, c_n = \sum_{k=1}^{n} \mathbf{E}\gamma_k^2$, and the Lindeberg condition is satisfied, for any

$\tau > 0$,

$$\lim_{n \to \infty} \sum_{k=1}^{n} \int_{|x|>\tau} x^2 dP\{c_n^{-1/2}\gamma_k < x\} = 0. \tag{5.4.5}$$

Then

$$\lim_{n \to \infty} P\{c_n^{-1/2} \sum_{k=1}^{n} \gamma_k < z\} = (2\pi)^{-1/2} \int_{-\infty}^{z} \exp(-y^2/2)dy.$$

Proof. It is obvious that

$$|E \exp\{is \sum_{p=1}^{n} c_n^{-1/2}\gamma_p\} - \exp\{-0.5s^2 \sum_{p=1}^{n} c_n^{-1}E\gamma_p^2\}|$$

$$\leq \sum_{k=1}^{n} |E \exp\{-0.5s^2 c_n^{-1} \sum_{p=1}^{k-1} E\gamma_p^2 + isc_n^{-1/2} \sum_{p=k}^{n} \gamma_p\}$$

$$- E \exp\{-0.5s^2 \sum_{p=1}^{k} c_n^{-1}E\gamma_p^2 + is \sum_{p=k+1}^{n} c_n^{-1/2}\gamma_p\} \leq I_1 + I_2 + I_3,$$

where

$$I_1 = \sum_{k=1}^{n} E|E(\exp\{isc_n^{-1/2}\gamma_k\} - isc_n^{-1/2}\gamma_k - 0.5c_n^{-1}(is\gamma_k)^2 - 1/\rho_k^{(n)}|,$$

$$I_2 = \sum_{k=1}^{n} |\exp\{-0.5s^2 c_n^{-1/2}E\gamma_k^2\} - 1 + 0.5s^2 c_n^{-1}E\gamma_k^2|,$$

$$I_3 = 0.5s^2 \sum_{k=1}^{n} E|E(c_n^{-1}\gamma_k^2/\rho_k^{(n)}) - c_n^{-1}E\gamma_k^2|.$$

According to (5.4.4), $\lim_{n \to \infty} I_3 = 0$ for every bounded s. For I_1 ad I_2 we have the following estimates:

$$I_1 = \sum_{k=1}^{n} E|\int_{|x|<\tau} (e^{isx} - isx - 0.5(isx)^2 - 1)$$

$$\times dP\{c_n^{-1/2}\gamma_k < x/\rho_k^{(n)}\} + \int_{|x|\geq\tau} (e^{isx} - isx$$

$$- 0.5(isx)^2 - 1)dP\{c_n^{-1/2}\gamma_k < x/\rho_k^{(n)}\}|$$

$$\leq |s|^3(\tau/3) \sum_{k=1}^{n} c_n^{-1}E\gamma_k^2 + 2s^2 \sum_{k=1}^{n} \int_{|x|\geq\tau} x^2$$

$$\times dP\{c_n^{-1/2}\gamma_k < x\}.$$

Since $c_n = \sum_{k=1}^{N} \mathbf{E}\gamma_k^2$ then, choosing sufficiently small τ and using (5.4.5), we obtain $\lim_{n\to\infty} I_1 = 0$. Similarly

$$J_2 \le \sum_{k=1}^{n}(0.5s^2c_n^{-1}\mathbf{E}\gamma_k^2)^2 \le s^4(\tau^2/4)\sum_k c_n^{-1}\mathbf{E}\gamma_k^2$$
$$+ (s^4/4)\sum_k \int_{|x|\ge\tau} x^2 d\mathbf{P}\{c_n^{-1/2}\gamma_k < x\}.$$

Therefore, $\lim_{n\to\infty} I_2 = 0$. Theorem 5.4.3 is proved.

Theorem 5.4.4. *For any finite s and some sequence of constants a_n let*

$$\lim_{n\to\infty}\sum_{k=1}^{n} \mathbf{E}|\mathbf{E}(\exp\{is\gamma_k a_n^{-1/2}\}/\rho_k^{(n)}) - 1|^2 = 0; \tag{5.4.6}$$

$$\lim_{n\to\infty}\sum_{k=1}^{n} \mathbf{E}|\mathbf{E}(\exp\{is\gamma_k a_n^{-1/2}\}/\rho_k^{(n)} - \mathbf{E}\exp(is\gamma_k a_n^{-1/2})| = 0. \tag{5.4.7}$$

Then

$$\lim_{n\to\infty}|\mathbf{E}\exp\{is(f_n - \mathbf{E}f_n)a_n^{-1/2}\} - \exp[\sum_{k=1}^{n}(\mathbf{E}\exp\{is\gamma_k a_n^{-1/2}\} - 1)]| = 0.$$

Proof. As in the proof of Theorem 5.4.3, we obtain

$$|\mathbf{E}\exp\{is(f_n - \mathbf{E}f_n)a_n^{-1/2}\} - \exp[\sum_{k=1}^{n}(\mathbf{E}\exp$$

$$\times \{is\gamma_k a_n^{-1/2}\} - 1)]) \le \sum_{k=1}^{n} \mathbf{E}|\mathbf{E}\{\exp[isa_n^{-1/2}\gamma_k]$$

$$- \exp[\mathbf{E}\exp\{is\gamma_k a_n^{-1/2}\} - 1]/\rho_k^{(n)}\}|$$

$$\le \sum_{k=1}^{n} \mathbf{E}|\mathbf{E}(\exp\{isa_n^{-1/2}\gamma_k\}/\rho_k^{(n)})$$

$$- \exp[\mathbf{E}(\exp\{is\gamma_k a_n^{-1/2}\}/\rho_k^{(n)} - 1]|$$

$$+ \sum_{k=1}^{n} \mathbf{E}|\exp[\mathbf{E}(\exp\{is\gamma_k a_n^{-1/2}\}/\rho_k^{(n)} - 1]$$

$$- \exp[\mathbf{E}\exp\{is\gamma_k a_n^{-1/2}\} - 1]|.$$

From this, using (5.4.6) and (5.4.7), we have the assertion of Theorem 5.4.4.

Theorem 5.4.5. *Let the sequence of martingale differences $\xi_{in}, i = \overline{1,n}$ be given and $\beta_k = (e^{is\nu_{kn}} - is\nu_{kn} - 1), k = \overline{1,n}$, where $\nu_{kn} = \xi_{kn} X(\sum_{l=k}^{n} \mathbf{E}_l \xi_{ln}^2 < c), c > 1$ for some $c > 1$,*

$$\text{plim}_{n\to\infty} \sum_{k=1}^{n} |\mathbf{E}\beta_k|^2 = 0, \qquad s \in [-S, S]; \qquad (5.4.8)$$

$$\text{plim}_{n\to\infty} [\sum_{k=1}^{n} \mathbf{E}_k \alpha_k - \alpha_n(s)] = 0, \qquad s \in [-S, S]; \qquad (5.4.9)$$

where $\alpha_k = (e^{is\xi_{kn}} - is\xi_{kn} - 1)$, and $s > 0, \alpha_n(s)$ are some nonrandom functions

$$\text{plim}_{n\to\infty} \sum_{k=1}^{n} \mathbf{E}_k \xi_{kn}^2 = 1. \qquad (5.4.10)$$

Then for any $s \in [-S, S]$,

$$\lim_{n\to\infty} [\mathbf{E} \exp\{is \sum_{k=1}^{n} \xi_{kn}\} - \exp\{\alpha_n(s)\}] = 0. \qquad (5.4.11)$$

Proof. Consider the expression

$$f_n(s) := \mathbf{E}[\exp\{is \sum_{k=1}^{n} \nu_{kn} - \sum_{k=1}^{n} \mathbf{E}_k \beta_k\} - 1]$$

$$= \sum_{l=1}^{n} \mathbf{E}[\exp\{is\nu_{ln}\} - \exp\{\mathbf{E}_l \beta_l\}]\theta_l, \qquad (5.4.12)$$

where $\theta_l = \exp\{is \sum_{k=l+1}^{n} \nu_{kn} - \sum_{k=l}^{n} \mathbf{E}_k \beta_k\}, \sum_{n+1}^{n} \equiv 0$.
  It is obvious that

$$|\theta_l| \le \exp\{0.5 \sum_{k=l}^{n} s^2 \mathbf{E}_k \nu_{kn}^2\} = \exp\{0.5s^2$$

$$\times \sum_{k=l}^{n} \mathbf{E}_k \xi_{kn}^2 X(\sum_{l=k}^{n} \mathbf{E}_l \xi_{ln}^2 < C) \le \exp\{0.5s^2 C\}. \qquad (5.4.13)$$

Then the statement (5.4.12) can be written in the form

$$f_n(s) = \sum_{l=1}^{n} \mathbf{E}[1 + \mathbf{E}_l \beta_l - \exp(\mathbf{E}_l \beta_l)]\theta_l.$$

Using this equality and (5.4.8), we obtain

$$\lim_{n\to\infty} \sup_{s\in[-S,S]} |f_n(s)| = 0. \qquad (5.4.14)$$

Represent $f_n(s)$ in the form,

$$f_n(s) = \mathbf{E}[\exp\{is \sum_{k=1}^{n} \nu_{kn} - \sum_{k=1}^{n} \mathbf{E}_k \beta_k\} - 1] X(A_n)$$

$$+ \mathbf{E}[\exp\{is \sum_{k=1}^{n} \nu_{kn} - \sum_{k=1}^{n} \mathbf{E}_k \beta_k\} - 1] X(\overline{A}_n),$$

$$(5.4.15)$$

where

$$A_n = \{\omega : |\sum_{k=1}^{n} \mathbf{E}_k \xi_{kn}^2 - 1| + |\sum_{k=1}^{n} \mathbf{E}_k \beta_k - \alpha_n(s)|$$

$$+ |\sum_{k=1}^{n} (\mathbf{E}_k \alpha_k - \mathbf{E}\alpha_k)| < \varepsilon \}.$$

It is obvious that

$$\sum_{k=1}^{n} \nu_{kn} - \sum_{k=1}^{n} \xi_k = \sum_{k=1}^{n} \xi_{kn} X(\sum_{l=k}^{n} \mathbf{E}_l \xi_{ln}^2 \geq c),$$

$$(5.4.16)$$

$$|\sum_{k=1}^{n} (\mathbf{E}_k \beta_k - \mathbf{E}_k \alpha_k)| \leq 0.5 s^2 \sum_{k=1}^{n} \mathbf{E}_k \xi_{kn}^2 X(\sum_{l=k}^{n} \mathbf{E}_l \xi_{ln}^2 \geq c).$$

$$(5.4.17)$$

Let us prove the following auxiliary assertion.

Lemma 5.4.1. *For the variables ξ_{kn},*

$$\mathrm{plim}_{n \to \infty} [\sum_{k=1}^{n} \nu_{kn} - \sum_{k=1}^{n} \xi_{kn}] = 0.$$

$$(5.4.18)$$

Proof. Using the relation (5.4.16) we have

$$\mathbf{E} \exp\{is \sum_{k=1}^{n} \xi_{kn} X(\sum_{l=k}^{n} \mathbf{E}_l \xi_{ln}^2 \geq c)\}$$

$$= \mathbf{E} \exp\{is \sum_{k=1}^{n} \xi_{kn} X(\sum_{l=k}^{n} \mathbf{E}_l \xi_{ln}^2 \geq c)\}$$

$$\times X(\sum_{l=1}^{n} \mathbf{E}_l \xi_{ln}^2 \geq c) + \mathbf{E} X(\sum_{l=1}^{n} \mathbf{E}_l \xi_{ln}^2 \geq c) + 1.$$

Since the condition (5.4.10) is fulfilled, then setting $c > 1$ from this correlation we obtain equality (5.4.18).

Similarly the following lemma holds.

Lemma 5.4.2. *For variables ξ_{kn} the equality*

$$\text{plim}_{n\to\infty} \sup_{s\in[-S,S]} |\sum_{k=1}^{n}(\mathbf{E}_k\beta_k - \mathbf{E}_k\alpha_k)| = 0. \qquad (5.4.19)$$

holds. Using Lemma 5.4.1 and Lemma 5.4.2, and also the conditions (5.4.9) and (5.4.10), we have for any $\varepsilon > 0$

$$\lim_{n\to\infty} \mathbf{E}X(\overline{A}_n) = 0. \qquad (5.4.20)$$

Granting this equality and (5.4.13), we find from the relation (5.4.15)

$$f_n(s) = \mathbf{E}[\exp\{is \sum_{k=1}^{n}\xi_{kn} - \alpha_{kn}^{(s)}\} - 1] + 0(1). \qquad (5.4.21)$$

It is obvious that $\overline{\lim}_{n\to\infty} |\alpha_n(s)| \leq 0.5s^2$ according to the condition (5.4.10). Therefore, the equality (5.4.1) follows from the expression (5.4.21). Theorem 5.4.5 is proved.

Corollary 5.4.2. *If for any $\varepsilon > 0$,*

$$\text{plim}_{n\to\infty} \sum_{k=1}^{n} \mathbf{E}_k\xi_{kn}^2 X(\mathbf{E}_k|\xi_{kn}| > \varepsilon) = 0, \qquad (5.4.22)$$

or

$$\text{plim}_{n\to\infty} \sum_{k=1}^{n} |\mathbf{E}_k\alpha_k|^2 = 0, \qquad s \in [-S, S],$$

and

$$\text{plim}_{n\to\infty} \sum_{k=1}^{n} \mathbf{E}_k\alpha_k - \alpha_n(s) = 0, \qquad s \in [-S, S],$$

$$\text{plim}_{n\to\infty} \sum_{k=1}^{n} \mathbf{E}_k\xi_{kn}^2 = 1,$$

then for any $s \in [-S, S]$,

$$\lim_{n\to\infty} |\mathbf{E}\exp\{is \sum_{k=1}^{n}\xi_{kn}\} - \exp(\alpha_n(s))| = 0.$$

The proof follows from the simple inequality

$$\sum_{k=1}^{n} |\mathbf{E}_k\beta_k|^2 \leq s^2 \sum_{k=1}^{n} \mathbf{E}_k\xi_{kn}^2 X(\mathbf{E}_k|\xi_{kn}| > \varepsilon) + \varepsilon s \sum_{k=1}^{n} \mathbf{E}_k\xi_{kn}^2$$

and inequality (5.4.17).

Corollary 5.4.3. *If for any $\varepsilon > 0$,*

$$\text{plim}_{n\to\infty} \sum_{k=1}^{n} \mathbf{E}_k \xi_{kn}^2 X(\mathbf{E}_k|\xi_{kn}| > \varepsilon) = 0,$$

and for almost every x,

$$\text{plim}_{n\to\infty}[G_n(x,\omega) - K_n(x)] = 0, \qquad (5.4.23)$$
$$\text{plim}_{n\to\infty} G_n(+\infty,\omega) = 1, \qquad (5.4.24)$$

where $G_n(x,\omega) = \sum_{k=1}^{n} \int_{-\infty}^{x} y^2 d\mathbf{P}\{\xi_{kn} < y/\sigma_{kn}\}$, K_n are nonrandom functions, then for finite s

$$\lim_{n\to\infty}[\mathbf{E}\exp\{is\sum_{k=1}^{n}\xi_{kn}\} - \exp\{\int(e^{isx} - isx - 1)x^{-2}dK_n(x)\}] = 0.$$

The proof follows from the fact, that under the conditions (5.4.23) and (5.4.24),

$$\text{plim}_{n\to\infty} \sum_{k=1}^{n} \mathbf{E}_k \alpha_k - \int(e^{isx} - isx - 1)x^{-2}dK_n(x) = 0.$$

Note that condition (5.4.22) is equal to one of the following conditions:

$$\text{plim}_{n\to\infty} \sum_{k=1}^{n}(\mathbf{E}_k\xi_{kn}^2)^{1+\delta} = 0, \quad \delta > 0,$$
$$\text{plim}_{n\to\infty} \sup_{k=\overline{1,n}} \mathbf{E}_k|\xi_{kn}| = 0,$$
$$\text{plim}_{n\to\infty} \sum_{k=1}^{n} \mathbf{E}_k\xi_{kn}^2 = 1.$$

By using limit theorems for accompanying infinite divisible laws for sums of independent random variables, we obtain the following assertion.

Theorem 5.4.6. *Let the conditions of Corollary 5.4.3 be fulfilled. The distribution functions of the random variables $\sum_{k=1}^{n}\xi_{kn}$ weakly converge to the limiting one if and only if there exists such a nondecreasing function $K(x)$ of bounded variation, that $K_n(x) \Rightarrow K(x)$. The logarithm of the characteristic function of the limit distribution is equal to*

$$\int(e^{itx} - itx - 1)x^{-2}dK(x).$$

Note, that in the general case, $K_n(x) \neq MG_n(x,\omega)$ under the conditions of Corollary 5.4.3, and $\alpha_n(s) \neq \sum_{k=1}^n \mathbf{E}\alpha_k$ under the conditions of Theorem 5.4.5.

If the condition is satisfied,

$$\lim_{A \to \infty} \overline{\lim_{n \to \infty}} \, \mathbf{E} \sum_{k=1}^n \mathbf{E}_k \xi_{kn}^2 X(\sum_{k=1}^n \mathbf{E}_k \xi_{kn}^2 > A) = 0, \qquad (5.4.25)$$

then we can assume that $K_n(x) = \mathbf{E}G_n(x,\omega), \alpha_n(s) = \sum_{k=1}^n \mathbf{E}\alpha_k$ in these assertions.

It is obvious that Theorem 5.4.5 can be formulated in the following form.

Theorem 5.4.7. *Let the condition (5.4.25) be satisfied, and*

$$\lim_{n \to \infty} \sum_{k=1}^n \mathbf{E}|\mathbf{E}_k \beta_k|^2 = 0, \quad s \in [-S, S],$$

$$\text{plim}_{n \to \infty} \sum_{k=1}^n (\mathbf{E}_k \alpha_k - \alpha_n(s)) = 0, \quad s \in [-S, S],$$

$$\text{plim}_{n \to \infty} \sum_{k=1}^n \mathbf{E}_k \xi_{kn}^2 = 1,$$

where $\alpha_n(s)$ are some nonrandom functions.
Then for any $s \in [-S, S]$,

$$\lim_{n \to \infty} [\mathbf{E} \exp\{is \sum_{k=1}^n \xi_{kn}\} - \exp\{\sum_{k=1}^n \mathbf{E}\alpha_k\}] = 0.$$

We show that for the function $\alpha_n(s)$ in Theorem 5.4.5 and Corollary 5.4.2 the following equality holds,

$$\alpha_n(s) = \int (e^{isx} - isx - 1)x^{-2} d\tilde{K}_n(x) + 0(1), \qquad (5.4.26)$$

where $\tilde{K}_n(x)$ is the nondecreasing function of bounded variation. We consider the expression

$$\mathbf{E}(\sum_{k=1}^n \mathbf{E}_k \alpha_k) X(\sum_{k=1}^n \mathbf{E}_k \xi_{kn}^2 \leq c), \quad c > 1.$$

It is obvious, that according to conditions (5.4.9) and (5.4.10), this expression is equal to

$$\int (e^{isx} - isx - 1)x^{-2} d\tilde{G}_n(x) = \alpha_n(s) + 0(1),$$

where

$$\tilde{G}_n(x) = \mathbf{E}\sum_{k=1}^{n}\int_{-\infty}^{x} y^2 d\mathbf{P}\{\xi_{kn} < y/\sigma_{kn}\}X(\sum_{k=1}^{n}\mathbf{E}_k\xi_{kn}^2 \le c).$$

From this, (5.4.26) follows.

We can take the function $\sum_{k=1}^{n}\mathbf{E}\alpha_k$ in the capacity of $\alpha_n(s)$ in Theorem 5.4.5 and Corollary 5.4.2.

For these functions we formulate Corollary 5.4.2 in the following form.

Corollary 5.4.4. *If*

$$\text{plim}_{n\to\infty}\sum_{k=1}^{n}|\mathbf{E}_k\alpha_k^2| = 0, \quad s \in [-S, S],$$

$$\text{plim}_{n\to\infty}\sum_{k=1}^{n}(\mathbf{E}_k\alpha_k - \mathbf{E}\alpha_k) = 0, \quad s \in [-S, S],$$

$$\text{plim}_{n\to\infty}\sum_{k=1}^{n}\mathbf{E}_k\xi_{kn}^2 = 1,$$

then for any $s \in [-S, S]$,

$$\lim_{n\to\infty}[\mathbf{E}\exp\{is\sum_{k=1}^{n}\xi_{kn}\} - \exp\{\sum_{k=1}^{n}\mathbf{E}\alpha_{kn}\}] = 0.$$

We can extend the method of the proof of the preceding theorems and corollaries to the case, when the variances of random variables ξ_{kn} do not exist.

Theorem 5.4.8. *Let sequence of the martingale differences* $\xi_{in}, i = \overline{1, n}$ *be given,*

$$\text{plim}_{n\to\infty}\sum_{k=1}^{n}|\mathbf{E}_k\alpha_k|^2 = 0, \quad s \in [-S, S], \tag{5.4.27}$$

$$\text{plim}_{n\to\infty}[\sum_{k=1}^{n}\mathbf{E}_k\alpha_k - \alpha_n(s)] = 0, \quad s \in [-S, S], \tag{5.4.28}$$

where $\alpha_k = (\exp(is\xi_{kn}) - is\xi_{kn} - 1), \alpha_n(s)$ *are some nonrandom functions,*

$$\text{plim}_{n\to\infty}\sum_{k=1}^{n}\mathbf{E}_k(1 + |\xi_{kn}|)(1 + \xi_{kn}^{-2})^{-1} = 1. \tag{5.4.29}$$

Then for any $s \in [-S, S]$,

$$\lim_{n \to \infty} [\mathbf{E} \exp\{is \sum_{k=1}^n \xi_{kn}\} - \exp\{\alpha_n(s)\}] = 0. \tag{5.4.30}$$

Proof. We consider the random variables

$$\nu_{kn} = \xi_{kn} X(\sum_{l=k}^n \mathbf{E}_l \xi_{ln}^2 (1 + \xi_{ln}^2)^{-1} < c), \quad k = \overline{1, n}, \quad c > 1.$$

Let $\beta_k = (\exp(is\nu_{kn}) - is\nu_{kn} - 1)$. If

$$|(e^{isx} - isx - 1) - (e^{isy} - isy - 1)| \leq 2|s||x - y|,$$

then

$$\sum_{k=1}^n |\mathbf{E}_k \beta_k|^2 \leq 2 \sum_{k=1}^n |\mathbf{E}_k \alpha_k|^2 + 2 \sum_{k=1}^n |\mathbf{E}_k \beta_k - \mathbf{E}_k \alpha_k|^2$$

$$\leq 2 \sum_{k=1}^n |\sum_k \alpha_k|^2 + 4|s| \sum_{k=1}^n \mathbf{E}_k |\xi_{kn}|$$

$$\times X(\sum_{l=k}^n \mathbf{E}_l \xi_{ln}^2 (1 + \xi_{ln}^2)^{-1} \leq c).$$

By using the condition (5.4.29) as in the proof of Lemma 5.4.2, we obtain

$$\text{plim}_{n \to \infty} \sum_{k=1}^n \mathbf{E}_k |\xi_{kn}| X(\sum_{l=k}^n \mathbf{E}_l \xi_{ln}^2 (1 + \xi_{ln}^2)^{-1} \geq c) = 0.$$

Therefore, taking into account condition (5.4.27), we find

$$\text{plim}_{n \to \infty} \sum_{k=1}^n |\mathbf{E}_k \beta_k|^2 = 0. \tag{5.4.31}$$

Later, as in Theorem 5.4.5, we consider the equality

$$f_n(s) := \sum_{l=1}^n \mathbf{E}[\exp\{is\nu_{in}\} - \exp\{\mathbf{E}_l \beta_l\}]\theta_l.$$

It is obvious that from

$$|e^{isx} - isx - 1| = |(e^{isx} - isx - 1)(1 + x^{-2})|(1 + x^2)^{-1} x^2 \leq (2|sx| + 0.5s^2) x^2 (1 + x^2)^{-1}$$

we find

$$|\theta_l| \leq \exp\{(2 + 0.5s^2)\mathbf{E}_k[\xi_{kn}^2 X(\sum_{l=k}^{n}\mathbf{E}_l\xi_{ln}^2$$

$$\times (1 + \xi_{ln}^2)^{-1} < c)[1 + \xi_{kn}^2 X(\sum_{l=k}^{n}\mathbf{E}_l\xi_{ln}$$

$$\times (1 + \xi_{ln}^2)^{-1} < c]^{-1}]\} \leq \exp\{(2 + 0.5s^2)c\}.$$

Later the proof is analogous to that of Theorem 5.4.5.

Corollary 5.4.5. *If*

$$\mathrm{plim}_{n\to\infty} \sum_{k=1}^{n}(\mathbf{E}_k\xi_{kn}^2(1 + \xi_{kn}^2)^{-1})^2 = 0$$

and the conditions (5.4.28) and (5.4.29) are fulfilled, then equality (5.4.30) holds.

Analogously, we can formulate the other corollaries of the preceding paragraph.

Let the random variables $\xi_{in}, i = \overline{1,n}$ be given. Assume that $\mathbf{E}_i\xi_{in}, i = \overline{1,n}$ do not exist. Then changing the proof of Theorem 5.4.8 respectively we can prove the following assertion.

Theorem 5.4.9. *Let $\alpha_k = (e^{is\xi_{kn}} - 1)$,*

$$\mathrm{plim}_{n\to\infty} \sum_{k=1}^{n}|\mathbf{E}_k\alpha_k|^2 = 0, \quad s \in [-S, S], \tag{5.4.32}$$

$$\mathrm{plim}_{n\to\infty}[\sum_{k=1}^{n}\mathbf{E}_k\alpha_k - \alpha_n(s)] = 0, \quad s \in [-S, S], \tag{5.4.33}$$

where $\alpha_n(s)$ are some nonrandom functions,

$$\mathrm{plim}_{n\to\infty} \sum_{k=1}^{n}\mathbf{E}_k(1 + |\xi_{kn}|^{-1})^{-1} = 1. \tag{5.4.34}$$

Then for any $s \in [-S, S]$,

$$\lim_{n\to\infty}[\mathbf{E}\exp\{is\sum_{k=1}^{n}\xi_{kn}\} - \exp\{\alpha_n(s)\}] = 0. \tag{5.4.35}$$

The proof of this theorem is analogous to that of Theorem 5.4.7, but we use the inequalities

$$|e^{isx} - 1| \leq (2 + |s|)(1 + |x|^{-1})^{-1},$$

$$|(e^{isx} - 1) - (e^{isy} - 1)| \leq (2 + |s|)(1 + |x - y|^{-1})^{-1},$$

and we consider the random variables

$$\nu_{kn} = \xi_{kn} X(\sum_{l=k}^{n} \mathbf{E}_l(1 + |\xi_{ln}|^{-1})^{-1} < c), \quad k = \overline{1, n}, \quad c > 1.$$

Assume

$$\sum_{i=1}^{n} \xi_{in} = \sum_{i=1}^{n} [\nu_{in} + \gamma_{in}],$$

where

$$\gamma_{in} = \mathbf{E}_{i-1} \sum_{k=1}^{n} \mathbf{E}_k \xi_{kn} X(|\xi_{kn}| < \tau) - \mathbf{E}_i \sum_{k=1}^{n} \mathbf{E}_k \xi_{kn}(|\xi_{kn}| < \tau),$$

$$\nu_{in} = \xi_{in} - \mathbf{E}_i \xi_{in} X(\xi_{in} < \tau).$$

Theorem 5.4.10.   *Let* $\beta_k = (e^{is(\nu_{kn} + \gamma_{kn})} - is\gamma_{kn} - 1)$,

$$\text{plim}_{n \to \infty} \sum_{k=1}^{n} [\mathbf{E}_k \beta_k - \beta_n(s)] = 0, \quad s \in [-S, S]; \tag{5.4.36}$$

$$\text{plim}_{n \to \infty} \sum_{k=1}^{n} |\mathbf{E}_k \beta_k|^2 = 0, \quad s \in [-S, S]; \tag{5.4.37}$$

where $\beta_n(s)$ *are some nonrandom functions,*

$$\text{plim}_{n \to \infty} \sum_{k=1}^{n} \mathbf{E}_k (1 + \nu_{kn}^{-2})^{-1} = 1, \tag{5.4.38}$$

and for some $h > 0$,

$$\lim_{n \to \infty} \mathbf{P}\{\sum_{k=1}^{n} \mathbf{E}_k \gamma_{kn}^2 > h\} = 0.$$

Then for any $s \in [-S, S]$,

$$\lim_{n \to \infty} [\mathbf{E} \exp\{is \sum_{k=1}^{n} \xi_{kn}\} - \exp\{\beta_n(s)\}] = 0. \tag{5.4.39}$$

Proof.   We introduce the random variables

$$\mu_{kn} = \nu_{kn} X(\sum_{l=k}^{n} \mathbf{E}_l(1 + \nu_{ln}^{-2})^{-1} < c),$$

$$\kappa_{kn} = \gamma_{kn} X(\sum_{l=k}^{n} \mathbf{E}_l \gamma_{ln}^2 \le h), \quad c > 1$$

and apply Theorem 5.4.8.

According to the construction of the variables μ_{kn} we find the inequality for the functions β_k,

$$|\mathbf{E}_k\beta_k| \le c(s)[\mathbf{E}_k(1+\nu_k^{-2})+\mathbf{E}_k\gamma_k^2].$$

Therefore, instead of condition (5.4.34), we can use condition (5.4.38).

The further proof is analogous to that of Theorem 5.4.7.

It is obvious that condition (5.4.36) is fulfilled if for any $\varepsilon > 0$,

$$\text{plim}_{n\to\infty} \sum_{k=1}^{n}(\mathbf{E}_k(1+\nu_k^{-2})^{-1})^2 = 0,$$

$$\text{plim}_{n\to\infty} \sum_{k=1}^{n}\mathbf{E}_k\gamma_k^2 X(\mathbf{E}_k\gamma_k^2 > \varepsilon) = 0.$$

Instead of conditions (5.4.37), we can consider the conditions

$$\text{plim}_{n\to\infty} \sum_{k=1}^{n}\int_{-\infty}^{x}(1+y^{-2})d\mathbf{P}\{\nu_k + \gamma_k < y/\sigma_{kn}\} = G(x),$$

$$\text{plim}_{n\to\infty} \sum_{k=1}^{n}\mathbf{E}_k(\nu_k + \gamma_k)(1+(\nu_k + \gamma_k)^2)^{-1} = \gamma,$$

where $G(x)$ is a nonrandom function of bounded variation, and γ is a non-random number.

Then

$$\lim_{n\to\infty} \beta_n(s) = is\gamma + \int(e^{isx} - isx(1+x^2)^{-1} - 1)(1+x^{-2})dG(x).$$

For $x = 0$, the integral expression is equal to

$$(e^{isx} - isx(1+x^2)^{-1} - 1)(1+x^{-2}) = -S^2/2.$$

Let us prove the central limit theorem for sums of martingale differences.

Theorem 5.4.11. *Let the sequence of the martingale differences $\xi_{in}, i = \overline{1,n}$ be given, for any $\tau > 0$,*

$$\text{plim}_{n\to\infty} \sum_{k=1}^{n}\mathbf{E}_k\xi_{kn}^2 X(|\xi_{kn}| > \tau) = 0, \qquad (5.4.40)$$

$$\text{plim}_{n\to\infty} \sum_{k=1}^{n}\mathbf{E}_k\xi_{kn}^2 = 1. \qquad (5.4.41)$$

Then [37]

$$\lim_{n \to \infty} \mathbf{P}\{\sum_{k=1}^{n} \xi_{kn} < x\} = (2\pi)^{-1/2} \int_{-\infty}^{x} \exp(-y^2/2) dy. \qquad (5.4.42)$$

The proof follows from inequalities (see the notations of Theorem 5.4.5).

$$\sum_{k=1}^{n} |\mathbf{E}_k \beta_k|^2 \leq 0.5\varepsilon s^2 \sum_{k=1}^{n} \mathbf{E}_k \xi_{kn}^2 + 2\varepsilon^{-2} \sum_{k=1}^{n} \mathbf{E}_k$$

$$\times \xi_{kn}^2 X(|\xi_{kn}| > \varepsilon), \qquad |\sum_{k=1}^{n} \mathbf{E}_k \alpha_k + 0.5 s^2 \sum_{k=1}^{n} \mathbf{E}_k \xi_{kn}^2|$$

$$\leq \varepsilon 6^{-1} |s|^3 \sum_{k=1}^{n} \mathbf{E}_k \xi_{kn}^2 + s^2 \sum_{k=1}^{n} \mathbf{E}_k \xi_{kn}^2 X(|\xi_{kn}| > \varepsilon).$$

Using Theorem 5.4.7, we prove the central limit theorem for sums of martingale differences, without the assumption that their variances exist.

Theorem 5.4.12. *Let the sequence of martingale differences $\xi_{in}, i = \overline{1,n}, n \in N$ be given, for any $\varepsilon > 0$,*

$$\text{plim}_{n \to \infty} \sum_{k=1}^{n} \mathbf{E}_k | \times_{kn} |^3 (1 + \xi_{kn}^2)^{-1} X(|\xi_{kn}| > \varepsilon) = 0,$$

$$\text{plim}_{n \to \infty} \sum_{k=1}^{n} \mathbf{E}_k (1 + \xi_{kn}^{-2})^{-1} = 1.$$

Then the equality (5.4.42) holds.

The proof follows from the inequalities (see the notations of Theorem 5.4.5),

$$\sum_{k=1}^{n} |\mathbf{E}_k \alpha_k|^2 \leq (2 + s^2) \sum_{k=1}^{n} (\mathbf{E}_k (1 + \xi_{kn}^{-2})^{-1})^2$$

$$\leq (2 + s^2) \sum_{k=1}^{n} \mathbf{E}_k (1 + \xi_{kn}^{-2})^{-1} [X(|\xi_{kn}| > \varepsilon) + \varepsilon],$$

$$\sum_{k=1}^{n} \mathbf{E}_k \alpha_k = \int_{|x|>\epsilon} (e^{isx} - isx - 1)(1 + x^{-2}) dG_n(x, \omega)$$

$$+ \int_{|x| \leq \epsilon} (e^{isx} - isx - 1)(1 + x^{-2}) dG_n(x, \omega),$$

where

$$G_n(x, \omega) = \sum_{k=1}^{n} \mathbf{E}(1 + \xi_{kn}^{-2})^{-1} X(\xi_{kn} < x), \quad \varepsilon > 0.$$

If $\xi_{in}, i = \overline{1,n}, n \in N$ is an arbitrary sequence of random variables series with finite expectation, then they can be represented in the form of martingale differences, subtracting $\mathbf{E}_i \xi_{in}$. Then the variables $\xi_{in} - \mathbf{E}_i \xi_{in}$ are the martingale differences and such assertion holds.

Theorem 5.4.13. *Let* $\nu_{in} = \xi_{in} - \mathbf{E}_i \xi_{in}$,

$$\text{plim}_{n \to \infty} \sum_{i=1}^{n} (\mathbf{E}_i \xi_{in} - \mathbf{E}\xi_{in}) = 0,$$

and for any $\varepsilon > 0$,

$$\text{plim}_{n \to \infty} \sum_{i=1}^{n} \mathbf{E}_i \nu_{in}^2 X(|\nu_{in}| > \varepsilon) = 0,$$

$$\text{plim}_{n \to \infty} \sum_{i=1}^{n} \mathbf{E}_i \nu_{in}^2 = 1.$$

Then

$$\lim_{n \to \infty} \mathbf{P}\{\sum_{i=1}^{n} (\xi_{in} - \mathbf{E}\xi_{in}) < x\} = (2\pi)^{-1/2} \int_{-\infty}^{x} \exp(-0.5y^2) dy.$$

Let the sequence of martingale differences $\xi_{in}, i = \overline{1,n}$ be given $\alpha_k = (\exp(is\xi_{kn}) - 1)$. In the previous theorems, we used the following condition for the proof of limit theorems:

$$\text{plim}_{n \to \infty} [\sum_{k=1}^{n} \mathbf{E}_k \alpha_k - \alpha_n(s)] = 0, \quad s \in [-S, S],$$

where $\alpha_n(s)$ are some nonrandom functions.

Now we weaken this condition, using the fact that $\alpha_n(s)$ is a random function.

The distribution with the characteristic function in the form:

$$\mathbf{E} \exp\{i\gamma(\omega)s + \int (e^{isx} - isx(1+x^2)^{-1} - 1)(1+x^{-2}) dG(x, \omega)\},$$

is the randomized infinite divisible distribution, where $\gamma(\omega)$ is a random variable, $G(x, \omega)$ is a random function, which is the nondecreasing function of bounded variance with fixed ω. We extend Theorem 5.4.5.

Theorem 5.4.14. *Let for any* $s \in [-S, S]$ *and the events* $A_{kn}, k = \overline{1, n}$:

$$\text{plim}_{n \to \infty} \sum_{k=1}^{n} |\mathbf{E}_k(\alpha_k/A_{kn})/^2 = 0, \tag{5.4.43}$$

where $\alpha_k = (\exp(is\nu_k) - 1), \nu_k = \xi_{kn} - \mathbf{E}_k(\xi_{kn}/A_{kn})$;

$$\text{plim}_{n \to \infty} [\sum_{k=1}^{n} \mathbf{E}_k(\alpha_k/A_{kn}) - \alpha_n(s, \omega)] = 0, \tag{5.4.44}$$

where $\alpha_n(s,\omega)$ are some B_n-measurable random functions $(B_n = U_{k=1}^n A_{kn})$,

$$\text{plim}_{n\to\infty}[\sum_{k=1}^n \mathbf{E}_k(\xi_{kn}^2/A_{kn}) - \xi_n] = 0, \qquad (5.4.45)$$

where ξ_n are some bounded B_n-measurable random variables.
    Then for any $s \in [-S, S]$,

$$\lim_{n\to\infty} [\mathbf{E}\exp\{is\sum_{k=1}^n \xi_{kn}\} - \mathbf{E}\exp\{\alpha_n(s,\omega) + is\sum_{k=1}^n \mathbf{E}_k(\xi_{kn}/A_{kn})\}] = 0. \quad (5.4.46)$$

Proof. Let $\mu_{kn} = \nu_{kn}(\sum_{l=k}^n \mathbf{E}_l(\nu_{ln}^2/A_{ln}) < c), c > 0, \beta_k = (e^{is\mu_{kn}} - is\mu_{kn} - 1)$. We consider the following statement:

$$f_n(s) := \mathbf{E}[\exp\{is\sum_{k=1}^n \mu_{kn} - \sum_{k=1}^n \mathbf{E}_k\beta_k/B_n]$$
$$\times \exp\{\alpha_n(s,\omega)\} - \exp\{\alpha_n(s,\omega)\}.$$

Using the proof of Theorem 5.4.5, we get the assertion of this theorem.

It is obvious that if

$$[\sum_{k=1}^n \mathbf{E}(\xi_{kn}/A_{kn}), \sum_{k=1}^n \int_{-\infty}^x y^2 d\mathbf{P}\{\xi_{kn} - \mathbf{E}_k$$
$$\times (\xi_{kn}/A_{kn}) < y/A_{kn}, \sigma_{kn}\}] \Rightarrow [\gamma(\omega), G(x,\omega)],$$

then under the conditions of Theorem 5.4.14,

$$\lim_{n\to\infty} \mathbf{E}\exp\{is\sum_{k=1}^n \xi_{kn}\} = \mathbf{E}\exp\{i\gamma(\omega)s + \int(e^{isx} - isx - 1)x^{-2}dG(x,\omega)\}.$$

We can easily extend theorems which were proved in this chapter to the sums of vector martingale differences. We extend Theorem 5.4.5.

For every n let the random vectors $\overrightarrow{\xi_{in}}, i = \overline{1,n}$, with values in space R_m be given, and $\mathbf{E}_i\xi_{in} = 0$, where $\mathbf{E}_i$ is the conditional expectation with respect to the minimal σ-algebra σ_{in} generated by the random vectors $\overrightarrow{\xi_{i+1n}}, \ldots, \overrightarrow{\xi_{nn}}$. By $\text{plim}_{n\to\infty}(\Xi_n - T_n) = 0$, where Ξ_n is a random of finite order, and T_n is a nonrandom one, we shall regard as the limits in probability of differences of corresponding elements of matrices Ξ_n and T_n.

Theorem 5.4.15. *Let for any $\varepsilon > 0$,*

$$\mathrm{plim}_{n\to\infty} \sum_{k=1}^{n} \mathbf{E}_k \|\overrightarrow{\xi_{kn}}\|^2 X(\mathbf{E}_k \|\overrightarrow{\xi_{kn}}\| > \varepsilon) = 0,$$

and for almost every Borel set B,

$$\mathrm{plim}_{n\to\infty}[G_n(B\backslash\{0\},\omega) - K_n(B\backslash\{0\})] = 0,$$
$$\mathrm{plim}_{n\to\infty} G_n(R_m,\omega) = 1,$$

where

$$G_n(B,\omega) = \sum_{k=1}^{n} \int_B \|y^2\| d\mathbf{P}(\overrightarrow{\xi_{kn}} < \overrightarrow{y}/\sigma_{kn}\},$$

$K_n(B)$ *are nonrandom finite measures on R_m;*

$$\lim_{\delta\downarrow 0}\mathrm{plim}_{n\to\infty}[\sum_{k=1}^{n} \int_{\|x\|<\delta} xx' d\mathbf{P}\{\overrightarrow{\xi_{kn}} < \overrightarrow{x}/\sigma_{kn}\} - T_n] = 0,$$

where T_n is a nonnegative-definite nonrandom matrix.
  Then for any finite $\overrightarrow{s} \in R_m$,

$$\lim_{n\to\infty}[\mathbf{E}\exp\{i(\sum_{k=1}^{n} \overrightarrow{\xi_{kn}}, \overrightarrow{s})\} - \exp\{-0.5(T_n\overrightarrow{s}, \overrightarrow{s})$$

$$+ \int_{R_m\backslash\{0\}} (e^{i(\overrightarrow{s},\overrightarrow{x})} - i(\overrightarrow{s}, \overrightarrow{x}) - 1)\|\overrightarrow{x}\|^{-2}K_n(dx)\}] = 0.$$

The proof of this theorem is based on that of Theorem 5.4.5 and Corollary 5.4.2, but instead of formula (5.4.12), we take formula

$$f_n(s) := \mathbf{E}[\exp\{i(\overrightarrow{s}, \sum_{k=1}^{n} \overrightarrow{\nu_{kn}}) - \sum_{k=1}^{n} \mathbf{E}_k\beta_k\} - 1]$$

$$= \sum_{l=1}^{n} \mathbf{E}[\exp\{i(\overrightarrow{s}, \overrightarrow{\nu_{ln}})\} - \exp\{\mathbf{E}_l\beta_l\}]\theta_l,$$

where

$$\theta_l = \exp\{i(\overrightarrow{s}, \sum_{k=l+1}^{n} \overrightarrow{\nu_{kn}}) - \sum_{k=l}^{n} \mathbf{E}_K\beta_k\},$$

$$\beta_l = (e^{i(\overrightarrow{s},\overrightarrow{\nu_{ln}})} - i(\overrightarrow{s},\overrightarrow{\nu_{ln}}) - 1),$$

$$\nu_{kn} = \xi_{kn} X(\sum_{l=k}^{n} \|\overrightarrow{E_{ln}}\|^2 < c), \quad k = \overline{1,n}.$$

To accomplish the proof, we use the limit theorem for sums of independent infinitesimal random vectors.

Theorem 5.4.16. *Let $\overrightarrow{\xi_{1n}}(\omega),\ldots,\overrightarrow{\xi_{nn}}(\omega)$ be infinitesimal and independent for each n, the values of the random vectors in the finite Euclidean space R_m. Then for the convergence of distributions of the random vectors $\sum_{k=1}^{n}\overrightarrow{\xi_{kn}}(\omega) - \overrightarrow{a_n}$, ($\overrightarrow{a_n}$ is some sequence of nonrandom vectors) it is necessary and sufficient that $G_n(B\backslash\{0\}) \Rightarrow G(B\backslash\{0\})$, where $G(B)$ is a finite measure on R_m, $B \in \mathfrak{B}$, and $\mathfrak{B}$ is a σ-algebra of Borel sets on R_m,*

$$\lim_{\delta\downarrow 0}\lim_{n\to\infty}\sum_{k=1}^{n}\int_{||x||<\delta}\overrightarrow{x}\,\overrightarrow{x}'\mu_{kn}(d\overrightarrow{x}) = s,$$

$$\lim_{n\to\infty}[\overrightarrow{a_n} - \sum_{k=1}^{n}\int_{||x||<\tau}\overrightarrow{x}\mu_{kn}(d\overrightarrow{x})] = \overrightarrow{a},$$

where $||s|| < \infty, \tau > 0$, and s is a nonnegative-definite matrix of order m.

The characteristic function of the limit distribution is

$$\exp\{\int_{R_m-\{0\}}(e^{i(\overrightarrow{q};\overrightarrow{x})} - 1 - i(\overrightarrow{q},\overrightarrow{x})(1+||\overrightarrow{x}||^2)^{-1}$$
$$\times (1+||\overrightarrow{x}||^{-2})G(d\overrightarrow{x}) + i(\overrightarrow{a},\overrightarrow{q}) + i\int_{R_m-\{0\}}(s,\overrightarrow{x})$$
$$\times ||x||^{-2}G(d\overrightarrow{x}) - 0.5(s\overrightarrow{q},\overrightarrow{q})\},$$

where $\overrightarrow{q} \in R^m$.

Proof. Necessity. If the distribution of the random vectors $\sum_{k=1}^{n}\overrightarrow{\xi_{1n}}(\omega) - \overrightarrow{a_n}$ weakly converge to the limit vector, then according to Theorem 5.3.2,

$$\sup_n \sum_{k=1}^{n}\int(1+||x||^{-2})^{-1}\mu_{kn}(dx) < \infty, \qquad (5.4.47)$$

where μ_{kn} is the distribution of the vector $\overrightarrow{\xi_{kn}} - \overrightarrow{\alpha_{kn}}$,

$$\overrightarrow{\alpha_k} = \int_{||\overrightarrow{x}||<\tau}\overrightarrow{x}\,d\mathbf{P}\{\overrightarrow{\xi_{kn}} < \overrightarrow{x}\}.$$

By using (5.4.47) and Theorem 5.3.3, we obtain for any real vector $\overrightarrow{s} \in R_m$

and $\delta > 0$,

$$\mathbf{E}\exp\{i(\sum_{k=1}^{n}\overrightarrow{\xi_{kn}} - \overrightarrow{a_n}, \overrightarrow{s})\} = \exp\{\int_{R_m\backslash\{x:\|x\|<\delta\}} (e^{i(\overrightarrow{s},\overrightarrow{x})} - 1 - i(\overrightarrow{s},\overrightarrow{x})$$

$$\times (1+\|\overrightarrow{x}\|^2)^{-1}G_n(d\overrightarrow{x}) + i(\overrightarrow{a_n} - \sum_{k=1}^{n}$$

$$\times \int_{\|\overrightarrow{x}\|<\tau} \overrightarrow{x}\mu_{kn}(d\overrightarrow{x}), \overrightarrow{s}) + \int_{R_m\backslash\{\overrightarrow{x}:\|\overrightarrow{x}\|<\delta\}} (\overrightarrow{s},\overrightarrow{x})\|\overrightarrow{x}\|^{-2}$$

$$\times G_n(d\overrightarrow{x}) + \int_{\|\overrightarrow{x}\|<\delta} (e^{i(\overrightarrow{s},\overrightarrow{x})} - 1 - i(\overrightarrow{s},\overrightarrow{x})(1+\|\overrightarrow{x}\|^2)^{-1}$$

$$\times (1+\|x\|^{-2})G_n(d\overrightarrow{x})\} + 0(1). \tag{5.4.48}$$

According to the Helly theorem from the sequence $G_n(B)$, we can easily choose the subsequence $G_n(B)$ which is weakly convergent to some finite measure $G(B)$.

Then

$$f_n(s) := \lim_{n'\to\infty} \mathrm{Re}\ln \mathbf{E}\exp\{i\sum_{k=1}^{n'}\overrightarrow{\xi_{kn'}}$$

$$-\overrightarrow{a'_n}, \overrightarrow{s})\} = \int_{R_m\backslash\{0\}} (\cos(\overrightarrow{s},\overrightarrow{x}) - 1)(1+\|x\|^{-2})$$

$$\times G(d\overrightarrow{x}) - \lim_{\delta\downarrow 0}\lim_{n\to\infty} (T_{n',\delta}\overrightarrow{s},\overrightarrow{s}),$$

where $T_{n',\delta} = \sum_{k=1}^{n}\int_{\|\overrightarrow{x}\|<\delta} \overrightarrow{x}\,\overrightarrow{x}'\mu_{kn}(d\overrightarrow{x})$ is a nonnegative-definite matrix of order n and $\sup_{n',\delta}\|T_{n',\delta}\| < \infty$.

The limit on the right-hand side of (5.4.48) exists and $\lim_{\delta\downarrow 0}\lim_{n'\to\infty} \times T_{n',\delta} = T_1$. Obviously,

$$\lim_{n'\to\infty} (\overrightarrow{a'_n} - \sum_{k=1}^{n}\int_{\|\overrightarrow{x}\|<\tau} \overrightarrow{x}\mu_{kn}(d\overrightarrow{x}))$$

$$= \overrightarrow{a}, \overrightarrow{a} \in R_m, \quad \lim_{\delta\downarrow 0}\lim_{n\to\infty}\int_{R_m\backslash\{\overrightarrow{x}:\|x\|<\delta\}} (\overrightarrow{s},\overrightarrow{x})\|\overrightarrow{x}\|^{-2}G_n(d\overrightarrow{x})$$

$$= \int_{R_m\backslash\{0\}} (\overrightarrow{s},\overrightarrow{x})\|\overrightarrow{x}\|^{-2}G(d\overrightarrow{x}),$$

where

$$
g(s) := \lim_{n' \to \infty} \mathbf{E} \exp\{i \sum_{k=1}^{n'} \overrightarrow{\xi_{kn'}} - \overrightarrow{a_{n'}}, \overrightarrow{s})\}
$$

$$
= \exp\Big\{ \int_{R_m \setminus \{0\}} (e^{i(\overrightarrow{s}, \overrightarrow{x})} - 1 - i(\overrightarrow{s}, \overrightarrow{x})
$$

$$
\times (1 + \|x\|^2)^{-1})(1 + \|\overrightarrow{x}\|^{-2})G(dx) + i(\overrightarrow{a}, \overrightarrow{s})
$$

$$
+ i \int_{R_m \setminus \{0\}} (\overrightarrow{s}, \overrightarrow{x})\|\overrightarrow{x}\|^{-2}G(dx)\} - 0.5(T_1 \overrightarrow{s}, \overrightarrow{s})\}.
$$

$$(5.4.49)$$

Let us show that the function $G(B)$, vector $\overrightarrow{a}$, and matrix T_1 are uniquely defined by the formula (5.4.48).

Let $\overrightarrow{s} = \overrightarrow{q} + \sqrt{\varepsilon}\,\overrightarrow{\eta}$, where η is a random normally distributed $N(\overrightarrow{0}, 1)$ vector, $\varepsilon > 0$. Then

$$
(\partial^3/\partial \varepsilon^3)\mathbf{E} \ln g(\overrightarrow{q} + \sqrt{\varepsilon}\,\overrightarrow{\eta}) = 2^{-3} \int e^{i(\overrightarrow{q}, \overrightarrow{x})} \exp(-0.5\varepsilon\|\overrightarrow{x}\|^2)
$$

$$
\times \|\overrightarrow{x}\|^4 (1 + \|\overrightarrow{x}\|^2)G(d\overrightarrow{x}) = \int e^{i(\overrightarrow{q}, \overrightarrow{x})} v(d\overrightarrow{x}),
$$

where $v(B) = 2^{-3} \int_B \exp(-0.5\varepsilon\|\overrightarrow{x}\|^2)\|\overrightarrow{x}\|^4(1 + \|\overrightarrow{x}\|^2)G(d\overrightarrow{x})$.

From this formula, by virtue of the uniqueness of the inversion formula for Fourier cosinuse-transform of measures on finite- dimension spaces,it obtains that $G(B)$ is uniquely defined by $g(s)$.

But then vector $\overrightarrow{a}$ and matrix T_1 are also uniquely defined. Consequently,

$$
G_n(B) \Rightarrow G(B), T_{n,\delta} \to T,
$$

$$
\lim_{n \to \infty} (\overrightarrow{a_n} - \sum_{k=1}^{n} \int_{\|\overrightarrow{x}\| < \tau} \overrightarrow{x} \mu_{kn}(\overrightarrow{dx})) = \overrightarrow{a}.
$$

The proof of sufficiency is obvious.

Theorem 5.4.16 is proved.

Theorem 5.4.17. *If for any $\tau > 0$,*

$$
\mathrm{plim}_{n \to \infty} \sum_{k=1}^{n} \mathbf{E}_k \|\overrightarrow{\xi_{kn}}\|^2 X(\|\overrightarrow{\xi_{kn}}\| > \tau) = 0,
$$

$$
\mathrm{plim}_{n \to \infty} \sum_{k=1}^{n} \mathbf{E}_k \overrightarrow{\xi_{kn}} \overrightarrow{\xi'_{kn}} = T_m,
$$

$$(5.4.50)$$

where T_m is a nonnegative-definite matrix, then

$$
\lim_{n \to \infty} \mathbf{P}\{\sum_{k=1}^{n} \overrightarrow{\xi_{kn}} \in B\} = (2\pi)^{-m/2} \det T_m^{-1/2} \int_B \exp\{-0.5(T_m^{-1}\overrightarrow{x}, \overrightarrow{x})\} \prod_{i=1}^{n} dx_i,
$$

where $B \in \mathfrak{B}$, and $\mathfrak{B}$ is a Borel σ-algebra on R_m.

We note that we can substitute condition (5.4.50) by the condition

$$\text{plim}_{n\to\infty} \sum_{k=1}^{n} \overrightarrow{\xi_{kn}}\overrightarrow{\xi'_{kn}} = T_m. \qquad (5.4.51)$$

To prove the equivalence of conditions, we must consider the expression

$$\mathbf{E} \exp\{-q| \sum_{k=1}^{n}(\overrightarrow{s},\overrightarrow{\xi_{kn}})^2 - \sum_{k=1}^{n}\mathbf{E}_k(\overrightarrow{s},\overrightarrow{\xi_{kn}})]\}$$

$$\times X(\sum_{k=1}^{n}\mathbf{E}_k(\overrightarrow{s},\overrightarrow{\xi_{kn}})^2 \leq c).$$

By using the Lindeberg condition, we find that this statement converges to 1. Therefore, the conditions (5.4.50) and (5.4.51) are equivalent.

For the proof of the limit theorems for the sums of martingale differences, we use the conditions, which reduce to convergence in probability of sums of some random functions to the sums of some nonrandom function. If these conditions are not satisfied, then under the specific assumption, we can prove that the limit random variable is represented in the form of the Ito integral, which is the solution of some stochastic differential equation. Let $\xi_{in}, i = \overline{1,n}, n \in N$ be the sequences of martingale differences and with probability 1, $\mathbf{E}_i\xi_{in}^2 > 0$. Then

$$\sum_{k=1}^{n}\xi_{kn} = \sum_{k=1}^{n}\xi_{kn}(\mathbf{E}_k\xi_{kn}^2)^{-1/2}(\mathbf{E}_k\xi_{kn}^2)^{1/2}.$$

If for the variables $\alpha_k = \xi_{kn}(\mathbf{E}_k\xi_{kn}^2 n)^{-1/2}$, the Lindeberg condition is fulfilled, then according to the preceding theorem, the finite-dimensional distributions of random processes $W_n(t) := \sum_{k\leq tn}\alpha_k, 0 \leq t \leq 1$ converge to finite-dimensional distributions of the Brownian motion process $W(t)$.

Let $(\mathbf{E}_k\xi_{kn}^2 n)^{1/2} = \xi_n(t)$, where $k/n < t \leq (k+1)/n$. Then if $\omega_n, \xi_n \Rightarrow \omega, \xi$ and with probability 1, $\int_0^1 \xi^2(t)dt < \infty$,

$$\lim_{h\downarrow 0} \overline{\lim_{n\to\infty}} \sup_{|t_1-t_2|\leq h} \mathbf{E}|\xi_n(t_1) - \xi_n(t_2)| = 0,$$

where $\xi(t)$ is a measurable random process nonforestalling with respect to the stream of σ-algebras, generated by the random process $W(t)$, then, according to the definition of stochastic integral,

$$\sum_{k=1}^{n}\xi_{kn} \Rightarrow \int_0^1 \xi(t)dW(t).$$

We define the stochastic difference equation for the process $\xi(t)$.
Note that

$$\sum_{l=1}^{k} \xi_{ln} = \eta_n(t), \qquad k/n < t \le (k+1)/n. \qquad (5.4.52)$$

Suppose that the process $\xi_n(s)$ is represented in the form of some functional $f_n(s, \eta_n(s))$ of process $\eta_n(t)$. Then formula (5.4.52) takes the form

$$\eta_n(t) = \int_0^t f_n(s, \eta_n(s))dW_n(s).$$

If $f_n(t, \cdot)$ converge on the process $\eta_n(\cdot)$ realizations to limit $f(t, \cdot)$, $f(t, \cdot)$ is a continuous function, and equation $\eta(t) = \int_0^t f(s, \eta(s))dW(s)$ with probability 1 has an unique strong solution, then $\eta_n(t) \Rightarrow \eta(t)$, and $\eta(t)$ is a solution of the equation $d\eta(t) = f(t, \eta(s))dW(t), \eta(0) = 0$. In this case,

$$\sum_{k=1}^{n} \xi_{kn} \Rightarrow \int_0^1 f(t, \eta(s))dW(t) = \eta(1).$$

§5 Limit Theorems for the Sums of Martingale Differences in Nonclassical Situations

Let a sequence of martingale differences $\xi_{in}, i = \overline{1, n}, n \in N$ be given. We are concerned with conditions under which the distribution of the sum $\sum_{i=1}^{n} \xi_{in}$ can be approximately substituted by the distribution of the sum $\sum_{i=1}^{n} \eta_{in}$ for larger n, where $\eta_{in}, i = \overline{1, n}, n \in N$ is a sequence of martingale differences which can be in particular the independent variables.

Theorem 5.5.1. *Let two independent sequences of martingale differences ξ_{in}, $i = \overline{1, n}$ and $\eta_{in}, i = \overline{1, n}, n \in N$ be given and the following conditions hold:*

$$\text{plim}_{n \to \infty} \sum_{k=1}^{n} (\sigma_k^2 - \delta_k^2) = 0, \qquad (5.5.1)$$

where $\sigma_k^2 = E_k \xi_{kn}^2, \delta_K^2 = E_k \eta_{kn}^2$ for some $h > 0$,

$$\lim_{n \to \infty} P\{\sum_{k=1}^{n} \sigma_k^2 > h\} = 0 \qquad (5.5.2)$$

Lindeberg condition hold: for any $\varepsilon > 0$,

$$\text{plim}_{n \to \infty} \sum_{k=1}^{n} \int_{|x|>\varepsilon} x^2 |d(P\{\xi_{kn} < x/\sigma_{kn}\}$$
$$- P\{\eta_{kn} < x/\sigma_{kn}\})| = 0, \qquad (5.5.3)$$

where $\sigma_{kn} = \sigma(\xi_{k+1n}, \ldots, \xi_{nn}, \eta_{k+1n}, \ldots, \eta_{nn})$.
  Then

$$\lim_{n \to \infty} [\mathbf{E} \exp\{is \sum_{k=1}^{n} \xi_{kn} - \mathbf{E} \exp\{is \sum_{k=1}^{n} \eta_{kn}\}] = 0. \qquad (5.5.4)$$

Proof. Set

$$\nu_{kn} = \xi_{kn} X(\sum_{l=k}^{n} \mathbf{E}_l \xi_{ln}^2 < c),$$

$$\mu_{kn} = \eta_{kn} X(\sum_{l=k}^{n} \mathbf{E}_l \eta_{ln}^2 < c), \quad k = \overline{1, n}, c > 0.$$

Consider the equality

$$f_n(s) := \mathbf{E}[\exp\{is \sum_{k=1}^{n} \nu_{kn} + 0.5s^2 \sum_{k=1}^{n} (\mathbf{E}\nu_{kn}^2$$

$$- \mathbf{E}_k \mu_{kn}^2)\} - \exp\{is \sum_{k=1}^{n} \mu_{kn}\}]$$

$$= \sum_{l=1}^{n} \mathbf{E} \exp\{is \sum_{k=1}^{l-1} \mu_{kn}\} \mathbf{E}_l \exp\{is\nu_{ln}$$

$$+ 0.5s^2(\mathbf{E}_l \nu_{ln}^2 - \mathbf{E}_l \mu_{ln}^2)\} - \mathbf{E}_l \exp\{is\mu_{ln}\}] \exp$$

$$\times is \sum_{k=l+1}^{n} (\mu_{kn} + \nu_{kn}) + 0.5s^2 \sum_{k=l+1}^{n} [\mathbf{E}_k \nu_{kn}^2 - \mathbf{E}_k \mu_{kn}^2]\}.$$
$$\qquad (5.5.5)$$

In the same manner, we obtain the proof of inequality (5.4.15),

$$\sum_{k=l+1}^{n} \mathbf{E}_k \nu_{kn}^2 \leq c, \quad \sum_{k=l+1}^{n} \mathbf{E}_k \mu_{kn}^2 \leq c, \quad l = \overline{0, n-1}. \qquad (5.5.6)$$

Now using equality (5.5.5) and the inequalities

$$|e^{isx} - isx - 1| \leq 0.5s^2 x^2,$$
$$|e^{isx} - isx - 1 - (isx)^2/2| \leq 6^{-1}|isx|^3,$$

we find

$$f_n(s) :\leq \sum_{l=1}^{n} \exp(s^2 c)\mathbf{E}|\mathbf{E}_l[\exp(is\nu_{ln}) - is\nu_{ln}$$
$$- 1 - 0.5(is\nu_{ln})^2 - \exp(is\mu_{ln}) + is\mu_{ln}$$
$$+ 1 + 0.5(is\mu_{ln})^2] + \mathbf{E}_l[-\exp(is\nu_{ln})$$
$$- 0.5s^2(\nu_{ln}^2 - \mu_{ln}^2) + \exp\{is\nu_{ln} + 0.5s^2$$
$$\times (\mathbf{E}_l\nu_{ln}^2 - \mathbf{E}_l\mu_{ln}^2)\}]| \leq \mathbf{E}\{\beta_n + 6^{-1}|s|^3\varepsilon$$
$$\times \sum_{l=1}^{n}[(\mathbf{E}_l\nu_{ln}^2 + \mathbf{E}_l\mu_{ln}^2) + | - \mathbf{E}_l e^{is\nu_{ln}}$$
$$\times [1 - \exp(0.5s^2(\mathbf{E}_l\nu_{ln}^2 - \mathbf{E}_l\mu_{ln}^2) + 0.5s^2$$
$$\times (\mathbf{E}_l\nu_{ln}^2 - \mathbf{E}_l\mu_{ln}^2)] + \mathbf{E}_l(e^{is\nu_{ln}} - 1)0.5s^2$$
$$\times (\mathbf{E}_l\nu_{ln}^2 - \mathbf{E}_l\mu_{ln}^2)|]\},$$

where

$$\beta_n = \sum_{l=1}^{n} \int_{|x|>\epsilon} x^2 |d(\mathbf{P}\{\nu_{ln} < x/\sigma_{ln}\} - \mathbf{P}\{\mu_{ln} < x/\sigma_{ln}\})|.$$

Together with (5.5.6) this yields

$$|f_n(s)| \leq c_1(s) \sum_{l=1}^{n} \mathbf{E}(\mathbf{E}_l\mu_{ln}^2 - \mathbf{E}_l\nu_{ln}^2)^2 + c^2(s)$$
$$\times \sum_{i=1}^{n} \mathbf{E}[\mathbf{E}_l\mu_{ln}^2|\mathbf{E}_l\mu_{ln}^2 - \mathbf{E}_l\nu_{ln}^2|] + \beta_n + \varepsilon c_3(s), \quad (5.5.7)$$

where $\sup_{|s|\leq s}[c_1(s) + c_2(s) + c_3(s)] < \infty$.
Obviously,

$$\sum_{l=1}^{n}(\mathbf{E}_l\mu_{ln}^2 - \mathbf{E}_l\nu_{ln}^2)^2 \leq \varepsilon c + c\beta_n,$$
$$\sum_{l=1}^{n} \mathbf{E}_l\mu_{ln}^2|\mathbf{E}_l\mu_{ln}^2 - \mathbf{E}_l\nu_{ln}^2| \leq \varepsilon c + c\beta_n.$$

Taking into account these inequalities and (5.5.7), we obtain $\lim_{n\to\infty}|f_n(s)| = 0$ for all $|s| \leq s$, if

$$\lim_{n\to\infty} \mathbf{E}\beta_n = 0. \quad (5.5.8)$$

The random variable β_n is bounded by the constant c, and provided that $c > h, \sum_{k=1}^{n}\sigma_k^2 \leq h, \sum_{k=1}^{n}\delta_k^2 \leq h$, we have

$$\beta_n = \sum_{k=1}^{n} \int_{|x|>\epsilon} x^2 |d(\mathbf{P}\{\xi_{kn} < x/\sigma_{kn}\} - \mathbf{P}\{\eta_{kn} < x/\sigma_{kn}\})|.$$

Thus the conditions (5.5.2), (5.5.3) imply (5.5.8). Hence,

$$\lim_{n\to\infty} |f_n(s)| = 0, \quad s \in [-S, S].$$

Thereby using condition (5.5.1) as in the proof of Theorem 5.4.5, we arrive at (5.5.4). The theorem is proved.

In particular, we can assume in Theorem 5.4.17 that the random variables $\eta_{in}, i = \overline{1,n}$ for each n are independent and do not depend on the random variables $\xi_{in}, i = \overline{1,n}$. Then, evidently $\delta_k^2 = \mathbf{E}\eta_{kn}^2, \mathbf{P}\{\eta_{kn} < x/\sigma_{kn}\} = \mathbf{P}\{\eta_{kn} < x\}$. Assuming in addition that η_{kn} are normally distributed $N(0, \delta_k^2)$, we derive from (5.5.4) the central limit theorem for the sums $\sum_{k=1}^{n} \xi_{kn}, n \in N$.

§6 Limit Theorems for Generalized U Statistics

Let $\xi_1^{(n)}, \ldots, \xi_n^{(n)}, n \in N$ be a sequence of series of random variables, $f_{i_1, \ldots i_k}(x_1, \ldots, x_n)$ Borel functions of real arguments, $i_1, \ldots, i_k$ are different integer numbers from 1 to n. Consider the random variables

$$\nu_n := \sum_{(i_1, \ldots, i_k)} f_{i_1, \ldots, i_k}^{(n)} (\xi_{i_1, \ldots, i_k}^{(n)}),$$

where sum is taken over all samples $(i_1 < \cdots < i_k)$ of numbers $1, 2, \ldots, n$.

If the random variables $\xi_i^{(n)}, i = \overline{1,n}$ do not depend on n and have identical distributions, $f_{i_1, \ldots, i_k}^{(n)}(x_1, \ldots, x_k) \equiv f(x_1, \ldots, x_k)$, then the random variable ν_n is called a U-statistic, and the function f is called the kernel of the U-statistic. For differently distributed random variables $\xi_i^{(n)}$, the random variables ν_n are called generalized U-statistics.

In order to prove the limit theorems for the random variables ν_n, we apply the above-mentioned limit theorems.

Theorem 5.6.1. *Let $\mathbf{E}\gamma_s^2, s = \overline{1,n}$ exist, where $\gamma_s = \mathbf{E}_{(s-1)}\nu_n - \mathbf{E}_{(s)}\nu_n, \mathbf{E}_{(s)}$ is the conditional expectation with respect to the minimal σ-algebra generated by the random variables $\xi_{s+1}^{(n)}, \ldots, \xi_n^{(n)}$,*

$$\lim_{n\to\infty} \sum_{s=1}^{n} \mathbf{E}\gamma_s > 0, \tag{5.6.1}$$

$$\text{plim}_{n\to\infty} c_n^{-1} \sum_{s=1}^{n} [\mathbf{E}\gamma_s^2 - \mathbf{E}(\gamma_s^2/\rho_s^{(n)})] = 0, \tag{5.6.2}$$

where $\rho_s^{(n)}$ is the minimal σ-algebra generated by the random variables $\xi_l^{(n)}, l \neq s, l = \overline{1,n}, c_n = \mathbf{V}(\nu_n - \mathbf{E}\nu_n)$, and the Lindeberg condition be satisfied: for any $\tau > 0$,

$$\text{plim}_{n\to\infty} \sum_{s=1}^{n} \int_{|x|>\tau} x^2 d\mathbf{P}\{c_n^{-1/2}\gamma_s < x/\rho_n^{(s)}\} = 0. \tag{5.6.3}$$

Then

$$L(c_n^{-1/2}(\nu_n - \mathbf{E}\nu_n)) \Rightarrow N(0,1). \tag{5.6.4}$$

If for some sequence of constants a_n

$$\lim_{n\to\infty} a_n^{-2} \sum_{s=1}^{n} \mathbf{E}\gamma_s^2 = 0, \tag{5.6.5}$$

then

$$\mathrm{plim}_{n\to\infty} a_n^{-1}[\nu_n - \mathbf{E}\nu_n] = 0. \tag{5.6.6}$$

We denote by $(i_1, \ldots, i_k)_s$ the sample of k numbers from $1, 2, \ldots, n$, where only one number is equal to $s, s = \overline{1, n}$. It is obvious that

$$\gamma_s = \mathbf{E}_{(s-1)} \sum_{(i_1,\ldots,i_k)_s} f_{i_1,\ldots,i_k}^{(n)}(\xi_{i_1}^{(n)}, \ldots, \xi_{i_k}^{(n)})$$

$$- \mathbf{E}_{(s)} \sum_{(i_1,\ldots,i_k)_s} f_{i_1,\ldots,i_k}^{(n)}(\xi_{i_1}^{(n)}, \ldots, \xi_{i_k}^{(n)}), \tag{5.6.7}$$

where the summation is taken over all possible samples $(i_1, \ldots, i_k)_s$.

The number of terms in (5.5.7) may be infinite as $n \to \infty$. Therefore, we can obtain the conditions when $\mathrm{plim}_{n\to\infty} b_{ns}^{-1}[\gamma_s - \mathbf{E}(\gamma_s/\xi_s^{(n)})] = 0$, where b_{ns} is a sequence of constant.

Theorem 5.6.2. *Let conditions (5.6.1) and (5.6.3) be fulfilled, there exists such sequence of constants b_{ns} that*

$$\lim_{h\to\infty} \lim_{n\to\infty} c_n^{-1} \sum_{s=1}^{n} b_{ns}^2 \mathbf{E}(b_{ns}^{-1}\gamma_s)^2/|b_{ns}^{-1}\gamma_s|$$

$$> h]\mathbf{P}\{|b_{ns}^{-1}\gamma_s| > h\} = 0, \tag{5.6.8}$$

$$\mathrm{plim}_{n\to\infty} b_{ns}^{-1}[\gamma_s - \mathbf{E}(\gamma_s/\xi_s^{(n)})] = 0,$$

$$\overline{\lim_{n\to\infty}} c_n^{-1} \sum_{s=1}^{n} b_{ns}^2 < \infty. \tag{5.6.9}$$

Then assertion (5.6.4) holds.

Theorem 5.6.3. *Let the random variables $\xi_1, \xi_2, \ldots, \xi_n, \ldots$ be independent, identically distributed, $f_{i_1,\ldots,i_k}(x_1, \ldots, x_k) \equiv f(x_1, \ldots, x_k)$ functions $f(x_1, \ldots, x_k)$ are symmetric and*

$$\mathrm{Var}\{\mathbf{E}[f(\xi_1, \xi_2, \ldots, \xi_k)/\xi_1]\} \neq 0.$$

Then

$$L\{[\sum_{(i_1,\ldots,i_k)} f(\xi_{i_1},\ldots,\xi_{i_k}) - C_n^k \mathbf{E}f$$

$$\times (\xi_1,\ldots,\xi_k)][n(C_{n-1}^{k-1})^2 \mathbf{V}\{\mathbf{E}[f$$

$$\times (\xi_1,\ldots,\xi_k)/\xi_1]\}]^{-1/2}\} \Rightarrow N(0,1),$$

where $A_n^k = n(n-1)\ldots(n-k+1), C_n^k = A_n^k(k!)^{-1}$.

First of all, let us prove the auxiliary assertion.

Lemma 5.6.1. *If the conditions of Theorem 5.6.3 are satisfied, then*

$$\mathbf{E}(\nu_n - \mathbf{E}\nu_n)^2 \le Cn(C_{n-1}^{k-1})^2,$$

where $C > 0$ *is a constant.*

Proof. On the basis of formula (5.6.7), it is obvious that

$$\mathbf{E}(\nu_n - \mathbf{E}\nu_n)^2 = \sum_{s=1}^n \mathbf{E}\gamma_s^2 \le 2\sum_{s=1}^n (\sum_{(i_1,\ldots,i_k)}$$

$$\times [\mathbf{E}f_{i_1,\ldots,i_k}^2(\xi_{i_1},\ldots,\xi_{i_k})^{1/2}]^2 \le Cn(C_{n-1}^{k-1})^2.$$

Lemma 5.6.2. *If the conditions of Theorem 5.6.3 are satisfied, then when* $k \ge 2$

$$\mathbf{E}(\nu_n - \mathbf{E}\nu_n)^2 = n[(C_{n-1}^{k-1})^2 \mathbf{V}\{\mathbf{E}[f(\xi_1,\ldots,\xi_k)/\xi_1]\} + 0(1). \tag{5.6.10}$$

Proof. Obviously,

$$\mathbf{E}(\nu_n - \mathbf{E}\nu_n)^2 = \sum_{s=1}^n \mathbf{E}\gamma_s^2 = \sum_{s=1}^n [\mathbf{E}(\gamma_s/\xi_s))^2$$

$$+ \mathbf{E}(\mathbf{E}(\gamma_s/\xi_s))^2]. \tag{5.6.1}$$

According to Lemma 5.6.1 and formula (5.6.7), we obtain

$$\mathbf{E}(\gamma_s - \mathbf{E}(\gamma_s/\xi_s))^2 \le C(n-1)(C_{n-2}^{k-2})^2, \tag{5.6.12}$$

$$\mathbf{E}(\mathbf{E}(\gamma_s/\xi_s))^2 = (C_{n-1}^{k-1})^2 \mathbf{V}\{\mathbf{E}[f(\xi_1,\ldots,\xi_k)/\xi_1]\}. \tag{5.6.13}$$

Since

$$\mathbf{E}(\gamma_s - \mathbf{E}(\gamma_s/\xi_s))^2 [\mathbf{E}(\mathbf{E}(\gamma_s/\xi_s))^2]^{-1} \le (n-1)$$

$$\times [C_{n-2}^{k-2}]^2 [C_{n-1}^{k-1}]^{-2} = k(n^{-1} + 0(1)), \tag{5.6.14}$$

if $k \geq 3$, then by using formula (5.6.12), (5.6.13), and (5.6.11), we obtain the assertion of Lemma 5.6.2, and the proof of Theorem 5.6.3. Take $n(C_{n-1}^{k-1})^2 \mathbf{V}\{\mathbf{E}[f(\xi_1, \ldots, \xi_k)/\xi_1]\}$ for the normalized constants C_n. On the basis of Lemma 5.6.1 and formula (5.6.14), we obtain

$$C_n^{-1} \sum_{s=1}^{n} \{\mathbf{E}(\gamma_s - \mathbf{E}(\gamma_s/\xi_s))^2 + [\mathbf{E}(\gamma_s - \mathbf{E}(\gamma_s/\xi_s))^2$$
$$\times \mathbf{E}\gamma_s^2]^{1/2}\} \leq C[n(C_{n-1}^{k-1})^2]^{-1} n\{(n-1)$$
$$\times C_{n-2}^{k-2})^2 + [(n-1)(C_{n-2}^{k-2})^2(C_{n-1}^{k-1})^2]^{1/2}\}, \quad k \geq 2.$$

Therefore, condition (5.6.9) holds, and condition (5.6.1) is fulfilled for equality (5.6.10). Prove that condition (5.6.8) holds.

We rewrite ν_n as

$$\nu_n = \sum_{(i_1, \ldots, i_k)} [\theta_{i_1, \ldots, i_k} + \kappa_{i_1, \ldots, i_k}],$$

where

$$\theta_{i_1, \ldots, i_k} = f_{i_1, \ldots, i_k}(\xi_{i_1}, \ldots, \xi_{i_k}) X(|f_{i_1, \ldots, i_k}(\xi_{i_1}, \ldots, \xi_{i_k})| \leq C),$$
$$\kappa_{i_1, \ldots, i_k} = f_{i_1, \ldots, i_k}(\xi_{i_1}, \ldots, \xi_{i_k}) X(|f_{i_1, \ldots, i_k}(\xi_{i_1}, \ldots, \xi_{i_k})| > C),$$
$$C > 0.$$

Obviously, the condition (5.6.8) is fulfilled for the sums

$$\sum_{(i_1, \ldots, i_k)} [\theta_{i_1, \ldots, i_k} - \mathbf{E}\theta_{i_1, \ldots, i_k}].$$

By using the proof of Lemma 5.6.1, we obtain

$$\mathbf{E}\Big(\sum_{(i_1, \ldots, i_k)} [\kappa_{i_1, \ldots, i_k} - \mathbf{E}\kappa_{i_1, \ldots, i_k}]\Big)^2 \leq Cn$$
$$\times (kC_{n-1}^{k-1}[\mathbf{E}f^2(\xi_1, \ldots, \xi_k)X(|f(\xi_1, \ldots, \xi_k)| > C)]^{1/2})^2.$$

This sum divided by $n(kC_{n-1}^{k-1})^2$ tends to zero as $C \to \infty$, because the variables $\xi_1, \ldots, \xi_n, \ldots$ are independent, identically distributed, and there exists $\mathbf{E}f^2(\xi_1, \ldots, \xi_n)$. So all the conditions of Theorem 5.6.1 are fulfilled.

Theorem 5.6.3 is proved.

§7 Central Limit Theorem for Some Functionals of Random Walk

Let $\xi_1, \xi_2, \ldots, \xi_k, \ldots$ be independent real random variables. The sequence of random variables $\{S_k, k = 0, 1, \ldots\}$, where $S_0 = 0, S_n = S_{n-1} + \xi_n, n \geq 0$, is referred to as the random walk, and ξ_n is the nth step of the walk.

We prove the central limit theorem for the sums $\sum_{k=1}^{n} \sin S_k$ by using the assertion of Theorem 5.4.11.

Theorem 5.7.1. *If*

$$|\mathbf{E}\exp(i\xi_p)| \le \varphi < 1/2, \quad p = 1, 2, \ldots, \tag{5.7.1}$$
$$\sup_{p \ge 1} |\mathbf{E}\exp(i2\xi_p)| \le \varphi_1 < 1, \tag{5.7.2}$$

then

$$\lim_{n \to \infty} \mathbf{P}\{(\sum_{k=1}^{n}\sin S_k)[\mathbf{V}\sum_{k=1}^{n}\sin S_k]^{-1/2} < x\} = (2\pi)^{-1/2}\int_{-\infty}^{x} e^{-y^2/2}dy.$$
$$\tag{5.7.3}$$

Proof. We present the sum $\sum_{k=1}^{n}\sin S_k$ in the following form:

$$\sum_{k=1}^{n}(\sin S_k - \mathbf{E}\sin S_k) = \sum_{k=1}^{n}\{\text{Im}\,e^{iS_k} - \mathbf{E}\,\text{Im}\,e^{iS_k}\} = \sum_{p=1}^{n}\gamma_p,$$

where

$$\gamma_p = \text{Im}\exp(iS_{p-1})(\exp(i\xi_p) - \mathbf{E}\exp(i\xi_p))$$

$$\times \sum_{k=p+1}^{n}\prod_{l=p+1}^{k}\varphi_l, \quad \varphi_l = \mathbf{E}\exp(i\xi_l), \quad \sum_{k=n+1}^{n}\prod_{l=n+1}^{k}\varphi_l = 1.$$

The variables $\gamma_p, p = \overline{1,n}$ are martingale differences. The assertion of Theorem 5.4.11 is valid for these variables, and we verify the fulfillment conditions (5.4.40) and (5.4.41).

It is obvious that

$$\mathbf{E}|\gamma_p|^2 \le [\mathbf{E}|\exp(iS_{p-1})|^2\mathbf{E}|\exp(i\xi_p) - \mathbf{E}\exp(i\xi_p)|^2$$

$$\times |\sum_{k=p+1}^{n}\prod_{l=p+1}^{k}\varphi_l|^2 \le c(1 - \varphi)^{-2}. \tag{5.7.4}$$

Lemma 5.7.1. *If (5.7.1) and (5.7.2) hold, then*

$$\lim_{n \to \infty} n^{-1}\sum_{p=1}^{n}\mathbf{E}\gamma_p^2 > 0. \tag{5.7.5}$$

Proof. Let $\varphi_p(2) := \mathbf{E}\exp(i2\xi_p)$.

It is obvious that

$$\mathbf{E}\gamma_p^2 = \frac{1}{2}[\mathbf{E}(\exp(i\xi_p) - \mathbf{E}\exp(i\xi_p))^2(\sum_{l=p+1}^{n}\prod_{l=p+1}^{k}\varphi_l)^2$$

$$- \mathrm{Re}(\prod_{s=1}^{p-1}\varphi_s(2)\mathbf{E}(\exp(i\xi_p) - \mathbf{E}\exp(i\xi_p))^2$$

$$\times \sum_{k=p+1}^{n}\prod_{l=p+1}^{k}\varphi_l)^2)].$$

Denote

$$\alpha_p = (\mathrm{Re}\prod_{s=1}^{p-1}\varphi_s(2)\mathbf{E}(\exp(i\xi_p) - \mathbf{E}\exp(i\xi_p))^2$$

$$\times (\sum_{k=p+1}^{n}\prod_{l=p+1}^{k}\varphi_l)^2).$$

Since $|\alpha_p| \le \varphi_1^{p-1}2(1-\varphi)^{-2}$, then $\sum_{p=1}^{n}|\alpha_p| \le 2(1-\varphi_1)^{-1}(1-\varphi^2)^{-1}$. Therefore, $\underline{\lim}_{n\to\infty} n^{-1}\sum_{p=1}^{n}\alpha_p = 0$,

$$n^{-1}\sum_{p=1}^{n}\mathbf{E}\gamma_p^2 = n^{-1}\sum_{p=1}^{n}\frac{1}{2}\mathbf{E}|\exp(i\xi_p) - \mathbf{E}\exp(i\xi_p)|^2$$

$$\times |1 + \sum_{p+1}^{n}\prod_{l=p+1}^{k}\varphi_l|^2 + 0(1).$$

It is obvious that $\underline{\lim}_{n\to\infty} n^{-1}\sum_{p=1}^{n}\mathbf{E}\gamma_p^2 > 0$, if $|1 + \sum_{k=p+1}^{n}\prod_{l=p+1}^{k}\varphi_l|^2 \ge \delta > 0$, and that $|1 + \sum_{k=p+1}^{n}\prod_{l=p+1}^{k}\varphi_l| \ge 1 - \varphi(1-\varphi)^{-1} > 0$, if $\varphi < 1/2$.

Lemma 5.7.1 is proved.

By using (5.7.5), we easily obtain the validity of condition (4.4.40) in Theorem 5.4.11. Prove that condition (5.4.41) is also satisfied.

It is obvious that

$$c_n^{-1}\sum_{p=1}^{n}\mathbf{E}_p\gamma_p^2 = 1 - \frac{1}{2}c_n^{-1}\sum_{p=1}^{n}\mathrm{Re}\exp(2iS_{p-1})$$

$$\times R_p + 0(1), \qquad R_p = \mathbf{E}(\exp(i\xi_p) - \mathbf{E}\exp(i\xi_p))^2(\sum_{k=p+1}^{n}\prod_{l=p+1}^{k}\varphi_l)^2,$$

where $\mathbf{E}_p$ is the conditional expectation with respect to the minimal σ-algebra generated by the random variables $\xi_i, i < p$,

$$\sum_{k=1}^{n}\mathrm{Re}(\exp(i2S_{k-1})R_k - \mathbf{E}(i2S_{k-1})R_k) = \sum_{p=1}^{n}\tilde{\gamma}_p,$$

where

$$\tilde{\gamma}_p = \mathrm{Re}((\exp(i2S_{p-1}))(\exp(i2\xi_p) - \mathbf{E}\exp(i2\xi_p))$$

$$\times (R_p + \sum_{k=p+1}^{n} R_k \prod_{l=p+1}^{k} \mathbf{E}\exp(i2\xi_l)).$$

The sequence $\tilde{\gamma}_p$ is a sequence of martingale differences $|\tilde{\gamma}_p| \leq c_1 < \infty$, since $|R_p| \leq c < \infty$, and the condition (5.7.2) is fulfilled.

Therefore,

$$\mathrm{plim}_{n\to\infty} n^{-1} \sum_{p=1}^{n} \tilde{\gamma}_p = 0.$$

Since

$$\lim_{n\to\infty} \frac{1}{2} c_n^{-1} \sum_{p=1}^{n} \mathbf{E}\,\mathrm{Re}(\exp(i2S_{p-1}))R_p = 0,$$

then the condition (5.4.41) is satisfied for the variables $\gamma_p c_n^{-1/2}$.

Since

$$|\sum_{k=1}^{n} \mathbf{E}\sin S_k c_n^{-1/2}| \leq |(1 + \sum_{k=1}^{n} \prod_{s=1}^{k} \varphi_s) c_n^{-1/2}| \to 0, \quad n \to \infty,$$

we obtain the validity of formula (5.7.3).

Theorem 5.6.1 is proved.

We notice that condition (5.7.2) can be replaced by the conditions

$$|\mathbf{E}\exp(i\xi_p)| \leq \varphi < 1$$

and

$$|1 + \sum_{k=p+1}^{n} \prod_{l=p+1}^{k} \varphi_l| \geq \delta > 0, \quad p = 1, 2, \ldots.$$

In addition, Theorem 5.7.1 can be extended to the sums in the form $\sum_{k=1}^{n} \times f(S_k)$, where $f(x) = \sum_{k=1}^{m} c_k \exp(ikx)$, or $f(s) = \int_{-\infty}^{\infty} \exp(isx)p(x)dx$, where $p(x)$ is an absolutely integrable function, and c_k are some constants.

§8 Limit Theorems for Sums of Random Variables Connected in a Markov Chain

An enormous amount of work has been devoted to the study of limit theorems for sums of random variables connected in a nonhomogeneous Markov chain. As a rule, the central limit theorem for such sums has been proved with the help of an approach proposed by Bernstein. The idea in this approach is very simple, but the proof of the theorems is cumbersome. Dobrushin was apparently the first one to apply limit theorems for sums of martingale differences to the proof of the limit theorems.

Here we use a different approach which consists of the direct reduction of sums of random variables to sums of martingale differences. With this approach it is simple to prove a known central limit theorem for sums of bounded random variables connected in a Markov chain.

Theorem 5.8.1. *Suppose that for each value of n the random variables ξ_{ns}, $s = \overline{1,n}$ are connected in a Markov chain with phase spaces (X_{ns}, A_{ns}) measurable transition functions $Q_{ns}(x, B)$, and initial distributions $\mu_{n1}(B)$, and that*

$$\sup_n \sup_{s=\overline{1,n}} \sup_{x_1,x_2,B} |Q_{ns}(x_1, B) - Q_{ns}(x_2, B)| < 1, \qquad (5.8.1)$$

$$\mathbf{E}\xi_{ns} = o, \quad |\xi_{ns}| < c < \infty, \quad n, s = 1, 2, \ldots, \qquad (5.8.2)$$

$$\lim_{n\to\infty} [\mathbf{V}\sum_{k=1}^{n} \xi_{nk}]n^{-1/2} = \infty. \qquad (5.8.3)$$

Then

$$\lim_{n\to\infty} \mathbf{P}\{\sum_{k=1}^{n} \xi_{kn}(\mathbf{V}\sum_{k=1}^{n} \xi_{kn})^{-1/2} < x\} = (2\pi)^{-1/2} \int_{-\infty}^{x} e^{-y^2/2} dy.$$

Proof. Obviously, $\sum_{s=1}^{n}(\xi_{sn} - \mathbf{E}\xi_{sn}) = -\sum_{s=1}^{n}\gamma_{sn}$, where

$$\gamma_{sn} = \mathbf{E}_{s-1}(\sum_{k=s-1}^{n} \xi_{kn}) - \mathbf{E}_s(\sum_{k=s-1}^{n} \xi_{kn})$$

$$= \sum_{k=s-1}^{n} \iint y[Q_{s-1}^k(dy, \xi_{(s-1)n}) - Q_{s-1}^k(dy, x)]Q_1^{s-1}(dx) - \xi_{sn}$$

$$- \sum_{k=s+1}^{n} \iint y[Q_s^k(dy, \xi_{sn}) - Q_s^k(dy, x)]Q_1^{s-1}(dx).$$

with $\mathbf{E}_s$ being the conditional expectation for the fixed minimal τ-algebra of events with respect to which the random variables $\xi_{ln}, l = \overline{1, s-1}$ are measurable, and with $Q_p^k(A, x) = \mathbf{P}\{\xi_{kn} \in A|\xi_{pn} = x\}$.

Using (5.8.1), we get $|\gamma_{sn}| \le c_1$. Thus, the $\sum_{s=1}^{n}\gamma_{sn}$ are sums of bounded martingale differences. The central limit theorem holds for them if the following well-known conditions hold:

$$\text{plim}_{n\to\infty} c_n^{-1}\sum_{k=1}^{n}(\mathbf{E}_k\gamma_{kn}^2 - \mathbf{E}\gamma_{kn}^2) = 0, \qquad (5.8.4)$$

and for any $\tau > 0$,

$$\mathrm{plim}_{n \to \infty} \, c_n^{-1} \sum_{k=1}^{n} \mathbf{E}_k \gamma_{kn}^2 X(c_n^{-1/2} |\gamma_{kn}| > \tau) = 0, \qquad (5.8.5)$$

where $c_n = \sum_{k=1}^{n} \mathbf{E}\gamma_{kn}^2$.

Since the variables γ_{kn} are bounded and condition (5.8.3) holds, we have (5.8.5). Let us prove (5.8.4). Obviously, $\mathbf{E}_k \gamma_{kn}^2 - \mathbf{E}\gamma_{kn}^2 = f_k(\xi_{k-1})$, where f_k is a Borel bounded function. Then

$$\mathbf{E}(c_n^{-1} \sum_{k=1}^{n} (\mathbf{E}_k \gamma_{kn}^2 - \mathbf{E}\gamma_{kn}^2))^2 = c_n^{-2} \sum_{k,p=1}^{n}$$

$$\times \mathbf{E} f_k(\xi_{k-1n}) f_p(\xi_{p-1n}) \le c_n^{-2} \sum_{p \ge l} \alpha^{p-l},$$

where

$$\alpha = \sup_n \sup_{s=\overline{1,n}} \sup_{x_1, x_2, B} |Q_{ns}(x_1, B) - Q_{ns}(x_2, B)|.$$

Using this relation and (5.8.1)–(5.8.3), we get (5.8.4). The theorem is proved.

CHAPTER 6

LIMIT THEOREMS OF THE LAW OF LARGE NUMBERS AND CENTRAL LIMIT THEOREM TYPES FOR RANDOM DETERMINANTS

Mainly, exact formulas for the distributions of the determinants of random matrices are either unknown or very complex. For this reason, the proofs of limit theorems for random determinants are of great interest. Besides, as a rule, there are a lot of applied problems that involve determinants of high orders.

In this chapter we deal with the following problems: choosing constants a_n and b_n to find the conditions of convergence of the distribution of a sequence of random variables $b_n^{-1}(\ln|\det \Xi_n| - a_n), n = 1, 2, \ldots$, where $\Xi_n = (\xi_{ij}^{(n)})$ is a random square matrix of the order n and the constants are properly chosen to the normal or degenerated distribution. All matrices are assumed to be real.

§1 Limit Theorems of the Law of Large Numbers Type for Random Determinants

Consider a sequence of matrices Ξ_n such that $\mathbf{E}\ln^2|\det \Xi_n|$ exist.
Denote

$$\gamma_k = \mathbf{E}[\ln|\det \Xi_n|/\rho_{k-1}^{(n)}] - \mathbf{E}[\ln|\det \Xi_n|/\rho_k^{(n)}],$$

where $\rho_k^{(n)}$ is the minimal σ-algebra with respect to which the random vector rows $\overrightarrow{\xi}_l, l = \overline{k+1, n}$ are measurable.

It follows from the proof of Theorem 5.4.1 that if for some sequence of constants c_n,

$$\lim_{n \to \infty} c_n^{-2} \sum_{k=1}^{n} \mathbf{E}\gamma_k^2 = 0, \tag{6.1.1}$$

then

$$\mathrm{plim}_{n \to \infty} c_n^{-1}(\ln|\det \Xi_n| - \mathbf{E}\ln|\det \Xi_n|) = 0, \tag{6.1.2}$$

170

and the proof of Theorem 5.4.2 implies that if there exist $\mathbf{E}\gamma_k^4$ and that for some sequence of constants

$$\sum_{n=1}^{\infty}(\sum_{k=1}^{n}\sqrt{c_n^{-4}\mathbf{E}\gamma_k^4})^2 < \infty, \tag{6.1.3}$$

then with probability 1,

$$\lim_{n\to\infty} c_n^{-1}(\ln|\det\Xi_n| - \mathbf{E}\ln|\det\Xi_n|) = 0. \tag{6.1.4}$$

Consider some particular cases, when conditions (6.1.1) and (6.1.3) are fulfilled.

Suppose that the vectors $\overrightarrow{\xi_l}, l = \overline{1,n}$ for each n are independent, that there exist $\mathbf{E}\ln^2|\det\Xi_n|$ and such Borel functions $f_k(x_1,\ldots,x_n)$ that

$$\lim_{n\to\infty} c_n^{-2}\sum_{k=0}^{n-1}\mathbf{E}\ln^2|\alpha_k| = 0, \tag{6.1.5}$$

where

$$\alpha_k = [\xi_{k+11}\Xi_{k+11} + \cdots + \xi_{k+1n}\Xi_{k+1n}][f_k(\Xi_{k+11},\ldots,\Xi_{k+1n})]^{-1},$$

Ξ_{ij} are the cofactors of the elements ξ_{ij} of the matrix Ξ_n. Then (6.1.2) holds.

The proof follows from the inequality

$$\mathbf{E}\gamma_k^2 = \mathbf{E}\{\mathbf{E}[\ln|\alpha_k|/\rho_{k-1}^{(n)}] - \mathbf{E}[\ln|\alpha_k|/\rho_k^{(n)}]\}^2 \le 2\mathbf{E}\ln^2|\alpha_k|.$$

Theorem 6.1.1. *For each n let the random variables $\xi_{pl}^{(n)}, p,l = \overline{1,n}$ be independent and distributed according to the stable laws with the characteristic functions*

$$\mathbf{E}\exp\{it\xi_{pl}^{(n)}\} = \exp\{-c_{pl}^{(n)}|t|^{\alpha_p}\}, c_{pl}^{(n)} \ge 0, \tag{6.1.6}$$

$$\sum_{l=1}^{n} c_{pl}^{(n)} \ne 0, \quad p = \overline{1,n}, \quad 0 \le \alpha_p \le 2;$$

$$\lim_{n\to\infty} c_n^{-2}\sum_{k=1}^{n}\mathbf{E}\ln^2|\eta_k| = 0, \tag{6.1.7}$$

where $\eta_k, k = \overline{1,n}$ are independent random variables, distributed according to the stable laws with the characteristic functions $\exp\{-|t|^{\alpha_k}\}$. Then the statement (6.1.2) is true.

Proof. Condition (6.1.6) implies (see §2, Chapter 4) that $\mathbf{E}\ln^2|\det\Xi_n|$ exists. Consequently, choosing f_k of the form

$$f_k = (\sum_{i=1}^{n} c_{k+1i}^{(n)}|\Xi_{k+1i}|^{\alpha_k+1})^{1/\alpha_k+1}$$

and using (6.1.5), we come to the statement of the Theorem 6.1.1.

Theorem 6.1.2. *For each n let the random vector rows $\overrightarrow{\xi_l}$ of the matrix Ξ_n be stochastically independent and distributed according to the normal law with nondegenerated covariance matrices and finite mathematical expectations. Then with probability 1,*

$$\lim_{n \to \infty} n^{-1}(\ln|\det \Xi_n| - \mathbf{E}\ln|\det \Xi_n|) = 0.$$

Proof. Obviously,

$$\mathbf{E}\gamma_k^4 = \mathbf{E}[\mathbf{E}\{\ln|\eta_{nk} + a_{nk}|/\rho_{k-1}^{(n)}\} - \mathbf{E}\{\ln|\eta_{nk} + a_{nk}|/\rho_k^{(n)}\}]^4,$$

where

$$\eta_{nk} = b_{nk}^{-1}\sum_{l=1}^{n}(\xi_{kl}^{(n)} - \mathbf{E}\xi_{kl}^{(n)})\Xi_{kl},$$

$$a_{nk} = b_{nk}^{-1}\sum_{l=1}^{n}(\mathbf{E}\xi_{kl}^{(n)})\Xi_{kl},$$

$$b_{nk} = (\sum_{l,p=1}^{n}\Xi_{kl}\Xi_{kp}r_{klp})^{1/2},$$

$$r_{klp} = \mathbf{E}\xi_{kl}^{(n)}\xi_{kp}^{(n)} - \mathbf{E}\xi_{kl}^{(n)}\mathbf{E}\xi_{kp}^{(n)}.$$

The random variables η_{nk} are distributed according to the normal law $N(0,1)$ and η_{np} is independent of a_{np}. Therefore,

$$\mathbf{E}\gamma_k^4 \le 16\mathbf{E}\ln^4|(\eta_{nk} + a_{nk})/f(a_{nk})|,$$

where

$$f(u) = \begin{cases} 1, & |u| < 1; \\ u, & |u| \ge 1; \end{cases}$$

as $|a_{nk}| < 1$, then $\sup_n \sup_{k=\overline{1,n}} \mathbf{E}\ln^4|\eta_{nk} + a_{nk}| < \infty$ and as $|a_{nk}| \ge 1$,

$0 < \varepsilon < 1$, we have

$$
\mathbf{E}\ln^4|\eta_{nk}|a_{nk}|^{-1} + 1| = (2\pi)^{-1/2}\mathbf{E}\int_{|x|/|a_{nk}|<\varepsilon} \ln^4|x|a_{nk}|^{-1}
$$

$$
+ 1|\exp(-x^2/2)dx + (2\pi)^{-1/2}\mathbf{E}\int_{L_1} \ln^4|1 + x/|a_{nk}||
$$

$$
\times \exp(-x^2/2)dx + (2\pi)^{-1/2}\mathbf{E}\int_{L_2} \ln^4|1 + x/|a_{nk}||
$$

$$
\times \exp(-x^2/2)dx \le -\ln^4(1-\varepsilon) + (2\pi)^{-1/2}
$$

$$
\times \int_{|y|\ge\varepsilon, |1+y|\le 1} \ln^4|1 + y|\exp(-a_{nk}^2 y^2/2)|a_{nk}|dy
$$

$$
+ \mathbf{E}|1 + \eta_{nk}/|a_{nk}||^4 \le -\ln^4(1-\varepsilon) + \varepsilon^{-2}\int_{|y|\le 1} \ln|y|dy
$$

$$
+ \mathbf{E}|1 + \eta_{nk}/|a_{nk}||^4 < \infty,
$$
$$
L_1 = \{|x| \ge \varepsilon|a_{nk}|, |1 + x/|a_{nk}|| \le 1\},
$$
$$
L_2 = \{|x| \ge \varepsilon|a_{nk}|, |1 + x/|a_{nk}|| \ge 1\}.
$$

From this and (6.1.3) the statement of Theorem 6.1.2 follows.

Analogous statements are true in the case when the elements of the matrix Ξ_n are some measurable functions of the random variables $\eta_i, i = \overline{1, N_n}$.

Let us divide the set of these variables into nonintersecting sets $K_i, i = \overline{1, m_n}$. Let $\beta_l^{(n)}$ be the smallest σ- algebra with respect to which the elements of the sets $K_i, i = \overline{1, m_n}$ are measurable.

We present $\ln|\det\Xi_n|$ in the form

$$
\ln|\det\Xi_n| - \mathbf{E}\ln|\det\Xi| = \sum_{l=1}^{m_n}\gamma_l,
$$

where

$$
\gamma_l = \mathbf{E}\{\ln|\det\Xi_n|/\beta_{l-1}^{(n)}\} - \mathbf{E}\{\ln|\det\Xi_n|/\beta_l^{(n)}\}.
$$

If for such sums (6.1.1) and (6.1.3) hold, then either do (6.1.2) or (6.1.4), respectively.

Theorem 6.1.3. *Let $H_{m_n n}$ be random $m_n \times n$ matrices $m_n \ge n$, which column vectors $\overrightarrow{\xi_l}, l = \overline{1, m_n}$ for each n are independent, B_n be a sequence of positive-definite nonrandom matrices of the order n and for some sequence of constants c_n,*

$$
\lim_{n\to\infty} c_n^{-2}\sum_{s=1}^{m_n}\mathbf{E}\ln^2(1 + (B_n^{-1}\overrightarrow{\xi_s}, \overrightarrow{\xi_s})) = 0. \tag{6.1.8}
$$

Then

$$\text{plim}_{n\to\infty} c_n^{-1}[\ln\det(B_n + H_{m_n n}H'_{m_n n}) - \mathbf{E}\ln\det(B_n + H_{m_n n}H'_{m_n n})] = 0.$$
(6.1.9)

Proof. It follows from §5, Chapter 2 that

$$\det(B_n + H_{m_n n}H'_{m_n n}) = \det(B_n + \sum_{l\neq s} K_n^l)[1 + (R_s\vec{\xi_s}, \vec{\xi_s})], \qquad (6.1.10)$$

where

$$R_s = (B_n + \sum_{l\neq s} K_n^l)^{-1}, \qquad K_n^l = (\xi_{il}\xi_{jl}).$$

Obviously,

$$(R_s\vec{\xi_s}, \vec{\xi_s}) \le (B_n^{-1}\vec{\xi_s}, \vec{\xi_s}). \qquad (6.1.11)$$

By (6.1.10) and (6.1.11) $\mathbf{E}\ln\det(B_n + H_{m_n n}H'_{m_n n})$ exists.

Let $\rho_k^{(n)}, k = \overline{0,n}$ be the minimal σ-algebra with respect to which the vectors $\vec{\xi_l}, l = \overline{k+1, m_n}$ are measurable. Then

$$\ln\det(B_n + C_n) - \mathbf{E}\ln\det(B_n + C_n) = \sum_{k=1}^{m_n} \mu_k,$$

where

$$\mu_k = \mathbf{E}\{\ln\det(B_n + C_n)/\rho_{k-1}^{(n)}\}$$
$$- \mathbf{E}\{\ln\det(B_n + C_n)/\rho_k^{(n)}\},$$
$$C_n = H_{m_n n}H'_{m_n n}.$$

Using (6.1.10) and (6.1.1) we obtain

$$\mathbf{E}\mu_k^2 \le \mathbf{E}\ln^2(1 + (B_n^{-1}\vec{\xi_s}, \vec{\xi_s})).$$

Thus Theorem 6.1.3 is proved.

The technique of the proof of Theorem 6.1.3 based on (6.1.10) is called the method of random matrices perturbation.

§2 The Perturbation Method

We use the notations introduced in the previous paragraph.

In accordance with the proof of Theorem 5.4.3, if

$$\underline{\lim}_{n \to \infty} \sum_{k=1}^{m_n} \mathbf{E}\gamma_k^2 > 0, \tag{6.2.1}$$

$$\lim_{n \to \infty} c_n^{-1} \sum_{k=1}^{m_n} \mathbf{E}|\mathbf{E}\gamma_k^2 - \mathbf{E}(\gamma_k^2/\theta_k^{(n)})| = 0, \tag{6.2.2}$$

where $\theta_s^{(n)}$ is the smallest σ-algebra with respect to which the random elements of sets $K_i, i \neq s, i = \overline{1, m_n}$ are measurable (see Chapter 6, §1); $c_n = \sum_{k=1}^{n} \mathbf{E}\gamma_k^2$ and satisfy the Lindeberg condition: for every $\tau > 0$,

$$\lim_{n \to \infty} \sum_{k=1}^{n} \int_{|x|>\tau} x^2 d\mathbf{P}\{\gamma_k c_n^{-1/2} < x\} = 0, \tag{6.2.3}$$

then

$$\lim_{n \to \infty} \mathbf{P}\{c_n^{-1}[\ln|\det \Xi_n| - \mathbf{E}\ln|\det \Xi_n|] < x\}$$
$$= (2\pi)^{-1/2} \int_{-\infty}^{x} \exp(-y^2/2) dy.$$

In many concrete cases it is easy to verify condition (6.2.3), for example, by using the Ljapunov condition. It is significantly more difficult to verify condition (6.2.2). In this paragraph we consider one special case of random matrices for which conditions (6.2.2) and (6.2.3) are satisfied.

Consider a sequence of random matrices $\Xi_n = I + B_n + HH'_n$, where B_n are nonnegative-definite matrices of order n, and $H_n = (\xi_{ij}^{(m)})$ are random matrices of dimension $n \times m_n, m_n \geq n$.

Theorem 6.2.1. *For each* $n = 1, 2, \ldots,$ *suppose the random elements of the matrix* $H_n = (\xi_{ij}^{(m)})$ *are independent,*

$$\lim_{h \downarrow 0} \overline{\lim}_{n \to \infty} \sup_{i=\overline{1,m_n}} \mathbf{E}\ln^2(1 + h\sum_{j=1}^{n}\xi_{ji}^2) = 0, \tag{6.2.4}$$

$\mathbf{E}\ln\det \Xi_n$ *and* $\mathrm{Var}\ln\det \Xi_n$ *exist, and*

$$\lim_{n \to \infty} m_n^{-1}\mathrm{Var}\ln\det \Xi_n > 0, \tag{6.2.5}$$

there exists such $0 < \tau < \infty$ *that*

$$\lim_{n \to \infty} \sup_{j=\overline{1,m_n}} \sum_{i=1}^{n}\alpha_{ij}^2 = 0,$$

where

$$\alpha_{ij} = \int_{|x|<\tau} x \, d\mathbf{P}\{\xi_{ij} < x\}$$

and

$$\lim_{n\to\infty} \sqrt{m_n} \sup_{i=\overline{1,n}, j=\overline{1,m_n}} \int x^2(1+x^2)^{-1} d\mathbf{P}\{\xi_{ij} - \alpha_{ij} < x\} = 0. \qquad (6.2.6)$$

Then

$$\lim_{n\to\infty} \mathbf{P}\{[\ln \det \Xi_n - \mathbf{E} \ln \det \Xi_n][\mathbf{V} \ln \det \Xi_n]^{-1/2}$$

$$< x\} = (2\pi)^{-1/2} \int_{-\infty}^{x} \exp(-y^2/2) dy. \qquad (6.2.7)$$

Proof. Obviously,

$$\ln \det \Xi_n - \mathbf{E} \ln \det \Xi_n = \sum_{k=1}^{m_n} \gamma_k,$$

where

$$\gamma_k = \mathbf{E}\{\ln \det \Xi_n / \rho_{k-1}^{(n)}\} - \mathbf{E}\{\ln \det \Xi_n / \rho_k^{(n)}\},$$

$\rho_k^{(n)}$ is the smallest σ-algebra with respect to which the random elements $\xi_{lp}^{(n)}, l = \overline{1,n}, p = \overline{k+1,m}$ of the matrix H_m are measurable.

In view of Theorem 5.4.3, the theorem will be proved if we can show that conditions (6.2.2) and (6.2.3) hold. For this we write the matrix $H_n H_n'$ in the form $H_n H_n' = \sum_{s=1}^{m_n} K_n^s$, where $K_n^s = (\xi_{is}\xi_{js})$ is an $n \times n$ matrix. We use the formula (see Chapter 2, §5)

$$\ln \det(I + C) - \ln \det(I + \tilde{C}) = \ln[1 + (\tilde{R}\vec{\xi_s}, \vec{\xi_s})], \qquad (6.2.8)$$

where C and $\tilde{C}$ are nonnegative-definite $n \times n$ matrices such that $C - \tilde{C} = K_n^s, \tilde{R} = (I + \tilde{C})^{-1}$, and $\vec{\xi_s} = (\xi_{1s}, \ldots, \xi_{ns})'$.

Using (6.2.8), we obtain for the γ_k the expression

$$\gamma_s = \mathbf{E}\{\log \det \Xi_n - \log \det(\Xi_n - K_n^s)|\rho_{s-1}^{(n)}\}$$

$$- \mathbf{E}\{\log \det \Xi_n - \log \det(\Xi_n - K_n^s)|\rho_s^{(n)}\}$$

$$= \mathbf{E}\{\log[1 + (R_s \vec{\xi_s}, \vec{\xi_s})]|\rho_{s-1}^{(n)}\} \qquad (6.2.9)$$

$$- \mathbf{E}\{\log[1 + (R_s \vec{\xi_s}, \vec{\xi_s})]|\rho_s^{(n)}\},$$

where $R_s = (\Xi_n - K_n^s)^{-1} := (r_{ij}^s)$.

Noting (6.2.9), we have

$$\mathbf{E}|\mathbf{E}\gamma_k^2 - \mathbf{E}[\gamma_k^2|\sigma_k^{(n)}]| = \mathbf{E}|\mathbf{E}[\mathbf{E}\{\log[1 + (R_k\overrightarrow{\xi_k}, \overrightarrow{\xi_k})]|\rho_{k-1}^{(n)}\}$$
$$- \mathbf{E}\{\log[1 + (R_k\overrightarrow{\xi_k}, \overrightarrow{\xi_k})]|\rho_k^{(n)}\}]^2$$
$$- \mathbf{E}\{[\mathbf{E}\log(1 + (R_k\overrightarrow{\xi_k}, \overrightarrow{\xi_k}))|\rho_k^{(n)}]]^2|\sigma_k^{(n)}\}|, \quad (6.2.10)$$

where $\sigma_k^{(n)}$ is the smallest σ-algebra with respect to which random elements $\xi_{ij}, i = \overline{1, n}, j \neq k, j = \overline{1, m_n}$ of matrix H_m are measurable.

The following integral representation is valid for

$$\ln[1 + (R_k\overrightarrow{\xi_k}, \overrightarrow{\xi_k})]$$
$$= -\int_1^A [\int_0^\infty \exp\{-y(R_k\overrightarrow{\xi_k}, \overrightarrow{\xi_k}) - yx\}dy]dx$$
$$+ \ln[A + (R_k\overrightarrow{\xi_k}, \overrightarrow{\xi_k})]$$
$$= -\mathbf{E}\{\int_1^A [\int_0^\infty \exp\{i\sqrt{y}\sum_{l=1}^n \eta_l\xi_{lk} - yx\}dy]dx|R_k,$$
$$\overrightarrow{\xi_k}\} + \ln[A + (R_k\overrightarrow{\xi_k}, \overrightarrow{\xi_k})], \quad (6.2.11)$$

where $\overrightarrow{\eta} = (\eta_1, \ldots \eta_n)$ is a random vector which for fixed matrix R_k distributed according to the normal law $N(0, 2R_k)$, and A is a constant greater than 1.

By substituting (6.2.11) into (6.2.10), we obtain

$$\mathbf{E}|\mathbf{E}\gamma_k^2 - \mathbf{E}(\gamma_k^2|\sigma_k^{(n)})| \leq 2\mathbf{E}|\mathbf{E}\int_1^A \int_1^A dx_1 dx_2$$
$$\times \int_0^\infty \int_0^\infty \exp\{i\sum_{l=1}^n (\sqrt{y_1}\eta_l + \sqrt{y_2}\eta_l^{k-1})\xi_{lk} - y_1 x_1 - y_2 x_2\}$$
$$\times dy_1 dy_2 - \mathbf{E}\{\int_1^A \int_1^A dx_1 dx_2 \int_0^\infty \int_0^\infty \exp\{i\sum_{l=1}^n (\sqrt{y_1}\eta_l$$
$$+ \sqrt{y_2}\eta_l^{k-1})\xi_{lk} - y_1 x_1 - y_2 x_2\}dy_1 dy_2|\rho_k, \overrightarrow{\rho_k}\}|$$
$$+ 2\mathbf{E}|\mathbf{E}\int_1^A \int_1^A dx_1 dx_2 \int_0^\infty \int_0^\infty \exp\{i\sum_{l=1}^n (\sqrt{y_1}\eta_l + \sqrt{y_2}\eta_l^k)\xi_{lk}$$
$$- y_1 x_1 - y_2 x_2\}dy_1 dy_2 - \mathbf{E}\{\int_1^A \int_1^A dx_1 dx_2 \int_0^\infty \int_0^\infty \exp\{i$$
$$\times \sum_{l=1}^n (\sqrt{y_1}\eta_l + \sqrt{y_2}\eta_l^k)\xi_{lk} - y_1 x_1 - y_2 x_2\}dy_1 dy_2|\rho_k, \overrightarrow{\rho_k}\}| + \delta_k, \quad (6.2.12)$$

where $\delta_k = 4\mathbf{E}\ln^2[1 + A^{-1}(\vec{\xi_k}, \vec{\xi_k})]$, $\vec{\eta}^{(k-1)} = (\eta_l^{(k-1)}, l = \overline{1,n})$ is the random vector, which for fixed matrices $\tilde{R}_k$ ad R_k is independent of the matrix H and the vector $\vec{\eta}$ and distributed according to the normal law $N(0, 2\tilde{R}_k)$, matrix $\tilde{R}_k$ is equal to $(I + B_n + \sum_{l=1}^{n} \tilde{K}_n^l + \sum_{s=k+1}^{m_n} K_n^s)^{-1}$, the matrices independent of the matrices K_n^s and distributed as the matrix K_n^l, $\bar{p}_{k-1}$ is the smallest σ-algebra with respect to which random elements of the matrices $\tilde{K}_n^l, l = k, m_n$ are measurable.

Let us prove a lemma.

Lemma 6.2.1. *For each n, let the random elements $\xi_{ij}^{(n)}$ of the matrix $H_n = (\xi_{ij}^{(n)})$ be independent and infinitesimal, and conditions (6.2.4) and (6.2.6) hold. Then*

$$\mathbf{E} \int_1^A \int_1^A dx_1 dx_2 \int_0^\infty \int_0^\infty \exp\{i \sum_{l=1}^n (\sqrt{y_1}\eta_l + \sqrt{y_2}\eta_l^{k-1})\xi_{lk}$$

$$- y_1 x_1 - y_2 x_2\} dy_1 dy_2 = \mathbf{E} \int_1^A \int_1^A dx_1 dx_2$$

$$\times \int_0^\infty \int_0^\infty \exp\{\sum_{l=1}^n \int (\exp[i(\sqrt{y_1}\eta_l + \sqrt{y_2}\eta_l^{k-1})u] - 1)$$

$$\times d\mathbf{P}\{\xi_{lk} - \alpha_{lk} < u\} - y_1 x_1 - y_2 x_2\} dy_1 dy_2 + 0(1).$$

$$(6.2.13)$$

Proof. From condition (6.2.4) if follows that

$$\lim_{h \to \infty} \overline{\lim}_{n \to \infty} \mathbf{P}\{\sum_{j=1}^n \xi_{ij}^2 \geq h\} = 0,$$

and so the random variables μ_{ij} are infinitesimal. But

$$\mathbf{P}\{|\nu_{11}(\sum_{j=1}^n \xi_{ij}^2)^{1/2}|^2 < x\} = \mathbf{P}\{|\sum_{j=1}^n \mu_{ji}\nu_{ji} + \sum_{j=1}^n \alpha_{ji}\nu_{ji}|^2 < x\},$$

where $\mu_{ij} = \xi_{ij} - \alpha_{ij}$, the $\nu_{ji}(i, j = 1, 2, \dots)$ are independent random variables distributed according to the normal law $N(0, 1)$ that do not depend on the random variables ξ_{ji}, and by virtue of (6.2.14) we obtain

$$\lim_{h \to \infty} \overline{\lim}_{n \to \infty} \mathbf{P}\{|\sum_{j=1}^n \xi_{ij}\nu_{ij}| \geq h\} = 0. \qquad (6.2.14)$$

By using (6.2.14) and (5.3.19), we obtain

$$\sup \sum_{j=1}^n \mathbf{E}(\mu_{ij}^2(1 + \mu_{ij}^2)^{-1}) < \infty. \qquad (6.2.15)$$

Noting

$$\mathbf{E}(\sum_{l=1}^{n} \eta_l \alpha_{lk})^2 = \mathbf{E}(R_k \overrightarrow{\alpha_k}, \overrightarrow{\alpha_k}) \le (\overrightarrow{\alpha_k}, \overrightarrow{\alpha_k}),$$

$$\mathbf{E}(\sum_{l=1}^{n} \eta_l^{k-1} \alpha_{lk})^2 = \mathbf{E}(\tilde{R}_k \overrightarrow{\alpha_k}, \overrightarrow{\alpha_k}) \le (\overrightarrow{\alpha_k}, \overrightarrow{\alpha_k}),$$

where $\overrightarrow{\alpha_k} = (\alpha_{lk}, l = \overline{1,n})$ and (6.2.6), we find that

$$\mathbf{E} \int_1^A \int_1^A dx_1 dx_2 \int_0^\infty \int_0^\infty \exp\{i \sum_{l=1}^{n} (\sqrt{y_1} \eta_l + \sqrt{y_2} \eta_l^{k-1})$$

$$\times \, \xi_{lk} - y_1 x_1 - y_2 x_2\} dy_1 dy_2$$

$$= \mathbf{E} \int_1^A \int_1^A dx_1 dx_2 \int_0^\infty \int_0^\infty dy_2 dy_1 \exp\{i \sum_{l=1}^{n} (\sqrt{y_1} \eta_l$$

$$+ \sqrt{y_2} \eta_l^{k-1})(\xi_{lk} - \alpha_{lk}) - y_1 x_1 - y_2 x_2\}.$$

On the basis of this relation, (6.2.15) and Theorem 5.3.4, we have (6.2.13). Lemma 6.2.1 is proved.

We write

$$f_l = \int (\exp\{i\sqrt{y_1} \eta_l + \sqrt{y_2} \eta_l^{k-1})u\} - 1) d\mathbf{P}\{\mu_{lk} < u\}.$$

Assume that y_1 and y_2 are some finite fixed numbers, $y_1 \ge 0, y_2 \ge 0$.

Lemma 6.2.2. *For each fixed y_1 and y_2*

$$\text{plim}_{n \to \infty} |\sum_{l=1}^{n} (f_l - \mathbf{E}\{f_l | R_k, \tilde{R}_k\})| = 0. \tag{6.2.16}$$

Proof. After some simple calculations, we obtain

$$\mathbf{E}\{|\sum_{l=1}^{n}(f_l - \mathbf{E}\{f_l | R_k, \tilde{R}_k\})|^2 | R_k, \tilde{R}_k\}$$

$$= \sum_{p,l=1}^{n} \iint \exp[-(y_1 r_{ll}^k + y_2 \tilde{r}_{ll}^{k-1})u^2 - (y_1 r_{pp}^k + y_2 \tilde{r}_{pp}^{k-1})v^2$$

$$- 2(y_1 r_{lp} + y_2 r_{lp}^{k-1})uv] dF_{lk}(u) dF_{pk}(v)$$

$$- \sum_{p,l=1}^{n} [\int \exp[-(y_1 r_{ll}^k + y_2 \tilde{r}_{ll}^{k-1})u^2]$$

$$\times dF_{lk}(u) \int \exp[-(y_1 r_{pp}^k + y_2 \tilde{r}_{pp}^{k-1})v^2] dF_{pk}(v)],$$

$$F_{pk}(v) = \mathbf{P}\{\mu_{pk} < v\}.$$

Let $\theta_{lp} = y_1 r_{lp}^k + y_2 \tilde{r}_{lp}^{k-1}$. We represent this sum in the following form:

$$2|\sum_{p,l=1}^{n} \iint [\rho_{lp}(u,v)]dF_{lk}(u)dF_{pk}(v)|$$

$$+ 2|\sum_{p,l=1}^{n} \iint \theta_{lp}uv(1+u^2)^{-1}(1+v^2)^{-1}dF_{lk}(u)dF_{pk}(v)$$

$$= I_1 + I_2, \tag{6.2.17}$$

where

$$I_1 = 2\sum_{l=1}^{n} \int [\sum_{p=1}^{n} \int \rho_{lp}(u,v)(1+u^2)u^{-2}dG_p^{(n)}(u)]$$

$$\times (1+v^2)v^{-2}dG_l^{(n)}(v),$$

$$\rho_{lq}(u,v) = \mathbf{E}(\exp\{i(\sqrt{y_2}\eta_l + \sqrt{y_1}\eta_l^{k-1})u\}$$
$$- i(\sqrt{y_1}\eta_l + \sqrt{y_2}\eta_l^{k-1})u(1+u^2)^{-1} - 1)(\exp\{i(\sqrt{y_1}\eta_l$$
$$+ \sqrt{y_2}\eta_l^{k-1})v\} - i(\sqrt{y_1}\eta_l + \sqrt{y_2}\eta_l^{k-1})v(1+v^2)^{-1} - 1).$$

If $u = 0 (v = 0)$, the expression $\rho_{lp}(u,v)$ is equal to

$$- 0.5u^2 \mathbf{E}(\sqrt{y_1}\eta_l + \sqrt{y_2}\eta_{l-1}^{k-1})^2(\exp\{i\sqrt{y_1}\eta_l + \sqrt{y_2}\eta_l^{k-1})v\}$$
$$- i(\sqrt{y_1}\eta_l + \sqrt{y_2}\eta_l^{k-1})v(1+v^2)^{-1} - 1),$$

$$- 0.5v^2 \mathbf{E}(\sqrt{y_1}\eta_l + \sqrt{y_2}\eta_l^{k-1})^2(\exp\{i(\sqrt{y_1}\eta_l + \sqrt{y_2}\eta_l^{k-1})u\}$$
$$- i(\sqrt{y_1}\eta_l + \sqrt{y_2}\eta_l^{k-1})u(1+u^2)^{-1} - 1),$$

$$G_p^{(n)}(u) = \int_{-\infty}^{u} y^2(1+y^2)^{-1}dF_{pk}(y),$$

$$I_2 = 2\sum_{p,l=1}^{n} \iint \theta_{pl}uv(1+u^2)^{-1}(1+v^2)^{-1}dF_{lk}(u)dF_{pk}(v).$$

Since $\sup_{j=\overline{1,n}} \sum_{i=1}^{n}[(r_{ij}^k)^2 + (\tilde{r}_{ij}^k)^2] \leq 2$ for every $\varepsilon > 0$, there are just a finite number c_ε of the $(r_{ij}^k)^2, (\tilde{r}_{ij}^k)^2, i = \overline{1,n}$ greater than ε (c_ε is independent of n and j); all the rest are less than ε. Obviously,

$$\sup_{u,v} |\rho_{lp}(u,v)| \leq [1 + c(y_1 + y_2)]^2.$$

Suppose for definiteness that

$$(r_{lq}^k)^2 > \varepsilon, \quad (\tilde{r}_{lq}^k)^2 > \varepsilon, \quad q = \overline{1,c_\varepsilon},$$
$$(r_{lq}^k)^2 \leq \varepsilon, \quad (\tilde{r}_{lq}^k)^2 \leq \varepsilon, \quad q = \overline{c_\varepsilon + 1, n},$$

for all $l = \overline{1,n}$. Otherwise we can renumber the r_{lq}^k. Noting the fact that $\lim_{n\to\infty} \sup_{p=\overline{1,n}} G_p^{(n)}(+\infty) = 0$, we conclude that $\lim_{n\to\infty} I_1 = 0$.

Let us prove a lemma.

Lemma 6.2.3 [101].

$$\sup_{n} \sup_{i=\overline{1,n}} \sum_{j=1}^{n} |\mathbf{E}\mu_{ij}(1+\mu_{ij}^2)^{-1}| < \infty. \qquad (6.2.18)$$

Proof. Obviously,

$$\sum_{j=1}^{n} |\mathbf{E}\mu_{ij}(1+\mu_{ij}^2)^{-1}| \leq \sum_{j=1}^{n} |\int_{|x|<\tau} x dF_{ij}(x)$$

$$- \int_{|x|<\tau} x^3(1+x^2)^{-1}dF_{ij}(x) + \int_{|x|\geq\tau} x(1+x^2)^{-1}dF_{ij}(x)|$$

$$\leq \sum_{j=1}^{n} |\int_{|x|<\tau} x dF_{ij}(x)| + \sum_{j=1}^{n} \mathbf{E}\mu_{ij}^2(1+\mu_{ij}^2)^{-1}(1+\tau^{-1}). \qquad (6.2.19)$$

Let us prove that

$$\sup_{n} \sum_{i=1}^{n} |\int_{|x|<\tau} x dF_{ij}(x)| < \infty.$$

Consider the following expression:

$$\sum_{j=1}^{n} |\int_{|x|<\tau} x dF_{ij}(x)| \leq \sum_{j=1}^{n} |\int_{|x|<\tau} (x - a_{ij})dF_{ij}(x)|$$

$$+ \sum_{j=1}^{n} |\int_{|x-\alpha_{ij}|<\tau, |x|\geq\tau} (x - \alpha_{ij})dF_{ij}(x)$$

$$- \int_{|x-\alpha_{ij}|\geq\tau, |x|<\tau} (x - \alpha_{ij})dF_{ij}(x)| \leq \sum_{j=1}^{n} |\alpha_{ij}|$$

$$\times \int_{|x|\geq\tau} dF_{ij}(x) + \tau \sum_{j=1}^{n} \int_{|x|\geq\tau} dF_{ij}(x)$$

$$+ 2(1+\tau^2)\tau^{-1} \sum_{j=1}^{n} \mathbf{E}\mu_{ij}^2(2+\mu_{ij}^2)^{-1}. \qquad (6.2.20)$$

For every $0 < \varepsilon < \tau$ beginning with some $n \geq n_0, \sup_n \sup_{j=\overline{1,n}} |\alpha_{ij}| < \varepsilon$. Therefore, as $n \geq n_0$,

$$\int_{|x|\geq\tau} dF_{ij}(x) \leq \int_{|x-\alpha_{ij}|\geq\tau-\varepsilon} dF_{ij}(x) \leq [1+(\tau-\varepsilon)^2](\tau-\varepsilon)^{-2}\mathbf{E}\mu_{ij}^2(1+\mu_{ij}^2)^{-1}.$$

From this inequality and (6.2.20) it follows that

$$\sup_{n} \sum_{j=1}^{n} |\int_{|x|<\tau} x dF_{ij}(x)| \leq c \sum_{j=1}^{n} \mathbf{E}\mu_{ij}^2(1+\mu_{ij}^2)^{-1},$$

where $c > 0$ is some constant.

The statement of Lemma 6.2.3 follows from (6.2.19) and (6.2.15).

By using Lemma 6.2.3, it is easy to show that $\lim_{n\to\infty} I_2 = 0$ for all fixed y_1, y_2.

We need another lemma.

Lemma 6.2.4. *For all fixed $y_1, y_2(y_1, y_2 \geq 0)$,*

$$\mathrm{plim}_{n\to\infty} \left| \sum_{l=1}^{n} (\mathbf{E}\{f_l/R_k, \tilde{R}_k\} - \mathbf{E}f_l) \right| = 0.$$

Proof. From §5, Chapter 2, we obtain the formula

$$r_{pp} - r_{pp}^s = -(\sum_{i=1}^{n} r_{ip}^{ks}\xi_{is})^2 [1 + (R_{ks}\overrightarrow{\xi}_s, \overrightarrow{\xi}_s)]^{-1}, \qquad (6.2.21)$$

where r_{pp}^{ks} are the elements of the matrix $R_{ks} = (I + B_n + \sum_{p\neq k,s} K_n^p)^{-1}$.

In the same way, equality is true for $\tilde{r}_{pp}^k - \tilde{r}_{pp}^{ks}$. It is obvious that

$$\mathbf{E}\{f_l/R_k, \tilde{R}_k\} - \mathbf{E}f_l = \sum_{s\neq k} (\mathbf{E}\{g_l/\Delta_{s-1}\} - \mathbf{E}\{g_l/\Delta_s\})$$
$$+ \mathbf{E}\{\varphi_l/R_k, \tilde{R}_k\} - \mathbf{E}\varphi_l,$$

where Δ_s is a minimal σ-algebra with respect to which the elements of matrices $K_n^p, \tilde{K}_n^p, p = \overline{s+1, m_n}, \varphi_l = f_l - g_l$ are measurable, $g_l = \int_{|x|<\tau} \exp\{-(y_1 r_{ll}^k + y_2 r_{ll}^k)u^2\} - 1]dF_{kl}(y)$.

Using this equality, we obtain

$$\mathbf{E}[\sum_{l=1}^{n} (\mathbf{E}\{f_l/R_k, \tilde{R}_k\} - \mathbf{E}f_l)]^2 \leq 2 \sum_{s\neq k} [\sum_{l=1}^{n} (\mathbf{E}\{g_l/\Delta_{s-1}\}$$

$$- \mathbf{E}\{g_l/\Delta_s\})]^2 + L_n = 2 \sum_{s\neq k} [\sum_{l=1}^{n} (\mathbf{E}\{(\exp\{-(y_1 r_{ll}^k$$

$$+ y_2\tilde{r}_{ll}^k)\mu_{lk}^2\} - \exp\{-y_1 r_{ll}^{ks} + y_2\tilde{r}_{ll}^{ks})\mu_{lk}^2\}X(|\mu_{lk}| < \tau)/\Delta_{s-1}\}$$

$$- \mathbf{E}\{[\exp\{-(y_1 r_{ll}^k + y_2\tilde{r}_{ll}^k)\mu_{lk}^2\} - \exp\{-(y_1 r_{ll}^{ks}$$

$$+ y_2\tilde{r}_{ll}^{ks})\mu_{lk}^2\}]X(|\mu_{lk}| < \tau)/\Delta_s\})]^2 + L_n$$

$$\leq 4 \sum_{s\neq k} \mathbf{E}[\sum_{l=1}^{n} \{y_1|r_{ll}^k - r_{ll}^{ks}| + y_2|\tilde{r}_{ll}^k - r_{ll}^{ks}|\}$$

$$\times \mathbf{E}\mu_{kl}^2(1 + \mu_{kl}^2)^{-1}(\tau^2 + 1)]^2 + L_n, \qquad (6.2.22)$$

where

$$L_n = 2\mathbf{E}[\sum_{l=1}^{n} \int_{|x|>\tau} (\exp\{-(y_1 r_{ll}^k) + y_2 \tilde{r}_{ll}^k)x^2\} - 1)dF_{kl}(x)]^2.$$

It is obvious that

$$L_n \leq (1 + \tau^{-2}) \int_{|x|>\tau} \sum_{l=1}^{n} x^2 (1 + x^2)^{-1} dF_{kl}(x).$$

Therefore, since (6.2.15) is true, it follows

$$\lim_{\tau \to \infty} \overline{\lim}_{n \to \infty} L_n = 0. \qquad (6.2.23)$$

Using (6.2.21), we obtain

$$\sum_{l=1}^{n} |r_{ll}^k - r_{ll}^{ks}| \leq (R_{ks}^2 \overrightarrow{\xi}_s, \overrightarrow{\xi}_s)[1 + (R_{ks} \overrightarrow{\xi}_s, \overrightarrow{\xi}_s)]^{-1} \leq 1$$

$$- [1 + \sum_{i=1}^{n} (1 + \lambda_i)^{-1} y_i^2]^{-1},$$

where $y_i = \sum_{p=1}^{n} h_{ip} \xi_{ps}$, h_{ip} are the components of eigenvectors of the matrix $B_n + \sum_{p \neq s} K_p$, and λ_i are its eigenvalues.

If $\lambda_i \geq 0$, then $\sum_{l=1}^{n} |r_{ll}^s - r_{ll}^{ks}| \leq 1$.

Analogously, we obtain $\sum_{l=1}^{n} |r_{ll}^k - r_{ll}^{ks}| \leq 1$. Considering these inequalities and conditions (6.2.7) and (6.2.23), we can prove Lemma 6.2.4.

Without any changes in the proof of Lemma 6.2.4 (except for trivial cases), we prove that

$$\text{plim}_{n \to \infty} |\sum_{l=1}^{n} (\mathbf{E}\{f_l/\rho_k, \overline{\rho_k}\} - \mathbf{E}f_l)| = 0$$

for all fixed y_1 and y_2.

By using Lemmas 6.2.4, 6.2.2, 6.2.1, and condition (6.2.4) from (6.2.12), we obtain

$$\lim_{n \to \infty} \sup_{k=\overline{1,m_n}} \mathbf{E}|\mathbf{E}\gamma_k^2 - \mathbf{E}(\gamma_k^2/\sigma_k^{(n)}) = 0.$$

As soon as (6.2.5) holds, (6.2.2) is valid. Validity of the condition (6.2.3) is obvious. Therefore (6.2.7) holds, and the theorem is proved. Suppose that for elements of each column of matrix H_n the condition of Ljapunov is true. Let us prove the next theorem.

Theorem 6.2.2. *For each n let the random elements of matrix $H_n = (\xi_{ij}^{(n)})$ be independent, $\mathbf{E}\xi_{ij}^{(n)} = 0, \operatorname{Var}\xi_{ij}^{(n)} = \sigma_{ij}^2, \sup_n \sup_{ij} n\sigma_{ij}^2 < \infty$, for certain $0 < \delta < 1$,*

$$\sup_n \sup_{i=\overline{1,n}, j=\overline{1,m_n}} \mathbf{E}|\xi_{ij}\sqrt{n}|^{2+\delta} < \infty, \qquad (6.2.24)$$

$$\sup_n \sup_{i=\overline{1,n}, j=\overline{1,m_n}} \mathbf{E}|(\xi_{ij}^2 - \sigma_{ij}^2)n|^{2+\delta} < \infty,$$

$$\lim_{n\to\infty} c_n > 0, \qquad (6.2.25)$$

where

$$c_n = \sum_{s=1}^n [1 + \mathbf{E}\sum_{i=1}^n r_{ii}^s \sigma_{is}^2]^{-1}[\sum_{i=1}^n \mathbf{E}r_{ii}^s \tilde{r}_{ii}^s \mathbf{E}(\xi_{is}^2 - \sigma_{is}^2)^2$$
$$+ \sum_{i\neq j} \mathbf{E}r_{ij}^s \tilde{r}_{ij}^s \sigma_{is}^2 \sigma_{js}^2],$$

where $\tilde{r}_{ij}^s$ are the elements of the matrix $(I + B_n + \sum_{p=1}^{s-1} \tilde{K}_n^p + \sum_{p=s+1}^{mn} K_n^p)^{-1}$, the matrices $\tilde{K}_n^p$ are independent of each other, do not depend on matrices K_n^p, and are distributed like the matrices K_n^p,

$$\lim_{n\to\infty} m_n n^{-1} < \infty. \qquad (6.2.26)$$

Then

$$\lim_{n\to\infty} \mathbf{P}\{[\ln\det\Xi_n - \mathbf{E}\ln\det\Xi_n]c_n^{-1/2} < x\}$$
$$= (2\pi)^{-1/2} \int_{-\infty}^x \exp(-y^2/2)dy. \qquad (6.2.27)$$

Proof. From (6.2.9) it follows that

$$\ln\det\Xi_n - \mathbf{E}\ln\det\Xi_n = \sum_{s=1}^{m_n}[\mathbf{E}\{\ln[1 + (R_s\overrightarrow{\xi}_s, \overrightarrow{\xi}_s)]/\rho_{s-1}^{(n)}\}$$

$$- \mathbf{E}\{\ln[1 + (R_s\overrightarrow{\xi}_s, \overrightarrow{\xi}_s)]/\rho_s^{(n)}\}] = \sum_{s=1}^{m_n}[\mathbf{E}\{\ln[1$$

$$+ \eta_s^{(n)}]/\rho_{s-1}^{(n)}\} - \mathbf{E}\{\ln[1 + \eta_s^{(n)}/\rho_s^{(n)}\}], \qquad (6.2.28)$$

where

$$\eta_s^{(n)} = [1 + \sum_{i=1}^n r_{ii}^s \sigma_{is}^2]^{-1}(\sum_{i=1}^n r_{ii}^s(\xi_{is}^2 - \sigma_{is}^2) + \sum_{i\neq j} r_{ij}^s \xi_{is}\xi_{js}).$$

According to (6.2.24) and $(R_s \vec{\xi}_s, \vec{\xi}_s) \leq (\vec{\xi}_s, \vec{\xi}_s)$,

$$\mathbf{E}|\eta_s^{(n)}|^{2+\delta} \leq 2^{1+\delta} \mathbf{E}|\sum_{i=1}^{n} r_{ii}^s(\xi_{is}^2 - \sigma_{is}^2)|^{2+\delta}$$

$$+ 2^{1+\delta}\mathbf{E}|\sum_{i \neq j} r_{ij}^s \xi_{is}\xi_{js}|^{2+\delta} = 2^{1+\delta}\mathbf{E}(\sum_{i,j=1}^{n} r_{ij}^s r_{ii}^s$$

$$\times (\xi_{is}^2 - \sigma_{is}^2)(\xi_{js}^2 - \sigma_{js}^2)) \sum_{i=1}^{n}(r_{ii}^s)^\sigma |\xi_{is}^2 - \sigma_{is}^2|^\sigma$$

$$+ 2^{1+\sigma}\mathbf{E}(\sum_{j \neq i, p \neq l} r_{ij}^s r_{pl}^s \xi_{is}\xi_{jp}\xi_{ps}\xi_{ls})(\sum_{i=1}^{n} \xi_{is}^2)^\sigma. \tag{6.2.29}$$

For $|r_{ij}| \leq 1$, $n^{-1}\sum_{i,j} r_{ij}^2 \leq 1$, then from (6.2.29) by using (6.2.26), if n is sufficiently large, we obtain

$$\mathbf{E}|\eta_s^{(n)}|^{2+\sigma} \leq 2^{1+\sigma}\sum_{i=1}^{n} \mathbf{E}|\xi_{is}^2 - \sigma_{is}^2|^{2+\delta}$$

$$+ 8\mathbf{E}\sum_{i \neq j \neq l} r_{ij}r_{il}\xi_{js}\xi_{ls}\sum_{i=1}^{n}|\xi_{is}|^{2\sigma}$$

$$+ 2^{1+\delta}\sum_{i \neq j}(r_{ij}^s)^2\xi_{is}^2\xi_{js}|\sum_{i \neq j} r_{ij}^s\xi_{is}\xi_{js}|^\delta \leq c_1 n^{-1-\delta}.$$

Hence, using (6.2.26) we have

$$\lim_{n \to \infty} \sum_{s-1}^{m_n} \mathbf{E}|\eta_s^{(n)}|^{2+\delta} = 0. \tag{6.2.30}$$

But then by using (6.2.28), it follows that

$$\text{plim}_{n \to \infty}[\ln \det \Xi_n - \mathbf{E}\ln \det \Xi_n - \sum_{s=1}^{m_n}\{\mathbf{E}[\eta_s^{(n)}$$

$$- 0.5(\eta_s^{(n)})^2/\rho_{s-1}^{(n)}] - \mathbf{E}[\eta_s^{(n)} - 0.5(\eta_s^{(n)})^2/\rho_s^{(n)}]\} = 0. \tag{6.2.31}$$

Denote

$$\delta_s = \mathbf{E}[\eta_s^{(n)} - 0.5(\eta_s^{(n)})^2/\rho_{s-1}^{(n)}] - \mathbf{E}[\eta_s^{(n)} - 0.5(\eta_s^{(n)})^2/\rho_s^{(n)}], \quad s = \overline{1, m_n}.$$

It is easy to check that $\mathbf{E}[\delta_s/\sigma_s] = 0, \mathbf{E}[\delta_s\delta_p/\sigma_s = 0], s < p$, where σ_s is a minimal σ-algebra, with respect to which the vector $\vec{\xi}_s$ is measured. Besides due to the strength of (6.2.30),

$$\lim_{n \to \infty} \sum_{s=1}^{m_n} \mathbf{E}|\delta_s|^{2+\delta} = 0.$$

If we prove that

$$\lim_{n \to \infty} \sum_{s=1}^{m_n} \mathbf{E}|\mathbf{E}(\delta_s^2/\sigma_s) - \mathbf{E}\delta_s^2| = 0, \qquad (6.2.32)$$

and then having (6.2.31), the theorem will be proved.

Since $\mathbf{E}|\eta_s^{(n)}|^{2+\delta} \le cn^{-1-\delta}$ (n is sufficiently large), then it is obvious that

$$\mathbf{E}(\delta_s^2/\sigma_s) = \mathbf{E}\{[1 + \sum_{i=1}^{n} r_{ii}^s \sigma_{is}^2]^{-1}[1 + \sum_{i=1}^{n} \tilde{r}_{ii}^s \sigma_{is}^2]^{-1}$$

$$\times \{\sum_{i=1}^{n} r_{ii}^s \tilde{r}_{ii}^s \mathbf{E}(\xi_{is}^2 - \sigma_{is}^2)^2 + \sum_{i \ne j} r_{ij}^s \tilde{r}_{ij}^s \sigma_{is}^2 \sigma_{js}^2\}$$

$$/\rho_{s-1}, \tilde{\rho}_{s-1}\} + \varepsilon_n n^{-1},$$

where ε_n is a random variable such that $\lim_{n \to \infty} \mathbf{E}|\varepsilon_n| = 0$, and $\tilde{\rho}_{s-1}$ is a minimal σ-algebra with respect to which the matrices $\tilde{K}_n^p, p = \overline{1, s-1}$ are measurable.

Lemma 6.2.5.

$$\mathbf{P} \lim_{n \to \infty} [\sum_{i=1}^{n} r_{ii}^s \sigma_{is}^2 - \sum_{i=1}^{n} \mathbf{E}r_{ii}^2 \sigma_{is}^2] = 0.$$

Proof.   Just as in Lemma 6.2.4 we have:

$$\sum_{i=1}^{n} r_{ii}^s \sigma_{is}^2 - \sum_{i=1}^{n} \mathbf{E}r_{ii}^s \sigma_{is}^2 = \sum_{p \ne s}\{\mathbf{E}[\sum_{i=1}^{n} r_{ii}^s \sigma_{is}^2/\Delta_{p-1}]$$

$$- \mathbf{E}[\sum_{i=1}^{n} r_{ii}^s \sigma_{is}^2/\Delta_p]\} = \sum_{p \ne s}\{\mathbf{E}[\sum_{i=1}^{n} (r_{ii}^s$$

$$- r_{ii}^{ps})\sigma_{is}^2/\Delta_{p-1}] - \mathbf{E}[\sum_{i=1}^{n} (r_{ii}^s - r_{ii}^{ps})\sigma_{is}^2/\Delta_p]\}.$$

From this and by using (6.2.21), we have

$$\mathbf{E}(\sum_{i=1}^{n} r_{ii}^s \sigma_{is}^2 - \sum_{i=1}^{n} \mathbf{E}r_{ii}^s \sigma_{is}^2)^2 \le \sum_{p \ne s} \max_i \sigma_{is}^4 \mathbf{E}[\sum_{i=1}^{n} (r_{ii}^s - r_{ii}^{ps})]^2.$$

Then, $\sum_{i=1}^{n} |r_{ii}^s - r_{ii}^{ps}| \le 1$, the statement of Lemma 6.2.5 is true.
   Analogously, we prove that

$$\text{plim}_{n \to \infty} n \sum_{i=1}^{n} [r_{ii}^s \tilde{r}_{ii}^s - \mathbf{E}r_{ii}^s \tilde{r}_{ii}^s]\mathbf{E}(\xi_{is}^2 - \sigma_{is}^2)^2 = 0. \qquad (6.2.33)$$

Lemma 6.2.6.

$$\mathrm{plim}_{n \to \infty} \, n \sum_{i,j=1}^{n} [r_{ij}^{s} \tilde{r}_{ij}^{s} - \mathbf{E} r_{ij}^{s} \tilde{r}_{ij}^{s}] \sigma_{is}^{2} \sigma_{js}^{2} = 0.$$

Proof. Denote $R_s = (r_{ij}^s), \tilde{R}_s = (\tilde{r}_{ij}), G = (\sigma_{is}^2 \delta_{ij})$. Introduce the following,

$$\mathrm{Tr}(R_s G)(\tilde{R}_s G) - \mathbf{E}\,\mathrm{Tr}(R_s G)(\tilde{R}_s G) = \sum_{p \neq s} \{\mathbf{E}[\mathrm{Tr}(R_s G)$$

$$\times (\tilde{R}_s G)/\Delta_{p-1}] - \mathbf{E}\{\mathrm{Tr}(R_s G)(\tilde{R}_s G)/\Delta_p]\} = \sum_{p \neq s} \{\mathbf{E}[\mathrm{Tr}(R_s G)$$

$$\times (\tilde{R}_s G) - \mathrm{Tr}\, R_{sp} G \tilde{R}_{sp} G/\Delta_{p-1}] - \mathbf{E}[\mathrm{Tr}\, R_s G \tilde{R}_s G$$

$$- \mathrm{Tr}\, R_{sp} G \tilde{R}_{sp} G/\Delta_p]\},$$

where $R_{sp} = (\Xi_n - K_n^p - K_n^s)^{-1}, \Delta_{p-1}$ is a minimal σ-algebra with respect to which the elements of matrices K_n^l and $\tilde{K}_n^l, l = \overline{p+1, m_n}$ are measurable.
We shall use

$$\mathrm{Tr}\, B(R_s - R_{sp}) = -(R_{sp} B R_{sp} \overrightarrow{\xi}_p, \overrightarrow{\xi}_p)[1 + (R_{sp} \overrightarrow{\xi}_p, \overrightarrow{\xi}_p)]^{-1}, \qquad (6.2.34)$$

where B is a square matrix of order n.
The proof of this formula follows from the equality

$$\mathrm{Tr}\, B R_s = \frac{\partial}{\partial z} \ln \det[Bz + \Xi_n - K_n^s]_{z=0}$$

and the formula (6.2.8).
By the formula (6.2.34), we obtain

$$|\,\mathrm{Tr}\, R_s G \tilde{R}_s G - \mathrm{Tr}\, R_{sp} G \tilde{R}_{sp} G| = |\,\mathrm{Tr}\, G R_s G[\tilde{R}_s - \tilde{R}_{sp}]$$

$$+ \mathrm{Tr}\, G \tilde{R}_{sp} G[R_s - R_{sp}]| \leq \max_{i=\overline{1,n}} \sigma_{is}^4.$$

As

$$\sum_{i,j=1}^{n} r_{ij}^s \tilde{r}_{ij}^s \sigma_{is}^2 \sigma_{js}^2 = \mathrm{Tr}\, R_s G \tilde{R}_s G,$$

then

$$\lim_{n \to \infty} n^2 \mathbf{E} |\,\mathrm{Tr}\, R_s G \tilde{R}_s G - \mathbf{E}\,\mathrm{Tr}\, R_s G \tilde{R}_s G|^2$$

$$\leq \lim_{n \to \infty} n^2 \sum_{p \neq s} (\max_{i=\overline{1,n}} \sigma_{is}^4)^2 = 0.$$

Lemma 6.2.6 is proved.

If Lemma 6.2.5, and 6.2.6, and the statements (6.2.40) are true, then

$$\lim_{n\to\infty} n\mathbf{E}|\mathbf{E}(\delta_s^2/\sigma_s) - \mathbf{E}\delta_s^2| = 0.$$

Consequently, (6.2.32) is fulfilled. Note that $\sum_{s=1}^{n} \mathbf{E}\delta_s^2 = c_n$ and $0 < \lim_{n\to\infty} c_n < \infty$ (see (6.2.25)). Theorem 6.2.2 is proved.

The proofs of Theorems 6.2.1 and 6.2.2 are examples of the application of the perturbation method for random determinants. Analogous statements can be formulated for matrices like $(I + i\Xi_n), (I + \Xi_n^2)$, where Ξ is a symmetric random matrix, which diagonal elements and all above the diagonal are independent. We should also use the following formula of perturbation of matrices

$$\ln\det(I + i\Xi_n) - \ln\det(I + i\Xi_n^k) = \ln(1 + i\xi_{kk} + ((I + i\Xi_n^k)^{-1}\overrightarrow{\xi_k}, \overrightarrow{\xi_k})).$$

§3 The Orthogonalization Method

Let $\Xi_n = (\xi_{ij})$ be a symmetric square random matrix of order n. Represent $\det\Xi_n^2$ in the following form (see Chapter 4, §1):

$$\det\Xi_n^2 (n!)^{-1} = \prod_{k=1}^{n} \gamma_{n-k+1}, \qquad (6.3.1)$$

where

$$\gamma_n = n^{-1}\sum_{i=1}^{n} \xi_{1i}^2,$$

$$\gamma_{n-k+1} = \sum_{j=k}^{n}(n-k+1)^{-1}(\sum_{L_k} \xi_{kp_1} t_{p_1p_2}^{(1)} t_{p_2p_3}^{(2)} \cdots t_{p_{k-1}j}^{(k-1)})^2,$$

$$k = \overline{2,n};$$

the summation is taken over the following sets of indices,

$$L_k = \{p_1 = \overline{1,n}, p_2 = \overline{2,n}, p_{k-1} = \overline{k-1,n}\};$$

$t_{ij}^{(k)}, i, j = \overline{k,n}$ are the elements of an orthogonal matrix T_k that are measurable with respect to the smallest σ-algebra generated by the random variables $\xi_{pi}, i = \overline{1,n}, p = \overline{1,k}$; the first column of matrix $T_k, k = \overline{2,n}$ is equal to the vector

$$\{\sum_{L_k} \xi_{kp_1} t_{p_1p_2}^{(1)} t_{p_2p_3}^{(2)} \cdots t_{p_{k-1}j}^{(k-1)}(n-k+1)^{-1/2}\gamma_{n-k+1}^{-1/2}, j = \overline{k,n}\},$$

$$\{\xi_{1j}/\sqrt{\gamma_n}, j = \overline{1,n}\}, \quad k = 1,$$

if $\gamma_{n-k+1} \neq 0$, and to an arbitrary nonrandom real unit vector if $\gamma_{n-k+1} = 0$; the first column of the matrix T_1 is equal to the vector

$$n^{-1/2}\{\xi_{1j}\gamma_n^{-1/2}, j = \overline{1,n}\},$$

if $\gamma_n \neq 0$, and to an arbitrary nonrandom real unit vector if $\gamma_n = 0$. We point out that the elements of the random matrix T_k (with the exception of the elements of the first column) have to be random variables that are measurable with respect to the smallest σ-algebra generated by the random variables $\xi_{pi}, i = \overline{1,n}, p = \overline{1,k}$, such that the matrix T_k is orthogonal.

Let us clarify formula (6.3.1). Let $\gamma_1 \neq 0$. Then

$$\frac{\det \Xi_n^2}{n!} = \gamma_n[(n-1)!]^{-1} \det \begin{bmatrix} \frac{\xi_{11}}{\sqrt{\gamma_n n}} & \frac{\xi_{12}}{\sqrt{\gamma_n n}} & \cdots & \frac{\xi_{1n}}{\sqrt{\gamma_n n}} \\ \xi_{21} & \xi_{22} & \cdots & \xi_{2n} \\ \cdots & \cdots & \cdots & \cdots \\ \xi_{n1} & \xi_{n2} & \cdots & \xi_{nn} \end{bmatrix}.$$

By multiplying both sides of this equation by $\det T_1^2$, we have

$$\det \Xi_n^2 (n!)^{-1} = \gamma_n[(n-1)!]^{-1} \det(\eta_{ij})^2, \quad \eta_{ij} = \sum_{k=1}^n \xi_{ik} t_{kj}^{(1)}, \quad i,j = \overline{2,n}.$$

Suppose that

$$\det \Xi_n^2 (n!)^{-1} = [\prod_{k=1}^m \gamma_{n-k+1}] \det H_m^2 [(n-m)!]^{-1},$$

where

$$H_m = (h_{ij})_{i,j=m+1}^n, \quad h_{ij} = \left(\sum_{p_1=\overline{1,n}; p_m=\overline{m,n}} \xi_{ip_1} t_{p_1 p_2}^{(1)} \cdots t_{p_m j}^{(m)} \right).$$

By multiplying both sides of this equation by $\det T_{m+1}^2$, we obtain

$$\det \Xi_n^2 (n!)^{-1} = [\prod_{k=1}^{m+1} \gamma_{n-k+1}] \det H_{m+1}^2 [(n-m-1)!]^{-1}.$$

Continuing this process, we arrive at formula (6.3.1).

Beginning with sufficiently large values of m, one can apply limit theorems to the variables γ_m, since γ_m is the sum of m positive random variables. But it does not appear possible to apply the limit theorems to the variables $\gamma_1, \ldots, \gamma_m$. And this is the defect of the method since the random variables $\gamma_i, i = \overline{1,m}$ may have an essential influence on the behavior of $\det \Xi_n^2$ for large values of n. For example, if $\gamma_1 = 0$ with a certain positive probability p, then deriving lower bounds for $\det \Xi_n^2$ larger than zero for sufficiently large n is out of the question. In order to overcome this difficulty, we replace the distribution of the normalized random determinant $\det \Xi_n$ for sufficiently large n by the distribution of the normalized determinant which last m rows are normally distributed. Then the random variables $\gamma_i, i = \overline{1,m}$ (m does not depend on n) will have a continuous limit distribution. Let us prove the following statement (to simplify some formulas, the index (n) will be dropped from the random variables $\xi_{ij}^{(n)}$).

Theorem 6.3.1. *Suppose that the elements $\xi_{ij}^{(n)}, i, j = \overline{1,n}$ of the random matrices Ξ_n are independent for each value of n, $\mathbf{E}\xi_{ij}^{(n)} = 0$, $\operatorname{Var}\xi_{ij}^{(n)} = 1, i, j = \overline{1,n}, n = 1, 2, \ldots,$*

$$\sup_n \sup_{i,j=\overline{1,n}} \mathbf{E}|\xi_{ij}^{(n)}|^{4+\delta} < \infty, \quad \delta > 0. \tag{6.3.2}$$

Then

$$\sum_{i=1}^n \xi_{1i}\alpha_{1i} \sim \sum_{i=1}^n \nu_{1i}\alpha_{1i}, \tag{6.3.3}$$

where

$$\alpha_{1i} = \begin{cases} A_{1i}(\sum_{i=1}^n A_{1i}^2)^{-1/2}, if & \sum_{i=1}^n A_{1i}^2 \neq 0, \\ n^{-1/2}, & if \quad \sum_{i=1}^n A_{1i}^2 = 0, \end{cases}$$

ν_{1i} are mutually independent random variables independent of the random variables ξ_{ij} that are normally distributed $N(0,1)$, and A_{ij} is the cofactor of the element ξ_{ij} of matrix Ξ_n.

Proof. By the central limit theorem, relation (6.3.3) holds if the following Lindeberg condition is satisfied: for some positive $\tau > 0$,

$$\lim_{n\to\infty} \mathbf{E}[\sum_{i=1}^n \int_{|x|>\tau} x^2 d\mathbf{P}\{\xi_{1i}\alpha_{1i} < x/\alpha_{1i}\}] = 0.$$

It follows from the Ljapunov condition that

$$\lim_{n\to\infty} \mathbf{E} \sum_{i=1}^n |\xi_{1i}\alpha_{1i}|^{2+\delta} = 0. \tag{6.3.4}$$

Indeed, by the familiar proof of the central limit theorem, we have

$$|\mathbf{E}\exp\{is\sum_{k=1}^{n}\xi_{1k}\alpha_{1k}\} - \mathbf{E}\exp\{is\sum_{k=1}^{n}\nu_{1k}\alpha_{1k}\}|$$

$$\leq \sum_{k=1}^{n}\mathbf{E}|\mathbf{E}(\exp\{is\xi_{1k}\alpha_{1k}\}/\alpha_{1k}) - \mathbf{E}(\exp\{is\nu_{1k}\alpha_{1k}\}/\alpha_{1k}|$$

$$\leq c\sum_{k=1}^{n}\mathbf{E}|\xi_{1k}\alpha_{1k}|^{2+\delta}.$$

Therefore, the relation (6.3.3) holds since the random variables $\sum_{k=1}^{n}\nu_{1k}\alpha_{1k}$ are normally distributed $N(0,1)$.

Since (6.3.2) is satisfied, condition (6.3.4) will hold if

$$\lim_{n\to\infty}\sum_{k=1}^{n}\mathbf{E}|\alpha_{1k}|^{2+\delta} = 0. \tag{6.3.5}$$

Note that

$$\sum_{k=1}^{n}A_{1i}^{2} = \mathbf{E}\{\det A_{n}^{2}|\Xi_{n}\},$$

where the matrix A_n is obtained form Ξ_n by replacing its first row by a vector $\vec{\nu}$ that does not depend on the matrix Ξ_n and is normally distributed $N(0,1)$.

Using this as well as formula (2.5.10), we have

$$|\alpha_{11}| = \begin{cases} [\mathbf{E}\{\det A_{n}^{2}(\det A_{11})^{-2}|\Xi_{n}\}]^{-1/2} \\ \quad = [1 + (B_{11}^{-1}\xi_{1},\xi_{1})]^{-1/2}, & \text{if } \sum_{i=1}^{n}A_{1i}^{2} \neq 0, \\ \\ n^{-1/2}, & \text{if } \sum_{i=1}^{n}A_{1i}^{2} = 0, \end{cases} \tag{6.3.6}$$

where $B_{kk} = \Xi_{kk}\Xi_{kk}'$, the matrix Ξ_{kk} is obtained from Ξ_{n} by deleting its kth row and kth column,

$$\xi_{k} = (\xi_{1k}, \dots, \xi_{k-1k}, \xi_{k+1k}, \dots, \xi_{nk}).$$

We shall assume in this formula and the subsequent expressions that the random variables ξ_{ij} have the variances n^{-1}, since all elements of Ξ_{n} appearing in the numerator and denominator of the fraction α_{1s} may be divided by $n^{-1/2}$.

We obtain similar formulas for the variables $\alpha_{1i}, i \neq 1$. To this end, we have to interchange the ith and first columns in the matrix A_n and then place the first column in A_{1i} after its $(i-1)$th column. Thus,

$$
|\alpha_{1i}| = \begin{cases} [1 + (\tilde{B}_{ii}^{-1}\overrightarrow{\eta}_i, \overrightarrow{\eta}_i)^{-1/2}, & \text{if } \sum_{i=1}^{n} A_{1i}^2 \neq 0, \\[2em] n^{-1/2}, & \text{if } \sum_{i=1}^{n} A_{1i}^2 = 0, \end{cases}
$$

the vector η_i is equal to the vector $(\xi_{2i}, \ldots, \xi_{ni})$; $\tilde{B}_{ii} = \tilde{\Xi}_{ii}\tilde{\Xi}'_{ii}$, and the matrix $\tilde{\Xi}_{ii}$ results from Ξ_{ii} by placing its first column after its $(i-1)$th column.

This formula is identical to (6.3.6). Without loss of generality, we may assume that

$$
|\alpha_{1i}| = [1 + (B_{ii}^{-1}\xi_i, \xi_i)]^{-1/2}, \qquad \text{if } \sum_{i=1}^{n} A_{1i}^2 \neq 0,
$$

$$
|\alpha_{1i}| = n^{-1/2}, \qquad \text{if } \sum_{i=1}^{n} A_{1i}^2 = 0.
$$

It is evident that $(B_{kk}^{-1}\xi_k, \xi_k) \geq (R_k(t)\xi_k, \xi_k)$, where $R_k(t) = (It + B_{kk})^{-1}$; t is some real positive constant.

Using this inequality, we obtain

$$
\mathbf{E} \sum_{i=1}^{n} |\alpha_{1i}|^{2+\delta} \leq n^{-\delta/2}\mathbf{P}\{\sum_{i=1}^{n} A_{1i}^2 = 0\}
$$

$$
+ \mathbf{E} \sum_{i=1}^{n} \alpha_{1i}^2 [1 + (R_i(t)\xi_i, \xi_i)]^{-\delta/2}. \tag{6.3.7}
$$

We proved the following inequality,

$$
\mathbf{E}[(R_i(t)\xi_i, \xi_i) - \sum_{s \neq i, s=1}^{n} r_{ss}^{(i)}(t)\xi_{si}^2]^4 \leq cn^{-2}t^{-4}, \tag{6.3.8}
$$

where $r_{ss}^{(i)}(t)$ are elements of the matrix $R_i(t) = (r_{pl}^{(i)}(t)), p, l \neq i, p, l = \overline{1, n}$ and c is a positive constant. Clearly,

$$
\mathbf{E}(\sum_{p \neq l; p, l \neq i} r_{pl}^{(i)}(t)\xi_{pi}\xi_{li})^4
$$

$$
= \mathbf{E} \sum_{\substack{\mu \neq \nu, p \neq l, k \neq m, q \neq s \\ (\mu, \nu, p, l, k, m, q, s \neq i)}} r_{\mu\nu}^{(i)} r_{pl}^{(i)} r_{km}^{(i)} r_{qs}^{(i)} \xi_{\mu i}\xi_{\nu i}\xi_{pi}\xi_{li}\xi_{ki}\xi_{mi}\xi_{qi}\xi_{si}.
$$

This expression is equal to one of the expressions

$$cn^{-4} \sum_{\mu \neq \nu, p \neq \nu, \mu \neq q, p \neq q} r_{\mu\nu}^{(i)} r_{p\nu}^{(i)} r_{\mu q}^{(i)} r_{pq}^{(i)}.$$

Since this expression, wherein c is a positive constant, does not exceed

$$n^{-4} \tilde{c} [\sum (r_{\mu\nu}^{(i)})^2 \sum (r_{p\nu}^{(i)})^2 \sum (r_{\mu q}^{(i)})^2 \sum (r_{pq}^{(i)})^2]^{1/2} \leq (\operatorname{Tr} R_i^2(t))^2 \tilde{c} n^{-4}$$

and

$$\operatorname{Tr} R_i^2(t) = \sum_k (t + \lambda_i)^{-2} \leq nt^{-2},$$

where the λ_i are the eigenvalues of the matrix B_{ii}, (6.3.8) is valid.

Applying the inequality [142]

$$\mathbf{E} |\sum_{i=1}^n \xi_i|^{2+\delta} \leq cn^{\delta/2} \sum_{i=1}^n \mathbf{E} |\xi_i|^{2+\delta},$$

where $0 \leq \delta < 1, \xi_i, i = \overline{1, n}$ are independent random variables, and $\mathbf{E}\xi_i = 0, \mathbf{E}|\xi_i|^{2+\delta} < \infty$, we obtain

$$\mathbf{E} |\sum_{s \neq i, s=1}^n r_{ss}^{(i)}(t)\xi_{si}^2 - n^{-1} \operatorname{Tr} R_i(t)|^{2+\delta} \leq cn^{-1-\delta/2} t^{-2-\delta}. \tag{6.3.9}$$

We also need the next lemma.

Lemma 6.3.1. *For all positive t and $i = \overline{1, n}$,*

$$\mathbf{E}[n^{-1} \operatorname{Tr} R_i(t) - n^{-1}\mathbf{E} \operatorname{Tr} R_i(t)]^4 \leq cn^{-2} t^{-4}. \tag{6.3.10}$$

Proof. Clearly, (see Chapter 5, §4)

$$\operatorname{Tr} R_i(t) - \mathbf{E} \operatorname{Tr} R_i(t) = \sum_{k=1}^n \tilde{\gamma}_k,$$

where

$$\tilde{\gamma}_k = \mathbf{E}_{k-1} \operatorname{Tr} R_i(t) - \mathbf{E}_k \operatorname{Tr} R_i(t),$$

and $\mathbf{E}_k$ is the conditional expectation under the fixed smallest σ-algebra of events generated by the random columns $\eta_s, s = (k+1), \dots, n-1$ of the matrix Ξ_{ii}. We express the matrix B_{ii} in the form $B_{ii} = \sum_{s=1}^{n-1} K_n^s$, where

$K_n^s = (\eta_{is}\eta_{js})$ are $(n-1)$th order square matrices, and the η_{ij} are the elements of Ξ_{ii}. From Chapter 2, §5, it follows that

$$\text{Tr } R_{is}(t) - \text{Tr } R_i(t) = -(d/dt)\ln[1 + (R_{is}(t)\eta_s, \eta_s)]$$
$$= (R_{is}^2(t)\eta_s, \eta_s)[1 + (R_{is}(t)\eta_s, \eta_s)]^{-1},$$

where

$$R_{is}(t) = (It + B_{ii} - K_n^s)^{-1}.$$

This formula leads to

$$|\text{Tr } R_{is}(t) - \text{Tr } R_i(t)| \le t^{-1}. \tag{6.3.11}$$

Applying Theorem 5.4.2 and (6.3.11), we obtain (6.3.10), namely,

$$\mathbf{E}|n^{-1}\text{Tr } R_i(t) - n^{-1}\mathbf{E}\text{Tr } R_i(t)|^4 \le c(\sum_{s=1}^{n} n^{-2}[\mathbf{E}\tilde{\gamma}_s^4]^{1/2})^2$$

$$\le cn^{-4}(\sum_{s=1}^{n}\{\mathbf{E}[\mathbf{E}_{k-1}(\text{Tr } R_i(t) - \text{Tr } R_{is}(t))$$

$$- \mathbf{E}_k(\text{Tr } R_i(t) - \text{Tr } R_{is}(t))]^4\}^{1/2})^2 \le c_1 n^{-2}t^{-4}. \tag{6.3.12}$$

From (6.3.8), (6.3.9), and (6.3.12), it follows, for any $\tau > 0$ and $\varepsilon > 0$, that

$$\lim_{n \to \infty} \sum_{i=1}^{n} \mathbf{P}\{|(R_i(t)\xi_i, \xi_i) - n^{-1}\mathbf{E}\text{Tr } R_i(t)| > \varepsilon\} = 0.$$

According to this relation and the fact that $\sum_{i=1}^{n} \alpha_{1i}^2 = 1$, we find, for any $\tau > 0$ and $\varepsilon > 0$, that

$$\mathbf{E}\sum_{i=1}^{n} \alpha_{1i}^2[1 + (R_i(t)\overrightarrow{\xi_i}\,\overrightarrow{\xi_i})]^{-\delta/2} \le \max_{i=\overline{1,n}}[1 + n^{-1}\mathbf{E}\text{Tr } R_i(t) - \varepsilon]^{-\delta/2} + 0(1). \tag{6.3.13}$$

Find the limit of the following expression $n^{-1}\mathbf{E}\text{Tr } R_i(t)$. For simplicity, we shall consider instead of it the value $n^{-1}\mathbf{E}\text{Tr } R(t)$. It is obvious that $\text{Tr } R(t) = \sum_{k=1}^{n} r_{kk}(t)$, where $r_{kk}(t)$ are the elements of the matrix $R(t)$ (parameter t will be omitted where it is possible). For r_{kk} we have the following formula (Chapter 2, §5)

$$r_{kk} = [t + \sum_{l=1}^{n} \xi_{kl}^2 - \sum_{i,j=1}^{n} r_{ij}^k(\sum_{l,p=1}^{n} \xi_{il}\xi_{kl}\xi_{jp}\xi_{kp})]^{-1}; \tag{6.3.14}$$

r_{ij}^k are the elements of matrix $R_k(t) = (It + H_k H_k')^{-1}$, the matrix H_k is obtained from the matrix Ξ_n by substitution of the elements of kth row by zeroes; $r_{kk}^k = 0, r_{ik}^k = 0, i = \overline{1,n}$.

Since the matrix H_k does not depend on the variables $\xi_{kl}, l = \overline{1,n}$, so

$$\mathbf{E}(\sum_{i,j=1}^n r_{ij}^k(\sum_{l \neq p} \xi_{il}\xi_{kl}\xi_{jp}\xi_{kp}))^2 = \sum_{l \neq p} n^{-2}$$

$$\times \mathbf{E}(\sum_{i,j=1}^n r_{ij}^k \xi_{il}\xi_{jp})^2 \leq^{-2} \mathbf{E} \sum r_{ij}^k r_{\nu\mu}^k \xi_{il}\xi_{jl}$$

$$\times \xi_{\nu l}\xi_{\mu p} = n^{-2}\mathbf{E}\,\mathrm{Tr}(R^k(t)H_k H_k')^2 \leq n^{-1}.$$

Using this inequality and the fact that $\mathrm{plim}_{n \to \infty} \sum_{l=1}^n \xi_{kl}^2 = 1$, we obtain from (6.3.14), that

$$\mathbf{E}r_{kk} = \mathbf{E}[t + 1 - \sum_{l=1}^n \xi_{kl}^2(\sum_{i,j=1}^n r_{ij}^k \xi_{il}\xi_{jl})]^{-1} + 0(1). \qquad (6.3.15)$$

Denote by $K_l^k = (\xi_{il}\xi_{jl})$ the square matrices of order n whose kth row and kth column have zero elements. Represent (6.3.15) in the form

$$\mathbf{E}r_{kk} = \mathbf{E}[t + 1 - \sum_{l=1}^n \xi_{kl}^2(\sum_{i,j=1}^n r_{ij}^k \xi_{il}\xi_{jl})]^{-1} + 0(1).$$

We set $R_k^p(t) := (t + \sum_{l \neq p} K_l^k)^{-1} = (r_{ij}^k(p))$ and consider the equation

$$\mathrm{Tr}\, R_k(t)K_l^k = \mathrm{Tr}\, R_k^l(t)K_l^k + \mathrm{Tr}(R_k(t) - R_l^k(t))K_l^k. \qquad (6.3.16)$$

Obviously,

$$\mathrm{plim}_{n \to \infty}[\mathrm{Tr}\, R_k^l(t)K_l^k - \sum_{i=1}^n r_{ii}^k(l)\xi_{il}^2] = 0. \qquad (6.3.17)$$

Making use of the formula for the differences of resolvents (Chapter 2,§5), we obtain

$$\mathrm{Tr}(R_k(t) - R_k^l(t))K_l^k = -(R_k^l(t)K_l^k R_k^l(t)\overrightarrow{\xi_l}, \overrightarrow{\xi_l})$$

$$\times [1 + (R_k^l(t)\overrightarrow{\xi_l}, \overrightarrow{\xi_l})]^{-1} = -(\sum_{i,p} r_{ip}^k(l)\xi_{il}\xi_{pl})$$

$$\times (\sum_{j,q} r_{qj}^k(l)\xi_{ql}\xi_{jl})[1 + (R_k^l(t)\overrightarrow{\xi_l}, \overrightarrow{\xi_l})]^{-1},$$

where $\overrightarrow{\xi_l} = (\xi_{1l}, \ldots, \xi_{nl})$.

From this formula, (6.3.16), and (6.3.17), the equation

$$\lim_{n \to \infty} \mathbf{E} | \operatorname{Tr} R_k(t) K_l^k - [\sum_{i=1}^{n} r_{ii}^k(l)\xi_{il}^2 - (\sum_{i=1}^{n} r_{ii}^k(l)\xi_{il}^2)^2 [1 + \sum_{i=1}^{n} r_{ii}^k(l)\xi_{il}^2]| = 0$$

follows.

Due to this limit correlation, (6.3.15) takes the following form,

$$n^{-1}\mathbf{E} \operatorname{Tr} R_i(t) = [t + 1 - n^{-1}\mathbf{E} \operatorname{Tr} R_i^k(t) \\ + (n^{-1}\mathbf{E} \operatorname{Tr} R_i^k(t))^2 [1 + n^{-1}\mathbf{E} \operatorname{Tr} R_i^k(t)]^{-1}]^{-1} + 0(1).$$
(6.3.18)

By using formula (6.2.21), we obtain

$$\lim_{n \to \infty} \mathbf{E}|r_{ii}^k(l) - r_{ii}^k| = 0.$$
(6.3.19)

According to the formula (2.5.11), for $p \neq k$,

$$\mathbf{E}|r_{pp} - r_{pp}^k| \le t^{-1} \sum_{l=1}^{n} n^{-1} \sigma_{lk}^2 \mathbf{E}(\sum_{i \neq k} r_{ip}^k \xi_{il})^2 \le t^{-1}n^{-1}c\mathbf{E} \sum_{i=1}^{n} r_{ip}^k b_{pi}, \quad (6.3.20)$$

where b_{pi} are the elements of matrix $R_k(t)H_k H_k'$.

If $\sum_{i=1}^{n}(r_{ip}^k)^2 \le t^{-2}, \sum_{i=1}^{n} b_{pi}^2 \le 1$, it follows from (6.3.20), that for any $t > 0$,

$$\lim_{n \to \infty} \mathbf{E}|r_{pp} - r_{pp}^k| = 0.$$
(6.3.21)

By using (6.3.19) and (6.3.21), we transform (6.3.18) to the following form: $m_n(t) = [t + [1 + m_n(t)]^{-1}]^{-1} + 0(1)$, where $m_n(t) = n^{-1}\mathbf{E} \operatorname{Tr} R_i(t)$. Solving this equation, we find that $m_n(t) = 2(1+0(1))[\{t^2+4t(1+0(1))\}^{1/2}+t]^{-1}$. By letting t go to zero, we obtain (6.3.5) from (6.3.13). Theorem 6.3.1 is proved.

We use the denotations introduced at the beginning of this paragraph.

Theorem 6.3.2. *For each value n let the random elements $\xi_{ij}^{(n)}, i, j = \overline{1, n}$ of matrix Ξ_n be independent, $\mathbf{E}\xi_{ij}^{(n)} = 0, \mathbf{V}\xi_{ij}^{(n)} = 1$, for some $\delta > 0$,*

$$\sup_{n} \sup_{i,j=\overline{1,n}} \mathbf{E}|\xi_{ij}^{(n)}|^{4+\delta} < \infty.$$

Then for any sequence of constants $c_n \to \infty$, as $n \to \infty$,

$$c_n^{-1} \ln[(n!)^{-1} \det \Xi_n^2] \sim c_n^{-1}\{\sum_{k=1}^{n}(\gamma_k - 1) - 0.5 \sum_{k=1}^{n}(\gamma_k - 1)^2\}.$$
(6.3.22)

Proof. Denote by A_{nk} the matrix, whose vector rows are equal to the corresponding vector rows of matrix Ξ_n, except for the last k vector rows, which are independent, do not depend on matrix Ξ_n, and are distributed according to the normal law $N(0,1)$.

By using Theorem 6.3.1, we obtain

$$\lim_{n\to\infty} [\mathbf{E}\exp\{ic_n^{-1}s\ln((n!)^{-1}\det\Xi_n^2)\}$$
$$- \mathbf{E}\exp\{ic_n^{-1}s\ln((n!)^{-1}\det A_{n1}^2)\}] = 0.$$

In this expression, it is assumed that $\exp(i\pm\infty) = 0$.

Applying Theorem (6.3.1) successively m times, we find that

$$\lim_{n\to\infty} [\mathbf{E}\exp\{ic_n^{-1}s\ln((n!)\det\Xi_n^2)\}$$
$$- \mathbf{E}\exp\{ic_n^{-1}s\ln((n!)^{-1}\det A_{nm}^2)\}] = 0. \qquad (6.3.23)$$

The correlation (6.3.23) enables us to consider the matrix Ξ_n instead of matrix A_{nm}. According to matrix A_{nm}, we build variables γ_k. We shall consider that they are defined by formula (6.3.1). Obviously, for any $0 < \varepsilon < 1$ and some $\delta > 0$,

$$\mathbf{P}\{|\gamma_k - 1| < \varepsilon, k = \overline{m,n}\} \geq \mathbf{P}\{\sum_{k=m}^{n} |\gamma_k - 1|^{2+\delta}$$

$$< \varepsilon^{2+\delta}\} \geq 1 - \sum_{k=m}^{n} \mathbf{E}|\gamma_k - 1|^{2+\delta}\varepsilon^{-2-\delta}. \qquad (6.3.24)$$

Let $\gamma_{n-k+1} = \sum_{j=k}^{n}(n-k+1)^{-1}\eta_j^2; \eta_j = \sum_{p=1}^{n}\xi_{kp}\theta_{pj}$, where the random vectors $\overrightarrow{\theta_j} = (\theta_{pj}, p = \overline{1,n}), j = \overline{k,n}$ are orthonormalized and do not depend on random variables ξ_{kp}. Obviously,

$$\gamma_{n-k+1} = \sum_{p,l=1}^{n} a_{pl}\xi_{kp}\xi_{kl},$$

$$a_{pl} = \sum_{j=k}^{n}(n-k+1)^{-1}\theta_{pj}\theta_{lj}.$$

For some $\delta_1 < \delta$, we have

$$\mathbf{E}|\gamma_{n-k+1} - 1|^{2+\delta_1} \leq 2^{1+\delta_1}\mathbf{E}|\sum_{p\neq l}\xi_{kp}\xi_{kl}a_{pl}|^{2+\delta_1}$$

$$+ 2^{1+\delta_1}\mathbf{E}|\sum_{p=1}^{n}(\xi_{kp}^2 - 1)a_{pp}|^{2+\delta_1} \leq 2^{1+\delta_1}$$

$$\times [\mathbf{E}|\sum_{p\neq l}\xi_{kp}\xi_{kl}a_{pl}|^4]^{(2+\delta_1)/4} + 2^{1+\delta_1}\mathbf{E}|\sum_{p=1}^{n}(\xi_{kp}^2 - 1)a_{pp}|^{2+\delta_1}. \qquad (6.3.25)$$

As in the proof of (6.3.8), we obtain

$$\mathbf{E}[\sum_{p\neq l}\xi_{kp}\xi_{kl}a_{pl}]^4 \leq c_1(n-k+1)^{-2}.$$

Using this inequality and (6.3.25), we find that

$$\mathbf{E}|\gamma_{n-k+1}-1|^{2+\delta_1} \leq c_4(n-k+1)^{-1-\delta_1/2}. \tag{6.3.26}$$

Equations (6.3.24) and (6.3.26) imply that

$$\mathbf{P}\{|\gamma_k-1|<\varepsilon, k=\overline{m,n}\} \geq 1 - c_4\varepsilon^{-2-\delta}\sum_{k=m}^{n}k^{-1-\delta_1/2}. \tag{6.3.27}$$

Consider the following expression:

$$\ln((n!)^{-1}\det A_{nm}^2) = \sum_{k=m}^{n}\ln(\gamma_k-1+1)+\sum_{k=1}^{m}\ln\gamma_k. \tag{6.3.28}$$

At the beginning, we estimate the first sum. From (6.3.27) for $0 < \varepsilon_1 < \varepsilon < 1$,

$$\mathbf{P}\{|\sum_{k=m}^{n}\ln(\gamma_k-1+1)-\sum_{k=m}^{n}(\gamma_k-1)+0.5$$
$$\times\sum_{k=m}^{n}(\gamma_k-1)^2|<\varepsilon\} \geq \mathbf{P}\{|\sum_{k=m}^{n}\ln(\gamma_k-1+1)$$
$$-\sum_{k=m}^{n}(\gamma_k-1)+0.5\sum_{k=m}^{n}(\gamma_k-1)^2|$$
$$<\varepsilon/\sum_{k=m}^{n}|\gamma_k-1|^{2+\delta_1}<\varepsilon_1\}\mathbf{P}\{\sum_{k=m}^{n}|\gamma_k-1|^{2+\delta}<\varepsilon_1\}$$
$$\geq 1 - \varepsilon_1^{-1}c\sum_{k=m}^{n}k^{-1-\delta/2}. \tag{6.3.29}$$

The variables $\gamma_k, k=\overline{1,m}$ are independent and distributed according to the $X_k^2, k=\overline{1,m}$ distributions because the last m, vector rows of matrix A_{nm} are independent, do not depend on Ξ_n and are distributed according to the normal law $N(0,I)$ (Chapter 2,§1). Therefore, for every fixed m, $\text{plim}_{n\to\infty} c_n^{-1}\sum_{k=1}^{m}\ln\gamma_k = 0$. Since these variables do not influence the form of the limit distribution in (6.3.23), we shall write $\ln\gamma_k, k=\overline{1,n}$ instead of

$\gamma_k - 1 - (\gamma_k - 1)^2/2$. Equations (6.3.29) and (6.3.23) imply that for any fixed $\varepsilon > 0$,

$$\lim_{n \to \infty} \mathbf{P}[|c_n^{-1} \ln((n!)^{-1} \det \Xi_n^2) - c_n^{-1}\{\sum_{k=1}^{n}(\gamma_k - 1)$$

$$- 0.5 \sum_{k=1}^{n}(\gamma_k - 1)^2\}| < \varepsilon] \geq 1 - \varepsilon_1^{-1}c \sum_{k=m}^{\infty} k^{-1-\delta/2}, \quad 0 < \varepsilon_1 < \varepsilon,$$

where c is some constant. Letting m go to infinity, we come to the assertion of Theorem (6.3.2).

Corollary 6.3.1. *If the conditions of Theorem 6.3.2 are fulfilled then*

$$\mathrm{plim}_{n \to \infty}(c_n')^{-1} \ln[(n!)^{-1} \det \Xi_n^2] = 0, \tag{6.3.30}$$

where c_n' is any sequence satisfying the condition $\lim_{n \to \infty} c_n/\ln n = \infty$.

The proof follows from the following simple remarks: the variables $\gamma_k - 1$ are noncorrelated, $\mathbf{E}(\gamma_k - 1)^2 \leq ck^{-1}$, where $c > 0$ is a constant.
From (6.3.30) we obtain the inequality

$$\lim_{n \to \infty} \mathbf{P}\{\det \Xi_n^2 \geq n!e^{-\varepsilon c_n'}\} = 1, \tag{6.3.31}$$

which is true for any $\varepsilon > 0$ when the conditions of Theorem 6.3.2 are satisfied.

§4 Logarithmic Law

By putting some supplementary bounds on the elements of a random matrix, we shall prove the asymptotic independence of the random variables γ_i introduced in the preceding section. It enables us to prove the central limit theorem for random determinants.

Theorem 6.4.1. *For each n let the random elements $\xi_{ij}^{(n)}, i, j = \overline{1, n}$ of matrix Ξ_n be independent, $\mathbf{E}\xi_{ij}^{(n)} = 0, \mathbf{V}\xi_{ij}^{(n)} = 1, \mathbf{E}[\xi_{ij}^{(n)}]^4 = 3$ for some $\delta > 0$,*

$$\sup_{n} \sup_{i,j=\overline{1,n}} \mathbf{E}|\xi_{ij}^{(n)}|^{4+\delta} < \infty.$$

Then

$$\lim_{n \to \infty} \mathbf{P}\{[\ln \det \Xi_n^2 - \ln(n-1)!](2\ln n)^{-1/2} < x\} = (2\pi)^{-1/2} \int_{-\infty}^{x} \exp(-y^2/2)dy. \tag{6.4.1}$$

Proof. It follows from Theorem 6.3.2 that

$$(2\ln n)^{-1/2}\ln[(n!)^{-1}\det \Xi_n^2] \sim (2\ln n)^{-1/2}\{\sum_{k=1}^{n} k^{-1/2}$$

$$\times [(\gamma_k - 1)\sqrt{k}] - 0.5\sum_{k=1}^{n}[(\gamma_k - 1)^2 - 2k^{-1}] - \sum_{k=1}^{n} k^{-1}\}. \tag{6.4.2}$$

Let $\sigma_k^{(n)}$ be a minimal σ-algebra with respect to which the random vector rows $\overrightarrow{\xi_l}, l \neq k, l = \overline{1,n}$ of the matrix Ξ_n are measurable. Since $\mathbf{E}\xi_{ij} = 0, \mathbf{V}\xi_{ij} = 1, \mathbf{E}\xi_{ij}^4 = 3$, it is easy to verify that $\mathbf{E}\{(\gamma_k - 1)\sqrt{k}]^2/\sigma_k^{(n)}\} = 2, \mathbf{E}|(\gamma_k - 1)\sqrt{k}|^{2+\delta} \leq c < \infty$. Hence, for the variables γ_k, Theorem 5.4.3 is true. From Theorem 5.4.3 we obtain

$$\lim_{n\to\infty} \mathbf{P}\{(2\ln n)^{-1/2}[\sum_{k=1}^{n} k^{-1/2}[(\gamma_k - 1)\sqrt{k}]]$$

$$< z\} = (2\pi)^{-1/2}\int_{-\infty}^{z} \exp(-y^2/2)dy. \tag{6.4.3}$$

By making use of inequality (6.3.26), we find that

$$\mathbf{E}|\sum_{k=1}^{n}[(\gamma_k - 1)^2 - 2k^{-1}]|^{1+\delta'} \leq c\sum_{k=1}^{n} k^{-1-\delta'/2} \leq c_1 < \infty.$$

Therefore,

$$\text{plim}_{n\to\infty}(2\ln n)^{-1/2}\sum_{k=1}^{n}[(\gamma_k - 1)^2 - 2k^{-1}] = 0. \tag{6.4.4}$$

From (6.4.2)–(6.4.4), the validity of (6.4.1) follows. Theorem 6.4.1 is proved.

By using Theorem 6.3.2, we prove the following important corollary.

Corollary 6.4.1. *If the conditions of Theorem 6.3.2 are satisfied, then*

$$\lim_{n\to\infty} \mathbf{P}\{\text{sign}\det \Xi_n = +1\} = 1/2,$$

$$\lim_{n\to\infty} \mathbf{P}\{\text{sign}\det \Xi_n = -1\} = 1/2. \tag{6.4.5}$$

Proof. It follows from (6.3.1) that $(n!)^{-1}\det \Xi_n = \prod_{k=1}^{n-1}\sqrt{\gamma_{n-k+1}}\delta_1$, where $\gamma_{n-k}, k = \overline{1, n-1}$ are defined by formulas (6.3.1), $\delta_1 = \sum_L \xi_{np_1}t_{p_1p_2}^{(1)}t_{p_2p_3}^{(2)}\cdots$ $t_{p_{n-1}n}^{(n-1)}, L = \{p_1 = \overline{1,n}, \ldots, p_{n-1} = \overline{n-1,n}\}$.

Choose the orthogonal matrices T_i in such a way that $\det T_i = 1$.

It follows from (6.3.30) that $\text{plim}_{n\to\infty} \sqrt{n!} \prod_{k=1}^{n-1} \sqrt{\gamma_{n-k+1}} = +\infty$. Therefore the sign $\det \Xi_n$ depends on the sign of the value δ_1. But in the proof of Theorem 6.3.2, the replacement of matrix Ξ_n by matrix A_{mn} is justified. In matrix A_{mn} the last vector rows are independent and distributed normally. Then the limit distribution of δ_1 is standard normal. Thus Corollary 6.4.1 is proved.

Note that in Theorem 6.4.1 the condition $\mathbf{E}[\xi_{ij}^{(n)}]^4 = 3$ is essential. If this condition is not introduced, then $\mathbf{E}\{[(\gamma_k - 1)\sqrt{k}]^2/\sigma_k^{(n)}\} \neq 0$. After some simple calculations it is easy to find that

$$\mathbf{E}\{[(\gamma_{n-k+1} - 1)\sqrt{n - k + 1}]^2/\sigma_{n-k+1}^{(n)}\} = \sum_{p=1}^{n}(n - k + 1)^{-1}$$

$$\times [\mathbf{E}(\xi_{kp}^2 - 1)^2 - 2](\sum_{j=k}^{n} \theta_{pj}^2)^2 + 2,$$

where $\theta_{p_1 j} = \sum t_{p_1 p_2}^{(1)} t_{p_2 p_3}^{(2)} \ldots t_{p_{k-1} j}^{(k-1)}$ and the summation is taken over the set of indices $p_2 = \overline{2, n}, \ldots, p_{k-1} = \overline{k - 1, n}$. Obviously, Theorem 6.4.1 may be generalized.

Theorem 6.4.2. *For each value n let the random elements $\xi_{ij}^{(n)}, i, j = \overline{1, n}$ of matrix Ξ_n be independent, $\mathbf{E}\xi_{ij}^{(n)} = 0, \text{Var}\,\xi_{ij}^{(n)} = 1$, for some $\delta > 0$,*

$$\sup_{n} \sup_{i,j=\overline{1,n}} \mathbf{E}|\xi_{ij}|^{4+\delta} < \infty, \text{plim}_{n\to\infty} \eta_n(\ln n)^{-1/2} = C.$$

Then

$$\lim_{n\to\infty} \mathbf{P}\{[\ln \det \Xi_n^2 - \ln(n - 1)!](2\ln n + \eta_n)^{-1/2}$$

$$< x\} = (2\pi)^{-1/2} \int_{-\infty}^{x} \exp(-y^2/2)dy,$$

where $\eta_n = 0.5 \sum_{k=1}^{n}(n - k + 1)^{-1} \sum_{p=1}^{n}[\mathbf{E}(\xi_{kp}^2 - 1)^2 - 2] \sum_{j=k}^{n} \theta_{pj}^2)^2$.

It is easy to show that the estimations are valid,

$$\min_{k,p}[\mathbf{E}(\xi_{kp}^2 - 1)^2 - 2] \sum_{k=1}^{n} k^{-1} \leq \eta_n \leq \max_{k,p}[\mathbf{E}(\xi_{kp}^2 - 1)^2 - 2] \sum_{k=1}^{n} k^{-1}.$$

§5 The Central Limit Theorem for the Determinants of Random Matrices of Finite Order

The study of the limit distribution of determinants like

$$\det(\sum_{k=1}^{n} \xi_{ik}\xi_{jk})_{i,j=\overline{1,m}} \qquad (6.5.1)$$

(ξ_{ij} are random variables, m does not depend on $n, n \to \infty$) may be applied to the study of the joint distribution of random variables $\nu_{ij}^{(n)} := (a_{ni}b_{nj})^{-1}$ $\sum_{k=1}^{n} \xi_{ik}\xi_{jk}, i,j = \overline{1,m}$, where a_{ni} and b_{nj} are normalizing constants. If under $n \to \infty, \{\nu_{ij}^{(n)}, i,j = \overline{1,m}\} \Rightarrow \{\nu_{ij}, i,j = \overline{1,m}\}$, then

$$\det(\sum_{k=1}^{n} \xi_{ik}\xi_{jk}) \prod_{i=1}^{n} a_{ni}^{-1}b_{ni}^{-1} \Rightarrow \det(\nu_{ij})_{i,j=\overline{1,m}}.$$

But with such an approach, one cannot obtain the central limit theorem.

Show that the variables 6.5.1 are generalized U-statistics, and prove the central limit theorem.

Obviously,

$$\nu_n := \det(\sum_{k=1}^{n} \xi_{ik}\xi_{jk}) = \sum_{\alpha_i \neq \alpha_j} \det[\xi_{i\alpha_j}\xi_{j\alpha_i}]_{i,j=1}^{m}. \qquad (6.5.2)$$

Suppose that the random vectors $(\xi_{1p}, \ldots, \xi_{mp}), p = \overline{1,n}$ are independent, denote $\gamma_s = \mathbf{E}_{(s-1)}\nu_n - \mathbf{E}_{(s)}\nu_n$, $\mathbf{E}_{(s)}$ is the conditional mathematical expectation under the fixed minimal σ-algebra in respect to which random vectors $(\xi_{1p}, \ldots, \xi_{np}), p = \overline{s+1,n}$ are measurable.

By Theorem 5.5.1 we obtain the following statements.

Corollary 6.5.1. *For each n let the random vectors $(\xi_{1p}, \ldots, \xi_{m,p}), p = \overline{1,n}$ be independent and $\mathbf{E}\gamma_s^2$ exist, $s = \overline{1,n}, n = 1, 2, \ldots$,*

$$\underline{\lim}_{n \to \infty} \sum_{s=1}^{n} \mathbf{E}\gamma_s^2 > 0,$$

$$\lim_{n \to \infty} c_n^{-1} \sum_{s=1}^{n} \mathbf{E}|\mathbf{E}\gamma_s^2 - \mathbf{E}(\gamma_s^2/\sigma_s^{(n)})| = 0,$$

where $\sigma_s^{(n)}$ is minimal σ-algebra with respect to which random vectors $(\xi_{1p}, \ldots, \xi_{mp}), p \neq s, c_n = \sum_{s=1}^{n} \mathbf{E}\gamma_s^2$ are measurable and the Lindeberg condition holds: for all $\tau > 0$

$$\lim_{n \to \infty} \sum_{s=1}^{n} \int_{|x|>\tau} x^2 d\mathbf{P}\{c_n^{-1/2}\gamma_\ell < x\} = 0.$$

Then

$$\lim_{n\to\infty} \mathbf{P}\{c_n^{-1/2}(\nu_n - \mathbf{E}\nu_n) < z\} = (2\pi)^{-1/2} \int_{-\infty}^{z} \exp(-y^2/2)dy.$$

Theorem 6.5.1. *For each n let the random variables $\xi_{kp}, k = \overline{1,m}, p = \overline{1,n}$ be independent, identically distributed, $\mathbf{E}\xi_{kp}^4$ exist, for some $\delta > 0$,*

$$\mathbf{E}|\xi_{11}|^{4+\delta} < \infty.$$

Then

$$\lim_{n\to\infty} \mathbf{P}\{[\nu_n - A_n^m(\sigma^{2m} + m\sigma^{2(m-1)}a^2)$$

$$\times [nm^2(A_{n-1}^{m-1})^2 \operatorname{Var}\xi]^{-1/2} < z\} = (2\pi)^{-1/2} \int_{-\infty}^{z} \exp(-y^2/2)dy,$$

where

$$\sigma^2 = \mathbf{E}\xi_{11}^2 - (\mathbf{E}\xi_{11})^2, \quad a^2 = (\mathbf{E}\xi_{11})^2,$$

$$\xi = \xi_{11}^2[\sigma^{2(m-1)} + (m-1)\sigma^{2(m-2)}a^2] - \xi_{11}\sum_{i=2}^{m}\xi_{1i}a^2\sigma^{2(m-2)},$$

$$A_n^m = n(n-1)\ldots(n-m+1), m \geq 2.$$

Proof. Obviously, it follows from (6.5.2) that

$$\mathbf{E}\nu_n = A_n^m \mathbf{E}\det[\xi_{ij}\xi_{ji}]_{i,j=1}^m = A_n^m \det[(1-\delta_{ij})a^2$$

$$+ \delta_{ij}(\sigma^2 + a^2)] = (\sigma^{2m} + m\sigma^{2(m-1)}a^2)A_n^m,$$

$$\gamma_s = \mathbf{E}_{(s-1)}\nu_n^{(s)} - \mathbf{E}_{(s)}\nu_n^{(s)},$$

where $\nu_n^{(s)} = \sum_s \det[\xi_{i\alpha_i}\xi_{j\alpha_j}]_{i,j=1}^m$, the summation is taken over all $\alpha_i = \overline{1,m}, \alpha_i \neq \alpha_j$, and one of the indices α_i is equal to s.

After simple calculations, we find that

$$\mathbf{E}[\nu_n^{(s)}/\sigma_s^{(n)}] = mA_{n-1}^{m-1}\mathbf{E}\{\det[\xi_{ij}\xi_{ji}]_{i,j=1}^m/\xi_{1i}, i = \overline{1,m}\},$$

$$\mathbf{E}\{\det[\xi_{ij}\xi_{ji}]_{i,j=1}^m/\xi_{1i}, i = \overline{1,m}\} = \xi.$$

It is easy to check that $\mathbf{E}(\nu_n - \mathbf{E}\nu_n)^2 = \sum_{s=1}^n \mathbf{E}\gamma_s^2 \leq n(mA_{n-1}^{m-1})^2c, c > 0$ is some constant,

$$\operatorname{plim}_{n\to\infty} \mathbf{E}_{(s-1)}\nu_n^{(s)}[mA_{n-1}^{m-1}]^{-1} = \xi,$$

$$\operatorname{plim}_{n\to\infty} \mathbf{E}_{(s)}\nu_n^{(s)}[mA_{n-1}^{m-1}]^{-1} = \mathbf{E}\xi.$$

Further, the proof is analogous to the proof of corollary 5.5.2. We obtain analogous assertions for the random permanents $\operatorname{per}(\sum_{k=1}^n \xi_{ik}\xi_{jk})_{i,j=\overline{1,m}}$.

CHAPTER 7

ACCOMPANYING INFINITELY DIVISIBLE
LAWS FOR RANDOM DETERMINANTS

The aim of this chapter is to give a general law for the method of proving limit theorems. The method is essentially connected with perturbation and orthogonalization methods. Besides, there also exists the integral representation method. Due to its importance, it is examined in the next chapter.

Briefly, the topic of this chapter is formulated as follows: to find the conditions for the convergence and the general form of the limit laws of distribution for sequences of the random variables $b_n^{-1}\{\ln|\det \Xi_n| - a_n\}$, where a_n, b_n are normalization constants. In general, the answer is obviously trivial—the limit law can be arbitrary. To obtain nontrivial theorems for the general form of the limit laws, it is required that the entries of the matrices Ξ_n are independent and have some identical properties.

§1 Perturbation Method and Accompanying Infinitely Divisible Laws for Random Determinants

Suppose that the random variables γ_l defined in Theorem 6.2.1 are satisfying the conditions of Theorem 5.4.4 or Theorem 5.4.5. Then

$$\lim_{n\to\infty} \left| \mathbf{E}\exp\left\{isc_n^{-1}\sum_{k=1}^{m_n}\gamma_k\right\} - \exp\left\{\sum_{k=1}^{m_n}(\mathbf{E}\exp[is\gamma_k c_n^{-1}] - 1)\right\} \right| = 0. \quad (7.1.1)$$

(see notations of Theorem 6.2.1). Formula (7.1.1) gives the opportunity to prove an important statement. Denote

$$K_n(x) = \sum_{k=1}^{m_n}\int_{-\infty}^{x} y^2 \mathbf{P}\{\gamma_k c_n^{-1} < x\}.$$

Theorem 7.1.1. *If the conditions of Theorem 5.4.5 hold, then for weak convergence of distribution functions of random variables* $[\ln\det\Xi_n - \mathbf{E}\ln\det\Xi_n]$

[**V** ln det Ξ]$^{-1/2}$ *to the limit function it is necessary and sufficient that there exists a nondecreasing function of bounded variation* $K(x)$ *such that* $K_n(x) \Rightarrow K(x)$. *The characteristic function of the limit distribution function is equal to*

$$\exp[\int (e^{isx} - isx - 1)x^{-2}dK(x)], \qquad (7.1.2)$$

$$((e^{isx} - isx - 1)x^{-2} = s^2/2, \qquad as \quad x = 0.$$

Proof.   Since $\mathbf{E}\gamma_k = 0$, then, obviously,

$$\sum_{k=1}^{m_n}(\mathbf{E}\exp\{is\gamma_k c_n^{-1}\} - 1) = \int (e^{isx} - isx - 1)x^{-2}dK_n(x). \qquad (7.1.3)$$

The sufficiency follows from the Helly theorem. Let us prove the necessity. Since $\sup_n K_n(+\infty) \leq 1$, then $K_n(x)$ is the weakly compact sequence. Choose from the sequence $K_n(x)$ a subsequence $K_{n'}(x)$ such that $K_{n'}(x) \Rightarrow K_1(x)$, where $K_1(x)$ is a nondecreasing function of bounded variation. Prove that $K_n(x) \Rightarrow K_1(x)$. Suppose the contrary. Then there exists a subsequence $K_{n''}(x)$ such that $K_{n''}(x) \Rightarrow K_2(x)$ and $K_1(\tilde{x}) \neq K_2(\tilde{x})$ at least at one point of the continuity $\tilde{x}$ of functions $K_1(x)$ and $K_2(x)$. By using the Helly theorem and (7.1.3), we obtain $\int (e^{isx} - isx - 1)x^{-2}dK_1(x) = \int (e^{isx} - isx - 1)x^{-2}dK_2(x)$. Taking the second derivative by s in this equality, we obtain $\int e^{isx}dK_1(x) = \int e^{isx}dK_2(x)$. Hence $K_1(\tilde{x}) = K_2(\tilde{x})$. This leads to a contradiction. Then, $K_n(x) \Rightarrow K_1(x)$. Theorem 7.1.1 is proved.

Consider the case when $\mathbf{E}$ ln det Ξ_n and $\mathbf{V}$ ln det Ξ_n do not exist, using formula

$$\ln \det \Xi_n = \sum_{k=1}^{m_n} \ln[1 + (R_k \overrightarrow{\xi}_k, \overrightarrow{\xi}_k)], \qquad (7.1.4)$$

where $R_k = (I + B_n + \sum_{s=k+1}^{n} K_n^s)^{-1}, K_n^s = (\xi_{is}\xi_{js})$ are the square matrices of n-order, $\overrightarrow{\xi}_s = (\xi_{1s}, \ldots \xi_{ns})$.

If the conditions of Theorem 6.2.1 hold then the quantities $(R_k \overrightarrow{\xi}_k, \overrightarrow{\xi}_k)$ are asymptotically independent.

Denote $c_n^{-1}(\ln[1 + (R_k \overrightarrow{\xi}_k, \overrightarrow{\xi}_k)] - a_k) = \nu_k, \alpha_k = \int_{|x|<\tau} x d\mathbf{P}\{\nu_k < x/\sigma_k^{(n)}\}$, $\tau > 0$ is an arbitrary constant, $\sigma_k^{(n)}$ is a minimal σ-algebra with respect to which all the random vectors $\overrightarrow{\xi}_s, s \neq k, s = \overline{1, m_n}$ are measured, and c_n, a_k is some sequence of constants.

The following statement holds.

Theorem 7.1.2. *If for each n the random vectors $\vec{\xi}_l, l = \overline{1, m_n}$ are indepen-dent, for any finite s,*

$$\lim_{n \to \infty} \sum_{k=1}^{m_n} \mathbf{E}|\mathbf{E}(\exp\{is(\nu_k - \alpha_k)\}/\sigma_k^{(n)}) - 1|^2 = 0; \qquad (7.1.5)$$

$$\lim_{n \to \infty} \sum_{k=1}^{m_n} \mathbf{E}|\mathbf{E}(\exp\{is(\nu_k - \alpha_k)\}/\sigma_k^{(n)})$$
$$- \mathbf{E}\exp\{is(\nu_k - \alpha_k)\}| = 0. \qquad (7.1.6)$$

Then for any finite s,

$$\lim_{n \to \infty} |\mathbf{E}\exp\{isc_n^{-1}[\ln \det \Xi_n - \sum_{k=1}^{m_n} a_k] - is\sum_{k=1}^{m_n} \alpha_k\}$$
$$- \exp\big[\sum_{k=1}^{m_n}(\mathbf{E}\exp\{is(\nu_k - \alpha_k)\} - 1)\big]| = 0.$$

Proof. Consider the inequality

$$|\mathbf{E}\exp\{is\sum_{k=1}^{m_n}(\nu_k - \alpha_k)\} - \exp[\sum_{k=1}^{m_n}(\mathbf{E}\exp\{is(\nu_k - \alpha_k)\}$$
$$- 1)]| \leq \sum_{p=1}^{m_n}|\mathbf{E}\exp\{\sum_{k=1}^{p-1}(\mathbf{E}\exp[is(\nu_k - \alpha_k)] - 1)$$
$$+ is\sum_{k=p}^{m_n}(\nu_k - \alpha_k)\} - \mathbf{E}\exp\{\sum_{k=1}^{p}(\mathbf{E}\exp[is(\nu_k - \alpha_k)]$$
$$- 1) + is\sum_{k=p+1}^{m_n}(\nu_k - \alpha_k)\}| \leq \sum_{p=1}^{m_n}\mathbf{E}|\mathbf{E}(\exp\{is(\nu_p$$
$$- \alpha_p)\}/\sigma_p^{(n)}) - \exp[\mathbf{E}\exp\{is(\nu_p - \alpha_p)\}]|$$
$$\leq \sum_{p=1}^{m_n}\mathbf{E}|\mathbf{E}(\exp\{is(\nu_p - \alpha_p)\}/\sigma_p^{(n)}) - 1]|$$
$$+ \sum_{p=1}^{m_n}\mathbf{E}|\exp[\mathbf{E}(\exp\{is(\nu_p - \alpha_p)\}/\sigma_p^{(n)})$$
$$- 1] - \exp[\mathbf{E}\exp\{is(\nu_p - \alpha_p)\}]|.$$

From this, (7.1.5), and (7.1.6), the statement of the theorem follows.

According to Lemma 5.3.1,

$$\sum_{k=1}^{m_n} \mathbf{E}|\mathbf{E}[\exp\{is(\nu_k - \alpha_k)\}/\sigma_k^{(n)}] - 1|^2$$

$$\leq c(s) \sum_{k=1}^{m_n} \mathbf{E}[1 + (\nu_k - \alpha_k)^2]^{-1}/\sigma_k^{(n)}\}]^2, \qquad (7.1.7)$$

where $c(s)$ is some bounded constant, while all values s are finite.

Rewrite the condition (7.1.6),

$$\sum_{k=1}^{m_n} \mathbf{E}|\mathbf{E}[\exp\{is(\nu_k - \alpha_k)\}/\sigma_k^{(n)}) - \mathbf{E}\exp\{is(\nu_k - \alpha_k)\}|$$

$$\leq \int |e^{isx} - isx(1 + x^2)^{-1} - 1|x^{-2}(1 + x^2)$$

$$\times \mathbf{E}d \sum_{k=1}^{m_n} |\int_{-\infty}^{x} y^2(1 + y^2)^{-1} d\mathbf{P}\{\nu_k - \alpha_k < y/\sigma_k^{(n)}\}$$

$$- \int_{-\infty}^{x} y^2(1 + y^2)^{-1} d\mathbf{P}\{\nu_k - \alpha_k < y\}| + |s| \sum_{k=1}^{m_n} \mathbf{E}|\mathbf{E}\{(\nu_k$$

$$- \alpha_k)[1 + (\nu_k - \alpha_k)^2]^{-1}/\sigma_k^{(n)}\} - \mathbf{E}\{(\nu_k - \alpha_k)$$

$$\times [1 + (\nu_k - \alpha_k)^2]^{-1}\}|. \qquad (7.1.8)$$

Thus, by using (7.1.7) and (7.1.8), we conclude that Theorem 7.1.2 is true if instead of the conditions (7.1.5) and (7.1.6) the following condition holds:

$$\sup_n \sum_{k=1}^{m_n} \mathbf{E}(\mathbf{E}[(\nu_k - \alpha_k)^2[1 + (\nu_k - \alpha_k)^2]^{-1}/\sigma_k^{(n)}])^2 < \infty, \qquad (7.1.9)$$

for almost all x

$$\text{plim}_{n\to\infty} \sum_{k=1}^{m_n} |\int_{-\infty}^{x} y^2(1 + y^2)^{-1} d\mathbf{P}\{\nu_k - \alpha_k < y/\sigma_k^{(n)}\}$$

$$- \int_{-\infty}^{x} y^2(1 + y^2)^{-1} d\mathbf{P}\{\nu_k - \alpha_k < y\} = 0,$$

$$\lim_{n\to\infty} \sum_{k=1}^{m_n} \mathbf{E}|\mathbf{E}((\nu_k - \alpha_k)[1 + (\nu_k - \alpha_k)^2]^{-1}/\sigma_k^{(n)})$$

$$- \mathbf{E}(\nu_k - \alpha_k)[1 + (\nu_k - \alpha_k)^2]^{-1}| = 0. \qquad (7.1.10)$$

Corollary 7.1.1. *If for each n the vectors $\overrightarrow{\xi}_s, s = \overline{1, m_n}$ are independent and the conditions (7.1.9) and (7.1.10) hold, then for the weak convergence of* *the distribution functions of the random variables $c_n^{-1}[\ln \det \Xi_n - \sum_{k=1}^{m_n} a_k] - \sum_{k=1}^{m_n} \alpha_k$ to the limit function it is necessary and sufficient that there exists a nondecreasing function of bounded variation $G(x)$, such that $G_n(x) \Rightarrow G(x)$, where*

$$G_n(x) = \sum_{k=1}^{m_n} \int_{-\infty}^{x} y^2 (1+y^2)^{-1} d\mathbf{P}\{\nu_k - \alpha_k < y\},$$

and

$$\lim_{n \to \infty} \sum_{k=1}^{m_n} \mathbf{E}(\nu_k - \alpha_k)[1 + (\nu_k - \alpha_k)^2]^{-1} = \gamma,$$

where γ is a constant (bounded).

The characteristic function of the limit distribution function equals

$$f(t) = \exp\{is\gamma t + \int (e^{isx} - isx(1+x^2)^{-1} - 1)x^{-2}(1+x^2)dG(x)\}. \quad (7.1.11)$$

Proof. The sufficiency follows from the Helly theorem. Prove the necessity. Obviously,

$$\sum_{k=1}^{m_n} (\mathbf{E}\exp\{is(\nu_k - \alpha_k)\} - 1) = \int (e^{isx} - isx(1+x^2)^{-1} - 1)dG_n(x) + \gamma_n,$$

where

$$\gamma_n = \sum_{k=1}^{m_n} \mathbf{E}(\nu_k - \alpha_k)[1 + (\nu_k - \alpha_k)^2]^{-1}.$$

The conditions (7.1.9) holds. Then from the sequence of functions $G_n(x)$ one can choose a subsequence $G_{n'}(x)$ such that $G_{n'}(x) = G_1(x)$, where $G_1(x)$ is a nondecreasing function of bounded variation.

To finish the proof we show that the characteristic function (7.1.11) one-to-one defines $G(x)$ and γ.

Consider the expression

$$\int_{t-1}^{t+1} \ln f(z)dz - 2\ln f(t) = -2\int e^{itx}(1 - (\sin x)/x)dG(x). \quad (7.1.12)$$

Set

$$\mathbf{V}(u) = 2\int_{-\infty}^{u} (1 - (\sin x)/x^{-2}(1+x^2)dG(x).$$

It is obvious that for all $x, 0 < c_1 \leq (1 - (\sin x)/x)x^{-2}(1+x^2)^{-1} \leq c_2 < \infty$. From this it follows that $\mathbf{V}(u)$ is a nondecreasing function of bounded variation. Function $\varphi(t) = \int_{t-1}^{t+1} \ln f(z)dz - 2\ln f(t)$ one-to-one defines the function $\mathbf{V}(u)$ at the points of continuity and $\mathbf{V}(u)$ one-to-one defines $G(u)$ according to the formula

$$G(x) = \int_{-\infty}^{x} y^2(1+y^2)^{-1}(1-\sin y)/y)^{-1}d\mathbf{V}(y).$$

Thus, the function $G(x)$ is one-to-one determined by the function $f(t)$, and therefore γ is also one-to-one determined by the function $f(t)$. Corollary 7.1.1 is proved.

We can apply Theorems 7.1.1 and 7.1.2 to the determinants of the random matrices of the form $\Xi_n = \sum_{k=1}^{n} \eta_k^{(n)} T_k$, where $\eta_k^{(n)}, k = \overline{1,n}$ are independent positive random variables for every n, $T_k = (\xi_{ik}^{(n)}\xi_{jk}^{(n)})$ are the random square matrices of n order, the random variables $\xi_{ij}^{(n)}, i, j = \overline{1,n}$ are independent for every n and do not depend on $\eta_k, k = \overline{1,n}$. If $\mathbf{E}\ln\det(I+\Xi_n)$ does not exist, we shall use the formula (7.1.4):

$$\ln\det(I+\Xi_n) = \sum_{k=1}^{n}\ln[1+\eta_k(R_k\vec{\xi}_k, \vec{\xi}_k)], \qquad (7.1.13)$$

where $R_k = (I + \sum_{s=k+1}^{n}\eta_s T_s)^{-1}, \vec{\xi}_k = (\vec{\xi}_{1k}, \ldots, \vec{\xi}_{nk})$.
Consider the difference

$$\ln\det(I+\Xi_n) - \sum_{k=1}^{n}\ln[1+\eta_k n^{-1}\mathbf{E}\operatorname{Tr}R_k]$$

$$= \sum_{k=1}^{n}\ln[1 + \{(R_k\vec{\xi}_k, \vec{\xi}_k) - n^{-1}\mathbf{E}\operatorname{Tr}R_k\}\eta_k$$

$$\times [1 + \eta_k n^{-1}\mathbf{E}\operatorname{Tr}R_k]^{-1}].$$

If we require the validity of one of the conditions, for any $\varepsilon > 0$ and some $\delta > 0$,

$$\lim_{n\to\infty}\mathbf{P}\{\sum_{k=1}^{n}|(R_k\vec{\xi}_k, \vec{\xi}_k) - n\mathbf{E}\operatorname{Tr}R_k|^{2+\delta}(n^{-1}\mathbf{E}SR_k)^{-2-\delta} > \varepsilon\} = 0,$$

$$(7.1.14)$$

or if the constants c_n are appropriately chosen,

$$\operatorname{plim}_{n\to\infty} c_n^{-1}\sum_{k=1}^{m_n}\ln[1 + \{(R_k\vec{\xi}_k, \vec{\xi}_k) - n^{-1}$$

$$\times \mathbf{E}\operatorname{Tr}R_k\}\eta_k[1 + \eta_k n^{-1}\mathbf{E}\operatorname{Tr}R_k]^{-1}] = 0, \qquad (7.1.15)$$

then in the first case,

$$\ln \det(I + \Xi) \sim \sum_{k=1}^{n} \ln[1 + \eta_k n^{-1} \mathbf{E} \operatorname{Tr} R_k],$$

and in the second case,

$$c_n^{-1} \ln \det(I + \Xi) \sim c_n^{-1} \sum_{k=1}^{n} \ln[1 + \eta_k n^{-1} \mathbf{E} \operatorname{Tr} R_k]. \tag{7.1.16}$$

Note that the condition (7.1.14) holds if the random variables $\xi_{ij}^{(n)}$ satisfy the following conditions:

$$\mathbf{E}\xi_{ij}^{(n)} = 0, \quad \mathbf{V}\xi_{ij}^{(n)} = n^{-1}\sigma_{ij}^2, \quad \sup_{i,j} \sigma_{i,j}^2 < \infty,$$

$$\sup_{i,j,n} \mathbf{E}[n^{1/2}\xi_{i,j}]^4 < \infty, \quad \lim_{n \to \infty} n^{-1}\mathbf{E} \operatorname{Tr} R_n > 0$$

[see the proof of (6.3.12), (6.3.8), and (6.3.9)].

The random variables $\ln[1 + \eta_k n^{-1} \mathbf{E} \operatorname{Tr} R_k]$ are independent, and therefore, the limit theorems for the sums of the independent random variables can be applied to (7.1.16).

Let us give some examples of the random matrices to which the proofs of the Theorems 7.1.1 and 7.1.2 can be applied.

1) Let $H_n = (\xi_{i,j}^{(n)})$ by symmetric random matrices; then according to the perturbation formulas for the determinants (see Chapter 2, §5),

$$\ln \det(I + i\Xi_n) = \sum_{k=1}^{n} \ln[1 + i\xi_{kk} + (\mathbf{P}_k \overrightarrow{\nu}_k, \overrightarrow{\nu}_k)],$$

where $\mathbf{P}_k = (\delta_{pl} + i\xi_{pl})_{p,l=k+1,n}^{-1}$, $\overrightarrow{\nu}_k = (\xi_{kk+1}, \ldots, \xi_{kn})$.

If $H_n = (h_{ij})_{i,j=1}^{n}$ are dominant matrices, then

$$\ln |\det H| = \sum_{k=1}^{n} \ln |h_{kk} - (Q\overrightarrow{h}_k, \overrightarrow{p}_k)|,$$

where $Q_k = (h_{ij})_{i,j=k+1,n}^{-1}$, $\overrightarrow{h}_k = (h_{kk+1}, \ldots, h_{kn})$, $\overrightarrow{p}_k = (h_{k+1k}, \ldots, h_{nk})$.

2) Let B_n be nonrandom matrices of n order, $\det B_n \neq 0$, and Ξ_n be random matrices of n order. Consider the random variables $\det B_n^{-1} \det(B_n + \Xi) = \det(I + B_n^{-1}\Xi_n) = \prod_{k=1}^{n}(1 + \lambda_k)$, where λ are the eigenvalues of the matrix

$B_n^{-1}\Xi_n$. Obviously, if $\text{plim}_{n\to\infty}\sum_{k=1}^n |\lambda_k|^s = 0$, where $s > 0$ is an integer, then

$$\text{plim}_{n\to\infty}\{\ln|\det(I + B_n^{-1}\Xi)| - \sum_{p=1}^{s-1}(-1)^{p+1}p^{-1}\text{Tr}(B_n^{-1}\Xi_n)^p\} = 0. \quad (7.1.17)$$

If the elements $\xi_{ij}^{(n)}, i, j = \overline{1, n}$ of the matrix Ξ_n are independent, then we apply the limit theorems for the polynomial functions of the independent random variables (see Chapter 5) to the random variables

$$\sum_{p=1}^{s-1}(-1)^{p+1}p^{-1}\text{Tr}(B_n^{-1}\Xi_n)^p.$$

To check the condition $\text{plim}_{n\to\infty}\sum_{k=0}^n |\lambda_k|^s = 0$, we use the Shur theorem (see Chapter 3). Indeed, we represent the matrix A^m, where $m > 0$ is an integer, in the form $A^m = USU^*$, where U is a unitary matrix and S is an upper triangular matrix, on the diagonal of which are the eigenvalues λ_i^m. Then

$$\sum_{i=1}^m |\lambda_i|^{2m} \le \text{Tr } A^m(A')^m.$$

Theorem 7.1.3. *For each n let the random entries $\xi_{ij}^{(n)}, i, j = \overline{1, n}$ be independent, $\text{plim}_{n\to\infty}\sum_{i=1}^n |\lambda_i|^3 = 0$,*

$$\lim_{n\to\infty} \mathbf{P}\{[\text{Tr}(B_n^{-1}\Xi_n) - \mathbf{E}\,\text{Tr}(B_n^{-1}\Xi_n)]$$
$$\times [\mathbf{V}\,\text{Tr }B_n^{-1}\Xi_n]^{-1/2} < z\} = (2\pi)^{-1/2}\int_{-\infty}^z \exp(-y^2/2)dy; \quad (7.1.18)$$

$$\text{plim}_{n\to\infty}[\text{Tr}(B_n^{-1}\Xi_n)^2 - \mathbf{E}\,\text{Tr}(B_n^{-1}\Xi_n)^2]$$
$$\times \mathbf{V}\,\text{Tr }B_n^{-1}\Xi)^{-1/2} = 0, \quad (7.1.19)$$

then

$$\lim_{n\to\infty} \mathbf{P}\{[\ln|\det(B_n + \Xi_n)| - \ln\det B_n - \mathbf{E}\,\text{Tr}(B_n^{-1}\Xi_n)$$
$$+ 0.5\mathbf{E}\,\text{Tr}(B_n^{-1}\Xi_n)^2](\mathbf{V}\,\text{Tr }B_n^{-1}\Xi_n)^{-1/2} < z\} = (2\pi)^{-1/2}$$
$$\times \int_{-\infty}^z \exp(-y^2/2)dy.$$

To check the conditions (7.1.18) and (7.1.19), we use the theorems of Chapter 5.

Note, that Theorem 7.1.3 can be proved also with the method of integral representation. Such proof is more complicated, but it does not need the condition:

$$\text{plim}_{n\to\infty} \sum_{i=1}^{n} |\lambda_i|^3 = 0.$$

As it was noted in Chapter 6, §5, the study of the limit distributions for the random determinants

$$\det \left(\sum_{k=1}^{n} \xi_{ik}\xi_{jk} \right)_{i,j=\overline{1,m}},$$

as $n \to \infty$, where m does not depend on n, can be reduced to the study of the joint limit distribution of random variables $a_{ni}^{-1}b_{nj}^{-1} \sum_{k=1}^{n} \xi_{ik}\xi_{jk}, i,j = \overline{1,m}$, where a_{ni}, b_{ni} are normalizing constants.

In this case the limit distributions are quite complicated and with their help it is difficult to describe the general form of the limit laws of distribution. Let us try to find the general form of the limit laws of distribution and the conditions of convergence to them for the random variables,

$$b_n^{-1}[\det \left(\sum_{k=1}^{n} \xi_{ik}\xi_{jk} \right)_{i,j=\overline{1,m}} - a_n],$$

when the constants a_n and b_n are appropriately chosen.

The proof of the following theorem is analogous to the proof of the necessity of Theorem 6.2.1. The notations are the same as in Corollary 6.5.1.

Theorem 7.1.4. *For each n let the random vectors $\overrightarrow{\xi}_p = (\xi_{1p}, \dots, \xi_{mp}), p = \overline{1,n}$ be independent and $\mathbf{E}\gamma_s^2, s = \overline{1,n}, n = 1, 2, \dots$ exist,*

$$\varlimsup_{n\to\infty} \sum_{s=1}^{n} \mathbf{E}\gamma_s^2 > 0, \quad \lim_{h\to\infty} \lim_{n\to\infty} c_n^{-1} \sum_{s=1}^{n} \mathbf{E}[b_{ns}^{-1}$$
$$\times \gamma_s^2/b_{ns}|\gamma_s| \geq h]\mathbf{P}\{b_{ns}|\gamma_s| \geq h\} = 0,$$
$$\text{plim}_{n\to\infty} b_{ns}^{-1}[\gamma_s - \mathbf{E}(\gamma_s/\sigma_s^{(n)})] = 0, \quad (7.1.20)$$
$$\varlimsup_{n\to\infty} c_n^{-1} \sum_{s=1}^{n} b_{ns}^2 < \infty,$$

where b_{ns} is some sequence of constants, and $\sigma_s^{(n)}$ is a minimal σ-algebra with respect to which the random vectors $\overrightarrow{\xi}_p, p \neq s$ are measurable, $c_n = \sum_{s=1}^{n} \mathbf{E}\gamma_s^2$.

Then for all $s \in [-S, S]$, where $S > 0$ is an arbitrary constant,

$$\lim_{n\to\infty} |\mathbf{E}\exp\{isc_n^{-1} \sum_{k=1}^{n} \gamma_k\} - \exp\left[\sum_{k=1}^{n}(\mathbf{E}\exp\{isc\gamma_k c_n^{-1}\} - 1)\right]| = 0. \quad (7.1.21)$$

By using (7.1.21) we can formulate the statement analogous to Theorem 7.1.1.

From all the conditions of Theorem 7.1.4 the most difficult one to check is (7.1.20). On the basis of (5.5.5) and (5.5.6) we can formulate a simple statement which implies the validity of this condition. Let

$$\gamma_s - \mathbf{E}[\gamma_s/\sigma_s^{(n)}] = \sum_{k=s+1}^{n} \{\mathbf{E}(\gamma_s/\Delta_{k-1}) - \mathbf{E}(\gamma_s/\Delta_k)\},$$

where Δ_k is a minimal σ-algebra with respect to which the random vectors $\vec{\xi}_p, p = \overline{1, s-1}, \overline{s+1, k}, k \geq s+1$ are measurable. If

$$\lim_{n \to \infty} b_{ns}^{-2} \sum_{k=s+1}^{n} \mathbf{E}[\mathbf{E}(\gamma_s/\Delta_{k-1}) - \mathbf{E}(\gamma_s/\Delta_k)]^2 = 0, \qquad (7.1.22)$$

then we obtain (7.1.20).

Obviously (7.1.22) holds if the random variables $\xi_{k,p}, k = \overline{1, m}, p = \overline{1, n}$ are independent, are identically distributed and $\mathbf{E}\xi_{kp}^4$ exist and $b_{ns} = mA_{n-1}^{m-1}$ (see Corollary 6.5.2).

§2 The Method of Orthogonaliztion and Accompanying Infinitely Divisible Laws

Let $\Xi_n = (\xi_{ij}^{(n)})$ be a square random matrix of order n,

$$\nu_n = \begin{cases} \det \Xi_n \left[\sum_{i=1}^{n} \Xi_{1i}^2\right]^{-1/2}, & \text{if } \sum_{i=1}^{n} \Xi_{1i}^2 \neq 0, \\ 0, & \text{if } \sum_{i=1}^{n} \Xi_{1i}^2 = 0. \end{cases}$$

From the proof of Theorem 6.3.1 this statement follows.

Theorem 7.2.1. *For each n let the entries $\xi_{ij}^{(n)}, i, j = \overline{1, n}$ of the random matrices Ξ_n be independent,*

$$\mathbf{E}\xi_{ij}^{(n)} = 0, \qquad \mathbf{V}\xi_{1j}^{(n)} = \sigma_{1j}^2, \qquad \mathbf{V}\xi_{kj}^{(n)} = 1, \quad k = \overline{2, n}, \quad j = \overline{1, n},$$

$$\sup_{n} \sup_{i=\overline{1,n}} \sigma_{1i}^2 < \infty, \qquad for some \quad \delta > 0,$$

$$\sup_{n} \sup_{j=\overline{1,n}, k=\overline{2,n}} \mathbf{E}|\xi_{kj}|^{4+\delta} < \infty.$$

Then $\sum_{i=1}^{n} \xi_{1i}\alpha_{1i} \sim \sum_{l=1}^{n} \gamma_{1l}\alpha_{1l}$, where $\gamma_{1l}^{(n)} l = \overline{1,n}$ *are independent random variables for each* n, *which do not depend on* $\xi_{ij}^{(n)}$, *are distributed according to the infinitely divisible laws and characteristic functions of which are equal to*

$$\exp\{\mathbf{E}\exp(i\xi_{1l}) - 1\}, \quad l = \overline{1,n}.$$

To prove the theorem for any $\varepsilon > 0$, we find

$$\lim_{n\to\infty} \mathbf{P}\{\sup_{i=\overline{1,n}} |\alpha_{1i}| > \varepsilon\} = 0,$$

and use the limit theorems for the sums of the infinitesimal independent random variables (see Chapter 5).

Let $\Xi_n = (\xi_{ij}^{(n)})$ be the square real random matrices of order n. By using the method of orthogonalization, we obtain (see Chapter 6, §3)

$$\det \Xi_n^2 \prod_{k=1}^{n} c_k^{-1} = \prod_{k=1}^{n} \gamma_{n-k+1}, \qquad (7.2.1)$$

where $\gamma_n = c_n^{-1}\sum_{i=1}^{n}\xi_{1i}^2, \gamma_{n-k+1} = \sum_{j=k}^{n} c_{n-k+1}^{-1}(\sum_k \xi_{Lkp_1} t_{p_1}^{(1)} t_{p_2 p_3}^{(2)} \cdots t_{p_{k-1}j}^{(k-1)})^2, k = \overline{2,n}, t_{ij}^{(k)}, i, j = \overline{k,n}$ are the entries of the orthogonal matrix T_k which are measurable with respect to the minimal σ-algebra induced by the random variables $\xi_{pi}, i = \overline{1,n}, p = \overline{1,k}$, and the first column vector of the matrix T_k is equal to the vector

$$\{\sum_{L_k} \xi_{kp_1} t_{p_1 p_2}^{(1)} t_{p_2 p_3}^{(2)} \cdots t_{p_{k-1}j}^{(k-1)}(\gamma_{n-k+1})^{-1/2}, j = \overline{k,n}\}, \quad k = \overline{2,n},$$

$$\{\xi_{1j}/\sqrt{\gamma_n}, j = \overline{1,n}\}, \quad k = 1,$$

if $\gamma_{n-k+1} \neq 0$ and to the arbitrary real independent vector of unity length, if $\gamma_{n-k+1} = 0, c_n$ are some normalizing constants.

From formula (7.2.1) one can see that the random variables γ_{n-k+1} are equal to the nonnegative-definite square forms

$$\sum_{i,j=1}^{n} \xi_{ki}\xi_{kj}\theta_{ijk}^{(n)}, \qquad (7.2.2)$$

where the random variables $\theta_{ijk}^{(n)}$ do not depend on the random variables $\xi_{ki}, i = \overline{1,n}$.

Suppose that the minimal σ-algebra with respect to which the random variables $\theta_{ijk}^{(n)}$ are measurable is fixed, that the elements ξ_{ki} are infinitesimal and centered by their cut mathematical expectations. Therefore, if for the square forms (7.2.2) the conditions of the Theorem 5.3.4 hold, then $\gamma_{n-k+1} \sim \sum_{i,j=1}^{n} \theta_{ijk}^{(n)} \eta_{ik}\eta_{jk}$, where $\eta_{ik}, i = \overline{1,n}$ are independent random variables which do not depend on $\theta_{ijk}^{(n)}$ and are distributed according to the infinitely divisible law.

§3 Central Limit Theorem for Random Permanents

Let $\Xi_n = (\xi_{ij})$ be square random matrices of order n. Suppose that $\mathbf{E}\ln^2|\operatorname{per}\Xi_n|$ exist. Denote $\gamma_k = \mathbf{E}[\ln|\operatorname{per}\Xi_n|/\sigma_{k-1}] - \mathbf{E}[\ln|\operatorname{per}\Xi_n|/\sigma_k]$, where σ_k is a minimal σ-algebra with respect to which all column vectors of matrix Ξ_n, beginning with $k+1$, are measurable. By using the Theorem 5.4.3, we set that if for the variables γ_k the conditions of the Theorem 5.4.3 hold, then

$$\lim_{n\to\infty} \mathbf{P}\{[\ln|\operatorname{per}\Xi_n| - \mathbf{E}\ln|\operatorname{per}\Xi_n|][\mathbf{V}\ln|\operatorname{per}\Xi_n|]^{-1/2}$$
$$< z\} = (2\pi)^{-1/2}\int_{-\infty}^{z}\exp(-y^2/2)dy, \tag{7.3.1}$$

but if for the variables γ_k the conditions of Theorem 5.4.4 hold, then for any finite s,

$$\lim_{n\to\infty} |\mathbf{E}\exp\{isc_n^{-1}[\ln|\operatorname{per}\Xi_n| - \mathbf{E}\ln|\operatorname{per}\Xi_n|]\}$$
$$- \exp\Big[\sum_{k=1}^{n}(\mathbf{E}\exp\{is\gamma_k c_n^{-1}\} - 1)\beta ig]| = 0, \tag{7.3.2}$$

where c_n is some sequence of constants.

Let us find some simple conditions which satisfy (7.3.1).

Theorem 7.3.1. *For each n let the entries $\xi_{ij}^{(n)}$ of the random matrix Ξ_n be independent, with probability 1,*

$$0 < c \le \inf_{i,j,n}\xi_{ij}^{(n)} \le \sup_{i,j,n}\xi_{ij}^{(n)} \le d < \infty,$$
$$\mathbf{E}\xi_{ij}^{(n)} = a, \quad i,j,n = 1,2,\ldots, \inf_{i,j,n}\mathbf{V}\xi_{ij}^{(n)} > 0. \tag{7.3.3}$$

Then

$$\lim_{n\to\infty} \mathbf{P}\{c_n^{-1/2}[\ln\operatorname{per}\Xi_n - \mathbf{E}\ln\operatorname{per}\Xi_n] < z\} = (2\pi)^{-1/2}\int_{-\infty}^{z}\exp(-y^2/2)dy.$$

where $c_n = \sum_{i,j=1}^{n}\sigma_{ij}^2 a^{-2}n^{-2}, \sigma_{ij}^2 = \mathbf{V}\xi_{ij}^{(n)}$.

Proof. Obviously $\gamma_s = \mathbf{E}[\ln(1+\alpha_s)/\sigma_{s-1}] - \mathbf{E}[\ln(1+\alpha_s)/\sigma_s]$, where $\alpha_s = \sum_{i=1}^{n}(\xi_{is}-a)T_{is}(\sum_{i=1}^{n}aT_{is})^{-1})$, and T_{is} is a permanent of the matrix obtained from the matrix Ξ_n by deleting the ith row and the sth column. By using (7.3.3), we check that

$$\lim_{n\to\infty}\sum_{s=1}^{n}\mathbf{E}|\alpha_s|^3 = 0. \tag{7.3.4}$$

Consequently,

$$\text{plim}_{n\to\infty}[\ln \text{ per } \Xi_n - \mathbf{E}\ln \text{ per } \Xi_n - \sum_{s=1}^{n}\beta_s + 0.5\sum_{s=1}^{n}\nu_s] = 0, \qquad (7.3.5)$$

where $\beta_s = \mathbf{E}(\alpha_s/\sigma_{s-1}) - \mathbf{E}(\alpha_s/\sigma_s), \nu_s = \mathbf{E}(\alpha_s^2/\sigma_{s-1}) - \mathbf{E}(\alpha_s^2/\sigma_s)$. From (7.3.4) we obtain $\lim_{n\to\infty}\sum_{s=1}^{n}\mathbf{E}|\beta_s|^3 = 0$.

Thus, for the variables β_s the condition (5.4.5) holds. Let us show that the condition (5.4.4) holds for them. To do this, we make the following transformation:

$$\sum_{s=1}^{n}\mathbf{E}|\mathbf{E}(\beta_s^2/\delta_s) - \mathbf{E}\beta_s^2| = \sum_{s=1}^{n}\mathbf{E}|\mathbf{E}((\mathbf{E}\alpha_s/\sigma_{s-1})^2/\delta_s)$$

$$- \mathbf{E}(\mathbf{E}(\alpha_s/\sigma_{s-1}))^2| = \sum_{s=1}^{n}\mathbf{E}|\mathbf{E}\{\sum_{i=1}^{n}T_{is}\tilde{T}_{is}\sigma_{is}^2 a^{-2}$$

$$\times (\sum_{i=1}^{n}T_{is})^{-1}(\sum_{i=1}^{n}\tilde{T}_{is})^{-1}/\sigma_{s-1}\tilde{\sigma}_{s-1}\} - \mathbf{E}\sum_{i=1}^{n}T_{is}\tilde{T}_{is}\sigma_{is}^2$$

$$\times a^{-2}(\sum_{i=1}^{n}T_{is})^{-1}(\sum_{i=1}^{n}\tilde{T}_{is})^{-1}|, \qquad (7.3.6)$$

where δ_s is a minimal σ-algebra with respect to which all vector columns of the matrix Ξ, except for the sth column, are measurable, $\tilde{T}_{is}$ is a permanent obtained from T_{is} by substituting the entries $\xi_{ij}, i = \overline{1, s-1}, j = \overline{1, n}$ for $\tilde{\xi}_{ij}$ which do not depend on Ξ_n and are distributed like the variables ξ_{ij}, and $\tilde{\sigma}_{s-1}$ is a minimal σ-algebra with respect to which the random variables $\tilde{\xi}_{ij}, i = \overline{1, s-1}, j = \overline{1, n}$ are measurable.

As in the proof of Theorem 2.4.1 for any $\varepsilon > 0$,

$$\lim_{n\to\infty} \sup_{i=\overline{1,n}} \mathbf{P}\{\sup_{j\neq i}[|T_{js}/T_{is} - 1| + |\tilde{T}_{js}/\tilde{T}_{is} - 1| > \varepsilon\} = 0. \qquad (7.3.7)$$

Transform (7.3.6) to the following form:

$$\sum_{s=1}^{n}\mathbf{E}|\mathbf{E}(\beta_s^2/\delta_s) - \mathbf{E}\beta_s^2| \leq \sum_{i,s=1}^{n}\sigma_{is}^2 a^{-2}n^{-2}$$

$$\times |\mathbf{E}\{[n^{-1}\sum_{j=1}^{n}T_{js}/T_{is}]^{-1}[n^{-1}\sum_{j=1}^{n}\tilde{T}_{js}/\tilde{T}_{is}]^{-1}/\sigma_{s-1}, \tilde{\sigma}_{s-1}\}$$

$$- \mathbf{E}[n^{-1}\sum_{j=1}^{n}T_{js}/T_{is}]^{-1}[n^{-1}\sum_{j=1}^{n}\tilde{T}_{js}/\tilde{T}_{is}]^{-1}|.$$

Hence, taking into consideration that $c \leq T_{js}/T_{is} \leq d, c \leq \tilde{T}_{js}/\tilde{T}_{is} \leq d$ and (7.3.7), we obtain

$$\lim_{n \to \infty} \sum_{s=1}^{n} \mathbf{E}|\mathbf{E}(\beta_s^2/\delta_s) - \mathbf{E}\beta_s^2| = 0.$$

Consequently for the variables β_s, the condition (5.4.4) holds.
   Analogously, we prove that

$$\sum_{s=1}^{n} \mathbf{E}\beta_s^2 = \sum_{i,s=1}^{n} \sigma_{is}^2 a^{-2} n^{-2} + 0(1).$$

According to Theorem 5.4.3,

$$\lim \mathbf{P}\{c_n^{-1/2} \sum_{s=1}^{n} \beta_s < z\} = (2\pi)^{-1/2} \int_{-\infty}^{z} \exp(-y^2/2)dy. \qquad (7.3.8)$$

With the help of simple transformations, we obtain

$$\mathbf{E}\Big(\sum_{s=1}^{n} \nu_s\Big)^2 = \sum_{s=1}^{n} \mathbf{E}\nu_s^2 \lambda eq \sum_{s=1}^{n} 4\mathbf{E}\alpha_s^4 \leq c_1 n^{-1},$$

c_1 is some constant.
   By taking into consideration this inequality and (7.3.8), we obtain the statement of Theorem 7.3.1.

CHAPTER 8

INTEGRAL REPRESENTATION METHOD

When solving the problems of science and engineering it is necessary to know the limit distribution of the random variables $\det(I + \Xi_n)$, where $\Xi_n = (\xi_{ij})$ is a random square matrix of order n. This chapter solves the following problem: by a suitable choice of constants a_n, to find the conditions of the convergence and the general form of all the limit distributions for the sequence of the random variables

$$\ln |\det(I + \Xi_n)| - a_n, \operatorname{sign} \det(I + \Xi_n)$$

where the entries $\xi_{ij}, i \geq j$ of the matrix Ξ_n are independent, asymptotically constant, and $\lim_{h \to \infty} \overline{\lim}_{n \to \infty} \mathbf{P}\{\operatorname{Tr} \Xi_n \Xi_n' \geq h\} = 0$.

§1 Limit Theorem for the Random Analytical Functions

For the sequence of the random processes (vectors) $\xi_n(t), \eta_n(t)$ let us introduce the notation $\xi_n(t) \sim \eta_n(t)$ denoting that for any set of time values $t_1, \ldots t_s$

$$\lim_{n \to \infty} [\mathbf{E} \exp\{i \sum_{k=1}^{s} \theta_k \xi_n(t_k)\} - \mathbf{E} \exp\{i \sum_{k=1}^{s} \theta_k \eta_n(t_k)\}] = 0,$$

where θ_k are real parameters.

If the random processes $\xi_n(t)$ and $\eta_n(t)$ are complex, then by $\xi_n(t) \sim \eta_n(t)$ we mean that

$$[\operatorname{Re} \xi_n(t), \operatorname{Im} \xi_n(t)] \sim [\operatorname{Re} \eta_n(t), \operatorname{Im} \eta_n(t)].$$

A complex-valued function of a real variable is said to be analytical if it can be expanded in a Taylor series in the neighborhood of any point of the real line.

We call the function of the type $\xi_n(t) = \sum_{k=1}^{m_n} \xi_{kn} t^k, t \in [a, b]$, where $\xi_{kn}, k = \overline{1, m_n}$ are random variables, the random analytical function. In this formula, m_n can be equal to infinity. In this case we consider that the series converge with probability 1 for all $t \in [a, b]$. Let us prove the following statement.

218

Theorem 8.1.1. *Let* $\xi_n(t) = \sum_{k=1}^{m_n} \xi_{kn} t^k, \eta_n(t) = \sum_{k=1}^{m_2(n)} \eta_{kn} t^k$ *be random analytical functions given on the interval* (c, d), *for the arbitrary constant* $0 < c < \infty, t \in (c, d)$

$$\sup_{n} \sup_{\xi_{kn}: |\xi_{kn}| \le c} | \sum_{k=1}^{m_1(n)} \xi_{kn} t^k | \le c_1 < \infty,$$

$$\sup_{n} \sup_{\eta_{kn}: |\eta_{kn}| \le c} | \sum_{k=1}^{m_2(n)} \eta_{kn} t^k | \le c_2 < \infty,$$

$$\lim_{h \to \infty} \overline{\lim_{n \to \infty}} \, \mathbf{P}\{ \sup_{k=\overline{1,m_1(n)}} |\xi_{kn}| \ge h \} = 0,$$

$$\lim_{h \to \infty} \overline{\lim_{n \to \infty}} \, \mathbf{P}\{ \sup_{k=\overline{1,m_2(n)}} |\eta_{kn}| \ge h \} = 0, \tag{8.1.1}$$

and also any convergent sequence of the joint conditional moments of the random functions $\xi_n(t_i), i = \overline{1,m}$ *and* $\eta_n(t_i), i = \overline{1,m}, m = 1, 2, \dots$ *converge to an analytical function with respect to* $t_i \in (c, d), i = \overline{1,m}, m = 1, 2, \dots,$ *under the condition that* $\sup_k |\xi_{kn}| \le c, \sup_k |\eta_{kn}| \le c$, *where* $c > 0$ *is an arbitrary constant,* $\xi_n(t) \sim \eta_n(t)$ *for all* $t \in [a, b] \subset [c, d]$. *Then* $\xi_n(t) \sim \eta_n(t)$, *for all* $t \in (c, d)$.

Proof.   Let us consider the moments

$$m_n(t_i, s_i, i = \overline{1,k}) = \mathbf{E}[\prod_{i=1}^{k} \xi_n^{s_i}(t_i) / \sup_{k=\overline{1,m_1(n)}} |\xi_{kn}| \le c],$$

$$p_n(t_i, s_i, i = \overline{1,k}) = \mathbf{E}[\prod_{i=1}^{k} \eta_n^{s_i}(t_i) / \sup_{k=\overline{1,m_2(n)}} |\eta_{kn}| \le c],$$

where $c > 0$ is some constant.

Obviously, the functions $m_n(t_i, s_i, i = \overline{1,k}), p_n(t_i, s_i, i = \overline{1,k})$ are analytical with respect to t_i, s_i for all $k = 1, 2, \dots$.

Using condition (8.1.1), we can choose a sequence n' such that $m_{n'}(t_i, s_i, i = \overline{1,k}) \to m(t_i, s_i, i = \overline{1,k}), p_{n'}(t_i, s_i, i = \overline{1,k}) \to p(t_i, s_i, i = \overline{1,k})$ as $n' \to \infty$ for all $t_i \in (c, d)$ and $m(t_i, s_i, i = \overline{1,k}) \equiv p(t_i, s_i, i = \overline{1,k})$ for all $t_i \in [a, b]$. Since the functions $m(t_i, s_i, i = \overline{1,k})$ and $p(t_i, s_i, i = \overline{1,k})$ are analytic for all $t_i \in (c, d)$, this identity is valid for all $t_i \in (c, d)$. Consequently,

$$m_n(t_i, s_i, i = \overline{1,k}) = p_n(t_i, s_i, i = \overline{1,k}) + 0(1).$$

Hence, $\xi_n(t) \sim \eta_n(t)$ for all $t \in (c, d)$, under the condition that $\sup_k |\xi_{kn}| \le c, \sup_k |\eta_{kn}| \le c$. From condition 8.1.1 and from the fact that c is arbitrary the statement of Theorem 8.1.1 follows.

§2 Limit Theorems for Random Determinants

Let us call the random variables $\xi_{ij}^{(n)}, i,j = \overline{1,n}$ asymptotically constant if there can be found non-random numbers $a_{ij}^{(n)}$, such that for all $\varepsilon > 0$,

$$\lim_{n \to \infty} \sup_{k,l=\overline{1,n}} \mathbf{P}\{|\xi_{kl}^{(n)} - a_{kl}^{(n)}| \geq \varepsilon\} = 0. \qquad (8.2.1)$$

We can choose the medians $m_{kl}^{(n)}$ of the variables $\xi_{kl}^{(n)}$ as a_{kl}. Indeed, if the probability to find ξ_{kl} in some interval is more than $1/2$, then $m_{kl}^{(n)}$ belongs to this interval. For each $\varepsilon > 0$, we can find n_ε such that if $n > n_\varepsilon$, we get

$$\sup_{k,l=\overline{1,n}} \mathbf{P}\{|\xi_{kl}^{(n)} - a_{kl}^{(n)}| > \varepsilon\} < 1/2,$$

$$\inf_{k,l=\overline{1,n}} \mathbf{P}\{|\xi_{kl}^{(n)} - a_{kl}^{(n)}| \geq \varepsilon\} > 1/2.$$

Then, as $n \geq n_\varepsilon, \sup_{k,l=\overline{1,n}} |m_{kl}^{(n)} - a_{kl}^{(n)}| \leq \varepsilon$. Consequently, the medians $m_{kl}^{(n)}$ can be chosen as $a_{kl}^{(n)}$.

The random vectors $\vec{\xi}_{nk}, k = \overline{1,n}, n = 1, 2, \ldots$ are called asymptotically constant if the constant vectors $\vec{a}_{nk}, k = \overline{1,n}$ can be found such that for all $\varepsilon > 0$,

$$\lim_{n \to \infty} \sup_{k=\overline{1,n}} \mathbf{P}\{(\vec{\xi}_{nk} - a_{nk}, \vec{\xi}_{nk} - \vec{a}_{nk}) \geq \varepsilon\} = 0. \qquad (8.2.2)$$

We call the variables $\xi_{kl}^{(n)} - a_{kl}^{(n)}$ and the vectors $\vec{\xi}_{nk} - \vec{a}_{nk}$ satisfying the (8.2.1) and (8.2.2) infinitesimal.

Let us consider the random variables $\nu_{ij}^{(n)} = \xi_{ij}^{(n)} - a_{ij}^{(n)} - \rho_{ij}^{(n)}$, where $\rho_{ij}^{(n)} = \int_{|x|<\tau} x dF_{ij}(x + a_{ij}^{(n)}, \tau > 0$ is an arbitrary constant, and $F_{ij}(x) = \mathbf{P}\{\xi_{ij}^{(n)} < x\}$. The square matrix $B_n; = (b_{ij}^{(n)})$ is composed of $b_{ij}^{(n)} := \rho_{ij}^{(n)} + a_{ij}^{(n)}$.

Theorem 8.2.1. *If for each n, the vectors $(\xi_{ij}^{(n)}, \xi_{ji}^{(n)})i \geq j, i,j = \overline{1,n}$ are independent and the column vectors and row vectors of the matrix $\Xi_n = (\xi_{ij}^{(n)})$ are asymptotically constant,*

$$\lim_{h \to \infty} \lim_{n \to \infty} \mathbf{P}\{|\sum_{i=1}^{n} \nu_{ii}^{(n)}| + \sum_{i,j=1}^{n} (\nu_{ij}^{(n)})^2 \geq h\} = 0; \qquad (8.2.3)$$

$$\sup_{n}[|\operatorname{Tr} B_n| + \operatorname{Tr} B_n B_n'] < \infty, \qquad (8.2.4)$$

then

$$\det(I + \Xi_n) \sim \det(I + B_n) \prod_{i>j,i,j=1}^{n} (1 - \nu_{ij}^{(n)}\nu_{ji}^{(n)}) \prod_{i=1}^{n}(1 + \nu_{ii}^{(n)}). \qquad (8.2.5)$$

Proof. Using Theorem 1.4.1, we obtain

$$\det(I + \alpha_t \Xi)^{-k} = \mathbf{E} \exp\{i\alpha_t \sum_{s=1}^{k}((\Xi - \Xi')\vec{\xi}_s, \vec{\eta}_s)$$

$$- \alpha_t \sum_{s=1}^{k}(\Xi\vec{\xi}_s, \vec{\xi}_s) - \alpha_t \sum_{s=1}^{k}(\Xi\vec{\eta}_s, \vec{\eta}_s)\},$$

where $\alpha_t = t[q + 0.5|\operatorname{Tr}(\Xi + \Xi')| + 0.25\operatorname{Tr}(\Xi + \Xi')^2]^{-1}$, and $0 \le t < 1, q \ge 1, \vec{\xi}_s, \vec{\eta}_s$ are independent random vectors distributed by the normal law $N(0.0.5I)$.

Let us consider the following transformation,

$$\mathbf{E} \exp\{i\theta_1 \sum_{s=1}^{k}((\Xi - \Xi')\vec{\xi}_s, \vec{\eta}_s) + i\theta_2 \sum_{s=1}^{k}[(\Xi\vec{\xi}_s, \vec{\xi}_s)$$

$$+ (\Xi\vec{\eta}_s, \vec{\eta}_s)] + i\theta_3 \operatorname{Tr}\Xi - \theta_4 \operatorname{Tr}(\Xi + \Xi')^2\}$$

$$= \mathbf{E} \exp\{\sum_{p=1}^{n} \nu_{pp}^{(n)} y_{pp} + i\sum_{p>l}(\nu_{pl}^{(n)} y_{pl} + \nu_{lp}^{(n)} y_{lp})$$

$$+ i\sum_{p=1}^{n} b_{pp}^{(n)} y_{pp} + i\sum_{p>l}(b_{pl}^{(n)} y_{pl} + b_{lp}^{(n)} y_{lp})\}, \qquad (8.2.6)$$

where

$$y_{pp} = \theta_2 \sum_{s=1}^{k} \tilde{\xi}_{ps}^2 + \theta_2 \sum_{s=1}^{k} \eta_{ps}^2 + \theta_3 + \sqrt{\theta_4}\zeta_{pp}, \quad p = \overline{1, r},$$

$$y_{pl} = \theta_1 [\sum_{s=1}^{k} \tilde{\xi}_{ps}\eta_{ls} - \sum_{s=1}^{k} \tilde{\xi}_{ls}\eta_{ps}]$$

$$+ \theta_2 [\sum_{s=1}^{k} (\tilde{\xi}_{ps}\tilde{\xi}_{ls} + \eta_{ps}\eta_{ls})] + \sqrt{\theta_4}(\zeta_{pl} + \zeta_{lp}),$$

$p > l, p, l = \overline{1, n}, \theta_4 \ge 0$, the random variables $\zeta_{pl}, p, l = 1, 2, \ldots$ are independent from one another, do not depend on Ξ and the vectors $\vec{\xi}_s, \vec{\eta}_s . s = \overline{1, k}$ and are distributed according to the normal law $N(0, 0, 5)$, and $\tilde{\xi}_{ps}, \eta_{ls}$ are the components of the vectors $\vec{\xi}_s, \vec{\eta}_s$. We rewrite (8.2.6) in the form,

$$\mathbf{E} \prod_{p=1}^{n}(1+\alpha_{pp}) \prod_{p>l}(1+\alpha_{pl}) \exp\{i\sum_{p=1}^{n} b_{pp}^{(n)} y_{pp} + i\sum_{p>l}(b_{pl}^{(n)} y_{pl} + b_{lp}^{(n)} y_{lp})\}, \quad (8.2.7)$$

where

$$\alpha_{pp} = \mathbf{E}[\exp(i\nu_{pp}^{(n)} y_{pp})/y_{pp}] - 1,$$

$$\alpha_{pl} = \mathbf{E}[\exp\{i\nu_{pl}^{(n)} y_{pl} + \nu_{lp}^{(n)} y_{lp}\}/y_{pl}, y_{lp}] - 1, \quad p \ne l.$$

From the conditions (8.2.3) and the fact that row vectors and column vectors of the matrix Ξ are asymptotically constant, we obtain [see Theorem 5.3.2 and the proof of (6.2.15)]:

$$\sup_n \sum_{p=1}^{n} \mathbf{E}[1 - (1 + \nu_{pp}^2)^{-1}] < \infty, \tag{8.2.8}$$

$$\sup_n \sum_{p \neq l} \mathbf{E}[1 - (1 + \nu_{pl}^2)^{-1}] < \infty, \tag{8.2.9}$$

$$\lim_{n \to \infty} \sup_{p=\overline{1,n}} \sum_{l=1}^{n} \mathbf{E}[1 - (1 + \nu_{pl}^2)^{-1}] + \mathbf{E}[1 - (1 + \nu_{lp}^2)^{-1}]) = 0. \tag{8.2.10}$$

Let

$$F_1(x) = \mathbf{P}\{\nu_{pl}^{(n)} < x\},$$
$$F_2(x) = \mathbf{P}\{\nu_{lp}^{(n)} < x\},$$
$$F(x, y) = \mathbf{P}\{\nu_{pl}^{(n)} < x. \nu_{lp}^{(n)} < y\}, \varepsilon, \tau > 0$$

be arbitrary constants. Consider the following inequalities:

$$|\alpha_{pl}| \leq |y_{pl}|\varepsilon + |y_{lp}|\varepsilon + 2\mathbf{P}\{|\nu_{pl}^{(n)}| \geq \varepsilon\} + 4\mathbf{P}\{|\nu_{lp}^{(n)}| \geq \varepsilon\}, \quad p \neq l, \tag{8.2.11}$$

$$
\begin{aligned}
|\alpha_{pl}| = &\left| \int_{|x|<\tau,|z|<\tau} (\exp(ixy_{pl} + izy_{lp}) - 1 - ixy_{pl} \right. \\
&- izy_{lp})dF(x, z) + \int_{\overline{\{x,z:|x|<\tau,|z|<\tau\}}} (\exp(ixy_{pl} \\
&+ izy_{lp}) - 1)dF(x, z) + iy_{pl} \int_{|x|<\tau} xdF_1(x) \\
&+ iy_{lp} \int_{|z|<\tau} zdF_2(z) - iy_{pl} \int_{|x|<\tau,|z|\geq\tau} xdF(x, z) \\
&- iy_{lp} \int_{|x|>\tau,|z|<\tau} zdF(x, z)\left| \leq y_{pl}^2 \int_{|x|<\tau} x^2 dF_1(x) \right. \\
&+ y_{pl}^2 \int_{|z|<\tau} z^2 dF_2(z) + 4 \int_{|x|\geq\tau} dF_1(x) + 2 \int_{|x|\geq\tau} dF_2(x) \\
&+ |y_{lp}|\left| \int_{|x|<\tau} xdF_1(x) \right| + |y_{lp}|\left| \int_{|x|<\tau} xdF_2(x) \right| \\
&+ |y_{pl}|\tau \int_{|x|\geq\tau} dF_1(x) + |y_{lp}|\tau \int_{|x|\geq\tau} dF_2(x), \quad p \neq l.
\end{aligned}
\tag{8.2.12}
$$

Since for sufficiently large n, $|\rho_{pl}| < \tau/2$,

$$\left| \int_{|x|<\tau} x \, dF_{pl}(x) - \int_{|x+\rho_{pl}|<\tau} x \, dF_{pl}(x) \right|$$

$$\leq \int_{\tau/2<|x|<3\tau/2} |x| \, dF_{pl}(x) \leq (3\tau/2) \int_{|x|>\tau/2} dF_{pl}(x),$$

$$\left| \int_{|x+\rho_{pl}|<\tau} x \, dF_{pl}(x) \right| = \left| \int_{|x|<\tau} (x) - \rho_{pl}) \, dF_{pl}(x - \rho_{pl}) \right|$$

$$= |\rho_{pl} \int_{|x|>\tau} dF_{pl}(x - \rho_{pl})| \leq (\tau/2) \int_{|x|>\tau/2} dF_{pl}(x).$$

Therefore, for sufficiently large n,

$$\left| \int_{|x|<\tau} dF_i(x) \right| \leq 2\tau \int_{|x|>\tau/2} dF_i(x), \quad i = \overline{1,2}. \tag{8.2.13}$$

It is obvious that

$$\int_{|x|<\tau} x^2 \, dF_i(x) \leq (1+\tau^2) \int x^2 (1+x^2)^{-1} \, dF_i(x), \tag{8.2.14}$$

$$\int_{|x|>\tau} dF_i(x) \leq (1+\tau^{-2}) \int x^2 (1+x^2)^{-1} \, dF_i(x), \quad i = \overline{1,2}.$$

Making use of (8.2.11)–(8.2.14), we get

$$\mathbf{E} \sum_{p>l} (|\alpha_{pl}|^2 + |\mathbf{E}\alpha_{pl}|^2) \leq 2 \sum_{p\geq l} \mathbf{E}|\alpha_{pl}|^2$$

$$\leq \sum_{p>l} \mathbf{E}(|y_{pl}|\varepsilon + |y_{lp}|\varepsilon + 2\mathbf{P}\{|\nu_{pl}^{(n)}| > \varepsilon\}$$

$$\times \left[\int x^2 (1+x^2)^{-1} \, dF_1(x) \{y_{pl}^2(1+\tau^2) + 4(1+\tau^{-2}) \right.$$

$$+ |y_{pl}|(8\tau^{-1}(4+\tau^2) + \tau^{-1} + \tau)\} + \int x^2 (1+x^2)^{-1} \, dF_2(x)$$

$$\left. \times \{y_{lp}^2 + 2(1+\tau^{-2}) + |y_{lp}|(8\tau^{-1}(4+\tau^2) + \tau + \tau^{-1})\} \right].$$

Since the variables $\nu_{pl}^{(n)}$ are infinitesimal, and by taking into account (8.2.8) and (8.2.9), we obtain

$$\lim_{n\to\infty} \mathbf{E} \sum_{p>l} (|\alpha_{pl}|^2 + |\mathbf{E}\alpha_{pl}|^2) = 0.$$

Analogously,

$$\lim_{n\to\infty} \mathbf{E} \sum_{p=1}^{n} (|\alpha_{pp}^2| + |\mathbf{E}\alpha_{pp}|^2) = 0.$$

From these two limit relations, it follows that

$$\lim_{n\to\infty} \mathbf{E} \sum_{p\geq l}(|\alpha_{pl}|^2 + |\mathbf{E}\alpha_{pl}|^2) = 0. \tag{8.2.15}$$

We need the inequality

$$\mathbf{E}|\sum_{p\geq l}(\alpha_{pl} - \mathbf{E}\alpha_{pl})|^2 = \mathbf{E}(\sum_{p\geq l}(\alpha_{pl} - \mathbf{E}\alpha_{pl}))$$

$$\times (\sum_{s\geq q}(\overline{\alpha}_{sq} - \mathbf{E}\overline{\alpha}_{sq})) = \sum_{p\geq l, s\geq l}(\alpha_{pl} - \mathbf{E}\alpha_{pl})$$

$$\times (\overline{\alpha}_{sl} - \mathbf{E}\overline{\alpha}_{sl}) + \sum_{p\geq l, p\geq q}(\alpha_{pl} - \mathbf{E}\alpha_{pl})(\overline{\alpha}_{pq} - \mathbf{E}\overline{\alpha}_{pq})$$

$$+ \sum_{p\geq l, l\geq s}(\alpha_{pl} - \mathbf{E}\alpha_{pl})(\overline{\alpha}_{ls} - \mathbf{E}\overline{\alpha}_{ls}) + \sum_{p\geq l, s\geq p}(\alpha_{pl} \tag{8.2.16}$$

$$- \mathbf{E}\alpha_{pl})(\overline{\alpha}_{sp} - \mathbf{E}\overline{\alpha}_{sp}) \leq \sum_{l=1}^{n}(\sum_{p=1}^{n}\sqrt{\mathbf{E}|\alpha_{pl}|^2})^2$$

$$+ \sum_{p=1}^{n}(\sum_{l=1}^{n}\sqrt{\mathbf{E}|\alpha_{pl}|^2})^2 + 2\sum_{l=1}^{n}\sqrt{\mathbf{E}|\alpha_{pl}|^2}$$

$$\times (\sum_{p=1}^{n}\sqrt{\mathbf{E}|\alpha_{lp}|^2}).$$

From the inequalities (8.2.11) and (8.2.12), it is easy to obtain the estimation

$$\mathbf{E}(\alpha_{pl})^2 \leq c(k,\tau)[\mathbf{E}\{1 - (1 + \nu_{lp}^2)^{-1}\} + \mathbf{E}\{1 - (1 + \nu_{pl}^2)^{-1}\}].$$

From this and from (8.2.8)–(8.2.10) and (8.2.16), it follows that

$$\lim_{n\to\infty} \mathbf{E}|\sum_{p\geq l}(\alpha_{pl} - \mathbf{E}_{pl})|^2 = 0. \tag{8.2.17}$$

By using (8.2.15) and (8.2.17), we obtain that (8.2.7) is equal to [see the proof of (5.3.20)]

$$\prod_{p=1}^{n}(1 + \mathbf{E}\alpha_{pp}) \prod_{p>l}(1 + \mathbf{E}\alpha_{pl}) \exp\{i\sum_{p=1}^{n}b_{pp}^{(n)}y_{pp}$$

$$+ i\sum_{p>l}(b_{pl}^{(n)}y_{pl} + b_{lp}^{(n)}y_{lp})\} + 0(1) = \mathbf{E}\exp\{i\sum_{p=1}^{n}\nu_{pp}^{(n)}$$

$$\times \overline{y}_{pp} + i\sum_{p>l}(\nu_{pl}^{(n)}\overline{y}_{pl} + \nu_{lp}\overline{y}_{lp} + i\sum_{p=1}^{n}b_{pp}^{(n)}y_{pp}$$

$$+ i\sum_{p>l}(b_{pl}^{(n)}y_{pl} + b_{lp}^{(n)}y_{lp})\} + 0(1), \tag{8.2.18}$$

where the random vectors $(\overline{y}_{pl}, \overline{y}_{lp}), p \geq l, p, l = 1, 2, \ldots$ are independent, do not depend on y_{pl}, ν_{pl}, and are distributed like the vectors $(y_{pl}, y_{lp}), p \geq l$.

Rewrite (8.2.18) in the form

$$
\mathbf{E} \exp\{i\theta_1 \sum_{s=1}^{k}[\sum_{p>l}(\nu_{pl} - \nu_{lp})(\overline{\xi}_{ps}\overline{\eta}_{ls} - \overline{\xi}_{ls}\overline{\eta}_{ps})
$$

$$
+ \sum_{p>l}(b_{pl} - b_{lp})(\tilde{\xi}_{ps}\eta_{ls} - \tilde{\xi}_{ls}\eta_{ps})] + i\theta_2
$$

$$
\times \sum_{s=1}^{k}[\sum_{p>l}(\nu_{pl} + \nu_{lp})(\overline{\xi}_{ps}\overline{\xi}_{ls} + \overline{\eta}_{ps}\overline{\eta}_{ls})
$$

$$
+ \sum_{p=1}^{n}\nu_{pp}(\overline{\xi}_{ps}^2 + \overline{\eta}_{ps}^2) + \sum_{p>l}(b_{pl} + b_{lp})(\tilde{\xi}_{ps}\tilde{\xi}_{ls} + \eta_{ps}\eta_{ls})
$$

$$
+ \sum_{p=1}^{n}b_{pp}(\tilde{\xi}_{ps}^2 + \eta_{ps}^2)] + i\theta_3 \operatorname{Tr} \Xi - \theta_4 \operatorname{Tr}(\Xi + \Xi')^2\},
$$

where $\overline{\xi}_{ps}, \overline{\eta}_{ps}$ are independent random variables, not depending on Ξ, ξ_{ps}, η_{ps}, $p, s = 1, 2, \ldots$, and are distributed like $\tilde{\xi}_{ps}, \eta_{ps}$. Hence,

$$
\{\sum_{s=1}^{k}((\Xi - \Xi')\overrightarrow{\xi}_s, \overrightarrow{\eta}_s), \sum_{s=1}^{k}\{(\Xi\overrightarrow{\xi}_s, \overrightarrow{\xi}_s)
$$

$$
+ (\Xi\overrightarrow{\eta}_s, \overrightarrow{\eta}_s)\}, \operatorname{Tr}\Xi, \operatorname{Tr}(\Xi + \Xi')^2\}
$$

$$
\sim \{\sum_{s=1}^{k}[\sum_{p>l}(\nu_{pl} - \nu_{lp})(\overline{\xi}_{ps}\overline{\eta}_{ls} - \overline{\xi}_{ls}\overline{\eta}_{ps})
$$

$$
+ \sum_{p>l}(b_{pl} - b_{lp})(\tilde{\xi}_{ps}\eta_{ls} - \tilde{\xi}_{ls}\eta_{ps})], \tag{8.2.19}
$$

$$
\sum_{s=1}^{k}[\sum_{p>l}(\nu_{pl} + \nu_{lp})(\overline{\xi}_{ps}\overline{\xi}_{ls} + \overline{\eta}_{ps}\overline{\eta}_{ls})
$$

$$
+ \sum_{p>l}(b_{pl} + b_{lp})(\tilde{\xi}_{ps}\tilde{\xi}_{ls} + \eta_{ps}\eta_{ls})
$$

$$
+ \sum_{p=1}^{n}\nu_{pp}(\overline{\xi}_{ps}^2 + \overline{\eta}_{ps}^2) + \sum_{p=1}^{n}b_{pp}(\xi_{ps}^2 + \eta_{ps}^2)],
$$

$$
\operatorname{Tr}\Xi, \operatorname{Tr}(\Xi + \Xi')^2\}.
$$

Further, we use the following easily verified statement.

Let ξ_n, η_n be some sequences of the random variables $\xi_n \sim \eta_n$,

$$
\sup_n \mathbf{E}|\xi_n|^{1+\beta} < \infty, \qquad \sup_n \mathbf{E}|\eta_n|^{1+\beta} < \infty,
$$

for some $\beta > 0$. Then $\mathbf{E}\xi = \mathbf{E}\eta + 0(1)$.

By using Lemma 1.4.1, we obtain

$$\mathbf{E}[\exp\{-\alpha_t \sum_{s=1}^{k}(\Xi\vec{\xi}_s, \vec{\xi}_s) - \alpha_t \sum_{s=1}^{k}(\Xi\vec{\eta}_s, \vec{\eta}_s)\}]^{\beta} \leq \exp\{t\beta[1 - t\beta]^{-1}\},$$

where $1 < \beta < t^{-1} - 1$.

It is easy to verify that the inequality holds if the corresponding random variable from the right-hand side of relation (8.2.19) is substituted for $\sum_{s=1}^{k}\{(\Xi\vec{\xi}_s, \vec{\xi}_s) + (\Xi\vec{\eta}_s, \vec{\eta}_s)\}$. Therefore, taking into account everything said above and (8.2.19), we find

$$\mathbf{E}\det(I + \alpha_t\Xi)^{-k} = \det(I + \alpha_l B)^{-k}\mathbf{E}\exp\{i\alpha_t \sum_{s=1}^{k}[\sum_{p>l}\nu_{pl}$$

$$- \nu_{lp})(\bar{\xi}_{ps}\bar{\eta}_{ls} - \bar{\xi}_{ls}\bar{\eta}_{ls})] - \alpha_t \sum_{s=1}^{k}[\sum_{p>l}(\nu_{pl} + \nu_{lp})$$

$$\times (\bar{\xi}_{ps}\bar{\xi}_{ls} + \bar{\eta}_{ps}\bar{\eta}_{ls}) + \sum_{p=1}^{n}\nu_{pp}(\bar{\xi}_{ps}^2 + \bar{\eta}_{ps}^2)]\} + 0(1).$$

From this expression and after simple operations of integration, it follows that

$$\mathbf{E}\det(I + \alpha_t\Xi)^{-k} = \mathbf{E}(\det(I + \alpha_t B)\prod_{p=1}^{n}(1 + \alpha_t\nu_{pp})$$

$$\times \prod_{p>l}(1 - \alpha_t^2\nu_{pl}^{(n)}\nu_{lp}^{(n)})]^{-k} + 0(1). \tag{8.2.20}$$

On the basis of Lemma 1.4.1, $\det(I + \alpha_t\Xi) \geq \exp(-t(1 - t)^{-1}), 0 \leq t < 1$. Therefore, by the Carleman Theorem the one-dimensional distributions of the random processes $\det(I + \alpha_t\Xi)$ are one-to-one restored by the moments (8.2.20). Thus, for all fixed $0 \leq t < 1$,

$$\det(I + \alpha_t\Xi) \sim \det(I + \alpha B)\prod_{p=1}^{n}(1 + \alpha_t\nu_{pp})\prod_{p>l}(1 - \alpha_t^2\nu_{pl}^{(n)}\nu_{lp}^{(n)}). \tag{8.2.21}$$

It is easy to verify that the random processes $\det(I + \alpha_t\Xi)$ and $\det(I + \alpha_t B)\prod_{p=1}^{n}(1 + \alpha_t\nu_{pp})\prod_{p>l}(1 - \alpha_t^2\nu_{pl}\nu_{lp})$ satisfy all the conditions of Theorem 8.1.1 and are analytical for all $0 \leq t < \infty$. Therefore, (8.2.21) holds for each fixed $0 \leq t < \infty$. Setting $t = q$ in (8.2.21), we get

$$\alpha_q = [1 + (2q)^{-1}|\operatorname{Tr}(\Xi + \Xi')| + (4q)^{-1}\operatorname{Tr}(\Xi + \Xi')^2]^{-1}.$$

Letting q go to infinity and using conditions (8.2.3), (8.2.4), we come to the statement of Theorem 8.2.1.

Corollary 8.2.1. *Let the conditions of Theorem 8.2.1 hold. Then*

$$\det(I + \Xi_n) \sim \det(I + B_n) \prod_{i>j}(1 - \nu_{ij}^2) \prod_{i=1}^{n}(1 + \nu_{ii})$$

if Ξ_n are symmetric,

$$\det(I + \Xi_n) \sim \det(I + B_n)$$

if for each n the entries $\xi_{pl}, p, l = \overline{1,n}$, are independent, and

$$\det(I + \Xi_n) \sim \det(I + B_n) \prod_{i>j}(1 + \nu_{ij}^2) \prod_{i=1}^{n}(1 + \nu_{ii})$$

if the matrices Ξ_n are antisymmetric.

Note that in the latter case, $\det(I + \Xi_n) = \det(I + \Xi_n \Xi_n')^{1/2}$, and the matrix $\Xi\Xi_n'$ is nonnegative-definite. For the matrices presented in the form $I + C_n$, where C is a nonnegative-definite matrix, the proofs of the theorems are considerably simplified, since instead of the cumbersome integral representation given by Theorem 1.4.1, we can use formula (1.4.3). On the basis of this formula and Theorem 8.2.1, we obtain the following statement.

Corollary 8.2.2. *For each n let the random variables $\xi_{ij}^{(n)}, i, j = \overline{1,n}$ be independent, and the column vectors and row vectors of the matrix $\Xi_n = (\xi_{ij}^{(n)})$ be asymptotically constant,*

$$\lim_{h\to\infty} \varlimsup_{n\to\infty} \mathbf{P}\{\sum_{i,j=1}^{n}(\nu_{ij}^{(n)})^2 \geq h\} = 0,$$

$$\sup_n \operatorname{Tr} B_n B_n' < \infty.$$

Then

$$\det(I + \Xi_n \Xi_n') \sim \det(I + B_n B_n') \prod_{i,j=1}^{n}(1 + (\nu_{ij}^{(n)})^2).$$

We can easily extend Theorem 8.2.1 to the case of several determinants of the random matrices, and also to the complex random matrices. For example, for the determinants of the complex random matrices $\det(I + \Xi_{1n} + i\Xi_{2n})$, where the matrices Ξ_{1n}, Ξ_{2n} are real, the scheme of applying the proof of

Theorem 8.2.1 is as follows. Instead of $\det(I+\Xi_{1n}+i\Xi_{2n})$ we consider $\det(I+\Xi_{1n}+t\Xi_{2n})$, where t is an arbitrary real parameter. Denote

$$\gamma_{n1} = -0.5t\det(I+\Xi_{1n}+t\Xi_{2n})+0.5t\det(I+\Xi_{1n}-t\Xi_{2n}),$$
$$\gamma_{2n} = 0.5\det(I+\Xi_{1n}+t\Xi_{2n})+0.5\det(I+\Xi_{1n}-t\Xi_{2n}).$$

We use the proof of Theorem 8.2.1 for the random variables γ_{n1} and γ_{n2} under certain conditions on the matrices Ξ_{1n} and Ξ_{2n}. Considering the moments of the cut random variables γ_{n1} and γ_{n2} (we put the conditions on the entries of matrices Ξ_{1n} and Ξ_{2n}), we establish that these moments are analytical functions with respect to t. They can be extended to the whole complex plane and, in particular, we can assume that $t = i$. Then $\gamma_{n1} = \operatorname{Im}\det(I+\Xi_{1n}+i\Xi_{2n})$, $\gamma_{n2} = \operatorname{Re}\det(I+\Xi_{1n}+i\Xi_{2n})$.

Theorem 8.2.2. *For each n let the vectors $(\xi_{ij}^{(1)}, \xi_{ji}^{(1)}, \xi_{ij}^{(2)}, \xi_{ij}^{(2)})$, $i \geq j, i, j = \overline{1,n}$ be independent, $\xi_{ij}^{(1)}$ and $\xi_{ij}^{(2)}$ be the entries of the matrices Ξ_{1n} and Ξ_{2n}, and the column vectors and row vectors of the matrix $\Xi_{pn}, p = \overline{1,2}$ be asymptotically constant,*

$$\lim_{h\to\infty}\overline{\lim_{n\to\infty}}\mathbf{P}\{|\sum_{i=1}^{n}\nu_{ii}^{(1)}|+|\sum_{i=1}^{n}\nu_{ii}^{(2)}|+\sum_{i,j=1}^{n}[(\nu_{ij}^{(1)})^2+(\nu_{ij}^{(2)})^2]\geq h\}=0,$$

where $\nu_{ij}^{(p)} = \xi_{ij}^{(p)} - a_{ij}^{(p)} - \rho_{ij}^{(p)}, \rho_{ij}^{(p)} = \int_{|x|>\tau} x dF_{ij}^{(p)}(x+a_{ij}^{(p)})$, τ is an arbitrary constant, and the nonrandom variables $a_{ij}^{(p)}$ are such that for any $\varepsilon > 0$,

$$\lim_{n\to\infty}\sup_{i,j=\overline{1,n}}\mathbf{P}\{|\xi_{ij}^{(p)}-a_{ij}^{(p)}|>\varepsilon\}=0,\quad p=\overline{1,2},$$
$$\sup_{n}[|\operatorname{Tr}B_{n1}|+|\operatorname{Tr}B_{n2}|+\operatorname{Tr}(B_{n1}B_{n1}'+B_{n2}B_{n2}')]<\infty,$$

where $B_{np} = (b_{ij}^{(p)}), b_{ij}^{(p)} = \rho_{ij}^{(p)}+a_{ij}^{(p)}$. Then

$$\det(I+\Xi_{n1}+i\Xi_{n2}) \sim \det(I+B_{n1}+iB_{n2})$$
$$\times \prod_{p>l}[1-(\nu_{lp}^{(1)}+i\nu_{lp}^{(2)})(\nu_{pl}^{(1)}+i\nu_{pl}^{(2)})]\prod_{p=1}^{n}(1+\nu_{pp}^{(1)}+i\nu_{pp}^{(2)}).$$

Corollary 8.2.3. *If the conditions of Theorem 8.2.2 hold, then*

$$|\det(I+\Xi_{n1}+i\Xi_{n2})| \sim |\det(I+B_{n1}+iB_{n2})|$$
$$\times \prod_{p>l}[(1-\nu_{pl}^{(1)}\nu_{lp}^{(1)}+\nu_{pl}^{(2)}\nu_{lp}^{(2)})^2+(\nu_{pl}^{(2)}\nu_{lp}^{(1)}+\nu_{pl}^{(1)}\nu_{lp}^{(2)})^2]^{1/2}$$
$$\times \prod_{p=1}^{n}[(1+\nu_{pp}^{(1)})^2+(\nu_{pp}^{(2)})^2]^{1/2}.$$

If in addition to the conditions of Theorem 8.2.2 the matrices Ξ_{n1}, Ξ_{n2} are symmetric, then $\det(I + \Xi_{n1} + i\Xi_{n2}) \sim \det(I + B_{n1} + iB_{n2}) \prod_{p>l}[1 - (\nu_{pl}^{(1)} + i\nu_{pl}^{(2)})^2] \prod_{p=1}^n (1 + \nu_{pp}^{(1)} + i\nu_{pp}^{(2)})$.

If in addition to the conditions of Theorem 8.2.2 the matrices Ξ_{n1}, Ξ_{n2} are antisymmetric, then $\det(I + \Xi_{n1} + i\Xi_{n2}) \sim \det(I + B_{n1} + iB_{n2}) \prod_{p>l}[1 + (\nu_{pl}^{(1)} + i\nu_{lp}^{(2)})^2]$.

Note that by proving the limit theorems for the random determinants $\det(I + i\Xi_n)$, where Ξ_n are the symmetric matrices, we can use the integral transformation (1.4.4).

If we exclude the condition that column vectors and row vectors of the matrix Ξ_n are asymptotically constant from Theorem 8.2.1, then on the basis of the proof of Theorem 8.2.1, it follows that $\det(I + \Xi_n) \sim \det(I + H_n)$, where $H_n = (\eta_{ij})$ is a random matrix with the entries η_{ij} being distributed by the infinitely divisible law with the characteristic functions

$$\mathbf{E} \exp\{is\eta_{pl} + iq\eta_{lp}\} = \exp\{isb_{pl} + iqb_{lp} + \mathbf{E}\exp[is\nu_{pl} + iq\nu_{lp}] - 1\}.$$

If the eigenvalues λ_i of the symmetric matrix Ξ_n satisfy the inequality $|\lambda_i| < 1$, then

$$|\ln\det(I+\Xi_n) - \sum_{i=1}^{s}(-1)^{i+1}i^{-1}\operatorname{Tr}\Xi_n^i| \le \operatorname{Tr}|A|^{s+1}\max_{i=\overline{1,n}}|\ln(1-|\lambda_i|)|, \quad (8.2.22)$$

where $|A| = H\operatorname{diag}(|\lambda_i|, i = \overline{1,n})H'$, and the orthogonal matrix H and eigenvalues λ_i are the solution of the equation

$$\Xi_n = H\operatorname{diag}(\lambda_i, i = \overline{1,n})H'.$$

If

$$\operatorname{plim}_{n\to\infty} c_n \operatorname{Tr}|A|^{s+1}\max_{i=\overline{1,n}}|\ln(1-|\lambda_i|)| = 0, \quad (8.2.23)$$

and when the constants c_n are appropriately chosen, then

$$\operatorname{plim}_{n\to\infty} c_n[\ln\det(I+\Xi_n) - \sum_{i=1}^{s}(-1)^{i+1}i^{-1}\operatorname{Tr}A^i] = 0.$$

Thus, the analysis of the limit distributions of the random determinants is reduced to proving the limit theorems for the traces of degrees of the random matrices.

It is extremely difficult to verify the conditions $|\lambda_i| < 1$ and (8.2.23). According to Theorems 8.2.1 and 8.2.2, it is possible to choose ξ_{ij} such that for any finite integer $s > 0, \operatorname{plim}_{n\to\infty}\operatorname{Tr}(\Xi\Xi')^s \ne 0$, and therefore it is unreasonable to expand the expression $\ln|\det(I + \Xi_n)|$ in series. The conditions

(8.2.22) or (8.2.23) are not necessary for the method of integral representations. But it is necessary that the random variables $|\det(I + \Xi_n)|$ be bounded in probability. The next chapter examines the method of the reduction of proof of the limit theorems for random determinants to that for functionals of random functions, which in some cases, being the generalization of the method of integral representations, does not demand that the random variables $\|\det(I + \Xi_n)\|$ be bounded in probability and that the conditions of the (8.2.23) type be held at the same time.

Let us show that in the cases when random variables $|\det(I + \Xi_n)|$ are bounded in probability, we ca use the method of integral representations.

Theorem 8.2.3. *Let Ξ_n be random square matrices of order n, and for each integer $k = 1, 2, \ldots$ and any $c > 0$,*

$$\lim_{n \to \infty} \mathbf{P}\{\sum_{s=1}^{2k}(\Xi_n \Xi_n' \overrightarrow{\eta}_s, \overrightarrow{\eta}_s) - 2k \operatorname{Tr} \Xi_n \Xi_n < x / \operatorname{Tr}(\Xi_n \Xi_n')^2 \le c\} = F_k(c, x),$$

(8.2.24)

where $F_k(c, x)$ is a distribution function depending on the two parameters c and k, and η_s are the independent random vectors not depending on the matrix Ξ_n, distributed by the normal law $N(0; 0.5I)$.

Then, for the sequence of the random variables,

$$\xi_n = \exp(\operatorname{Tr} \Xi_n \Xi_n') \det(I + \Xi_n \Xi_n')^{-1}, n = 1, 2, \ldots$$

for all $c > 0$ and $k = 1, 2, \ldots$,

$$\lim_{n \to \infty} \mathbf{E}[\xi_n^k / \operatorname{Tr}(\Xi_n \Xi_n')^2 \le c] = \int e^{-x} dF_k(c, x). \qquad (8.2.25)$$

The proof follows from the inequality

$$|\exp(\operatorname{Tr} \Xi_n \Xi_n') \det(I + \Xi_n \Xi_n')^{-1} - 1| \le \operatorname{Tr}(\Xi_n \Xi_n')^2 \exp[\operatorname{Tr} \Xi_n \Xi_n']^{1/2}$$

and from formula (1.4.3).

Corollary 8.2.4. *For each n let the random entries $\xi_{ij}^{(n)}, i, j = \overline{1, n}$ of the matrices Ξ_n be independent, and let $\mathbf{E}\xi_{ij}^{(n)} = 0$, $\operatorname{Var} \xi_{ij}^{(n)} = \sigma_{ij}^2 n^{-3/2}$, $\sup_{i,j,n} \sigma_{ij}^2 < \infty$, for some $\delta > 0$, $\sup_{i,j,n} \mathbf{E}|\xi_{ij}^{(n)} n^{-3/4}|^{4+\delta} < \infty$ exist.*
    Then

$$\operatorname{plim}_{n \to \infty}[\ln \det(I + \Xi_n \Xi_n') - \operatorname{Tr} \Xi_n \Xi_n'$$

$$+ 0.5 \sum_{p,l,s=1}^{n} \sigma_{ps}^2 \sigma_{ls}^2 n^{-3}] = 0.$$

Proof. Let us represent the random variables $(\Xi_n \Xi_n' \vec{\eta_s}, \vec{\eta_s}) - \text{Tr}\,\Xi_n \Xi_n'$ in the form

$$(\Xi_n \Xi_n' \vec{\eta_s}, \vec{\eta_s}) - \text{Tr}\,\Xi_n \Xi_n' = \sum_{i=1}^{n} a_{ii}(\eta_{is}^2 - 0.5) + \sum_{i \neq j} a_{ij} \eta_{is} \eta_{js}, \qquad (8.2.26)$$

where $a_{ij} = \sum_{k=1}^{n} \xi_{ik} \xi_{jk}$.

Then

$$\begin{aligned}
\mathbf{E}\xi_n^{-k} = \mathbf{E}[(\mathbf{E}\{\exp[-\sum_{i=1}^{n} a_{ii}(\eta_{is}^2 - 0.5) &- \sum_{ii \neq j} a_{ij}\eta_{is} \\
\times \eta_{is}]/\Xi_n\})^{-1}/\text{Tr}(\Xi_n \Xi_n')^2 &\leq c]^k \mathbf{P}\{\text{Tr}(\Xi_n \Xi_n')^2 \leq c\} \\
+ \mathbf{E}[\xi_n^{-k}/\text{Tr}(\Xi_n \Xi_n')^2 &> c]\mathbf{P}\{\text{Tr}(\Xi_n \Xi_n')^2 > c\}.
\end{aligned} \qquad (8.2.27)$$

If on the basis of Corollary 5.2.1 the conditions of Corollary 8.2.3 for fixed random variables a_{ij} hold, we get

$$\sum_{i=1}^{n} a_{ii}(\eta_{is}^2 - 0.5) - \sum_{i \neq j} a_{ij}\eta_{is}\eta_{js}$$

$$\sim \sum_{i=1}^{n} a_{ii}\nu_i - \sum_{i \neq j} a_{ij}\eta_{is}\eta_{js}, \qquad s = \overline{1, 2k}, \qquad (8.2.28)$$

where the random variables $\nu_i, i = \overline{1, n}$ are independent of one another, do not depend on the random variables a_{ij}, η_{is}, and are distributed by the normal law $N(0; 0.5)$. It is obvious that $\text{plim}_{n \to \infty} \sum_{i=1}^{n}(a_{ii}^2 - \mathbf{E}a_{ii}^2) = 0$. Therefore,

$$\sum_{i=1}^{n} a_{ii}\nu_i \sim \nu_i[\sum_{i=1}^{n}(\sum_{k=1}^{n} n^{-3/2}\sigma_{ik})^2]^{1/2}. \qquad (8.2.29)$$

By using Corollary 5.2.1 again, under the condition that η_{is} are fixed, we obtain

$$\sum_{i \neq j} a_{ij}\eta_{is}\eta_{js} \sim \sum_{i \neq j}(\sum_{k=1}^{n} \nu_{ik}\nu_{jk})\eta_{is}\eta_{js}, \qquad (8.2.30)$$

where for each n the random variables $\nu_{ij}, i, j = \overline{1, n}$ are independent, do not depend on Ξ and η_{is}, ν_i, and are distributed by the normal law $N(0, n^{-3/2}\sigma_{ij}^2)$.

Consider the characteristic functions

$$\mathbf{E}\exp\{i\theta \sum_{p \neq l}^{n}(\sum_{s=1}^{n} \nu_{ps}\nu_{ls})\sum_{q=1}^{2k} \eta_{pq}\eta_{lq}\}$$

$$= \mathbf{E}\prod_{s=1}^{n} \mathbf{E}[\exp\{i\theta \sum_{p \neq l} \nu_{ps}\nu_{ls} \sum_{q=1}^{2k} \eta_{pq}\eta_{lq}\}/\eta_{pq}, \quad p = \overline{1, n}, q = \overline{1, 2k}]$$

$$= \mathbf{E}\prod_{s=1}^{n} \det[\delta_{pl} - i\theta n^{-3}\sigma_{ps}\sigma_{ls} \sum_{q=1}^{2k} \eta_{pq}\eta_{lq}(1 - \delta_{pl})]^{-1/2}. \qquad (8.2.31)$$

We need the following theorem by Gershgorin.

Let $A_n = (a_{ij})$ be the random square matrix of order n and the $\lambda_i, i = \overline{1,n}$ its eigenvalues. Then the eigenvalues λ_i satisfy the relations

$$|a_{ii} - \lambda_i| \le \sum_{j \ne i, j = \overline{1,n}} |a_{ij}|, \quad i = \overline{1,n}.$$

According to this theorem, the moduli of the eigenvalues of the matrices $(n^{-3}\sigma_{ps}\sigma_{ls}\sum_{q=1}^{2k} \eta_{pq}\eta_{lq}(1 - \delta_{pl}))$ tend to zero in probability. Taking into account the large numbers law, we find from (8.2.31) that

$$\mathbf{E}\exp\{i\theta \sum_{p \ne l}^{n}(\sum_{s=1} \nu_{ps}\nu_{ls}) \sum_{q=1}^{2k} \eta_{pq}\eta_{lq}\}$$

$$= \exp\{-(\theta^2/4)k \sum_{s=1}^{n}\sum_{p \ne l} n^{-3}\sigma_{ps}^2\sigma_{ls}^2\} + 0(1).$$

$$(8.2.32)$$

It is easy to show that

$$\mathrm{plim}_{n \to \infty}[\mathrm{Tr}(\Xi_n\Xi_n')^2 - \sum_{i \ne j} n^{-3}\sigma_{ik}^2\sigma_{jk}^2] = 0 \qquad (8.2.33)$$

By using the (8.2.28)–(8.2.30), (8.2.32), and (8.2.33), we get

$$\mathbf{E}[\xi_n^k \,\mathrm{Tr}(\Xi_n\Xi_n')^2 \le c] = \mathbf{E}\exp\{-\nu_1[k\sum_{s=1}^{n}\sum_{p,l} n^{-3}$$

$$\times \sigma_{ps}^2\sigma_{ls}^2]^{1/2}\} + \varepsilon_n(c) = \exp\{0.5k \sum_{p,l,s=1}^{n} n^{-3}\sigma_{ps}^2\rho_{ls}^2\}$$

$$+ \varepsilon_n'(c), \quad \lim_{c \to \infty} \mathrm{plim}_{n \to \infty}[|\varepsilon_n(c)| + |\varepsilon_n'(c)|] = 0.$$

From this and from (8.2.27), letting c go to infinity, we come to the statement of Theorem 8.2.3.

With the help of the method of integral representations, we can prove limit theorems for random determinants not only of the form $\det(I + \Xi_n)$ but also $\det(B_n + \Xi_n)$ where B_n are some nonrandom matrices. Let us consider a particular case.

Theorem 8.2.4. *For each n the random entries $\xi_{ij}^{(n)}, i, j = \overline{1,n}$ of the matrices $\Xi_n = (\xi_{ij}^{(n)})$ be independent, $a_{ij}^{(n)}$ be nonrandom values such that for each $\varepsilon > 0$,*

$$\lim_{n \to \infty} \sup_{k,l=\overline{1,n}} \mathbf{P}\{|\xi_{kl}^{(n)} - a_{kl}^{(n)}| > \varepsilon\} = 0,$$

$$\nu_{ij}^{(n)} = \xi_{ij}^{(n)} - a_{ij}^{(n)} - \rho_{ij}^{(n)},$$

$$\rho_{ij}^{(n)} = \int_{|x|<\tau} x dP\{\xi_{ij}^{(n)} - a_{ij}^{(n)} < x\},$$

and $B_n = (b_{ij}^{(n)})$, $C_n = (b_{ij}^{(n)} + c_{ij}^{(n)})$, $c_{ij}^{(n)} = \rho_{ij}^{(n)} + a_{ij}^{(n)}$, $\tau > 0$ *be an arbitrary constant,* $B_n, n = 1, 2, \ldots$ *nonrandom square matrices, there exist inverse matrices* $R_n := (C_n C_n')^{-1} = (r_{ij}^{(n)})$ *satisfying the conditions* $\sum_{j=1}^{n} r_{ij}^2 \leq c, i = \overline{1, n}, c > 0$ *a constant,*

$$\lim_{n \to \infty} \lim_{n \to \infty} \mathbf{P}\{|\operatorname{Tr} H_n C_n^{-1}| + \operatorname{Tr}[H_n C_n^{-1} + (H_n C_n^{-1})']^2 + \sum_{i,j=1}^{n} \nu_{ij}^2 \geq h\} = 0, \qquad (8.2.34)$$

where $H_n = (\nu_{ij}^{(n)})$, *the vector rows and vector columns of this matrix are infinitesimal,*

$$\lim_{h \to \infty} \overline{\lim_{n \to \infty}} \mathbf{P}\{|\sum_{p,l=1}^{n} \tilde{c}_{pl} \nu_{pl}| + \sum_{p,l=1}^{n} \tilde{c}_{pl}^2 \nu_{pl}^2 \geq h\} = 0,$$

$\tilde{c}_{pl}$ *the entries of the matrix* C_n^{-1}. *Then*

$$\det(B_n + \Xi_n)/\det C_n \sim \prod_{p,l=1}^{n} (l + \tilde{c}_{pl} \nu_{pl}).$$

Proof. By using Theorem 1.4.1, we get [see the (8.2.6)]

$$\mathbf{E} \det(I + \alpha_t H_n C_n^{-1})^{-K}$$

$$= \mathbf{E} \exp\{i\alpha_t \sum_{s=1}^{k} (H_n C_n^{-1} \vec{\xi_s}, \vec{\eta_s})$$

$$- i\alpha_t \sum_{s=1}^{k} (H_n C_n^{-1} \vec{\xi_s}, \vec{\xi_s})$$

$$- \alpha_t \sum_{s=1}^{k} (H_n C_n^{-1} \vec{\eta_s}, \vec{\eta_s})$$

$$- \alpha_t \sum_{s=1}^{k} (H_n C_n^{-1} \vec{\eta_s}, \vec{\eta_s})\},$$

where

$$\alpha_t = t\{q + |\operatorname{Tr} H_n C_n^{-1}| + 0.25 \operatorname{Tr}[H_n C_n^{-1} + (H_n C_n^{-1})']^2$$

$$+ |\sum_{p,l=1}^{n} \nu_{pl} c_{pl}| + \sum_{p,l=1}^{n} \nu_{pl}^2 \tilde{c}_{pl}^2\}, \quad q \geq 1, \quad 0 \leq t < 1,$$

$\overrightarrow{\xi_s}, \overrightarrow{\eta_s}, s = \overline{1,k}$ are independent random vectors distributed by the normal law $N(0,0,5I)$ ($\overrightarrow{\xi_s} \overrightarrow{\eta_s}$ do not depend on the entries of the matrix H_n).

Denote $C_n^{-1}\overrightarrow{\xi_s} = \overrightarrow{\widetilde{\xi_s}}, C_n^{-1}\overrightarrow{\eta_s} = \overrightarrow{\widetilde{\eta_s}}$. Consider the characteristic function

$$\mathbf{E}\exp\{i\theta_i \sum_{s=1}^{k}(H_n\overrightarrow{\widetilde{\xi_s}}, \overrightarrow{\eta_s}) + i\theta_2 \sum_{s=1}^{k}(H_n\overrightarrow{\widetilde{\eta_s}}, \overrightarrow{\xi_s})$$

$$+ i\theta_3 \sum_{s=1}^{k}(H_n\overrightarrow{\widetilde{\xi_s}}, \overrightarrow{\xi_s}) + i\theta_4 \sum_{s=1}^{k}(H_n\overrightarrow{\widetilde{\eta_s}}, \overrightarrow{\eta_s}) + i\theta_5 \operatorname{Tr} H_n C_n^{-1}$$

$$- \theta_6 \operatorname{Tr}(H_n C_n^{-1} + (H_n(C_n^{-1})')')^2\} = \mathbf{E}\exp\{i\sum_{p,l=1}^{n} \nu_{pl} y_{pl}\}, \qquad (8.2.35)$$

where

$$y_{pl} = \theta_1 \sum_{s=1}^{k} \widetilde{\xi_{ps}}\eta_{ls} + \theta_2 \sum_{s=1}^{k} \widetilde{\eta_{ps}}\xi_{ls} + \theta_3 \sum_{s=1}^{k} \widetilde{\xi_{ps}}\xi_{ls}$$

$$+ \theta_4 \sum_{s=1}^{k} \widetilde{\eta_{ps}}\eta_{ls} + \theta_5 \widetilde{C}_{pl} + \sqrt{\theta_6} \sum_{j=1}^{n} \widetilde{c_{pj}}\zeta_{jl},$$

$\theta_6 \geq 0, \zeta_{ij}, i, j = 1, 2, \ldots$ are independent random variables, not dependent on the variables $\xi_{ij}, \eta_{ij}, \widetilde{\xi}_{ij}, i, j = 1, 2, \ldots$, are distributed by the normal law $N(0, 1/2)$.

We represent (8.2.35) in the following form:

$$\mathbf{E}\prod_{p,l=1}^{n}(1 + \alpha_{pl}), \qquad (8.2.36)$$

where $\alpha_{pl} = \mathbf{E}[\exp\{i\nu_{pl}^{(n)}y_{pl}\}/y_{pl}] - 1$.

From the condition (8.2.34), it follows that [see (8.2.9)]

$$\sup_{n} \sum_{p,l=1}^{n} \mathbf{E}\nu_{pl}^2(1 + \nu_{pl}^2)^{-1} < \infty,$$

$$\lim_{n \to \infty} \sup_{p=\overline{1,n}} \sum_{l=1}^{n}[\mathbf{E}\nu_{pl}^2(1 + \nu_{pl}^2)^{-1}$$

$$+ \mathbf{E}\nu_{lp}^2(1 + \nu_{lp}^2)^{-1}] = 0. \qquad (8.2.37)$$

Therefore, in the same manner as in the proof of (8.2.15), we get

$$\lim_{n \to \infty} \sum_{p,l=1}^{n} \mathbf{E}|\alpha_{pl}|^2 = 0.$$

Since $\sum_{j=1}^n r_{ij}^2 \le C, \sum_{j=1}^n \tilde{c}_{ij}^2 = r_{ij} \le \sqrt{c}, i = \overline{1,n}$, by using the proof of Lemma 7.2.2 and (8.2.37), we find

$$\lim_{n\to\infty} \mathbf{E}|\sum_{p,l=1}^n \alpha_{pl} - \sum_{p,l=1}^n \mathbf{E}[\alpha_{pl}/\zeta_{ij}, i,j = \overline{1,n}]| = 0. \qquad (8.2.38)$$

It is easy to check this relation by considering the expression

$$\mathbf{E}|\sum_{p,l=1}^n (\alpha_{pl} - \beta_{pl})|^2 = \sum_{p,l,s,q,} \mathbf{E}(\alpha_{pl} - \beta_{pl})(\overline{\alpha_{sq} - \beta_{sq}}),$$

where $\beta_{pl} = \mathbf{E}[\alpha_{pl}/\zeta_{ij}, i,j = \overline{1,n}]$.

If $p \ne s,q$ and $l \ne s,q$, then from Lemma 6.2.2 it follows that this sum tends to zero as $n \to \infty$; if the two indices p,s or l,q coincide, then

$$\sum_{p,q} \mathbf{E}(\alpha_{pl} - \beta_{pl})(\overline{\alpha_{pq} - \beta_{pq}}) \le c \sum_{p=1}^n [\sum_{l=1}^n G_{pl}^{(n)}(+\infty))(\sum_{q=1}^n G_{pq}^{(n)}(+\infty)) \to 0,$$

since $\lim_{n\to\infty} \sup_{p=\overline{1,n}} \sum_{l=1}^n G_{pl}^n(+\infty) = 0, G_{pl}^{(n)}(+\infty) = \mathbf{E}\nu_{pl}^2(1 + \nu_{pl}^2)^{-1}$.

According to (8.2.38) and the proof of Theorem 8.2.1,

$$\mathbf{E}\det(I + \alpha_t H_n C_n^{-1})^{-k} = \mathbf{E}\prod_{p,l=1}^n [\mathbf{E}\exp\{i\alpha_t\nu_{pl}[\tilde{\xi}_{pl}\eta_{l1}$$

$$- \tilde{\eta}_{p1}\xi_{l1}] - \alpha_t\nu_{pl}[\tilde{\xi}_{p1}\xi_{l1} + \tilde{\eta}_{p1}\eta_{l1}]\}]^k + 0(1)$$

$$= \mathbf{E}\prod_{p,l=1}^n (1 + \alpha_t\tilde{c}_{pl}\nu_{pl})^{-k} + 0(1).$$

Theorem 8.2.4 is proved.

§3 Method of Integral Representations and Accompanying Infinitely Divisible Laws

From the Theorem 8.2.1 it follows that the distribution of the random determinant under increasing order towards infinity and under some conditions tends to the distribution of the product of some independent random variables. This fact gives an opportunity to find not only the general form of the limit distributions for random determinants, but the conditions of convergence to them.

Denote

$$\ln|1 + \nu_{ij}^{(n)}\nu_{ji}^{(n)}| = \gamma_{ij}^{(n)}, \quad i \ne j,$$

$$\ln|1 + \nu_{ii}^{(n)}| = \gamma_{ii}^{(n)}, \qquad \beta_{ij}^{(n)} = \int_{|x|<\tau_1} x dP\{\gamma_{ij}^{(n)} < x\},$$

$$T_n(x) = \sum_{i\ge j} \int_{-\infty}^x y^2(1+y^2)^{-1}dP\{\gamma_{ij}^{(n)} - \beta_{ij}^{(n)} < y\}, \quad \tau_1 > 0$$

is an arbitrary constant.

Theorem 8.3.1. *If the conditions of Theorem 8.2.1 hold and* $\det(I + B_n) \neq 0, n = 1, 2, \ldots,$ *then for the convergence of the random variables* $\ln |\det(I + \Xi_n)| - a_n$ *by a suitable choice of the constants* a_n *such that* $\sup_n |a_n| \leq \infty$, *it is necessary and sufficient that there exists the nondecreasing function* $T(x)$ *of bounded variation such that* $T_n(x) \Rightarrow T(x)$ *and* $\ln |\det(I + B_n)| - a_n + \int x (1 + x^2)^{-1} dT_n(x) - \sum_{i>j} \beta_{ij} \to \gamma$ *where* γ *is a finite constant. The logarithm of the characteristic function of the limit law has the form*

$$i\gamma t + \int (\exp(itx) - itx(1 + x^2)^{-1} - 1)(1 + x^2)x^{-2} dT(x).$$

Proof. Necessity. For each n the random variables $\gamma_{ij}^{(n)}, i \geq j, i, j = \overline{1, n}$ are independent and infinitesimal. Therefore, according to Theorem 5.3.2, we obtain

$$\lim_{n \to \infty} \sum_{i \geq j, i, j = \overline{1, n}} |f_{ij}(t) - 1|^2 = 0,$$

where $f_{ij}(t) = \mathbf{E} \exp\{it(\gamma_{ij}^{(n)} - \beta_{ij}^{(n)})\}$. Hence,

$$\mathbf{E} \exp\{it[\ln |\det(I + \Xi_n)| - a_n - \sum_{i \geq j} \beta_{ij}^{(n)}]\}$$

$$= \exp\{\sum_{i \geq j} \mathbf{E}[\exp\{it(\gamma_{ij}^{(n)} - \beta_{ij}^{(n)})\} - 1]$$

$$+ it \ln |\det(I + B_n)| - it a_n\} + 0(1).$$

Further we need to use the proof of Theorem 7.1.1. The necessity of the conditions of the theorem is proved. Sufficiency is obtained in the same way as in Theorem 7.1.1. Theorem 8.3.1 is proved.

By using Theorem 8.3.1, we can find the necessary and sufficient conditions of the convergence of the distributions of the random determinants to the normal, degenerate laws, to Puasson law. The theory of such limit theorems is developed in the paper [101].

Let us prove the limit theorems for the joint distribution of the determinant sign and its module. The most suitable transformation for this is the following one which in what follows will be called M-transformation (or the Mellin transformation),

$$f_n(k, t) = \mathbf{E}[\det(I + \Xi_n)/|\det(I + \Xi_n)|]^k |\det(I + \Xi_n)|^{it}, \qquad (8.3.1)$$

where $k = 0, \pm 1, \pm 2, \cdots - \infty < t < +\infty$.

In the case $\det(I + \Xi_n) = 0$, we assume $\det(I + \Xi_n) : |\det(I + \Xi_n)| = 0, |\det(I + \Xi_n)|^0 = 1$. We can find the distribution $\det(I + \Xi_n)$ by the function

$f_n(k,t)$ by using the formulas

$$\mathbf{E}[\exp\{it \ln |\det(I + \Xi_n)|\}/\det(I + \Xi_n)$$
$$\geq 0]\mathbf{P}\{\det(I + \Xi_n)$$
$$\geq 0\} = (f_n(0,t) + f_n(1,t))/2,$$
$$\mathbf{E}[\exp\{it \ln |\det(I + \Xi_n)|\}/\det(I + \Xi_n) < 0]\mathbf{P}\{\det(I + \Xi_n)$$
$$< 0\} = (f_n(0,t) - f_n(1,t))/2$$

and the inversion formula for the characteristic functions. Denote

$$K_n^{(1)}(x) = \sum_{p>l} \int_{-\infty}^{x} y^2(1 + y^2)^{-1} d\mathbf{P}\{\ln |1 - \gamma_{lp}^{(n)}\gamma_{pl}^{(n)}|$$

$$- \beta_{pl}^{(n)} < y, \gamma_{lp}^{(n)}\gamma_{pl}^{(n)} \leq 1\} + \sum_{p=1}^{n} \int_{-\infty}^{x} y^2(1 + y^2)^{-1}$$

$$\times d\mathbf{P}\{\ln |1 + \gamma_{pp}^{(n)}| - \beta_{pp}^{(n)} < y, \gamma_{pp}^{(n)} \geq -1\},$$

$$K_n^{(2)}(x) = \sum_{p>l} \int_{-\infty}^{x} y^2(1 + y^2)^{-1} d\mathbf{P}\{\gamma_{lp}^{(n)}\gamma_{pl}^{(n)} < y\}$$

$$+ \sum_{p=1}^{n} \int_{-\infty}^{x} y^2(1 + y^2)^{-1} d\mathbf{P}\{\gamma_{pp}^{(n)} < y\}.$$

Theorem 8.3.2. *If the conditions of Theorem 8.2.1 hold and $\det(I + B_n) \neq 0$, $n = 1, 2, \ldots$, then for the distribution functions of the random variables $\det(I + \Xi_n)e^{-a_n}, n = 1, 2, \ldots$ to be weakly convergent to the limit distribution function under some suitable choice of constants $a_n(\sup_n |a_n| < \infty)$, it is necessary and sufficient that*

$$\ln |\det(I + B_n)| - a_n + \int x(1 + x^2)^{-1} dT_n^{(1)} x - \sum_{i \geq j} \beta_{ij} \to \gamma$$

(γ is a finite constant) and that there exist the decreasing functions $T_n^{(1)}(x)$, $T_n^{(2)}(x)$ of bounded variation such that $T_n^{(1)}(x) \Rightarrow T^{(1)}(x), T_n^{(2)}(x) \Rightarrow T^{(2)}(x)$. The random limit variable is distributed analogously to $\xi_1 \xi_2$. The M-transformation of the variables ξ_1 and ξ_2 is equal to

$$\mathbf{E}\xi_1^k \xi_2^{it} = [\text{sign} \det(I + B_n)]^k \exp\{it\gamma + \int [e^{itx}$$

$$- itx(1 + x^2)^{-1} - 1](1 + x^2)x^{-2} dT^{(1)}(x) + \int_1^{\infty} (e^{ity}(-1)^k$$

$$- 1)(1 + y^2)y^{-2} dT^{(2)}(y)\}, \quad k = 0, 1.$$

Proof. If in formula (8.3.1) k is even, then the proof coincides with that of Theorem 8.3.1. If k is odd, the by using Theorem 8.2.1, we obtain

$$f_n(k,t)\exp\{-\sum_{p\geq l}\beta_{pl}\} = \prod_{p>l} f_{pl} \prod_p f_{pp} |\det(I+B)|^{it}(\operatorname{sign}\det(I+B_n))^k + 0(1),$$

where

$$f_{pl} = \mathbf{E}|1 - \nu_{pl}^{(n)}\nu_{lp}^{(n)}|^{it} \exp\{-it\beta_{pl}\} \operatorname{sign}[1 - \nu_{pl}^{(n)}\nu_{lp}^{(n)}],$$
$$f_{pp} = \mathbf{E}|1 + \nu_{pp}^{(n)}|^{it} \exp\{-it\beta_{pp}\} \operatorname{sign}[1 + \nu_{pp}^{(n)}], \quad p > l, \quad p,l = \overline{1,n}.$$

Obviously, as $0 < \tau < 1$,

$$|f_{pl} - 1|^2 = |\mathbf{E}\exp\{it\ln|1 - \nu_{pl}^{(n)}\nu_{lp}^{(n)}| - it\beta_{pl}\} - 1$$
$$+ \int_{|x|>\tau} \exp\{it\ln|1 - x| + it\beta_{pl}\}d\mathbf{P}\{\nu_{pl}^{(n)}\nu_{lp}^{(n)} < x\}$$
$$+ \int_{|x|>\tau} \exp\{it\ln|1 - x| - it\beta_{pl}\}\operatorname{sign}[1 - x]d\mathbf{P}\{\nu_{pl}^{(n)}\nu_{lp}^{(n)}$$
$$<x\}|^2 \leq 2|\mathbf{E}\exp\{it\ln|1 - \nu_{pl}^{(n)}\nu_{lp}^{(n)}| - it\beta_{pl}\} - 1|^2$$
$$+ 8(1+\tau^{-2})[1 - \mathbf{E}\{1 + (\nu_{pl}^{(n)}\nu_{lp}^{(n)})^2\}^{-1}]^2. \tag{8.3.2}$$

We have an analogous inequality for the variables f_{pp}.

From the proof of Theorem 8.3.1, it follows that

$$\lim_{n\to\infty}\{\sum_{p>l}|\mathbf{E}\exp\{it\ln|1 - \nu_{pl}^{(n)}\nu_{lp}^{(n)}| - it\beta_{pl}\} - 1|^2$$
$$+ \sum_{p=1}^n |\mathbf{E}\exp\{it\ln|1 + \nu_{pp}^{(n)}| - it\beta_{pp}\} - 1|^2\} = 0. \tag{8.3.3}$$

It is evident that

$$1 - \mathbf{E}\{1 + (\nu_{pl}^{(n)}\nu_{lp}^{(n)})^2\}^{-1} \leq 2\mathbf{E}(\nu_{pl}^{(n)})^2[1 + (\nu_{pl}^{(n)})^2]^{-1}$$
$$+ 4\mathbf{E}(\nu_{lp}^{(n)})^2[1 + (\nu_{lp}^{(n)})^2]^{-1}.$$

Since the variables $\nu_{pl}^{(n)}$ are infinitesimal and

$$\sup_n \sum_{p,l} \mathbf{E}(\nu_{pl}^{(n)})^2[1 + (\nu_{pl}^{(n)})^2]^{-1} < \infty,$$

then

$$\lim_{n\to\infty}\{\sum_{p>l}[1-\mathbf{E}\{1+(\nu_{pl}^{(n)}\nu_{lp}^{(n)})^2\}^{-1}]^2+\sum_{p=1}^n[1-\mathbf{E}\{1+(\nu_{pp}^{(n)})^2\}^{-1}]^2\} = 0. \tag{8.3.4}$$

Making use of (8.3.3) and (8.3.4), we obtain from (8.3.2),

$$\lim_{n\to\infty} \sum_{p\geq l} |f_{pl} - 1|^2 = 0.$$

Consequently,

$$\prod_{p>l} f_{pl} \prod_{p} f_{pp} = \exp\{\sum_{p\geq l}(f_{pl} - 1)\} + 0(1)$$

$$= \exp\{\sum_{p>l}\int_{1-x\geq 0} [\exp\{it \ln |1 - x| - it\beta_{pl}\} - 1]$$

$$\times dP\{\nu_{pl}^{(n)}\nu_{lp}^{(n)} < x\} + \sum_{p=1}^{n}\int_{1-x\geq 0} [\exp\{it \ln |1 - x|$$

$$- it\beta_{pl}\} - 1]dP\{-\nu_{pp}^{(n)} < x\} + \sum_{p>l}\int_{1-x<0} [\exp\{it$$

$$\times \ln |1 - x| - it\beta_{pl}\}(-1)^k - 1]dP\{\nu_{pl}^{(n)}\nu_{lp}^{(n)} < x\} \qquad (8.3.5)$$

$$+ \sum_{p=1}^{n}\int_{1-x<0} [\exp\{it \ln |1 - x| - it\beta_{pl}\}(-1)^k - 1]$$

$$\times dP\{-\nu_{pp}^{(n)} < x\}\} + 0(1) = \exp\int(e^{ity} - ity(1 + y^2)^{-1}$$

$$- 1)(1 + y^{-2})dT_n^{(1)}(y) + \int_1^{\infty} (\exp\{it \ln(1 - y) - it\varepsilon_n\}(-1)^k$$

$$- 1)(1 + y^{-2})dT_n^{(2)}(y) + it\int y^{-1}dT_n^{(1)}(y)\} + 0(1),$$

where ε_n is some number such that $\varepsilon_n \to 0$ as $n \to \infty$.

From formula (8.3.5) follows the sufficiency of the conditions of the theorem. Let us prove the necessity. Since random variables $\sum_{i\geq j} \gamma_{ij}^{(n)} - a_n$ converge by a suitable choice of constants a_n, then (see Chapter 5) $\sup_n T_n^{(1)}(+\infty) < \infty$. Let us choose the weakly convergent subsequence of the functions $T_n^{(1)}(x) \Rightarrow T^{(1)}(x)$, $T_n^{(2)}(x) \Rightarrow T^{(2)}(x)$. On the basis of (8.3.5),

$$\prod_{p>l} f_{pl} \prod_{p} f_{pp} \exp\{-it\int y^{-1}dT_n^{(1)}(y)\}$$

$$\to \exp\{\int(e^{ity} - ity(1 + y^2)^{-1} - 1)(1 + y^{-2})dT^{(1)}(y)$$

$$+ \int_1^{\infty} (e^{ity}(-1)^k - 1)(1 + y^{-2})dT^{(2)}(y)\} \qquad (8.3.6)$$

Let us show that the transformation (8.3.6) one-to-one defines the functions $T^{(1)}(y)$ and $T^{(2)}(y)$. (Denote it by $f(k,t)$. By using (7.1.12) we obtain

$$\int_{s-1}^{s+1} \ln f(k,t)dt - 2\ln f(k,s) = -2\int e^{isx}$$

$$\times (1 - x^{-1}\sin x)(1+x^2)x^{-1}\{dT^{(1)}(x) + \theta(x)(-1)^k dT^{(2)}(x)\},$$

$$\theta(x) = 1, \quad x \geq 1, \quad \theta(x) = 0, \quad x < 1.$$

Hence, as in the proof of Corollary 7.1.1, we obtain that the functions $T^{(1)}(x)$ and $T^{(2)}(x)$ are one-to-one restored by the function $f(k,t)$ at almost all points x. Theorem 8.3.2 is proved.

Corollary 8.3.1. *Let the conditions of Theorem 8.3.2 hold and the nondecreasing function of bounded variation $T^{(2)}(x)$ exist such that $T_n^{(2)}(x) \Rightarrow T^{(2)}(x)$. Then*

$$\lim_{n\to\infty} [\mathbf{P}\{\det(I+\Xi_n)\det(I+B_n) \geq 0\} + \mathbf{P}\{\det(I$$

$$+\Xi_n)\det(I+B_n) > 0\}] = 1 + \exp[-2\int_1^\infty (1+y^{-2})dT^{(2)}(y)].$$

We obtain the proof from Theorem 8.3.2 and from the following equality, $\mathbf{P}\{\xi > 0\} + \mathbf{P}\{\xi \geq 0\} = \{1 + \mathbf{E}\operatorname{sign}\xi\}$, where ξ is some random variable.

Now we come to the proof of the limit theorems for the determinants of the complex random matrices. We use the following transformation:

$$\mathbf{E}|\det H_n|^{it}\exp\{k[\arg\det H_n]\}, -\infty < t < \infty, \quad k = 0, \pm 1, \pm 2, \ldots,$$

where H_n is a complex random matrix.

To simplify the proofs and statements of the theorems, we consider the complex symmetric matrices $I + i\Xi_n$, where Ξ_n is a real symmetric random matrix.

Theorem 8.3.3. *For each n let the random entries $\xi_{ij}^{(n)}$,, $i \geq j$, $i,j = \overline{1,n}$ of the matrix Ξ_n by independent, the row vectors of the matrix Ξ_n be asymptotically constant,*

$$\lim_{h\to\infty}\lim_{n\to\infty}\mathbf{P}\{|\sum_{i=1}^n \nu_{ii}| + \sum_{i,j=1}^n (\nu_{ij}^{(n)})^2 \geq h\} = 0,$$

where $\nu_{ij} = \xi_{ij} - a_{ij} - \rho_{ij}, \rho_{ij} = \int_{|x|<\tau} xdF_{ij}^{(n)}(x + a_{ij}^{(n)})$, τ is an arbitrary constant, and the nonrandom variables $a_{ij}^{(n)}$ are such that for any $\varepsilon > 0$,

$$\lim_{n\to\infty}\sup_{i,j=\overline{1,n}}\mathbf{P}\{|\xi_{ij}^{(n)} - a_{ij}^{(n)}| > \varepsilon\} = 0, \quad \sup_n[|\operatorname{Tr}B_n| + \operatorname{Tr}B_n B_n'] < \infty,$$

where $B_n = (b_{ij}), b_{ij} = \rho_{ij} + a_{ij}$.

Then in order for the distribution functions of the random vectors $\{\ln |\det (I+i\Xi_n)| - a_n, \arg \det(I+i\Xi_n) - b_n\}$ to converge weakly to the limit distribution function with a suitable choice of constants $a_n (\sup_n |a_n| < \infty)$, it is necessary and sufficient that there exist the nondecreasing functions of bounded variation $G_1(x)$ $G_2(x)$ such that $G_{n1}(x) \Rightarrow G_1(x), G_{n2}(x) \Rightarrow G_2(x), 0.5 \ln \det(I + B_n^2) - a_n \to \gamma_1, \int x^{-1} dG_{n2}(x) + \arg \det(I + iB_n) - b_n \to \gamma_2$, where γ_1 and γ_2 are finite numbers,

$$G_{n1}(x) = \sum_{p>l} \int_{-\infty}^{x} y^2 (1+y^2) dP\{\nu_{pl}^{(n)} < y\},$$

$$G_{n2}(x) = \sum_{p=1}^{n} \int_{-\infty}^{x} y^2 (1+y^2)^{-1} dP\{\nu_{pp}^{(n)} < y\}.$$

The distribution function of the limit random variables are one-to-one defined by the following transformation

$$m(t,k) := \exp\{it\gamma_1 + ik\gamma_2 + \int [(1+x^2)^{it} - 1](1+x^2)$$

$$\times x^{-2} dG_1(x) + \int [(1+ix)^k (1+x^2)^{it/2-k/2} - ikx(1+x^2)^{-1}$$

$$- 1](1+x^{-2}) dG_2(x)\}, \quad k = 0, \pm 1, \pm 2, \ldots \tag{8.3.7}$$

Proof. By using Corollary 8.2.3, we get

$$\mathbf{E} \exp\{it \ln |\det(I + i\Xi_n)| - a_n) + ik(\arg \det(I + i\Xi_n)$$

$$- b_n)\} = \prod_{p=1}^{n} f_{pp} \prod_{p>1} f_{pl} \exp\{it\delta_{1n} + ik\delta_{2n}\} + 0(1),$$

where $f_{pp} = \mathbf{E}(1 + i\nu_{pp})^k (1 + \nu_{pp}^2)^{(it-k)/2}, f_{pl} = \mathbf{E}(1 + \nu_{pl}^2)^{it}, \delta_{1n} = \ln \det(I + B_n^2) - a_n, \delta_{n2} = \arg \det(I + iB_n) - b_n$.

It is obvious that for any finite t,

$$\lim_{n \to \infty} \sup_{p=\overline{1,n}} |f_{pp}| = 1,$$

$$\lim_{n \to \infty} \sup_{p,l=\overline{1,n}} |f_{pl}| = 1,$$

$$|f_{pl} - 1| = |\int [(1+x^2)^{it} - 1]dF_{pl}(x)| \leq 2\int_{|x| \geq \tau} dF_{pl}(x)$$

$$+ |t| \int_{|x| < \tau} x^2 dF_{pl}(x) \leq [|t|(1+\tau^2) + 2(1+\tau^{-2})]$$

$$\times \mathbf{E}\nu_{pl}^2 [1 + \nu_{pl}^2]^{-1}, \tag{8.3.8}$$

$$|f_{pp} - 1| = |\int [(1+ix)^k (1+x^2)^{(it-k)/2} - 1]dF_{pp}(x)|$$

$$\leq \int |(1+x^2)^{it/2} - 1|dF_{pp}(x) + \int_{|x| < \tau} |(1+ix)^k (1+x^2)^{-k/2}$$

$$- ikx - 1|dF_{pp}(x) + 2\int_{|x| \geq \tau} dF_{pp}(x)$$

$$+ |k|\tau| \int_{|x| < \tau} dF_{pp}(x)|. \tag{8.3.9}$$

After some simple calculations, as $|x| < \tau$,

$$|(1+ix)^k (1+x^2)^{-k/2} - ikx - 1| \leq |(1+ix)^k - ikx - 1|$$

$$+ (1 + |kx|)|(1+x^2)^{k/2} - 1| \leq x^2[\tau^{-2}(1+\tau)^k$$

$$+ (1 + |k|\tau)\tau^{-2}(1+\tau^2)^k]. \tag{8.3.10}$$

For sufficiently large n [see (8.2.13)],

$$|\int_{|x| < \tau} dF_{pp}(x)| \leq 2\tau \int_{|x| > \tau/2} dF_{pp}(x). \tag{8.3.11}$$

Using the inequalities

$$\int_{|x| < \tau} x^2 dF_{pl}(x) \leq (1+\tau^2) \int x^2 (1+x^2)^{-1} dF_{pl}(x),$$

$$\int_{|x| > \tau} dF_{pl}(x) \leq (1+\tau^{-2}) \int x^2 (1+x^2)^{-1} dF_{pl}(x),$$

(8.3.11), (8.3.10), and (8.3.9), we obtain

$$|f_{pp} - 1| \leq \{|t|(1+\tau^2) + (1+\tau^2)\tau^{-2}$$

$$+ (1+\tau^2)((1+\tau)^k \tau^{-2} + (1 + |k|\tau)(1+\tau^2)^k \tau^{-2})$$

$$+ 2|k|(4+\tau^2)\}[1 - \mathbf{E}\{1 + \nu_{pp}^2\}^{-1}]. \tag{8.3.12}$$

On the basis of (8.3.12) and (8.3.8), $\lim_{n\to\infty}\sum_{p\geq l}|f_{pl}-1|^2 = 0$. Consequently,

$$\mathbf{E}\exp\{it(\ln|\det(I+i\Xi_n)|-a_n)+ik(\arg\det(I+i\Xi_n)-b_n)\}$$

$$=\exp[\sum_{p\geq l}(f_{pl}-1)]+0(1)=\exp\{it\delta_{1n}+ik\delta_{2n}+\int[(1+x^2)^{it}-1]$$

$$\times(1_x^{-2})dG_{n1}(x)+\int[(1+ix)^k(1+x^2)^{-k/2}(1+x^2)^{it/2}$$

$$-ikx(1+x^2)^{-1}-1](1+x^{-2})dG_{n2}(x)\}+0(1).$$

According to the conditions of this theorem, $\sup_n G_{n1}(+\infty)<+\infty, \sup_n G_{n2}$ $(+\infty)<+\infty$. Therefore, the sets of the functions $\{G_{n1}(x)\}$ and $\{G_{n2}(x)\}$ will be weakly compact. To complete the proof of the theorem, we need to show that the formula (8.3.7) one-to-one defines $G_1(x)$ and $G_2(x)$. It follows from the equality

$$\int_{t-1}^{t+1}\ln m(z,0)dz-2\ln m(t,0)=\int(1+x^2)^{it}dV(x),$$

where

$$\mathbf{V}(u)=-2\int_{-\infty}^{u}[1-\sin\ln(1+x^2)[\ln(1+x^2)]^{-1}$$

$$\times(1+x^{-2})d[G_1(x)+G_2(x)],$$

$$\ln m(0,k+2)-\ln m(0,k)-\ln m(0,k-2)$$

$$=\int\exp[ik\arg(1+ix)]dW(x),$$

$$W(u)=-\int_{-\infty}^{u}[1+2\sin^2(\arg(1+ix))](1+x^{-2})dG_2(x).$$

By using the fact that the distribution function concentrated on $(-\pi,\pi)$ and is one-to-one defined by its Fourier coefficients, from these two equalities, we conclude that formula (8.3.7) one-to-one defines $G_1(x)$ and $G_2(x)$. Theorem (8.3.3) is proved.

Corollary 8.3.2. *Let the conditions of Theorem 8.3.3 hold. Then in order that*

$$\text{plim}_{n\to\infty}[\ln|\det(I+i\Xi_n)|-0.5\ln\det(I+B_n^2)]=\sigma^2,$$

$$\lim_{n\to\infty}\mathbf{P}\{\arg\det(I+i\Xi_n)-\arg\det(I+iB_n)$$

$$-\int x^{-1}dG_{n2}(x)<z\}=\int_{-\infty}^{z}p(x)dx,$$

where

$$p(x) = \begin{cases} (2\pi c)^{-1/2} \sum_{-\infty}^{\infty} \exp\{-0.5c^{-1}(x+2k\pi)^2\}, & x \in (-\pi, \pi), \\ 0, & x \in (-\infty, \pi] \cup [\pi, +\infty), \quad c > 0, \end{cases}$$

it is necessary and sufficient that for any fixed $\tau > 0$, the conditions

$$\sum_{p,l=1}^{n} \mathbf{P}\{|\nu_{pl}| \geq \tau\} \to 0,$$

$$\sum_{p>l} \{\int_{|x|<\tau} x^2 dF_{pl}(x) - (\int_{|x|<\tau} x dF_{pl}(x)^2\} \to \sigma^2,$$

$$\sum_{p=1}^{n} \{\int_{|x|<\tau} x^2 dF_{pp}(x) - (\int_{|x|<\tau} x dF_{pp}(x))^2\} \to c \qquad (8.3.13)$$

hold.

Proof. For the assertion of Corollary 8.3.2 to hold in formula (8.3.7), we set

$$G_{n1}(x) = \begin{cases} 0, x \leq 0, \\ \sigma^2, x > 0, \end{cases} \qquad G_2(x) = \begin{cases} 0, x \leq 0, \\ c, x > 0. \end{cases}$$

This is easy to verify by using the summation Poisson formula [34, p. 710] $\sum_{-\infty}^{\infty} \varphi(x + 2k\lambda) = \pi\lambda^{-1} \sum_{-\infty}^{\infty} f(n\pi/\lambda) \exp\{in(\pi/\lambda)x\}$, where f is the probability density with integrable characteristic function $\varphi; \lambda, x$ are the real parameters. The conditions $G_{n1}(x) \Rightarrow G_1(x), G_{n2}(x) \Rightarrow G_2(x)$ are equivalent to the conditions of (8.3.13) (see [101]). Corollary 8.3.2 is proved.

Let us consider the determinants of the random matrices $I + \Xi_n \Xi_n'$. For such matrices the calculations are considerably simplified; the conditions of the convergence to the limit laws of distribution have the same form as the conditions of the convergence of the sums of independent infinitesimal random variables.

Denote $G^n(x) = \sum_{p,l=1}^{n} \int_0^x y(1+y)^{-1} d\mathbf{P}\{(\nu_{pl}^{(n)})^2 < y\}$.

Theorem 8.3.4. *If for each n the random entries $\xi_{pl}^{(n)}, p, l = \overline{1, n}$ of the matrix $\Xi_n = (\xi_{pl}^{(n)})$ are independent, the row vectors and column vectors of the matrix Ξ_n are asymptotically constant, and $\sup_n \operatorname{Tr} B_n B_n' < \infty$, then for the convergence of the random variables $\kappa_n := \ln \det(I + \Xi_n \Xi_n') - \ln \det(I + B_n B_n')$, it is necessary and sufficient that there exists a nondecreasing function $G(x)$ of bounded variation such that $G_n(x)$ converges weakly to $G(x)$. The logarithm of the Laplace transformation of the limit law is equal to*

$$\varphi(t) = \int_0^\infty [(1+u)^{-t} - 1](1 + u^{-1}) dG(u).$$

Proof. Necessity. From the convergence of the distributions and the fact that $\sup_n \operatorname{Tr} B_n B'_n < \infty$, it follows that $\lim_{h\to\infty} \overline{\lim}_{n\to\infty} \mathbf{P}\{\sum_{p,l=1}^n (\nu_{pl}^{(n)})^2 \geq h\} = 0$. Consequently (see Chapter 5), $\sup_n G_n(+\infty) < +\infty$.

Denote

$$\mu_{kn} = \mathbf{E}\det(I + \Xi_n \Xi'_n)^{-k/2}$$

$$= \mathbf{E}\exp\{i \sum_{p,l=1}^n \nu_{pl}\beta_{pl} + i \sum_{p,l=1}^n b_{pl}\beta_{pl}\}$$

$$= \mathbf{E}\prod_{p,l=1}^n (\alpha_{pl} + 1)\exp\{i \sum_{p,l=1}^n b_{pl}\beta_{pl}, \qquad (8.3.14)$$

$$\alpha_{pl} = \mathbf{E}\{\exp(i\mu_{pl}\beta_{pl}/\beta_{pl}\} - 1,$$

where $\beta_{pl} = 2\sum_{m=1}^k \eta_{pm}\zeta_{lm}$; η_{pm}, ζ_{lm} are independent among themselves and of the random variables ν_{pl} random variables distributed by the normal $N(0,1)$ law. (In formula (8.3.14) we use the integral representations for determinants [see Chapter 1].

Since the row vectors and column vectors of the matrix Ξ_n are asymptotically constant, $\sum_{p=1}^n \mathbf{E}\nu_{pl}^2(1+\nu_{pl}^2)^{-1} \to 0$, $\sum_{p=1}^n \mathbf{E}\nu_{lp}^2(1+\nu_{lp}^2)^{-1} \to 0(n \to \infty)$. By using these expressions as in the proof of Theorem 8.2.1, we obtain

$$\lim_{n\to\infty} \sum_{p,l=1}^n \mathbf{E}|\alpha_{pl}|^2 = 0, \quad \operatorname{plim}_{n\to\infty} |\sum_{p,l=1}^n (\alpha_{pl} - \mathbf{E}\alpha_{pl}| = 0.$$

Consequently, (8.3.14) can be transformed to

$$\lim_{n\to\infty} \mathbf{E}\exp(-k\kappa_n) = \lim_{n\to\infty} \exp\{\varphi_n(k)\}, \qquad (8.3.15)$$

where $\varphi_n(k) = \int_0^\infty [(1+u)^{-k} - 1](1+u^{-1})dG_n(u)$.

Obviously, $\exp(\varphi_n(t))$ is a completely monotone function and $\varphi_n(0) = 0$. Therefore $\exp(\varphi_n(t)) = \mathbf{E}\exp(-t\gamma_n)$, where $\gamma_n \geq 0$ is some random variable. From the convergence of variables κ_n the convergence of the variables γ_n, and consequently, the convergence of $\exp(\varphi_n(t))$ to some completely monotone function $m(t)$ follow. Therefore, on the basis of Lemma 2.2.1 in (8.3.15), k can be replaced by any $t > 0$. Since the set $\{G_n(x)\}$ is weakly compact, from the convergence of $\varphi_n(t)$ follows the weak convergence of $G_n(x)$ to the nondecreasing function $G(x)$ of bounded variation whose uniqueness follows from the equality

$$\int_{t-1}^{t+1} \ln m(z)dz - 2\ln m(t) = \int_0^\infty (1+x)^{-t}dv(x),$$

where $v(x) = \int_0^x (1 + u - (1+u)^{-1})[[\ln(1+u)]^{-1} - 2](1+u^{-1})dG(u)$. The necessity is proved. The proof of the sufficiency is based on the Helly theorem. Theorem 8.3.4 is proved.

Corollary 8.3.3. *If the conditions of Theorem 8.3.4 hold, then for the convergence of the random variables* χ_n, *it is necessary and sufficient that*

$$\sum_{p,k=1}^{n} (1 - \mathbf{P}\{[\xi_{kp}^{(n)} - \alpha_{kp}]^2 < x\}) \Rightarrow L(x), x > 0; \qquad (8.3.16)$$

$$\lim_{\varepsilon \to 0} \overline{\lim_{n \to \infty}} \sum_{p,k=1}^{n} \{\int_{|x|<\varepsilon} x^2 dF_{kp}(x) - (\int_{|x|<\varepsilon} x dF_{kp}(x))^2\}$$

$$= \lim_{\varepsilon \to 0} \lim_{n \to \infty} \sum_{p,k=1}^{n} \{\int_{|x|<\varepsilon} x^2 dF_{kp}(x)$$

$$- (\int_{|x|<\varepsilon} x dF_{kp}(x))^2\} = G^2, \qquad (8.3.17)$$

the function $L(x)$ *is not decreasing in the interval* $(0, +\infty), L(+\infty) = 0$, *for any finite* $\delta > 0 \int_0^\delta x dl(x) < \infty$ *(zero is excluded from the domain of integration). The logarithm of the Laplace transformation of the limit law is equal to*

$$\varphi(t) = - \int_0^\infty [(1+u)^{-t} - 1] dL(u) - t\sigma^2, \quad t \geq 0. \qquad (8.3.18)$$

Proof. The theorem is proved if we show that under the fulfillment of the condition of infinitesimalness, (8.3.16) and (8.3.17) are equivalent to $G_n(x) \Rightarrow G(x)$. The last condition is equivalent to

$$\sum_{p,k=1}^{n} (1 - \mathbf{P}\{[\xi_{kp}^{(n)} - a_{kp}]^2 < x\}) \Rightarrow \int_x^\infty (1 + y^{-1}) dG(y), \quad x > 0,$$

$$\lim_{\varepsilon \to 0} \overline{\lim_{n \to \infty}} \sum_{p,k=1}^{n} \int_{0<x<\varepsilon} x(1+x)^{-1} d\mathbf{P}\{\nu_{pk}^{(n)} < x\}$$

$$= \lim_{\varepsilon \to 0} \lim_{n \to \infty} \sum_{p,k=1}^{n} \int_{0<x<\varepsilon} x(1+x)^{-1} d\mathbf{P}\{\nu_{pk}^{(n)} < x\}$$

From this we obtain (8.3.16) and (8.3.17) [see 142].

Corollary 8.3.4. *If the random variables* $\xi_{pl}, p, l = 1, 2, \ldots$ *are independent and identically distributed with the distribution functions* $F(x)$, *then for the convergence of the random variables,*

$$\kappa_n := \ln \det[I + ((\xi_{ij} - a_n)c_n^{-1})((\xi_{ij} - a_n)c_n^{-1})'],$$

with a suitable choice of constants a_n, c_n, *it is necessary and sufficient for* $F(x)$ *to belong to the domain of attraction of the stable law with characteristic index*

$\alpha(0 < \alpha \leq 2)$. *The Laplace transformation of the limit distribution function equals*

$$\exp\{-d[B(t, \alpha/2)]^{-1}\}, \quad t \geq 0,$$

$$d = \begin{cases} c\Gamma(1 - \alpha/2)\Gamma(\alpha/2), & 0 < \alpha < 2 \\ \sigma^2, & \alpha = 2 \end{cases}$$

The constants a_n and c_n are chosen so that $a_n = c_n \int_{|x|<\tau} x\, dF(c_n x), \tau > 0$ is an arbitrary constant and the following condition holds,

$$\lim_{n \to \infty} n^2[1 - \mathbf{P}\{\xi_{ij}^2 > c_n x\}] = cx^{-\alpha/2}, \quad x > 0 \, (0 < \alpha < 2), \qquad (8.3.19)$$

for all $\varepsilon > 0$,

$$n^2 \int_{|x| \geq \varepsilon} dF(c_n x) \to 0,$$

$$n^2 \left[\int_{|x| < \varepsilon} x^2 dF(c_n x) - \left(\int_{|x| < \varepsilon} x\, dF(c_n x) \right)^2 \right] \to \sigma^2 \, (\alpha = 2) \quad (8.3.20)$$

Proof. From Corollary 8.3.4 it follows that the row vectors and the column vectors of the matrix $(\xi_{ij} - a_n)c_n^{-1}$ are infinitesimal in case of the convergence of the variables κ_n. The condition (8.3.16) is equivalent to the condition (8.3.19) [101] as $0 < \alpha < 2$ and to the condition (8.3.20) as $\alpha = 2$ [101].

The Laplace transformation of the limit distribution function is obtained from the following expression:

$$-\int_0^\infty [(1 + u)^{-t} - 1] dL(u) - t\sigma^2 = 0.5ac \int_0^\infty [(1 + x)^{-t} - 1]$$
$$\times x^{-\alpha/2 - 1} dx - t\sigma^2 = c\Gamma(1 + \alpha/2)\Gamma(t + \alpha/2)[\Gamma(t)]^{-1} - t\sigma^2$$
$$= -d[B(t, \alpha/2)].$$

Corollary 8.3.4 is proved.

§4 Limit Theorems of the General Form for Random Determinants

Let the square matrices $\Xi_n = (\xi_{ij}^{(n)})$ of order n be given. If for each n the vectors $(\xi_{ij}^{(n)}, \xi_{ij}^{(n)}), i \geq j, i, j = \overline{1, n}$ are independent and the conditions (8.2.3) and (8.2.4) hold, then [see (8.2.9)]

$$\sup_n \sum_{p \geq l} \mathbf{E}\nu_{pl}^2(1 + \nu_{pl}^2)^{-1} < \infty.$$

By using this condition and the fact that the variables ν_{pl} are infinitesimal, it is easy to prove that the set of the vectors $\{(\nu_{ij}, \nu_{ji}), i > j, i, j = \overline{1,n}\}$ can be divided into $2n$ noncutting sets $R'_{in}, R''_{in}, i = \overline{1,n}$ so that the vectors $\overrightarrow{\mu_i}$, composed of the elements of each set, will be infinitesimal, with the set R'_{in} including only the entries of the ith row vector of the matrix $\Xi_n + \Xi'_n$, and with the set R''_{in} including only the entries of the ith column vectors of the matrix $\Xi_n + \Xi'_n$.

Denote the set of values of the index i of the variables $(\nu_{ij}, \nu_{ji}) \in U^n_{p=1} R'_{pn}$ by T_{jn} and the set of values of the index j of the variables $(\nu_{ij}, \nu_{ji}) \in U^n_{p=1} R''_{pn}$ by K_{in} (suppose that $\sum_{i \in \varnothing} = 0$).

We shall use the notations of §1 of this chapter in the following theorem.

Theorem 8.4.1. *If for each n the vectors $(\xi^{(n)}_{ij}, \xi^{(n)}_{ji}), i \geq j$ are independent and asymptotically constant and if conditions (8.2.3) and (8.2.4) hold, then*

$$\det(I + \Xi_n) \sim \prod_{i=1}^n (1 + \nu_{ii}^{(n)}) \det[\delta_{ij} - \delta_{ij} \sum_{p \in T_i \cup K_i} \nu_{pi}\nu_{ip} + b_{ij}]. \qquad (8.4.1)$$

Proof. Denote $\beta_{pp} = \mathbf{E}\alpha_{pp}$,

$$\beta_{pl} = \begin{cases} \mathbf{E}(\alpha_{pl}/\eta_{ls}, \tilde{\xi}_{ls}, s = \overline{1,k}), & (\nu_{lp}, \nu_{pl}) \in R'_{ln} \\ \mathbf{E}(\alpha_{pl}/\eta_{ps}, \tilde{\xi}_{ps}, s = \overline{1,k}), & (\nu_{lp}, \nu_{pl}) \in R''_{ln}, p \neq l \end{cases}$$

As in the proof of Theorem 8.2.1, we determine that (8.2.15) holds. Since the vectors $\overrightarrow{\mu_i}$ are infinitesimal,

$$\lim_{n \to \infty} \sup_{i=\overline{1,n}} \sum_{\nu_{ij}, \nu_{ji} \in R'_{in} R''_{in}} \mathbf{E}\mu_{ij}^2 (1 + \mu_{ij}^2)^{-1} = 0 \qquad (8.4.2)$$

By using a limit, we obtain

$$\lim_{n \to \infty} \mathbf{E}|\sum_{p \geq l} (\alpha_{pl} - \beta_{pl})|^2 \leq \lim_{n \to \infty} \left\{ \sum_{p=1}^n \mathbf{E}|\alpha_{pp}|^2 \right.$$
$$\left. + \sum_{p=1}^n \mathbf{E}\left[\sum_{(\nu_{lp}\nu_{pl} \in R'_{in} \cup R''_{in})} |a_{lp}| \right]^2 \right\}$$

Hence, from (8.4.2),(8.2.8), and (8.2.9) it follows that

$$\lim_{n \to \infty} \mathbf{E}|\sum_{p \geq l} (\alpha_{pl} - \beta_{pl})|^2 = 0.$$

Further,

$$
\mathbf{E}\det(I + \alpha_t\Xi)^{-k} = \mathbf{E}\exp\Big\{ i\alpha_t\sum_{s=1}^{k}((B - B')\vec{\xi_s}, \vec{\varepsilon_s})
$$

$$
- \alpha_t\sum_{s=1}^{k}(B\vec{\xi_s}, \vec{\xi_s}) - \alpha_t\sum_{s=1}^{k}(B\vec{\eta_s}, \vec{\eta_s} + \sum_{s=1}^{k}\sum_{p=1}^{n}\sum_{l\in T_p}[i
$$

$$
\times \alpha_t\eta_{lsp}(\nu_{lp}\tilde{\xi}_{ps} - \nu_{lp}\tilde{\xi}_{ps}) + i\alpha_t\xi_{lsp}(\nu_{lp}\eta_{ps} - \nu_{pl}\eta_{pl})
$$

$$
- \alpha_t\eta_{lsp}\eta_{ps}(\nu_{pl} + \nu_{lp}) - \alpha_t\tilde{\xi}_{lsp}\tilde{\xi}_{ps}(\nu_{pl} + \nu_{lp})] \tag{8.4.3}
$$

$$
+ \sum_{s=1}^{k}\sum_{l=1}^{n}\sum_{p\in K_l}[i\alpha_t\overline{\eta}_{psl}(\nu_{lp}\tilde{\xi}_{ls} - \nu_{pl}\tilde{\xi}_{ls}
$$

$$
+ i\alpha_t\xi_{psl}(\nu_{pl}\eta_{ls} - \nu_{lp}\eta_{ls}) - \alpha_t\overline{\eta_{psl}}\eta_{ls}(\nu_{pl} + \nu_{lp})
$$

$$
- \alpha_t\overline{\xi}_{psl}\tilde{\xi}_{ls}(\nu_{pl} + \nu_{lp})]\Big\} \prod_{p=1}^{n}(1 + \alpha_t\nu_{pp})^{-k} + 0(1),
$$

where the random variables $\eta_{lsp}, \xi_{lsp}, l, s, p = 1, 2, \ldots$ are independent, do not depend on the variables $\nu_{pl}, \eta_{ls}, \xi_{ls}$ and are distributed by the normal law $N(0, 1/2)$. From (8.4.3), after some simple calculations, we obtain $\mathbf{E}\det(I + \alpha_t\Xi)^{-k} = \mathbf{E}\det[\delta_{ij} - \delta_{ij}\alpha_t^2\sum_{p\in T_i\cup K_i}\nu_{pi}\nu_{ip} + \alpha_tb_{ij}]^{-k}\prod_{p=1}^{n}(1 + \alpha_t\nu_{pp})^{-k} + 0(1)$. From this, (8.4.1) follows. Theorem 8.4.1 is proved.

Corollary 8.4.1. *If in addition to the conditions of Theorem 8.4.1,*

$$
\lim_{n\to\infty}[|\operatorname{Tr} B_n| + \operatorname{Tr} B_nB_n'] = 0, \tag{8.4.4}
$$

then

$$
\det(I + \Xi_n) \sim \prod_{i=1}^{n}(1 + \nu_{ii})\prod_{i=1}^{n}(1 - \sum_{p\in T_i\cup K_i}\nu_{pi}\nu_{ip}).
$$

Corollary 8.4.2. *If in addition to the conditions of Theorem 8.4.1, condition (8.4.4) holds and the vectors $(\xi_{ik}, i = \overline{1, k}), (\xi_{ki}, i = \overline{1, k}), i = \overline{1, n}$ are asymptotically constant, then the sets T_{in}, K_{in} can be chosen so that*

$$
\det(I + \Xi_n) \sim \prod_{i=1}^{n}(1 + \nu_{ii})\prod_{i=1}^{n}(1 - \sum_{j=i+1}^{n}\nu_{ji}\nu_{ij}).
$$

Let us consider the symmetric matrices $\Xi_n = (\xi_{ij})$. Analogously to Theorem 8.4.1, we introduce the vectors $\vec{\mu_i}$, composed of the elements of the set $L_n = \{\nu_{ij}, i < j, i, j = \overline{1, n}\}$.

Theorem 8.4.2. *If for each n the random variables $\xi_{ij}^{(n)}, i \geq j, i, j = \overline{1,n}$ of the symmetric matrix $\Xi_n = (\xi_{ij}J)$ are independent and asymptotically constant, $\sup_n \operatorname{Tr} B_n^2 < \infty$, then*

$$\det(I + i\Xi_n) \sim \det(I + iB_n + \Lambda_n) \prod_{k=1}^{n}(1 + i\nu_{kk}), \qquad (8.4.5)$$

where $\Lambda_n = \operatorname{diag}\{\sum_{l \in T_p \cup K_p} \nu_{pl}^2, p = \overline{1,n}\}$.

Proof. Suppose that the distribution functions of the random variables on the left-hand side of the relation (8.4.5) converge weakly to the limit distribution function. Then from the inequalities $|\det(I + i\Xi_n)| \geq [1 + \sum_{i,j=1}^{n} \xi_{ij}^2]^{1/2}$, $\sup_n \operatorname{Tr} B_n^2 < \infty$, the boundedness of the variables $\sum_{i,j=1}^{n} \nu_{ij}^2$ in probability follows. Hence, $\sup_n \sum_{p,l=1}^{n} \mathbf{E}\nu_{pl}^2(1 + \nu_{pl}^2)^{-1} < \infty$.

By using the proof of Theorems 8.2.2 and 8.4.1, we obtain (8.4.5). The relation (8.4.5) is proved analogously if the distribution functions of the random variables in the right-hand side are weakly convergent. Theorem 8.4.2 is proved.

Now let us pass over to the limit theorems for the determinants of the random matrices $1 + \Xi_n \Xi_n'$.

Theorem 8.4.3. *If for each n the random entries $\xi_{ij}^{(n)}, i, j = \overline{1,n}$ of the matrices $\Xi_n = (\xi_{ij}^{(n)})$ are independent and asymptotically constant, $\sup_n \operatorname{Tr} B_n B_n' < \infty$, then*

$$\det(I + \Xi_n \Xi_n') \sim \prod_{j=1}^{n}\left(1 + \sum_{i \in T_{jn}} \nu_{ij}^2\right)$$

$$\times \det[\delta_{ij}(1 + \sum_{j \in k_{in}} \nu_{ij}^2) + c_{ij}],$$

where $c_{ij} = \sum_{k=1}^{n} b_{ik}b_{jk}(1 + \sum_{i \in T_{kn}} \nu_{ik}^2)^{-1}$. The sets K_{in} and T_{in} are defined as in Theorem 8.4.2, with the only difference that instead of the set L_n we consider the set $\{\nu_{ij}, i, j = \overline{1,n}\}$.

Proof. Since $\det(I + \Xi_n \Xi_n' \geq 1 + \sum_{p,l=1}^{n} \xi_{pl}^2$ and $\sup_n \operatorname{Tr} B_n B_n' < \infty$, in case of the convergence of the distributions of the random variables $\det(I + \Xi_n \Xi_n')$, we therefore get $\sup_n G_n(+\infty) < \infty$, where

$$G_n(x) = \sum_{p,l=1}^{n} \int_{-\infty}^{x} y^2(1 + y^2)^{-1} d\mathbf{P}\{\nu_{pl}^{(n)} < y\}.$$

By using (8.3.14) as in the proof of Theorem 8.4.1, we obtain

$$\mu_{kn} := \mathbf{E}\exp\left\{-\sum_{p=1}^{n}[\sum_{q=1}^{k}\eta_{pq}^2]\sum_{l \in K_{pn}}\nu_{pl}^2\right.$$

$$\left. -\sum_{l=1}^{n}[(\sum_{q=1}^{n}\zeta_{lq}^2)\sum_{p \in T_{ln}}\nu_{lp}^2] + i\sum_{p,l=1}^{n} b_{pl}\beta_{pl}\right\} + 0(1).$$

This implies the statement of Theorem 8.4.3.

Corollary 8.4.3. *Let the conditions of Theorem 8.4.3 hold. Then* $\det(I + \Xi_n\Xi'_n) \sim \prod_{j=1}^{n}(1 + \sum_{i\in T_{jn}} \nu_{ij}^2)(1 + \sum_{i\in K_{jn}} \nu_{ji}^2)$, *if* $\lim_{n\to\infty} \operatorname{Tr} B_n B'_n = 0$; $\det(I + \Xi_n\Xi'_n) \sim \det[I + \operatorname{diag}(\sum_{i=1}^{n} \nu_{ij}^2, j = \overline{1,n}) + B_n B'_n]$ *if the row vectors of the matrix* Ξ_n *are asymptotically constant.*

Theorem 8.4.4. *For each n let the random variables $\xi_{pl}^{(n)}, p, l = \overline{1,n}$ be independent, have the finite variance of $\sigma_{pl}^2, \sup_n \sum_{p,l=1}^{n} \sigma_{pl}^2 < \infty, \sup_n \operatorname{Tr} B_n B'_n < \infty$, where $B_n = \mathbf{E}\Xi_n$ and the Lindeberg condition holds, for each $\tau > 0$,*

$$\lim_{n\to\infty} \sum_{p,l=1}^{n} \int_{|x|>\tau} x^2 dF_{pl}(x) = 0,$$

$$F_{pl}(x) = \mathbf{P}\{\nu_{pl}^{(n)} < x\},$$

$$\nu_{pl}^{(n)} = \xi_{pl}^{(n)} - \mathbf{E}\xi_{pl}^{(n)}.$$

Then

$$\operatorname{plim}_{n\to\infty}[\det(I + \Xi_n\Xi'_n) - \prod_{j=1}^{n}(1 + 0.5 \sum_{i\in T_{jn}} \sigma_{ij}^2)$$

$$\times \det[\delta_{ij}(1 + 0.5 \sum_{j\in k_{in}} \sigma_{ij}^2) + d_{ij}]] = 0,$$

where $d_{ij} = \sum_{k=1}^{n} \mathbf{E}\xi_{ik}^{(n)}\mathbf{E}\xi_{jk}^{(n)}[1 + 0.5\sum_{i\in T_{kn}} \sigma_{ik}^2]^{-1}$, the set $\{\xi_{ij}^{(n)}, i, j = \overline{1,n}\}$ is divided into the sets R'_{in}, R''_{in}, so that

$$\lim_{n\to\infty} \sup_{j=\overline{1,n}} \sum_{\xi_{ij}\in R'_{in}\cup R''_{in}} \sigma_{ij}^2 = 0, \tag{8.4.6}$$

and the sets T_{in}, K_{in} are defined by the sets $R'_{in}, R''_i n$ as in Theorem 8.4.3.

Proof. Denote $\alpha_{pl} = \mathbf{E}\{\exp(i\nu_{pl}\beta_{pl})/\beta_{pl}\} - 1, \varphi = \sum_{p,l=1}^{n} \theta_{pl}^2,$

$$\theta_{pl}^2 = \begin{cases} \sum_{q=1}^{k} \eta_{pq}^2, & \text{if} \quad \xi_{pl} \in R''_{ln}, \\ \sum_{q=1}^{k} \zeta_{lq}^2, & \text{if} \quad \xi_{pl} \in R'_{ln}, \end{cases}$$

$$P_n = \sum_{p,l=1}^{n} |\alpha_{pl}|^2 + |\sum_{p,l=1}^{n} (\alpha_{pl} + 0.5\sigma_{pl}^2\beta_{pl}^2)|$$

$$+ |\sum_{p,l=1}^{n} 0.5\sigma_{pl}^2(\beta_{pl}^2 - \theta_{pl}^2)|.$$

Since $|\alpha_{pl}| \le \beta_{pl}^2 \sigma_{pl}^2/2, |\alpha_{pl}| \le |\beta_{pl}|\varepsilon + 2\mathbf{P}\{|\nu_{pl}| \ge \varepsilon\}, \varepsilon > 0$ and by the Lindeberg condition, the variables ν_{pl} are infinitesimal,

$$\lim_{n \to \infty} \sum_{p,l=1}^{n} \mathbf{E}|\alpha_{pl}|^2 = 0. \qquad (8.4.7)$$

By using the condition (8.4.6), we obtain

$$\lim_{n \to \infty} |\sum_{p,l=1}^{n} 0.5\sigma_{pl}^2(\beta_{pl}^2 - \theta_{pl}^2)| = 0. \qquad (8.4.8)$$

It is obvious that

$$|\alpha_{pl} + 0.5\sigma_{pl}^2\beta_{pl}^2| = \int_{|x| \le \varepsilon} (\exp\{ix\beta_{pl}\} - 1 - ix\beta_{pl}$$

$$0.5(ix\beta_{pl})^2)dF_{pl}(x) + \int_{|x|>\varepsilon} (0.5x^2\beta_{pl}^2 + \exp\{ix\beta_{pl}\} - 1$$

$$ix\beta_{pl})dF_{pl}(x) \le |\beta_{pl}|^3(\varepsilon/6)\int_{|x| \le \varepsilon} x^2 dF_{pl}(x)$$

$$+ \beta_{pl}^2 \int_{|x|>\varepsilon} x^2 dF_{pl}(x).$$

From this inequality and by using the Lindeberg condition, we obtain

$$\lim_{n \to \infty} \mathbf{E}|\sum_{p,l=1}^{n} (\alpha_{pl} + 0.5\sigma_{pl}^2\beta_{pl}^2)| = 0. \qquad (8.4.9)$$

From the (8.4.7)–(8.4.9), it follows that $\lim_{n \to \infty} \mathbf{E}P_n = 0$. Therefore,

$$\mu_{kn} = \mathbf{E}\exp\{-\sum_{p,l=1}^{n} \sigma_{pl}^2\theta_{pl}^2 + i\sum_{p,l=1}^{n} \beta_{pl}\mathbf{E}\xi_{pl}^{(n)}\} + 0(1).$$

From this it follows the assertion of Theorem 8.4.4.

§5 Limit Theorems for the Determinants of Random Matrices with Dependent Random Elements

Let us consider the sequence of series of the random variables $\xi_{ij}^{(n)}, i, j = \overline{1,n}, n = 1, 2, \ldots$. Let $\Xi_n = (\xi_{ij}^{(n)}), \mathbf{E}\Xi_n = (\mathbf{E}\xi_{ij}^{(n)})$ be the square matrices of order n, and $r_{ij}(n)$ is the covariance function of the random variables $\xi_{ii}^{(n)}$ and $\xi_{jj}^{(n)}$. Suppose that $\mathbf{E}\xi_{ij}^{(n)} = a_{ij}^{(n)}$, and $\operatorname{Var}\xi_{ij}^{(n)} = \sigma_{ij}^2(n)$ exist.

Theorem 8.5.1. *If the matrices Ξ_n are nonnegative-definite and*

$$\lim_{n\to\infty} \sum_{i,j=1}^{n} [2\sigma_{ij}^2(n) + r_{ij}(n)] = 0, \qquad (8.5.1)$$

then $\operatorname{plim}_{n\to\infty}[\det(I + \Xi_n)^{-1} - \det(I + \mathbf{E}\Xi_n)^{-1}] = 0$.

Proof. By using the integral representation for determinants, we obtain (Chapter 1, §5)

$$\mathbf{E}\det(I + \Xi_n)^{-1/2} = \mathbf{E}\exp\{-\sum_{i,j=1}^{n} \xi_{ij}^{(n)}\eta_i\eta_j\},$$

$$\mathbf{E}\det(I + \Xi_n)^{-1} = \mathbf{E}\exp\{-\sum_{i,j=1}^{n} \xi_{ij}^{(n)}(\eta_i\eta_j + \zeta_i\zeta_j)\},$$

where $\eta_i, \zeta_i, i = 1, 2, \ldots$ are independent, not depending on the $\xi_{ij}^{(n)}$ random variables, and distributed by the normal law $N(0, 1/2)$. From these formulas, by using (8.5.1) and the fact that the matrix $\mathbf{E}\Xi_n$ is nonnegative-definite, we get $\operatorname{plim}_{n\to\infty}[\det(I+\Xi_n)^{-1/2} - \det(I+\mathbf{E}\Xi_n)^{-1/2}] = 0$. It implies the assertion of Theorem 8.5.1.

In the same way, prove that $\operatorname{plim}_{n\to\infty}[\det(I+i\Xi_n)^{-1} - \det(I+i\mathbf{E}\Xi_n)^{-1}] = 0$ if the matrices Ξ_n are symmetric and the condition (8.5.1) holds.

From formulas (8.2.6) and Theorem 8.2.1, we obtain the following statement.

Theorem 8.5.2. *Let Ξ_n be the square random matrices, $n = 1, 2, \ldots$ and*

$$\operatorname{plim}_{n\to\infty}[|((\Xi_n - \mathbf{E}\Xi_n)\vec{\xi}, \vec{\eta})| + |((\Xi_n - \mathbf{E}\Xi_n)\vec{\xi}, \vec{\xi})|] = 0, \qquad (8.5.2)$$

where $\vec{\xi}, \vec{\eta}$ are the random vectors not depending on the matrix Ξ_n and distributed by the normal law $N(0, I)$.

Then for each $0 \le t < 1$ and $q \ge 1$,

$$\operatorname{plim}_{n\to\infty}|\det(I + \alpha_t\Xi_n)^{-1} - \det(I + \alpha_t\mathbf{E}\Xi_n)^{-1}| = 0,$$

where $\alpha_t = t[q + |\operatorname{Tr}\Xi_n| + 0.25\operatorname{Tr}(\Xi + \Xi')^2]^{-1}$.

If in addition to condition (8.5.2),

$$\sup_{n}[|\operatorname{Tr}\mathbf{E}\Xi_n| + \operatorname{Tr}\mathbf{E}\Xi_n\mathbf{E}\Xi_n'] < \infty,$$

$$\lim_{h\to\infty}\overline{\lim_{n\to\infty}}\,\mathbf{P}\{|\sum_{i=1}^{n}(\xi_{ii} - \mathbf{E}\xi_{ii})|$$

$$+ \sum_{i,j=1}^{n}(\xi_{ij} - \mathbf{E}\xi_{ij})^2 \ge h\} = 0,$$

hold, then $\text{plim}_{n\to\infty} |\det(I + \Xi_n) - \det(I + \mathbf{E}\Xi_n)| = 0.$

Analogously we obtain the following assertion.

Let the random variables $\xi_{ij}^{(n)}$ be such that for each integer $k > 0$,

$$\{((\Xi_n - \Xi_n')\overrightarrow{\xi_s}, \overrightarrow{\xi_s}), (\Xi_n\overrightarrow{\xi_s}, \overrightarrow{\xi_s}), (\Xi_n\overrightarrow{\eta_s}, \overrightarrow{\eta_s}), \quad s = \overline{1,k}\}$$
$$\sim \{((H_n - H_n')\overrightarrow{\xi_s}, \overrightarrow{\xi_s}), (H_n\overrightarrow{\xi_s}, \overrightarrow{\xi_s}), (H_n\overrightarrow{\eta_s}, \overrightarrow{\eta_s}), \quad s = \overline{1,k}\},$$
$$(8.5.3)$$

where $H_n = (h_{ij})$ is a random matrix with the entries $(h_{ij}, h_{ji}), i > j, i, j = \overline{1,n}$ for each n being independent and not depending on the vectors $\overrightarrow{\xi_s}, \overrightarrow{\eta_s}, s = \overline{1,k}$.

For the entries of the matrices Ξ_n and H_n, conditions (8.2.3) and (8.2.4) hold. Then $\det(I + \Xi_n) \sim \det(I + H_n)$.

To verify condition (8.5.3), the limit theorems for sums of dependent random variables can be used.

CHAPTER 9

THE CONNECTION BETWEEN THE CONVERGENCE OF RANDOM DETERMINANTS AND THE CONVERGENCE OF FUNCTIONALS OF RANDOM FUNCTIONS

In this chapter, the random determinant is represented in the form of the functional of the random function, and limit theorems for the functionals of the random functions are proved.

§1 The Method of Integral Representations and Limit Theorems for Functionals of Random Functions

Let B_n be nonrandom matrices, Ξ_n random square matrices, and B_n^{-1} exist. Then according to Theorem 1.4.1, for every integer $k > 0$,

$$
\mathbf{E} \det(I + \gamma_t \Xi B^{-1})^{-k} = \mathbf{E} \exp \left\{ i\gamma_t \sum_{s=1}^{k} ((\Xi B^{-1} \right.
$$

$$
- (\Xi B^{-1})') \vec{\xi}_s, \vec{\eta}_s) - \gamma_t \sum_{s=1}^{k} (\Xi B^{-1} \vec{\xi}_s, \vec{\xi}_s)
$$

$$
\left. - \gamma_t \sum_{s=1}^{k} (\Xi B^{-1} \vec{\eta}_s, \vec{\eta}_s) \right\}, \tag{9.1.1}
$$

where

$$
\gamma_t = t[q + 0.5 \operatorname{Tr}(\Xi B^{-1} + (\Xi B^{-1})') + 0.25 \operatorname{Tr}(\Xi B^{-1} + (\Xi B^{-1})')^2)]^{-1},
$$

and $0 \le t < 1$, $q > 1$, $\vec{\xi}_s, \vec{\eta}_s$, $s = \overline{1,k}$ are independent random vectors distributed by the normal law $N(0,0,5I)$.

255

If B_n are positive-definite matrices, formula (9.1.1) can be represented in the form,

$$\mathbf{E}\det(I + \gamma_t \Xi B^{-1})^{-k} = \mathbf{E}\exp\left\{i\gamma_t \sum_{s=1}^{k}((\Xi - \Xi')\vec{x}_s, \vec{y}_s)\right.$$

$$\left. - \gamma_t \sum_{s=1}^{k}(\Xi\vec{x}_s, \vec{x}_s) - \gamma_t \sum_{s=1}^{k}(\Xi\vec{y}_s, \vec{y}_s)\right\}, \qquad (9.1.2)$$

where $\vec{x}_s$, $\vec{y}_s$ are independent random vectors (not depending on Ξ), distributed by the normal law $N(0, 0, 5B^{-1})$.

Denote $x_{ns}(u) = x_{ks}$, $y_{ns}(u) = y_{ks}$, $k/n \leq u < (k+1)/n$. We introduce the random functionals $\xi_n(x_{ns}(\cdot), y_{ns}(\cdot)) := (\Xi x_s, \vec{y}_s)$.

Then (9.1.2) takes the form

$$\mathbf{E}\det(I + \gamma_t \Xi B^{-1})^{-k} = \mathbf{E}\exp\left\{i\gamma_{tn}\sum_{s=1}^{n}[\xi_n(x_{ns}(\cdot), y_{ns}(\ cdot))\right.$$

$$\left. - \xi_n(y_{ns}(\cdot), x_{ns}(\cdot))] - \gamma_{tn}\sum_{s=1}^{k}[\xi_n(x_{ns}(\cdot), x_{ns}(\cdot))\right.$$

$$\left. + \xi_n(y_{ns}(\cdot), y_{ns}(\cdot))]\right\},$$

$$\gamma_{tn} = t[q + 2\mathbf{E}\{\xi_n(x_{ns}(\cdot), x_{ns}(\cdot))/\xi_n(\cdot, \cdot)\}$$
$$+ \mathbf{E}\{\xi_n^2(x_{ns}(\cdot), y_{ns}(\cdot)/\xi_n(\cdot, \cdot)\}]^{-1}.$$

We suppose that for every k pairs of random functions

$$x_{ns}(u), \ y_{ns}(u), \ 0 \leq u \leq 1, \ s = \overline{1, k}$$
$$\{\xi_n(x_{ns}(\cdot), y_{ns}(\cdot)), \xi_n(x_{ns}(\cdot), x_{ns}(\cdot)), \xi_n(y_{ns}(\cdot), y_{ns}(\cdot)),$$
$$\gamma_{tn}, s = \overline{1, k}\} \Rightarrow \{\xi(x_s(\cdot), y_s(\cdot)), \xi(x_s(\cdot), x_s(\cdot)),$$
$$y_s(\cdot)), \delta_t, s = \overline{1, k}\}, \qquad (9.1.3)$$

where $\delta_t = t[q + 2\mathbf{E}\{\xi(x_s(\cdot), x_s(\cdot))/\xi(\cdot, \cdot)\} + \mathbf{E}\{\xi^2(x_s(\cdot), y_s(\cdot))/\xi(\cdot, \cdot)\}]^{-1}$, the random functionals $\xi(\cdot, \cdot)$ do not depend on the random functions $(x_s(u), y_s(u))$ and are given on a set of the values of the functions $x_s(u)$ and $y_s(u)$; they are such that the variables in the right-hand side of Formula (9.1.3) are random. Then

$$\lim_{n\to\infty}\mathbf{E}\det(I + \gamma_t \Xi B^{-1})^{-k} = \mathbf{E}\exp\left\{i\delta_t\sum_{s=1}^{n}[\xi(x_s(\cdot),\right.$$

$$\left. y_s(\cdot)) - \xi(y_s(\cdot), x_s(\cdot))] - \delta_t\sum_{s=1}^{k}[\xi(x_s(\cdot), x_s(\cdot))\right.$$

$$\left. + \xi(y_s(\cdot), y_s(\cdot))]\right\}. \qquad (9.1.4)$$

Different particular cases can be given when (9.1.3) holds. We shall consider one of them. We introduce the random step functions,

$$\nu_n(u,v) = \xi_{ij}^{(n)}, \qquad r_n(u,v) = b_{ij}^{(-1)},$$
$$i/n \le u < (i+1)/n, \qquad j/n \le v < (j+1)/n, \quad 0 \le u, \quad v \le 1,$$

($\xi_{ij}^{(n)}$ are the entries of the matrix Ξ_n; $B_{ij}^{(-1)}$ are the entries of the matrix B_n^{-1}).

Theorem 9.1.1. *Let B_n be positive-definite matrices, $\nu_n(u,v) \Rightarrow \nu(u,v)$, $0 \le u$, $v \le 1$, $\nu(u,v)$ be some measurable random function,*

$$\overline{\lim_{n \to \infty}} \sup_{u,v} \mathbf{E}[\nu^2(u,v) + \nu_n^2(u,v,)] < \infty,$$

$$\lim_{h \downarrow 0} \overline{\lim_{n \to \infty}} \sup_{|u_1-u_2| \le h, |v_1,v_2| \le h} \mathbf{E}\,|\nu_n(u_1,v_1) - \nu_n(u_2,v_2)| = 0,$$

$$\lim_{n \to \infty} r_n(u,v) = r(u,v),$$

where $r(u,v)$ is a continuous function on $[0,1] \times [0,1]$. Then

$$\lim_{n \to \infty} \mathbf{E} \det(I + \gamma_t \Xi_n B_n^{-1})^{-k} = \mathbf{E}\left\{ \mathbf{E}\left[\exp\left\{ i\delta_t \int_0^1 \int_0^1 [\nu(u,v)] \right.\right.\right.$$
$$- \nu(v,u)]\xi(u)\eta(v)\,du\,dv - \delta_t \left[\int_0^1 \int_0^1 \nu(u,v)\xi(u)\xi(v)\,du\,dv \right.$$
$$\left.\left.\left. + \int_0^1 \int_0^1 \nu(u,v)\eta(u)\eta(v)\,du\,dv \right] \right\}/F \right]^k \right\},$$

$\xi(u)$, $\eta(u)$ are independent Gaussian processes (not depending on $\nu(u,v)$) with zero mean and covariance functions $r(u,v)$, F is a minimal σ-algebra, with respect to which the random processes $\xi(u)$, $\eta(u)$, $0 \le u \le 1$ are measurable, $\delta_t = t[q + \int_0^1 \nu(u,u)du + 0.25 \int_0^1 \int_0^1 (\nu(u,v) + \nu(v,u))^2\,du\,dv]^{-1}$.

Use formulas (9.1.3), (9.1.4), and the following lemma to prove Theorem 9.1.1.

Lemma 9.1.1. *Let the measurable random processes $\xi_n(x)$, $\xi(x)$ be given on $[0,1]$, and*

$$\overline{\lim_{n \to \infty}} \sup_t \mathbf{E}|\xi_n(t)| < \infty,$$

$$\lim_{h \downarrow 0} \overline{\lim_{n \to \infty}} \sup_{|t-s| \le h} \mathbf{E}\,|\xi_n(t) - \xi_n(s)| = 0.$$

Then $\int_0^1 \xi_n(x)dx \Rightarrow \int_0^1 \xi(x)\,dx$.

Apply Theorem 9.1.1 to skew-symmetric random matrices. Denote

$$a_{pl}^{(n)} = [b_{pp}^{(n)} b_{ll}^{(n)} - (b_{lp}^{(n)})^2]^{\frac{1}{2}},$$

$$f_{pl}(y) = \int (\exp(ixy) - 1) dF_{pl}(x), \qquad F_{pl}(x) = \mathbf{P}\{\nu_{pl}^{(n)} < x\},$$

$$\nu_{pl}^{(n)} = a_{pl}^{(n)} \zeta_{pl}^{(n)} - \gamma_{pl}^{(n)}, \qquad \gamma_{pl}^{(n)} = \int_{|x|<\tau} x d\mathbf{P}\{a_{pl}^{(n)} \zeta_{pl}^{(n)} < x\},$$

$\tau > 0$ is an arbitrary constant.

Theorem 9.1.2. *Let B_n be positive-definite matrices, $\Xi_n = (\xi_{ij}^{(n)})$ skew-symmetric matrices, for each n random entries $\xi_{pl}^{(n)}$, $l > p$, p, $l = \overline{1,n}$ independent, $\xi_{pl}^{(n)} a_{pl}^{(n)}$, p, $l = \overline{1,n}$ infinitesimal.*
   Then for the convergence of the random variables $\eta_n := \det B_n \det(B_n^{-1} + \Xi_n)$, it is necessary and sufficient that for all $k = 1, 2, \ldots$, there exist a limit

$$m_k := \lim_{n \to \infty} \mathbf{E} \exp\left\{ \sum_{p>l} [f_{pl}(z_{pl}^k) + i\gamma_{pl} z_{pl}^k] \right\},$$

$$\lim_{h \to \infty} \overline{\lim_{n \to \infty}} \, \mathbf{P}\{\eta_n \geq h\} = 0, \tag{9.1.5}$$

where $z_{pl}^k = a_{pl}^{-1} \sum_{m=1}^k [x_{pm} y_{lm} - x_{lm} y_{pm}]$, the vectors $\{x_{pm}, p = \overline{1,n}\}$, $\{y_{pm}, p = \overline{1,n}\}$, $m = 1, 2, \ldots$ are independent, do not depend on the matrices Ξ_n, and are distributed by the normal law $N(0, 1/2B_n^{-1})$.
   A limit distribution function for the variables η_n^{-1} is one-to-one determined by the moments m_k.

Proof. *Sufficiency.* Obviously, $\ln \det(B_n + \Xi_n) - \ln \det B_n = \ln \det(I + K_n)$, where $K_n = (\sum_{p,l=1}^n t_{ip} \zeta_{pl}^{(n)} t_{jl} (\mu_i \mu_j)^{-\frac{1}{2}}$, t_{ij} and μ_i are components of the eigenvectors and eigenvalues of the matrix B_n, respectively, defined by the equation $B_n = T \Lambda T'$, $T = (t_{ij})$, $\Lambda = (\mu_i \delta_{ij})$. Therefore, $\eta_n \geq 1 + \text{Tr} \, K_n K_n'$. From condition (9.1.5) it follows that for each sequence α_n ($\alpha_n \to 0$, as $n \to \infty$),

$$\text{plim}_{n \to \infty} \alpha_n \sum_{i>j} a_{ij}^{(n)} \xi_{ij}^{(n)} \kappa_{ij} = 0, \tag{9.1.6}$$

where $\kappa_{ij} = \sum_{l,p=1}^n \eta_{lp}(t_{pi} t_{lj} - t_{pj} t_{li})(\mu_p \mu_l)^{-\frac{1}{2}} (a_{pl}^{(n)})^{-1}$, η_{lp}, l, $p = 1, 2, \ldots$ are the independent random variables distributed by the normal law $N(0, 1)$.
   It is evident that we can find such a sequence C_{n^2} of Borel sets of Euclidean space R_{n^2} that $\lim_{n \to \infty} \mathbf{P}\{(\eta_{lp}, l, p = \overline{1,n}) \in C_{n^2}\} = 1$, and with the fixed $(\eta_{lp}, l, p = \overline{1,n}) \in C_{n^2}$ (9.1.6) will hold. Therefore, as in the proof of (5.3.19), taking into account that $\mathbf{E} \kappa_{ij}^2 \leq 1$, we obtain

$$\sup_n \mathbf{E} \sum_{p>l} \nu_{pl}^2 [1 + \nu_{pl}^2]^{-1} < \infty. \tag{9.1.7}$$

By the integral representation (1.4.3),

$$\mathbf{E}\eta_n^{-k} = \mathbf{E}\exp\left\{\sum_{p>l}[\nu_{pl}z_{pl}^k + \gamma_{pl}z_{pl}^k]\right\}.$$

Since the variables $\xi_{pl}^{(n)}a_{pl}^{(n)}$ are infinitesimal and (9.1.7) holds,

$$\lim_{n\to\infty}\mathbf{E}\sum_{p>l}|f_{lp}(z_{lp}^k)|^2 = 0.$$

This implies the sufficiency of the conditions of Theorem 9.1.2.

Necessity. Since the distributions η_n converge to the eigenfunction, (9.1.5) holds. The rest of the proof is analogous to that of the sufficiency. It is evident that $\eta_n^{-1} \le 1$. Therefore the Carleman series for the moments m_k diverges. Hence, the limit distribution function for the variable η_n^{-1} is one-to-one, determined by the moments m_k. Theorem 9.1.2 is proved.

Corollary 9.1.1. *If in addition to the conditions of Theorem 9.1.2,*

$$\mathop{\mathrm{plim}}_{n\to\infty}[f_{pl}(z_{pl}^k) - \mathbf{E}f_{pl}(z_{pl}^k)] = 0, \quad k = 1, 2, \dots,$$

$$\sup_n \det B_n^{-1}\det(B_n + \Gamma_n) < \infty,$$

where $\Gamma_n = (\gamma_{pl}^{(n)})$ is a square matrix of order n, then for the convergence of the random values $\ln\det(B_n + \Xi_n) - \ln\det(B_n + \Gamma_n)$, it is necessary and sufficient that there exists a nondecreasing function $G(x)$ of bounded variation such that $G_n(x) \Rightarrow G(x)$, where

$$G_n(x) = \sum_{p>l}\int_0^x y(1+y)^{-1}d\mathbf{P}\{\nu_{lp}^2 < y\}, \quad x \ge 0.$$

The Laplace transformation of the limit distribution function is equal to

$$\exp\left\{\int_0^\infty [(1+x^2)^{-t} - 1](1+x^{-1})dG(x)\right\}.$$

Denote

$$\eta_n(t, s) = B_{kl}^{(n)}(2a_{kk}^{(n)})^{-1}, \quad k/n \le t < (k+1)/n, \quad l/n \le s < (l+1)/n.$$

Corollary 9.1.2. *In addition to the conditions of Theorem 9.2.1 for all $p, l = \overline{1, n}$, let*

$$n^2\int_{-\infty}^x y^2(1+y^2)^{-1}d\mathbf{P}\{\nu_{pl} < y\} \Rightarrow G(x); \tag{9.1.8}$$

$$\lim_{n\to\infty}n^2\left[\gamma_{pl} + \int y(1+y^2)^{-1}d\mathbf{P}\{\nu_{pl} < y\}\right] = \gamma; \tag{9.1.9}$$

$$\lim_{n\to\infty}r_n(t, s) = r(t, s), \tag{9.1.10}$$

where $r(t,s)$ is continuous in the domain $0 \le t \le s \le 1$, γ is an arbitrary constant, and $G(x)$ is a nondecreasing function of bounded variation.

Then for each integer $k > 0$,

$$\lim_{n \to \infty} \mathbf{E} \exp\{-k \ln[\det B_n^{-1} \det(B_n + \Xi_n)]\}$$

$$= \mathbf{E} \exp\left\{ \iint_{0<t<s<1} [i\gamma\eta_k(t,s) + \int (\exp\{ix\eta_k(t,s)\} \right.$$

$$\left. - ix\eta_k(t,s)(1+x^2)^{-1} - 1)(1+x^{-2})dG(x)]dt\,ds \right\},$$

$$(9.1.11)$$

where $\eta_k(t,s) = \sum_{m=1}^{k}[\xi_m(t)\eta_m(s) - \xi_m(s)\eta_m(t)]$, $\xi_m(s)$, $\eta_m(t)$ are independent Gaussian processes, identically distributed with zero mean and covariance functions $r(t,s)$.

Proof. Denote

$$\varphi_{pl}(t) = \int (e^{itx} - itx(1+x^2)^{-1} - 1)(1+x^{-2})dG_n(x),$$

$$G_n(x) = \int_{-\infty}^{x} (1+y^{-2})^{-1}d\mathbf{P}\{\nu_{pl} < y\},$$

$$\delta_{pl} = \int y(1+y^2)^{-1}d\mathbf{P}\{\nu_{pl} < y\}.$$

From conditions (9.1.8) and (9.1.9) it follows that

$$\mathbf{E} \left| \sum_{p \le l} [\{i\gamma_{pl}z_{pl}^k + f_{pl}(z_{pl}^k)\} - n^{-1}\{\gamma z_{pl}^k + f(z_{pl}^k)\}] \right|$$

$$\le \sup_{p,l=\overline{1,n}} [\mathbf{E} |n^2(\gamma_{pl} + \delta_{pl})z_{pl}^k - \gamma z_{pl}^k|$$

$$+ \mathbf{E} |n^2\varphi_{pl}(z_{pl}^k) - f(z_{pl}^k)|] \to 0 \quad (n \to \infty), \qquad (9.1.12)$$

where $f(t) = \int (e^{itx} - itx(1+x^2)^{-1} - 1)(1+x^{-2})dG(x)$.

Consider the Gaussian processes $\xi_m^{(n)}(x) = a_{pl}^{-1}x_{pm}$, $\eta_m^{(n)}(y) = a_{pl}^{-1}y_{pm}$ for $p/n \le x < (p+1)/n$; $l/n \le y < (l+1)/n$. Making use of condition (9.1.10), we obtain $\xi_m^{(n)}(x) \Rightarrow \xi_m(x)$, $\eta_m^{(n)}(y) \Rightarrow \eta_m(y)$, $m = 1, 2, \ldots$, $0 \le x, y \le 1$. Let $\eta_k^{(n)}(t,s) = z_{pl}^k$, $p/n \le t < (p+1)/n$; $l/n \le s < (l+1)/n$. Then $\eta_k^{(n)}(t,s) \Rightarrow \eta_k(t,s)$, $\lim_{h \downarrow 0} \overline{\lim}_{n \to \infty} \sup_{|t'-t''|\le h, |s'-s''|\le h} \mathbf{E}|\eta_k^{(n)}(t',s') - \eta_k^{(n)}(t'',s'')| = 0$. Hence, from (9.1.12) and Lemma 9.1.1, it follows that

$$\lim_{n \to \infty} \mathbf{E} \exp(-k\eta_n) = \mathbf{E} \exp\left\{ \iint_{0<t<s<1} [i\gamma\eta_k(t,s) + f(\eta_k(t,s))]\,dt\,,ds \right\}.$$

Corollary 9.1.2 is proved.

Now by using (9.1.11) and by giving the concrete form to the function $G(x)$, we can find the condition of the convergence of the variables $\ln \det(B_n + \Xi_n) - \ln \det B_n$ to some given distributions. In particular, as

$$G(x) = \begin{cases} \sigma^2, & x > 0 \\ 0, & x \le 0, \end{cases}$$

and the conditions of Corollary 9.1.2 hold:

$$\lim_{n \to \infty} \mathbf{E} \exp\{k \ln[\det(B_n + \Xi_n) \det B_n^{-1}]\}$$

$$= \mathbf{E} \exp\left\{ \iint_{0 < t < s < 1} [i\gamma \eta_k(t, s) - 0.5\sigma^2 \eta_k^2(t, s)] \, dt \, ds \right\}.$$

We shall consider this case in more detail. We prove the following statement.

Theorem 9.1.3. *For every n, let the random variables $\xi_{pl}^{(n)}, l > p, l, p = \overline{1, n}$, be independent with zero expectations and finite variances σ_{pl}^2, $\sup_n \sum_{p > l} \sigma_{pl}^2 a_{pl}^2 < \infty$, let there exist $\lim_{n \to \infty} r_n(t, s) = r(t, s)$, where $r_n(t, s) = 0.5 b_{pl} \sigma_{pl}$ if $p/n \le t < (p+1)/n;\ l/n \le s < (l+1)/n$ and the function $r(t, s)$ be continuous in the domain $0 \le t \le s \le 1$.*
Then in order that for every integer $k > 0$,

$$\lim_{n \to \infty} \mathbf{E} \exp\{-k \ln \det(I + B_n^{-1}\Xi_n)\}$$

$$= \mathbf{E} \exp\left\{ -0.5 \iint_{0 < x < y < 1} \left[\sum_{m=1}^{k} (\xi_m(x)\eta_m(y) \right. \right.$$

$$\left. \left. - \xi_m(y)\eta_m(x)) \right]^2 \, dx \, dy \right\}$$

and the random variables $\xi_{pl}^{(n)} a_{pl}^{(n)}$ be infinitesimal, it is sufficient and for symmetric variables $\xi_{pl}^{(n)}$ also necessary that the Lindeberg condition hold for every $\tau > 0$,

$$\lim_{n \to \infty} \sum_{p > l} \int_{|x| > \tau} x^2 dF_{pl}(x) = 0,$$

$$F_{pl}(x) = \mathbf{P}\{\xi_{pl}^{(n)} a_{pl}^{(n)} < x\}.$$

Proof. *Sufficiency.* From the Lindeberg condition it follows that the variables $\xi_{pl}^{(n)} a_{pl}^{(n)}$ are infinitesimal. Therefore, as in the proof of Theorem 5.3.4, we get

$$\underset{n \to \infty}{\text{plim}} \sum_{p > l} [|f_{pl}(z_{pl}^k)|^2 + |f_{pl}(z_{pl}^k) + 0.5(z_{pl}^k)^2 a_{pl}^2 \sigma_{pl}^2|] = 0.$$

By using this and on the basis of Corollary 9.1.2, we obtain sufficiency of the conditions of Theorem 9.1.3.

Necessity. Since the variables $\xi_{pl}^{(n)} a_{pl}^{(n)}$ are infinitesimal,

$$\operatorname*{plim}_{n \to \infty} \left| \ln \mathbf{E} \exp(-k\eta_n) - \sum_{p > l} f_{pl}(z_{pl}^k) \right| = 0.$$

Consequently,

$$\lim_{n \to \infty} \mathbf{E} \left[\exp\left\{ \sum_{p > l} f_{pl}(z_{pl}^k) \right\} - \exp\left\{ -0.5 \sum_{p > l} (z_{pl}^k)^2 \sigma_{pl}^2 a_{pl}^2 \right\} \right] = 0.$$

The expression under the sign of expectation is nonnegative, therefore

$$\lim_{n \to \infty} \sum_{p > l} \mathbf{E}[0.5\sigma_{pl}^2 a_{pl}^2 (z_{pl}^k)^2 - f_{pl}(z_{pl}^k)] = 0.$$

Hence, for every $\varepsilon > 0$,

$$0.5k \sum_{p > l} [b_{pp} b_{ll} - (b_{pl})^2] \sigma_{pl}^2 - \sum_{p > l} \mathbf{E} \int_{|x| \le \epsilon} \operatorname{Re}(1$$
$$- \exp\{isxz_{pl}^k\}) dF_{pl}(x) = \sum_{p > l} \mathbf{E} \int_{|x| > \epsilon} \operatorname{Re}(1$$
$$- \exp\{isxz_{pl}^k\}) dF_{pl}(x) + 0(1).$$

The integrand on the right-hand side does not exceed $x^2 \varepsilon^{-2}$ and does not exceed $x^2 (z_{pl}^k)^2$ on the left-hand side. Consequently,

$$\sum_{p > l} \int_{|x| > \epsilon} x^2 dF_{pl}(x) \le \varepsilon^{-2} k^{-1} \sum_{p > l} \sigma_{pl}^2 a_{pl}^2.$$

By letting k go to infinity, we obtain the Lindeberg condition. Theorem 9.1.3 is proved.

Analogous statements can be proved not only for the random determinants considered above, but also for $\det(B_n + \Xi_n \Xi_n')$, $\det(B_n + iH_n)$, where B_n are nonnegative-definite matrices, Ξ_n and H_n are random matrices, and H_n are symmetric.

§2 The Spectral Functions Method of Proving Limit Theorems for Random Determinants

Let Ξ_n be a complex random matrix. Denote its eigenvalues by λ_i, $i = \overline{1, n}$. From Chapter 3, it follows that λ_i can be chosen to be random variables. By the spectral function of matrix Ξ_n is meant the expression

$$\mu_n(x, y) = c_n^{-1} \sum_{k=1}^{n} F(x - \operatorname{Re}(\lambda_k/b_n))F(y - \operatorname{Im}(\lambda_k/b_n)), \qquad (9.2.1)$$

where $F(x) = 0$ as $x < 0$, and $F(x) = 1$ as $x \geq 0$, and c_n, b_n are some nonrandom numbers. If $c_n = n$, then $\mu_n(x, y)$ is a normalized spectral function of the matrix $\Xi_n b_n^{-1}$. If the eigenvalues λ_i of the matrix Ξ_n are real, then the normalized spectral function of the matrix Ξ_n takes the form

$$\mu_n(x) = n^{-1} \sum_{i=1}^{n} F(x - \lambda_i).$$

Obviously, the distribution functions are the realization of the function $\mu_n(x)$.

By using (9.2.1), a random determinant can be represented in the form

$$c_n^{-1} \ln \det(\Xi_n b_n^{-1}) = \iint \ln(x + iy)d\mu_n(x, y), \qquad (9.2.\ 2)$$

assuming that there exists an integral on the right-hand side of the equation. Let us show the way to represent logarithms of the modules of random determinants by the spectral functions of symmetric matrices.

Let Ξ_n be a real random matrix, λ_i, $i = \overline{1, n}$ the eigenvalues of the matrix $\Xi_n \Xi_n' \exp(-2a_n)$. Denote the spectral function by

$$\nu_n(x) = (2b_n)^{-1} \sum_{i=1}^{n} F(x - \lambda_i)\varphi(\lambda_i), \qquad (9.2.3)$$

where $\varphi(x)$ is a continuous function on $(-\infty, \infty)$.

Then

$$[\ln | \det \Xi_n| - a_n]b_n^{-1} = \int_0^{\infty} \ln x(\varphi(x))^{-1}d\nu_n(x), \qquad (9.2.4)$$

under the condition that there exists the integral in the right-hand side of the equation.

Let us prove the following theorem.

Theorem 9.2.1. *Let the function* $f(x)(-\infty < x < \infty)$ *be continuous and bounded on the whole straight line* R_1, $\mu_n(x)$ *be the normalized spectral functions of the symmetric random matrices* Ξ_n, $\mu_n(x) \Rightarrow \mu(x)$ *on some everywhere dense set* C *of the straight line* R_1, $\mu_n(-\infty) \Rightarrow \mu(-\infty)$, $\mu_n(+\infty) \Rightarrow \mu(+\infty)$, *where* $\mu(x)$ *is a random distribution function. Then*

$$\int f(x)d\mu_n(x) \Rightarrow \int f(x)d\mu(x).$$

Proof. Let $a < 0$, $b < 0$ and $a, b \in c$. Denote

$$I_1 = \left| \int_{-\infty}^a f(x)d\mu(x) - \int_{-\infty}^a f(x)d\mu_n(x) \right|,$$

$$I_2 = \left| \int_b^\infty f(x)d\mu(x) - \int_b^\infty f(x)d\mu_n(x) \right|,$$

$k = \sup_x |f(x)|$, then

$$I_1 \le k[\mu(a) - \mu(-\infty)] + k[\mu_n(a) - \mu_n(-\infty)],$$
$$I_2 \le k[\mu(+\infty) - \mu(b)] + k[\mu_n(+\infty) - \mu_n(b)].$$

Since $\mathbf{E}|\mu(a) - \mu(-\infty)| = \mathbf{E}\mu(a) - \mathbf{E}\mu(-\infty) \to 0$ when $a \to -\infty$, $\mathbf{E}|\mu(+\infty) - \mu(b)| = \mathbf{E}\mu(+\infty) - \mathbf{E}\mu(b) \to 0$, when $b \to \infty$, $\lim_{n\to\infty} \mathbf{E}|\mu_n(a) - \mu_n(-\infty)| = \mathbf{E}\mu(a) - \mathbf{E}\mu(-\infty)$ and $\lim_{n\to\infty} \mathbf{E}|\mu_n(+\infty) - \mu_n(b)| = \mathbf{E}\mu(+\infty) - \mathbf{E}\mu(b)$, then the variables $\mathbf{E}I_1$ and $\mathbf{E}I_2$ can be made as small as desired if $|a|, b$, and n are chosen sufficiently large.

Let us divide the interval (a, b) by the points x_n so that for all

$$x \in (x_{k-1}, x_k] \, |f(x) - f(x_k)| < \varepsilon, \quad k = \overline{1, m_\varepsilon},$$

where $\varepsilon > 0$ is an arbitrary constant number. Denote $f_\varepsilon(x) = f(x_k)$, $x \in (x_{k-1}, x_k]$. The points x_k can be chosen so that they belong to the set C. It is evident that

$$\left| \int_a^b f(x)d\mu_n(x) - \int_z^b f_\varepsilon(x)d\mu_n(x)\beta iggr \right| \le \varepsilon[\mu_n(b) - \mu_n(a)]; \tag{9.2.5}$$

$$\left| \int_a^b f(x)d\mu(x) - \int_a^b f_\varepsilon(x)d\mu(x) \right| \le var\varepsilon[\mu(b) - u(a)]. \tag{9.2.6}$$

From the convergence of the finite-dimensional distributions of the random functions $\mu_n(x)$, it follows that

$$\lim_{n\to\infty} \mathbf{E} \exp\left\{ is \sum_{k=0}^{m-1} f_\varepsilon(x_k)[\mu_n(x_{k+1}) - \mu_n(x_k)] \right\}$$

$$= \mathbf{E} \exp\left\{ is \sum_{k=0}^{m-1} f_\varepsilon(x_k)[\mu(x_{k+1}) - \mu(x_k)] \right\}.$$

Hence by using (9.2.5) and (9.2.6), we obtain $\int_a^b f(x)d\mu_n(x) \Rightarrow \int_a^b f(x)d\mu(x)$. Taking into account that $\lim_{n\to\infty}(\mathbf{E}I_1 + \mathbf{E}I_2) = 0$, we get the statement of Theorem 9.2.1.

Theorem 9.2.1 allows the following generalization.

Theorem 9.2.2. *Let function $f(x)$ be continuous on the straight line R_1; $\mu_n(x) \Rightarrow \mu(x)$ on some everywhere dense set C of the straight line R_1, $\mu_n(-\infty) \Rightarrow \mu(-\infty)$, $\mu_n(+\infty) \Rightarrow \mu(+\infty)$ for some $\alpha > 0$ $\sup_n \mathbf{E} \int |f(x)|^{1+\alpha} \times d\mu_n(x) < \infty$. Then $\int f d\mu_n \Rightarrow \int f d\mu$.*

Proof. Obviously, $\int f d\mu_n = \int_{|f(x)|<A} f d\mu_n + \int_{|f(x)|\geq A} f d\mu_n$. From Theorem 9.2.1, we get $\int_{|f(x)|<A} f d\mu_n \Rightarrow \int_{|f(x)|<A} f d\mu$ for any fixed $A > 0$. Applying the inequalities

$$\int_{|f(x)|\geq A} f d\mu_n \leq A^{-\alpha} \int |f|^{1+\alpha} d\mu_n,$$

$$\mathbf{E} \int |f|^{1+\alpha} d\mu \leq \sup_n \mathbf{E} \int |f(x)|^{1+\alpha} d\mu_n(x)$$

and letting A go to infinity, we prove Theorem 9.2.2.

Corollary 9.2.1. *Let the Ξ_n be the symmetric random matrices $\mu_n(x)$ their normalized spectral functions, $\mu_n(x) \Rightarrow \mu(x)$ on some everywhere dense set C of the straight line R_1, $\mu_n(-\infty) \Rightarrow \mu(-\infty)$, $\mu_n(+\infty) \Rightarrow \mu(+\infty)$ for some $\alpha > 0$ $\sup_n n^{-1}\mathbf{E}\operatorname{Tr}|\ln|\Xi_n||^{1+\alpha} < \infty$. Then*

$$n^{-1}\ln|\det \Xi_n| \Rightarrow \int \ln|x| d\mu(x).$$

We prove analogous statements for the spectral functions (9.2.1) and the random determinants (9.2.2). Let us consider the spectral functions (9.2.3). The following statement extends to them.

Corollary 9.2.2. *Let $\nu_n(x) \Rightarrow \nu(x)$ on some everywhere dense set C of the straight line R_1, $\nu_n(+\infty) \Rightarrow \nu(+\infty)$, $P\{\nu_n(+\infty) \leq h\} = 1$, where $h > 0$ is some constant number, $\nu(x)$ is a random nondecreasing function, for some $\alpha > 0$,*

$$\sup_n \mathbf{E} \int_0^\infty |\ln x/\varphi(x)|^{1+\alpha} d\nu_n(x) < \infty.$$

Then $b_n^{-1}[\ln|\det \Xi_n| - a_n] \Rightarrow \int_0^\infty [\ln x/\varphi(x)]d\nu(x)$.

Thus, to prove the limit theorems for the random determinants, it is necessary to find the conditions for the convergence of finite-dimensional distributions of random spectral functions. It is convenient to prove limit theorems for $\mu_n(x)$, $\nu_n(x)$ with the help of the Stieltjes transformation:

$$\int (x - z)^{-1} d\mu_n(x) = n^{-1}\operatorname{Tr}(\Xi - zI)^{-1}, \qquad (9.2.7)$$

where z is a complex number. $\text{Im}\, z \neq 0$, Ξ is a symmetric random matrix and $\mu_n(x)$ is its normalized spectral function.

In addition to (9.2.7), it is possible to use the Stieltjes transformation

$$\int (1 + itx)^{-1} d\mu_n(x) = n^{-1} \text{Tr}(I + it\Xi)^{-1} \qquad (9.2.8)$$

It is obvious that for analytical functions $\varphi(x)$,

$$\int (1 + itx)^{-1} d\nu_n(x) = \text{Tr}(2b_n)^{-1}(I + it\Xi_n \Xi_n'$$
$$\times \exp(-2n^{-1}a_n))^{-1} \varphi(\Xi_n \Xi_n' \exp(-2n^{-1}a_n)). \qquad (9.2.9)$$

Denote $\eta_n(t) = \int(1 + itx)^{-1} d\mu_n(x)$, $\xi_n(z) = \int (x - z)^{-1} d\mu_n(x)$, $\text{Im}\, z \neq 0$. Let us show that the inversion formula at the points of stochastic continuity x_1 and x_2 of the function $\mu_n(x)$ has the form

$$\mathbf{P}\{\mu_n(x_2) - \mu_n(x_1) < u\} = \lim_{\varepsilon \to 0} \mathbf{P}\{\pi^{-1} \int_{x_2}^{x_1} \text{Im}\, \xi_n(y + i\varepsilon)\, dy < u\}. \qquad (9.2.10)$$

Thus,

$$\pi^{-1} \int_{x_1}^{x_2} \text{Im}\, \xi_n(y + i\varepsilon)\, dy = \pi^{-1} \int \text{arctg}((x - x_2)/\varepsilon) d\mu_n(x) - \pi^{-1}$$
$$\int \text{arctg}((x - x_1)/e) d\mu_n(x), \lim_{\delta \to 0} \mathbf{P}\{|\mu_n(x_2 + \delta) - \mu_n(x_2 - \delta)| + \mu_n(x_1 + \delta)$$
$$-\mu_n(x_1 - \delta)| < \varepsilon'\} = 1,\ \varepsilon' > 0.$$

Obviously,

$$\pi^{-1} \int \text{arctg}((x - x_2)/\varepsilon) d\mu_n(x) = \pi^{-1} \int_{-\infty}^{x_2 - \delta} \text{arctg}((x - x_2)/\varepsilon) d\mu_n(x)$$
$$+\pi^{-1} \int_{x_2 + \delta}^{\infty} \text{arctg}((x - x_2)/\varepsilon) d\mu_n(x) + \pi^{-1} \int_{x_2 - \delta}^{x_2 + \delta} \text{arctg}((x - x_2)/\varepsilon) d\mu_n(x).$$

The latter integral tends in probability to zero as $\delta \to 0$, and the sum of the first two integrals tends in probability to $\mu_n(x_2) - 1$ as $\delta \to 0$ and $\varepsilon \to 0$ (δ is chosen to be equal to $\sqrt{\varepsilon}$). Formula (9.2.10) is proved.

If $\mu_n(x)$ is a noneigen spectral random function, i.e., $\mu_n(+\infty) - \mu_n(-\infty) < 1$ with probability 1, then formula (9.2.10) will give one-to-one the function $\mu_n(x)$ at the points of its stochastic continuity if with probability 1 $\mu_n(-\infty) = \xi\ (\mu_n(+\infty) = \eta)$, where $\xi(n)$ is some random variable.

Analogously, we obtain the inversion formula for finite-dimensional distributions of the function $\mu_n(x)$

$$\mathbf{P}\{\mu_n(x_2^k) - \mu_n(x_1^k) < u_k, k = \overline{1,m}\} = \lim_{\varepsilon \to 0} \mathbf{P}\{\pi^{-1}$$

$$\times \int_{x_1^k}^{x_2^k} \operatorname{Im}\xi_n(y + i\varepsilon)\, dy < u_k, \ k = \overline{1,m}\},$$

where $x_1^k, x_2^k, k = \overline{1,m}$ are points of the stochastic continuity of the function $\mu_n(x)$.

For the transformation $\eta_n(t)$, it is possible to write the inversion formula on the basis of formula (9.2.10). Let $\int (x - is)^{-1} d\mu_n(x) = -(is)^{-1}\eta_n(s^{-1}) = m_n(s)$, where s is a real parameter. Function $m_n(s)$ is analytical for all $s \neq 0$. Therefore it allows analytical continuity on z, $\operatorname{Im} z \neq 0$ and using formula (9.2.10). The inversion formula at the points of stochastic continuity of the function $\mu_n(x)$ has the form

$$\mathbf{P}\{\mu_n(x_2) - \mu_n(x_1) < u\}$$

$$= \lim_{\varepsilon \to 0} \mathbf{P}\left\{ \pi^{-1} \int_{x_1}^{x_2} \operatorname{Im}(y + i\varepsilon)^{-1}\eta_n(-i(y + i\varepsilon)^{-1}\, dy < u \right\}.$$

$$(9.2.11)$$

By the notation $\mu_n(x) \simeq > \mu(x)$ is meant the convergence of the finite-dimensional distributions of the random spectral functions $\mu_n(x)$ to the finite-dimensional distributions of random spectral function $\mu(x)$ at the points of the stochastic continuity of the latter function.

Let us prove the following theorem.

Theorem 9.2.3. *Let $\mu_n(x)$ be the sequence of the random spectral functions and with probability 1, $\lim_{h \to -\infty} \sup_n \mathbf{E}\mu_n(h) = 0$. Then, in order that $\mu_n(x) \simeq \to \mu(x)$, where $\mu(x)$ is some random spectral function, it is necessary and sufficient that $\xi_n(z) \Rightarrow \xi(z)$, $\operatorname{Im} z \neq 0$. Finite-dimensional distributions $\mu(x)$ are expressed by $\xi(z)$ according to (9.2.10).*

Proof. The necessity follows from Theorem 9.2.1. Let us prove the sufficiency. Consider the joint moments of the finite-dimensional distributions of the random functions.

$$\mathbf{E} \prod_{k=1}^m \xi_n(z_k) = \int \cdots \int \prod_{k=1}^m (x_k - z_k)^{-1} d\mathbf{E} \prod_{k=1}^m \mu_n(x_k), \quad \operatorname{Im} z_k \neq 0 \quad (9.2.12)$$

By setting some z_n equal to one another on the left-hand side of this equation and by choosing the corresponding m, we obtain an arbitrary joint moment of the finite-dimensional distributions of the random functions $\zeta_n(z)$. From equation (9.2.12), it follows that the distribution functions $\mathbf{E} \prod_{k=1}^m \mu_n(x_k)$ converge to the limit distribution function $F(x_k, k = \overline{1,m})$ at the points of its continuity in the force of the well-known theorems for multidimensional functions of distribution and their Fourier and Stieltjes transformations. Thus, the joint moments of the finite-dimensional distributions of the functions $\mu_n(x)$ converge on some everywhere dense set to the joint moments of the finite-dimensional distributions of some random functions $\mu(x)$. It is evident that the realizations of the function $\mu(x)$ are the distribution functions. Hence, $\xi(z) \approx \int (x - z)^{-1} d\mu_n(x)$; therefore, formula (9.2.10) is valid for $\mu(x)$.

The proof of Theorem 9.2.3 can be applied to the other transformations of the random functions $\mu_n(x)$. Since in the future we mainly use the transformation $\eta_n(t)$, let us formulate a corollary from Theorem 9.2.3.

Corollary 9.2.3. *Let $\mu_n(x)$ be the sequence of the random spectral functions and with probability 1 $\lim_{h \to -\infty} \sup_n E\mu_n(h) = 0$. Then*

a) in order that $\mu_n(x) \simeq \to \mu(x)$, where $\mu(x)$ is some random spectral function, it is necessary and sufficient that $\eta_n(t) \Rightarrow \eta(t)$;

b) in order that at every point of the continuity of the nonrandom distribution function $\mu(x)$ $\mathrm{plim}_{n \to \infty} \mu_n(x) = \mu(x)$, it is necessary and sufficient that for every t, $\mathrm{plim}_{n \to \infty} \eta_n(t) = \eta(t)$ where $\eta(t) = \int (1 + itx)^{-1} d\mu(x)$.

Corollary 9.2.4. *Let $\mu_n(x)$ and $\lambda_n(x)$ be the sequences of the random spectral functions given on a common probability space, and with probability 1, $\lim_{h \to \infty} \sup_n \mu_n(h) = 0$, $\lim_{h \to -\infty} \sup_n \lambda_n(h) = 0$, $m_n(t) = \int (1 + itx)^{-1} d\mu_n$, $p_n(t) = \int (1 + itx)^{-1} d\lambda_n$. Then*

a) in order that $\mu_n(x) \sim \lambda_n(x)$ on some everywhere dense set C, it is necessary and sufficient that $m_n(t) \sim p_n(t)$, $-\infty < t < \infty$;

b) in order that $\mathrm{plim}_{n \to \infty}[\mu_n(x) - \lambda_n(x)] = 0$ for all x from some everywhere dense set C, it is necessary and sufficient that for each t, $\mathrm{plim}_{n \to \infty}[m_n(t) - p_n(t)] = 0$.

Theorem 9.2.4. *Let $\mu_n(x)$ be the sequence of random spectral functions and with probability 1, $\lim_{h \to -\infty} \sup_n \mathbf{E}\mu_n(h) = 0$. In order that with probability 1 at every point of continuity of some nonrandom distribution function $\mu(x)$ whose Stieltjes transformation equals $\eta(t) = \int (1 + itx)^{-1} d\mu(x)$, $\lim_{n \to \infty} \mu_n(x) = \mu(x)$, it is necessary and sufficient that with probability 1 for every t, $\lim_{n \to \infty} \eta_n(t) = \eta(t)$.*

Proof. The necessity of the condition is obvious. Let us prove the sufficiency. With probability 1, the functions $t\eta_n(t) - t\eta(t)$ are equipotentially continuous. Therefore, for any $\varepsilon > 0$ and bounded $T > 0$, there exist numbers t_i, $i = \overline{1, m_\varepsilon}$ such that with probability 1, $\sup_{|t| \le T} |t| \, |\eta_n(t) - \eta(t)| \le \varepsilon + \max_{i=\overline{i, m_\varepsilon}} |t_i| \, , |\eta_n(t_i) - \eta(t_i)|$. Therefore, with probability 1 for any bounded $T > 0$, $\lim_{n \to \infty} \sup_{|t| \le T} t \, |\eta_n(t) - \eta(t)| = 0$. Theorem 9.2.4 is proved.

Sometimes, from the convergence of the random determinants, the convergence of the random spectral functions follows. Let us consider the so-called logarithmic transformation.

$$n^{-1}\ln\det(Iz+\Xi) = \int \ln(z+x)d\mu_n(x), \quad \operatorname{Im} z \neq 0.$$

The inversion formula of this transformation at points of the stochastic continuity $\mu_n(y)$ is

$$\mathbf{P}\{\mu_n(y) < u\} = \lim_{\varepsilon \to 0} \mathbf{P}\{\operatorname{Im} n^{-1}\ln\det(I(-y+i\varepsilon)+\Xi) < u\}. \qquad (9.2.13)$$

Let us briefly describe the following approach to the study of limit theorems for random determinants. It is obvious that for the symmetric matrices, $\ln\det(I+i\Xi_n) = \int_0^1 [1 - \operatorname{Tr}(1+it\Xi_n)^{-1}t^{-1}\,dt$. Therefore, if there exists $\mathbf{E}\ln\det(I+i\Xi_n)$, then $\ln\det(I+i\Xi_n) - \mathbf{E}\ln\det(I+i\Xi_n)^{-1} = \int_0^1 [\mathbf{E}\operatorname{Tr}(I+it\Xi_n)^{-1} - \operatorname{Tr}(I+it\Xi_n)^{-1}t^{-1}\,dt$. Denote $\eta_n(t)v\mathbf{E}\operatorname{Tr}(I+it\Xi_n)^{-1} - \operatorname{Tr}(I+it\Xi_n)^{-1}$. If $\eta_n(t) \Rightarrow \eta(t)$ and

$$\lim_{h\to 0}\lim_{n\to\infty}\sup_{|t'-t''|\le h}\mathbf{E}|\eta_n(t')(t')^{-1} - \eta_n(t'')(t'')^{-1}| = 0, \quad \sup_{0\le t\le 1}\mathbf{E}|t|^{-1}|\eta(t)| < \infty,$$

then

$$\ln\det(I+i\Xi_n) - \mathbf{E}\ln\det(I+i\Xi_n) \Rightarrow \int_0^1 \eta(t)t^{-1}\,dt.$$

§3 The Canonical Spectral Equation

The peculiar feature of the normalized spectral functions of the symmetric random matrices with independent entries on the diagonal and above is their convergence to some nonrandom function of distribution under the condition that the dimension of the matrices is increasing. Let us give an exact wording and prove this fact. Let $\Xi_n = (\xi_{ij}^{(n)})$ be a symmetric random matrix and $\mu_n(x)$ its normalized spectral function.

Theorem 9.3.1. *If for every n the vectors $\vec{\xi}_i = (\xi_{ii}^{(n)}, \xi_{ii+1}^{(n)}, \ldots \xi_{in}^{(n)})$, $i = \overline{1,n}$ are independent, random values $\xi_{ij}^{(n)}$, $i,j,n = 1,2,\ldots$ are given on a common probability space, there exists a limit $\lim_{n\to\infty} n^{-1}\mathbf{E}\operatorname{Tr}(I+it\Xi_n)^{-1} = m(t)$ and the function $m(t)$ is continuous at zero, then with probability 1, $\lim_{n\to\infty}\mu_n(x) = \mu(x)$ at every point of continuity of the nonrandom function $\mu(x)$, whose Stieltjes transformation is equal to $\int(1+itx)^{-1}d\mu_n = m(t)$.*

Proof. Denote $\gamma_k = \mathbf{E}[\operatorname{Tr} R_t/\rho_{k-1}] - \mathbf{E}[\operatorname{Tr} R_t/\rho_k]$, $k = \overline{1,n}$, where $R_t = (I+it\Xi_n)^{-1}$, ρ_k is a minimal σ-algebra with respect to which the random vectors $\vec{\xi}_i$, $i = \overline{k+1,n}$ are measurable.

Let the matrix Ξ^k be obtained from the matrix Ξ_n by replacing the entries of the kth row and kth column by zeroes, $R_t^k = (I + it\Xi_n^k)^{-1} = (r_{ij}^k)$. It is obvious that $r_{ij}^k = 0$, if $i \neq k$, $r_{kk}^k = 1$. From §5 of Chapter 2 it follows that

$$\operatorname{Tr} R_t - \operatorname{Tr} R_t^k = t \frac{d}{dt} \ln \det[I + it R_t^k (\Xi_n - \Xi_n^k)]$$

$$= t \frac{d}{dt} \ln[1 + it\xi_{kk}^{(n)} + t^2 (R_t^k \vec{\theta}_k, \vec{\theta}_k)], \qquad (9.3.1)$$

where $\vec{\theta}_k = (\xi_{k1}, \ldots, \xi_{kk-1}, 0, \xi_{kk+1}, \ldots, \xi_{kn})$.

By using (9.3.1), we find

$$|\operatorname{Tr} R_t - \operatorname{Tr} R_t^k| = |it\xi_{kk}^{(n)} + t^2 (R_t^k \vec{\theta}_k, \vec{\theta}_k) + t^2 ((R_t^k)^2 \vec{\theta}_k, \vec{\theta}_k)|$$

$$\times |1 + it\xi_{kk}^{(n)} + t^2 (R_t^k \vec{\theta}_k, \vec{\theta}_k)|^{-1}$$

$$\leq 2 + |t^2 ((R_t^k)^2 \vec{\theta}_k, \vec{\theta}_k)| \, |1 + it\xi_{kk}^{(n)} + t^2 (R_t^k \vec{\theta}_k, \vec{\theta}_k)|^{-1}$$

$$\leq 2 + t^2 \sum_{k=1}^n y_k^2 (1 + t^2 \lambda_k^2)^{-1} [1 + t^2 \sum_{k=1}^n y_k^2 (1 + t^2 \lambda_k^2)^{-1}]^{-1} \leq 3,$$

where λ_k are the eigenvalues of the matrix Ξ^k, $y_k = (\vec{h}_k, \vec{\theta}_k)$, $\vec{h}_k$ are the eigenvectors of the matrix Ξ^k.

From this inequality and the proof of Theorem 5.4.2, it follows that with probability 1 for any t, $\lim_{n\to\infty}[n^{-1} \operatorname{Tr} R_t - n^{-1} \mathbf{E} \operatorname{Tr} R_t] = 0$. Therefore, from Theorem 9.2.4, the statement of Theorem 9.3.1 follows.

Theorem 9.3.2. *For every n, let the random entries $\xi_{ij}^{(n)}$, $i \geq j$, $i, j = \overline{1,n}$ of the matrix $\Xi_n = (\xi_{ij}^{(n)} - \alpha_{ij}^{(n)})$ be independent, infinitesimal, $\alpha_{ij} = \int_{|x|<\tau} x\, dP\{\xi_{ij} < x\}, \tau > 0$ an arbitrary constant, $K_n(u, v, z) \Rightarrow K(u, v, z)$, where $K_n(u, v, z) = n \int_{-\infty}^z y^2 (1 + y^2)^{-1} dP\{\xi_{ij} - \alpha_{ij} < y\}$; $i/n \leq u < (i + 1)/n$; $j/n \leq v < (j + 1)/n$, the $K(u, v, z)$ a nondecreasing function with bounded variation on z and continuous on u and v in the domain $0 \leq u, v \leq 1$ for some $\alpha > 0$,*

$$\sup_n n^{-1} \mathbf{E} \operatorname{Tr} |\ln(I + i\Xi_n)|^{1+\alpha} < \infty. \qquad (9.3.2)$$

Then

$$\operatorname{plim}_{n \to \infty} n^{-1} \ln \det(I + i\Xi_n) = \int \ln(1 + ix) dF(x), \qquad (9.3.3)$$

where $F(x)$ is a distribution function whose Stieltjes transformation equals

$$\int (1 + itx)^{-1} dF(x) = \lim_{a \downarrow 0} \int_0^1 \int_0^1 x\, dG_a(x, z, t)\, dz, \qquad (9.3.4)$$

$G_a(x, z, t)$ *is a distribution function on* x $(0 \le x \le 1, 0 \le z \le 1, -\infty < t <$
$\infty)$, *satisfying the equation at the points of continuity*

$$G_a(x, z, t) = \mathbf{P}\{[1 + t^2 \xi_a(G_a(\cdot, \cdot, t), z)]^{-1} < x\}, \qquad (9.3.5)$$

$\xi_a(G_a(\cdot, \cdot, t), z)$ *is a random functional whose Laplace transformation of one-dimensional distributions equals*

$$\mathbf{E} \exp\{-s\xi(G(\cdot, \cdot, t), z)\} = \exp\left\{ \int_0^1 \int_0^1 \left[\int_0^\infty (\exp\{-syx^2(1 \right. \right.$$
$$\left. \left. + a|x|)^{-2}\} - 1)(1 + x^{-2})dK(v, z, x)dG(y, v, t)\, dv \right\}, \quad \alpha > 0, s \ge 0.$$

The solution of Eq. (9.3.5) exists and is unique in the class L of the functions $G(x, z, t)$, which are distribution functions on x $(0 \le x \le 1)$ for any fixed $0 \le z \le 1$, $-\infty < t < \infty$ and such that for any integer $k > 0$ and z functions $\int x^k dG_a(x, z, t)$ are analytical[1] on t (excluding, perhaps, point zero).

The solution of the equation (9.3.5) can be found by means of the method of successive approximations.

Proof. Denote $R_t = (I + it\Xi_n)^{-1} = (r_{pl}(t))$. To simplify formulas, we note r_{pl} instead of $r_{pl}(t)$. From §5 of Chapter 2 it follows that $r_{kk}(t) = [1 + it\nu_{kk}^{(n)} + t^2(R_t^k \vec{\nu}_k, \vec{\nu}_k)]^{-1}$, where $\vec{\nu}_k = (\nu_{1k}, \ldots, \nu_{k-1k}, 0, \nu_{k+1k}, \ldots, \nu_{nk})$, $\nu_{ij} = \xi_{ij} - \alpha_{ij}$, $R_t^k = (r_{ij}^k) = (I + it\Xi_n^k)^{-1}$.

Let us prove that

$$\operatorname*{plim}_{n \to \infty} \left| \sum_{i \ne j}^n r_{ij}^k \nu_{ki} \nu_{kj} \right| = 0, \qquad (9.3.6)$$

supposing that $\nu_{kk} = 0$. It is clear that the matrix R_t^k can be represented in the form $R_t^k = B_n + iC_n - iD_n$, where $B_n = (b_{ij})$, $C_n = (c_{ij})$, $D_n = (d_{ij})$ are nonnegative-definite matrices, whose entries satisfy the conditions $|b_{ij}| \le 1$, $|c_{ij}| \le 1$, $|d_{ij}| \le 1$, $\sum_{i=1}^n b_{ij}^2 \le 1$, $\sum_{i=1}^n c_{ij}^2 \le 1$, $\sum_{i=1}^n d_{ij}^2 \le 1$, $i, j = \overline{1, n}$. To prove (9.3.6) it is sufficient to show that

$$\operatorname*{plim}_{n \to \infty} \sum_{i \ne j} s_{ij} \nu_{ik} \nu_{jk} = 0, \qquad (9.3.7)$$

where $s_{ij} = b_{ij} + c_{ij} + d_{ij}$.

From the condition $K_n(u, v, z) \Rightarrow K(u, v, z)$, it follows that

$$\sup_n \sum_{i=1}^n \mathbf{E}\nu_{ik}^2(1 + \nu_{ik}^2)^{-1} < \infty. \qquad (9.3.8)$$

---

[1] We call a complex function of the real variable analytical on the interval (a, b), provided that it can be decomposed into Tailor's converging series in some vicinity of every point of the interval (a, b).

Hence, according to Theorem 5.3.4, (the matrix R_t^k is fixed),

$$
\mathbf{E}\exp\left\{-q\sum_{i=1}^{n}s_{ii}\nu_{ik}^2 - s\sum_{i,j=1}^{n}s_{ij}\nu_{ik}\nu_{jk}\right\}
$$

$$
= \mathbf{E}\exp\left\{\sum_{l=1}^{n}\mathbf{E}\exp\{i\sqrt{s}\,\nu_{lk}\eta_l - q\nu_{lk}^2\} - 1/\eta_l)\right\} + 0(1)
$$

$$
= \mathbf{E}\exp\left\{\sum_{l=1}^{n}\int(\exp\{i\sqrt{s}\,x\eta_l - qx^2\} - i\sigma qrts\,x\eta_l(1+x^2)^{-1} - 1)\right.
$$

$$
\left. \times (1+x^{-2})dG_l^{(n)}(x) + i\sum_{l=1}^{n}\sqrt{s}\,\eta_l\mathbf{E}\nu_{lk}(1+\nu_{lk}^2)^{-2}\right\} + 0(1),
$$
(9.3.9)

where $(\eta_1,\ldots,\eta_n)$ is a normally distributed vector with the zero vector of means and with the covariance matrix $2(s_{ij})$, $q, s \geq 0$, $G_l^{(n)}(x) = \int_{-\infty}^{x}y^2(1+y^2)^{-1}dP\{\nu_{lk} < y\}$. Denote $\mathbf{E}[\exp\{i\sqrt{s}\,\nu_{lk}\eta_l - q\nu_{lk}^2s_{ll}\} - 1/\eta_l] = f_l$. Let us show that

$$
\operatorname*{plim}_{n\to\infty}\sum_{l=1}^{n}(f_l - \mathbf{E}f_l) = 0.
$$

We consider the expression

$$
\varepsilon_n := \sum_{p,l=1}^{n}\iint \mathbf{E}[\theta_l(x) - \mathbf{E}\theta_l(x)][\overline{\theta_p(y) - \mathbf{E}\theta_p(y)}]
$$

$$
\times dG_l^{(n)}(x)dG_p^{(n)}(y),
$$

where $\theta_l(x) = (\exp\{i\sqrt{s}\,x\eta_l - qx^2s_{ll}\} - i\sqrt{s}\,x\,eta_l(1+x^2)^{-1} - 1)(1+x^{-2})$. The following inequality is valid for ε_n:

$$
\varepsilon_n \leq \sum_{l=1}^{n}\int\left|\sum_{p=l}^{n}\int \rho_{lp}(x,y)dG_p^{(n)}(x)\right|dG_l^{(n)}(y),
$$
(9.3.10)

where $\rho_{lp}(x,y) = \mathbf{E}(\theta_l(x) - \mathbf{E}\theta_l(x))(\overline{\theta_p(y) - \mathbf{E}\theta_p(y)})$.

Since $\sup_{j=\overline{1,n}}\sum_{i=1}^{n}s_{ij}^2 \leq 3$, for any $\varepsilon > 0$, there exists the finite number of the variables s_{ij}^2 being greater than $\varepsilon > 0$, and the rest being less than ε. Clearly, $\sup_{x,y}|\rho_{lp}(x,y)| \leq c(s,q)$, where $c(s,q)$ is a constant depending only on s and q and bounded for any finite s and q. Suppose that for all $l = \overline{1,n}$ $s_{lk} > \varepsilon, k = \overline{1,m_\varepsilon}, s_{lk} \leq \varepsilon, k = \overline{m_\varepsilon + 1, n}$, where m_ε is some constant not depending on n. The random variables $\theta_l(x), \theta_p(y)$ are asymptotically independent as $s_{lp} \to 0$. Therefore, there exist the variables A_ε and B_ε such that $|\rho_{lp}(x,y)| < c(\varepsilon)$, $p = \overline{m_\varepsilon + 1, n}$ as $|x| \le A_\varepsilon$, $|y| \le B_\varepsilon$ and $c(\varepsilon) \to 0$, $A_\varepsilon \to \infty$, $B_\varepsilon \to \infty$, as $\varepsilon \to \infty$. Consequently,

$$
\begin{aligned}
\varepsilon_n \le{} & \sum_{l=1}^{n} \int_{|y| \ge A_\varepsilon} \left| \sum_{p=1}^{n} \int dG_p^{(n)}(x) \right| dG_l^{(n)}(y) c(s,q) \\
& + \sum_{l=1}^{n} \int_{|y| > A_\varepsilon} \left| \sum_{p=1}^{m_\varepsilon} dG_p^{(n)}(x) c(s,q) + c(\varepsilon) \cdot \sum_{p=m_\varepsilon+1}^{n} \int_{|x| \le B_\varepsilon} dG_p^{(n)}(x) \right. \\
& + \left. \sum_{p=m_\varepsilon+1}^{n} \int_{|x| > B_\varepsilon} c(s,q) dG_p^{(n)}(x) \right| dG_l^{(n)}(y).
\end{aligned}
$$

Hence, taking into account (9.3.8) and $\lim_{\varepsilon \to 0} \overline{\lim}_{n \to \infty} \sum_{p=1}^{n} [G_l^{(n)}(+\infty) - \int_{|y| < A_\varepsilon} dG_l^{(n)}(y)] = 0$, $\lim_{n \to \infty} \sup_{p=\overline{1,n}} G_p^{(n)}(+\infty) = 0$, we get $\lim_{n \to \infty} \varepsilon_n = 0$ for any fixed s and q.

Let us show that

$$
\operatorname*{plim}_{n \to \infty} \sum_{l=1}^{n} \eta_l \mathbf{E} \nu_{lk} (1 + \nu_{lk}^2)^{-1} = 0. \tag{9.3.11}
$$

According to Lemma 6.2.3, $\sup_n \sum_{l=1}^{n} |\mathbf{E}\nu_{lk}(1 + \nu_{lk}^2)^{-1}| < \infty$. Hence, taking into account that

$$
\begin{aligned}
\mathbf{E} \left(\sum_{l=1}^{n} \eta_l \mathbf{E}\nu_{lk}(1 + \nu_{lk}^2)^{-1} \right)^2 ={} & \sum_{l,p=1}^{n} r_{pl} \mathbf{E}\nu_{lk}(1 + \nu_{lk}^2)^{-1} \\
& \times \mathbf{E}\nu_{pk}(1 + \nu_{pk}^2)^{-1} \le \sum_{l=1}^{n} |\mathbf{E}\nu_{lk}(1 + \nu_{lk}^2)^{-1}| \sup_{p=\overline{1,n}} \left[\sum_{p=1}^{n} r_{pl}^2 \right]^{\frac{1}{2}} \\
& \times \left[\sum_{p=1}^{n} |\mathbf{E}\nu_{pk}(1 + \nu_{pk}^2)^{-1}|^2 \right]^{\frac{1}{2}},
\end{aligned}
$$

we obtain the expression (9.3.11).

Using it and $\lim_{n \to \infty} \varepsilon_n = 0$ from (9.3.9) and Theorem 5.3.4, we find (matrices R_t^k are fixed)

$$
\begin{aligned}
\mathbf{E} \exp \left\{ -q \sum_{i=1}^{n} s_{ii} \nu_{ik}^2 - s \sum_{i,j=1}^{p} s_{ij} \nu_{ik} \nu_{jk} \right\} \\
= \exp \left\{ \sum_{l=1}^{n} (\mathbf{E} \exp\{-s\nu_{lk}^2 s_{ll} - q\nu_{lk}^2 s_{ll}\} - 1) \right\} + 0(1) \\
= \mathbf{E} \exp \left\{ -\sum_{l=1}^{n} (s + q) \nu_{lk}^2 s_{ll} \right\} + 0(1).
\end{aligned}
$$

Equation (9.3.6) follows from this relation. According to §5 of Chapter 2,
$r_{pp} - r_{pp}^k = -t^2 (\sum_{i=1}^n r_{ip}^k \nu_{ik})^2 [1 + it\nu_{kk} + (R_s^k \overrightarrow{\nu_k}, \overrightarrow{\nu_k})]^{-1}$, $p \neq k$.
    Obviously,

$$\left| \sum_{i=1}^n r_{ip}^k \nu_{ik} \right|^2 \leq 2 \left(\sum_{i=1}^n \operatorname{Re} r_{ip}^k \nu_{ik} \right)^2 + 2 \left(\sum_{i=1}^n \operatorname{Im} r_{ip}^k \nu_{ik} \right)^2.$$

On the basis of the proof of Theorem (5.3.4),

$$\mathbf{E} \exp \left\{ iq \sum_{l=1}^n \operatorname{Re} r_{ll}^k \nu_{kl} \right\} = \mathbf{E} \exp \left\{ \sum_{l=1}^n [\exp\{ iq \operatorname{Re} r_{lp}^k \nu_{lk} \} - 1/r_{lp}^k] \right\} + 0(1).$$

Hence, as in the proof of (9.3.7) and by taking into account $\sum_{l=1}^n (\operatorname{Re} r_{lp}^k)^2 \leq 1$,
we obtain $\operatorname{plim}_{n \to \infty} \sum_{l=1}^n \nu_{lk} \omega p \operatorname{Re} r_{lp}^k = 0$.
    Analogously, we prove $\operatorname{plim}_{n \to \infty} \sum_{l=1}^n \operatorname{Im} r_{lp}^k \nu_{lk} = 0$. Consequently,

$$\lim_{n \to \infty} \sup_{p = \overline{1,n}} \mathbf{E} |r_{pp} - r_{pp}^k| = 0. \tag{9.3.12}$$

Now let us show that

$$\operatorname{plim}_{n \to \infty} \sum_{j=1}^n \nu_{kj}^2 , |r_{jj}^k - r_{jj}| = 0. \tag{9.3.13}$$

Suppose that ν_{kj} and r_{jj} are independent of one another. According to Theorem 5.3.4,

$$\mathbf{E} \exp \left\{ is \sum_{j=1}^n \nu_{kj}^2 |r_{jj}^k - r_{jj}| \right\} = \mathbf{E} \exp \left[\sum_{j=1}^n \int \exp\{ isx^2 |r_{jj}^k \right.$$
$$\left. - r_{jj}| \} - 1)(1 + x^{-2}) dG_j^{(n)}(x) \right] + 0(1).$$

Let us consider the inequalities

$$\mathbf{E} \left| \sum_{j=1}^n \int (\exp\{ isx^2 |r_{jj}^k - r_{jj}| \} - 1)(1 + x^{-2}) dG_j^{(n)}(x) \right|$$
$$\leq \sup_{j = \overline{1,n}} \mathbf{E} |r_{lj}^k - r_{jj}| (|s|c + |s|c\tau) + 2 \sum_{j=1}^n \int_{x^2 > \tau} dG_j^{(n)}(x).$$

Hence, by choosing sufficiently large τ and by using (9.3.12), we obtain
(9.3.13). From Lemma 6.2.4 it follows that

$$\operatorname{plim}_{n \to \infty} \left| \sum_{p=1}^n [\mathbf{E} \exp\{ -s\nu_{kp}^2 r_{pp}^k \} - 1/r_{pp}^k] - \mathbf{E}[\exp\{ -s\nu_{kp}^2 r_{pp}^k \} - 1] \right| = 0. \tag{9.3.14}$$

Let us introduce the random variables $\tilde{\nu}_{ij} = \nu_{ij}[1 + a |\nu_{ij}|]^{-1}$, where $a > 0$ is an arbitrary constant number. Denote $\widetilde{\Xi}_n = (\tilde{\nu}_{ij})$, $\widetilde{R}_t = (I + it\widetilde{\Xi}_n)^{-1} = (\tilde{r}_{ij})$. We consider the difference

$$n^{-1}\mathbf{E}\operatorname{Tr} R_t - n^{-1}\mathbf{E}\operatorname{Tr} \widetilde{R}_t = n^{-1}\sum_{s-1}^{n}\{\mathbf{E}\operatorname{Tr} \widetilde{R}_t^{s-1} - \mathbf{E}\operatorname{Tr} \widetilde{R}_t^s\},$$

where $\widetilde{R}_t^{(s-1)} = (I + it\widetilde{\Xi}_n^{(s-1)})^{-1}$, and the matrix $\widetilde{\Xi}_n$ is obtained from the matrix Ξ_n by replacing the variables $\nu_{ij}, j = \overline{1,s}$ by the variables $\tilde{\nu}_{ij}$.

Using (9.3.1), we obtain

$$n^{-1}\mathbf{E}\operatorname{Tr} R_t - n^{-1}\mathbf{E}\operatorname{Tr} \widetilde{R}_t = n^{-1}\sum_{s=1}^{n}\mathbf{E}\Big[t\frac{d}{dt}\ln[1 + it\nu_{kk}$$

$$+ t^2(\widetilde{R}_{tk}\vec{\nu}_k, \vec{\nu}_k)] - t\frac{d}{dt}\ln[1 + it\nu_{kk} + t^2(\widetilde{R}_{tk}\overrightarrow{\tilde{\nu}}_k, \overrightarrow{\tilde{\nu}}_k)]\Big],$$

where $\widetilde{R}_{tk} = (I + it\widetilde{\Xi}_{nk}^{(k-1)})^{-1}$ and the matrix $\widetilde{\Xi}_{nk}^{(k-1)}$ is obtained from the matrix $\widetilde{\Xi}_n^{(k-1)}$ by replacing the entries $\xi_{ki}, i = \overline{1,n}$ by zeros.

From this formula, we get (see the estimation of the resolvent difference in the proof of Theorem 9.3.1)

$$\lim_{a \to 0} n^{-1}|\mathbf{E}\operatorname{Tr} R_t - \mathbf{E}\operatorname{Tr} \widetilde{R}_t| = 0. \qquad (9.3.15)$$

Denote $\tilde{p}_{ss} = \operatorname{Re}\tilde{r}_{ss}$, $\tilde{q}_{ss} = \operatorname{Im}\tilde{r}_{ss}$, $\tilde{p}_{ii}^k = \operatorname{Re}\tilde{r}_{ii}^k$, $\tilde{q}_{ii}^k = \operatorname{Im}\tilde{r}_{ii}^k$,[2] where $\tilde{r}_{ij}^k$ are the entries of the matrix $\widetilde{R}_{tk}$. From the formula for the variables r_{kk} and from (9.3.7), it follows that

$$\tilde{q}_{kk} = -t^2\sum_{i=1}^{n}\tilde{q}_{ii}^k\tilde{\nu}_{ik}^2\Big[1 + t^2\sum_{i=1}^{n}\tilde{p}_{ii}^k\tilde{\nu}_{ik}^2$$

$$+ t^4\Big(\sum_{i=1}^{n}\tilde{q}_{ii}^k\nu_{ik}^2\Big)^2\Big]^{-1} + \varepsilon_n, \qquad (9.3.16)$$

where ε_n are some random variables and $\lim_{n\to\infty}\mathbf{E}|\varepsilon_n| = 0$ for all $|t| \leq T$. Taking into account that $\mathbf{E}|\tilde{q}_{ii} - \tilde{q}_{ii}^k| \to 0$ as $n \to \infty$ and $i \neq k$, $i = \overline{1,n}$, by using (9.3.12) we find $\sup_k \mathbf{E}|\tilde{q}_{kk}| \leq t^2 a^{-2}\sup_k \mathbf{E}|q_{kk}| + 0(1)$. The random functions $\tilde{q}_{kk}(t, w)$ are analytical on t (excluding, perhaps, point zero). Indeed, $r_{kk} = \sum_{l=1}^{n}(1 + it\lambda_l)^{-1}h_{kl}^2$, where h_{kl} are the components of the eigenvectors of the matrix Ξ_n, and λ_l are eigenvalues of the matrix Ξ_n. Hence, it follows

---

[2] The functions $\tilde{p}_{ss}$ and $\tilde{q}_{ss}$ are obtained from the functions p_{ss} and q_{ss} by replacing ν_{pl} and $\tilde{\nu}_{pl}$.

that $q_{kk}(t, w)$ is an analytical function for all t, excluding, perhaps, point zero. According to (9.3.16), and under the condition that $t^2 a^{-2} < 1$,

$$\lim_{n \to \infty} \sup_{k=\overline{1,n}} \mathbf{E}|\tilde{q}_{kk}| = 0.$$

Consequently, $\lim_{n \to \infty} \sup_{k=\overline{1,n}} \mathbf{E}|\tilde{q}_{kk}| = 0$ for all finite values t, since $\tilde{q}_{kk}(t, w)$, is an analytical function, $q_{kk}(o, w) = 0$. It means that $\tilde{r}_{kk} = \tilde{p}_{kk} + \varepsilon_n = \left[1 + t^2 \sum_{i=1}^{n} p_{ii}^k \nu_{ik}^2\right]^{-1} + \delta_n$, where ε_n and δ_n are complex random variables such that $\lim_{n \to \infty} \mathbf{E}(|\varepsilon_n| + |\delta_n|) = 0$.

Let $G_{na}(x, z, t) = \mathbf{P}\{\tilde{r}_{kk}(t) < x\}$ for $k/n \leq z < (k+1)/n$. From this equation, (9.3.14), and (9.3.13) we obtain

$$G_{na}(x, z, t) = \mathbf{P}\{[1 + t^2 \xi_{na}(G_{na}(\cdot, \cdot, t), z)]^{-1} < x\} + 0(1), \qquad (9.3.17)$$

where $\xi_{na}(G_{na}(\cdot, \cdot, t), z)$ is a random functional, whose Laplace transformation of the one-dimensional distributions equals

$$\mathbf{E} \exp\{-s\xi_{na}(G_{na}(\cdot, \cdot, t), z)\}$$
$$= \exp\left\{\int_0^1 \int_0^1 \left[\int_0^\infty (\exp\{-sx^2 y(1 + a|x|)^{-2}\} - 1)(1 + x^{-2})\right.\right.$$
$$\left.\left. \times dK(v, z, x)\right] dG_{na}(y, v, t)\, dv\right\} + 0(1).$$

On the basis of this formula and (9.3.17), the function $G_n(x, z, t)$ for the sufficiently large n can be considered uniformly continuous on z and t, for $|t| < T$. The functions $G_{na}(x, z, t)$ are nondecreasing and of bounded variation on x, equipotentially continuous on t and z on some everywhere dense set C for $|t| \leq T$, $0 \leq z \leq 1$, $T > 0$ is an arbitrary constant number. Therefore, there exists the subsequence $G'_{na}(x, z, t)$ of the sequence $G_{na}(x, z, t)$ with weak convergence to some function $G_a(x, z, t)$, and $G_a(x, z, t)$ satisfies Eq. (9.3.5) at the points of the continuity. If we prove that Eq. (9.3.5) has a unique solution for all functions from class L, we thus prove that $G_{na} \Rightarrow G_a$.

We suppose that there exist two subsequences $G_{n'}$ and $G_{n''}$ which are weakly convergent to the solutions $G_1(x, z, t)$ and $G_2(x, z, t)$ of equation (9.3.5). Denote $\xi_a(G_1(\cdot, \cdot, t), z) = \eta_1(z, t)$, $\xi_a(G_2(\cdot, \cdot, t), z) = \eta_2(z, t)$. Introduce the functionals

$$\tilde{\xi}_a(\theta_1(\cdot, \cdot, t), z), \quad \xi_a(\theta_2(\cdot, \cdot, t), z), \qquad (9.3.18)$$

given on the set of bounded nonnegative random functions $\Theta(w, z, t)$, with the distribution functions being continuous on z for $0 \leq z \leq 1$. The two-dimensional distributions of the functionals (9.3.18) are determined as follows:

$$\mathbf{E} \exp\{-s\tilde{\xi}_a(\theta_1(\cdot, \cdot, t), z) - q\tilde{\xi}_a(\theta_2(\cdot, \cdot, t), z)\}$$
$$= \exp\left\{\int_0^1 \mathbf{E}\left[\int_0^\infty (\exp\{-x^2(1 + a|x|)^2[s\theta_1(w, v, t)\right.\right.$$
$$\left.\left. + q\theta_2(w, v, t)]\} - 1)(1 + x^{-2})dK(v, z, x)\right] dv\right\}, \quad s, q \geq 0.$$
$$(9.3.19)$$

To make sure of the fact that such functionals exist, it is necessary to consider underlimiting random variables

$$\sum_{i=1}^{n} \tilde{\nu}_{ik}^2 \theta_{ii}^{(1)}(w, i/n, t), \quad \sum_{i=1}^{n} \nu_{ik}^2 \theta_{ii}^{(2)}(w, i/n, t), \tag{9.3.20}$$

where the random variables $\theta_{ii}^{(p)}, i = \overline{1,n}$ are independent for every $p = \overline{1,2}$ and n, do not depend on the random variables $\tilde{\nu}_{ik}^2$, and are distributed analogously to the random variables $\theta_p(w, i/n, t)$.

Let $\theta_1^{(1)}(w, z, t)$ and $\theta_2^{(1)}(w, z, t)$, for fixed z and t, have the distributions $G_1(x, z, t)$, $G_2(x, z, t)$, respectively, and arbitrary joint distribution. We consider the system of the functional random equations

$$[1 + t^2 \tilde{\xi}_a(\theta_1^{(1)}(\cdot, \cdot, t), z)]^{-1} = \theta_1^{(2)}(w, z, t),$$
$$[1 + t^2 \tilde{\xi}_a(\theta_2^{(1)}(\cdot, \cdot, t), z)]^{-1} = \theta_2^{(2)}(w, z, t). \tag{9.3.21}$$

The random variables $\theta_1^{(2)}(w, z, t)$ and $\theta_2^{(2)}(w, z, t)$ have the joint distribution that, in general, differs from the joint distributions of the variables $\theta_1^{(1)}(w, z, t)$, $\theta_2^{(1)}(w, z, t)$. From the equations (9.3.21), (9.3.20), and (9.3.19), it follows that

$$\sup_{0 \le z < 1} \mathbf{E} |\theta_1^{(2)}(w, z, t) - \theta_2^{(2)}(w, z, t)| \le t^2 c$$

$$\times \sup_{0 \le z \le 1} \mathbf{E} |\theta_1^{(1)}(w, z, t) - \theta_2^{(1)}(w, z, t)|, \tag{9.3.22}$$

$$c = \sup_{0 \le z \le 1} \int_0^1 \int x^2 (1 + a|x|)^{-2} (1 + x^{-2}) dK(v, z, x) \, dv.$$

Analogously, for the random functions $\theta_1^{(2)}(w, z, t)$, $\theta_2^{(2)}(w, z, t)$

$$\sup_{0 \le z \le 1} \mathbf{E} |\theta_1^{(3)}(w, z, t) - \theta_2^{(3)}(w, z, t)| \le t^2 c$$

$$\times \sup_{0 \le z \le 1} \mathbf{E} |\theta_1^{(2)}(w, z, t) - \theta_2^{(2)}(w, z, t)|,$$

where $\theta_1^{(3)}(w, z, t), \theta_2^{(3)}(w, z, t)$ are random functions with one-dimensional distributions coinciding with the function distributions $\theta_1^{(1)}(w, z, t)$, $\theta_2^{(1)}(w, z, t)$, respectively, and their joint distributions, in general, do not coincide. After n steps, we obtain $\sup_{0 \le z \le 1} \mathbf{E} |\theta_1^{(n)}(w, z, t) - \theta_2^{(n)}(w, z, t)| \le (t^2 c)^n$. Passing to the limit on n, we get $\tilde{G}_1(x, z, t) = G_2(x, z, t)$ for all $0 \le x \le 1$, $0 \le z \le 1$ and $t^2 < c^{-1}$. Since the functions $\int x^k dG_p(x, z, t)$ are analytical on t for all finite $t \ne 0$, $G_{na}(x, z, t) \Rightarrow G_a(x, z, t)$ for all finite t.

Thus, $G_{na}(x,z,t) \Rightarrow G_a(x,z,t)$, and $G_a(x,z,t)$ is a unique solution of Eq. (9.3.5). But then there exists $\lim_{n\to\infty} n^{-1}\mathbf{E}\,\mathrm{Tr}\,R_t = m(t)$, and the function $m(t)$ is continuous at zero. Consequently, from Theorem 9.3.1 with probability 1, $\lim_{n\to\infty}\mu(x) = F(x)$ at every point of the continuity of the nonrandom spectral function $F(x)$, whose Stieltjes transformation equals $m(t) = \lim_{a\to 0}\int_0^1\int_0^1 dG_a(x,z,t)\,dz$. Hence, by using (9.3.2), we get (9.3.3).

Let us now prove that Eq. (9.3.5) can be solved by the method of successive approximations. For this purpose, we consider the system of the functional random equations

$$[1 + t^2\tilde{\xi}_a(\nu_n(\cdot,\cdot,t),z)]^{-1} = \nu_{n+1}(w,z,t),$$
$$[1 + t^2\tilde{\xi}_a(\nu_{n-1}(\cdot,\cdot,t),z)]^{-1} = \nu_n(w,z,t), \qquad (9.3.23)$$

where $\nu_0(w,z,t), -\infty < t < \infty$ is an arbitrary bounded nonnegative random variable, with the distribution function being continuous on z $(0 \le z \le 1)$. Obviously, $\nu_n(w,z,t)$ belongs to the class L.

By using underlimiting random variables (9.3.20), it is easy to show that

$$\sup_{0\le z\le 1}\mathbf{E}\,|\nu_{n+1}(w,z,t) - \nu_n(w,z,t)| \le t^2 c \sup_{0\le z\le 1}\mathbf{E}\,|\nu_n(w,z,t) - \nu_{n-1}(w,z,t)|.$$

Hence, $\sup_{0\le z\le 1}\mathbf{E}\,|\nu_{n+1}(w,z,t) - \nu_n(w,z,t)| \le (t^2 c)^n$. For $t^2 c < 1$, the series $\sum_{n=1}^\infty \mathbf{E}\,|\nu_{n+1}(w,z,t) - \nu_n(w,z,t)|$ converges, and therefore there exists the limit $\lim_{n\to\infty}\mathbf{P}\{\nu_{n+1}(w,z,t) < x\}$. By making use of the first equation in (9.3.23), we obtain

$$G_{n+1}(x,z,t) = \mathbf{P}\{[1 + t^2\xi_a(G_n(\cdot,\cdot,t),z)]^{-1} < x\},$$

where $G_{n+1}(x,z,t) = \mathbf{P}\{\nu_{n+1}(w,z,t) < x\}$, $n = 0,1,\ldots$, and the limit $G_n(x,z,t) \Rightarrow G(x,z,t)$ exists. Theorem 9.3.2 is proved.

Different conditions of the convergence of the functions $k_n(u,v,z)$ to the limit function were considered in Chapter 8. By using them, we find particular conditions of the convergence of the variables $n^{-1}\ln\det(I + i\Xi_n)$.

Corollary 9.3.1. *In Theorem 9.3.2, instead of the condition $K_n(u,v,z) \Rightarrow K(u,v,z)$, let $N_n(u,v,z) \Rightarrow N(u,v,z)$, $z > 0$, $0 \le u \le 1$, $0 \le v \le 1$ hold where $N_n(u,v,z) = n[1 - \mathbf{P}\{(\xi_{ij}^{(n)})^2 < z\}]$ for $i/n \le u < (i+1)/n$, $j/n \le v < (j+1)/n$; the function $N(u,v,z)$ is a nondecreasing one on z and continuous on v and u $(0 \le u,v \le 1)$, and for every $\varepsilon > 0$ and $\varepsilon < \delta < \infty$,*

$$-\sup_{0\le u,v\le 1}\int_\varepsilon^\delta z\,dN(u,v,z) < \infty,$$

$$\lim_{\varepsilon\to 0}\lim_{n\to\infty}\sigma_n^2(u,v,\varepsilon) = \lim_{\varepsilon\tau 0}\overline{\lim_{n\to\infty}}\,\sigma_n^2(u,v,\varepsilon) = \sigma^2(u,v),$$

where $\sigma_n^2(u,v,\varepsilon) = n\left\{\int_{|x|<\varepsilon} x^2\,d\mathbf{P}\{\xi_{ij} < x\} - \biggl(\int_{|x|<\varepsilon} x\,d\mathbf{P}\{\xi_{ij} < x\}\biggr)^2\right\},$

$i/n \le u < (i+1)/n$, $\quad j/n \le v < (j+1)/n$, $\quad \sigma^2(u,v)$

is bounded and continuous in the domain $0 \leq u, v \leq 1$.

Then (9.3.3) holds, and the function $F(x)$ is determined by Eq. (9.3.5), with the random functional given by the following Laplace transformation,

$$
\mathbf{E} \exp\{-s\xi_a(G_a(\cdot, \cdot, t), z)\} = \exp\left\{-s \int_0^1 \sigma^2(v, z)\right.
$$

$$
\times \int_0^1 y dG_a(y, v, t)\, dv - \int_0^1 \int_0^1 \left[\int_0^\infty (\exp\{-sx(1 + a\sqrt{x})^{-2}\} - 1)\right.
$$

$$
\left.\times\, dN(v, z, x)\right] dG_a(y, v, t)\, dv \Bigg\}.
$$
$$(9.3.24)$$

Corollary 9.3.2. *If in addition to the conditions of Corollary 9.3.1, $N(u, v, z) \equiv 0$, then $\operatorname{plim}_{n\to\infty} \ln \det(I + i\Xi_n) = \int \ln(1 + ix)dF(x)$, where $F(x)$ is a distribution function with the Stieltjes transformation being equal to $\int (1 + itx)^{-1} dF(x) = \int_0^1 c(x, t)\, dx$ and $c(x, t)$ satisfying the equation*

$$
c(x, t) = \left[1 + t^2 \int_0^1 \sigma^2(x, y) c(y, t)\, dy\right]^{-1}. \tau ag 9.3.25
$$

The solution of Eq. (9.3.25) exists, is unique in the class of the functions $c(x, t)$, analytical on t, continuous on $x[0, 1]$, and it can be found with the method of successive approximations.

Proof. From (9.3.24) it follows that Eq. (9.3.5) has the form

$$
G(x, z, t) = \mathbf{P}\left\{\left[1 + t^2 \int_0^1 \sigma^2(v, z) \int_0^1 y dG(y, v, t)\,, dv\right]^{-1} < x\right\},
$$

provided the conditions of Corollary (9.3.2) hold. Putting $\int_0^1 y dG(y, v, t) = c(v, t)$, we obtain (9.3.25). Obviously, it can be solved with the method of successive approximaitons. Corollary 9.3.2 is proved.

Let Q_n be square matrices of order n, with all entries being equal to 1.

Corollary 9.3.3. *Let the random entries ξ_{ij}, $i \geq j$, $i, j = 1, 2, \ldots$ of the matrices $\Xi_n = (\xi_{ij})$ be independent and identically distributed, and there exist the constants a_n and c_n such that*

$$
\lim_{n\to\infty} n[1 - \mathbf{P}\{(\xi_{ij} - a_n)^2 > c_n x\}] = cx^{-\alpha},
$$

$$
x > 0, \quad 0 < \alpha < 1, \quad c > 0, \tag{9.3.26}
$$

for some $\beta > 0$,

$$
\sup_n n^{-1}\mathbf{E}\operatorname{Tr}|\ln(I + i\Xi_n)|^{1+\beta} < \infty.
$$

Then,

$$\operatorname*{plim}_{n\to\infty} n^{-1} \ln \det(I + ic_n^{-\frac{1}{2}}(\Xi_n - a_n Q_n))$$

$$= \int \ln(1 + ix) dF(x),$$

the Stieltjes transformation of the distribution function $F(x)$ equals $\int (1 + itx)^{-1} dF(x) = \int_0^1 x dG(x,t)$, and $G(x,t)$ satisfies the equation

$$G(x,t) = \mathbf{P}\left\{ \left[1 + t^2 \eta \left[\int_0^1 y^\alpha dG(y,t) \right]^{\frac{1}{\alpha}} \right]^{-1} < x \right\}, \tag{9.3.27}$$

where η is a random variable distributed by a stable law, whose Laplace transformation is equal to $\mathbf{E}\exp(-s\eta) = \exp(-s^\alpha h)$, $h = c\Gamma(1-\alpha)$, $s \geq 0$. The solution of equation (9.3.27) exists and is unique in the class of the distribution functions $G(x,t)$ on x, $0 \leq x \leq 1$ for any fixed t, and such that the functions $\int_0^1 y^{\frac{\alpha}{2}} dG(y,t)$ are analytical on t for all $t \neq 0$.

Proof. Let us check the validity of the conditions of Corollary 9.3.1 by setting $\xi_{ij}^{(n)} = c_n^{-\frac{1}{2}}(\xi_{ij} - a_n)$. Obviously, $N(u,v,x) \equiv cx^{-\alpha}$, $x > 0$, $0 < \alpha < 1$, $c > 0$, the random variables $\xi_{ij}^{(n)}$ are infinitesimal, and by (9.3.26)

$$\lim_{\varepsilon \to 0} \overline{\lim_{n\to\infty}} \, \sigma_n^2(u,v,\varepsilon) \leq \lim_{\varepsilon\to 0} 2\varepsilon^2 \, \overline{\lim_{n\to\infty}} \int_{|x|<\varepsilon} d\mathbf{P}\{[\xi_{ij}^{(n)}]^2 < x\} = 0.$$

Under these conditions, the Laplace transformation (9.3.24) has the form $\exp\{-s^\alpha c\Gamma(1-\alpha)\int_0^\infty y^\alpha dG(y,t)\}$. By using it, we obtain (9.3.27).

Let us prove that the solution of this equation is unique. Denote $m(t) = \int_0^1 x^\alpha dG(x,t)$. Then $m(t)$ satisfies the equation

$$m(t) = \mathbf{E}[1 + t^2 \eta m^{\frac{1}{\alpha}}(t)]^{-\alpha}. \tag{9.3.28}$$

Suppose that there exist two solutions $G_1(x,t)$ and $G_2(x,t)$. Then, $m_i(t) = \int_0^1 x^\alpha dG_i(x,t)$, $i = 1, 2, \dots$ must be different. Obviously, that $1 \geq m(t) \geq [1 + t^2 A]^{-\alpha} P\{\eta < A\}$. Hence, $|m_1(t) - m_2(t)| \leq t^2 c |m_1(t) - m_2(t)|$.

Since the functions $m_i(t)$, $i = 1, 2$ are analytical, we conclude that the solution of Eq. (9.3.27) is unique. Corollary (9.3.3) is proved.

Let us consider some examples.

1) *The Wigner semicircle law.* Let us set $\sigma^2(x,y) \equiv \sigma^2$ in equation (9.3.25). Then $c(x,t) = [1 + t^2\sigma^2 \int_0^1 c(y,t) dy]^{-1}$, but the Stieltjes transformation just equals $\int_0^1 c(x,t) dx$. Therefore, we obtain the equation $m(t) = [1 + t^2 m(t)]^{-1}$ for it. The solution of this equation has the form $m(t) = 2t^2\sigma^2[1 + \sqrt{1 + 4t^2\sigma^2}]^{-1}$. Converting this Stieltjes transformation, we obtain

$$F'(x) = \begin{cases} (2\pi\sigma)^{-1}\sqrt{4\sigma^2 - y^2}, & |y| \leq 2\sigma, \\ 0 & |y| > 2\sigma. \end{cases}$$

2) Random entries of the matrix Ξ_n belong to the domain of attraction of the stable law with the exponent $\alpha = \frac{1}{2}$. In this case, Eq. (9.3.28) has the form $m(t) = \mathbf{E}[1 + t^2 c\xi^{-2} m^2(t)]^{\frac{1}{2}}$, where ξ is a normally $N(0,1)$ distributed random variable, $c > 0$ is a constant. After making some simple transformations, we get

$$m(t) = \exp(t^2 m^2(t)/2) \int_{t\sqrt{cm(t)}}^{\infty} \sqrt{2/\pi} \exp(-x^2/2)\, dx.$$

Differentiating with respect to t, we obtain the following equation,

$$m'(t) = [tm^3(t)c - \sqrt{c}\, m(t)][1 - t^2 cm^2(t) + \sqrt{c}t]^{-1},$$

with the initial condition $m(0) = 1$.

Note that the uniqueness of the solution of Eq. (9.3.5) can be proved in another way. In this equation, we proceed to the limit as $a \downarrow 0$. Such limit exists, if

$$G(x, z, t) = \mathbf{P}\{[1 + t^2 \xi(G(\cdot, \cdot, t), z)]^{-1} < x\}$$

has the unique solution in the class of functions L. Let us transform this equation,

$$\int x^k dG(x, z, t) = \mathbf{E} \int_0^{\infty} \exp\{-t^2 y\xi(G(\cdot, \cdot, t), z)\} y^{k-1}$$

$$\times e^{-y} dy[(k-1)!]^{-1} = \int_0^{\infty} \exp\left\{\int_0^1 \int_0^1 \beta_{iggl}\left[\int_0^{\infty} (\exp\{-t^2 yx^2\} - 1)\right.\right.$$

$$\left.\left.\times (1 + x^{-2}) dK(v, z, x)\right] dG(y, v, t)\, dv\right\} y^{k-1} e^{-y}\, dy[(k-1)!]^{-1}.$$

We multiply both sides of the equation by $(-s)^k (k!)^{-1}$ and sum over k from 1 to ∞. Then

$$-1 + m(s, t, z) = \int_0^{\infty} \exp\left\{\int_0^1 \left[\int_0^{\infty} (m(t^2 yx^2, t, v) - 1)\right.\right.$$

$$\left.\left.\times (1 + x^{-2}) dK(v, z, x)\right] dv\right\} \sum_{k=1}^{\infty} y^{k=1}[(k-1)!k!]^{-1} e^{-y}\, dy,$$

where $m(s, t, z) = \int_0^1 e^{-sx} dG(x, z, t)$, $s \geq 0$.

According to the paper [104]

$$\sum_{k=1}^{\infty} (-s)^k y^{k-1}[(k-1)!k!]^{-1} = \frac{\partial}{\partial y} J_0(2\sqrt{sy})$$

$$= -(2\pi)^{-1} \int_0^{2\pi} \sin(2\sqrt{sy} \sin u)(s/y)^{\frac{1}{2}} \sin u\, du,$$

where $J_0(z)$ is the Bessel function.

Thus, we get the equation

$$m(s,t,z) - 1 = \int_0^\infty \exp\left\{ \int_0^1 \left[\int_0^\infty (m(t^2 yx^2, t, v) - 1)(1 + x^{-2}) \right. \right.$$
$$\left. \left. \times\, dK(v,z,x) \right] dv \right\} \frac{\partial}{\partial y} J_0(\sqrt{sy}) e^{-y}\, dy. \tag{9.3.29}$$

We give simple conditions when the solution of (9.3.29) is unique.

Let $\sup_{u,v} \int_0^\infty x^2 dK(v, u, x) < \infty$ and there exist two solutions $m_1(s,t,z)$ and $m_2(s,t,z)$ of this equation. Denote $u(s,t,z) = |m_1(s,t,z) = m_2(s,t,z)|$. From Eq. (9.3.29), we get

$$u(s,t,z) \le c \int_0^\infty \int_0^1 \int_0^\infty u(t^2 yx^2, t, v)(1 + x^{-2}) dK(v, z, x) dv s e^{-y}\, dy,$$

where $c = \pi^{-1} \int_0^{2\pi} \sin^2 u\, du$.

Hence,

$$u(s,t,z) \le t^2 \int_0^\infty \int_0^1 \int_0^\infty u(t^2 yx^2, t, v)(1 + x^{-2})\, dK(v, z, x)\, dv$$
$$\times \int_0^\infty \int_0^\infty y(1 + x^2)\, dK(v, z, x) e^{-y}\, , dy.$$

as $t^2 < \sup_v [\int_0^\infty \int_0^\infty y(1+x^2)\, dK(v, z, x) e^{-y}\, dy]^{-1} u(s,t,z) \equiv 0$. From the fact that the function $u^2(s,t,z)$ is analytical with respect to t as $t \ne 0$, it follows that $u(s,t,z) \equiv 0$ for all $s \ge 0$, $-\infty < t < \infty$, $0 \le z \le 1$. Thus, we conclude that

$$m_1(s,t,z) = m_2(s,t,z).$$

§4 The Wigner Semicircle Law

Let $\Xi_n = (\xi_{ij}^{(n)})_{i,j=1}^n$, $n = 1, 2, \ldots$, be symmetric matrices, $\mu_n(x) = n^{-1} \sum_{\lambda_{in} < x} 1$, where λ_{in}, $i = \overline{1, n}$, are the eigenvalues of Ξ_n, and the random variables $\xi_{ij}^{(n)}$, $i, j = \overline{1, n}$, $n = 1, 2, \ldots$, given on a common probability space.

The semicircle law is any assertion which states that the normalized spectral funciton $\mu_n(x)$ converges, with probability 1 or in probability, to the nonrandom spectral function $\mu(x)$ whose density has the semicircle form:

$$\mu'(x) = \begin{cases} (2\pi\sigma)^{-1}\sqrt{4\sigma^2 - x^2}, & |x| \le 2\sigma, \\ 0, & |x| > 2\sigma, \ \sigma > 0. \end{cases}$$

Wigner [167] was the first one to prove that $\lim_{n\to\infty} \mathbf{E}\mu_n(x) = \mu(x)$, under the following restrictions on the entries of Ξ_n : $\xi_{ij}^{(n)} = \eta_{ij} n^{-\frac{1}{2}}$, the η_{ij} for $i \geq j$, $i,j = 1,2,\ldots$, are independent and symmetric, $\mathbf{E}\eta_{ij}^2 = \sigma^2$, and η_{ij} have bounded moments of all orders. Later, Grenander [105] showed that $p - \lim_{n\to\infty} \mu_n(x) = \mu(x)$ under the same restrictions on the random variables ξ_{ij}.

It was proved in [4,5] that if η_{ij}, $i \geq j$, $i,j = 1,2,\ldots$, are independent, $\mathbf{E}\eta_{ii}^2 < \infty$, $\mathbf{E}\eta_{ij}^4 < \infty$ for $i \neq j$, and $\mathbf{E}\eta_{ij} = 0$, then $p - \lim_{n\to\infty} \mu_n(x) = \mu(x)$; if, in addition, $\mathbf{E}\eta_{ii}^4 < \infty$ and $\mathbf{E}\eta_{ij}^6 < \infty$, $i \neq j$, then $\lim_{n\to\infty} \mu_n(x) = \mu(x)$ with probability 1. At the "level of rigor of physics," a proof of the semicircle law was proposed in [132] by means of perturbation theory, under the assumption that η_{ij}, $i \geq j$, $i,j = 1,2,\ldots$, are independent, indentically distributed, and have finite variance. Then Pastur [140] proved that $p - \lim_{n\to\infty} \mu_n(x) = \mu(x)$, under the assumptions that the η_{ij}, $i \geq j$, $i,j = 1,2,\ldots$, are independent, that $\mathbf{E}\eta_{ij} = 0$, $\mathrm{Var}\,\eta_{ij} = \sigma^2$, and that for any $\tau > 0$ and $j = \overline{1,n}$, the Lindeburg condition

$$\lim_{n\to\infty} \sum_{i=1}^{n} \int_{|x|>\tau} x^2 d\mathbf{P}\{\xi_{ij}^{(n)} < x\} = 0$$

is satisfied.

The author [47] succeeded in proving that $\lim_{n\to\infty} \mu_n(x) = \mu(x)$ with probability 1 under the same assumptions. In another study, [50], the author proved that under the conditions $\mathbf{E}\xi_{ij}^{(n)} = 0$, and $\mathrm{Var}\,\xi_{ij}^{(n)} = \sigma^2/n$, the random variables $\xi_{ij}^{(n)}$, $i \geq j$, $i,j = \overline{1,n}$, are independent for every n, $\lim_{n\to\infty} \mu_n(x) = \mu(x)$ with probability 1, if and only if for any $\tau > 0$, we have

$$\lim_{n\to\infty} n^{-1} \sum_{i,j=1}^{n} \int_{|x|>\tau} x^2 d\mathbf{P}\{\xi_{ij}^{(n)} < x\} = 0. \qquad (9.4.1)$$

Theorem 9.4.1. *If the random variables $\xi_{ij}^{(n)}$, $i \geq j$, $i,j = \overline{1,n}$ are independent for each n, $\mathbf{E}\xi_{ij}^{(n)} = 0$, and $\mathrm{Var}\,\xi_{ij}^{(n)} = \sigma^2/n$, and $0 < \sigma^2 < \infty$, then $\lim_{n\to\infty} \mu_n(x) = \mu(x)$ with probability 1 if and only if condition (9.4.1) is satisfied.*

Proof. *Sufficiency.* It is obvious that (see the notations introduced in the proof of Theorem 9.3.1),

$$\mathbf{E}\left|(R_t^k \overrightarrow{\nu_k}, \overrightarrow{\nu_k}) - \sum_{i=1}^{n} r_{ii}^k (\nu_{ik})^2\right|^2$$

$$= \sigma^4 n^{-2} \sum_{i,j=1}^{n} \mathbf{E}[r_{ij}^k]^2 \to 0, \qquad (9.4.2)$$

as $n \to \infty$, $\overrightarrow{\nu_k} = (\nu_{1k}, \ldots, \nu_{nk})$.

Therefore,

$$n^{-1}\mathbf{E}\operatorname{Tr} R_t = n^{-1} \sum_{k=1}^{n} \mathbf{E}[1 + it\nu_{kk}^{(n)} + t^2(R_t^k \overrightarrow{\nu_k}, \overrightarrow{\nu_k})]^{-1}$$

$$= n^{-1} \sum_{k=1}^{n} \mathbf{E}[1 + t^2 \sum_{i=1}^{n} r_{ij}^k \nu_{ik}^2]^{-1} + 0(1)$$

$$= n^{-1} \sum_{k=1}^{n} \mathbf{E}\exp\{-\gamma t^2 \sum_{i=1}^{n} r_{ii}^k \nu_{ik}^2\} + 0(1), \qquad (9.4.3)$$

where γ is a random variable with density e^{-x}, $x \geq 0$, not depending on Ξ_n. Denote $\alpha_i^k = \mathbf{E}[\exp(-\gamma t^2 r_{ii}^k \nu_{ik}^2) - 1/r_{ii}^k, \gamma]$. Using the inequalities

$$|\alpha_i^k| \leq |\gamma r_{ii}^k|\varepsilon + 2\mathbf{P}\{\nu_{ik}^2 > \varepsilon\},$$

$$|\alpha_i^k| \leq |\gamma r_{ii}^k|\sigma^2 n^{-1},$$

we get $\lim_{n\to\infty} \sum_{i=1}^{n} \mathbf{E}|\alpha_i^k|^2 = 0$. Hence, Eq. (9.4.3) is equal to

$$n^{-1}\mathbf{E}\operatorname{Tr} R_t - n^{-1} \sum_{k=1}^{n} \mathbf{E}\exp\left\{\sum_{i=1}^{n} \alpha_i^k\right\} + 0(1).$$

On the basis of this equality

$$|n^{-1}\mathbf{E}\operatorname{Tr} R_t - n^{-1} \sum_{k=1}^{n} \mathbf{E}[1 + n^{-1}t^2\sigma^2 \operatorname{Tr} R_t^k]^{-1}|$$

$$= n^{-1}\left|\sum_{k=1}^{n} \mathbf{E}\exp\left\{\sum_{i=1}^{n} \alpha_i^k\right\} - \sum_{k=1}^{n} \mathbf{E}\exp\left\{-\gamma t^2\sigma^2 n^{-1} \sum_{i=1}^{n} r_{ii}^k\right\}\right|$$

$$+ 0(1) \leq n^{-1} \sum_{k=1}^{n} \mathbf{E}\left|\sum_{i=1}^{n} \alpha_i^k + n^{-1}\gamma t^2\sigma^2 \sum_{i=1}^{n} r_{ii}^k\right|$$

$$+ 0(1) \leq n^{-1} \sum_{k,i=1}^{n} \mathbf{E}\int_{|x|\leq\varepsilon} |e^{-\gamma t^2 r_{ii}^k x^2} - 1 + \gamma t^2 r_{ii}^k x^2|dF_{ik}(x)$$

$$+ n^{-1} \sum_{k,i=1}^{n} \mathbf{E}\int_{|x|>\varepsilon} |e^{-\gamma t^2 r_{ii}^k x^2} - 1 + \gamma t^2 r_{ii}^k x^2|dF_{ik} + 0(1)$$

$$\leq n^{-1}t^4 \sup_{k,i=\overline{1,n}} \mathbf{E}\gamma^2|r_{ii}^k|^2\varepsilon^2 \sum_{i,k=1}^{n} \int_{|x|\leq\varepsilon} x^2 dF_{ik}(x)$$

$$+ n^{-1}2t^2 \sup_{k,i=\overline{1,n}} \mathbf{E}|\gamma r_{ii}^k| \sum_{i,k=1}^{n} \int_{|x|>\varepsilon} x^2 dF_{ik}(x) + 0(1).$$

From this, and taking into account Lindeberg condition (9.4.1), we find

$$n^{-1}\mathbf{E}\operatorname{Tr} R_t = n^{-1}\sum_{k=1}^{n}\mathbf{E}[1 + n^{-1}t^2\sigma^2\operatorname{Tr} R_t^k]^{-1} + 0(1). \qquad (9.4.4)$$

From Theorem 9.3.1, it follows that with probability 1,

$$\lim_{n\to\infty} n^{-1}[\operatorname{Tr} R_t^k - \mathbf{E}\operatorname{Tr} R_t^k] = 0,$$

and using formula (9.3.1), we get

$$\lim_{n\to\infty} n^{-1}|\mathbf{E}\operatorname{Tr} R_t - \mathbf{E}\operatorname{Tr} R_t^k| = 0.$$

Therefore, (9.4.4) can be written in the form

$$n^{-1}\mathbf{E}\operatorname{Tr} R_t = [1 + n^{-1}t^2\sigma^2\mathbf{E}\operatorname{Tr} R_t]^{-1} + 0(1).$$

Having solved this square equation, we get

$$n^{-1}\mathbf{E}\operatorname{Tr} R_t = 2\left[1 + \sqrt{1 + 4t^2\sigma^2}\right]^{-1} + 0(1).$$

From this, using the inversion formula for the Stieltjes transfomations and Theorem 9.3.1, we get the sufficiency of the conditions of Theorem 9.4.1. Let us prove the necessity.

Using (9.4.2), we find

$$r_{kk} = \left[1 + t^2\sum_{i=1}^{n} r_{ii}^k \nu_{ik}^2\right]^{-1} + \varepsilon_n, \qquad (9.4.5)$$

where $\lim_{n\to\infty}\mathbf{E}|\varepsilon_n| = 0$. Denote $p_{kk} = \operatorname{Re} r_{kk}$, $q_{kk} = \operatorname{Im} r_{kk}$, $p_{ii}^k = \operatorname{Re} r_{ii}^k$, $q_{ii}^k = \operatorname{Im} r_{ii}^k$. By force of these notations,

$$q_{kk} = -t^2\sum_{i=1}^{n} q_{ii}^k \nu_{ik}^2\left[\left(1 + t^2\sum_{i=1}^{n} p_{ii}^k \nu_{ik}^2\right)^2 + \left(\sum_{i=1}^{n} q_{ii}^k \nu_{ik}^2\right)^2 biggr]^{-1} + \varepsilon_n.$$

From this equality, we get

$$\mathbf{E}|q_{kk}| \le t^2\sigma^2 n^{-1}\sum_{i=1}^{n}\mathbf{E}|q_{ii}^k| + \mathbf{E}|\varepsilon_n|. \qquad (9.4.6)$$

From §5 of Chapter 2, it follows that $\mathbf{E}\,|q_{ii} - q_{ii}^k| \le n^{-1}t^2\sigma^2$, $i \ne k$. Using this inequality from (9.4.6), we have

$$\mathbf{E}|q_{kk}| \le t^2\sigma^2 n^{-1}\sum_{i=1}^{n}\mathbf{E}|q_{ii}| + 0(1) \le t^2\sigma^2 \max_{i=\overline{1,n}}\mathbf{E}|q_{ii}| + 0(1). \qquad (9.4.7)$$

Suppose that $\alpha_n = \max_{i=\overline{1,n}} \mathbf{E}|q_{ii}|$ does not tend to zero as $n \to \infty$ and $z^2\sigma^2 < 1$. By assuming that on the left-hand side of inequality (9.4.7) the value of k is such that with $\mathbf{E}|q_{kk}| = \alpha_n$, we come to the inequality $1 \le t^2\sigma^2 + 0(1)$; and that is impossible for large $n > n_0$ since $t^2\sigma^2 < 1$. Thus, $\alpha_n = 0$ for $t^2\sigma^2 < 1$, $n > n_0$. Let us take into account that $q_{kk} = \mathrm{Im}\sum_{s=1}^n (1 + it\lambda_s)^{-1}h_{sk}^2$, where λ_s are the eigenvalues of the matrix Ξ and h_{s_k} are the components of the eigenvector $\overrightarrow{h_s}$ corresponding to the eigenvalue λ_k. Obviously, the function is analytical on t, maybe excluding the point zero. Therefore, $\alpha_n \to 0$ as $n > n_0$, and for all values of t. But then (9.4.5) is

$$\mathbf{E}r_{kk} = \mathbf{E}\left[1 + t^2\sum_{i=1}^n p_{ii}^k\nu_{ik}^2\right]^{-1} + 0(1).$$

Using this correlation, analogous to the proof of sufficiency, we find

$$n^{-1}\mathbf{E}\,\mathrm{Tr}\,R_t = n^{-1}\sum_{k=1}^n \mathbf{E}\left[1 + t^2\sum_{i=1}^n p_{ii}^k\nu_{ik}^2\right]^{-1} + 0(1)$$

$$= n^{-1}\sum_{k=1}^n \mathbf{E}\exp\left\{-\gamma t^2\sum_{i=1}^n \tilde{p}_{ii}\nu_{ik}^2\right\} + 0(1)$$

$$= n^{-1}\sum_{k=1}^n \mathbf{E}\exp\left\{\sum_{i=1}^n \alpha_i^k\right\} + 0(1), \qquad (9.4.8)$$

where

$$\alpha_i^k = \mathbf{E}[\{\exp(-\gamma t^2\tilde{p}_{ii}\nu_{ik}^2) - 1\}/p_{ii}, \gamma].$$

We suppose that the variables $\tilde{p}_{ii}$, $i = \overline{1,n}$ do not depend on the values ν_{ik}, i, $k = \overline{1,n}$ and the distributed as well as the variables p_{ii} in this formula. Since the normalized spectral function of the matrix Ξ_n converges to the semicircle law,

$$n^{-1}\mathbf{E}\,\mathrm{Tr}\,R_t = [1 + t^2\sigma_{n-1}^2\mathbf{E}\,\mathrm{Tr}\,R_t]^{-1} + 0(1) = 2\left[1 + \sqrt{1 + 4t^2\sigma^2}\right]^{-1} = 0(1).$$
$$(9.4.9)$$

Using that $\alpha_n \equiv 0$ as $n > n_0$ from (9.4.9), we get

$$n^{-1}\mathbf{E}\,\mathrm{Tr}\,R_t = n^{-1}\mathbf{E}\sum_{k=1}^n \left[1 + t^2\sigma_{n-1}^2\sum_{i=1}^n \tilde{p}_{ii}\right]^{-1} + 0(1)$$

$$= n^{-1}\sum_{k=1}^n \mathbf{E}\exp\{-n^{-1}\gamma t^2\sigma^2\sum_{i=1}^n \tilde{p}_{ii}\} + 0(1).$$

From this equality and (9.4.8), it follows that

$$\beta_n := n^{-1}\sum_{k=1}^n \mathbf{E}\left[\exp\left\{\sum_{i=1}^n \alpha_i^k\right\} - \exp\left\{-\gamma n^{-1}t^2\sigma^2\sum_{i=1}^n \tilde{p}_{ii}\right\}\right] = 0. \quad (9.4.10)$$

Since the expression under the sign of the mathematical expectation is non-negative,

$$\beta_n \geq \mathbf{E}C_n\left[n^{-1}\sum_{i,k=1}^{n}\{\alpha_i^k + \gamma t^2\sigma^2 n^{-1}\tilde{p}_{ii}\}\right], \qquad (9.4.11)$$

where $C_n = \exp\{-\gamma n^{-1}t^2\sigma^2\sum_{i=1}^{n}\tilde{p}_{ii}\}$. It is obvious that $\inf_n C_n > C > 0$ as $\gamma \leq a < \infty$, where $a > 0$ is an arbitrary constant. Therefore, by (9.4.10),

$$\lim_{n\to\infty} n^{-1}\mathbf{E}\sum_{i,k=1}^{n}\{\alpha_i^k + \gamma t^2\sigma^2 n^{-1}\tilde{p}_{ii}\}$$

$$= \lim_{n\to\infty} n^{-1}\mathbf{E}\sum_{i,k=1}^{n}\{[1 + t^2\tilde{p}_{ii}\nu_{ik}^2]^{-1} - 1 + t^2\sigma^2 n^{-1}\tilde{p}_{ii}\} = 0.$$

Hence,

$$n^{-1}\sum_{i=1}^{n}\mathbf{E}\sum_{k=1}^{n} n^{-1}t^2\sigma^2 p_{ii} - n^{-1}\sum_{i=1}^{n}\mathbf{E}\sum_{k=1}^{n}\int_{|x|\leq\epsilon}[1 - (1 + t^2\tilde{p}_{ii}x^2)^{-1}]dF_{ik}$$

$$(x) = n^{-1}\sum_{i=1}^{n}\mathbf{E}\sum_{k=1}^{n}\int_{|x|>\epsilon}[1 - (1 + t^2\tilde{p}_{ii}x^2)^{-1}]dF_{ik}(x) + 0(1).$$

The expression under the integral sign on the right-hand side does not exceed $x^2\epsilon^{-2}$, and that on the left-hand side is not less than $t^2\tilde{p}_{ii}x^2$. Therefore,

$$\mathbf{E}\sum_{i=1}^{n} n^{-1}\tilde{p}_{ii}\sum_{k=1}^{n}\int_{|x|>\epsilon} x^2 dF_{ik}(x) = 2\sigma^2\epsilon^{-2}t^{-2} + 0(1).$$

Obviously,

$$\mathbf{E}\tilde{p}_{ii} \geq [1 + t^2\sigma_{n-1}^2\mathbf{E}\operatorname{Tr} R_t]^{-1} + 0(1) = 2\left[1 + \sqrt{1 + 4t^2\sigma^2}\right]^{-1} + 0(1),$$

then

$$n^{-1}\sum_{i,k=1}^{n}\int_{|x|>\epsilon} x^2 dF_{ik}(x) \leq \sigma^2\epsilon^{-2}t^{-2}\left(1 + \sqrt{1 + 4t^2\sigma^2}\right) + 0(1). \qquad (9.4.12)$$

By choosing sufficiently large t and approximating n to infinity, the right-hand side of the inequality (9.4.12) can be made infinitesimal. Consequently, for any $\epsilon > 0$,

$$\lim_{n\to\infty} n^{-1}\sum_{i,k=1}^{n}\int_{|x|>\epsilon} x^2 dF_{ik}(x) = 0.$$

Theorem 9.4.1 is proved.

From Theorem 9.3.2 we can get the conditions with respect to which the Wigner semicircle law is valid, with the entries of the matrix Ξ_n not having the finite variances. We shall use the notations of Theorem 9.3.2 in the theorem proved below.

Theorem 9.4.2. *For every value of n, let the random entries* $\xi_{ij}^{(n)}$, $i \geq j$, $i, j = \overline{1, n}$ *of the matrix* $\Xi_n = (\xi_{ij}^{(n)} - \alpha_{ij}^{(n)})$ *be independent and identically distributed,*

$$\lim_{\varepsilon \to 0} \lim_{n \to \infty} \sigma_n^2(u, v, \varepsilon) = \lim_{\varepsilon \uparrow 0} \lim_{n \to \infty} \sigma_n^2(u, v, \varepsilon) = \sigma^2, \ (0 \leq u, v \leq 1), \tag{9.4.13}$$

$$\lim_{h \to \infty} \overline{\lim_{n \to \infty}} \sup_{i = \overline{1, n}} \mathbf{P} \left\{ \sum_{j=1}^{n} (\xi_{ij}^{(n)} - \alpha_{ij}^{(n)})^2 \geq h \right\} = 0. \tag{9.4.14}$$

Then with probability 1, $\lim_{n \to \infty} \mu_n(x) = \mu(x)$, *where*

$$\frac{d\mu(x)}{dx} = \begin{cases} \frac{1}{2\pi\sigma} \sqrt{4\sigma^2 - x^2}, & \text{if } |x| \leq 2\sigma, \\ 0, & \text{if } |x| > 2\sigma, \end{cases}$$

and the variables $\xi_{ij}^{(n)}$ *are infinitesimal if and only if for every fixed* $\varepsilon > 0$,

$$\lim_{n \to \infty} n \mathbf{P} \{ \xi_{11}^2 > \varepsilon \} = 0. \tag{9.4.15}$$

Proof. *Sufficiency.* From the conditions (9.4.15), the validity of the condition $N_n(u, v, t) \Rightarrow 0$ and the infinitesimality of the variables $\xi_{ij}^{(n)}$, $i, j = \overline{1, n}$ follow. Now it is not difficult to make sure of the fact that

$$n^{-1} \mathbf{E} \operatorname{Tr} R_t = n^{-1} \sum_{k=1}^{n} \mathbf{E}[1 + n^{-1} t^2 \sigma^2 \operatorname{Tr} R_t^k]^{-1} + 0(1).$$

From this, and by using the proof of Theorem 9.4.1, we get the sufficiency of the conditions of Theorem 9.4.2.

Necessity. From Theorem 9.3.1 and 9.3.2, we can choose a subsequence $\mu_{n'}(x)$ such that with probability 1 at every point of the continuity of the nonrandom spectral function $\mu(x) \lim_{n' \to \infty} \mu_{n'}(x) = \mu(x)$, Stieltjes transformation of the function $\mu(x)$ is equal to

$$\int (1 + itx)^{-1} d\mu(x) = \int_0^1 x dG(x, t) = \left[1 + \sigma^2 t^2 \int_0^1 x dG(x, t) \right]^{-1},$$

where $G(x, t)$ is a distribution function on x, satisfying the equation

$$G(x, t) = \mathbf{P} \{ [1 + t^2 \xi(G(\cdot, t))]^{-1} < x \}$$

at the points of continuity. To prove this, we use the fact that the main condition (9.3.8) follows from (9.4.13) and (9.4.14).

Here $\xi(G(\cdot, t))$ is a random functional whose Laplace transformation of one-dimensional distributions is equal due to (9.4.13),

$$\mathbf{E}e^{-s\xi(G(\cdot,t))} = \exp\left\{ \int_0^1 \int_0^\infty (e^{-syx} - 1)dN(y)dG(x,t) \right.$$
$$\left. - \frac{s\sigma^2}{2} \int_0^1 xdG(x,t) \right\}.$$

Then

$$\mathbf{E}[1 + t^2\xi(G(\cdot,t))]^{-1} = \left[1 + \sigma^2 t^2 \int_0^1 xdG(x,t)\right]^{-1}.$$

Hence

$$\mathbf{E}\exp\left\{ \int_0^\infty \int_0^1 (e^{-\gamma t^2 zx} - 1)dN(z)dG(x,t) \right\} = 1,$$

where $\gamma \geq 0$ is a random variable with density e^{-x}, $x \geq 0$.

From condition (9.4.14) it follows that $G(x,t) \neq 0$ for almost all values x and the finite t. Therefore, $N(z) \equiv 0$, $z > 0$. Consequently, (9.4.15) holds. Theorem 9.4.2 is proved.

§5 The General Form of Limit Spectral Functions

In the general case, the entries of the random matrix Ξ_n have an arbitrary form and their expectations may not be equal to zero. If the random entries of the matrix Ξ_n are independent, then from Theorem 9.3.1 it follows that the normalized spectral function of such a matrix can be approximately replaced by a nonrandom spectral function. The problem of finding the general form of the limit nonrandom spectral function is of interest. Without any additional assumptions for random variables, this problem seems trivial since any distribution function can be limiting. We shall make the following assumptions which are not too bounded and are confirmed by numerous problems: the random variables $\nu_{ij}^{(n)}$, $i, j = \overline{1, n}$ are infinitesimal and the vector rows of the random matrices $(a_{ij} + \nu_{ij})$ are asymptotically constant. By making such general assumptions, it is possible to get a formula for the Stieltjes transformations of the limit spectral functions and to show the general form of limit distributions of the random determinants. The normalized spectral function of the matrix $(a_{ij} + \nu_{ij})$ for large n approaches the normalized spectral function of the matrix $(a_{ij} + \gamma_{ij})$, where γ_{ij}, $i > j$ are infinitely divisible independent random variables with the characteristic functions

$$\exp\{\mathbf{E}\exp(is\nu_{ij}) - 1\}.$$

Theorem 9.5.1. *For every n, let the entries $\xi_{ij}^{(n)}$, $i > j$, $i, j = \overline{1, n}$ of the matrix $\Xi_n = (\xi_{ij}^{(n)})$ be independent and asymptotically constant, i.e., there* *exist such constants* $a_{ij}^{(n)}$ *that for any* $\varepsilon > 0$,

$$\lim_{n \to \infty} \sup_{i,j=\overline{1,n}} \mathbf{P}\{|\xi_{ij}^{(n)} - a_{ij}^{(n)}| > \varepsilon\} = 0,$$

the norms of the vector rows $(\nu_{ij}^{(n)})$ *are bounded in probability,*

$$\nu_{ij}^{(n)} = \xi_{ij}^{(n)} - b_{ij}^{(n)}, \ b_{ij}^{(n)} = a_{ij}^{(n)} + \int_{|x|<\tau} x dP\{\xi_{ij}^{(n)} - a_{ij}^{(n)} < x\},$$

$\tau > 0$ *is an arbitrary constant.*

$$\sup_n \sup_{j=\overline{1,n}} \sum_{i=1}^{n} b_{ij}^2 < \infty.$$

Then

$$n^{-1}\mathbf{E}\operatorname{Tr}(I + it(\xi_{ij}^{(n)}))^{-1} = n^{-1}\mathbf{E}\operatorname{Tr}(I + it(b_{ij}^{(n)} + \gamma_{ij}^{(n)}))^{-1} + 0(1),$$

where, for every n, γ_{ij}, $i \geq j$, $i,j = \overline{1,n}$ *are independent, not depending on the* Ξ_n, *and are distributed according to the infinitely divisible laws with characteristic function* $\exp\{\mathbf{E}\exp(is\nu_{ij}^{(n)}) - 1\}$.

Proof. Denote $R_t^s = (I + it\Xi_n^s)^{-1}$ where the matrix Ξ_n^s is obtained from the matrix Ξ_n by replacing the entries $\nu_{ij}^{(n)}$, $i = \overline{1,s}$, $j = \overline{1,n}$ by the entries γ_{ij}. We consider the difference

$$n^{-1}\mathbf{E}\operatorname{Tr} R_t^0 - n^{-1}\mathbf{E}\operatorname{Tr} R_t^n = \sum_{s=1}^{n}[n^{-1}\mathbf{E}\operatorname{Tr} R_t^{s-1}$$
$$- n^{-1}\mathbf{E}\operatorname{Tr} R_t^s](R_t^0 = (I + it\Xi_n)^{-1}).$$

Let $\tilde{R}_t^s = (I + it\tilde{\Xi}_n^s)^{-1}$, where the matrix $\tilde{\Xi}_n^s$ is obtained from the matrix Ξ_n^s by replacing the entries γ_{ki}, $i = \overline{1,n}$ by zeros. By using the formula for the resolvent difference (see Theorem 9.3.1), we find

$$n^{-1}\mathbf{E}\operatorname{Tr} R_t^0 - n^{-1}\mathbf{E}\operatorname{Tr} R_t^n = n^{-1}\sum_{s=1}^{n}\{n^{-1}\mathbf{E}\operatorname{Tr} R_t^{\cdot}$$

$$- n^{-1}\mathbf{E}\operatorname{Tr} \tilde{R}_t^s - (-n^{-1}\mathbf{E}\operatorname{Tr} r_t^s - n^{-1}\mathbf{E}\operatorname{Tr} \tilde{R}_t^s)\} = n^{-1}\sum_{s=1}^{n}\mathbf{E}\delta_s,$$
$$\tag{9.5.1}$$

where

$$\delta_s = t\frac{d}{dt}\ln\{[1 + it(\nu_{ss} + b_{ss}) + t^2(\tilde{R}_t^s(\overrightarrow{\nu}_s + \overrightarrow{b}_s),$$

$$(\overrightarrow{\nu}_s + \overrightarrow{b}_s))][1 + it(\nu_{ss} + b_{ss}) + t^2(\tilde{R}_t^s(\overrightarrow{\gamma}_s + \overrightarrow{b}_s),$$

$$(\overrightarrow{\gamma}_s + \overrightarrow{b}_s))]^{-1}\},$$

$$\overrightarrow{\nu}_s = (\nu_{s1}, \ldots, \nu_{ss-1}, 0, \nu_{ss+1}, \ldots, \nu_{sn}),$$

$$\overrightarrow{\gamma}_s = (\gamma_{s1}, \ldots, \gamma_{ss-1}, 0, \gamma_{ss+1}, \ldots, \gamma_{sn}),$$

$$\overrightarrow{b}_s = (b_{s1}, \ldots, b_{ss-1}, 0, b_{ss+1}, \ldots, b_{sn}).$$

According to Theorem 5.3.4, $\lim_{n\to\infty} \mathbf{E}|\delta_s| = 0$. From this and from (9.5.1), follows the statement of Theorem 9.5.1.

Theorem 9.5.2. *For every n, let the entries $\xi_{ij}^{(n)}$, $i \geq j$, $i,j = \overline{1,n}$ of the matrix $\Xi_n = (\xi_{ij}^{(n)})$ be independent,*

$$\mathbf{E}\xi_{ij} = b_{ij}, \qquad \mathrm{Var}\,\xi_{ij} = \sigma_{ij}^2, \qquad \lim_{n\uparrow\infty}\sup_{ij=\overline{1,n}} \sigma_{ij}^2 = 0,$$

$$\sup_n \sup_{j=\overline{1,n}} \sum_{i=1}^{n}[\sigma_{ij}^2 + b_{ij}^2] < \infty,$$

and Lindeberg condition holds; for any $\tau > 0$,

$$\lim_{n\to\infty} n^{-1}\sum_{i,j=1}^{n}\int_{|x|>\tau} x^2 d\mathbf{P}\{\nu_{ij}^{(n)} < x\} = 0 (\nu_{ij} = \xi_{ij} - \mathbf{E}\xi_{ij}).$$

Then

$$n^{-1}\mathbf{E}\,\mathrm{Tr}(I + it(\xi_{ij}^{(n)}))^{-1} = n^{-1}\mathbf{E}\,\mathrm{Tr}(I + it(b_{ij}^{(n)} + \gamma_{ij}^{(n)}))^{-1} + 0(1), \quad (9.5.2)$$

where for every n, $\gamma_{ij}^{(n)}$, $i \geq j$, $i,j = \overline{1,n}$ are independent, not depending on Ξ_n, and are distributed by the normal law $N(0, \sigma_{ij}^2)$.

Proof. We use Eq. (9.5.1). As in the proof of Theorem 9.4.1, we obtain

$$\lim_{n\to\infty} \mathbf{E}\left|(\tilde{R}_t^s(\overrightarrow{\nu}_s + \overrightarrow{b}_s), (\overrightarrow{\nu}_s + \overrightarrow{b}_s))\right.$$

$$\left. - \sum_{p\neq s}r_{pp}^s\nu_{ps}^2 - (\tilde{R}_t^s\overrightarrow{b}_s, \overrightarrow{b}_s)\right| = 0,$$

$$\lim_{n\to\infty} \mathbf{E}|(\tilde{R}_t^s(\overrightarrow{\gamma}_s + \overrightarrow{b}_s), (\overrightarrow{\gamma}_s + \overrightarrow{b}_s)) - \sum_{p\neq s}r_{pp}^s\gamma_{ps}^2 - (\tilde{R}_t^s\overrightarrow{b}_s, \overrightarrow{b}_s)| = 0.$$

The difference in (9.5.1) can be approximately replaced by the following:

$$n^{-1}\mathbf{E}\operatorname{Tr}R_t^0 - n^{-1}\mathbf{E}\operatorname{Tr}R_t^n = n^{-1}\sum_{s=1}^{n}\mathbf{E}\Delta_s + 0(1),$$

where

$$\Delta_s = t\frac{d}{dt}\ln\left\{\left[1 + itb_{ss} + t^2\sum_{p\neq s}r_{pp}^s\nu_{ps}^2 + t^2(\tilde{R}_t^s\overrightarrow{b}_s, \overrightarrow{b}_s)\right]\right.$$
$$\left.\times[1 + itb_{ss} + t^2\sum_{p\neq s}r_{pp}^s\gamma_{ps}^2 + t^2(\tilde{R}_t^s\overrightarrow{b}_s, \overrightarrow{b}_s)]^{-1}\right\}.$$

Taking into account the Lindeberg condition as in the proof of the sufficiency of the conditions of Theorem 9.4.1, we find that $\mathbf{E}|\Delta_s| \to 0$ as $n \to \infty$. Theorem 9.5.2 is proved.

In the general case it is difficult to indicate an equation to satisfy the Stieltjes transformation (9.5.2). If $b_{ij}^{(n)} = b_i\delta_{ij}$, then, obviously, we can find the equations of the (9.3.5) type, since the formula

$$n^{-1}\mathbf{E}\operatorname{Tr}[I + it(b_i\delta_{ij}) + it(\gamma_{ij}^{(n)})]^{-1}$$
$$= n^{-1}\sum_{s=1}^{n}\mathbf{E}[1 + it(b_s + \gamma_{ss}) + (R_t(s)\overrightarrow{\gamma}_s, \overrightarrow{\gamma}_s)]^{-1}$$

is valid where the matrix $R_t(s) = (I + it\Xi_n(s))^{-1}$ is obtained from the matrix $(b_i\delta_{ij} + \gamma_{ij})$ by replacing the entries b_s, γ_{si}, $i = \overline{1, n}$ by zeros.

The functional equation for (9.5.2) can be deduced if the variables $\gamma_{ij}^{(n)}$ are distributed by the normal law $N(0, \sigma^2 n^{-1})$. Denote $m_n(t) = n^{-1}\mathbf{E}\operatorname{Tr}(I + it(\xi_{ij}^{(n)}))^{-1}$.

Corollary 9.5.1. *If, in addition to the conditions of Theorem 9.5.2, $\sigma_{ij}^2 = \sigma^2/n$, $0 < \sigma^2 < \infty$, then $m_n(t) = n^{-1}\sum_{s=1}^{n}[1 + it\lambda_s + \sigma^2 t^2 m_n(t)]^{-1} + 0(1)$, where λ_s are the eigenvalues of the matrix $(b_{ij}^{(n)})$.*

Proof. Under the conditions of Corollary 9.5.1, the distribution of the matrix $(\gamma_{ij}^{(n)})$ is invariant with respect to the orthogonal transformation, therefore,

$$m_n(t) = n^{-1}\mathbf{E}\operatorname{Tr}(I + it(\lambda_s\delta_{ps}) + it(\gamma_{ps}))^{-1}$$
$$= n^{-1}\sum_{s=1}^{n}\mathbf{E}[1 + it(\lambda_s + \gamma_{ss}) + (R_t(s)\overrightarrow{\gamma}_s, \overrightarrow{\gamma}_s)]^{-1}.$$

Further on, we use the proof of Theorem 9.4.1.

Let $\lambda_n(x) = n^{-1}\sum_{s=1}^n F(x - \lambda_s)$. Obviously, if $\lambda_n(x) \Rightarrow \lambda(x)$, then $m_n(t) \to m(t)$, and $m(t)$ satisfies the equation [141]

$$m(t) = \int [1 + itx + t^2\sigma^2 m(t)]^{-1} d\lambda(x). \tag{9.5.3}$$

Theorem 9.5.3. *Let the condition of Theorem 9.5.1 hold and*

$$\lim_{n\to\infty} \sup_{i,j=\overline{1,n}} \sqrt{n}\mathbf{E}\nu_{ij}^2(1 + \nu_{ij}^2)^{-1} = 0,$$

then

$$n^{-1}\mathbf{E}\operatorname{Tr}(I + it(\xi_{ij}^{(n)}))^{-1} = \mathbf{E}\frac{\partial}{\partial\theta}\ln\det[I\theta - it$$

$$\times \operatorname{diag}\{\xi_{sn} \cdot (\operatorname{Re} r_{pp}^{(s)}(\theta), p = \overline{s,n}) + i\xi_{sn}(\operatorname{Im} r_{pp}^{(s)}(\theta),$$

$$p = \overline{s,n}), s = \overline{1,n}\} + itB_n]_{\theta=1} + 0(1), \tag{9.5.4}$$

$\xi_{sn}(\operatorname{Re} r_{pp}^{(s)}(\theta), \pi = \overline{s,n})$, $\xi_{sn}(\operatorname{Im} r_{pp}^{(s)}(\theta), p = \overline{s,n})$, $s = \overline{1,n}$ *are independent for every n random variables,* $r_{pp}^{(s)}(\theta)$ *are entries of the matrix* $\tilde{T}_n^s = (I\theta + it\tilde{H}_n^s)^{-1}$, *the matrix* $\tilde{H}_n^s$ *is obtained from the matrix* Ξ_n *by replacing the entries* ν_{ij}, $i = \overline{1,s-1}$, $\nu_{sj} + b_{sj}$, $j = \overline{1,n}$ *by zeros,* $B_n = (b_{ij})$, $\theta \geq 1$, H_n^s *is a matrix obtained from the matrix* Ξ_n^s *by replacing the entries* ν_{ij}, $i = \overline{1,s-1}$, $j = \overline{1,n}$ *by zeros,* $T_n^s = (I\theta + itH_n^s)^{-1}$.

The characteristic function of the random variables $\xi_{sn}(\cdot)$, $\frac{\partial}{\partial\theta}\xi_{sn}(\cdot)$ *equals*

$$\mathbf{E}\exp\Big\{i\beta_1\xi_{sn}(\operatorname{Re} r_{pp}^{(s)}(\theta), p = \overline{s,n}) + i\beta_2\xi_{sn}(\operatorname{Im} r_{pp}^{(s)}(\theta), p = \overline{s,n})$$

$$+ i\beta_3\frac{\partial}{\partial\theta}\xi_{sn}(\operatorname{Re} r_{pp}^{(s)}(\theta), p = \overline{s,n}) + i\beta_4\frac{\partial}{\partial\theta}\xi_{sn}$$

$$\times (\operatorname{Im} r_{pp}^{(s)}(\theta), p = \overline{s,n})\Big\} = \exp\Big\{\sum_{k=s}^n\Big[\mathbf{E}\int \exp\{ix[\beta_1 \operatorname{Re} r_{kk}^{(s)}(\theta)$$

$$+ \beta_2 \operatorname{Im} r_{kk}^{(s)}(\theta) + \beta_3\frac{\partial}{\partial\theta}\operatorname{Re} r_{kk}^{(s)}(\theta) + \beta_4\frac{\partial}{\partial\theta}\operatorname{Im} r_{kk}^{(s)}(\theta)]\} - 1\Big]$$

$$\times d\mathbf{P}\{\nu_{sk}^2 < x\}\Big\}. \tag{9.5.5}$$

Proof. We denote by K_n^s the matrix which is obtained from the matrix Ξ_n by replacing the entries ν_{ij}, $i = \overline{1,s-1}$, by zeros, and by replacing the sth diagonal entry $\nu_{ss} + b_{ss}$ by the value $it(\tilde{T}_n^s(\overrightarrow{\nu}_s + \overrightarrow{b}_s), (\overrightarrow{\nu}_s + \overrightarrow{b}_s)) + it(\tilde{T}_n^s\overrightarrow{b}_s, \overrightarrow{b}_s) + it\nu_{ss} + \theta$, $P_n^s := (I\theta + itK_n^s)^{-1}$, where

$$\overrightarrow{b}_s = (b_{s1}, \ldots, b_{ss-1}, 0, b_{ss+1}, \ldots, b_{sn}),$$

$$\overrightarrow{\nu}_s = (0, \ldots, 0, \nu_{ss+1}, \ldots, \nu_{sn}),$$

the number of zeros in the vector $\overrightarrow{\nu}_s$ equals s.

We consider the difference (see §5 of Chapter 2)

$$
\begin{aligned}
&\operatorname{Tr}(T_n^s - \tilde{T}_n^s) - \operatorname{Tr}(P_n^s - \tilde{T}_n^s) \\
&\quad = \operatorname{Tr}[(I\theta + itH_n^s)^{-1} - (I\theta + it\tilde{H}_n^s)^{-1}] - \operatorname{Tr}[(I\theta + itK_n^s)^{-1} \\
&\quad - (I\theta + it\tilde{H}_n^s)^{-1}] = \frac{\partial}{\partial\theta} \ln \det[I + it\tilde{T}_n^s(H_n^s - \tilde{H}_n^s)] \\
&\quad - \frac{\partial}{\partial\theta} \ln \det[I + it\tilde{T}_n^s(K_n^s - \tilde{H}_n^s)] = \frac{\partial}{\partial\theta} \ln\{1 + \theta^{-1}it \\
&\quad \times (\nu_{ss} + b_{ss}) + \theta^{-1}t^2(\tilde{T}_n^s(\overrightarrow{\nu}_s + \overrightarrow{b}_s),(\overrightarrow{\nu}_s + \overrightarrow{b}_s))\} \\
&\quad - \frac{\partial}{\partial\theta} \ln\{1 + \theta^{-1}it(\nu_{ss} + b_{ss}) + \theta^{-1}t^2(\tilde{T}_n^s(\overrightarrow{\nu}_s + \overrightarrow{b}_s),(\overrightarrow{\nu}_s + \overrightarrow{b}_s)) \\
&\quad - \theta^{-1}t^2(\tilde{T}_n^s\overrightarrow{b}_s,\overrightarrow{b}_s) + \theta^{-1}t^2(\tilde{T}_n^s\overrightarrow{b}_s,\overrightarrow{b}_s)\} = 0.
\end{aligned}
$$

Therefore,

$$
\begin{aligned}
n^{-1}\mathbf{E}\operatorname{Tr}(I + it\Xi_n)^{-1} = \mathbf{E}\frac{\partial}{\partial\theta}\ln\det[I\theta - it\,\operatorname{diag}\{(\tilde{T}_n^s \\
\times (\overrightarrow{\nu}_s + \overrightarrow{b}_s),(\overrightarrow{\nu}_s + \overrightarrow{b}_s)) - (\tilde{T}_n^s\overrightarrow{b}_s,\overrightarrow{b}_s)), s = \overline{1,n}\} + itB_n]_{\theta=1}.
\end{aligned}
$$

Using the proof of Theorem 9.3.2, we obtain

$$
(\tilde{T}_n^s(\overrightarrow{\nu}_s + \overrightarrow{b}_s),(\overrightarrow{\nu}_s + \overrightarrow{b}_s)) - (\tilde{T}_n^s\overrightarrow{b}_s,\overrightarrow{b}_s) \sim \sum_{p=s}^{n} \tilde{r}_{pp}^{(s)}\nu_{sp}^2,
$$

$$
\left(\operatorname{Re}\sum_{p=s}^{n} \tilde{r}_{pp}^{(s)}\nu_{sp}^2, \operatorname{Im}\sum_{p=s}^{n} \tilde{r}_{pp}^{(s)}\nu_{sp}^2, \frac{\partial}{\partial\theta}\operatorname{Re}\sum_{p=s}^{n} \tilde{r}_{pp}^{(s)}\nu_{sp}^2, \right.
$$

$$
\frac{\partial}{\partial\theta}\operatorname{Im}\sum_{p=s}^{n} \tilde{r}_{pp}^{(s)}\nu_{sp}^2) \sim (\xi_{sn}(\operatorname{Re} r_{pp}^{(s)}(\theta), p = \overline{s,n}),
$$

$$
\xi_{sn}(\operatorname{Im} r_{pp}^{(s)}(\theta), p = \overline{s,n}), \frac{\partial}{\partial\theta}\xi_{sn}(\operatorname{Re} r_{pp}^{(s)}(\theta), p = \overline{s,n}),
$$

$$
\left. \frac{\partial}{\partial\theta}\xi_{sn}(\operatorname{Im} r_{pp}^{(s)}(\theta), p = \overline{s,n}) \right).
$$

From this and from (9.5.5), (9.5.4) follows. Theorem 9.5.3 is proved.

Formula (9.5.4) is convenient for the analysis of the random perturbations of a determinant of a random matrix. It is clear that if under the conditions of Theorem 9.5.3 there exists

$$
\mathbf{E}|n^{-1}\ln\det(I + i\Xi_n)|^{1+\alpha}, \quad \alpha > 0,
$$

then

$$
n^{-1} \mathbf{E} \ln \det(I + i\Xi_n) = n^{-1} \int_0^1 t^{-1} [1 - \mathbf{E} \frac{\partial}{\partial \theta} \ln \det[I\theta
$$
$$
- it \operatorname{diag}\{\xi_{sn}(\operatorname{Re} r_{pp}^{(s)}(\theta), p = \overline{s,n}) + i\xi_{sn}(\operatorname{Im} r_{pp}^{(s)}(\theta),
$$
$$
p = \overline{s,n}), s = \overline{1,n}\} + itB_n]_{\theta=1}] \, dt + 0(1),
$$

and there is a determinant of the sum of the nonrandom matrix and of the random diagonal matrix with independent diagonal entries on the right-hand side of this equation.

§6 Normalized Spectral Functions of Symmetric Random Matrices with Dependent Random Entries

We prove limit theorems for the normalized spectral functions of random matrices with dependent random elements.

Let $\Xi_n = (\xi_{ij}^{(n)})$ be symmetric $n \times n$ random matrices on a probability space (Ω, F, P) and let

$$
\alpha(n, k, s) = \sup_{A \in A'_n, B \in A''_n} [\mathbf{P}(AB) - \mathbf{P}(A)\mathbf{P}(B)], \qquad (9.6.1)
$$

where A'_n is the smallest σ-algebra with respect to which the random variables ξ_{ij}, $1 \leq i \leq k$, $1 \leq j \leq n$ are measurable, and A''_n is the smallest σ-algebra with respect to which the random variables ξ_{ij}, $s \leq i \leq n$, $1 \leq j \leq n$ are measurable. Let $\mu_n(x) = n^{-1} \sum_{k=1}^n F(-\lambda_k + x)$, where the λ_k are the eigenvalues of the matrix Ξ_n.

Theorem 9.6.1. *If*

$$
\lim_{n \to \infty} n^{-1} [k_n^2 + n^2 \sup_{1 \leq k < s \leq n, |k-s|=k_n} \alpha(n, k, s)] = 0, \qquad (9.6.2)
$$

and

$$
\lim_{h \to -\infty} \sup_n \mathbf{E}\mu_n(h) = 0, \qquad (9.6.3)
$$

then for almost every x and any $\varepsilon > 0$,

$$
\lim_{n \to \infty} \mathbf{P}\{|\mu_n(x) - \mathbf{E}\mu_n(x)| > \varepsilon\} = 0. \qquad (9.6.4)
$$

Proof. Consider the Stieltjes transforms of the spectral functions $\mu_n(x)$ (see Theorem 9.3.1), where $\gamma_k = \mathbf{E}_{k-1} \operatorname{Tr}(I + it\Xi_n)^{-1} - \mathbf{E}_k \operatorname{Tr}(I + it\Xi_n)^{-1}$ and $\mathbf{E}_k$ is the conditional expectation for the fixed smallest σ-algebra with respect to which the random variables ξ_{ij}, $k \leq j \leq i \leq n$ are measurable. Let Ξ_{nk}^s be the matrix obtained from Ξ_n by replacing the entries ξ_{ij}, $k \leq i \leq s$, $1 \leq j \leq n$ by zeros. It follows from (9.3.1) that

$$|\operatorname{Tr}(I + it\Xi_n)^{-1} - \operatorname{Tr}(I + it\Xi_{nk}^k)^{-1}| \leq 2.$$

Therefore,

$$|\operatorname{Tr}(I + it\Xi_n)^{-1} - \operatorname{Tr}(I + it\Xi_{nk}^s)^{-1}| \leq (s - k)2. \qquad (9.6.5)$$

We represent γ_k in the form

$$\begin{aligned}
\gamma_k = {}& \mathbf{E}_{k-1}[\operatorname{Tr}(I + it\Xi_n)^{-1} - \operatorname{Tr}(I + it\Xi_{k-k_n}^{k+k_n})^{-1}] \\
& - \mathbf{E}_k[\operatorname{Tr}(I + it\Xi_n)^{-1} - \operatorname{Tr}(I + it\Xi_{k-k_n}^{k+k_n})^{-1}] \\
& + \mathbf{E}_{k-1}\operatorname{Tr}(I + it\Xi_{k-k_n}^{k+k_n})^{-1} - \mathbf{E}_k\operatorname{Tr}(I + it\Xi_{k-k_n}^{k+k_n})^{-1},
\end{aligned}$$

where the matrix $\Xi_{k-k_n}^{k+k_n}$ is obtained from Ξ_n by replacing the entries ξ_{ij}, $1 \leq i \leq k - k_n$ and $k + k_n \leq i \leq n$, $1 \leq j \leq n$ by zeros.

For γ_k it is easy to derive the inequality (see formula (9.3.1) and (9.6.5)),

$$|\gamma_k|^2 \leq c[(2k_n)^2 \sup_{1 \leq k < s \leq n, |k-s|=k_n} \alpha(n, k, s)n^2].$$

But (9.6.4) follows from this, with the use of (9.6.2) and (9.6.3). Theorem (9.6.1) is proved.

Note that if $\alpha(k_n) = c\exp(-\alpha_k k_n)$, where $\alpha > 1$ and $c > 0$, then (9.6.4) is valid for $k_n = \log n$.

CHAPTER 10

LIMIT THEOREMS FOR RANDOM
GRAM DETERMINANTS

Let H be $m_n \times n$ matrix, $m_n \geq n$. We find the general form of the limited expressions $n^{-1}\mathbf{E}\ln\det HH'$ as $n \to \infty$ by means of limit theorems for normalized spectral random functions.

§1 Spectral Equation for Gram Matrices

Let $H_n = \xi_{pl}^{(n)} + i\eta_{pl}^{(n)}$ be a complex $m_n \times n$ random matrix and $\mu_n(x)$ a normalized spectral function of the matrix $H_n H_n *$.

Theorem 10.1.1. *If for each n the vector columns of the matrix H_n are independent, the random variables $\xi_{pl}^{(n)}, \eta_{pl}^{(n)}, p, l, n = 1, 2, \ldots$ are given on a common probability space, there exists a limit $\lim_{n\to\infty} n^{-1}\mathbf{E}\operatorname{Tr}(I + itH_nH_n*)^{-1} = a(t)$, the function $a(t)$ is continuous in zero and the series $\sum_{n=1}^{\infty} n^{-4}m_n^2$ is convergent, then $\lim_{n\to\infty}\mu_n(x) = \mu(x)$ with probability 1 at each point of the continuity of the nonrandom spectral function $\mu(x)$, with the Stieltjes transformation being $\int(1 + itx)^{-1}d\mu(x) = a(t)$.*

Proof. Let us denote $\gamma_k = \mathbf{E}[\operatorname{Tr} R_t \mid \sigma_{k-1}] - \mathbf{E}[\operatorname{Tr} R_t \mid \sigma_k], k = \overline{1, m_n}$, where $R_t = (I + itH_nH_n*)^{-1}, \sigma_k$ is the smallest σ-algebra generated by the random vector-columns $\vec{\eta}_s, s = \overline{k + 1, m_n}$ of the matrix H_n.

It is obvious that $\operatorname{Tr} R_t - \mathbf{E}\operatorname{Tr} R_t = \sum_{k=1}^{m_n} \gamma_k$. By using the proof of Lemma 6.3.1, we get $\mathbf{E}\mid\sum_{k=1}^{m_n}\gamma_k\mid^4 \leq cm_n^2$, where $c > 0$ is a constant. Hence, on the basis of the Borel–Kantelli lemma and the proof of the Theorem 9.3.1, we come to the assertion of Theorem 10.1.1.

It is easy to show that Theorem 10.1.1 holds, provided the matrix $HH*$ can be represented in the form of $HH* = B_n + \sum_{s=1}^{m_n} \nu_s^{(n)} K_n^s$, where B_n are nonrandom square matrices of order n, $\nu_s^{(n)}$ are random variables, independent for each n and not depending on the matrices $K_n^s = \eta_s \vec{\eta}_s^*$.

Analogously, we get the following assertion.

297

Theorem 10.1.2. *If in Theorem 10.1.1 the condition* $\lim_{n\to\infty} n^{-2}m_n = 0$ *is satisfied instead of the convergence of the series* $\sum_{n=1}^{\infty} n^{-4}m_n^2$, *then for any* $\varepsilon > 0$ *at each point of the continuity of the function* $\mu(x)$

$$\lim_{n\to\infty} \mathbf{P}\{|\mu_n(x) - \mu(x)| > \varepsilon\} = 1.$$

Theorem 10.1.3. *For each* n, *let the random entries* $\xi_{pl} + i\eta_{pl}, p = \overline{1,n}, l = \overline{1,m_n}$ *of the matrix* $H_n = (\xi_{pl} + i\eta_{pl} - \alpha_{pl} - i\beta_{pl})$ *be independent, infinitesimal,* $\alpha_{pl} = \int_{|x|<\tau} x d\mathbf{P}\{\xi_{pl} < x\}, \beta_{pl} = \int_{|x|<\tau} x d\mathbf{P}\{\eta_{pl} < x\}, \tau > 0$ *an arbitrary constant,* $K_n(u,v,z) \Rightarrow K(u,v,z),$ *where* $K_n(u,v,z) = n \int_0^z y(1+y)^{-1} d\mathbf{P}\{\xi_{pl} - \alpha_{pl})^2 + (\eta_{pl} - \beta_{pl})^2 < y\}$ *as* $p/n \leq u < (p+1)/n, l/n \leq v < (l+1)/n$ *the function* $K(u,v,z)$ *nondecreasing, of bounded variation on* z *and continuous on* u *and* v *in the domain* $0 \leq u \leq 1, 0 \leq v \leq c, \lim_{n\to\infty} m_n/n = c,$ *where* c *is an absolute constant, and for some* $\alpha > 0$,

$$\sup_n n^{-1} \mathbf{E}\, \mathrm{Tr}\,|\ln H_n H_n*|^{1+\alpha} < \infty. \tag{10.1.1}$$

Then

$$\mathrm{plim}_{n\to\infty} n^{-1} \ln \det H_n H_n* = \int_0^\infty \ln x dF(x), \tag{10.1.2}$$

where $F(x)$ *is a distribution function whose Stieltjes transformation is*

$$\int_0^\infty (1+tx)^{-1} dF(x) = \lim_{\alpha\to 0} \int_0^1 \int_0^1 x dG_\alpha(x,z,t)dz, \tag{10.1.3}$$

where $G_\alpha(x,z,t)$ *is a distribution function on* $x(0 \leq x \leq 1, 0 \leq z \leq 1, 0 \leq t < \infty)$ *satisfying the equation at the point of continuity:*

$$G_\alpha(x,z,t) = \mathbf{P}\{[1 + t\xi_{1a}([1 + t\xi_{2a}(G_\alpha(\cdot,\cdot,t),\cdot)]^{-1}, z)]^{-1} < x\},^1 \tag{10.1.4}$$

the random functional ξ_{1a} *is given on the set of independent bounded random real continuous functions; the random functional* ξ_{2a} *is given on the set of distribution functions* $G_\alpha(x,z,t)$; *and these functionals are mutually independent.*

The Laplace transformations of the random functionals $\xi_{1a}(\cdot)$ *and* $\xi_{2a}(\cdot)$ *distributions are equal to*

$$\mathbf{E}\exp\{-s\xi_{2a}(G_a(\cdot,\cdot,t),z)\} = \exp\{\int_0^c \int_0^1 [\int_0^\infty (\exp(-x$$
$$\times (1 + a\sqrt{x})^{-2}sy) - 1)(1 + x^{-1})dK(v,z,x)]$$
$$\times dG_a(y,v,t)dv\},$$

$$\mathbf{E}\exp\{-s\xi_{1a}(h(\cdot,\cdot,t),z)\} = \exp\{\mathbf{E}\int_0^1 \int_0^\infty (\exp\{-sx$$
$$\times h(w,v,t)(1 + a\sqrt{x})^{-2}\} - 1)(1 + x^{-1})_d K(z,v,x)dv\},$$

---

[1] We shall call equation (10.1.4) a canonical spectral function.

where $h(w, v, t)$ is any arbitrary bounded real function, continuous on $v, t(0 \leq v \leq 1, 0 \leq t < \infty)$ not depending on $\xi_{1a}(\cdot), s \geq 0, a \geq 0$.

The solution of Eq. (10.1.4) exists and is unique in the class of functions $G_a(x, z, t)$ which are distribution functions on x and $\int_0^1 x^k dG(x, z, t)$ are functions analytical on t (excluding, perhaps, zero) for any integer $k > 0$. The solution of Eq. (10.1.4) exists and can be found by means of the method of successive approximates.

Proof. We find $\lim_{n\to\infty} n^{-1} \mathbf{E} \operatorname{Tr} R_t$. Obviously, $\operatorname{Tr} R_t = \sum_{k=1}^n r_{kk}(t)$, where $r_{kk}(t)$ are entries of matrix R_t. (Parameters t will be omitted in order to simplify the formulas.) From § 5 of Chapter 2 and also from (6.3.4), it follows that

$$r_{kk} = [1 + t \sum_{l=1}^{m_n} |\nu_{l_k}|^2 - t^2 \sum_{i,j=1}^n r_{ij}^k (\sum_{l,p=1}^n \bar{\nu}_{ie} \nu_{ke} \nu_{jp} \nu_{kp})]^{-1}, \qquad (10.1.5)$$

where $\nu_{ke} = \xi_{ke} - \alpha_{ke} + i(\eta_{ke} - \beta_{ke}), r_{ij}^k$ are the entries of the matrix $R_k(t) = (I + tH_k H_k *)^{-1}$, matrix H_k is obtained from the matrix H by replacing the kth row by zeros, the summation is taken over all $i, j = \overline{1, n}$, except for $i = j = k$.

Let us prove that

$$\operatorname{plim}_{n\to\infty} t^2 |\sum_{i,j=1}^n r_{ij}^k \sum_{e \neq p} \bar{\nu}_{ie} \nu_{ke} \nu_{jp} \nu_{kp}| = 0. \qquad (10.1.6)$$

It is evident that $\sum_{i,j=1}^n r_{ij}^k \sum_{e,p} \bar{\nu}_{ie} \nu_{ke} \nu_{jp} \bar{\nu}_{kp} = \sum_{e,p} \nu_{ke} \bar{\nu}_{kp} a_{ep}$ where $a_{ep} = \sum_{i,j=1} r_{ij}^k \bar{\nu}_{ie} \nu_{jp}$. After some simple transformations, we have

$$\sum_{p=1}^n |a_{ep}|^2 = \sum_{p=1}^n |\sum_{i,j} r_{ij}^k \bar{\nu}_{ie} \nu_{jp}|^2$$
$$= \sum_p \sum_{i,j} \sum_{s,q} r_{ij}^k \bar{\nu}_{ie} \bar{\nu}_{jp} r_{sq}^k \nu_{se} \bar{\nu}_{pq}$$
$$= \operatorname{Tr} R_k H_k H_k * R_k K_e$$
$$\leq \operatorname{Tr} R_k K_l \leq \operatorname{Tr} K_l (I + tK_l)^{-1}, \qquad (10.1.7)$$

where $K_l = (\nu_{il} \bar{\nu}_{jl})$ is an Hermitian matrix of order n. The Eigenvalues of the matrix K_l are equal to $\sum_{i=1}^n |\nu_{il}|^2, 0, \ldots, 0$. Therefore, we get $\sum_{p=1}^n |a_{lp}|^2 \leq t^{-1}$ from the equality (10.1.7). Analogously, $\sum_{l=1}^n |a_{lp}|^2 \leq t^{-1}$. Since the variables a_{lp} do not depend on the variables ν_{kl}, (10.1.6) holds. By using (10.1.6) we can represent (10.1.5) in the form

$$r_{kk} = [1 + t \sum_{l=1}^{m_n} |\nu_{kl}|^2 - t^2 \sum_{l=1}^{m_n} |\nu_{kl}|^2 \operatorname{Tr} R_k \bar{T}_l^k]^{-1} + \varepsilon_n, \qquad (10.1.8)$$

where ε_n are random variables such that $\lim_{n \to \infty} \mathbf{E}|\varepsilon_n| = 0$, $T_l^k = (\nu_{il}\bar{\nu}_{jl})$ are square matrices of order n, with the kth column and the kth row having zero elements.

Let us denote $R_p^k(t) = (I + t \sum_{l \neq p} T_l^k)^{-1} = (r_{ij}^k(p))$ and consider the equality

$$\text{Tr } R_k(t)\bar{T}_l^k = \text{Tr } R_l^k(t)\bar{T}_l^k + \text{Tr}(R_k(t) - R_e^k(t))\bar{T}_l^k. \tag{10.1.9}$$

Analogously, to the proof of (9.3.7), we get

$$\text{plim}_{n \to \infty}[\text{Tr } R_l^k \bar{T}_l^k - \sum_{i=1}^n r_{ii}^k(l)|\nu_{il}|^2] = 0. \tag{10.1.10}$$

Taking into account that (see § 5 of Chapter 2)

$$\text{Tr } C_n R_t - \text{Tr } C_n R_t^k = -t(R_t^k C_n R_t^k \vec{\nu}_k, \vec{\nu}_k)[1 + t(R_t^k \vec{\nu}_k, \vec{\nu}_k)]^{-1},$$

where $\vec{\nu}_k = (\nu_{ik}, i = \overline{1,n})$ and C_n is a square matrix of order n, we find

$$\text{Tr}(R^k(t) - R_l^k(t))\bar{T}_l^k = -t(R_l^k(t)\bar{T}_l^k R_l^k(t)\vec{\nu}_l, \vec{\nu}_l)$$
$$\times [1 + t(R_l^k(t)\vec{\nu}_l, \vec{\nu}_l)]^{-1} = -t(\sum_{i,j} r_{ij}^k(l)\nu_{il}\bar{\nu}_{jl})$$
$$\times (\sum_{p,q} r_{pq}^k(l)\nu_{pl}\bar{\nu}_{ql})[1 + t(R_l^k(t)\vec{\nu}_l, \vec{\nu}_l)]^{-1}. \tag{10.1.11}$$

From (10.1.10), (9.3.7), and (10.1.11), it follows that

$$\text{plim}_{n \to \infty}[\text{Tr } R_k(t)\bar{T}_l^k - \sum_{i=1}^n r_{ii}^k(l)|\nu_{il}|^2$$
$$+ (\sum_{i=1}^n r_{ii}^k(l)|\nu_{il}|^2)^2[1 + t\sum_{i=1}^n r_{ii}^k(l)|\nu_{il}|^2]^{-1}] = 0.$$

On the basis of this expression, (10.1.8) can be replaced by the formula

$$r_{kk} = [1 + t\sum_{l=1}^{m_n} |\nu_{kl}|^2\{1 + t\sum_{i=1}^n r_{ii}^k(l)|\nu_{il}|^2\}^{-1}]^{-1} + \varepsilon_n',$$
$$\lim_{n \to \infty} \mathbf{E}|\varepsilon_n'| = 0. \tag{10.1.12}$$

By using (6.2.21) analogously to the proof of (9.3.12), we obtain

$$\lim_{n \to \infty} \mathbf{E}|r_{ii}^k(l) - r_{ii}^k| = 0. \tag{10.1.13}$$

From formula (6.3.20), it follows that

$$r_{pp} - r_{pp}^k = -t^2(\sum_{l=1}^n a_l \nu_{kl})^2[1 + t\sum_{l=1}^n |\nu_{kl}|^2 - t^2(R_k(t)\vec{\mu}_k, \vec{\mu}_k)]^{-1}, \tag{10.1.14}$$

where

$$a_l = \sum_{i=l}^{n} r_{ip}^k \nu_{il}, \vec{\mu}_k = \{\mu_{kl}, l = \overline{1,n}\}, \mu_{kl} = \sum_{p=1}^{n} \nu_{kp}\vec{\nu}_{lp}.$$

It is evident that

$$\sum_{l=1}^{n} |a_l|^2 = \sum_{l=1}^{n} \sum_{i,j} r_{ip}^k \nu_{il} \vec{r}_{jp}^k \vec{\nu}_{jl} = \text{Tr } R_k(t) C_p R_k * (t) H_k H_k *,$$

where

$$C_p = (c_k \delta_{kl}), c_p = 1, c_l = 0, l \neq p, \text{Tr } R_k(t) C_p R_k * (t) H_k H_k *$$
$$= t^{-1} \text{Tr } C_p R_k * (t) - t^{-1} \text{Tr } C_p |R_k * (t)|^2.$$

Consequently, $\sum_{l=1}^{n} |a_e|^2 \leq 2t^{-1}$. But then $\text{plim}_{n \to \infty} |t \sum_{l=1}^{n} a_l \nu_{kl}| = 0$. Hence,

$$\lim_{n \to \infty} \mathbf{E} |r_{pp} - r_{pp}^k| = 0. \tag{10.1.15}$$

Taking into account (10.1.15), the expression (10.1.12) can be transformed to the form

$$r_{kk} = [1 + t \sum_{l=1}^{m_n} |\nu_{kl}|^2 \{1 + t \sum_{i=1}^{n} \tilde{r}_{ii} |\nu_{il}|^2\}^{-1}]^{-1} + \varepsilon_n''.$$

Here the random variables $\tilde{r}_{ii}$ do not depend on ν_{ij} and have the same distribution as the variables r_{kk}.

The further proof of this theorem is analogous to that of Theorem 9.3.2. Theorem 10.1.3 is proved.

Let us examine some particular cases of the convergence $K_n(u, v, z)$ to the limit laws.

Corollary 10.1.1. *Let* $N_n(u, v, z) \Rightarrow N(u, v, z), z > 0, 0 \leq u \leq 1, 0 \leq v \leq c$, *where* $N_n(u, v, z) = n[2 - \mathbf{P}\{(\text{Re } \nu_{ij}^{(n)})^2 < z\} - \mathbf{P}\{\text{Im } \nu_{ij}^{(n)})^2 < z\}], i/n \leq u < (i+1)/n, j/n \leq v < (j+1)/n$, *hold instead of the condition* $K_n((u, v, z) \Rightarrow K(u, v, z)$ *in Theorem 10.1.3; let the function* $N(u, v, z)$ *be nonincreasing on* z *and continuous on* u *and* v, $0 \leq u \leq 1, 0 \leq v \leq c$ *for any* $\varepsilon > 0$ *and* $\varepsilon < \delta < \infty$,

$$\sup_{0 \leq u, v \leq 1} [-\int_{\varepsilon}^{\delta} z_d |N(u, v, z)] < \infty,$$

$$\lim_{\varepsilon \to 0} \underline{\lim}_{n \to \infty} \sigma_n^2(u, v, \varepsilon) = \lim_{\varepsilon \to 0} \overline{\lim}_{n \to \infty} \sigma_n^2(u, v, \varepsilon) = \sigma^2(u, v),$$

$$\sigma_n^2(u, v, \varepsilon) = n\{\int_{|x| < \varepsilon} x^2 dP\{\text{Re } \nu_{ij} < x\} - (\int_{|x| < \varepsilon} x dP\{\text{Re } \nu_{ij} < x\})^2$$

$$+ \int_{|x| < \varepsilon} x dP\{\text{Im } \nu_{ij} < x\} - (\int_{|x| < \varepsilon} x dP\{\text{Im } \nu_{ij} < x\})^2\},$$

$$i/n \leq u < (i+1)/n,$$
$$j/n \leq v < (j+1)/n;$$

the function $\sigma^2(u, v)$ be bounded and continuous in the domain $0 \le u \le 1, 0 \le v \le c$; and the random variables $\xi_{pl}, \eta_{pl}, p = \overline{1, n}, l = \overline{1, m_n}$ be independent for any n.

Then (10.1.2) holds, and function $F(x)$ is defined by equation (10.1.4), with the random functional being given by the following Laplace transformation:

$$\mathbf{E} \exp\{ - s\xi_{2a}(G_a(\cdot, \cdot, t), z)\} = \exp\{-s \int_0^c \sigma^2(v, z)$$

$$\times \int_0^1 y dG_a(y, v, t) dv - \int_0^c \int_0^1 [\int_0^\infty (\exp\{-sxy(1 + \sqrt{x}a)^{-2}\}$$

$$- 1) dN(v, z, x)] dG_a(y, v, t) dv\}.$$

It is possible to write an analogous expression for $\xi_{1a}(\cdot)$.

Corollary 10.1.2. *If in addition to the conditions of Corollary 10.1.1, $N(u, v, z) \equiv 0$, then $\text{plim}_{n \to \infty} n^{-1} \ln \det H_n H_n* = \int_0^\infty \ln x dF(x)$, where $F(x)$ is a distribution function, whose Stieltjes transformation is*

$$\int_0^\infty (1 + tx)^{-1} dF(x) = \int_0^1 u(x, t) dx$$

and $u(x, t)$ is satisfying the equation

$$u(x, t) = [1 + t \int_0^c \sigma^2(x, y)[1 + t \int_0^1 u(z, t)$$

$$\times \sigma^2(z, y) dz]^{-1} dy]^{-1}. \qquad (10.1.16)$$

The solution of equation (10.1.16) exists, is unique in the class of functions $u(x, t) \ge 0$, analytical on t and continuous on $x(0 \le x \le 1)$, and can be found by means of the method of successive approximations.

§2 Limit Theorems for Random Gram Determinants with Identically Distributed Elements

Theorem 10.2.1. *Let the real and imaginary parts of the matrix $H_n = (\xi_{pl} + i\eta_{pl})$ be independent, identically distributed, and let the constants a_n and c_n exist such that $\lim_{n \to \infty} n[2 - \mathbf{P}\{(\xi_{ij} - a_n)^2 > c_n x\} - \mathbf{P}\{(\eta_{ij} - a_n)^2 > c_n x\}] = dx^{-\alpha}, x > 0, 0 < \alpha < 1, d > 0, \lim_{n \to \infty} m_n/n = c, 0 < c < \infty$ for some $\beta > 0$,*

$$\sup_n n^{-1} \mathbf{E} \operatorname{Tr} |\ln |\{c_n^{-1}(H_n - a_n Q_n)(H_n - a_n Q_n)*\}||^{1+\beta} < \infty.$$

Then

$$\text{plim}_{n \to \infty} n^{-1} \ln \det\{c_n^{-1}(H_n - a_n Q_n)(H_n - a_n Q_n)*\} =$$

$$= \int_0^\infty \ln x dF(x),$$

where Q is an $m_n \times n$ matrix, with the entries being equal to $1 + i$, the Stieltjes transformation of the distribution function $F(x)$ is equal to $\int_0^\infty (1 + tx)^{-1} dF(x) = \int_0^1 x dG(x,t)$, and $G(x,t)$ satisfies the equality

$$G(x,t) = \mathbf{P}\{[1 + ct\eta(\mathbf{E}\{1 + t\eta[\int_0^1 y^{\alpha/2}$$
$$\times dG(y,t)]^{2/\alpha}\}^{-1})^{2/\alpha}]^{-1} < x\}, \qquad (10.2.1)$$

where η is a random variable, distributed by the stable law, whose Laplace transformation is

$$\mathbf{E}\exp(-s\eta) = \exp(-s^\alpha h), \qquad h = d\Gamma(1 - \alpha)$$

The solution of equation (10.2.1) exists and is unique in the class of the functions $G(x,t)$, which are distribution functions on $x(0 \leq x \leq 1)$ for fixed t and are such that the functions $\int_0^1 y^{\alpha/2} dG(y,t)$ are analytical on t for all $t \neq 0$.

The proof is analogous to that of Corollary 9.3.3.

Note that with the help of Eq. (10.2.11) for the Stieltjes transformation $\int_0^1 x dG(x,t)$, we can get

$$\int_0^1 x dG(x,t) = \mathbf{E}\{1 + ct\eta[\mathbf{E}(1 + t\eta[g(t)]^{2/\alpha})^{-1}]^{2/\alpha}\}^{-1},$$

$$g(t) = \mathbf{E}[1 + ct\eta[\mathbf{E}(1 + t\eta[g(t)]^{2/\alpha})^{-1}]^{2/\alpha}]^{-\alpha/2}.$$

Analogous theorems can be proved for random matrices of the $V_n = \sum_{s=1}^{m_n} \nu_s^{(n)} K_n^s$ type, where $K_n^s = \vec{\eta}_s \vec{\eta}_s*, \vec{\eta}_s$ are vector columns of the matrix H_n and $\nu_s^{(n)}$ are random variables.

Let the conditions of Theorem 10.1.3 hold for the entries of the vectors $\vec{\eta}_s$; let the random variables $\nu_s^{(n)}$ be independent for any n, not depending on the matrix H, and $\mathbf{P}\{\nu_s^{(n)} < x\} \Rightarrow \mathbf{P}\{\nu < x\}$ for all $s = \overline{1,m}$, where ν is a random variable, for some $\alpha < 0 \sup_n n^{-1}\mathbf{E}\,\mathrm{Tr}\,|\ln|V_n||^{1+\alpha} < \infty$. Then

$$\mathrm{plim}_{n\to\infty} n^{-1}\ln|\det V_n| = \int_{-\infty}^\infty \ln|x| dF(x),$$

where $F(x)$ is a distribution function, whose Stieltjes transformation is

$$\int_0^\infty (1 + it(x))^{-1} d|F(x) = \lim_{a\to 0} \int_0^1 \int_0^1 (x + iy) dG_a(x,y,z,t) dz,$$

where $G_a(x,y,z,t)$ is a distribution function on x and y, $0 \leq x \leq 1, 0 \leq z \leq 1, 0 \leq y \leq 1, -\infty < t < \infty$ satisfying the equation at the points of continuity,

$$G_a(x,y,z,t) = \mathbf{P}\{\mathrm{Re}[1 + it\tilde{\xi}_{1a}(\nu[1 + it\nu\tilde{\xi}_{2a}(G_a(\cdot,\cdot,\cdot,t),\cdot)]^{-1}, z)]^{-1} < x$$
$$\mathrm{Im}[1 + it\tilde{\xi}_{1a}(\nu[1 + it\nu\tilde{\xi}_{2a}(G_a(\cdot,\cdot,\cdot,t),\cdot)]^{-1}, z)]^{-1} < y\}.$$

The random functional $\tilde{\xi}_{1a}(\cdot)$ and $\tilde{\xi}_{2a}(\cdot)$ are given as in Theorem 10.1.3. But now the random functions on which the functional $\tilde{\xi}_{1a}(\cdot)$ is given, are complex, therefore the following changes are necessary:

$$\mathbf{E}\exp\{is_1(\mathrm{Re}\,\tilde{\xi}_{1a}(\eta(\cdot,\cdot,t),z)+is_2\,\mathrm{Jm}\,\tilde{\xi}_{1a}(\eta(\cdot,\cdot,t),z)\}$$
$$=\exp\{\mathbf{E}\int_0^1\int_0^1\int_{-\infty}^{\infty}[\exp(ix^2(1+a|x|)^{-2}(S_1\,\mathrm{Re}\,\eta(w,v,t)$$
$$+S_2\,\mathrm{Jm}\,\eta(w,v,t))^{-1}-1](1+x^{-2})dK(v,z,x)dv\},$$

where $\eta(w,v,t)$ is an arbitrary bounded complex random function, continuous on $v,t,0\le v\le 1,-\infty<t<+\infty$ and not depending on $\tilde{\xi}_{1a}(\cdot)),a\ge 0,$

$$\mathbf{E}\exp\{is_1\,\mathrm{Re}\,\tilde{\xi}_{2a}(G_a(\cdot,\cdot,\cdot t),z)+is_2\,\mathrm{Im}\,\tilde{\xi}_{2a}(G_a(\cdot,\cdot,\cdot,t),z)\}$$
$$=\exp\{\int_0^c\int_0^1[\int_{-\infty}^{+\infty}[\exp(ix^2(1+a|x|)^{-2}(is_1u_1+is_2u_2))$$
$$-1](1+x^{-2})dK(v,z,x)]dG(u_1,u_2,v,t)dv\}.$$

Let us consider a particular case of the solution of Eq. (10.1.4) [128]. If in Eq. (10.1.16), $\sigma^2(x,y)\equiv\sigma^2$, then $u(x,t)\equiv u(t)=[1+ct\sigma^2(1+t\sigma^2u(t))^{-1}]^{-1}$. Solving this equation, we get

$$u(t)=[-[1+t\sigma^2(c-1)]$$
$$\pm\sqrt{(1+t\sigma^2(c-1))^2+4t\sigma^2}](2t\sigma^2)^{-1}.$$

The minus sign before the square root should be omitted, since the Stieltjes transformation in this case is always nonnegative. The function $u(t)$ is analytical for all $t\ne 0$. Let us continue it analytically on the whole complex surface. Then for the Stieltjes transformation $\int(x-z)^{-1}d\mu(x),\mathrm{Im}\,z\ne 0$, we obtain $\int(x-z)^{-1}d\mu(x)=(2\sigma^2)^{-1}[1-(1-\sigma^2(c-1)z^{-1})+[(1-z^{-1}\sigma^2(c-1))^2-z^{-1}4\sigma^2]^{1/2}].$

Using the inversion formula (9.2.10), we find

$$\mu(x_2)-\mu(x_1)=(2\sigma^2)^{-1}\lim_{\varepsilon\downarrow 0}\pi^{-1}\int_{x_1}^{x_2}\mathrm{Im}[-(1-(y+i\varepsilon)^{-1}\sigma^2(c-1))$$
$$+\{(1-(y+i\varepsilon)^{-1}\sigma^2(c-1))^2-(y+i\varepsilon)^{-1}4\sigma^2\}^{1/2}]dy$$
$$=\int_{x_1}^{x_2}p(x)dx,$$

where $p(x)=p_1(x)+p_2(x),$

$$p_1(x)=\begin{cases}(1-c)\delta(x),&0\le c\le 1,\\0,&c>1,\end{cases}$$

$$p_2(x)=\begin{cases}[4c\sigma^4-(x-(c+1)\sigma^2)^2]^{1/2}&(2\pi\sigma^2x)^{-1},\\&(x-(c+1)\sigma^2)^2\le 4\sigma^4c,\\0,&(x-(c+1)\sigma^2)^2>4\sigma^4c.\end{cases}$$

§3 Limit Spectral Functions

Let us examine the random matrices of the type $V_n = B_n + \sum_{s=1}^{m_n} \theta_s^{(n)} K_n^s$, where $K_n^s = \vec{h}_s \vec{h}_s *, h_s$ are vector columns of the matrix H_n, $\theta_S^{(n)}$ are real random variables, and B_n are Hermitian matrices of order n. For such matrices, Thoerem 10.1.1 can be applied, with the normalized spectral function of the matrix V_n being approximately replaced (if n values are large) by the nonrandom spectral function that satisfies the functional equations in some cases. Suppose that for every n, the random variables $\theta_s^{(n)}, v_{pl} := \xi_{pl} + i\eta_{pl}, p = \overline{1,n}, l = \overline{1,m_n}$ are independent, the variables $\xi_{pl}, \eta_{pl}, p = \overline{1,n}, l = \overline{1,m_n}$ are infinitesimal, $\lim_{n \to \infty} m_n |n^{-1} = c, 0 < c < \infty$,

$$\lim_{h \to \infty} \overline{\lim}_{n \to \infty} \sup_{l=\overline{1,m_n}, k=\overline{1,n}} \mathbf{P}\{\sum_{p=1}^n |v_{pl}|^2$$

$$+ \sum_{k=1}^{m_n} |v_{lk}|^2 \geq h\} = 0, \quad B_n = (\delta_{pl} b_l).$$

Analogous to the proof of Theorem (10.1.3), we get

$$r_{kk} \approx [1 + itb_k + it \sum_{l=1}^{m_n} |v_{kl}|^2 \{1 + it\theta_l^{(n)}$$

$$\times \sum_{i=1}^n \tilde{r}_{ii} |v_{il}|^2\}^{-1} \theta_l^{(n)}]^{-1} + \varepsilon_n, \tag{10.3.1}$$

where $\lim_{n \to \infty} \mathbf{E}|\varepsilon_n| = 0, b_k$ are the diagonal entries of the matrix B_n, the random variables $\tilde{r}_{ii}$ do not depend on the matrix V and are distributed analogously to the variables r_{kk}.

If the conditions of Corollary 10.1.2 hold for the entries of matrix K_n^s and in addition $\mathbf{P}\{\theta_s^{(n)} < x\} \Rightarrow \mathbf{P}\{\theta < x\}$, where θ is some random variable $b_n(x) \to b(x)$, where $b_n(x) = b_k, k|n^{-1} \leq x < (k+1)|n^{-1}, b(x)$ is a continuous function on $[0, 1]$, then $\lim_{n \to \infty} \mathbf{E}_n^{-1} \operatorname{Tr}(I + itV_n)^{-1} = \int_0^1 u(x,t) dx$, and $u(x,t)$ satisfies the equation

$$u(x,t) = [1 + itb(x) + it\mathbf{E} \int_0^c \sigma^2(x,y)\theta$$

$$\times [1 + it\theta \int_0^1 u(z,t)\sigma^2(z,y)_d|z]^{-1} dy]^{-1}. \tag{10.3.2}$$

The solution of Eq. (10.3.2) exists and is unique in the class of functions $\operatorname{Re} u(x,t) \geq 0$, analytical on t.

Theorem 10.3.1. *Let for every n the random variables $\theta_p^{(n)}, \eta_{pl}^{(n)}, \xi_{pl}^{(n)}$, $p = \overline{1,n}, l = \overline{1,m_n}$ be independent, asymptotically constant, i.e., there exist constants $a_{ij}^{(n)}$ such that for any $\varepsilon > 0$,*

$$\lim_{n \to \infty} \sup_{p=\overline{1,n}} \mathbf{P}\{|\xi_{pl} + i\eta_{pl} - a_{pl}^{(n)}| > \varepsilon\} = 0,$$

$$\lim_{h \to \infty} \sup_{i=\overline{1,n}} \mathbf{P}\{\sum_{j=1}^{n} |v_{ij}|^2 + |v_{ji}|^2 \geq h\} = 0,$$

where

$$v_{ij} = \xi_{ij} - i\eta_{ij} - b_{ij}, b_{ij} = a_{ij} + \int_{|x|<\tau} x dP\{\xi_{ij} - \operatorname{Re} a_{ij} < x\}$$

$$+ i \int_{|x|<\tau} x dP\{\eta_{ij} - \operatorname{Im} a_{ij} < x\}, \tau > 0$$

is being an arbitrary absolute constant, and $\lim_{n \to \infty} \sup_{j=\overline{1,m_n}} \sum_{i=1}^{n} |b_{ij}|^2 < \infty$.
Then

$$n^{-1}\mathbf{E}\operatorname{Tr}(I + itV_n)^{-1}$$

$$= n^{-1}\mathbf{E}\operatorname{Tr}[I + itB_n + it\sum_{s=1}^{m_n} \theta_s^{(n)}(\vec{b_s} + \vec{\gamma_s})(\vec{b_s} + \vec{\gamma_s})*]^{-1} + 0(1) \tag{10.3.3}$$

where $\vec{b_s} = (b_{ks}, k = \overline{1,n}), \vec{\gamma_s} = (\gamma_{ks}, K = \overline{1,n})$, the random variables γ_{ks}, $k = \overline{1,n}$, $s = \overline{1,m_n}$ for each n, are independent, do not depend on the variables $\theta_s^{(n)}$, and are distributed by the infinitely divisible laws with the characteristic functions $\exp\{\mathbf{E}\exp(is_1 \operatorname{Re} v_{kp} + is_2 \operatorname{Im} v_{kp}) - 1\}$.

Proof. Let

$$R_t^p = [I + itB_n + it\sum_{s=1}^{p} \theta_s^{(n)}(\vec{b_s} + \vec{\gamma_s})(\vec{b_s} + \vec{\gamma_s}) * + it\sum_{s=p+1}^{n} \theta_s^{(n)}K_n^s]^{-1}.$$

Let us consider the difference

$$n^{-1}\mathbf{E}\operatorname{Tr} R_t^0 - n^{-1}\mathbf{E}\operatorname{Tr} R_t^n = \sum_{p=1}^{m_n}[n^{-1}\mathbf{E}\operatorname{Tr} R_t^{p-1} - n^{-1}\mathbf{E}\operatorname{Tr} R_t^p].$$

Let

$$\tilde{R}_t^p = [I + itB_n + it\sum_{s=1}^{p-1} \theta_s^{(n)}(\vec{b_s} + \vec{\gamma_s})(\vec{b_s} + \vec{\gamma_s}) * + it\sum_{s=p+1}^{n} \theta_s^{(n)}K_n^s]^{-1}.$$

Using the formula for the resolvent difference (see Lemma 6.3.1), we find

$$n^{-1}\mathbf{E}\operatorname{Tr} R_t^0 - n^{-1}\mathbf{E}\operatorname{Tr} R_t^n = \sum_{p=1}^{m_n} \{n^{-1}\mathbf{E}\operatorname{Tr} R_t^{p-1}$$

$$- n^{-1}\mathbf{E}\operatorname{Tr} \tilde{R}_t^p - (n^{-1}\mathbf{E}\operatorname{Tr} R_t^p - n^{-1}\mathbf{E}\operatorname{Tr} \tilde{R}_t^p)\}$$

$$= n^{-1}\sum_{p=1}^{m_n} \mathbf{E}\delta_s,$$

$$\delta_s = t\frac{d}{dt}\ln[1 + it(\tilde{R}_t^p(\vec{b}_s + \vec{v}_s), (\vec{b}_s + \vec{v}_s))] - t\frac{d}{dt}\ln[1$$

$$+ it(\tilde{R}_t^p(\vec{b}_s + \vec{\gamma}_s), (\vec{b}_s + \vec{\gamma}_s)]. \tag{10.3.4}$$

Analogously, as in Theorem 5.3.4, we get $\lim_{n\to\infty} \mathbf{E}|\delta_s| = 0$ for any finite t. Theorem 10.3.1 is proved.

Theorem 10.3.2. *For any n, let the random variables $\theta_p^{(n)}, \eta_{pl}^{(n)}, \xi_{pl}^{(n)}, p = \overline{1,n}, l = \overline{1,m_n}$ be independent*

$$\mathbf{E}(\xi_{pl}^{(n)} + i\eta_{pl}^{(n)}) = b_{pl}^{(n)}, \mathbf{V}\xi_{pl}^{(n)} = \sigma_{pl}^2, \mathbf{V}\eta_{pl}^{(n)} = \rho_{pl}^2,$$

$$\lim_{n\to\infty} \sup_{p,l}(\sigma_{pl}^2 + \rho_{pl}^2) = 0, \quad \lim_{n\to\infty} m_n n^{-1} = c, \quad 0 < c < \infty,$$

$$\sup_n \sup_{j=\overline{1,n}} \sum_{i=1}^n (|b_{ij}^2| + |b_{ji}|^2|) < \infty,$$

$$\sup_n [\sup_{j=\overline{1,n}} \sum_{i=1}^{m_n}(\sigma_{ij}^2 + \rho_{ij}^2) + \sup_{j=\overline{1,m_n}} \sum_{i=1}^n (\sigma_{ji}^2 + \rho_{ji}^2)] < \infty$$

and the Lindeberg condition holds, for any $\tau > 0$,

$$\lim_{n\to\infty} n^{-1} \sum_{p=\overline{1,n},l=\overline{1,m_n}} \int_{|x|>\tau} x^2 d[\mathbf{P}\{\xi_{pl}^{(n)} - \mathbf{E}\xi_{pl}^{(n)} < x\}$$

$$+ \mathbf{P}\{\eta_{pl}^{(n)} - \mathbf{E}\eta_{pl}^{(n)} < x\}] = 0.$$

Then

$$n^{-1}\mathbf{E}\operatorname{Tr}(I + itV_n)^{-1} = n^{-1}\mathbf{E}\operatorname{Tr}[I + itB_n$$

$$+ it\sum_{s=1}^{m_n} \theta_s^{(n)}(\vec{b}_s + \vec{\gamma}_s)(\vec{b}_s + \vec{\gamma}_s)*]^{-1} + 0(1), \tag{10.3.5}$$

where $\operatorname{Re}\gamma_{ij}, \operatorname{Jm}\gamma_{ij}, i = \overline{1,n}, j = \overline{1,m_n}$ are independent for each n and distributed by the normal laws $N(0,\sigma_{ij}^2), N(0,\rho_{ij}^2)$.

The proof is analogous to that of Theorem 9.5.2, with (10.3.4) taken into account.

In some cases, using Eq. (10.3.5), we can find functional equations which the Stieltjes transformation of the normalized spectral function satisfies. For example, if $\vec{b}_s \equiv 0, \sigma_{ij}^2 = \rho_{ij}^2 \equiv \sigma_n^2$, then the distribution of the matrix $\sum_{s=1}^{m_n} \theta_s \vec{\gamma}_s \vec{\gamma}_s *$ and the matrix $T_n \sum_{s=1}^{m_n} \theta_s \vec{\gamma}_s \vec{\gamma}_s * T_n'$, where T_n is a unitary matrix, coincide. Therefore, Eq (10.3.5) has the following form

$$n^{-1}\mathbf{E}\operatorname{Tr}(I + itV_n)^{-1} = n^{-1}\mathbf{E}\operatorname{Tr}[I + it\operatorname{diag}(\beta_i, i = \overline{1,n})$$
$$+ it\sum_{s=1}^{m_n} \theta_s \vec{\gamma}_s \vec{\gamma}_s *]^{-1} + 0(1).$$

By using this equation and (10.3.1), we get the following statement.

Corollary 10.3.1.   *In addition to the conditions of Theorem 10.3.2, let* $\vec{b}_s \equiv 0, \sigma_{ij}^2 = \rho_{ij}^2 = \sigma^2/n, \mathbf{P}\{\theta_s^{(n)} < x\} \Rightarrow \mathbf{P}\{\theta < x\}$ *for all* $s = \overline{1, m_n}$, *where* θ *is some random variable,* $\lambda_n(x, B_n) \Rightarrow \lambda(x)$, *where* $\lambda_n(x, B_n)$ *is a normalized spectral function of the matrix* B_n.

Then

$$\lim_{n \to \infty} n^{-1}\mathbf{E}\operatorname{Tr}(I + itV_n)^{-1} = u(t)$$
$$u(t) = \int [1 + itx + ict\mathbf{E}\sigma^2\theta[1 + it\theta\sigma^2 u(t)]^{-1}]^{-1} d\lambda(x)$$

Suppose that $m_n = n, \theta_s^{(n)} \equiv 1, \sigma_{ij}^2 = \rho_{ij}^2 = \sigma^2/n$. Then

$$n^{-1}\mathbf{E}\operatorname{Tr}(I + itV_n)^{-1} = n^{-1}\mathbf{E}\operatorname{Tr}[I + itB_n$$
$$+ it(Q_n + \sigma^2\Xi_n)(Q_n^* + \sigma^2\Xi_n^*)]^{-1} + 0(1),$$

$$(10.3.6)$$

where $Q_n = (b_{ij}), \Xi_n = (\gamma_{ij})$ are square matrices of order n. The matrix Q_n can be represented in the form $Q_n = U_1 \wedge U_2$, where U_1, and U_2 are unitary matrices, $\Lambda = (\lambda_i \delta_{ij})$ is a diagonal matrix of the eigenvalues of the matrix $Q_n Q_n *$, and (10.3.5) is written in the following ways

$$n^{-1}\mathbf{E}\operatorname{Tr}(I + itV_n)^{-1} = n^{-1}\mathbf{E}\operatorname{Tr}[I + itU_1 * B_n U_1$$
$$+ it(\Lambda + \Xi_n)(\Lambda + \Xi_n*)]^{-1} + 0(1)$$

$$(10.3.7)$$

If the matrices $B \equiv 0$, then the functional equation for the Stieltjes transformation of a limiting spectral function can easily be derived from Eq (10.3.7).

CHAPTER 11

THE DETERMINANTS OF TOEPLITZ
AND HANKEL RANDOM MATRICES

In this chapter, the perturbation and orthogonalization methods are applied to the analysis of determinant distribution of Toeplitz and Hankel matrices. Besides, limit theorems of the law of large numbers and the central limit theorem type are proved: the general form of limiting distributions for determinants of random Toeplitz and Hankel matrices has been found.

§1 Limit Theorem of the Law of Large Numbers Type

By the Hankel random matrix of order n we mean the matrix $\Gamma_n = (\xi_{i+j})_{i,j=1,r}^{n}$ (where $\xi_i, i = \overline{2, 2n}$ are random complex variables), as well as the matrix of the kind $(u_n(w, \xi_i + \xi_j))$ (where ξ_i are random variables, and $u_n(w, t)$ is a random function such that $u_n(w, \xi_i + \xi_j)$ are random variables).

The square matrix $(\xi_{i-j})_{i,j=0}^{n-1}$ is called a random Toeplitz matrix of order n, where $\xi_i, i = 0, \pm 1, \ldots, \pm(n-1)$ are complex variables. The Toeplitz matrix will be Hermitian if it satisfies the conditions $\xi_{-p} = \overline{\xi_p}, p = \overline{0, n-1}$.

It is possible for the determinants of such matrices to formulate limit theorems of the law of large numbers type. Let there exist $\mathbf{E}\ln^2 |\det \Gamma_n|$ and $\gamma_k = \mathbf{E}[\ln |\det \Gamma_n|/\sigma_{k-1}^{(n)}] - \mathbf{E}[\ln |\det \Gamma_n|/\sigma_k^{(n)}]$, where $\sigma_k^{(n)}$ are the smallest σ- algebras, with respect to which the random variables $\xi_l, l = \overline{k+1, 2n}$ are measurable. If for a certain sequence of constants c_n,

$$\lim_{n \to \infty} c_n^{-2} \sum_{k=1}^{n} \mathbf{E}\gamma_k^2 = 0, \qquad (11.1.1)$$

then $\operatorname{plim}_{n \to \infty} c_n^{-1}(\ln |\det \Gamma_n| - \mathbf{E}\ln |\det \Gamma_n|) = 0$.

If $\mathbf{E}\gamma_k^2$ does not exist, then it is possible to use the conditions of Corollary 5.4.1.

Theorem 11.1.1. *Let $\Gamma_n = (u(\xi_i + \xi_j))$, where $u(x)$ is a nonrandom real even Borel function, for each n the random variables $\xi_i^{(n)}$ are independent*

$$\lim_{n \to \infty} c_n^{-2} \sum_{k=1}^{n} \mathbf{E}\{\mathbf{E}[|\ln(1 + \delta_n)|/\sigma_{k-1}^{(n)}]$$

$$- \mathbf{E}[\ln|1 + \delta_n|/\sigma_k^{(n)}]\}^2 = 0, \tag{11.1.2}$$

where c_n is some sequence of constant values, $\sigma_k^{(n)}$ is the smallest σ-algebra with respect to which the random variables ξ_l, $l = \overline{k+1, n}$ are measurable,

$$\delta_n = iu(0) + (R_k \vec{u_k}, \vec{u_k}), \vec{u_k}$$
$$= (u(\xi_k + \xi_1), \dots, u(\xi_k + \xi_{k-1}), 0, \dots, u(\xi_k + \xi_n)), R_k = (I + i\Gamma_n^k)^{-1}.$$

and Γ_n^k is obtained from the matrix Γ_n by replacing the entries of the kth vector row and of the kth vector column by zeros.
   Then

$$\text{plim}_{n \to \infty} c_n^{-1}\{\ln \det(I + \Gamma_n^2) - \mathbf{E}\ln\det(I + \Gamma_n^2)\} = 0.$$

Proof. Obviously, the matrix Γ_n is symmetric and

$$\ln \det(I + \Gamma_n^2) = 2|\ln \det(I + i\Gamma_n)|.$$

Condition (11.1.2) is obtained by using the formula

$$\ln \det(I + i\Gamma_n) - \ln \det(I + i\Gamma_n^k) = \ln(I + \delta_n).$$

Theorem 11.1.1 is proved.

   Note that (11.1.2) holds under $\sup_n \sup_{k=\overline{1,n}} |1 + \delta_n| \leq c, \lim_{n \to \infty} c_n^{-2}n = 0$.
   As has been shown in Corollary 5.4.1, it is possible to demand the existence not of $\mathbf{E}\ln^2 \det(I + \Gamma_n^2)$ but of $\mathbf{E}\ln^{1+\epsilon} \det(I + \Gamma_n^2), \epsilon > 0$. The condition of Theorem 11.1.1 can be weakened if we use limit random theorems for normalized spectral functions (see Chapter 9).

Theorem 11.1.2. *Let $\Gamma_n = (u(\xi_i + \xi_j)), u(x)$ be a nonrandom real even Borel function, the random variables $\xi_i^{(n)}, i = \overline{1, n}$ for each n be independent and given on a common probability space, and $\mu_n(x)$ be a normalized spectral function of the matrix Γ_n.*
   Then with probability 1 for almost all x, $\lim_{n \to \infty}[\mu_n(x) - F_n(x)] = 0$, where $F_n(x)$ is a nonrandom distribution function, whose Stieltjes transformation is

$$\int (1 + itx)^{-1} dF_n(x) = n^{-1}\mathbf{E}\,\text{Tr}(1 + it\Gamma_n)^{-1}.$$

   Let us now study Hankel matrices of the kind $\Gamma_n = (\xi_{i+j})$. Let $\theta_k = (\delta_{i,k-j}), i, j = \overline{1, n}$, where δ_{ij} is the Kronecker symbol.

Corollary 11.1.1. *For each n, let the random variables $\xi_i, i = \overline{2, 2n}$ of the matrix $\Gamma_n = (\xi_{i+j})$ be independent and given on the common probability space.*

$$\lim_{h \to -\infty} \sup_n \mathbf{E}\mu_n(h) = 0,$$

$$\lim_{n \to \infty} n^{-2} \sum_{k=2}^{n} \mathbf{E}\xi_k^2 k^2 + \sum_{k=n+1}^{2n} \mathbf{E}\xi_k^2 (2n - k)^2 = 0.$$

(11.1.3)

Then for almost all x, $\operatorname{plim}_{n\to\infty}[\mu_n(x) - F_n(x)] = 0$, where $F_n(x)$ is a nonrandom distribution function, whose Stieltjes transformation is

$$\int (1 + itx)^{-1} dF_n(x) = n^{-1}\mathbf{E}\operatorname{Tr}(I + it\Gamma_n)^{-1}.$$

Proof. Let us examine the sums $n^{-1}\operatorname{Tr}(I + it\Gamma_n)^{-1} - n^{-1}\mathbf{E}\operatorname{Tr}(I + it\Gamma_n)^{-1} = n^{-1}\sum_{k=2}^{2n}[\mathbf{E}\{\operatorname{Tr}(I + it\Gamma_n)^{-1}/\sigma_{k-1}^{(n)}\} - \mathbf{E}\{\operatorname{Tr}(I + it\Gamma_n)^{-1}/\sigma_k^{(n)}\}]$, where $\sigma_k^{(n)}$ is the smallest σ-algebra, with respect to which the random variables $\xi_l, l = \overline{k+1, 2n}$ are measurable. Let $\Gamma_n^{(k)}$ be a matrix, obtained from the matrix Γ_n by replacing the entries ξ_k by zeros. It is obvious that $\Gamma_n - \Gamma_n^k = \xi_k \theta_k$. Therefore, $\operatorname{Tr}(I + it\Gamma_n)^{-1} - \operatorname{Tr}(I + it\Gamma_n^k)^{-1} = t\frac{d}{dt}\ln\det(I + it\xi_k R_t^k \theta_k)$, where $R_t^k = (I + it\Gamma_n^k)^{-1}$.

By using the integral representation for determinants, we get (the variable ξ_k and the matrix R_t^k are fixed),

$$t\frac{d}{dt}\ln\det(I + it\xi_k R_t^k \theta_k) = -2t\frac{d}{dt}\ln\mathbf{E}\exp\{-it\xi_k$$

$$\times(\sqrt{R_t^k\theta_k}\sqrt{R_t^k\vec{\eta}, \vec{\eta}})\} = -2t\mathbf{E}\exp\{-it\xi_k(\sqrt{R_t^k\theta_k}$$

$$\times\sqrt{R_t^k\vec{\eta}, \vec{\eta}})\}\{-i\xi_k(\frac{d}{dt}t\sqrt{R_t^k\theta_k}\sqrt{R_t^k\vec{\eta}, \vec{\eta}})\}$$

$$\times\{\mathbf{E}\exp[-it\xi_k(\sqrt{R_t^k\theta_k}\sqrt{R_t^k\vec{\eta}, \vec{\eta}})]\}^{-1}$$

$$= -2it\xi_k\operatorname{Tr}\frac{d}{dt}\sqrt{R_t^k\theta_k}\sqrt{R_t^k}(I + it\xi_k\sqrt{R_t^k\theta_k}\sqrt{R_t^k})^{-1}$$

$$= -2it\xi_k\operatorname{Tr}\frac{d}{dt}t\theta_k(I + it\Gamma_n^k + it\xi_k\theta_k)^{-1},$$

where $\vec{\eta}$ is a random vector distributed by the normal law $N(0, 0.5I)$ not depending on ξ_k and the matrix R_t^k.

From this formula, it is easy to get

$$|t\frac{d}{dt}\ln\det(I + it\xi_k R_t^k \theta_k)| = |2t||\operatorname{Tr}\xi_k\theta_k[I + it\Gamma_n^k + it\xi_k\theta_k)^{-1}$$

$$- (I + it\Gamma_n^k + it\xi_k\theta_k)^{-1}it\Gamma_n^k(I + it\Gamma_n^k + it\xi_k\theta_k)^{-1}]|$$

$$= 2|\operatorname{Tr}\{it\xi_k\theta_k(I + it\Gamma_n^k + it\xi_k\theta_k)^{-1}\}^2|.$$

The eigenvalues of the matrix θ_k are equal to ± 1, and the modules of the eigenvalues of the matrix R_t are bounded by unity, therefore,

$$|t\frac{d}{dt}\ln\det(I + it\xi_k R_t^k\theta_k)| \leq 2|t\xi_k|k$$

The further proof is analogous to that of Theorem 9.3.1.

Analogous statements can be formulated for Toeplitz random matrix determinants.

§2 The Method of Integral Representations for Determinants of Toeplitz and Hankel Random Matrices

Let $\Gamma_n = (\xi_{i+j}^{(n)})$ be the Hankel matrix of order n. From § 5 of Chapter 1, we get the following integral representation:

$$\det(I + i\Gamma_n)^{-1/2} = \mathbf{E}[\exp\{-0.5i\sum_{k=2}^{2n}\xi_k^{(n)}$$
$$\times (\sum_{l+p=k}\eta_l\eta_p)\}/\xi_k^{(n)}, k = \overline{2, 2n}], \tag{11.2.1}$$

where $\eta_l, l = 1, 2, \ldots$ are independent of one another and of random variables. $\xi_k^{(n)}$ random variables, distributed according to the normal law $N(0, 1)$.

We get an analogous formula for the Toeplitz matrix $T_n = (\xi_{|i-j|}^{(n)})$,

$$\det(I + iT_n)^{-1/2} = \mathbf{E}[\exp\{-0.5i\sum_{k=0}^{n-1}\xi_k$$
$$\times (\sum_{|l-p|=k}\eta_l\eta_p)\}/\xi_k, k = \overline{0, n-1}]. \tag{11.2.2}$$

With the matrices $(I + \Gamma_n), (I + T_n)$ being badly conditioned (i.e., when their determinants can be equal to zero), the integral representations have the form

$$\det(I + \alpha_t\Gamma_n)^{-k} = \mathbf{E}[\exp\{-0.5\alpha_t\sum_{s=1}^{2k}(\Gamma_n\vec{\eta_s}, \vec{\eta_s})\}/\Gamma_n], \tag{11.2.3}$$

where $\vec{\eta_s}$ are independent random vectors, not depending on the matrix Γ_n and distributed according to the normal law $N(0, I)$, $\alpha_t = t[q + |\operatorname{Tr}\Gamma_n| + \operatorname{Tr}\Gamma_n^2]^{-1}, 0 \leq t < 1, q \geq 1$.

An analogous formula is valid for the matrix $(I + T_n)$.

Let us prove limit theorems for $\det(I + i\Gamma_n)$. These theorems can be easily extended to the integral representations (11.2.2), and (11.2.3).

Theorem 11.2.1. *For every n, let the random entries $\xi_k^{(n)}, k = \overline{2, 2n}$ of the matrices $\Gamma_n = (\xi_{i+j})$ be independent, the variables $\sqrt{n}\xi_k^{(n)}$ be infinitesimal,*

$$\lim_{h \to \infty} \overline{\lim}_{n \to \infty} P\{\sum_{k=2}^{2n} (\sqrt{n}\xi_k^{(n)} - \alpha_k)^2 \geq h\} = 0, \qquad (11.2.4)$$

where $\alpha_k = \int_{|x|<\tau} x \, dP\{\sqrt{n}\xi_k^{(n)} < x\}, \tau > 0$ is an arbitrary constant.

Then

$$\mathbf{E} \det(I + i\Gamma_n)^{-m/2}$$

$$= \mathbf{E}[\exp\{\sum_{k=2}^{2n} f_k(\sum_{s=1}^{m} \sum_{l+p=k} n^{-1}\eta_l^{(s)}\eta_p^{(s)}/2)$$

$$+ 0.5i \sum_{s=1}^{m} \sum_{l+p=k} \eta_l^{(s)}\eta_p^{(s)}/2n\}] + 0(1), \qquad (11.2.5)$$

where $\eta_l^{(s)}, l = \overline{1, n}$ are the components of the vector $\vec{\eta_s}$,

$$\rho_k = \alpha_k + \int x(1 + x^2)^{-1} dP\{\sqrt{n}\xi_k^{(n)} - \alpha_k < x\}$$
$$f_x(x) = \mathbf{E}\{\exp[ix(\xi_k\sqrt{n} - \alpha_k) - ix(\xi_k\sqrt{n} - \alpha_k)$$
$$\times [1 + (\xi_k\sqrt{n} - \alpha_k)^2]^{-1} - 1]\}.$$

Analogous statements hold for the values $|\det(I + i\Gamma_n)|, \det(I + iT_n)$, $|\det(I + iT_n)|$.

We shall need limit theorems for the sequence of the random variables $\gamma_k = \sum_{l+p=k} \eta_l\eta_p/2\sqrt{n}, k = \overline{2, 2n}$.

Let us consider the characteristic function of the two arbitrary variables $\gamma_k, \gamma_m, k < m$,

$$\mathbf{E}\exp\{-is_1\gamma_k - is_2\gamma_m\} = \det[I + is_1 n^{-1/2}\Gamma_k + is_2 n^{-1/2}\Gamma_m]^{-1/2},$$

where Γ_k is a square matrix of order n, with all entries being zero, except for entries with the indices p and l, $p + l = k$, and which are equal to 1. Find the limit of this determinant as $n \to \infty$. Note that the eigenvalues of the matrix $s_1\Gamma_k + s_2\Gamma_m$ are bounded for all bounded s_1 and s_2, $\mathrm{Tr}(s_1\Gamma_k + s_2\Gamma_m) = s_1 + s_2$. Therefore,

$$\lim_{n \to \infty} \det[I + is_1 n^{-1/2}\Gamma_k + is_2 n^{-1/2}\Gamma_m]^{-1/2}$$
$$= \lim_{n \to \infty} \exp\{-0.25(s_1^2 k + s_2^2 m)/n\}.$$

If k and m depend on n so that $\lim_{n \to \infty} k/n = p, \lim_{n \to \infty} m/n = q$, where p and d q are constants, then

$$\lim_{n \to \infty} \mathbf{E}\exp\{-is_1\gamma_k - is_2\gamma_m\} = \exp\{-0.25(s_1^2 p + s_2^2 q)\}.$$

Hence, we conclude that the random variables γ_k and $\gamma_m, k \neq m$ are asymptotically independent and the characteristic function of variables tends to $\exp\{-s^2 p/4\}$ if $\lim_{n \to \infty} k/n = p$.

Corollary 11.2.1. *For every n, let the random entries $\xi_k^{(n)}$, $k = \overline{2, 2n}$ of the matrices $\Gamma_n = (\xi_{i+j}^{(n)})$ be independent, the variables $\sqrt{n}\xi_k^{(n)}$ infinitesimal, for each $k = \overline{1, n}$,*

$$n \int_{-\infty}^{x} y^2(1 + y^2)^{-1} d\mathbf{P}\{\sqrt{n}\xi_k^{(n)} - d_k < x\} \Rightarrow G(x), \sup_n \operatorname{Tr} B_n^2 < \infty,$$

where $G(x)$ is a nondecreasing function of bounded variation, $B_n = (\rho_{i+j})$. Then the distribution of the random variable

$$\ln|\det(I + i\Gamma_n)| - \ln|\det(I + iB_n)|$$

converges weakly to the limit, whose Laplace transformation is

$$\exp\{2 \int [(1 - \exp(-x^2 s/2))x^{-2}s^{-1} - 1](1 + x^{-2})dG(x)\}.$$

Proof. Analogous to the proof (11.2.5) for each integer $m > 0$,

$$\mathbf{E}|\det(I + i\Gamma_n)|^{-m}$$

$$= \mathbf{E}\exp\{\sum_{k=2}^{2n} f_k(\sum_{s=1}^{m}\sum_{l+p=k} n^{-1}\eta_l^{(s)}\eta_p^{(s)}/2 - \sum_{s=m+1}^{2m}\sum_{l+p=k} \eta_l^{(s)}\eta_p^{(s)}/2n)$$

$$+ 0.25n^{-1}\sum_{k=2}^{2n} \rho_k \sum_{s=1}^{m}\sum_{l+p=k} \eta_l^{(s)}\eta_p^{(s)} - \sum_{s=m+1}^{2m}\sum_{l+p=k} \eta_l^{(s)}\eta_p^{(s)}]\}.$$

From the asymptotic independence of the variables γ_k and (11.2.6), it follows that

$$\operatorname{plim}_{n\to\infty} \sum_{k=2}^{2n}[f_k(\cdot) - \mathbf{E}f_k(\cdot)] = 0.$$

Using this as well as the limiting characteristic functions for the variables γ_k, we get

$$\mathbf{E}|\det(I + i\Gamma_n)|^{-m} = |\det(I + iB_n)|^{-m}\exp\{\int_0^1 \int [\exp(-x^2 mt/2) - 1]$$

$$\times (1 + x^{-2})dG(x)dt\}.$$

Hence, the methods of the proofs of §3 of Chapter 8 allow us to come to the assertion of Corollary 11.2.1.

Let us consider the Hankel random matrix determinants of the form $\det(B_n + i\Gamma_n)$, where B_n are positive-definite nonrandom matrices. Corollary 9.1.2 implies the following assertion.

Theorem 11.2.2. *For every* n, *let the random variables* $\xi_k^{(n)}$, $k = \overline{2,2n}$ *be independent, the variables* $\xi_k^{(n)}[\sum_{L_k} b_{pp}^2 b_{il}^2]^{1/2}$, $k = \overline{2,2n}$ *be infinitesimal, (the summation is taken over the domain* L_k: $p+l = k, l \leq p \leq n, 1 \leq l \leq n$,

$$n^2 \int_{-\infty}^{x} y^2(1+y^2)^{-1}dP\{\nu_p < y\} \Rightarrow G(x),$$

where $G(x)$ *is a nondecreasing function of bounded variation,*

$$\nu_p = a_k \xi_k^{(n)} - \int_{|x|<\tau} x dP\{\xi_k^{(n)} a_k < x\},$$

$$a_k = [\sum_{L_k} b_{pp}^2 b_{il}^2]^{1/2}$$

(where b_{pp} *are the entries of the matrix* B_n^{-1}),

$$n^2[\gamma_p + \int y(1+y^2)^{-1}dP\{\gamma_p < y\}] \to \gamma, \gamma_p = \int_{|x|<\tau} x dP\{\xi_k^{(n)} a_k < x\},$$

γ *is a certain constant,* $\lim_{n\to\infty} r_n(t,s) = r(t,s), r_n(t,s) = b_{kl}^{(n)}/2a_k^{(n)}$ *for* $k/n \leq t < (k+1)/n, l/n \leq s < (l+1)/n$, *and the function* $r(t,s)$ *is continuous in the domain* $0 \leq t \leq s \leq 1$.
    Then for each integer $k > 0$,

$$\lim_{n\to\infty} \mathbf{E} \exp\{-k \ln[\det(B_n + i\Gamma_n) \det B_n^{-1}]\} = \mathbf{E} \exp\{\int_0^1 i\gamma \xi_k(t)$$

$$+ \int (\exp(ix\xi_k(t)) - ix\xi_k(t)(1+x^2)^{-1} - 1)(1 + x^{-2})dG(x)dt\},$$

where

$$\xi_k(t) = \begin{cases} \sum_{m=1}^{k} \int_0^t \xi_m(t-y)\eta_m(y)dy, & 0 \leq t \leq 1/2 \\ \sum_{m=1}^{k} \int_{t-1/2}^{1/2} \eta_m(t-y)\xi_m(y)dy, & 1 \geq t \geq 1/2. \end{cases}$$

The Gaussian random processes $\xi_m(y), \eta_m(y), m = 1, 2, \ldots$ *are independent, have zero mean and covariance functions* $r(t,s)$.

§3 The Stochastic Analogue of the Szegö Theorem

Let $T = (\xi_{|i-j|})_{i,j=1}^n$ be a Toeplitz random matrix of order n, and $\xi_{|k|}$, $k = 0, \pm 1, \ldots$ be real random variables. It is obvious that $\operatorname{Tr} T^{(s)} = \sum \xi_{|i_1-i_2|} \xi_{|i_2-i_3|} \cdots \xi_{|i_s-i_1|}$, and the summation is taken over all values of the variables $i_p = \overline{1,n}, p = \overline{1,s}$. Assuming $\psi(i) = 1$ for $0 \leq i \leq n$ and $\psi(i) = 0$ for the other values of i, this sum can be written in the form

$$\operatorname{Tr} T^s = \sum \psi(j_1)\psi(j_1 + j_2)\ldots\psi(j_1 + \cdots + j_s)$$
$$\times \xi_{|j_2|}\xi_{|j_3|} \cdots \xi_{j_1 + \cdots + j_s|}, \qquad (11.3.1)$$

where the summation is taken over the values of the variables $j_k = 0, \pm 1, \cdots \pm (n-1)$.
    Szegö proved the following theorem [115].

Theorem 11.3.1. *Let $R_n = (r_{i-j})i, j = \overline{0,n}, r_k = (2\pi)^{-1} \int_{-\pi}^{\pi} \exp(ikx) f(x) dx$, where $f(x)$ is a real even measurable function with $\sum_{k=1}^{\infty} k|r_k| < \infty$.*
Then

$$\lim_{n\to\infty}\{\sum_{k=0}^{n}(\lambda_k^{(n)})^s - (2\pi)^{-1}(n+1)\int_{-\pi}^{\pi} f^s(x)dx\}$$

$$= -2\sum\{\max(0,l_1,l_1+l_2,\ldots,l_1+\cdots+l_{s-1}\}r_{l_1}r_{l_2}\cdots r_{l_{s-1}}r_{l_1+\cdots+l_{s-1}},$$
$$(11.3.2)$$

where $\lambda_k^{(n)}, k = \overline{1,n}$ are eigenvalues of the matrix R_n, the summation is taken over the domain $-\infty < l_1, l_2, \ldots, l_{s-1} < \infty, s \geq 0$ in any integer.

Let us extend this theorem to the case when the matrix R_n is random. We introduce the notation $\xi_n(x) = \sum_{l=-\infty}^{\infty} \xi_l \exp(ilx)$.

Theorem 11.3.2. *Let $\xi_n(x) \Rightarrow \xi(x), x \in (-\pi, \pi)$ with probability 1,*

$$\sum_{k=1}^{\infty} k|\xi_k| < \infty, \sup_{|x|<\pi} \sup_n \mathbf{E}|\xi_n(x)| < \infty \qquad (11.3.3)$$

$$\lim_{h\to 0}\lim_{n\to\infty}\sup_{t',t'',|t'-t''|<k} \mathbf{E}|\xi_n(t') - \xi_n(t'')| = 0 \qquad (11.3.4)$$

Then formula (11.3.2) holds, where we substitute $\xi(x)$ for the function $f(x)$, and we substitute the variables ξ_k for the variables r_k, and the limit is understood in the sense of convergence of distributions.

Proof. Obviously,

$$\sum_{j_1}\psi(j_1)\psi(j_1+j_2)\ldots\psi(j_1+\ldots+j_{s-1})$$

$$= n+1-\{\max(0, j_2, j_2+j_3, \ldots j_2+\ldots+j_s)$$
$$- \min(0, j_2, j_2+j_3, \ldots, j_2+\ldots+j_s)\}$$

(if the right-hand side of the equality is not positive, then we consider it to be equal to zero), $\xi_{|k|} = \xi_{|-k|}, k = 0$, the (11.3.1) holds, therefore,

$\text{Tr}\,T^s$

$$= (2\pi)^{-1}(n+1)\int_{-\pi}^{\pi} \xi_n^2(x)dx - \sum\{\max(0, j_2, j_2 + j_3, \ldots, j_2 + \cdots + j_3)$$

$$- \min(0, j_2, j_2 + j_3, \ldots, j_2 + \cdots + j_s)\}\xi_{j_2}\xi_{j_3}\ldots\xi_{j_2 + \cdots + j_s}$$

$$= (2\pi)^{-1}(n+1)\int_{-\pi}^{\pi} \xi_n^s(x)dx - 2\sum\max(0, l_1, l_1 + l_2, \ldots, l_1 + \cdots + l_{s-1})$$

$$\times r_{l_1}r_{l_2}r_{l_{s-1}}\cdots r_{l_{s-1}}r_{l_1+\cdots+l_{s-1}},$$

where the summation is taken over the domain $l_1, l_2, \ldots, l_{s-1} = 0, \pm 1, \pm 2, \ldots$,
$r_k = (2\pi)^{-1} \int_{-\pi}^{\pi} \exp(ikx)\xi_n(x)dx$.

Hence, using (11.3.3) and (11.3.4), we come to the assertion of Theorem 11.3.3.

If in addition to the conditions of Theorem 11.3.2 with probability 1, $\sup_{|x| \le \pi} |\xi(x)| \le E < \infty$, then for some sufficiently small $\varepsilon > 0$,

$$n^{-1} \ln \det[I - \varepsilon T] \Rightarrow (2\pi)^{-1} \int_{-\pi}^{\pi} \ln[1 + \varepsilon\xi(x)]dx.$$

Let us proceed to the examination of the Toeplitz random matrix of $T_n = (u(\xi_i - \xi_j))$ form. Let $u(x)$ be a real even measurable function, $\lambda_i, i = \overline{1, n}$ be the eigenvalue of the matrix T_n; the random variables $\xi_i, i = 1, 2, \ldots$ be independent, identically distributed with the distribution $F(x)$ function and $Eu(\xi_i - \xi_j)$ exist. Then, obviously,

$\lim_{n \to \infty} \mathbf{E} \operatorname{Tr}(n^{-1}T_n)^m$

$$= \int \cdots \int u(x_1 - x_2)u(x_2 - x_3) \ldots u(x_m - x_1)dF(x_1) \ldots dF(x_m). \tag{11.3.5}$$

The expression (11.3.5) can be simplified. Let us prove the following Theorem [107].

Theorem 11.3.3. *Let the random variables $\xi_i, i = 1, 2, \ldots$ be independent, identically distributed with distribution density $p(x)$ which is continuous on $(-\infty, \infty)$, there exist the integrals*

$$\int p^m(x)dx, \int u(x_1)u(x_2) \ldots u(x_{m-1})u(x_1 + \cdots + x_{m-1})\Pi_i dx_i := c_m.$$

Then

$$\lim_{\beta \downarrow 0} \lim_{n \to \infty} \beta \mathbf{E} \operatorname{Tr}(\beta n^{-1}T_n(\beta))^m = c_m \int p^m(y)dy, \tag{11.3.6}$$

where $T_n(\beta) = (u(\xi_i\beta^{-1} - \xi_j\beta^{-1}))$.

Proof. From Eq. (11.3.5), it follows that

$$\lim_{n \to \infty} \mathbf{E} \operatorname{Tr}(\beta n^{-1}T_n(\beta))^m = \int \cdots \int u(x_1 - x_2)u(x_2 - x_3)$$
$$\ldots u(x_m - x_1)p(\beta x_1) \ldots p(\beta x_m)\Pi_i dx_i.$$

After replacing the variables $x_k = \beta^{-1}s_m + \sum_{i=k}^{m-1} s_i, k = \overline{1, m-1}, x_m = \beta^{-1}s_m$, the expression takes the form

$$\lim_{n \to \infty} \beta \mathbf{E} \operatorname{Tr}(\beta n^{-1}T_n(\beta))^m$$

$$= \int \cdots \int u(s_1)u(s_2) \ldots u(s_{m-1})$$
$$\times u(s_1 + s_2 + \cdots + s_{m-1})p(s_m + \beta(s_{m-1} + \cdots + s_2 + s_1))$$
$$\times p(s_m + \beta(s_{m-1} + \cdots + s_2)) \ldots p(s_m)\Pi_i ds_i.$$

Proceeding to the limit as $\beta \downarrow 0$, we come to the assertion of Theorem 11.3.3.

Note that if the function $u(x)$ is the Fourier transformation of some measurable function $f(x)$: $u(x) = (2\pi)^{-1} \int \exp(ixy)f(y)dy$, then

$$\int \cdots \int \Pi_{i=1}^{m-1} u(x_i)u(x_1 + \cdots + x_{m-1})\Pi_i dx_i = \int f^m(y)dy,$$

and therefore

$$\lim_{\beta \downarrow 0} \lim_{n \to \infty} \beta \mathbf{E} \operatorname{Tr}(n^{-1}\beta \operatorname{Tr}(\beta))^m = \int f^m(y)dy \int p^m(y)dy. \qquad (11.3.7)$$

Let $\lambda_n(x, \beta) = \beta \sum F(x - n^{-1}\beta\lambda_i(\beta))$, where $\lambda_i(\beta)$ are eigenvalues of the matrix $T_n(\beta)$. Then from (11.3.7) it follows that

$$\lim_{\beta \downarrow 0} \lim_{n \to \infty} \int x^m d\lambda_n(x, \beta) = \int \int [f(y)p(x)]^m dydx. \qquad (11.3.8)$$

If the moments $\mu_n := \int [f(y)p(x)]^m dydx$ determine a distribution function uniquely, then from (11.3.8) we get $\lambda_n(x, \beta) \Rightarrow \lambda(x)$ as $n \to \infty$, $\beta \downarrow 0$, and the moments of the function $\lambda(x)$ are equal to $\int x^m d\lambda(x) = \int \int [t(y)p(x)]^m dydx$.

§4 The Method of Perturbation for Determinants of Some Toeplitz Random Matrices

Let $\Xi_n = \rho_{ij}(\xi_i - \xi_j)n_{i,j=1}^{-1/2}, \rho_{ij}(x), -\infty < x < \infty, i, j = 1, 2, \ldots$ be independent random functions, with identical one-dimensional distributions the random variables $\xi_i, i = 1, 2, \ldots$ do not depend on the random functions $\rho_{ij}(x)$ and identically distributed $\mathbf{E}\rho_{ij}(x) = 0, \mathbf{E}\rho_{ij}^2(x) = \sigma^2(x), \sup_x \sigma^2(x) < \infty$.

Theorem 11.4.1. *If the above-mentioned conditions for the matrix Ξ_n hold, then with probability 1, $\lim_{n \to \infty} \mu_n(x, \Xi_n) = \mu(x)$ at each point of continuity of the nonrandom spectral function $\mu(x)$, whose Stieltjes transformation is*

$$\int (1 + itx)^{-1} d\mu(x) = \int [1 + t^2 f(t, y)]^{-1} dF(y),$$

where $F(y) = P\{\xi_1 < y\}$, and $f(t, y)$ satisfies the functional equation

$$f(t, y) = \int \sigma^2(y - x)[1 + t^2 f(t, x)]^{-1} dF(x). \qquad (11.4.1)$$

The solution of equation (11.4.1) exists, is unique in the class of analytical functions on t, $\operatorname{Re} f \geq 0$, and can be found with the help of the method of successive approximations

Proof. By using limit theorems for spectral functions of symmetric random matrices, proved in § 2 of Chapter 9, we conclude that with probability 1 for every real t,

$$\lim_{n \to \infty} [n^{-1} \operatorname{Tr} R_t - n^{-1} \mathbf{E} \operatorname{Tr} R_t] = 0,$$

and

$$n^{-1}\mathbf{E}\operatorname{Tr} R_t = n^{-1}\sum_{k=1}^{n}\mathbf{E}[1 + itn^{-1/2}\rho(0) + n^{-1}t^2$$

$$\times \sum_{i,j\neq k} r_{ij}^{k}\rho_{ki}(\xi_k - \xi_j)\rho_{kj}(\xi_k - \xi_j)]^{-1}, \qquad (11.4.2)$$

where $R_t = (I + it\Xi_n)^{-1}, r_{ij}^{k}$ are entries of the matrix $(I + it\Xi_n^{k})^{-1}$, the matrix Ξ_n^{k} is obtained from the matrix Ξ_n by replacing the entries of the kth vector row and the kth vector column by zeros. Analogous to §4 of Chapter 9, we get

$$\operatorname{plim}_{n\to\infty}\left[\sum_{i,j\neq k} r_{ij}^{k}\rho_{ki}(\xi_k - \xi_j)\rho_{kj}(\xi_k - \xi_j) - \sum_{i\neq k} r_{ii}^{k}\sigma^2(\xi_k - \xi_i)n^{-1}\right] = 0.$$
$$(11.4.3)$$

From §3 of Chapter 9, it follows that $\lim_{n\to\infty}\mathbf{E}|r_{ii}^{k} - r_{ii}| = 0, i \neq k$. Therefore,

$$\operatorname{plim}_{n\to\infty}|\sum_{i=1}^{n} r_{ii}^{k}n^{-1}\sigma^2(\xi_k - \xi_i) - \sum_{i=1}^{n} r_{ii}n^{-1}\sigma^2(\xi_k - \xi_i)| = 0 \qquad (11.4.4)$$

Obviously, $\sum_{i=1}^{n} r_{ii}\sigma^2(\xi_k - \xi_i) - \mathbf{E}[\sum_{i=1}^{n} r_{ii}\sigma^2(\xi_k - \xi_i)/\xi_k] = \sum_{k=1}^{n}\gamma_k$ where $\gamma_p = \mathbf{E}[\sum_{i=1}^{n} r_{ii}\sigma^2(\xi_k - \xi_i)/\sigma_{p-1}] - \mathbf{E}\sum_{i=1}^{n} r_{ii}\sigma^2(\xi_k - \xi_i)/\sigma_p]$, σ_p is the smallest σ- algebra, with respect to which the random variables $\rho_{ij}(x), i, j = \overline{p+1, n}, \xi_i, i = \overline{p+1, n}, \xi_k$ are measurable. Let us represent $\gamma_p, p \neq k$ in the form:

$$\gamma_p = \mathbf{E}[\{\sum_{i\neq p} r_{ii}\sigma^2(\xi_k - \xi_i) - \sum_{i\neq p} r_{ii}^{p}\sigma^2(\xi_k - \xi_i)\}/\sigma_{p-1}]$$

$$- \mathbf{E}[\{\sum_{i\neq p} r_{ii}\sigma^2(\xi_k - \xi_i) - \sum_{i\neq p} r_{ii}^{p}\sigma^2(\xi_k - \xi_i)\}/\sigma_p]$$

$$+ (\mathbf{E}[r_{pp}\sigma^2(\xi_k - \xi_p)|\sigma_{p-1}] - \mathbf{E}[r_{pp}\sigma^2(\xi_k - \xi_p)/\sigma_p]).$$

Hence, it is not difficult to conclude that $\sup_p |\gamma_p| < \infty$. Hence,

$$\operatorname{plim}_{n\to\infty} n^{-1}\sum_{k=1}^{n}\gamma_p = 0.$$

By using (11.4.2)–(11.4.4), we get

$$n^{-1}\mathbf{E}\operatorname{Tr} R_t = \int[1 + t^2\mathbf{E}r_{11}\sigma^2(x - \xi_1)]^{-1}dF(x) + 0(1),$$

$$\mathbf{E}r_{kk}\sigma^2(z - \xi_k) = \int[1 + t^2\mathbf{E}r_{11}\sigma^2(x - \xi_1)]^{-1}\sigma^2(z - x)dF(x) + 0(1).$$

By letting $\mathbf{E}r_{kk}\sigma^2(z - \xi_k) = f_n(t, z)$, we get the functional equation

$$f_n(t, z) = \int [1 + t^2 f_n(t, x)]^{-1} \sigma^2(z - x) dF(x) + 0(1).$$

Hence, the assertion of Theorem 11.4.1 follows.

Theorem 11.4.2. *Let the conditions of the Theorem 11.4.1 hold and* $\mathbf{V} \ln |1 + f(t, \xi_1)| \neq 0$ *where* $f(t, x)$ *satisfies Eq. (11.4.1). Then*

$$\lim_{n \to \infty} \mathbf{P}\{[n\mathbf{V} \ln |1 + f(t, \xi_1)|]^{-1/2}[\ln |\det(I + i\Xi_n)|$$

$$-\mathbf{E} \ln |\det(I + i\Xi_n)|] < x\} = (2\pi)^{-1/2} \int_{-\infty}^{x} \exp(-y^2/2) dy$$

Proof. Let us represent the difference $\ln |\det(I + i\Xi_n)| - \mathbf{E} \ln |\det(I + i\Xi_n)|$ in the form of $\ln |\det(I + i\Xi_n)| - \mathbf{E} \ln |\det(I + i\Xi_n)| = \sum_{k=1}^{n} \gamma_k$, where

$$\gamma_k = \mathbf{E}[\ln |1 + n^{-1/2} i \rho_k(0)$$
$$+ n^{-1} \sum_{i,j \neq k} r_{ij}^k \rho_{ki}(\xi_k - \xi_j) \rho_{kj}(\xi_k - \xi_j)| / \sigma_{k-1}] - \mathbf{E}[\ln |1$$
$$+ n^{-1/2} i \rho_k(0) + n^{-1} \sum_{i,j \neq k} r_{ij}^k \rho_{ki}(\xi_k - \xi_j) \rho_{kj}(\xi_k - \xi_j)| \sigma_k],$$

and σ_k is the smallest σ-algebra, with respect to which the random functions and variables $\rho_{ij}(x), \xi_i, i, j = \overline{k+1, n}$ are measurable. By using the method of the proof of Theorem 11.4.1, we establish that $\mathbf{E}\gamma_k^2 = \mathbf{V} \ln |1 + f(1, \xi_1)| + 0(1)$ and the random variables $\gamma_k, k = \overline{1, n}$ are asymptotically independent. After some simple transformations using Theorem 5.4.3, we come to the assertion of Theorem 11.4.2.

Analogous statements are also valid for the variables $\rho_{ij}(x), \xi_i$ with different distributions.

CHAPTER 12

LIMIT THEOREMS FOR DETERMINANTS
OF RANDOM JACOBI MATRICES

Problems of theoretical physics and numerical analysis can be reduced to the determination of the distribution function $F(x)$ of the eigenvalues of a random Jacobi matrix

$$\Xi_n = (\xi_i \delta_{ij} + \eta_i \delta_{ij-1} + \zeta_i \delta_{ij+1}),$$

where $\delta_{ij} = \begin{cases} 0, & i \neq j, \\ 1, & i = j, \end{cases}$ is a Kronecker symbol.

A method for finding $F(x)$ was first worked out by Dyson [30]. The methods presented in this chapter for determining the limiting distributions of the determinants of random Jacobi matrices are based on the investigation of the Stieltjes transform of the normalized spectral function (n.s.f)

$$\mu_n(x) = n^{-1} \sum_{i=\overline{1,n}} F(x - \lambda_i),$$

where λ_i are the eigenvalues of the matrix Ξ_n.

§1 Limit Theorems of the Law of Large Numbers Type

Let Ξ_n be a random Jacobi matrix, and $\rho_k^{(n)}$ be the minimal σ-algebra with respect to which the random vectors ξ_l $l = \overline{1, k}$ of Ξ_n are measurable. Suppose that $\mathbf{E} \ln^2 |\det \Xi_n|$ exists. According to §1 in Chapter 6, if there is a sequence of constants a_n such that

$$\lim_{n \to \infty} a_n^{-2} \sum_{k=1}^{n} \mathbf{E} \gamma_k^2 = 0, \qquad (12.1.1)$$

then

$$\text{plim}_{n \to \infty} a_n^{-1} (\ln |\det \Xi_n| - \mathbf{E} \ln |\det \Xi_n|) = 0. \qquad (12.1.2)$$

321

For determinants of random Jacobi matrices, Theorems 6.1.1 and 6.1.2 hold. It is required that nonzero entries of the random matrices be distributed by the stable law. For the random Jacobi determinants, these assumptions can be considerably weakened.

Let us consider random Jacobi matrices of the form

$$\Xi_n = (\xi_i \delta_{ij} + \eta_i \delta_{ij-1} + \zeta_j \delta_{ij+1}).$$

We introduce the notation $b_i = \det B_i$, $c_{n-1} = \det c_{n-1}$, where

$$B_k = (\xi_i \delta_{ij} + \eta_i \delta_{ij-1} + \zeta_j \delta_{ij+1}), \quad i,j = \overline{1,k},$$
$$C_{n-k} = (\xi_i \delta_{ij} + \eta_i \delta_{ij-1} + \zeta_j \delta_{ij+1}), \quad i,j = \overline{k,n}.$$

Expending $\det \Xi_n$ by the entries of the first row, the second row, etc., we get

$$c_{n-1} = \xi_1 c_{n-2} - \eta_1 \zeta_1 c_{n-3},$$
$$c_{n-2} = \xi_2 c_{n-3} - \eta_2 \zeta_2 c_{n-4},$$
$$c_{n-k} = \xi_k c_{n-(k+1)} - \eta_k \zeta_k c_{n-(k-2)},$$
$$c_0 = \xi_n; \; C_1 \equiv 1; \; C_0 \equiv 0, \; k = \overline{1,n},$$

$$\det \Xi_n = - \eta_{k-1} \zeta_{k-1} b_{k-2} c_{n-(k+1)}$$
$$+ b_{k-1} c_{n-(k+1)} \xi_k - b_{k-1} c_{n-(k+2)} \eta_k \zeta_k;$$
$$b_0 \equiv 1; b_{-1} \equiv 0; \tag{12.1.3}$$

Theorem 12.1.1. *For each n, let the random vectors $(\xi_k, \eta_k, \zeta_k), k = \overline{1,n}$, be independent and with probability 1,*

$$\xi_k - |\eta_k \zeta_k| \geq 1, \; |\eta_{k-1} \zeta_{k-1}| \leq 1,$$
$$\xi_n \geq 1, \; k = \overline{1,n}, \; n = 1, 2, \dots; \tag{12.1.4}$$

$$\sup_n \sup_{k=\overline{1,n}} \mathbf{E} \ln^2(\xi_k \pm |\eta_k \zeta_k| \pm |\eta_{k-1} \zeta_{k-1}|) < \infty. \tag{12.1.5}$$

Then

$$\text{plim}_{n \to \infty} n^{-1}(\ln \det \Xi_n - \mathbf{E} \ln \det \Xi_n) = 0. \tag{12.1.6}$$

Proof. By using (12.1.4) and (12.1.5), we get $C_{n-i} \cdot C_{n-(i+1)}^{-1} \geq 1$ and $b_i \cdot b_{i-1}^{-1} \geq 1$, $i = \overline{1,n}$.
Consequently, $\det \Xi_n \geq 0$, and

$$\xi_k - |\eta_k \xi_k| - |\eta_{k-1} \xi_{k-1}| \leq \det \Xi_n (b_{k-1} C_{n-(k+1)})^{-1} \leq \xi_k + |\eta_k \xi_k| + |\eta_{k-1} \xi_{k-1}|.$$

Hence,

$$\begin{aligned}
\mathbf{E}\gamma_k^2 =&\, \mathbf{E}[\mathbf{E}[\ln(\det \Xi_n(b_{k-1}C_{n-k-1})^{-1})/\rho_{k-1}^{(n)}] \\
&- \mathbf{E}[\ln(\det \Xi_n(b_{k-1}C_{n-k-1})^{-1})/\rho_k^{(n)}]]^2 \\
\leq&\, 2\max\{\mathbf{E}\ln^2(\xi_k - |\eta_k\zeta_k| - |\eta_{k-1}\zeta_{k-1}|), \\
&\mathbf{E}\ln^2(\xi_k + |\eta_k\zeta_k| + |\eta_{k-1}\zeta_{k-1}|)\}.
\end{aligned}$$

From this inequality, (12.1.1), and (12.1.5) follows the assertion of Theorem 12.1.1.

Consider one particular case of Jacobi matrices $H_n(\lambda) = ((2 + \lambda\xi_i^{(n)})\delta_{ij} - \delta_{ij-1} - \delta_{ij+1})$, where λ is an arbitrary constant.

Corollary 12.1.1. *If for each n the random variables ξ_i, $i = \overline{1,n}$ are independent and nonnegative, and*

$$\sup_n \sup_{i=\overline{1,n}} [\mathbf{E}\ln^2 \xi_i^{(n)} + \mathbf{E}\ln^2(\xi_i^{(n)}\lambda + 2)] \leq c < \infty,$$

then (12.1.6) holds for the matrix $H_n(\lambda)$,

$$H_n(\lambda) = ((2 + \lambda\xi_i^{(n)})\delta_{ij} - \delta_{ij-1} - \delta_{ij+1}),$$

for any $0 < \lambda < \infty$.

Theorem 12.1.2. *Let the random variables ξ_i, $i = \overline{1,n}$ be nonnegative, independent, identically distributed, and $\mathbf{E}\ln\xi$ exist. Then for any $0 < \lambda < \infty$,*

$$\mathrm{plim}_{n\to\infty} n^{-1}(\ln\det\theta_n(\lambda) - \mathbf{E}\ln\det\theta_n(\lambda)) = 0,$$

where $\theta_n(\lambda) = ((2 + \lambda\xi_i)\delta_{ij} - \delta_{ij-1} - \delta_{ij+1})$.

We obtain the assertion of Theorem 12.1.2 by introducing the truncation method,

$$\gamma_k' = \begin{cases} \gamma_k, & \xi_i < n\delta, \\ 0, & \xi_i \geq n\delta, \end{cases} \qquad \gamma_k'' = \begin{cases} 0, & \xi_i < n\delta, \\ \gamma_k, & \xi_i \geq n\delta, \end{cases}$$

where δ is a constant, and by using limit theorems of the law of large numbers type for Borel functions of independent random values, proved in §3 of Chapter 5.

§2 The Dyson Equation

From (12.1.3) it follows that

$$\det \Xi_n = \Pi_{i=1}^n C_{n-i}C_{n-i-1}^{-1},$$

$$C_{n-i}C_{n-i-1}^{-1} = \xi_i - \eta_i\zeta_i \cdot \left(\frac{C_{n-i-1}}{C_{n-i-2}}\right)^{-1},$$

i.e., $\det \Xi_n$ can be represented as the product of continued fractions

$$C_{n-i}C_{n-i-1}^{-1} = \xi_i - \eta_k\zeta_k/(\xi_i - \eta_{i+1}\zeta_{i+1}/\cdots(\cdots\eta_n\zeta_n/\xi_n)\cdots).$$

If the vectors (ξ_i, η_i, ξ_i), $i = 1, 2, \ldots$ are identically distributed and independent, and the distribution functions of $C_{n-i}C_{n-i-1}^{-1}$ converge to the limiting function $F(x)$, then the latter satisfies the following integral equations:

$$F(x) = \int \cdots \int_{y_1 - y_2 y_3 z^{-1} < x} dF(z) d\Phi(y_1, y_2, y_3), \qquad (12.2.1)$$

where $\Phi(y_1, y_2, y_3) = \mathbf{P}(\xi_1 < y_1, \ \eta_1 < y_2, \ \zeta_1 < y_3)$.

The equation (12.2.1) and the like will be called the Dyson equations.

Theorem 12.2.1. *Let the random variables ξ_i, $i = 1, 2, \ldots$ of the matrices*

$$\Xi_n = ((2 + \xi_k)\delta_{ij} - \delta_{ij-1} - \delta_{ij+1})$$

be independent, nonnegative, and indentically distributed, the sequence of sums the $\sum_{k=1}^n \xi_k$, $n = 1, 2, \ldots$ tend in probability to infinity, and for some $\delta > 0$,

$$\mathbf{E}|\ln \xi_1|^{1+\delta} < \infty. \qquad (12.2.2)$$

Then

$$\text{plim}_{n\to\infty} n^{-1} \ln \det \Xi_n = \int_1^\infty \ln x dF(x),$$

where $F(x)$ satisfies the integral equation

$$F(x) = \int \int_{2+y-z^{-1}<x} dF(z) d\mathbf{P}\{\xi_1 < y\}. \qquad (12.2.3)$$

Proof. Obviously $\det \Xi_n \geq 0$, and $\mathbf{E}(\ln \det \Xi_n)$ exists. Consider the difference

$$\ln \det \Xi_n - \mathbf{E} \ln \det \Xi_n = \sum_{k=1}^n (\mathbf{E}[\delta_k^{(n)}/\sigma_{k-1}^{(n)}] - \mathbf{E}[\delta_k^{(n)}/\sigma_k^{(n)}]),$$

where $\delta_k^{(n)} = \ln[\det \Xi_n \cdot (b_{k-1}C_{n-k-1})^{-1}]$; we introduce the following quantities:

$$\nu_k^{(n)} = \begin{cases} \delta_k^{(n)} & \text{if } |\delta_k^{(n)}| \leq n\varepsilon, \\ 0 & \text{if } |\delta_k^{(n)}| > n\varepsilon, \end{cases}$$

$$\mu_k^{(n)} = \begin{cases} \delta_k^{(n)} & \text{if } |\delta_k^{(n)}| > n\varepsilon, \\ 0 & \text{if } |\delta_k^{(n)}| \leq n\varepsilon. \end{cases}$$

Clearly, $\delta_k^{(n)} = \nu_k^{(n)} + \mu_k^{(n)}$. By using (12.2.2), we get

$$\lim_{n \to \infty} n^{-1} \mathbf{E} | \sum_{k=1}^{n} (\mathbf{E}[\mu_k^{(n)} / \sigma_{k-1}^{(n)}] - \mathbf{E}[\mu_k^{(n)} / \sigma_k]|$$

$$\leq \lim_{n \to \infty} n^{-2} \sum_{k=1}^{n} \mathbf{E} |\mu_k^{(n)}|^{1+\delta} \cdot (n\varepsilon)^{-\delta} = 0; \qquad (12.2.4)$$

$$\mathbf{E}[n^{-1} \mathbf{E}(\sum_{k=1}^{n} \nu_k^{(n)} / \sigma_{k-1}^{(n)}) - n^{-1} \mathbf{E}(\sum_{k=1}^{n} \nu_k^{(n)} / \sigma_k^{(n)})]^2$$

$$\leq n^{-2} \sum_{k=1}^{n} \mathbf{E}(\nu_k^{(n)})^2 \leq \varepsilon n^{-2} \sum_{k=1}^{n} \mathbf{E} |\nu_k^{(n)}| \to 0, \varepsilon \to 0. \qquad (12.2.5)$$

On the basis of (12.2.4) and (12.2.5),

$$\text{plim}_{n \to \infty} n^{-1} (\ln \det \Xi_n - \mathbf{E} \ln \det \Xi_n) = 0.$$

Let us now find the limit of $n^{-1} \mathbf{E} \ln \det \Xi_n$. The fact that the sums $\sum_{k=1}^{n} \xi_k$ diverge in probability implies the existence of $\text{plim}_{n \to \infty} C_{n-i} C_{n-i-1}^{-1} = \xi$ (where ξ is some random variable) [110]. Taking this fact into account for the distribution function of the variable ξ, we obtain equation (12.2.3). Now we write

$$n^{-1} \mathbf{E} \ln \det \Xi_n = n^{-1} \sum_{k=1}^{n} \mathbf{E} \ln(C_{n-k} C_{n-k-1}^{-1}).$$

By using (12.2.2), we obtain

$$\lim_{n \to \infty} \mathbf{E} \ln(C_{n-k} C_{n-k-1}^{-1}) = \int_{1}^{\infty} \ln x \, dF(x), \quad \forall k : k \to \infty, \ n \to \infty.$$

This implies the assertion of Theorem 12.2.1.

Theorem 12.2.2. *Let the random vectors* (ξ_i, η_i, ζ_i) $i = 1, 2, \ldots,$ *whose components are the entries of the matrix* $\Xi_n = \{\xi_i \delta_{ij} + \eta_i \delta_{ij-1} + \zeta_i \delta_{ij+1}\}$ *be independent and identically distributed, with probability 1,*

$$\xi_k - |\eta_k \xi_k| > 1, \qquad |\eta_{k-1} \xi_{k-1}| \leq 1, \quad k = \overline{1, n}, \quad \xi_n \geq 1,$$

and for some $\delta > 0$,

$$\mathbf{E} |\ln(\xi_2 - |\eta_2 \zeta_2| - |\eta_1 \zeta_1|)|^{1+\delta} + \mathbf{E} |\ln(\xi_2 + |\eta_2 \zeta_2| + |\eta_1 \zeta_1|)|^{1+\delta} < \infty.$$

Then

$$\text{plim}_{n \to \infty} n^{-1} \ln \det \Xi_n = \int_{1}^{\infty} \ln x \, dF(x),$$

where $F(x)$ satisfies the integral equation

$$F(x) = \int \cdots \int_{y_1 - y_2 y_3 z^{-1} < x} dF(z) d\Phi(y_1, y_2, y_3)$$

$$\Phi(y_1, y_2, y_3) = \mathbf{P}\{\xi_1 < y_1, \ \eta_1 < y_1, \zeta_1 < y_3\}. \tag{12.2.6}$$

Proof. As in the proof of Theorem 12.2.1, we obtain

$$\mathrm{plim}_{n \to \infty}(n^{-1} \ln \det \Xi_n - n^{-1} \mathbf{E} \ln \det \Xi_n) = 0.$$

Equation (12.2.6) follows from the fact that under the condition $\xi_k - |\eta_k \xi_k| > 1$ the continued fraction $C_{n-1} C_{n-i-1}^{-1}$ converges as $n \to \infty$ [110, p.12] and $C_{n-i} C_{n-i-1}^{-1} > 1$. Thus, Theorem 1.2.2 is proved.

Theorems 12.2.1, and 12.2.2 can be easily generalized for the case when the distribution of the random matrices is periodic in the following sense: The random vectors $[(\xi_i, \eta_i, \zeta_i), \ i = \overline{(l-1) \cdot k + 1, \ lk}] \ l = 1, 2, \ldots,$ are independent and identically distributed. Under the condition that the continued fraction $C_{n-i} C_{n-i-1}^{-1}$ converges as $n \to \infty$, we get the following integral equation for the limit distribution function:

$$F(x) = \int \cdots \int dF(z) d\Phi(y_1^{(i)}, y_2^{(i)}, y_3^{(i)}), \tag{12.2.7}$$

where the integration is over the domain

$$y_1^{(1)} - y_2^{(1)} y_3^{(1)} / y_1^{(2)} - y_2^{(2)} y_3^{(2)} / \ldots (\ldots y_2^{(k-1)} y_3^{(k-1)} / z) \ldots) < x,$$

$$\Phi(y_1^{(i)}, y_2^{(i)}, y_3^{(i)}) = \mathbf{P}\{\xi_i < y_1^{(i)}, \eta_i < y_2^{(i)}, \zeta_i < y_3^{(i)}\}.$$

In Theorems 12.2.1, 12.2.2 we proved that the limit $\mathrm{plim}_{n \to \infty} n^{-1} \ln \det \Xi_n$ exists when the fractions $C_{n-i} C_{n-i-1}^{-1}$ converge. If we do not make this assumption, then we need conditions under which equations (12.2.3), (12.2.6), and (12.2.7) have a unique solution in the class of distribution functions. It is possible to approach the solution of this problem without using continued fractions. Suppose that Ξ_n is nonsymmetric and has a real spectrum.

Obviously,

$$n^{-1} \ln |\det \Xi_n| = \int_{-\infty}^{\infty} \ln |x| d\mu_n(x),$$

where $\mu_n(x)$ is the n.s.f. of Ξ_n. We consider the Stieltjes transform of $\mu_n(x)$,

$$\int (1 + itx)^{-1} d\mu_n(x) = n^{-1} \mathrm{Tr}(I + it\Xi_n)$$

and prove the limit theorems for $n^{-1} \ln \det \Xi_n$.

Let

$$\Xi_n = (\xi_i \delta_{ij} + \eta_i \delta_{ij-1} + \zeta_i \delta_{ij+1}),$$

where ξ_i, η_i, ζ_i are the real random variables, and $\eta_i \zeta_i > 0$. We find the expression for the resolvent $R_t = (I + it\Xi_n)^{-1}$ of such a matrix.

Clearly,

$$\text{Tr } R_t = n - t \frac{d}{dt} \ln \det(I + it\Xi_n).$$

Let $d_n(t) = \det(I + it\Xi_n)$. It follows from (12.1.3) that

$$d_n(t) = (1 + it\xi_1)d_{n-1}(t) + t^2 \eta_1 \zeta_1 d_{n-2}(t),$$

where d_{n-k} is the determinant of the matrix

$$D_{n-k} = I + it(\xi_i \delta_{ij} + \eta_i \delta_{ij-1} + \zeta_i \delta_{ij+1}), \quad i,j = \overline{k+1,n};$$

$d_0(t) = 1$. According to §5 in Chapter 2, $d_{n-1}(t)d_n^{-1}(t) = r_{11}^{(n)}(t)$, where $r_{11}^{(n)}(t)$ is an entry of the resolvent $(I+it\Xi_n)^{-1}$. Therefore, $r_{11}^{(n)}(t) = [1+it\xi_1 + t^2 r_{11}^{(n-1)}(t)]^{-1}$, where $r_{11}^{(n-1)}(t)$ is an entry of the matrix $((I+it\xi_{pl}), p,l = \overline{2,n})$, and ξ_{pl} are the entries of Ξ_n. Since

$$d_n d_0^{-1} = \Pi_{k=0}^{n-1} d_{n-k} d_{n-k-1}^{-1} = \Pi_{k=0}^{n-1}[r_{11}^{(k+1)}(t)]^{-1},$$

it follows that $\text{Tr } R_t = n - t \sum_{k=1}^{n} d \ln r_{11}^k/dt$. We give another recurrent relation. Let

$$\text{Tr } R_{n-k} = n - k + t \frac{d}{dt} \ln d_{n-k}(t), \qquad (12.2.8)$$

where

$$R_{n-k} = (I + it(\xi_i \delta_{ij} + \eta_i \delta_{ij-1} + \zeta_i \delta_{ij+1}))^{-1}, \quad i,j = \overline{k+1n}.$$

$$\text{Tr } R_t = n - t \sum_{k=1}^{n} \frac{d}{dt} \ln r_{11}^{(k)}$$

Using the expression for $\text{Tr } R_{n-k}$, we can write

$$\text{Tr } R_{n-k} - \text{Tr } R_{n-k-1} = -1 + t \frac{d}{dt} \ln r_{11}^{(k)}(t). \qquad (12.2.9)$$

Theorem 12.2.3. *If the random vectors (ξ_i, η_i, ζ_i), $i = 1, 2, \ldots$, whose components are the entries of the matrix $\Xi_n = (\xi_i \delta_{ij} + \eta_i \delta_{ij-1} + \zeta_i \delta_{ij+1})$, are independent and identically distributed, with probability 1, $\eta_i \zeta_i > 0$, $|\zeta_i| < C < \infty$, $|\eta_i| < C < \infty$, $i = \overline{1,n}$, then with probability 1,*

$$\lim_{n \to \infty} \mu_n(x) = \mu(x)$$

at each point of continuity of the nonrandom spectral function whose Stieltjes transform is

$$\int (1 + itx)^{-1} d\mu(x) = 1 - \iiiint (x_1 + ix_2)(x_3 + ix_4)^{-1} dG_t(x_p, \ p = \overline{1,4}),$$

where the distribution function $G_t(x_p, \ p = \overline{1,4})$ defined for $|t| < \infty$, $|x_p| \leq 2$, $x_3 \geq 0$ satisfies the integral equation

$$G_t(x_p, \ p = \overline{1,4}) = \int \cdots \int dF(z_i, \ i = \overline{1,3}) \, dG_t(s,q,u,v), \qquad (12.2.10)$$

the integration being over the domain

$$
\begin{aligned}
\{s,q,u,v: \ & - \operatorname{Re}[itz_1 + 2z_2 z_3 t^2 (u + iv) + t^3 z_2 z_3 (s + iq)] \\
& \times [1 + itz_1 + t^2 z_2 z_3 (u + iv)]^{-2} < x_1, \\
& - \operatorname{Im}[itz_1 + z_2 z_3 (2t^2 (u + iv) + t^3 (s + iq))] \\
& \times [1 + itz_1 + t^2 z_2 z_3 (u + iv)]^{-2} < x_2, \\
& \operatorname{Re}[1 + itz_1 + t^2 z_2 z_3 (u + iv)]^{-1} < x_3, \\
& \operatorname{Im}[1 + itz_1 + t^2 z_2 z_3 (u + iv)]^{-1} < x_4\}, \\
& u \geq 0, \ |x_p| \leq 2, \ p = \overline{1,4}; \\
& F(z_i, \ i = \overline{1,3}) = \mathbf{P}\{\xi_1 < z_1, \ \eta_1 < z_2, \ \zeta_1 < z_3\}.
\end{aligned}
\qquad (12.2.11)
$$

The solution of (12.2.10) exists, is unique in the class of distribution functions $G_t(x_p, \ p = \overline{1,4})$, depending on the parameter t, $(-\infty < t < \infty)$, and such that the integral $\int \cdots \int \Pi_{i=1}^{n} x_i^{k_p} dG_t(x_p, \ p = \overline{1,4})$ is analytical on t for any positive integers k_p, $p = \overline{1,4}$.

Proof. It is known that the eigenvalues of Ξ_n are real when $\eta_i \zeta_i > 0$; therefore, $n^{-1} \operatorname{Tr} R_t$ exists.

It follows from Chapter 9 that with probability 1,

$$\lim_{n \to \infty} n^{-1}(\operatorname{Tr} R_t - \mathbf{E} \operatorname{Tr} R_t) = 0,$$

and from (12.2.8) we obtain that

$$n^{-1} \mathbf{E} \operatorname{Tr} R_t = 1 - tn^{-1} \sum_{k=0}^{n-1} E \frac{d}{dt} \ln r_{11}^{(k)}(t).$$

Therefore, for the proof of the theorem, it is necessary to find the limit of the expression $\mathbf{E}[\frac{d}{dt} \ln r_{11}^{(k)}(t)]$ as $n \to \infty$ (the symbol (t) in $r_{11}^{(k)}(t)$ will be omitted in what follows). Obviously,

$$
\begin{aligned}
t \frac{d}{dt} r_{11}^{(k)} = & - [1 + it\xi_k + \eta_k \zeta_k t^2 r_{11}^{(k-1)}]^{-2} \\
& \times [it\xi_k + 2\eta_k \zeta_k (t^2 r_{11}^{(k-1)} + t^3 \frac{d}{dt} r_{11}^{(k)})].
\end{aligned}
$$

By using this expression, we get the following recurrent equation:

$$G_t^{(k)}(x_p, \ p = \overline{1,4})$$

$$= \mathbf{P}\{\operatorname{Re} t\frac{d}{dt}r_{11}^{(k)} < x_1, \quad \operatorname{Im} t\frac{d}{dt}r_{11}^{(k)} < x_2, \quad \operatorname{Re} r_{11}^{(k)} < x_3, \operatorname{Im} r_{11}^{(k)} < x_4\}$$

$$= \int \cdots \int dF(z_1, z_2, z_3)dG_t(x_p), \qquad (12.2.12)$$

where the integration is over the domain (12.2.11).

Since $\eta_i\zeta_i > 0$ with probability 1, Ξ_n can be replaced by the matrix $\tilde{\Xi}_n$,

$$\tilde{\Xi}_n = \{\xi_i\delta_{ij} + (\eta_i\zeta_i)^{1/2}\delta_{ij-1} + (\eta_i\zeta_i)^{1/2}\delta_{ij-1}\}, \quad i = \overline{1,n}.$$

Clearly, when $\det\Xi_n = \det\tilde{\Xi}_n$, and the eigenvalues of this matrix coincide, then

$$r_{11}^{(n)} = \sum_{s=1}^{n}(1 + it\lambda_s)^{-1}h_{s1}h_{s1}^{-1} = (H\Lambda H^{-1}\vec{x}, \vec{x}),$$

where h_{sk} and h_{sk}^{-1} are the entries of the matrices H and H^{-1}, respectively. The eigenvectors corresponding to the λ_s are orthogonal with the weight $(h_{sk}, \beta_s h_{sk}^{-1}) = 0$, where $\beta_s = \beta_1\Pi_{k=1}^{s-1}\eta_k\xi_k$ are the weight coefficients, and β_1 is arbitrary. By choosing $\beta_1 > 0$, we get, as in §2 in Chapter 9, that an upper bound for $|t\frac{d}{dt}\ln r_{11}^{(k)}(t)|$ is equal to 2.

We show that for all $|t| \leq T$,

$$\lim_{k\to\infty}\sup_{|x_i|\leq 2}|G_t^{(k+1)}(\cdot) - G_t^{(k)}(\cdot)| = 0, \qquad (12.2.13)$$

where $T > 0$ is any bounded number.

Suppose that $\tilde{r}_{11}^{(p)}$ and $\tilde{r}_{11}^{(p+1)}$ are random variables defined on the common probability space, and moreover, $\tilde{r}_{11}^{(p)} \approx r_{11}^{(p)}$, $\tilde{r}_{11}^{(p+1)} \approx r_{11}^{(p+1)}$, the joint distribution of the variables $\tilde{r}_{11}^{(p)}, \tilde{r}_{11}^{(p+1)}$, is arbitrary and belongs to some set of distributions K. The variables $\tilde{r}_{11}^{(p)}, r_{11}^{(p-1)}$ do not depend on $\xi_{p+1}, \eta_{p+1}, \zeta_{p+1}$. Let us consider the random recurrent equations

$$\tilde{r}_{11}^{(p)} \approx [1 + it\xi_{p+1} + t^2\eta_{p+1}\zeta_{p+1}\tilde{r}_1^{(p-1)}]^{-1},$$

$$\tilde{r}_{11}^{(p+1)} \approx [1 + it\xi_{p+1} + t^2\eta_{p+1}\zeta_{p+1}\tilde{r}_{11}^{(p)}]^{-1}. \qquad (12.2.14)$$

Obviously, $\tilde{r}_{11}^{(p+1)} \approx r_{11}^{(p+1)}$ and the variables $r_{11}^{(p)}, \tilde{r}_{11}^{(p)}$ have a certain joint distribution.

From (12.2.14), we obtain,

$$\mathbf{E}|\tilde{r}_{11}^{(p+1)}(t) - \tilde{r}_{11}^{(p)}(t)| \leq t^2 \mathbf{E}|\tilde{r}_{11}^{(p)}(t) - \tilde{r}_{11}^{(p-1)}(t)|.$$

By using (12.2.14) for $\tilde{r}_{11}^{(p+1)}$ and $\tilde{r}_{11}^{(p+2)}$, we find that (symbol t is ommited)

$$\mathbf{E}|\tilde{r}_{11}^{(p+2)} - \tilde{r}_{11}^{(p+1)}| \le t^4 \mathbf{E}|\tilde{r}_{11}^{(p)} - \tilde{r}_{11}^{(p-1)}(t)|,$$

where $\tilde{r}_{11}^{(p+2)} \approx r_{11}^{(p+2)}$, $\tilde{r}_{11}^{(p+1)} \approx r_{11}^{(p+1)}$ and the variables $\tilde{r}_{11}^{(p+2)}$ and $\tilde{r}_{11}^{(p+1)}$ have a certain joint distribution. Repeating this procedure $k - p$ times, we get

$$\mathbf{E}|\tilde{r}_{11}^{(k)} - \tilde{r}_{11}^{(k-1)}| \le t^{2(k-p)} \mathbf{E}|\tilde{r}_{11}^{(p)} - \tilde{r}_{11}^{(p-1)}|.$$

Hence, for a fixed number p, we find that as $|t| < 1$,

$$\lim_{k \to \infty} \mathbf{E}|\tilde{r}_{11}^{(k)} - \tilde{r}_{11}^{(k-1)}| = 0.$$

But then (12.2.13) holds for $|t| < 1$. Since the characteristic functions of the distributions $G_t^{(k-1)}(\cdot)$, and $\Gamma_t^{(k)}(\cdot)$ are analytical on t, (12.2.3) holds for all $|t| \le T$. Let $G_t^{(m')}(\cdot)$ be a subsequence of the sequence $G_t^{(m)}(\cdot)$, $m = k, k + 1, \ldots$, such that $G_t^{(m')} \Rightarrow G_t(\cdot)$. Then by (12.2.3), $G_t(\cdot)$ satisfies (12.2.10). We show that (12.2.10) has the unique solution in the class of functions defined under the conditions of the theorem. Suppose that there are two solutions $G_t^{(1)}(\cdot)$ and $G_t^{(2)}(\cdot)$ of (12.2.10). Instead of (12.2.10), we consider the pair of random functional equations

$$\nu_1^{(p)}(t) + i\nu_2^{(p)}(t) \approx [it\xi_1 + \eta_1\zeta_1[2t(\mu_3^{(p)}(t) + i\mu_4^{(p)}(t)) \\
+ t^3(\mu_1^{(p)}(t) + i\mu_2^{(p)}(t))]][1 + it\xi_1 + t^2\eta_1\zeta_1(\mu_3^{(p)}(t) + i\mu_4^{(p)}(t))]^{-2},$$
$$\nu_3^{(p)}(t) + i\nu_4^{(p)}(t) \approx [1 + it\xi_1 + t^2\eta_1\zeta_1(\mu_3^{(p)}(t) + i\mu_4^{(p)}(t))]^{-1}.$$
$$(12.2.15)$$

The vectors $(\nu_i^{(p)}(t), i = \overline{1,4})$ have the distribution $G_t^{(p)}(\cdot)$, $p = 1, 2$ and do not depend on ξ_1, η_1, ζ_1.

The joint distribution of the two vectors $(\mu_i^{(1)}, i = \overline{1,4})$ and $(\mu_i^{(2)}, i = \overline{1,4})$ has an arbitrary form. Let K denote the set of all joint distributions of these vectors. The joint distributions of the vectors $(\nu_i^{(1)}, i = \overline{1,4})$ and $(\nu_i^{(2)}, i = \overline{1,4})$ belong to some set $L \subset K$. From equations (12.2.15), we get for $|t| < 1$,

$$\int |\vec{x} - \vec{y}| dF_2(\vec{x}, \vec{y}) \le ct^2 \int |\vec{x} - \vec{y}| dF_1(\vec{x}, \vec{y}),$$

where $C > 0$ is a constant, $F_2(\vec{x}, \vec{y})$ is the distribution function of the vectors $\nu_i^{(1)}, \nu_i^{(2)}(t)$, $i = \overline{1,4}$, $F_1(\vec{x}, \vec{y})$ is the distribution function of the vectors $\mu_i^{(1)}(t), \mu_i^{(2)}(t)$ chosen in such a way that $F_1(\vec{x}, \vec{a}) = G_t^{(1)}(\vec{x})$, $F_1(\vec{a}, \vec{x}) = G_t^{(2)}(x)$, $a = (2, 2, 2, 2)$.

Substituting F_2 for F_1, we write

$$\int |\vec{x}, -\vec{y}| dF_3(\vec{x}, \vec{y}) \le ct^2 \int |\vec{x} - \vec{y}| dF_2(\vec{x}, \vec{y}),$$

where $F_3(\vec{x}, \vec{y})$ is a distribution function satisfying the condition $F_3(\vec{x}, \vec{a}) = G_t^{(1)}(\vec{x})$, $F_3(\vec{a}, \vec{x}) = G_t^{(2)}(\vec{x})$. Thus, we obtain the sequence of distribution functions $F_k(\vec{x}, \vec{y})$, $k = \overline{1, n}$ satisfying the inequality

$$\int |\vec{x} - \vec{y}| dF_{k+1}(\vec{x}, \vec{y}) \le ct^2 \int |\vec{x} - \vec{y}| dF_k(\vec{x}, \vec{y}).$$

Consequently,

$$\int |\vec{x} - \vec{y}| dF_n(\vec{x}, \vec{y}) \le (ct^2)^n \int |\vec{x} - \vec{y}| dF_1(\vec{x}, \vec{y}).$$

Letting n go to infinity, we get for $ct^2 < 1$, that

$$\lim_{n \to \infty} \int |\vec{x} - \vec{y}| dF_n(\vec{x}, \vec{y}) = 0.$$

Since the characteristic functions of these distributions are analytic on t, this identity holds for all finite t. Thus, Theorem 12.2.3 is proved.

If $\mathbf{E} \ln r_{11}^{(k)}(t)$ exist, then (12.2.10) can be written in the simplified form.

Corollary 12.2.1. *If in addition to the conditions of Theorem 12.2.3,*

$$\sup_n \sup_{k = \overline{1, n}, |t| < T} \mathbf{E} |\ln r_{11}^{(k)}(t)|^{1+\delta} \le C,$$

for some $\delta > 0$ and any bounded $T > 0$, then with probability 1,

$$\lim_{n \to \infty} \mu(x) = \mu(x)$$

at each point of continuity of the continuous nonrandom spectral function $\mu(x)$, whose Stieltjes transform is

$$\int (1 + itx)^{-1} d\mu(x) = 1 + t\frac{d}{dt} \int\int \ln(y_1 + iy_2) dG_t(y_1, y_2),$$

where the distribution function $G_t(y_1, y_2)$ (t is a parameter), given on the set $0 \le y_1 \le 1$, $|y_2| \ge 1$ satisfies the integral equation

$$G_t(y_1 + y_2) = \int\int\int dG_t(x_1, x_2) dF(z_1, z_2, z_3), \qquad (12.2.16)$$

where the integration domain is over the domain

$$\{x_1, x_2, \operatorname{Re}[1 + itz_1 + z_2z_3t^2(x_1 + x_2)]^{-1} < y_1,$$
$$\operatorname{Im}[1 + itz_1 + z_2z_3t^2(x_1 + ix_2)]^{-1} < y_2\},$$
$$0 \le x_1 \le 1, \quad |x_2| < 1.$$

The solution of equation (12.2.16) exists and is unique in the class of the distribution functions $G_t(x_1, x_2)$ depending on parameter t, $-\infty < t < \infty$ and satisfying the condition that $\int \int x_1^{k_1} x_2^{k_2} dG_t(x_1, x_2)$ is the analytical function on t for any positive integers k_1, k_2.

Corollary 12.2.2.   *If in addition to the condition of Theorem 12.2.3 or Corollary 12.2.1 there exists such $\delta > 0$ that*

$$\sup_n n^{-1} \sum_{k=1}^{n} \mathbf{E}|\ln|\lambda_k||^{1+\delta} \le C < \infty,$$

then

$$\operatorname{plim}_{n \to \infty} n^{-1} \ln |\det \Xi_n| = \int \ln |x| d\mu(x),$$

where $\mu(x)$ is defined in Theorem 12.2.3 or Corollary 12.2.1, respectively.

All the preceeding arguments are extended easily to random Jacobi matrices of the form

$$\Xi_n = \{\xi_i \delta_{ij} + \eta_i \delta_{ij-1} + \eta_i \delta_{ij+1}\}$$

whose pairs (ξ_i, η_i) of random variables are independent and identically distributed. In this case, for example, we replace Eq. (12.2.6) by the following:

$$G_t(y_1, y_2) = \int \cdots \int dG_t(x_1, x_2) d\mathbf{P}\{\xi_1 < z_1, \eta_1 < z_2,\}$$

where the integration is over the domain

$$\{x_1, x_2 : \operatorname{Re}[1 + itz_1 + t^2z_2^2(x_1 + ix_2)]^{-1} < y_1,$$
$$\operatorname{Im}[1 + itz_1 + t^2z_2^2(x_1 + ix_2)]^{-1} < y_2\},$$
$$0 \le x_1 \le 1, |x_2| \le 1. \tag{12.2.17}$$

We generalize Corollary 12.2.2 for the case when the random variables ξ_i and η_i are not identically distributed.

Corollary 12.2.3.   *Suppose that the pairs (ξ_i, η_i) of the entries of the random matrices*

$$H_n = (\xi_i \delta_{ij} + \eta_i \delta_{ij-1} + \eta_i \delta_{ij+1})$$

are independent, that there exists the limit

$$\lim_{n \to \infty} F_n(z_1, z_2, u) = F(z_1, z_2, u), \quad 0 \le u \le 1,$$

where $F_n(z_1, z_2, u) = \mathbf{P}\{\xi_i < z_1, \eta_i < z_2\}$, for $in^{-1} \leq u \leq (i+1)n^{-1}$, that $F(z_1, z_2, u)$ is a distribution function continuous on the parameter u on $[0,1]$, and that there exists a number $\delta > 0$ such that

$$\sup_n \sup_{k=\overline{1,n}, |t| \leq T} \mathbf{E}|\ln r_{11}^{(k)}(t)|^{1+\delta} \leq C < \infty.$$

Then with probability 1, the relation $\lim_{n \to \infty} \mu(x) = \mu(x)$ is valid at each point of continuity of the nonrandom spectral function $\mu(x)$ whose Stieltjes transform is

$$\int (1 + itx)^{-1} d\mu(x) = 1 + t\frac{d}{dt} \int \cdots \int \ln(y_1 + iy_2) dG_t(y_1, y_2, u) du,$$

where the distribution function $G_t(y_1, y_2, u)$ depends on two parameters u and t, $0 \leq u \leq 1$, $-\infty \leq t \leq \infty$, $0 \leq y_1 \leq 1$, $|y_2| \geq 1$, and satisfies the integral equation

$$G_t(y_1, y_2, u) = \int \cdots \int dG_t(y_1, y_2, u) dF(z_1, z_2, u),$$

with the integration is over the domain (12.2.17).

Since $r_{11}^{(k)}$ and $\frac{d}{dt} r_{11}^{(k)}$ can be represented as continued fractions, analogous methods can be used to prove limit theorems for certain random continued fractions.

§3 The Stochastic Sturm–Liouville Problem

Let us study the distribution of eigenvalues and eigenfunctions of the differential equation

$$u''(t) + (\xi(t) + \lambda)u(t) = 0; \qquad u(0) = u(1) = 0, \tag{12.3.1}$$

where $\xi(t)$ is a real, continuous and bottom-bounded random process defined on $[0, L]$.

Sometimes, instead of boundary conditions, we use the following conditions

$$u(0)\cos\alpha - u'(0)\sin\alpha = 0,$$
$$u(L)\cos\beta - u'(L)\sin\beta = 0.$$

In the cases when Eq. (12.3.1) can be approximately reduced to a difference equation in order to solve the stochastic Sturm–Liouville problem, it is necessary to use limit theorems for determinants of random Jacobi matrices.

Indeed, after replacing the second derivative in (12.3.1) by the difference of the second order $n^{-2}(u_{k+1} - 2u_k + u_{k-1})$, where $u_k = u(k/n)$, $k = \overline{0, n-1}$.

Then we obtain the difference equation equivalent to the linear homogeneous system of equations

$$\Xi_{n-1}(\lambda)\vec{u}_{n-1} = 0,$$

where

$$\Xi_{n-1}(\lambda) = \{(-2 + n^2(\xi(i/n) + \lambda))\delta_{ij} + \delta_{ij-1} + \delta_{ij+1}\},$$

$i, j = \overline{1, n-1}$. The matrix $-\Xi_{n-1}(0)$ is a nonnegative-positive definite matrix. Consider the random process $\lambda_n(x) = \sum_{i=1}^{n} \lambda_{in}^{-1} F(x - \lambda_{in})$, where $\lambda_{1n} \geq \lambda_{2n} \geq \cdots \geq \lambda_{nn}$ are the eigenvalues of the matrix $\Xi_n(0)$. It is obvious that

$$\int (1 + \lambda x)^{-1} d\lambda_n(x) = \frac{d}{dt} \ln \det \Xi_n(\lambda). \tag{12.3.2}$$

Let us prove the limit theorems for the determinants of random Jacobi matrices.

Theorem 12.3.1. *Let $\xi(t)$ be a measurable process on $[0, L]$ such that*

$$\mathbf{P}\{\inf_{0 \leq t \leq L} \xi(t) > 0\} = 1, \tag{12.3.3}$$

$$\lim_{h \to \infty} \mathbf{P}\{\sup_{0 \leq t \leq L} \xi(t) \geq h|\} = 0. \tag{12.3.4}$$

Then for all $\lambda \geq 0$,

$$n^{-1} \ln \det \Xi_n(\lambda) \Rightarrow \int_0^L \{\mathbf{E}[\exp\{-\frac{1}{2}\int_0^t (\xi(x) + \lambda)\omega^2(x)dx\}/\sigma]\}^{-2} dt, \tag{12.3.5}$$

as $n \to \infty$, where $\omega(x)$ is a Brownian motion process for $\xi(t)$, and σ is the minimal σ-algebra, with respect to which the process $\xi(x)$ is measurable, $x \in [0, L]$.

Proof.   Consider the matrices

$$L_n = (\delta_{ij} + \sum_{k=\max(i,j)}^{n} \nu_k),$$

where $\nu_i = n^{-2}(\xi(i/n) + \lambda)$. Subtract the second row from the first one, then the third row from the second one, and so on. Further, subtract the second column from the first one, then the third column from the second one, etc. Then $\det L_n = \det \Xi_n - \det \Xi_{n-1}(\det L_0 = 1)$ and $\det \Xi_n = \sum_{i=1}^{n} \det L_i$. Using the integral representation for the determinant (§5, Chapter 1), we get

$$\det L_k^{-1/2} = \mathbf{E}[\exp\left\{-0.5 \sum_{i=1}^{n} n^2 \nu_i \left(\sum_{p=1}^{k} \eta_p n^{-1/2}\right)^2 n^{-1}\right\}/\nu_i, i = \overline{1, n}],$$

$$\tag{12.3.6}$$

where $\eta_1, \eta_2, \ldots$ are independent $N(0,1)$-distributed variables that do not depend on the process $\xi(x)$. Let

$$\xi_n(x) = \xi(k\backslash n), \qquad w(x) = n^{-1/2} \sum_{i=1}^{k} \eta_i, \quad \frac{k}{n} \leq x \leq \frac{k+1}{n}.$$

Then

$$n^{-1}(\det \Xi_n(\lambda) - 1)$$
$$= n^{-1} \sum_{i=1}^{n} [\mathbf{E}\exp\{-0.5 \int_0^{i/n} (\xi_n(x) + \lambda)w_n^2(x)dx\}/\sigma]^{-2} = \int_0^1 \eta_n(t)dt,$$

where $\eta_n(t) = [\mathbf{E}(\exp\{-0.5 \int_0^{i/n} (\xi_n(x) + \lambda)w_n^2(x)dx\}/\sigma]^{-2}, \frac{i}{n} \leq t \leq \frac{i+1}{n}$. Let $h_2 > 0$ be some constant. If $\inf_{0 \leq t \leq L} \xi(t) > 0$, $\sup_{0 \leq t \leq L} \xi(t) \leq h_2$, then the ability of $\xi(x)$ and $w_n(x) \Rightarrow w(x)$ imply $\eta_n(t) \Rightarrow \eta(t)$ and

$$\lim_{h \to 0} \overline{\lim}_{n \to \infty} \sup_{|t'-t''| \leq h} \mathbf{E}\{|\eta_n(t') - \eta_n(t'')|/ \inf_{0 \leq t \leq L} \xi(t) > 0, \sup_{0 \leq t \leq L} \xi(t) \leq h_2\} = 0.$$

By using Lemma 9.2.1 and conditions (12.3.3) and (12.3.4), we get (12.3.5). A similar assertion is true for the finite sequence of random variables $\det \Xi_n(\lambda_k)$, $k = \overline{1, m}$. Theorem (12.3.1) is proved.

Theorem 12.3.2. *Under the assumptions of Theorem 12.3.1,*

$$\lambda_n(x) \overset{\sim}{\to} \lambda(x), \quad 0 \leq x < \infty,$$

where $\lambda(x)$ is nondecreasing, random process, bounded with probability 1, whose Stieltjes transform is

$$\int (1 + tx)^{-1}d\lambda(x) = \frac{d}{dt} \ln \int_0^1 \{\mathbf{E}[\exp\{-0.5 \int_0^y (t + \xi(x))w^2(x)dx\}/\sigma]\}^{-2}dy,$$
$$t > 0. \tag{12.3.7}$$

Proof. By using (12.3.2) we found

$$\int_0^\infty (1 + tx)^{-1}d\lambda(x) = -2n^{-1} \sum_{i=1}^{n} \{\mathbf{E}[\exp\{-0.5\eta_n \left(\tfrac{i}{n}\right)\}/\sigma\}^{-3}$$
$$\times \mathbf{E}[\exp\{-0.5\eta_n \left(\tfrac{i}{n}\right)\} \int_0^{i/n} \xi_n(x)w_n^2(x)dx/\sigma]$$
$$\times [n^{-1} + n^{-1} \sum_{i=1}^{n} \{\mathbf{E}[\exp\{-0.5\eta_n \left(\tfrac{i}{n}\right)\}/\sigma]\}^{-2}]^{-1},$$
$$\tag{12.3.8}$$

where

$$\eta_n(\frac{i}{n}) = \int_0^{i/n} (\xi_n(x) + \lambda)w_n^2(x)dx.$$

Since

$$1 \geq \mathbf{E}[\exp\{-0.5\eta_n(\tfrac{i}{n})\}/\sigma] \geq \exp\{-0.5\mathbf{E}[\eta_n(\tfrac{i}{n})/\sigma]\},$$

then from (12.3.8), we obtain

$$\lambda_n(+\infty) \leq n^{-1} \sum_{i=1}^n [\exp\{-0.5 \int_0^{i/n} \xi_n(x)dx\}]^{-3} \int_0^{i/n} \xi_n(x)dx.$$

This inequality and (12.3.4) imply

$$\lim_{h\to\infty} \overline{\lim_{n\to\infty}} \mathbf{P}\{\lambda_n(+\infty) \geq h\} = 0. \qquad (12.3.9)$$

By using the method of proving Theorem 12.3.1, we obtain (12.3.7).

We need the following assertion.

Lemma 12.3.1.   *Let $\lambda_n(x)$, $0 \leq x < \infty$ be a sequence of nondecreasing random processes,*

$$\lim_{h\to\infty} \sup_n \mathbf{P}\{\lambda_n(+\infty) \geq h\} = 0,$$

and for all t, $\eta_n(t) \Rightarrow \eta(t)$, where

$$\eta_n(t) = \int (1 + itx)^{-1} d\lambda_n(x).$$

Then $\lambda_n(x) \overset{\sim}{\to} \lambda(x)$, where $\lambda(x)$ is a nondecreasing random proccess of bounded variation, whose Stieltjes transform is

$$\eta(t) = \int (1 + itx)^{-1} d\lambda(x).$$

Proof.   Consider conditional moments

$$\mathbf{E}\{\Pi_{k=1}^m \int (1 + it_k x_k)^{-1} d\lambda_n(x_k)/\lambda_n(+\infty) \leq c\}$$
$$= \mathbf{E}\{\Pi_{k=1}^m \eta_n(t_k)/\lambda_n(+\infty) \leq c\},$$

where $c > 0$ is an arbitrary constant.

Passing on the subsequences, we obtain, for almost all x, that

$$\{\lambda_n(x), \lambda_n(+\infty) \leq c\} \Rightarrow \{\lambda(x), \lambda(+\infty) \leq c\}.$$

Letting $c \to \infty$, we come to the statement of Lemma 12.3.1.

Obviously, Lemma 12.3.1 holds for the transform $\int_0^\infty (1+tx)^{-1} d\lambda_n(x)$, $t \geq 0$. By using Lemma 12.3.1 and (12.3.9), we accomplish the proof of Theorem 12.3.2. In the same way, we shall prove the following assertion.

Theorem 12.3.3. *Let the random variables* $\xi_l^{(n)}$, $l = \overline{1,n}$ *of the matrix* $\Xi_n(\lambda) = \{(2 + n^{-2}(\xi_l^{(n)} + \lambda))\delta_{ij} + \delta_{ij-1} + \delta_{ij+1}\}$ *be independent and non-negative, there exist*

$$\mathbf{E}l\xi_l^{(n)} = a_l^{(n)}, \qquad \sup_n \sum_{i=1}^n a_i^{(n)} \le c, \qquad \lim_{n\to\infty} \varphi_n(t) = \varphi(t),$$

$$t \in [0,1], \qquad \lim_{h\to 0} \sup_{|t'-t''|\le h} |\varphi(t') - \varphi(t'')| = 0,$$

where $\varphi_n(t) = a_l^{(n)}n$, *for* $\frac{l}{n} \le t < \frac{l+1}{n}$ *the Lindeberg condition holds: for every* $\tau > 0$,

$$\lim_{n\to\infty} \sum_{l=1}^n \int_{x>\tau} x\, d\mathbf{P}\{\xi^{(n)}l > x\} = 0.$$

Then $\operatorname{plim}_{n\to\infty} \lambda_n(x) = \lambda(x)$ *at each point of continuity of the nondecreasing nonrandom bounded function* $\lambda(x)$, *whose Stieltjes transform is*

$$\int_0^\infty (1 + tx)^{-1} d\lambda(x) = \frac{d}{dt} \ln \int_0^1 [\mathbf{E}\exp\{-0.5\int_0^y [\varphi(x) + t]w^2(x)dx\}]^{-2} dy.$$

§4 The Sturm Oscillation Theorem

In order to find the limiting spectral functions of random Jacobi matrices, we have to invert their Stieltjes transform, which is the solution of the Dyson equation. Note that such an inversion, in general, is a very difficult task. The Sturm oscillation theorem [39] makes it possible to avoid this operation in some cases. We shall prove one of its generalizations.

Theorem 12.4.1. *Let* A_n *be a symmetric real matrix of order* n *and* $\det A_i$, $i = \overline{0,n}(\det A_0 = 1)$ *the sequence of its main minors,* $\det A_i \ne 0$, $i = \overline{0,n}$.

Then the number of negative eigenvalues of matrix A *is equal to the number of changes of the sign in the sequence* $\det A_i$, $i = \overline{0,n}$.

Proof. The quadratic form $(A\vec{x}, \vec{x})$ can be written in two ways:

$$(A\vec{x}, \vec{x}) = \sum_{i=1}^n \det A_{i-1}(\det A_i)^{-1}y_i^2,$$

where $y_j = \sum_{j=1}^n \beta_{ij}x_j$, β_{ij} are real numbers and in the form

$$(Ax, x) = \sum_{i=1}^n \lambda_i z_i^2,$$

where $\vec{z} = H\vec{x}$, λ_i are the eigenvalues of matrix A, and H is the matrix of the eigenvectors of matix A. Using these two representations of the $A(x, x)$ and the law of inertia for the quadratic forms, we obtain the assertion of Theorem 12.4.1.

From Theorem 12.4.1 such an assertion follows. Let $\Xi_n = (\xi_{ij}^{(n)})$ be a random real symmetric matrix of order n, $(\det \Xi_0 = 1)$ $\det \Xi_i, i = \overline{0, n}$ its main minors, $\mu_n(x)$ the normalized spectral function of Ξ_n and with probability 1 $\det \Xi_i \neq 0$. Then with probability 1,

$$\mu_n(x) = (2n^{-1}) \sum_{i=1}^{n} (1 - \text{sign}[\det(\Xi_{i-1} - Ix)(\det(\Xi_i - Ix))^{-1}. \qquad (12.4.1)$$

By using (12.4.1), we can approximately calculate $\mu_n(x)$.

If the random entries $\xi_{ij}^{(n)}, i \geq j$, $i, j = \overline{1, n}$ are independent and have continuous distributions for every n, $\lim_{h \to \infty} \sup_n \mathbf{E}\mu_n(h) = 0$, then (see §1, Chapter 9) with probability 1 for almost all x, $\lim_{n \to \infty} (\mu_n(x) - \mathbf{E}\mu(x)) = 0$, where

$$\mathbf{E}\mu_n(x) = n^{-1} \sum_{i=1}^{n} \mathbf{P}\{\det(\Xi_{i-1} - Ix)(\det(\Xi_i - Ix)) < 0\}. \qquad (12.4.2)$$

In general, we cannot obtain a functional equation for the limiting distributions of random variables $\det(\Xi_{i-1} - Ix) \det(\Xi_i - Ix)^{-1}$. It is possible to obtain the functional equation for the case, when Ξ_n are Jacobi matrices analogous to the one obtained in [§2, Chapter 12].

Let $\Xi_n = (\delta_{ij}\xi_i - \delta_{ij-1} - \delta_{ij+1})$ be a random Jacobi matrix, $n = m \cdot k$, where the random vectors $(\xi_{i+1}, \dots, \xi_{i+k})$, $i = 0, k, 2k, \dots$ formed by the entries of matrix Ξ_n are independent and identically distributed with continuous distribution functions $F(x_1, \dots x_n)$. Denote by K_λ the set of all limiting distributions for the converging sequences of the distribution functions

$$G_\lambda^{(n)}(x) := n^{-1} \sum_{i=1}^{n} \mathbf{P}\{\det(\Xi_{i-1} - I\lambda) \cdot (\det(\Xi_i - I\lambda))^{-1} < x\}.$$

Theorem 12.4.2. *With probability 1, for almost all x,*

$$\lim_{n \to \infty} \mu_n(x) = G_x(0)$$

for any function $G_x(z)$ from the set K_x satisfying the equation

$$G_x(u) = \int \cdots \int_{\mathcal{L}_x(u)} dF(y_1, \dots, y_n) dG_x(z),$$

where the integration is over the domain

$$\mathcal{L}_x(u) = y_1 - x - (y_2 - x - (\cdots - y_k - x - z^{-1})^{-1} \ldots)^{-1} < u.$$

Proof. The main minors of matrix $\Xi_n - I\lambda$ satisfy the recurrent equation

$$d_{n-k}(\lambda) = (\xi_k - \lambda)d_{n-1-k}(\lambda) - d_{n-2-k}(\lambda), \quad d_0 = 1.$$

From (12.4.2) and Theorem 12.2.1, we obtain

$$\lim_{n\to\infty} \mathbf{E}\mu_n(x) = \mu(x) = \lim_{n\to\infty} \int \cdots \int dF(y_1, \ldots y_n) dG_x^{(n)}(z),$$

where

$$G_x^{(n)}(z) = n^{-1} \sum_{k=1}^{n} \mathbf{P}\{\det(\Xi_{k-1} - Ix)(\det(\Xi_k - Ix))^{-1} < z\},$$

$$G_x^{(n)}(u) = \int \cdots \int_{\mathcal{L}_x(u)} dF(y_1, \ldots, y_n) dG_x^{(n)}(z) + 0(1).$$

This implies the assertion of Theorem 12.4.2.

§5 The Central Limit Theorem for Determinants of Random Jacobi Matrices

Let $\Xi_n = (\xi_i \delta_{ij} + \eta_i \delta_{ij-1} + \zeta_j \delta_{ij+1})$, σ_k be the minimal σ-algebra with respect to which the random variables $\xi_l, \eta_l, \zeta_l, \ l = \overline{k+1,n}$ are measurable. Suppose that $\mathbf{E}\ln^2|\det\Xi_n|$ exist. Using (12.1.3) we get

$$\ln|\det\Xi_n| - \mathbf{E}\ln|\det\Xi_n|$$

$$= \sum_{k=1}^{n}\{\mathbf{E}[\ln|-\eta_{k-1}\zeta_{k-1}\frac{d_{k-2}}{d_{k-1}} + \xi_k - \eta_k\zeta_k\frac{b_{n-(k+2)}}{b_{n-(k+1)}}|/\sigma_{k-1}] - \mathbf{E}[\ln|-\eta_{k-1}$$

$$\times \zeta_{k-1}d_{k-2}d_{k-1}^{-1} + \xi_k - \eta_k\zeta_k b_{n-(k+2)}b_{n-(k+1)}^{-1}|/\sigma_k]\}, \tag{12.5.1}$$

where

$$d_k = \det(\xi_i\delta_{ij} + \eta_i\delta_{ij-1} + \zeta_j\delta_{ij+1}) \ i,j = \overline{1,k},$$

$$b_{n-k} = \det(\xi_i\delta_{ij} + \eta_i\delta_{ij-1} + \zeta_j\delta_{ij+1}) \ i,j = \overline{k,n}.$$

For the sums (12.5.1), Theorems (5.4.5)–(5.4.7) are applicable. In some cases, by proving the central limit theorem for random Jacobi determinants, we can omit the condition (5.4.9).

Theorem 12.5.1. *Let the random variables ξ_i, $i = 1, 2, \ldots$ of the matrix $\Xi_n = ((2 + \xi_i)\delta_{ij} - \delta_{ij+1} - \delta_{ij-1})$ be independent, nonnegative, identically distributed, and let there exist*

$$\mathbf{E} \ln^2 \xi_1 < \infty. \tag{12.5.2}$$

Then

$$\lim_{n \to \infty} \mathbf{P}\{n^{-1/2}\sigma^{-1}[\ln \det \Xi_n - \mathbf{E} \ln \det \Xi_n] < x\}$$
$$= (2\pi)^{-1/2} \int_{-\infty}^{x} \exp(-y^2/2)dy,$$

where

$$\sigma^2 = \int_1^\infty \int_0^\infty \Big[\int_1^\infty \ln(-z^{-1} + 2 + u - x^{-1})dG(x)$$
$$- \int_0^\infty \int_1^\infty \ln(-z^{-1} + 2 + u - x^{-1})dG(x)dF(u) \Big]^2 dF(u)dG(z), \tag{12.5.3}$$

with the distribution function $G(z)$ satisfying the integral equation

$$G(x) = \iint_{2+y-z^{-1}<x, \; z \geq 1} dG(z)dF(y), \qquad F(y) = \mathbf{P}\{\xi_1 < y\}.$$

Proof. Using (12.5.1), we get

$$\ln \det \Xi_n - \mathbf{E} \ln \det \Xi_n = \sum_{k=1}^{n} \gamma_k,$$

where

$$\gamma_k = \mathbf{E}\{\ln[-(2 + \xi_k)d_{k-2}d_{k-1}^{-1} - b_{n-(k+2)}b_{n-(k+1)}^{-1}]/\sigma_{k-1}\}$$
$$- \mathbf{E}\{\ln[-(2 + \xi_k)d_{k-2}d_{k-1}^{-1} - b_{n-(k+2)}b_{n-(k+1)}^{-1}]/\sigma_k\},$$

σ_k is the minimal σ-algebra, with respect to which the random variables $\xi_{k+1}, \xi_{k+2}, \ldots$, are measurable,

$$d_k = ((2 + \xi_i)\delta_{ij} - \delta_{ij-1} - \delta_{ij+1}) \; i, j = \overline{1, k},$$
$$b_{n-k} = ((2 + \xi_i)\delta_{ij} - \delta_{ij-1} - \delta_{ij+1}), \quad i, j = \overline{k, n}.$$

The theorem will be proved, if we show that

$$\text{plim}_{n \to \infty} n^{-1} \sum_{k=1}^{n} \mathbf{E}[\gamma_k^2/\sigma_k] = \sigma^2, \tag{12.5.4}$$

$$\lim_{n \to \infty} n^{-1} \sum_{k=1}^{n} \mathbf{E}\gamma_k^2 = \sigma^2, \tag{12.5.5}$$

and for any $\tau > 0$,

$$\text{plim}_{n \to \infty} n^{-1} \sum_{k=1}^{n} \mathbf{E}[\gamma_k^2 \chi(|\gamma_k| > \tau\sqrt{n})/\sigma_k] = 0. \qquad (12.5.6)$$

Condition (12.5.2) implies $\mathbf{P}\{\xi_i = 0\} = 0$.

Therefore, the sequence of sums $\sum_{k=1}^{n} \xi_k$, $n = 1, 2, \dots$ tends to infinity in probability. But then (see Theorem 12.2.1),

$$\text{plim}_{n \to \infty}[b_{n-(k+1)} \cdot b_{n-(k+2)}^{-1}] = \theta_1,$$
$$\text{plim}_{n \to \infty}[d_{k-1} \cdot d_{k-2}^{-1}] = \theta_2,$$

for all k such that $(n - k) \to \infty$ as $n \to \infty$, where θ_1 and θ_2 are independent random variables with the distribution function $G(x)$. Since

$$|\gamma_k| \leq \mathbf{E}\ln^2 \xi_1 + \mathbf{E}\ln^2(2 + \xi_1) + \ln^2 \xi_k + \ln^2(2 + \xi_k),$$

$\lim_{n \to \infty, k \to \infty} \mathbf{E}\gamma_k^2 = \sigma^2$. It means that (12.5.4) and (12.5.6) hold.

Estimate the difference

$$b_{n-(k+1)}b_{n-(k-2)}^{-1} - b_{n-(s+1)}b_{n-(s+2)}^{-1}.$$

Since

$$b_{n-(k+1)}b_{n-(k+2)}^{-1} = 2 + \xi_{k+1} - b_{n-(k+3)}b_{n-(k+2)}^{-1},$$

we have

$$b_{n-(k+1)}b_{n-(k+2)}^{-1} = 2 + \xi_{k+1} - (\cdots - (2 + \xi_{s+1} - c_s)^{-1} \dots)^{-1},$$

where $c_s = b_{n-(s+1)}b_{n-(s+2)}^{-1}$. This implies

$$c_k - 2 + \xi_{k+1} - (\dots(2+\xi_{s+1})^{-1}\dots)^{-1} \leq c_s \prod_{i=k+1}^{s+1}(1+\xi_i)^{-1}, \quad s < k. \quad (12.5.7)$$

Hence, the random variables C_{kn} and C_{sn}, where $|k_n - s_n| \to \infty$ as $n \to \infty$, are asymptotically independent as $n \to \infty$. Consequently,

$$|\gamma_k| \leq \mathbf{E}\ln^2 \xi_1 + \mathbf{E}\ln^2(2 + \xi_1) + \ln^2 \xi_k + \ln^2(2 + \xi_k)$$

implies (12.5.5) (see the proof of Theorem 12.2.1). Hence, (12.5.3) follows and Theorem 12.5.1 is proved.

Using the inequality (12.5.7), we can prove that the random variables C_{kn} and C_{sn} ($|k_n - s_n| \to \infty$ as $n \to \infty$) are asymptotically independent in the case when $\xi_i^{(n)}$, $i = \overline{1, n}$ are nonnegative, independent and

$$\text{plim}_{n \to \infty} \Pi_{i=k_n}^{s_n}(1 + \xi_i^{(n)}) = \infty.$$

Therefore, Theorem 12.5.1 can be formulated in the general form.

Theorem 12.5.2. *Let the random variables $\xi_i^{(n)}$, $i = \overline{1,n}$ be independent, nonnegative for every n, and for some $\delta > 0$,*

$$\lim_{n \to \infty} \sum_{k=1}^{n} \mathbf{E}|\ln \xi_k^{(n)}|^{2+\delta} n^{-2-\delta} = 0;$$

for any $k_n, s_n < n$, $|k_n - s_n| \to \infty$, $\varepsilon > 0$,

$$\lim_{n \to \infty} \mathbf{P}\{|\sum_{i=k_n}^{s_n} \xi_i^{(n)}| \geq \varepsilon\} = 1, \quad s_n > k_n,$$

$$\underline{\lim}_{n \to \infty} n^{-1} \sum_{k=1}^{n} \mathbf{E}\gamma_k^2 > 0.$$

Then

$$\lim_{n \to \infty} \mathbf{P}\{[\ln \det \Xi_n - \mathbf{E} \ln \det \Xi_n]\left[\sum_{k=1}^{n} \mathbf{E}\gamma_k^2\right]^{-1/2} < x\}$$

$$= (2\pi)^{-1} \int_{-\infty}^{x} \exp(-y^2/2)dy.$$

§6 The Central Limit Theorem for Normalized Spectral Functions of Random Jacobi Matrices

One way to investigate the spectral properties of the finite difference analogue of the Schrödinger equation with a random potential is to study the spectral normalized distribution functions (n.d.f.'s) of a random Jacobi matrix [141]. In [146] the central limit theorem (c.l.t.) was proved for the spectrum of random Jacobi matrices whose entries are independent and identically distibuted and have distribution densities. The method of proving was based on the c.l.t. for homogeneous Markov chains.

In this section we prove a c.l.t. for "smoothed" n.d.f.'s of random Jacobi matrices

$$H_n = \{\xi_i\delta_{ij} + \eta_i\delta_{ij-1} + \eta_i\delta_{ij+1}\}, \quad i,j = 1, \ldots, n,$$

whose row vectors are independent and identically distributed.

Let

$$\mu_n(x) = \sum_{i=1}^{n} \mathcal{F}(x - \lambda_i)n^{-1}$$

be the n.d.f. of H_n, where the λ_i are its eigenvalues. In general, $\mathbf{E}\mu_n(x) - \mu_n(x) \to 0$ as $n \to \infty$ for almost all x, and it is difficult to find the points where the convergence fails, so we consider the "smoothed" n.d.f.

$$\tilde{\mu}_n(x) = a^{-1} \int (1 + y^2)^{-1} \mu_n(x + ay) dy$$

$$= (-a)^{-1} \int_{-\infty}^{x} \operatorname{Re} \operatorname{Tr}[(1 + iua^{-1})I_n + ia^{-1} H_n]^{-1} du, \quad a > 0.$$

Let

$$\theta_n(z_1, z_2) = \mu_n(z_1) - \mu_n(z_2)$$

$$= (-a)^{-1} \int_{z_1}^{z_2} \operatorname{Re} \operatorname{Tr}[(1 + iua^{-1})I_n + ia^{-1} H_n]^{-1} du.$$

We represent the difference $\theta_n(z_1, z_2) - \mathbf{E}\theta_n(z_1, z_2)$ as a sum of martingale differences and show that its finite-dimensional distributions tend to those of a Gaussian function as $n \to \infty$. Let $\gamma_k = \mathbf{E}_k \theta_n(z_2, z_2) - \mathbf{E}_{k-1} \theta_n(z_1, z_2)$, $k = 1, \ldots, n$, where $\mathbf{E}_k$ means that the expectation is taken for a fixed minimal σ-algebra with respect to which the random column vectors $\vec{h}_p$, $(p = k, \ldots, n)$ of H_n are measurable.

Theorem 12.6.1. *Suppose that the entries of the random matrix*

$$H_n = \{\xi_{ij}\delta_{ij} + \eta_i\delta_{ij-1} + \eta_i\delta_{ij+1}\}, \quad i,j = \overline{1,n}$$

are independent and identically distributed, and $\underline{\lim}_{n \to \infty} c_n n^{-1} > 0$. *Then*

$$\lim_{n \to \infty} P\{c_n^{-1/2} \sum_{k=1}^{n} \gamma_k < x\} = (2\pi)^{-1/2} \int_{-\infty}^{x} e^{-y^2/2} dy,$$

where $c_n = \sum_{1}^{n} \mathbf{E}\gamma_k^2$.

Proof. Let $R(u, a) = [(1 + iua^{-1})I + ia^{-1} H_n]^{-1}$. We represent

$$\theta_n(z_1, z_2) - \mathbf{E}\theta_n(z_1, z_2) = \sum_{i=1}^{n} \gamma_k.$$

It is clear that if the conditions

$$\operatorname{plim}_{n \to \infty} c_n^{-1} \sum_{k=1}^{n} \mathbf{E}_k \gamma_k^2 = 1, \tag{12.6.1}$$

$$\operatorname{plim}_{n \to \infty} c_n^{-1} \sum_{k=1}^{n} \mathbf{E}\gamma_k^2 \chi(|\gamma_k| C_n^{-1/2} > \tau) = 0, \quad \forall \tau > 0 \tag{12.6.2}$$

hold for $\{\gamma_k\}$, then the theorem will be proved. Let us represent γ_k in the form

$$\gamma_k = \mathbf{E}_{k=1}\theta_n(z_1, z_2) - \mathbf{E}_{k-1}\theta_n(z_1, z_2)$$

$$= (-an)^{-1}\int_{z_1}^{z_2}[\mathbf{E}_{k-1}[\operatorname{Tr}\operatorname{Re}R(u,a) - \operatorname{Tr}\operatorname{Re}R^k(u,a)]du$$

$$= (-an)^{-1}\int_{z_1}^{z_2}\{\mathbf{E}_{k-1}[\operatorname{Tr}\operatorname{Re}R(u,a) - \operatorname{Tr}\operatorname{Re}R^k(u,a)]$$

$$- \mathbf{E}_k[\operatorname{Tr}\operatorname{Re}R(u,a) - \operatorname{Tr}\operatorname{Re}R^k(u,a)]\}du,$$

where $R^k(u,a) = [(1 + iua^{-1})I_n + ia^{-1}H_n^k]^{-1}$ and H_n^k is the matrix obtained from H_n by replacing the entries of the kth row and the kth column by zeros. Using formula (2.5.13), we write

$$\gamma_k = (-an)^{-1}\int_{z_1}^{z_2}\{\mathbf{E}_{k-1}(d/du)\operatorname{Re}\ln[1 + ia^{-1}(u + \xi_k) + a^{-2}(R(u,a)$$

$$\times \vec{\xi}_k, \vec{\xi}_k)] - \mathbf{E}_k(d/du)\operatorname{Re}\ln[1 + ia^{-1}(u + \xi_k) + a^{-2}(R(u,a)\vec{\xi}_k, \vec{\xi}_k)]\}du.$$

We estimate the integrand:

$$\operatorname{Re}(d/dt)\ln[1 + ia^{-1}(u + \xi_k) + a^{-2}(R(u,a)\vec{\xi}_k, \vec{\xi}_k)]$$

$$= \operatorname{Re}[ia^{-1} + a^{-2}(R^2(u,a)\vec{\xi}_k, \vec{\xi}_k)]$$

$$\times [1 + ia^{-1}(u + \xi_k) + a^{-2}(R(u,a)\vec{\xi}_k, \vec{\xi}_k)]^{-1}$$

$$= \operatorname{Re}[ia^{-1} + a^{-2}\sum_{l=1}^{n}y_l^2[1 - a^{-2}(u + x_l)^2][1 + a^{-2}(u + \lambda_l)^2]^{-2}$$

$$+ 2ia^{-2}\sum_{l=1}^{n}y_l^2a^{-1}(u + \lambda_l)[1 + a^{-2}(u + \lambda_l)^2]^{-2}$$

$$\times [\sum_{l=1}^{n}y_l^2[1 + a^{-2}(u + \lambda_l)^2]^{-1}$$

$$- i\sum_{l=1}^{n}y_l^2a^{-2}(u + \lambda_l)[1 + a^{-2}(u + \lambda_l)]^{-1}]^{-1}]$$

$$\leq a^{-2} + a^{-1}(u + \lambda_k)(1 + a^{-2}(u + \lambda_k)^2)^{-1},$$

where $y_l = (\vec{\eta}_k, \vec{\xi}_k)$, h_k are the eigenvectors of H_n.

With the help of this inequality, it is not difficult to show that $\gamma_k \leq b$, where b is a constant. Let us now verify conditions (12.6.1) and (12.6.2) for $\{\gamma_k\}$. From the Lyapunov conditions and the boundedness of γ_k, it follows that

$$c_n^{-1-\delta}\sum_{k=1}^{n}\mathbf{E}_k|\gamma_k|^{2+\delta} \leq nb^2c_n^{-1-\delta} \leq n_{n\to\infty}^{-\delta} \to 0, \quad \delta > 0.$$

Let us check (12.6.1). We show that $\mathbf{E}_k \gamma_k^2$ and $\mathbf{E}_s \gamma_s^2$ are asymptotically independent, assuming for certainty that $k > s$. Let

$$d_n(u, a) = \det[(1 + iua^{-1})I_n + ia^{-1}H_n],$$

and write the decomposition [(see formula (12.5.1)]

$$d_{n-k}(u, a) = (1 + ia^{-1}(u + \xi_k))d_{n-k-1}(u, a) + a^{-2}\eta_k^{-2}d_{n-k-2}(u, a),$$

where

$$d_{n-k}(u, a)$$
$$= \det\{(1 + ia^{-1}(u + \xi_k))\delta_{ij} + ia^{-1}\eta_i\delta_{ij-1} + ia^{-1}\eta_i\delta_{ij+1}\}, \quad i, j = k \ldots n.$$

Then

$$\gamma_k = (-an)^{-1} \int_{z_1}^{z_2} [\mathbf{E}_{k-1}(d/du) \operatorname{Re} \ln d_n(u, a) - \mathbf{E}_k(d/du)$$

$$\operatorname{Re} \ln d_n(u, a)]du = (-an)^{-1} \int_{z_1}^{z_2} \{\sum_{k=1}^{n} [\mathbf{E}_{k-1} \operatorname{Re}(d/du)$$

$$\ln[(1 + ia^{-1}(u + \xi_k)) + a^{-2}\eta_k^2 r_{k-1}^{-1}(u, a)]$$

$$- \mathbf{E}_k \operatorname{Re}(d/du) \ln[1 + ia^{-1}(u + \xi_k)$$

$$+ a^{-2}\eta_k^2 r_{k-1}^{-1}(u, a)]]\}du, \qquad (12.6.3)$$

where $r_k(u, a) = d_{n-k}(u, a)d_{n-k-1}^{-1}(u, a)$. Below, we omit the parameters u and a of r_k and d_n for simplicity.

Using the representations

$$r_k = 1 + ia^{-1}(u + \xi_k) + a^{-2}\eta_k r_{k-1}^{-1}, \qquad (d/du)r_k = ia^{-1} + a^{-2}\eta_k^2 r_{k-1}^2 \frac{d}{du} r_{k-1},$$

we find that

$$G_{u,a}^{(k)}(x_p, \; p = 1, \ldots, 4, r_s) = \mathbf{P}\{\operatorname{Re} r_k < x_k,$$
$$\operatorname{Im} r_k < x_2, \operatorname{Re}(d/du)r_k < x_3, \operatorname{Im}(d/du)r_k < x_4 | \xi_i, \eta_i, \quad i = k, \ldots, n\}$$
$$+ \int \cdots \int_{\mathcal{L}} d\mathcal{F}(y_1, y_2)dG_{u,a}^{(k-1)} \quad (t_p, \; p = 1, \ldots, 4, r_s),$$

where $\mathcal{F}(y_1, y_2)$ is the distribution function of the entries of H_n, $G_{u,a}^{(k)}(x_p, \; p = 1, 4, r_s)$ is the conditional distribution function of H_n, and

$$L = \{t_p, \; p = 1, \ldots, 4, y_1, y_2 : \; \operatorname{Re}[1 + ia^{-1}(u + y_1)$$
$$+ a^{-2}y_2^2(t_1 + t_2)^{-1}] < x_1,$$
$$\operatorname{Im}[1 + iua^{-1}(u + y_1) + a^{-2}y_2^2(t_1 + it_2)^{-1}] < x_2,$$
$$\operatorname{Re}[ia^{-1} + a^{-1}y_2(t_1 + it_2)^{-2}(t_3 + it_2)^{-2}(t_3 + it_4)] < x_3,$$
$$\operatorname{Im}[ia^{-1} + a^{-2}y_2(t_1 + it_2)(t_3 + it_4)] < x_4\}.$$

Following the proof of the Theorem 12.2.3, we can show that

$$G_{u,a}^{(k)}(x_p, \ p = 1, \ldots, 4, r_s) \Rightarrow G_{u,a}(x_p, \ p = 1, \ldots, 4, r_s),$$
$$G_{u,a}(x_p, \ p = 1, \ldots, 4, r_s),$$
$$= \int \cdots \int_L dF(y_1, y_2) dG_{u,a}(t_p, \ p = 1, \ldots, 4, r_s),$$

$$(12.6.4)$$

and the solution of (12.6.4) exists and is unique in the class of distribution functions. But (12.6.4) coincides with equation (12.2.10) for the unconditional distribution functions, and the solution of that equation also exists and is unique. Consequently,

$$G_{u,a}(x_p, \ p = 1, \ldots, 4, r_s) = G_{u,a}(x_p, \ p = 1, \ldots, 4).$$

This means that r_k and r_s are asymptotically independent. But since $\mathbf{E}_k \gamma_k^2 = f(r_k)$, $\mathbf{E}_s \gamma_s^2 = f(r_s)$, and

$$G_{u,a}^{(k)}(x_p, \ p = 1, \ldots, 4, r_s) \Rightarrow G_{u,a}(x_p, \ p = 1, \ldots, 4, r_s),$$

(12.6.3) gives us that $\mathbf{E}_k \gamma_k^2$ and $\mathbf{E}_s \gamma_s^2$ are asymptotically independent as $k - s \to \infty$. Moreover, it was proved earlier that $\mathbf{E}_k \gamma_k^2$ and $\mathbf{E}_s \gamma_s^2$ are bounded. The combination of these conditions means that $R_{ks} \to 0$ as $|k - s| \to \infty$, where R_{ks} is the correlation function of $\mathbf{E}_k \gamma_k^2$ and $\mathbf{E}_s \gamma_s^2$. Then

$$\mathbf{E} n^{-2} \left(\sum_{k=1}^{n} \mathbf{E}_k \gamma_k^2 - \sum_{k=1}^{n} \mathbf{E} \gamma_k^2 \right) = n^{-2} \sum_{i,j=1}^{n} R_{ij}$$
$$- n^{-2} \sum_{|i-j| \geq m} R_{ij} + n^{-2} \sum_{|i-j| \leq m} R_{ij} = 0(m) + cmn^{-2} \to 0,$$

as $m, n \to \infty$. Thus, the condition (12.6.1) holds and Theorem 12.6.1 is proved.

CHAPTER 13

THE FREDHOLM RANDOM DETERMINANTS

Let Ξ_n be a square random matrix. We call the random function $\det(I + t\,\Xi_n)$, where t is a real or complex variable, the Fredholm random determinant of the matrix Ξ_n. Fredholm random determinants carry important information about random matrices. With their help, the limiting distributions for eigenvalues of the random matrices can be found. In this chapter, on the basis of the limit theorems for Fredholm random determinants, limit theorems for the eigenvalues of symmetric and nonsymmetric random matrices are proved.

We give a brief account of a scheme of proving limit theorems for the eigenvalues of random matrices. Let $\Xi_n = (\xi_{ij})$ be a random $n \times n$ matrix and $\lambda_i, i = \overline{1,n}$ its eigenvalues in the order of increasing moduli. To simplify the formulae, we assume that the eigenvalues $\lambda_i, i = \overline{1,n}$ are distinct. For λ_i, the following formulae hold:

$$
\begin{aligned}
\lambda_1 &= \operatorname{Tr} \Xi_n^s (\operatorname{Tr} \Xi_n^{s-1})^{-1} + \varepsilon_s, \\
\lambda_2 &= (\operatorname{Tr} \Xi_n^s - \lambda_1^s)(\operatorname{Tr} \Xi_n^{s-1} - \lambda_1^{s-1})^{-1} + \delta_s,
\end{aligned}
\tag{13.0.1}
$$

where $\varepsilon_s \to 0$ and $\delta_s \to 0$ as $s \to \infty$.

If we suppose that

$$
\overline{\lim}_{h \to \infty} \sup_n \mathbf{P}\{\operatorname{Tr} \Xi_n \Xi_n^* \geq h\} = 0,
\tag{13.0.2}
$$

then in (13.0.1) we can set

$$
\overline{\lim}_{s \to \infty} p - \overline{\lim}_{n \to \infty} [|\varepsilon_s| + |\delta_s|] = 0.
$$

Due to this relation, in order to study limit theorems for the eigenvalues λ_i, we need limit theorems for the distributions of the random vectors $(\operatorname{Tr} \Xi_n^s,$ $\operatorname{Tr} \Xi_n^{s-1})$ for any fixed integer $s \geq 0$. We mention that for symmetric random matrices, (13.0.1) also holds for multiple eigenvalues λ_i.

The investigation of the distribution of $\operatorname{Tr} \Xi_n^s$ is a very difficult problem, but under the condition (13.0.2) it can be reduced to the study of certain

347

sums of conditionally independent random variables. If the Ξ_n are symmetric matrices, then

$$\operatorname{Tr} \Xi_n^s = i^{-s}\big[(s-1)!\big]^{-1}(-1)^{s-1}(\partial^s/\partial t^s)\ln\det(I + it\,\Xi_n) \quad t = 0. \quad (13.0.3)$$

For $\det(I + it\Xi_n)$, there is an integral representation (see Chapter 1)

$$\det(I + it\Xi_n)^{-1} = \big[\mathbf{E}\exp\big(-it(\Xi_n\,\vec{\xi}_n,\,\vec{\xi}_n)/\Xi_n\big)\big]^2, \quad (13.0.4)$$

where $\vec{\xi}_n$ is a random vector that does not depend on Ξ and that is distributed according to the normal law $N(0, 2^{-1}I)$.

The expression (13.0.3) is equal to

$$\operatorname{Tr}\Xi_n^s = \mathbf{E}\Big[\varphi\big(\exp(-it(\Xi_n\,\vec{\xi}_n,\,\vec{\xi}_n)),(\Xi_n\,\vec{\xi}_n,\,\vec{\xi}_n)\big)$$
$$\times\,\big[\mathbf{E}\exp\big(-it(\Xi_n\,\vec{\xi}_n,\,\vec{\xi}_n)\big)\big]^{-s}/\Xi_n\Big],$$

where φ is a polynomial.

When we now consider the moments

$$\mathbf{E}\varphi^m\big[\mathbf{E}\exp(-it(\Xi_n\,\vec{\xi}_n,\,\vec{\xi}_n))\big]^l, \quad m,l = 1, 2, \ldots,$$

we can reduce the problem of finding the distributions of the $\operatorname{Tr}\Xi_n^s$ to the same problem as for the random variables $(\Xi_n\,\vec{\xi}_{nk},\,\vec{\xi}_{nk})$, $k = 1, 2, \ldots$, where the $\vec{\xi}_{nk}$ are independent random vectors that do not depend on Ξ_n and are distributed according to the normal law $N(0, 2^{-1}I)$. If the distributions of the $(\Xi_n\,\vec{\xi}_{nk},\,\vec{\xi}_{nk})$ approach those of the $(H_n\,\vec{\xi}_{nk},\,\vec{\xi}_{nk})$, where H_n are random matrices, then under certain hypotheses, it is easy to derive

$$\lambda_1(\Xi_n) \sim \lambda_1(H_n).$$

For non-symmetric random matrices Ξ_n, we have to use integral representations of the following type (see Chapter 1):

$$\det(I + \alpha_t\Xi_n)^{-1} = \mathbf{E}\big[\exp\big\{i\alpha_t((\Xi_n - \Xi_n')\,\vec{\xi},\,\vec{\eta})$$
$$- \alpha_t(\Xi_n\,\vec{\xi},\,\vec{\xi}) - \alpha_t(\Xi_n\,\vec{\eta},\,\vec{\eta})\big\}/\Xi_n\big].$$

§1 Fredholm Determinants of Symmetric Random Matrices

Suppose that random matrices Ξ_n satisfy the conditions of Theorem 8.4.3. It is clear that

$$\det(I + t\,\Xi_n\Xi_n') \sim \Pi_{j=1}^n(1 + t \sum_{i \in T_{jn}} \nu_{ij}^2)$$
$$\times \det[\delta_{ij}(1 + \sum_{j \in k_{in}} \nu_{ij}^2) + C_{ij}(t)], \quad t \geq 0, \tag{13.1.1}$$

where

$$C_{ij}(t) = t \sum_{k=1}^n b_{ik}b_{jk}(1 + t \sum_{i \in T_{kn}} \nu_{ik}^2)^{-1}.$$

To prove (13.1.1), we consider the moments $E\Pi_{s=1}^m \det(I + t_s\Xi_n\Xi_n^1)^{-k_s}$, where $t_s \geq 0$, $s = \overline{1,m}$ are arbitrary real numbers, m is a finite integer, and k_s are arbitrary positive integers.

Obviously,

$$E\Pi_{s=1}^n \det(I + t_s\Xi_n\Xi_n^1)^{-k_s} = E \exp\left\{ -\sum_{s=1}^m t_s \sum_{l=1}^{2k_s}(\Xi_n\Xi_n' \,\overrightarrow{\eta_{ls}}, \overrightarrow{\eta_{ls}}) \right\},$$

where η_{ls}, $l, s = 1, 2, \ldots$, are independent random vectors distributed normally $N(0, 0, 5I)$ which don't depend on the matrix Ξ_n. By applying the proof of Theorem 8.4.3 to this formula, we obtain (13.1.1). Analogously, we establish the following more general statement: if the conditions of Theorem 8.4.3 hold, then for integers s, m,

$$\frac{d^s}{dt^s} \det(I + t\Xi_n\Xi_n')^m \sim \frac{d^s}{dt^s}\left\{ \Pi_{j=1}^n\left(1 + t \sum_{i \in T_{jn}} \nu_{ij}^2\right)^m \right.$$
$$\left. \times \det\left[\delta_{ij}\left(1 + t \sum_{j \in K_{in}} \nu_{ij}^2\right) + C_{ij}(t)\right]^m. \tag{13.1.2}$$

Let us prove (13.1.2) as $s = 1$, $m = 1$. Using the integral representations for random determinants, we find

$$\frac{d}{dt} \det(I + t\Xi_n\Xi_n')$$
$$= E\big[\exp\{-t(\Xi_n\Xi_n' \,\overrightarrow{\eta}_{11}, \overrightarrow{\eta}_{11}) - t(\Xi_n\Xi_n' \,\overrightarrow{\eta}_{12}, \overrightarrow{\eta}_{12})\}$$
$$\times [(\Xi_n\Xi_n' \,\overrightarrow{\eta}_{11}, \overrightarrow{\eta}_{11}) + (\Xi_n\Xi_n' \,\overrightarrow{\eta}_{12}, \overrightarrow{\eta}_{12})]/\Xi_n]$$
$$\times \big\{ E[\exp\{-t(\Xi_n\Xi_n' \,\overrightarrow{\eta}_{11}, \overrightarrow{\eta}_{11}) - t(\Xi_n\Xi_n' \,\overrightarrow{\eta}_{12}, \overrightarrow{\eta}_{12})\}/\Xi_n]\big\}^{-2}.$$

Denote the numerator of the fraction by $\rho_{1n}(t)$, the denominator by $\rho_{2n}(t)$, and consider the moments

$$\mathbf{E}\Pi_{i=1}^{p}\rho_{1n}^{k_i}(t_i)\rho_{2n}^{m_i}(t_i) = \mathbf{E}\exp\Big\{-\sum_{i=1}^{p}t_i$$

$$\times \sum_{j=1}^{2(k_i+m_i)}[(\Xi_n\Xi_n'\,\overrightarrow{\eta}_{ij},\,\overrightarrow{\eta}_{ij}) + (\Xi_n\Xi_n'\,\overrightarrow{\tilde{\eta}}_{ij},\,\overrightarrow{\tilde{\eta}}_{ij})]\Big\}$$

$$\times \Pi_{i=1}^{p}\Pi_{j=1}^{m_i}[(\Xi_n\Xi_n'\,\overrightarrow{\eta}_{ij},\,\overrightarrow{\eta}_{ij}) + (\Xi_n\Xi_n'\,\overrightarrow{\tilde{\eta}}_{ij},\,\overrightarrow{\tilde{\eta}}_{ij})],$$

where $\overrightarrow{\tilde{\eta}}_{ij}$ are independent random vectors which do not depend on $\overrightarrow{\eta}_{ij}$ and Ξ_n and have the same distributions as the vectors $\overrightarrow{\eta}_{ij}$, $t_i \neq 0$.

Thus, the problem is reduced to the study of the finite number of random variables $(\Xi_n\Xi_n'\,\overrightarrow{\eta}_{ij},\,\overrightarrow{\eta}_{ij})$. Therefore, on the basis of the proof of Theorem 8.4.3, we obtain (13.1.2).

If the conditions of Theorem 8.4.3 hold, then

$$\frac{d^s}{dt^s}\ln\det(I + t\Xi_n\Xi_n')$$

$$\sim \frac{d}{dt^s}\mathrm{Ln}\Big\{\Pi_{j=1}^{n}\Big(1+t\sum_{i\in T_{in}}\nu_{ij}^2\Big)\det\Big[\delta_{ij}\Big(1+t\sum_{j\in K_{in}}\nu_{ij}^2\Big)+C_{ij}(t)\Big]\Big\}.$$

$$(13.1.3)$$

Formula (13.1.3) and the ones similar to it play a leading role in proving limit theorems for eigenvalues of the random matrices.

If the conditions of Corollary 8.4.3 hold, then

$$\frac{d^s}{dt^s}\ln\det(I + t\Xi_n\Xi_n')$$

$$\sim \frac{d^s}{dt^s}\Big\{\mathrm{Ln}\Big[\Pi_{j=1}^{n}\Big(1+t\sum_{i\in T_{jn}}\nu_{ij}^2\Big)\Big(1+t\sum_{i\in K_{jn}}\nu_{ji}^2\Big)\Big]\Big\}, \quad t \geq 0;$$

if the conditions of Corollary 8.4.3 hold, then

$$\frac{d^s}{dt^s}\ln\det(I + t\Xi_n\Xi_n')$$

$$\sim \frac{d^s}{dt^s}\Big\{\mathrm{Ln}\det\Big[I + tB_nB_n' + \mathrm{diag}\Big(t\sum_{i=1}^{n}\nu_{ij}^2,\ j = \overline{1,n}\Big)\Big]\Big\}, \quad t \geq 0;$$

if the conditions of Theorem 8.2.2 hold and Ξ_n is a symmetric matrix, then

$$\frac{d^s}{dt^s}\ln\det(I + it\Xi_n)$$

$$\sim \frac{d^s}{dt^s}\mathrm{Ln}\Big[\Pi_{p<l}(1+t^2\nu_{pl}^2)\prod_{p=1}^{n}(1+it\nu_{pl})\Big], \quad t \geq 0;$$

if the conditions of Theorem 8.4.2 hold, then

$$\frac{d^s}{dt^s} \ln \det(I + it\Xi_n)$$

$$\sim \frac{d^s}{dt^s} \text{Ln}\left[\det(I + itB_n + t^2\Lambda_n)\prod_{k=1}^{n}(1 + it\nu_{kk})\right], \quad t \geq 0,$$

where $\Lambda_n = \text{diag}\left(\sum_{l \in T_p \cup K_p} \nu_{pl}^2, \ p = \overline{1, n}\right)$.

Analogous statements can be formulated for Fredholm determinants of random Jacobi, Teoplitz, Hankel, and dominant matrices. (See Chapters 11 and 12.)

Note that with the help of Fredholm determinants, limit theorems for traces of powers of random matrices can be proved since the following formula holds:

$$\text{Tr}\,\Xi^k = i^{k-1}[(k-1)!]^{-1}\frac{d^k}{dt^k}\ln\det(I + it\Xi_k)_{t=0}.$$

§2 Limit Theorems for Eigenvalues of Symmetric Random Matrices

Let Ξ_n be the square random matrices of the order n. Arrange the eigenvalues of the matrix $\Xi_n\Xi_n'$ in nonincreasing order $\lambda_{1n} \geq \lambda_{2n} \geq \cdots \geq \lambda_{nn}$. Consider the random process $\lambda_n(x)$ equal to the sum of eigenvalues belonging to the semi-interval $[0, x)$. If with probability 1, $\text{Tr}\,\Xi_n\Xi_n' < \infty$, then

$$\int_0^\infty (1 + tx)^{-1}d\lambda_n(x) = d\ln\det(I + i\Xi_n\Xi_n')/dt := \eta_n(t).$$

Thus the proof of the limit theorems for spectral functions is reduced to that for Fredholm random determinants.

Theorem 13.2.1. *If* $\lim_{h\to\infty}\overline{Lim}_{n\to\infty}P\{\lambda_n(+\infty) \geq h\} = 0$, *then in order that* $\lambda_n(x) \xrightarrow{\sim} \lambda(x)$, $x \geq 0$, *where* $\lambda(x)$ *is a random function, nondecreasing and of bounding variation with probability 1, it is necessary and sufficient that* $\eta_n(t) \Rightarrow \eta(t)$, $t \geq 0$, *where* $\eta(t)$ *is some random function.*

Proof. Sufficiency.
Consider the following moments (see Theorem 9.2.3):

$$\mathbf{E}\left\{\Pi_{k=1}^m\eta_n(t_k)/\lambda_n(+\infty) \leq h\right\}$$

$$= \int\cdots\int\prod_{k=1}^m(1 + t_kx_k)^{-1}d\mathbf{E}\left\{\Pi_{k=1}^m\lambda_n(x_k)/\lambda_n(+\infty) \leq h\right\}, \quad (13.2.1)$$

where $t_k \geq 0$ are arbitrary real variables. Since $\lambda_n(+\infty) = \eta_n(0)$, from the convergence of $\eta_n(t) \Rightarrow \eta(t)$, $t \geq 0$ it follows that $\lambda_n(x) \xrightarrow{\sim} \lambda(x)$ under the condition that $\lambda_n(+\infty) \leq h$.

By letting h go to infinity, we obtain the assertion of Theorem 13.2.1. Necessity is evident.

Analogously, we prove the following assertion.

Theorem 13.2.2. *If* $\lim_{h\to\infty}\lim_{n\to\infty}\mathbf{P}\{\lambda_n(+\infty)\geq h\}=0$ *and* $\eta_n(t)\sim\zeta_n(t)$, $t\geq 0$, *where* $\zeta_n(t)=\int_0^\infty(1+tx)^{-1}d\mu_n(x)$, $\mu_n(x)$ *is some random function, nondecreasing and of bounded variation with probability 1, then for almost all values* x,

$$\lambda_n(x)\sim\mu_n(x).$$

Theorem 13.2.2 and formula (13.1.13) yield the following important assertion.

Theorem 13.2.3. *If for the matrix* Ξ_n *the conditions of Theorem 8.4.3 hold, then for almost all values* x, $\lambda_n(x)\sim\mu_n(x)$, $x\geq 0$, *where* $\mu_n(x)$ *is a nondecreasing random process, whose Stieltjes transform is*

$$\int_0^\infty(1+tx)d\mu_n(x)=\frac{d}{dt}\operatorname{Ln}\det[I+t\Lambda_{1n}]\{I'+t\Lambda_{2n}+tB_n(I+t\Lambda_{1n})^{-1}B_n'\}],$$

$$\Lambda_{1n}=\Big(\delta_{ij}\sum_{i\in Tjn}\nu_{ij}^2\Big),\qquad \Lambda_{2n}=\Big(\delta_{ij}\sum_{j\in K_{in}}\nu_{ij}^2\Big),\quad t\geq 0.$$

Let us show that convergence of finite-dimensional distributions of spectral functions $\lambda_n(x)$ implies the convergence of joint distributions of the eigenvalues.

Theorem 13.2.4. *If* $\lambda_n(x)\Rightarrow\lambda(x)$, $x\geq 0$, *where* $\lambda(x)$ *is a nondecreasing process and* $\lim_{h\to\infty}\overline{\operatorname{Lim}}_{n\to\infty}P\{\lambda_n(+\infty)\geq h\}=0$, *then for any integers* $k_1\geq k_2\geq\cdots\geq k_m\{\lambda_{k_{1n}},\lambda_{k_{2n}},\ldots,\lambda_{k_{mn}}\}\Rightarrow\{\lambda_{k_1},\lambda_{k_2},\ldots\lambda_{k_m}\}$, *where* $\lambda_1\geq\lambda_2\geq\ldots$ *are the moments of the steps of the process* $\lambda(x)$, m *is an integer.*

Proof. Obviously, $\Big[\sum_{i=1}^n\lambda_{in}^k\Big]^{1/k}=\lambda_{1n}+\varepsilon_1(n,h,k)$ under the condition that $\lambda_n(+\infty)\leq h$, where the random variable $\varepsilon_n(n,h,k)$ satisfies the relation $\lim_{k\to\infty}\operatorname{plim}_{n\to\infty}\varepsilon_1(n,h,k)=0$.

Clearly, $\lambda(x)$ is a step process, therefore for every fixed k,

$$\mathbf{P}\{\lambda_{1n}<x\}=\mathbf{P}\Big\{\Big[\sum_{i=1}^n\lambda_{in}^k\Big]^{1/k}<x/\lambda_n(+\infty)\leq h\Big\}\mathbf{P}\{\lambda_n(+\infty)\leq h\}$$

$$+\varepsilon_2(n,h,k)=\mathbf{P}\Big\{\Big[\int_0^h y^{k-1}d\lambda_n(y)\Big]^{1/k}<x/\lambda_n(+\infty)\leq h\Big\}$$

$$\times\mathbf{P}\{\lambda_n(+\infty)\leq h\}+\varepsilon_2(n,h,k)=\mathbf{P}\Big\{\Big[\int_0^h y^{k-1}d\lambda(y)\Big]^{1/k}$$

$$<x/\lambda(+\infty)\leq h\Big\}\mathbf{P}\{\lambda(+\infty)\leq h\}$$

$$+\varepsilon_3(n,h,k)=\mathbf{P}\{\lambda_1<x\}+\varepsilon_4(n,h,k),$$

where the variables $\varepsilon_i(n,h,k)$, $i=\overline{2,4}$ satisfy the relations

$$\lim_{h\to\infty}\operatorname{Lim}_{k\to\infty}\operatorname{Lim}_{n\to\infty}|\varepsilon_i(n,h,k)|=0,\quad i=\overline{2,4}.$$

Analogously,

$$\left[\sum_{i=1}^{n} \lambda_{in}^{k} - \left(\sum_{i=1}^{n} \lambda_{in}^{l}\right)^{k/l}\right]^{1/k} = \lambda_{2n} + \varepsilon_5(n, h, k, l),$$

where $\lambda_n(+\infty) \leq h$, $\lim_{k\to\infty} \overline{\lim}_{l\to\infty} \overline{\text{pLim}}_{n\to\infty} \varepsilon_5(n, h, k, l) = 0$. Hence,

$$\lim_{n\to\infty} \mathbf{P}\{\lambda_{1n} < x, \lambda_{2n} < y\} = \mathbf{P}\{\lambda_1 < x, \lambda_2 < y\}.$$

Note that the proved assertions hold when the eigenvalues λ_{in}, λ_i, $i = 1, 2, \ldots$ are multiple.

Continuing the arguments, we obtain

$$\lim_{n\to\infty} \mathbf{P}\{\lambda_{k_1 n} < x_1, \ldots, \lambda_{k_m n} < x_m\} = \mathbf{P}\{\lambda_{k_1} < x_1, \ldots, \lambda_{k_m} < x_m\}.$$

Theorem 13.2.4 is proved.

Analogously, we prove the following assertion.

Theorem 13.2.5. *If $\lambda_n(x) \sim \theta_n(x), x \geq 0$, where $\theta_n(x)$ are nondecreasing processes and $\lim_{h\to\infty} \overline{\text{Lim}}_{n\to\infty} \mathbf{P}\{\lambda_n(+\infty) \geq h\} = 0$, then for all numbers $k_1 \geq k_2 \geq \cdots \geq k_m\{\lambda_{k_1 n}, \lambda_{k_2 n}, \ldots, \lambda_{k_m n}\} \sim \{\theta_{k_1 n}, \theta_{k_2 n}, \ldots, \theta_{k_m n}\}$, where $\theta_{1n} \geq \theta_{2n} \geq \cdots$ are the moments of steps of the process, $\theta_n(x), m$ is an integer.*

From Theorems 13.2.3 and 13.2.5, we obtain the following assertion.

Corollary 13.2.1. *I) If in addition to the conditions of Theorem 13.2.3, $\lim_{n\to\infty} \operatorname{Tr} B_n B_n' = 0$, then for any integers $k_1, k_2, \ldots, k_m$,*

$$\{\lambda_{k_1 n}, \lambda_{k_2 n}, \ldots, \lambda_{k_m n}\} \sim \{\mu_{k_1 n}, \mu_{k_2 n}, \ldots, \mu_{k_m n}\},$$

where $\mu_{1n} \geq \cdots \geq \mu_{2n,n}$ are the variables $\sum_{i\in T_{jn}} \nu_{ij}^2, \sum_{i\in K_{jn}} \nu_{ij}^2$ arranged in nondecreasing order.

II) If in addition to the conditions of Theorem 13.2.3, the vector rows and vector columns of Ξ_n are asymptotically constant, then for any integers $k_1 > k_2 > \cdots k_m$,

$$\{\lambda_{k_1 n}, \ldots, \lambda_{k_m n}\} \sim \{\theta_{k_1 n}, \ldots, \theta_{k_m n}\},$$

where $\theta_{1n} \geq \cdots \geq \theta_{(n^2+n),n}$ are the variables ν_{ij}^2, $i, j = \overline{1, n}$, $\beta_{in}, i = \overline{1, n}$, arranged in increasing order. β_{in} are eigenvalues of the matrix B_n.

Theorem 13.2.6. *For every n let the random entries $\xi_{ij}^{(n)}$, $i, j = \overline{1, n}$ of the matrix $\Xi_n = (\xi_{ij}^{(n)})$ be independent; let the vector rows and vector columns of the matrix Ξ_n be asymptotically constant,*

$$\lim_{n\to\infty} \operatorname{Tr} B_n B_n' = 0,$$

$$\sum_{i,j=1}^{n} [1 - F_{ij}^{(n)}(z)] \Rightarrow K(z), \quad z \geq 0, \tag{13.2.2}$$

where $F_{ij}^{(n)}(z) = \mathbf{P}\{\nu_{ij}^2 < z\}$, the function $K(z)$ is continuous and bounded for all $z > 0$.

Then for all integers $k_1 > k_2 > \cdots > k_m > 0$ and the real numbers $x_m \geq x_{m-1} \geq \cdots \geq x_1 > 0$,

$$\lim_{n \to \infty} \mathbf{P}\{\lambda_{k_1 n} < x_1, \ldots, \lambda_{k_m n} < x_m\} = (-1)^m [(k_m - 1)!]^{-1}$$

$$\times \int_0^{x_1} \exp(-K(z_1)) \partial K(z_1) \Pi_{i=1}^{m-1} \{[(k_i - k_{i+1} - 1)!]^{-1}$$

$$\times \int_{z_i}^{x_{i+1}} [K(z_i) - K(z_{i+1})]^{k_i - k_{i+1} - 1} \partial K(z_{i+1})\} [K(z_m)]^{k_m - 1}. \tag{13.2.3}$$

Proof. Divide the intervals $(0, x_i)$, $i = \overline{1, m}$ by the points $0 = z_1^{(i)} < z_2^{(i)} < \ldots < z_{s_i}^{(i)} = x_i$, $z_{l+1}^{(i)} - z_l^{(i)} = \Delta z_i$, $l = \overline{1, s_i - 1}$. Let $\nu_p := \nu_{ij}^2$, $i, j = \overline{1, n}$, $p = \overline{1, n^2}$.

Let

$$\mathbf{P}\{\nu_{j_{n^2}} < z_{p_1 - 1}^{(1)}, \nu_{j_{n^2} - 1} < z_{p_1 - 1}^{(1)}, \ldots \nu_{j_{k_1} + 1} < z_{p_1 - 1}^{(1)},$$

$$\nu_{j_{k_1}} \in [z_{p_1}^{(1)}, \ z_{p_1 - 1}^{(1)}), \ z_{p_1}^{(1)} \leq \nu_{j_{k_1} - 1} < z_{p_2 - 1}^{(2)}, \ z_{p_1}^{(1)} \leq \nu_{j_{k_1} - 2}$$

$$< z_{p_2 - 1}^{(2)}, \ldots, z_{p_1}^{(1)} \leq \nu_{j_{k_2} + 1} < z_{p_2 - 1}^{(2)}, \ \nu_{j_{k_2}} \in [z_{p_2}^{(2)}, z_{p_2 - 1}^{(2)}),$$

$$\ldots, z_{p_m - 1}^{(m-1)} \leq \nu_{j_{k_m} + 1} < z_{p_m}^{(m)}, \ \nu_{j_{k_m}} \in [z_{p_m}^{(m)}, z_{p_m - 1}^{(m)}),$$

$$z_{p_m}^{(m)} \leq \nu_{j_{k_m} - 1}, \ldots, z_{p_m}^{(m)} \leq \nu_{j_1}\} = \mathbf{P}_{j_1 \ j_2 \ldots j_m}^{p_1 \ p_2 \ldots p_m},$$

where $j_1, \ldots, j_{n^2}$ is a transposition of $1, \ldots, n^2$ and p_i, $i = \overline{1, m}$ are integers.

Arrange the values ν_j $j = \overline{1, n^2}$ in increasing order $\mu_{1n} \geq \mu_{2n} \geq \cdots \geq \mu_{n^2 n}$. Using Corollary 13.2.1 (1), we obtain

$$\beta_n := \mathbf{P}\{\lambda_{k_1 n} < x_1, \ldots, \lambda_{k_m n} < x_m\} = \mathbf{P}\{\mu_{k_1 n} < x_1,$$

$$\ldots, \mu_{k_m n} < x_m\} + o(1) = \text{Lim}_{\Delta Z_l \to 0, l = \overline{1, m}} \sum \mathbf{P}_{j_1 \ j_2 \ldots j_{n^2}}^{p_1 \ p_2 \ldots p_m} + o(1),$$

where the summation is taken over all possible values of nonequal indices $j_{k_1}, \ldots, j_{k_m}$ and over all nonintersecting samples $\langle j_{k_1 - 1}, \ldots, j_{k_2 + 1} \rangle$, $\langle j_{k_m}, \ldots j_1 \rangle$ of numbers $1, 2, \ldots, n^2$ from which the numbers $j_{k_1}, \ldots, j_{k_m}$, $p_i = \overline{1, s_i}$, $i = \overline{1, m}$ are deleted.

It is obvious that

$$\beta_n = \mathrm{Lim}_{\Delta z_l \to 0,\, l=\overline{1,m}} \sum_{j_1,\ldots,j_{k_m}} \sum_{p_1=1}^{s_1} F_{j_{n2}}(Z_{p_1-1}^{(1)}$$

$$\ldots F_{j_{k_1+1}}(Z_{p_1}^{(1)}-1)[F_{j_{k_1}}(Z_{p_1}^{(1)}) - F_{j_{k_1}}(Z_{p_1}^{(1)}-1)]\Pi_{i=1}^{k-1}\Big\{ \sum_{p_{i+1}=1}^{s_i+1} T(Z_{p_i}^{(i)},$$

$$Z_{p_i+1}^{(i+1)}) \sum_{(j_{k_{i-1}},\ldots,j_{k_{i+1}-1})} [F_{j_{k_i}-1}(Z_{p_{i+1}-1}^{(i+1)})$$

$$- F_{j_{k_i}-1}(Z_{p_i}^{(i)})]\ldots[F_{j_{k_{i+1}}+1}(Z_{p_{i+1}-1}^{(i+1)}) - F_{j_{k_{i+1}}}(Z_{p_i}^{(i)})]$$

$$\times [F_{j_{k_{i+1}}}(Z_{p_{i+1}}^{(i+1)}) - F_{j_{k_{i+1}}}(Z_{p_{i+1}-1}^{(i+1)})]\Big\} \sum_{(j_{k_m-1},\ldots,j_1)}$$

$$\times [1 - F_{j_{k_m}-1}(Z_{p_m}^{(m)}]\ldots[1 - F_{j_1}(Z_{p_m}^{(m)})],$$

where $T(u,v) = \begin{cases} 1, & u \le v \\ 0, & u > v \end{cases}$ $F_i(u) = \mathbf{P}\{\nu_i < u\}$.

Consider the expression

$$\gamma_n := \sum_{(i_1,\ldots,i_m)} [F_{i_1}(x) - F_{i_2}(z)]\ldots[F_{im}(x) - F_{im}(z)]$$

$$= (m!)^{-1}\Big[\sum_{p=1}^{n^2}[F_p(x) - F_p(z)]\Big]^m + \theta_n(m),$$

where $\theta_n(m) = -\sum(p_1!\ldots p_{n^2}!)^{-1}[F_1(x) - F_1(z)]^{P_1}\ldots[F_{n^2}(x) - F_{n^2}(z)]^{P_{n^2}}$, summation is taken over all $p_i = \overline{1,n^2}$, $\sum_{i=1}^{n^2} p_i = m$, and among the numbers p_i, at least one is greater than or equals 2, $\langle i_1,\ldots,i_m\rangle$ is a sample of m numbers from $(1,2,\ldots,n^2)$.

After some simple transformations, we find

$$|\theta_n(m)| \le [(m-1)!]^{-1} \sup_{l=\overline{1,n^2}}[F_l(x) - F_l(z)]\Big[\sum_{p=1}^{n^2}[F_p(x) - F_p(z)\Big]^{m-1}.$$

Using this inequality, (13.2.2), and the and the fact that the values $\nu_i, i = \overline{1,n^2}$ are infinitesimal, we obtain $\lim_{n\to\infty}\theta_n(m) = 0$. Consequently,

$$\gamma_n = (m!)^{-1}[K(z) - K(x)]^m + O(1). \qquad (13.2.4)$$

Analogously,

$$F_{l_{n^2}}(x)\ldots F_{l_1}(x) = \exp(-K(x)) + \varepsilon_{l_{n^2}}\ldots l_1, \qquad (13.2.5)$$

where $(l_{n^2}, \ldots, l_1)$ is a transposition of the numbers $1, \ldots n^2$, and

$$\lim_{n \to \infty} \varepsilon_{l_{n^2} \ldots l_1}(n) = 0$$

uniformly by $l_{n^2, \ldots, l_1}$.

According to (13.2.4) and (13.2.5),

$$\beta_n = \mathrm{Lim}_{\Delta z_i \to 0, i=\overline{1,m}} \sum_{j_{k_1}, \ldots, j_{k_m}} \sum_{p_1=1}^{s_1} \exp[-K(z_{p_1-1}^{(1)})]$$

$$\times [F_{j k_1}(z_{p_1}^{(1)}) - F_{j k_1}(z_{p_1-1}^{(1)})]\Pi_{i=1}^{m-1} \sum_{p_{i+1}=1}^{s_{i+1}} [(k_i - k_{i+1} - 1)!]^{-1}$$

$$\times T(z_{p_i}^{(i)}, z_{p_{i+1}-1}^{(i+1)})[K(z_{p_i}^{(i)}) - K(z_{p_{i+1}-1}^{(i+1)})]^{k_i - k_{i+1} - 1}$$

$$\times [F_{j k_{i+1}}(z_{p_{i+1}}^{(i+1)}) - F_{j k_{i+1}}(z_{p_{i+1}}^{(i+1)})]][(k_m - 1)!]^{-1}$$

$$\times [K(z_{p_m}^{(m)})]^{k_m - 1} + o(1) = \mathrm{Lim}_{\Delta z_i \to 0, i=\overline{1,m}} \sum_{p_1=1}^{s_1} \exp[-K(z_{p_1-1}^{(1)})]$$

$$\times [K(z_{p_1-1}^{(1)}) - K(z_{p_1}^{(1)})]\Pi_{i=1}^{m-1}\Big\{ \sum_{p_{i+1}=1}^{s_i+1} [(k_i - k_{i+1} - 1)!]^{-1}$$

$$\times T(z_{p_i}^{(i)}, z_{p_{i+1}-1}^{(i+1)})[K(z_{p_i}^{(i)}) - K(z_{p_{i+1}-1}^{(i+1)})]^{k_i - k_{i+1} - 1}$$

$$\times [K(z_{p_{i+1}-1}^{(i+1)}) - K(z_{p_{i+1}}^{(i+1)})]\Big\}[(k_m - 1)!]^{-1}$$

$$\times [K(z_{p_m}^{(m)})]^{k_m - 1} + o(1).$$

Hence, since the function $K(z), z > 0$ is continuous and bounded, we come to the assertion of Theorem 13.2.6.

From formula (13.2.3) it follows that

$$\lim_{n \to \infty} \mathbf{P}\{\lambda_{kn} < x\} = -[(k-1)!]^{-1} \int_0^x \exp[-K(z)][K(z)]^{k-1} dK(z), \quad x > 0,$$

for every $k > m, 0 < x < y$,

$$\lim_{n \to \infty} \mathbf{P}\{\lambda_{kn} < x, \lambda_{mn} < y\} = [(m-1)!(k-m-1)!]^{-1}$$

$$\times \int_0^x \exp[-K(z_1)] dK(z_1)$$

$$\times \int_{z_1}^y [K(z_1) - K(z_2)]^{k-m-1}[K(z_2)]^{m-1} dK(z_2).$$

If function $K(z)$ is differentiable, then

$$\lim_{n \to \infty} \mathbf{P}\{\lambda_{kn} < x\} = \exp(-K(x)) \sum_{m=0}^{k-1} (m!)^{-1} K^m(x).$$

On the basis of Formula (13.2.3), we can cite different particular cases of convergence of the sums $\sum[1 - F_{ij}(x)]$ to a limiting function.

For symmetric random matrices Ξ_n, we shall study instead of $\lambda_n(x)$ the random functions of the form $\mu_n(x) = \sum_{i=1}^{n} F(x - \lambda_{in})\lambda_{in}^2$, where $\lambda_{1n} \geq \cdots \geq \lambda_{nn}$ are the eigenvalues of the matrix Ξ_n. The process $\mu_n(x)$ will have bounded variation if $\operatorname{Tr}\Xi_n^2 < \infty$.

For $\mu_n(x)$, we shall study the following Stieltjes transform,

$$\int (1 + itx)^{-1} d\mu_n(x) = d^2 \ln \det(I + it\Xi_n)/dt^2 := \eta_n(t).$$

From Theorems 13.2.2 and 13.2.4, the following assertions follow.

Corollary 13.2.2. *If* $\lim_{h\to\infty} \overline{\lim}_{n\to\infty} P\{\operatorname{Var} \mu_n(\cdot) \geq h\} = 0$, *then for* $\mu_n(x)$ $\simeq> \mu(x)$, *where* $\mu(x)$ *is some random function of bounded variation with probability 1, it is necessary and sufficient that* $\eta_n(t) \Rightarrow \eta(t)$, *where* $\eta(t)$ *is some random function.*

Corollary 13.2.3. *If* $\mu_n(x) \sim \theta_n(x)$, *where* $\theta_n(x)$ *are nondecreasing random processes and* $\overline{\lim}_{h\to\infty} \lim_{n\to\infty} P\{\operatorname{Var} \mu_n(\cdot) \geq h\} = 0$, *then for every integer* $k_1 \geq k_2 \geq \cdots \geq k_m > 0\{\lambda_{k_1 n}, \lambda_{k_2 n}, \ldots, \lambda_{k_m n}\} \sim \{\theta_{k_1 n}, \ldots \theta_{k_m n}\}$, *where* $\theta_{1n} \geq \theta_{2n} \geq \ldots$ *are moments of steps of process* $\theta_n(x)$.

Corollary 13.2.4. *If for the matrix* Ξ_n, *the conditions of Theorem 8.4.2 hold, then for almost all* x, $\mu_n(x) \sim \theta_n(x)$, *where* $\theta_n(x)$ *is a nondecreasing random process whose Stieltjes transform is*

$$\int (1 + itx)^{-1} d\theta_n(x) = \frac{d^2}{dt^2} \ln[\det(I + itB_n$$
$$+ t^2 \Lambda_n) \prod_{k=1}^{n} (1 + it\nu_{kk})],$$

where $\Lambda_n = \operatorname{diag}(\sum_{l \in T_p \cup K_p} \nu_{pl,p=\overline{1,n}}^2)$.

Corollary 13.2.5. *If in addition to the conditions of Corollary 13.2.4, the vector rows of the matrix* Ξ_n *are asymptotically constant, then for almost all* x, $\mu_n(x) \sim \gamma_n(x)$, *where* $\gamma_n(x) = \sum_{i=1}^{2(n^2+n)} F(x - \gamma_{in})\gamma_{in}^2$, $\gamma_{in} := \{\nu_{ij}^{(n)}, -\nu_{ij}^{(n)}, i > j, i,j = \overline{1,n}, \nu_{ii}^{(n)}, \beta_i^{(n)}, i = \overline{1,n}\}$, $\beta_i^{(n)}$ *are eigenvalues of the matrix* B_n.

Denote the function on the right side of formula (13.2.3) by

$$F(x_1, \ldots, x_m, k(\cdot)).$$

Theorem 13.2.7. *If for every* n, *the random variables* $\nu_{ij}^{(n)}, i \geq j, i,j = \overline{1,n}$ *are independent, the vector- rows of symmetric matrix* Ξ_n *are asymptotically* *constant* $\lim_{n\to\infty} \operatorname{Tr} B_n^2 = 0$, *and*

$$\sum_{i>j}[1 - \mathbf{P}\{|\nu_{ij}^{(n)}| < z\}] + \sum_i[1 - \mathbf{P}\{\nu_{ii}^{(n)} < z\}] \Rightarrow K_1(z),$$

$$\sum_{i>j}[1 - \mathbf{P}\{|\nu_{ij}^{(n)}| < z\}] + \sum_i[1 - \mathbf{P}\{-\nu_{ii}^{(n)} < z\}] \Rightarrow K_2(z),$$

where $K_1(z), K_2(z)$ *are continuous and bounded functions for every* $z > 0$, *then for all integers* $k_1 > k_2 > \cdots > k_m > 0$ *and the real numbers* $x_m \geq x_{m-1} \geq \cdots \geq x_1 > 0$,

$$\lim_{n\to\infty} \mathbf{P}\{\lambda_{k_1 n} < x_1, \ldots, \lambda_{k_m n} < x_m\}$$
$$= F(x_1, \ldots, x_m, K_1(\cdot)); \tag{13.2.6}$$
$$\lim_{n\to\infty} \mathbf{P}\{\lambda_{n-k_1,n} < -x_1, \ldots, \lambda_{n-k_m,n} < -x_m\}$$
$$= \sum_{l=0}^{m}(-1)^l \sum_{\langle i_1 \ldots i_l \rangle} F(x_{i_1}, \ldots, x_{i_l}, K_2(\cdot)), \tag{13.2.7}$$

where $\langle i_1, \ldots, i_l \rangle$ *is a sample of* l *numbers from* $1, 2, \ldots, m$, $\sum_{\langle i_0 \rangle} \equiv 1$.

Proof. It is evident that the limiting distributions of the eigenvalues $\lambda_{k_1 n}, \ldots,$ $\lambda_{k_m n}$ coincide with those of the corresponding members of the rank statistics $\mu_1^{(n)}, \mu_2^{(n)}, \ldots, \mu_{n(n+1)/2}^{(n)}$, constructed by the values $|\nu_{ij}|, \nu_{ii}, i > j, i, j = \overline{1, n}$. Thus, using the proof of Theorem 13.2.6, we obtain (13.2.6).

It is easy to show that

$$\mathbf{P}\{\lambda_{n-k_1,n} < -x_1, \ldots, \lambda_{n-k_m,n} < -x_m\}$$
$$= \mathbf{P}\{\theta_{k_1 n} > x_1, \theta_{k_2 n} > x_2, \ldots, \theta_{k_m n} > x_m\} + 0(1),$$

where $\theta_{k_1 n}, \ldots, \theta_{k_m n}$ are members of the rank statistics constructed by the values $|\nu_{ij}|, -\nu_{ii}, i > j, i, j = \overline{1, n}$.

Hence, (13.2.7) follows. Theorem 13.2.7 is proved.

From formulas (13.2.6) and (13.2.7), we may obtain limiting distributions for the maximal and minimal eigenvalues of the matrix Ξ_n,

$$\lim_{n\to\infty} \mathbf{P}\{\lambda_{1n} < x\} = -\int_0^x \exp(-K_1(z))dK_1(z), \quad x > 0,$$

$$\lim_{n\to\infty} \mathbf{P}\{\lambda_{n,n} < -x\} = 1 + \int_0^x \exp(-K_2(z))dK_2(z).$$

Now consider the Fredholm determinants of the beam of matrices $B_n + \lambda C_n$, where B_n is a positive-definite nonrandom matrix, and C_n is an antisymmetric random matrix. Obviously, the roots of the characteristic equation

$\det(B_n + \lambda C_n) = 0$ will be imaginary with zero real parts. Denote them by $i\lambda_{1n}, \ldots, i\lambda_{nn}$ where $\lambda_{1n} \geq \cdots \geq \lambda_{nn}$. Let $\gamma_n(x) = \sum_{i=1}^{n} F(x - \lambda_{in})\lambda_{in}^2$. Then

$$\int (1 + itx)^{-2} d\gamma_n(x) = d^2 \ln \det(B_n + tC_n)/dt^2. \tag{13.2.8}$$

By using Corollary 13.2.2, Theorem 9.1.3, and Corollary 9.1.2, we arrive at the following statements.

Theorem 13.2.8. *If the conditions of Corollary 9.1.2 hold, then $\gamma_n(x) \simeq>$ $\kappa(x)$, where $\kappa(x)$ is a random function having almost surely bounded variation, and the Stieltjes transform $\int (1 + itx)^{-2} d\kappa(x) = -d^2 \ln \eta(t)/dt^2, \eta(t)$ is a random function, given one-to-one by its moments*

$$\mathbf{E}\eta^{k_1}(t)\eta^{k_2}(t_2) \ldots \eta^{k_l}(t_l)\left(\frac{d^2}{dS_1^2}\eta(S_1)\right)^{l_1}$$

$$\times \left(\frac{d^2}{dS_2^2}\eta(S_2)\right)^{l_2} \ldots \left(\frac{d^2}{dS_p^2}\eta(S_p)\right)^{l_p}\left(\frac{d}{dq_1}\eta(q_1)\right)^{m_1}$$

$$\times \left(\frac{d}{dq_2}\eta(q_2)\right)^{m_2} \ldots \left(\frac{d}{dq_r}\eta(q_r)\right)^{m_2} = \mathbf{E}\prod_{i=1}^{p}\frac{d^2}{dS_i^2}$$

$$\times \prod_{i=1}^{r}\frac{d}{dq_i}\exp\left\{\iint_{0<z<y<1}[i\gamma\eta(z,y) + \int (e^{ix\eta(z,y)} - ix\eta(z,y)\right.$$

$$\times (1+x^2)^{-1} - 1)(1 + x^{-2})dG(x)]dzdy\},$$

$$\eta(z,y) = \sum_{i=1}^{l}t\sum_{\alpha=1}^{k_i}[\xi_\alpha^{(i)}(z)\eta_\alpha^{(i)}(z) - \xi_\alpha^{(i)}(y)\eta_\alpha^{(i)}(z)]$$

$$+ \sum_{j=1}^{p}s_j\sum_{\beta=1}^{l_j}[\bar{\xi}_\beta^{(j)}(z)\bar{\eta}_\beta^{(j)}(y) - \bar{\xi}_\beta^{(j)}(y)\bar{\eta}_\beta^{(j)}(z)]$$

$$+ \sum_{k=1}^{r}q_k\sum_{\gamma=1}^{m_k}[\tilde{\xi}_\gamma^{(k)}(z)\tilde{\eta}_\gamma^{(k)}(y) - \tilde{\xi}_\gamma^{(k)}(y)\tilde{\eta}^{(k)}(z)],$$

$$\xi_\alpha^{(i)}(z), \eta_\alpha^{(i)}(z), \bar{\xi}_\beta^{(j)}(z), \bar{\eta}_\beta^{(j)}(z), \tilde{\xi}_\gamma^{(k)}(z), \tilde{\eta}^{(k)}(y),$$

where $\alpha, \beta, i, j, k = 1, 2, \ldots$ are independent identically distributed Gaussian processes, with zero mean and covariance functions $r(t,s)$, k_i, l_i, m_i are positive integers, and t_i, s_i, q_i are real variables.

Theorem 13.2.9. *Let the conditions of Theorem 9.1.3 hold. Then in order that $\lambda_n(x) \simeq> \theta(x)$, where $\theta(x)$ is a random function of bounded variation* *whose Stieltjes transform is* $\int (1 + itx)^{-2} d\theta(x) = d^2 \ln \xi(t)/dt^2, \xi(t)$ *is a random function, given one-to-one by its moments*

$$\mathbf{E}\xi^{k_1}(t_1)\xi^{k_2}(t_2)\ldots \xi^{k_l}(t_l)(d^2\xi(s_1)/ds_1^2)^{l_1}$$

$$\times (d^2\xi(s_2)/ds_2^2)^{l_2}\ldots (d^2\xi(s_p)/ds_p^2)^{l_p}(d\xi(q_1)/dq_1)^{m_1}$$

$$\times (d\xi(q_2)/dq_2)^{m_2}\ldots (d\xi(q_r)/dq_r)^{m_r} = \mathbf{E}\prod_{i=1}^{p}(d^2/ds_i^2)$$

$$\times \prod_{i=1}^{r}(d/dq_i)\exp\{-0.5 \iint_{0<z<y<1} \eta^2(z,y)dzdy\},$$

and the random variables $\xi_{pl}^{(n)} a_{pl}^{(n)}$ *be infinitesimal, it is sufficient, and for symmetric variables* $\xi_{pl}^{(n)}$ *also necessary, that the Lindeberg condition hold: for every* $\tau > 0$,

$$\lim_{n\to\infty} \sum_{p>l,p,l=\overline{1,n}} \int_{|x|>\tau} x^2 d\mathbf{P}\{\xi_{pl}^{(n)} a_{pl}^{(n)} < x\} = 0.$$

Theorem 13.2.10 [50, p. 364]. *Let the conditions of Theorem 8.4.3 hold, and*

$$\lim_{n\to\infty} \mathbf{P}\{\sum_{i\in T_{jn}} \nu_{ij}^2 < x_j, j = \overline{1,s}\} = \prod_{j=1}^{s} F_j(x_j),$$

where $F_j(x), j = \overline{1,m}$ *are continuous functions,*

$$\mathrm{plim}_{n\to\infty} [\sum_{j=\overline{s,n}}\sum_{i\in T_{jn}} \nu_{ij}^2 + \sum_{j=1}^{n}\sum_{i\in K_{jn}} \nu_{ij}^2] = 0.$$

Then

$$\lim_{n\to\infty} \mathbf{P}\{\lambda_{k_1n} < x_1, \lambda_{k_2n} < x_2, \ldots, \lambda_{k_mn} < x_m\} = [(s - k_1)!$$

$$\times (k_{m-1})!]^{-1} \prod_{i=1}^{m-1}[(k_i - k_{i+1} - 1)!]^{-1} \int_0^{x_1}\int_{z_1}^{x_2}\cdots\int_{z_{m-1}}^{x_m} \text{per } c_s, \quad (13.2.9)$$

where $s \geq k_1 > k_2 > \cdots > k_m, 0 < x_1 \leq x_2 \leq \cdots \leq x_m, c_s = (c_{ij}), i,j = \overline{1,s}$ *are square matrices of the order* s,

$$c_{ij} = \begin{cases} F_j(z_2) - F_j(z_1), i = \overline{1, k_1 - k_2 - 1}, \\ F_j(z_{l+2}) - F_j(z_{l+1}), \\ \quad i = \overline{k_1 - k_{l+1} - l + 1, k_1 - k_{l+2} - l - 1}, l = \overline{1, m-2}, \\ 1 - F_j(z_m), i = \overline{k_1 - k_m - m + 2, k_1 - k_m}, \\ dF_j(z_i), i = \overline{k_1 - m + 1, k_1} \\ F_j(z_i), i = \overline{k_1 + 1, s}. \end{cases}$$

Proof. By using Theorem 13.2.6, we obtain

$$\lim_{n\to\infty} \mathbf{P}\{\lambda_{k_1 n} < x_1, \ldots, \lambda_{k_m n} < x_m\} = \sum_{[j_{k_1}\cdots j_{k_m}]}$$

$$\times \prod_{i=1}^{m-1} \sum_{(j_{k_i-1},\ldots,j_{k_{i+1}+1})} \sum_{(j_{k_m-1},\ldots,j_1)}$$

$$\times \int_0^{x_1} \int_{z_1}^{x_2} \cdots \int_{z_{m-1}}^{x_m} F_{js}(z_1)\ldots F_{j_{k_1+1}}(z_1) dF_{j_{k_1}}(z_1)$$

$$\times \prod_{i=1}^{m-1} \{[F_{j_{k_i-1}}(z_{i+1}) - F_{j_{k_i-1}}(z_i)]\ldots[F_{j_{k_{i+1}+1}}(z_{i+1}) - F_{j_{k_{i+1}+1}}(z_i)]$$

$$\times dF_{j_{k_i+1}}(z_{i+1})\}[1 - F_{j_{k_m-1}}(z_m)]\ldots[1 - F_{j_1}(z_m)].$$

(See the proof of Theorem 13.2.6.)

Hence we have formula (13.2.9). Theorem 13.2.10 is proved.

If the functions $F_j(z), j = \overline{1, m}$ are differentiable, then

$$\lim_{n\to\infty} \mathbf{P}\{\lambda_{k_1 n} < x_1, \lambda_{k_2 n} < x_2, \ldots, \lambda_{k_m n} < x_m\} = [(m - k_1)!$$

$$\times (k_m-1)!]^{-1} \prod_{i=1}^{m-1} [(k_i - k_{i+1} - 1)!]^{-1}$$

$$\times \int_0^{x_1} \int_{z_1}^{x_2} \cdots \int_{z_{m-1}}^{x_m} \operatorname{per} B_s dz_1 \ldots dz_m,$$

where $B_s = (b_{ij})$ are square matrices of the order s,

$$b_{ij} = \begin{cases} F_j(z_2) - F_j(z_1), i = \overline{1, k_1 - k_2 - 1}, \\ F_j(z_{l+2}) - F_j(z_{l+1}), \\ \quad i = \overline{k_1 - k_{l+1} - l + 1, k_1 - k_{l+2} - l + 1}, l = \overline{1, m - 2}, \\ 1 - F_j(z_m), i = \overline{k_1 - k_m - m + 2, k_1 - m}, \\ \dfrac{dF(z_i)}{dz_i}, i = \overline{k_1 - m + 1, k_1}, \\ F_j(z_i), i = \overline{k_1 + 1, s}. \end{cases}$$

§3 Fredholm Determinants of Nonsymmetric Random Matrices and Limit Theorems for Eigenvalues

As in §1 of this chapter, when the conditions of Theorem 8.2.1 for random matrices Ξ_n hold, we obtain

$$\det(I + z\Xi_n) \sim \det(I + zB_n) \prod_{i>j}(1 - z^2 \nu_{ij}^{(n)} \nu_{ji}^{(n)}) \prod_{i=1}^{n}(1 + z\nu_{ii}^{(n)}), \quad (13.3.1)$$

where z is a complex number.

If the conditions of Theorem 8.4.1 hold, then

$$\det(I + z\Xi_n) \sim \prod_{i=1}^{n}(1 + z\nu_{ii}^{(n)}) \det[\delta_{ij} - \delta_{ij} z^2 \sum_{p \in T_i \cup K_i} \nu_{pi}\nu_{ip} + zb_{ij}]. \quad (13.3.2)$$

Except for these relations we obtain the following:

$$\frac{d^s}{dt^s} \det(I + z\Xi_n) \sim \frac{d^s}{dt^s}\{\det(I + zB_n) \prod_{i>j}(1$$

$$- z^2\nu_{ij}^{(n)}\nu_{ji}^{(n)}) \prod_{i=1}^{n}(1 + z\nu_{ii}^{(n)}), \quad (13.3.3)$$

if the conditions of Theorem 8.2.1 hold and

$$\frac{d^s}{dt^s} \det(I + z\Xi_n) \sim \frac{d^s}{dt^s} \left\{ \prod_{i=1}^{n}(1 + z\nu_{ii}^{(n)}) \right.$$

$$\left. \times \det[\delta_{ij} - \delta_{ij} z^2 \sum_{p \in T_i \cup K_i} \nu_{pi}\nu_{ip} + zb_{ij}] \right\}, \quad (13.3.4)$$

if the conditions of Theorem 8.4.1 hold, where $s > 0$ is an arbitrary integer.

With the help of the limit theorems for Fredholm random determinants, we shall prove the limit theorems for eigenvalues of nonsymmetric random matrices.

Theorem 13.3.1. *Let the entries $\xi_{ij}^{(n)}$ of the random matrix Ξ_n satisfy the conditions of Theorem 8.2.1, let $\lambda_1, \ldots, \lambda_n$ be its eigenvalues arranged in increasing order of their moduli (or in increasing order of their arguments if the moduli of some eigenvalues coincide), $\beta_i, i = \overline{1,n}$ eigenvalues of matrix B_n. Then for any integers $k_1, k_2, \ldots, k_m \{\lambda_k, \lambda_{k_2}, \ldots, \lambda_{k_m}\} \sim \{\mu_{k_1}, \mu_{k_2}, \ldots, \mu_{k_m}\}$, where $\mu_1, \ldots, \mu_{n^2+n}$ are arranged in increasing order of their moduli (or arguments if the moduli of some random variables coincide) random variables $\beta_p, \nu_{pp}, p = \overline{1,n}, \sqrt{|\nu_{pl}\nu_{lp}|}i^{(1-\text{sign }\nu_{pl}\nu_{lp})/2} - \sqrt{|\nu_{pl}\nu_{lp}|}i^{(1-\text{sign }\nu_{pl}\nu_{lp})/2}, p \neq l.$*

Proof. Suppose that the moduli of every finite number of the first eigenvalues of the matrix Ξ_n do not coincide: $\underline{\lim}_{n\to\infty} \mathbf{P}\{\|\lambda_i| - |\lambda_j\| > 0\} = 1, i \neq j, i, j = \overline{1,n}, k$ is an integer. Then, using the formulas

$$\lambda_{1n} = \lim_{s\to\infty} [\text{Tr }\Xi_n^{s+1} / \text{Tr }\Xi_n^s],$$

$$\lambda_{2n} = \lim_{s\to\infty} [\text{Tr }\Xi_n^{s+1} - \lambda_{1n}^{s+1}][\text{Tr }\Xi_n^s - \lambda_{1n}^s]^{-1}$$

and the proof of Theorem 13.2.4, we come to the assertion of the theorem.

Let the moduli of some eigenvalues coincide. Consider the expression $\theta_s = \mu_{1n}^{-s} \sum_{i=1}^n \lambda_{in}^s$. On the basis of the proof of Theorem 8.2.1, it is easy to determine that for any integers $k_1, k_2, \ldots, k_s, p_1, \ldots, p_l$,

$$\{\operatorname{Tr} \Xi^{k_m}, m = \overline{1,s}, \mu_{p_i}, i = \overline{1,l}\} \sim \{\sum_i \mu_i^{k_m}, m = \overline{1,s}, \mu_{p_i}, i = \overline{1,l}\}.$$

Therefore, using the proof of Theorem 13.2.4, we obtain that for some sufficiently large s,

$$\varlimsup_{n \to \infty} \mathbf{P}\{h_2 \geq |\theta_{s'}| \geq h_1\} = 1 + 0(s'),$$

where h_1, h_2 are constants $(h_1 > 0, h_2 < \infty), 0(s) \to 0$ as $s \to \infty$. Hence, $\operatorname{plim}_{n \to \infty}[|\lambda_{1n}| - |\mu_{1n}|] = 0$. Find the argument of the eigenvalue λ_{1n}. Suppose that moduli of the first $2m$ eigenvalues of the matrix Ξ_n coincide and moduli of the rest of the eigenvalues are less than $|\lambda_{1n}|$. Denote the first $2m$ eigenvalues by $|\lambda_{1n}| \exp(i\varphi_s^{(n)}), s = \overline{1,2m}$. Then

$$\operatorname{Tr} \Xi_n^k |\mu_{1n}|^{-k} = \sum_{s=1}^{2m} \exp(ik\varphi_s) + \varepsilon_1(n,k)$$

$$\lim_{k \to \infty} \operatorname{plim}_{n \to \infty} \varepsilon_1(n,k) = 0.$$

Consider the expressions

$$\sum_{s=1}^n e^{ip\varphi_s^{(n)}} e^{ik\varphi_s^{(n)}} = \operatorname{Tr} \Xi_n^{k+p} |\mu_{1n}|^{-k-p} - \varepsilon_1(n, k+p). \qquad (13.3.5)$$

By multiplying (13.3.5) by the Fourier coefficients $c_p^{(k)}$ of the function $F(y)$, where $F(y) = 1$, if $y > 0$ and $F(y) = 0$, if $y \leq 0, y = z - x$ (z is fixed), and by summing over p, we obtain

$$\sum_{s=1}^{2m} e^{ik\varphi_s} F(z - \varphi_s) = \sum_{s=0}^N e^{ik\mu_s} F(z - \mu_s) + \varepsilon_2(n, k);$$

$$\lim_{k \to \infty} \operatorname{plim}_{n \to \infty} \varepsilon_2(n, k) = 0.$$

Choose the subsequence n' such that the distributions of the arguments of the first $2m$ eigenvalues of the matrix Ξ_n and the random variables $\mu_{in}, i = \overline{1,2m}$ converge to the distribution function of some random variables φ_i, μ_i. By passing to the limit in this inequality as n', we obtain

$$\sum_s e^{ik\varphi_s} F(z - \varphi_s) \approx \sum_s e^{ik\mu_s} F(z - \mu_s) + 0(k), 0(k) \to 0, k \to \infty.$$

Hence, if follows that $\varphi_s \approx \mu_s, s = 1, 2, \ldots$. Theorem 13.3.1 is proved.

If the conditions of Theorem 8.2.1 hold, then from Theorem 13.3.1 the statements follow that:

a) if $\lim_{n\to\infty}\sum_{i=1}^{n}|\beta_i| = 0$, then the eigenvalues of the matrix Ξ_n as $n \to \infty$ are located on the coordinate axes of the complex plane;

b) if $\lim_{n\to\infty}\sum_{i=1}^{n}|\beta_i| = 0, \mathrm{plim}_{n\to\infty}\sum_{i=1}^{n}|\nu_{ii}| = 0$, then the eigenvalues of the matrix Ξ_n as $n \to \infty$ are located on each axis of the complex plane symmetrically with respect to the origin;

c) if with probability 1, $\nu_{ij}\nu_{ji} \geq 0, i,j = \overline{1,n}$ and $\mathrm{plim}_{n\to\infty}\sum_{i=1}^{n}|\beta_i| = 0$, then $\mathrm{plim}_{n\to\infty}\mathrm{Im}\,\lambda_{in} = 0, i = \overline{1,n}$;

d) if with probability 1, $\nu_{ij}\nu_{ji} < 0, i,j = \overline{1,n}$ and $\mathrm{plim}_{n\to\infty}\sum_{i=1}^{n}(|\beta_i| + |\nu_{ii}|) = 0$, then $\mathrm{plim}_{n\to\infty}\mathrm{Re}\,\lambda_{in} = 0$. On the basis of Theorem 13.2.6, we can formulate the limit theorems for the distributions of the finite number of eigenvalues of the matrix Ξ_n.

§4 Fredholm Determinants of Random Linear Operators in the Hilbert Space

Let $(\Omega, B, \mathbf{P})$ be a probability space, H be the real separable Hilbert space, h the σ algebra of Borel sets in H, and (H, h) the measurable Hilbert space.

A mapping $H \times \Omega$ into H which for every fixed $x \in D$ is a measurable mapping $\{\Omega, B\}$ into $\{H, h\}$, i.e., for every $D' \in h\{\omega : A(x, w) \in D'\} \in B$ is called a random operator $A(\cdot, \omega)$, acting in H and defined on the set of nonrandom elements $D \in H$. Let D_H be a linear variety of random elements with values in the measurable space $\{H, h\}$ (see §1, Chapter 5). If for every $\xi \in D_h, \tilde{A}(\xi(\omega), \omega)$ is a random element with values in the measurable space $\{H, h\}$, then $\tilde{A}(\cdot, \omega)$ is a random operator with a random domain of definition. Its range is $R_{\tilde{A}} = \{\tilde{A}(\xi(\omega), \omega) : \xi(\omega) \in D_H\}$. An operator $\tilde{A}(\cdot, \omega)$ is called 'linear strong', if for every $\xi_1, \xi_2 \in D_H$ and the real numbers α, β such that $\alpha\xi_1 + \beta\xi_2 \in D_H, \tilde{A}(\alpha\xi_1 + \beta\alpha_2, \omega) = \alpha\tilde{A}(\xi_1, \omega) + \beta\tilde{A}(\xi_2, \omega)$ with probability 1 and 'linear weak', if for every $\xi_1, \xi_2 \in D_H$ and real numbers α, β such that $\alpha\xi_1 + \beta\xi_2 \in D_H, \tilde{A}(\alpha\xi_1 + \beta\xi_2, \omega) \approx \alpha\tilde{A}(\xi_1, \omega) + \beta\tilde{A}(\xi_2, \omega)$. Obviously, almost for all fixed ω a linear strong operator $\tilde{A}(\cdot, \omega)$ is a nonrandom one which operates in H. Therefore, many facts of the theory of linear operators in Hilbert space hold for $\tilde{A}(\cdot, \omega)$.

Since H is the separable Hilbert space, the operator norm of the bounded operator $\tilde{A}(\cdot, \omega)$ will be a random variable. For the bounded quite continuous operator $A(\cdot, \omega)$ we can choose its eigenvalues and eigenfunctions so that they are random numbers and random elements in (H, h), respectively.

Schur Lemma. *Let $\tilde{A}(\cdot, \omega)$ be a quite continuous bounded operator with the domain of definition $D_{\tilde{A}}$. Then there exists an orthonormal (for the general case, random) basis $\{\varphi_j, j = 1, 2, \ldots\}$, where the matrix of operator $\tilde{A}(\cdot, \omega)$ is triangular: $\tilde{A}\varphi_j = a_{j1}\varphi_1 + a_{j2}\varphi_2 + \cdots + a_{jj}\varphi_j, j = 1, 2, \ldots, a_{jj} = (A\varphi_j, \varphi_j) = \lambda_j(\tilde{A}), j = 1, 2, \ldots$, where $\lambda(\tilde{A})$ are random eigenvalues of the operator $\tilde{A}(\cdot, \omega)$.*

By a random Fredholm determinant of a kernel operator $\tilde{A}(\cdot,\omega)$, we mean the expression $\det(I - \mu\tilde{A}(\cdot,\omega)) = \prod_j(1 - \mu\lambda_j(\tilde{A}))$, where μ is an arbitrary complex number, $\lambda_j(\tilde{A})$ are random eigenvalues of operator $\tilde{A}(\cdot,\omega)$, and I is the identical operator.

Let $\tilde{A}_1(\cdot), \tilde{A}_2(\cdot)$ be kernel operators and $\varphi_n, n = 1, 2, \ldots$ an arbitrary random normalized basis of the space $D_{\tilde{A}}$, then with probability 1,

$$\det(I - \tilde{A}(\cdot,\omega)) = \lim_{n\to\infty} \det[\delta_{jk} - (\tilde{A}(\cdot,\omega)\varphi_j, \varphi_k)]_{j,k=1}^n;$$
$$\tag{13.4.1}$$

$$\det[(I - A_1(\cdot,\omega))(I - \tilde{A}_2(\cdot,\omega))] = \det(I - \tilde{A}_1(\cdot,\omega))\det(I - \tilde{A}_2(\cdot,\omega)).$$
$$\tag{13.4.2}$$

The proof of (13.4.1) and (13.4.2) follows from [103].

CHAPTER 14

THE SYSTEMS OF LINEAR ALGEBRAIC
EQUATIONS WITH RANDOM COEFFICIENTS

The simplest properties of solutions of linear algebraic systems with random coefficients are studied in this chapter. In some cases, the explicit formulas for solutions of these systems and their mean values and variances are found.

§1 The Systems of Normal Linear Algebraic Equations

By the system of linear random algebraic equations we mean the equality $\Xi \vec{x}(\omega) = \vec{\eta}(\omega)$, where $\Xi = (\xi_{ij})$ is a random matrix, $\vec{\eta}(\omega)$ is a random vector, and $\vec{x}(\omega)$ is a desired vector from some set D of random vectors whose dimension is the same as that of the vector $\vec{x}(\omega)$.

We call a system of equations $\Xi \vec{x}(\omega) = \vec{\eta}(\omega)$ normal if the entries of matrix Ξ or of vector $\vec{\eta}$ or if the entries of both of them are distributed according to the joint normal law.

The equation $\Xi \vec{x}(\omega) = \vec{\eta}(\omega)$ has a unique solution if Ξ_n is a square matrix and $\mathbf{P}\{\det \Xi = 0\} = 0$. If the square matrix Ξ_n and the vector $\vec{\eta}$ have the joint distribution density $p(z_n, x)$ then the distribution density of the solution of equations $\Xi \vec{x} = \vec{\eta}$ is equal to

$$\int p(z_n, z_n \vec{y})|\det z_n|dz_n, \qquad (14.1.1)$$

on the assumption that this integral exists.

If the vector $\vec{\eta}_n$ does not depend on the matrix Ξ_n and is distributed according to the nondegenerate normal law with parameters $\vec{a}_n, T_n$ and $\mathbf{P}\{\det \Xi_n = 0\} = 0$, then the distribution density of the solution $\vec{x}_n(\omega)$ of the system $\Xi_n \vec{x}_n(\omega) = \vec{\eta}_n(\omega)$ is equal to

$$p(\vec{y}_n) := (2\pi)^{-n/2} \det T_n^{-1/2} \mathbf{E} \det \Xi_n \exp\{-0.5$$
$$\times (T_n^{-1}(\Xi_n \vec{y}_n - \vec{a}_n), (\Xi_n \vec{y}_n - \vec{a}_n))\}. \qquad (14.1.2)$$

366

In this formula, we suppose the distribution of matrix Ξ_n to be such that the density $p(\overrightarrow{y}_n)$ exists. For example, we can require that

$$\mathbf{E}|\det \Xi_n| < \infty.$$

Let us obtain the distribution density of solutions of some systems of normal linear algebraic equations in an explicit form.

Theorem 14.1.1. *Let the vector η_n be normally distributed $N(0,1)$, the column vectors $\overrightarrow{\xi}_i = (\xi_{i1}, \ldots, \xi_{in}), i = \overline{1,n}$ of matrix Ξ_n be independent, do not depend on the vector $\overrightarrow{\eta}$, and be normally distributed $N(0, R_n)$ (the matrix R_n nondegenerate). Then the distribution density of the solution $\overrightarrow{x}_n$ of the system of equations $\Xi_n \overrightarrow{x}_n = \overrightarrow{\eta}_n$ is equal to*

$$p(\overrightarrow{y}_n) = \Gamma((n+1)/2) \det R_n^{1/2} \pi^{-(n+1)/2} [1 + (R_n \overrightarrow{y}_n, \overrightarrow{y}_n)]^{-(n+1)/2}. \quad (14.1.3)$$

Proof. From formula (14.1.2), it follows that

$$p(\overrightarrow{y}_n) = \int \det(X_n X_n')^{1/2} \exp\{-0.5 \operatorname{Tr}(X_n X_n' Y_n)$$
$$- 0.5 \operatorname{Tr}(X_n X_n' R_n^{-1})\} dX_n (2\pi)^{-(n+n^2/2)} \det R_n^{-n/2},$$

where $dX_n = \prod_{i,j=1}^n dx_{ij}, Y_n = (y_i y_j)$. After the change of variables $X_n = Z_n(Y_n + R_n^{-1})^{-1/2}, p(\overrightarrow{y}_n) = c \det(Y_n + R_n^{-1})^{-(n-1)/2} \det R_n^{-n/2}$, where c is a normalizing constant.

According to §1 in Chapter 6, $\det(Y_n + R_n^{-1}) = \det R_n^{-1}[1 + (R_n \overrightarrow{y}_n, \overrightarrow{y}_n)]$. Therefore, $p(\overrightarrow{y}_n) = c \det R_n^{1/2}[1 + (R_n \overrightarrow{y}_n, \overrightarrow{y}_n)]^{-(n+1)/2}$. We find the constant c from the condition $\int p(\overrightarrow{y}_n) d\overrightarrow{y}_n = 1$. Theorem 14.1.1 is proved.

By formula (14.1.3), we find

$$\int \cdots \int p(\overrightarrow{y}_n) \prod_{i=k+1}^n dy_i = p(\overrightarrow{y}_k). \quad (14.1.4)$$

In particular, $\int \cdots \int p(\overrightarrow{y}_n) \prod_{i=2}^n dy_i = p(y_1) = \pi^{-1}[1 + \sigma^2 y_n^2]^{-1}\sigma$. If the matrix Ξ in formula (14.1.2) has the distribution density $p(X_n)$, then using the generalized Wishart density (see §3, Chapter 1) for the distribution density of vector $x_n(\omega)$ we obtain

$$p(\overrightarrow{y}_n) = (2\pi)^{-n/2} \det T_n^{-1/2} c_{n,n} \int \exp\{-0.5(T_n^{-1}(\sqrt{Z_n} H$$
$$\times \overrightarrow{y}_n - a_n), (\sqrt{Z_n} H \overrightarrow{y}_n - \overrightarrow{a}_n))\} p(\sqrt{Z_n} H) \det \sqrt{Z_n} \mu(dH) dZ_n,$$

where Z_n is a nonnegative definite matrix, the integration is over all these matrices, $\mu(dH)$ is a normalized Haar measure on the group of orthogonal real matrices H, and $c_{nn} = \pi^{n^2/2 - n(n-1)/4}[\prod_{i=1}^{n} \Gamma((n+1-i)/2)]^{-1}$.

Let the system $(A_n + \Xi_n)\overrightarrow{x} = \overrightarrow{b} + \overrightarrow{\eta}$ be given, where A_n and $\overrightarrow{b}$ are the nonrandom matrix, and the vector $\Xi = (\xi_{ij})$ and $\overrightarrow{\eta} = (\eta_1, \ldots, \eta_n)$ is the random matrix and vector whose entries are independent and normally distributed $N(0,1)$. The distribution density of the vector $\overrightarrow{x}$ is equal to [see formula (14.1.2)]

$$p(\overrightarrow{y}) = (2\pi)^{-(n^2+n)/2} \int \exp\{-0.5(X\overrightarrow{y} - \overrightarrow{b}, X\overrightarrow{y} - \overrightarrow{b})\}$$
$$- 0.5 \operatorname{Tr}(X - A)(X - A)'\} |\det X| dX.$$

Hence,

$$p(\overrightarrow{y}) = (2\pi)^{-(n^2+n)/2} \int \exp\{-0.5 \operatorname{Tr} XX'(I + \overrightarrow{y}\,\overrightarrow{y}') + \operatorname{Tr} X(A'$$
$$+ \overrightarrow{y}\,\overrightarrow{b}') - 0.5(\overrightarrow{b}, \overrightarrow{b}) - 0.5 \operatorname{Tr} AA'\} |\det X| dX.$$

We write $(I + \overrightarrow{y}\,\overrightarrow{y}') = Q, (A' + \overrightarrow{y}\,\overrightarrow{b}') = \mathbf{P}, 0.5(\overrightarrow{b}, \overrightarrow{b}) - 0.5 \operatorname{Tr} AA' = \beta$. After the change of variables $X = Q^{-1/2}Y$,

$$p(\overrightarrow{y}) = (2\pi)^{-(n^2+n)/2} \int \exp\{-0.5 \operatorname{Tr} YY' + \operatorname{Tr} YQ^{-1/2}\mathbf{P} - \beta\}$$
$$\times |\det Y| dY \det Q^{-(n+1)/2} = (2\pi)^{-(n^2+n)/2} \int \exp\{-0.5 \operatorname{Tr}(Y$$
$$- Q^{-1/2}\mathbf{P})(Y \quad Q^{-1/2}\mathbf{P})' - \beta + 0.5 \operatorname{Tr} \mathbf{P}\mathbf{P}'Q^{-1}\}$$
$$\times |\det Y| dY \det Q^{-(n+1)/2}.$$

Hence,

$$p(\overrightarrow{y}) = (2\pi)^{-(n^2+n)/2} \int \exp\{-0.5 \operatorname{Tr} YY'\} |\det[Y + Q^{-1/2}\mathbf{P}]|$$
$$\times dY \exp[-\beta + 0.5 \operatorname{Tr} \mathbf{P}\mathbf{P}'Q^{-1}] \det Q^{-(n+1)/2}.$$

It follows from §3, Chapter 2, that $|\det(\Xi + Q^{-1/2}\mathbf{P})| \approx |\det(\Xi + \Lambda)|$, where Λ is a diagonal matrix of eigenvalues of the matrix $Q^{-1/2}\mathbf{P}\mathbf{P}'Q^{-1/2}$. We note that $\det Q = 1 + (\overrightarrow{y}, \overrightarrow{y})$. As a result, $p(\overrightarrow{y}) = (2\pi)^{-n/2}\mathbf{E}|\det \Xi + \Lambda| \exp(-\beta + 0.5 \operatorname{Tr} \mathbf{P}\mathbf{P}'Q^{-1})[1 + (\overrightarrow{y}, \overrightarrow{y})]^{-(n+1)/2}$.

Let us consider the systems of normal random algebraic equations with the symmetric matrix of the random coefficients. Let the system of equations $\Xi_n \overrightarrow{x} = \overrightarrow{\eta}$ be given, where Ξ_n is the random symmetric matrix which does not depend on the random normal $N(0, I)$ vector $\overrightarrow{\eta}$ and whose distribution density equals

$$(2\pi)^{-(n^2+n)/2}\exp(-0.5\,\mathrm{Tr}\,X^2).$$

The distribution density of vector $\overrightarrow{x}$ is equal to [see (14.1.2)]

$$p(\overrightarrow{y}) = (2\pi)^{-n^2/2-n}\int \exp[-\,\mathrm{Tr}\,X^2(I + \overrightarrow{y}\,\overrightarrow{y}')]|\det X|dX. \qquad (14.1.5)$$

The matrix $\overrightarrow{y}\,\overrightarrow{y}'$ has the eigenvalues $(\overrightarrow{y}, \overrightarrow{y}), 0, \ldots, 0$. Thus, (14.1.5) takes the form $p(\overrightarrow{y}) = (2\pi)^{-n^2/2-n}\int \exp[-\,\mathrm{Tr}\,X^2\Lambda]|\det X|dX, \Lambda = \mathrm{diag}(1 + \lambda,$ $1,\ldots,1)$ and is a matrix of order $n, \lambda = (\overrightarrow{y}, \overrightarrow{y})$. From this formula, we obtain

$$p(\overrightarrow{y}) = (2\pi)^{-n^2/2-n}(1+\lambda)^{-1/2}(2+\lambda)^{-(n+1)/2}$$

$$\times \int \exp(-\,\mathrm{Tr}\,X^2)\det[x_{ij}(1 + (2+\lambda)(1+\lambda)^{-1/2})\delta_{i1}\delta_{j1})]dX,$$

where $x_{ij} = x_{ji}, dX = \prod_{i\geq j} dx_{ij}$.

Let us consider the case, when Ξ is an upper triangular matrix and its entries $\xi_{ij}, i \leq j$ are independent and have a normal distribution $N(0, I)$, the vector $\overrightarrow{\eta}$ does not depend on the matrix Ξ and has a normal distribution $N(0, I)$. Using formula (14.1.2) for the distribution density of vector $\overrightarrow{x}$, we have

$$p(\overrightarrow{y}) = (2\pi)^{-n/2}\mathrm{E}\prod_{i=1}^{n}|\xi_{ii}|\exp\{-0.5\sum_{i,j=1}^{n}y_iy_j$$

$$\times \sum_{k=1}^{n}\xi_{ik}\xi_{jk}\} = (2\pi)^{-n/2}\prod_{i=1}^{n}[[1$$

$$+ \sum_{j=1}^{i-1}y_j^2]^{1/2} + y_i^2]^{-2}\prod_{i=1}^{n}[1 + \sum_{j=1}^{i-1}y_j^2]^{1/2}.$$

Similarly, we obtain the formula for $p(\overrightarrow{y})$ when ξ_{ij}, η_i can have different variances.

Let us find the distribution densities of the solution of systems of linear random algebraic equations whose entries are independent and distributed according to the symmetric stable law. Let the system of linear nonhomogeneous equations be given,

$$\Xi_{mn}\overrightarrow{x}_n = \overrightarrow{b}_m, \qquad (14.1.6)$$

where $\Xi_{mn} = (\xi_{ij}), i = \overline{1, m}, j = \overline{1, n}, m \leq n$ is a rectangular random matrix $\overrightarrow{x}_n = (x_1, \ldots, x_n), \overrightarrow{b}_m = (\eta_1, \ldots, \eta_n)$.

Suppose that the first components $x_k, k = \overline{1,m}$ of the vector $\overrightarrow{x}_n$ are unknown; we move the components $x_l, l = \overline{m+1, n}$ to the right-hand side and treat them as arbitrary constants. Then the system (14.1.6) takes the form $\Xi_{mn}\overrightarrow{x}_m = \overrightarrow{\gamma}_m$, where $\overrightarrow{\gamma}_m = \overrightarrow{b}_m - x_{m+1}\overrightarrow{\xi}_{m+1} - \cdots - x_n\overrightarrow{\xi}_n, \overrightarrow{\xi}_k = (\overrightarrow{\xi}_{1k}, \ldots, \xi_{mk}), k = \overline{m+1, n}$.

Theorem 14.1.2. *If the random variables $\xi_{ij}, \eta_k, i, k = \overline{1,m}, j = \overline{1,n}$ are independent indentically distributed according to a stable distribution law with the characteristic function $f(t, \alpha, c) = \exp(-c|t|^\alpha), 0 \le \alpha \le 2, c > 2$, then the random variables $x_k, k = \overline{1,m}$ are indentically distributed with the density function*

$$p(u; \alpha, \beta) = 2\beta^{-1}\int_0^\infty zp(zu\beta^{-1}; \alpha)p(z, \alpha)dz, \qquad (14.1.7)$$

where $p(z, \alpha)$ is the density of the postulated stable distribution, $\beta = (1 + |x_{m+1}|^\alpha + \cdots + |x_n|^\alpha)^{1/\alpha}$, and the ratios $x_k/x_l, k \ne l, k, l = \overline{1,m}$ have densities $p(u, \alpha, 1)$.

Proof. It is evident that $\mathbf{P}\{\det \Xi_{mm} = 0\} = 0$, by Cramer's formula with probability 1,

$$x_k = \left[\sum_{p=1}^m \eta_p\alpha_p - \sum_{l=m+1}^n x_l\left(\sum_{p=1}^n \xi_{pl}\alpha_p\right)\right]\left[\sum_{p=1}^n \xi_{pk}\alpha_p\right]^{-1}, \qquad (14.1.8)$$

where $\alpha_p = \Xi^{pk}(\sum_{p=1}^n |\Xi^{pk}|^\alpha)^{1/\alpha}$, and Ξ^{pk} is the cofactor of the entry ξ_{pk} of the matrix Ξ_{mm}.

The joint characteristic function of the numerator and the denominator in (14.1.8) has the form $\exp\{-c|t\beta|^\alpha\}\exp\{-c|\tau|^\alpha\}$. Hence, the density of the random variables x_k is equal to (14.1.7). Let us consider the ratios $x_k/x_l, k \ne l$. Using the Cramer's formula, we obtain $x_k/x_l = \sum_{p=1}^m \xi_{pl}t_{pl}/\sum_{p=1}^m \xi_{pk}t_{pl}$, where $t_{pl} = T^{pl}(\sum_{p=1}^m |T^{pl}|^\alpha)^{1/\alpha}, T^{pl}$ are the cofactors of the coefficients of the matrix Ξ_{mm}, the kth column vector of which is replaced by the vector $\overrightarrow{\gamma}_m$.

On the basis of the preceding theorem, we obtain the density of the random variable x_k/x_l. Theorem 14.1.2 is proved.

We note, that the explicit form of $p(u; \alpha, \beta)$ is known in the cases $p(u; 2, \beta) = \beta\pi^{-1}(u^2 + \beta^2)^{-1}, p(u; 1, \beta) = 2\beta\pi^{-2}[\ln|u| - \ln\beta][u^2 - \beta^2]^{-1}$.

§2 The Stochastic Method of Least Squares

Let us study the connection between the solutions of the random equations and the stochastic method of the least squares.

Let us consider the linear model of the multiple regression $\overrightarrow{y} = \Xi\overrightarrow{\theta} + \overrightarrow{\varepsilon}$, where $\overrightarrow{y} = (y_1, \ldots, y_n)$ is the vector of observation of some variable, $\overrightarrow{\varepsilon} =$

$(\varepsilon_1, \ldots, \varepsilon_n)$ is the vector of errors of observation, $\overrightarrow{\theta} = (\theta_1, \ldots, \theta_m)$ is the vector of unknown parameters, and Ξ is the random $(n \times m)$ matrix of the values of controllable variables and independent of the vector $\overrightarrow{\varepsilon}$.

The difference between this model and the generally accepted one is that the entries of the matrix of the planned variables Ξ are random values.

Suppose, that $\mathbf{E}\,\overrightarrow{\varepsilon} = 0, \mathbf{E}\,\overrightarrow{\varepsilon}\,\overrightarrow{\varepsilon}' = R$, R is a positive definite matrix. Then the estimation of the vector of parameters $\overrightarrow{\theta}$ by method of the least squares is equal to

$$\widehat{\overrightarrow{\theta}} = \lim_{\delta \downarrow 0} (\Xi' R^{-1} \Xi + I\delta)^{-1} \Xi' R^{-1} \overrightarrow{y}.$$

Consider the case, when $(\Xi' R^{-1} \Xi)^{-1}$ exists with probability 1. The estimation of $\overrightarrow{\theta}$ is equal to $\widehat{\overrightarrow{\theta}} = (\Xi' R^{-1} \Xi)^{-1} \Xi' R^{-1} \overrightarrow{y}$. This estimation will be unbiased, and its covariance matrix $K = \mathbf{E}(\Xi' R^{-1} \Xi)^{-1}$. Without loss of generality, we assume, that

$$K = \mathbf{E}(\Xi R_n^{-1} \Xi')^{-1}, \qquad (14.2.1)$$

where Ξ is a random $n \times m$ matrix.

Let us prove the following assertion.

Theorem 14.2.1. *Let the random matrix Ξ have the distribution density $p(x)$, $R_m > 0, m \geq n$, and the integral*

$$\int_{Z>0} Z^{-1} p(\sqrt{Z_n} H^{(n)} \sqrt{R_m}) \det Z^{(m-n-1)/2} dZ \mu(dH), \qquad (14.2.2)$$

be finite, where the integration is over the set N of nonnegative-definite matrices Z of nth order, dZ is the element of the Lebesgue measure of set N, and μ is the normalized Haar measure on the group G of orthogonal matrices of mth order, $H^{(n)} = (h_{ij}), i = \overline{1,n}, j = \overline{1,m}$.

Then $\|K\| < \infty$, and

$$K = c_{n,m} \det R_m^{n/2} \int_{Z>0} Z^{-1} p(\sqrt{Z} H^{(n)} \sqrt{R_m})$$
$$\times \det Z^{(m-n-1)/2} dZ \mu(dH) \qquad (14.2.3)$$

(constant $c_{n,m}$ is defined in Theorem 1.3.1).

Proof. It is obvious that $K = \int (X R_m^{-1} X')^{-1} p(X) dX$. After the change of variables $X = Y \sqrt{R_m}$, this integral takes the form $K = \int (YY')^{-1} p(Y\sqrt{R_m})$ $dY \det R_m^{n/2}$ (see the proof of Theorem 1.3.1). By using the generalized Wishart density, we obtain (14.2.3). If the integral is finite, then $\|K\| < \infty$. Theorem 14.2.1 is proved.

Corollary 14.2.1. *Let random matrix Ξ be given, $\operatorname{Tr} \Xi\Xi' \le c < \infty$ with probability 1, the bounded distribution density $p(X)$ of matrix Ξ exists, $R > 0$ and $m \ge n + 2$. Then $\|K\| < \infty$.*

Proof. From formula (14.2.2) it follows that

$$\int_{Z>0} \operatorname{Tr} Z^{-1} p(\sqrt{Z} H^{(n)} \sqrt{R_m}) \det Z^{(m-n-1)/2} dZ \mu(dH)$$

$$\le c_1 \int_{\operatorname{Tr} Z^2 \le c} \operatorname{Tr} Z^{-1} \det Z^{(m-n-1)/2} dZ$$

$$\le c_2 \int_{\operatorname{Tr} Z^2 \le c} \det Z^{-1/2} dZ$$

$$= c_3 \int_{\operatorname{Tr} XX' \le c} dX < \infty,$$

where X is a square matrix of order n.

Corollary 14.2.2 is proved.

Theorem 14.2.1 and Corollary 14.2.1 are valid for matrices $\mathbf{E}(\Xi' R^{-1} \Xi)^{-s}$, where $s \ge 1$ is an integer and

$$A = \left(\sum_{k=1}^{m} \int \xi_{ik}(x)\xi_{jk}(x) dF(x) \right)_{i,j=\overline{1,n}},$$

where $\xi_{ij}(x)$ are real measurable random processes, and $x \in (-\infty, \infty)$, $F(x)$ is some distribution function.

Note, that the matrix A is positive-semidefinite, and

$$\mathbf{E} \det A^{-s} \le \int \mathbf{E}\left\{ \det \left(\sum_{k=1}^{m} \xi_{ik}(x)\xi_{jk}(x) \right) \right\}^{-s} dF(x).$$

If the distribution density of a random matrix $(\sum_{k=1}^{m} \xi_{ik}(x)\xi_{jk}(x))$ exists for all $x \in (-\infty, \infty)$, then by using the method of proving Theorem 14.2.1, we obtain the condition under which $\mathbf{E} \det A^{-s}$ exists.

§3 Spectral Method for the Calculation of Moments of Inverse Random Matrices

In this section, the conditions of existence of mathematical expectation of inverse symmetric random matrices are given.

Theorem 14.3.1. *Let the boundary distribution density of the random real $n \times m$ matrix Ξ exist, $R_m > 0$, and for some $\varepsilon > 0$*

$$\sum_{i=1}^{n} \sup_{0 < \lambda_i < \varepsilon} \int_L \prod_{i \ne j} \lambda_j^{(m-n-1)/2} \lambda_i^{(m-n-2)/2}$$

$$\times p(U_n \sqrt{\Lambda} U_n' H^{(n)} \sqrt{R_m}) \prod_{i>j}(\lambda_i - \lambda_j) \mu(dH)\nu(dU) \prod d\lambda_j < \infty, \tag{14.3.1}$$

$$L = \{0 < \lambda_1 < \cdots < \lambda_n, U_{1i} > 0, i = \overline{1,n}\},$$

where U_n are orthogonal matrices of nth order, and $\nu(dU)$ is the normalized Haar measure on the group of matrices $U_n, \Lambda = (\lambda_i \delta_{ij})$.
   Then $\|K\| := \|\mathbf{E}(\Xi R_m^{-1} \Xi')^{-1}\| < \infty.$

Proof. Change the variables $Z = U_n \Lambda U_n'$ in the integral (14.2.2). Then

$$
\begin{aligned}
\text{Tr } K = & c_n c_{n,m} \det R_m^{n/2} \sum_{i=1}^n \int_{L, 0 < \lambda_i < \epsilon} \lambda_i^{-1} \prod_{j=1}^n \lambda_j^{(m-n-1)/2} \\
& \times p(U_n \sqrt{\Lambda_n} U_n' H^{(n)} \sqrt{R_m}) \mu(dH) \nu(dU) d\Lambda_n \\
& + c_n c_{n,m} \det R_m^{n/2} \sum_{i=1}^n \int_{L, \lambda_i \geq \epsilon} \lambda_i^{-1} \prod_{j=1}^n \lambda_j^{(m-n-1)/2} \\
& \times p(U_n \sqrt{\Lambda_n} U_n' H^{(n)} \sqrt{R_m}) \prod_{i>j} (\lambda_i - \lambda_j) \mu(dH) \nu(dU) d\Lambda_n.
\end{aligned}
$$

The constant c_n is defined in Theorem 3.1.1, $d\Lambda_n = \prod_{i=1}^n d\lambda_i$.
   The first sum in this expression is finite if the sum of (14.3.1) is finite. The second sum is equal to

$$
\sum_{i=1}^n \mathbf{E}\{\nu_i^{-1}/\nu_i \geq \epsilon\} \mathbf{P}\{\nu_i \geq \epsilon\} \leq n\epsilon,
$$

where $\nu_i, i = \overline{1,n}$ are the eigenvalues of the matrix $\Xi R_m \Xi'$. Theorem 14.3.1 is proved.

   Instead of the condition (14.3.1), we can use the following; for some $\epsilon > 0$,

$$
\begin{aligned}
\sup_{0 < \lambda < \epsilon} & \int_{0 < \lambda_2 < \cdots < \lambda_n, u_{1i} \geq 0, i = \overline{1,n}} \prod_{j=2}^n \lambda_j^{(m-n-1)/2} \lambda_1^{(m-n-1)/2} \\
& \times p(U_n \sqrt{\Lambda_n} U_n' H^{(n)} \sqrt{R_m}) \prod_{j>i} (\lambda_i - \lambda_j) \mu(dH) \nu(dU) \\
& \times \prod_{i=2}^n d\lambda_i < \infty. \quad\quad\quad (14.3.2)
\end{aligned}
$$

   A large number of examples of the existence of integrals (14.2.2), (14.3.1), and (14.3.2) is known. Let us consider the most important case when the entries of the matrix Ξ are distributed by the joint normal law.

Corollary 14.3.1. *If the entries of matrix Ξ are distributed according to joint nondegenerate normal law $R_m > 0$ and $m \geq n + 2$, then $\|K\| < \infty$.*

Proof. Since the entries of matrix Ξ are distributed according to joint nondegenerate normal law, then there exists $\delta > 0$, such that $p(x) \leq c\{-\delta \text{ Tr}(X - A)(X - A)'$, where $c > 0$ is an arbitrary constant, A is the $n \times m$ matrix. It is obvious, that there exists a constant $\delta_1 > 0$ such that $R - I\delta_1 > 0$. By using these unequalities, we obtain the following upper estimation for (14.3.2):

$$
\sup_{0 < \lambda_1 < \epsilon} c \int_{L_1} \prod_{j=2}^{n} \lambda_j^{(m-n-1)/2} \lambda_1^{(m-n-2)/2} \exp\{-\delta\delta_1
$$

$$
\times \sum_{i=1}^{n} \lambda_i - \delta \operatorname{Tr} AA' + \delta \operatorname{Tr}(U_n \sqrt{\Lambda} U_n' H^{(n)} \sqrt{R_m} A')
$$

$$
+ \delta \operatorname{Tr}(U_n \sqrt{\Lambda_n} U_n' H^{(n)} \sqrt{R_m} A')'\} \prod_{i>j} (\lambda_i - \lambda_j) \mu(dH)
$$

$$
\times \nu(dU) \prod_{i=2}^{n} d\lambda_i = \sup_{0 < \lambda_1 < \epsilon} c \int_{L_1} \prod_{i=2}^{n} \lambda_i^{(m-n-1)/2}
$$

$$
\times \lambda_1^{(m-n-2)/2} \exp\{-\delta\delta_1 \sum_{i=1}^{n} \lambda_i - \delta AA' + \delta \sum_{i=1}^{n} \sqrt{\lambda_i} b_i\}
$$

$$
\times \prod_{i>j} (\lambda_i - \lambda_j) \mu(dH) \nu(dU) \prod_{i=2}^{n} d\lambda_i,
$$

$$
L_1 = \{0 < \lambda_2, < \cdots < \lambda_n, \quad u_{1i} \geq 0, \quad i = \overline{1,n}\},
$$

where $\max_{i=\overline{1,n}} |b_i| \leq c_1 < \infty$.

Hence, since $\int \mu(dH)\nu(dU) = 1$ and $m \geq n + 2$, it follows that this integral is finite. Therefore the assertion of Corollary 14.3.1 holds.

Theorem 14.3.2. *If the entries of matrix Ξ are distributed according to the joint non-degenerate normal law, $R_m > 0$ and $m \geq n + 2s$, then*

$$
\|\mathbf{E}(\Xi R_m^{-1} \Xi')^{-s}\| < \infty,
$$

where $s \geq 1$ is an integer.

Let us perform the explicit calculation of matrix K. If the matrix Ξ has the distribution $p(XX')$, then formula (14.2.3) as $R_m = I$ has the form

$$
K = c_{n,m} \det R_m^{n/2} \int_{Z > 0} Z^{-1} p(Z) \det Z^{(m-n-1)/2} dZ. \tag{14.3.3}
$$

Let us calculate this integral for the case, when the entries of matrix Ξ are distributed according to the normal law.

Theorem 14.3.3. *If the random matrix Ξ has a distribution density $p(X) = (2\pi)^{-mn/2} \exp\{-0.5 \operatorname{Tr} XX'\}$ and $R = I, m \geq n+2$, then $K = (m-n-1)^{-1} I$.*

Proof. From §1, Chapter 3 it follows that $\Xi\Xi' = U\Lambda U'$ with probability 1, and the diagonal matrix Λ of the eigenvalues does not depend on the orthogonal matrix U; and $u_{1i} \geq 0, i = \overline{1,n}, \lambda_1 > \cdots > \lambda_n > 0$. Therefore,

$$K = \mathbf{E}(\Xi\Xi')^{-1} = \mathbf{E}(U\Lambda U')^{-1} = \mathbf{E}(U\mathbf{E}\Lambda^{-1})U' = I\delta.$$

$$\delta = (n!)^{-1}\Delta_{m,n} \int \cdots \int_{\lambda_i > 0} \lambda_1^{-1} \exp\{-0.5 \sum_{i=1}^{n} \lambda_i\}$$

$$\times \prod_{i=1}^{n} \lambda_i^{(m-n-1)/2} \prod_{i>j} |\lambda_i - \lambda_j| \prod_{i=1}^{n} d\lambda_i$$

$$= \Delta_{m,n} n^{-1} \int \cdots \int_{\lambda_1 > \cdots > \lambda_n > 0} \exp\{-0.5 \sum_{i=1}^{n} \lambda_i\}$$

$$\times \prod_{i=1}^{n} \lambda_i^{(m-n-3)/2} [\lambda_2 \ldots \lambda_n + \lambda_1 \lambda_3 \ldots \lambda_n + \cdots + \lambda_1$$

$$\ldots \lambda_{n-1}] \prod_{i>j} |\lambda_i - \lambda_j| \prod_{i=1}^{n} d\lambda_i$$

$$= \Delta_{m,n} \Delta_{m-2,n}^{-1} \mathbf{E} \frac{d}{dz} \det[Iz + HH']_{z=0}, \qquad (14.3.4)$$

where H is the $(m-2) \times n$ matrix, whose entries are independent and distributed according to the normal law $N(0,1)$,

$$\Delta_{m,n} = \pi^{n/2} [2^{mn/2} \prod_{i=1}^{n} \{\Gamma[(m+1-i)/2]\Gamma[(n+1-i)/2]\}]^{-1}.$$

It is obvious, that $\frac{d}{dz} \det[Iz + HH']_{z=0} = \sum_{k=1}^{n} \det \Xi_{(k)}\Xi'_{(k)}$, where the matrix Ξ_k is obtained from the matrix H by deleting the kth row vector. By using (2.1.3), we find

$$\mathbf{E}\frac{d}{dz} \det[Iz + HH']_{z=0} = n\mathbf{E} \det \Xi_{(1)}\Xi'_{(1)} = 2^{(n-1)m/2}$$

$$\times \prod_{i=1}^{n-1} \Gamma((m+1-i)/2)[2^{(m-2)(n-1)/2} \prod_{i=1}^{n-1} \Gamma((m-1-i)/2)]^{-1}.$$

By substituting this in (14.3.4), we obtain

$$\delta = 0.5\Gamma[(m-n-1)/2]\Gamma^{-1}[(m-n+1)/2] = (m-n-1)^{-1}.$$

Theorem 14.3.3 is proved.

We note that the assertion proved in this section can be extended to the random matrix Ξ whose column vectors have no continuous distribution or are nonrandom vectors. Let these be the last k column vectors of matrix Ξ. We denote them by $\overrightarrow{h}_s, s = \overline{1,k}$.

Then $\Xi\Xi' = \sum_{i=1}^{m-k} \overrightarrow{\xi}_i\,\overrightarrow{\xi}_i' + \sum_{s=1}^{k} \overrightarrow{h}_s\,\overrightarrow{h}_s'$, and the problem is reduced to the calculation of $\mathbf{E}(\tilde{\Xi}\tilde{\Xi}' + H)^{-1}$, where $\tilde{\Xi}$ is $(m-k) \times n$ matrix, $H = \sum_{s=1}^{k} \overrightarrow{h}_s\,\overrightarrow{h}_s'$. For these matrices we can apply all results of this section, because $\mathbf{E}(\tilde{\Xi}\tilde{\Xi}' + H)^{-1} \leq \mathbf{E}(\tilde{\Xi}\tilde{\Xi}')^{-1}$.

We obtain analogous assertions for the matrices $A = (\sum_{k=1}^{m} \int \xi_{ik}(x)\xi_{jk}(x)\, p(x)dx)_{i,j=1}^{n}$, where $\xi_i(x)$ are real measurable random processes, $x \in (-\infty, \infty)$, $p(x)$ is the distribution density.

It is not difficult to show, that $\mathbf{E}\,\mathrm{Tr}\,A^{-1} \leq n\mathbf{E}\lambda_1^{-1} \leq \int \mathbf{E}\tilde{\lambda}_1^{-1}(x)p(x)dx$, where $\tilde{\lambda}_1(x)$ is the minimal eigenvalue of the matrix

$$A(x) = \left(\sum_{k=1}^{m} \xi_{ik}(x)\xi_{jk}(x) \right).$$

On the basis of these formulas for matrix A, we can formulate assertions which are analogous to the preceding ones. For example, from Corollary 14.3.3 follows the following assertion: if the finite-dimensional distributions of random processes $\{\xi_{ik}(x), i = \overline{1,n}, k = \overline{1,m}\}$ are Gaussian nondegenerate and $m \geq n + 2s$, then $\|\mathbf{E}A^{-s}\| < \infty$.

CHAPTER 15

LIMIT THEOREMS FOR THE SOLUTION OF THE SYSTEMS OF LINEAR ALGEBRAIC EQUATIONS WITH RANDOM COEFFICIENTS

The distribution functions of the solutions of the systems of linear algebraic equations $\Xi_n \vec{x}_n = \vec{y}_n$, in general, have a cumbersome form; the order of these systems is large, therefore, the asymptotic behaviour of the solutions should be studied in increasing order of the system to infinity. A general form of the limit theorems for the solutions of the systems $\Xi_n \vec{x}_n = \vec{\eta}_n$ with independent random coefficients are given in this chapter.

§1 The Arctangent Law

Let us consider systems of linear algebraic equations $\Xi_n \vec{x}_n = \vec{\eta}_n$, where $\Xi_n = (\xi_{ij}^{(n)})$ is a real random square matrix of order n, and $\vec{\eta}_n = (\eta_1, \ldots, \eta_n)$ is a random vector. Let Ξ_{ij} be the cofactor of the entry ξ_{ij}. If $\det \Xi \neq 0$, then the solution of this system exists and equals $\vec{x}_n = \Xi_n^{-1} \vec{\eta}_n$; if $\det \Xi_n = 0$, then the solution cannot exist. Suppose, that the components $x_k^{(n)}$ of the vector x_n are equal to ∞, if $\det \Xi_n = 0$.

Theorem 15.1.1. *For every n, let the random variables ξ_{ij}, η_i, $i, j = \overline{1, n}$ be independent, $\mathbf{E}\xi_{ij} = 0$, $\mathbf{E}\eta_i = 0$, $\mathrm{Var}\,\xi_{ij} = \mathrm{Var}\,\eta_i = \sigma^2$, $0 < \sigma^2 < \infty$, $i, j = \overline{1, n}$, for some $\delta > 0 \sup_{n,i,j} \mathbf{E}[|\xi_{ij}|^{4+\delta} + |\eta_i|^{4+\delta}] < \infty$. Then for any $k \neq l$, $k, l = \overline{1, n}$,*

$$\lim_{n \to \infty} \mathbf{P}\{x_k^{(n)} < z\} = \lim_{n \to \infty} \mathbf{P}\{x_k^{(n)}/x_l^{(n)} < z\} = 2^{-1} + \pi^{-1} \arctan z.$$

Proof. Without loss of generality, we suppose that $\sigma^2 = 1$. It is obvious that $x_k^{(n)} = \sum_{i=1}^n \eta_i \alpha_{ik} \left[\sum_{i=1}^n \xi_{ik} \alpha_{ik}\right]^{-1}$, $\alpha_{ik} = \Xi_{ik} \left(\sum_{i=1}^n \Xi_{ik}^2\right)^{1/2}$, under the condition that $\det \Xi_n \neq 0$, $\sum_{i=1}^n \Xi_{ik}^2 \neq 0$. It follows from Theorem 6.3.2., that for any $\varepsilon > 0$,

$$\lim_{n \to \infty} \mathbf{P}\{\Xi_{ik}^2 \geq (n-1)! \exp(-\varepsilon c_n)\} = 1,$$

where c_n is an arbitrary sequence of the positive numbers, satisfying the condition $\lim_{n \to \infty} c_n / \ln n = \infty$. Therefore,

$$\lim_{n \to \infty} \mathbf{P}\{x_k < z\} = \lim_{n \to \infty} \mathbf{P}\{x_k < z/\Xi_{1k}^2 > \varepsilon\}. \tag{15.1.1}$$

By using Theorem 6.3.1 under the condition $\Xi_{1k}^2 > \varepsilon$, we obtain

$$x_k \sim \sum_{i=1}^n \nu_i \alpha_{ik} \left[\sum_{i=1}^n \nu_{ik} \alpha_{ik} \right]^{-1},$$

where ν_i, ν_{ik}, $i, k = 1, 2, \ldots$ are independent random variables which do not depend on Ξ_n, and the η_n have normal distribution $N(0, 1)$. From this, and from (15.1.1), we have

$$\lim_{n \to \infty} \mathbf{P}\{x_k^{(n)} < z\} = 1/2 + \pi^{-1} \operatorname{arctg} z.$$

Let us now consider the ratio $x_k x_l$, $k \neq l$. It is obvious that under the condition $\det \Xi_n \neq 0$, $\sum_{p=1}^n \Xi_{pl}^2 \neq 0$,

$$x_k^{(n)}/x_l^{(n)} = -\sum_{p=1}^n \xi_{pl}^{(n)} \alpha_{pl} / \sum_{p=1}^n \xi_{pk}^{(n)} \alpha_{pl}.$$

On the basis of this formula for $x_k^{(n)}/x_l^{(n)}$, the previous reasonings are valid. Theorem 15.1.1 is proved.

Theorem 15.1.2. *If the conditions of Theorem 15.1.1 are fulfilled, then for any finite k,*

$$\lim_{n \to \infty} \mathbf{P}\{x_{i_1}^{(n)} < y_1, \ldots, x_{i_k}^{(n)} < y_k\}$$
$$= \pi^{-(k+1)/2} \Gamma((k+1)/2)$$
$$\times \int_{-\infty}^{y_1} \cdots \int_{-\infty}^{y_k} (1 + z_1^2 + \cdots + z_k^2)^{-(k+1)/2} \prod_{i=1}^k dz_1,$$

where $i_1, \ldots, i_k$ are various arbitrary integer numbers from 1 to n.

Proof. Without loss of generality, we suppose that $i_1 = n - k, \ldots, i_k = n$. Otherwise, in the system of equations $\Xi \vec{x} = \vec{\eta}$, we carry out a corresponding transposition of the equations and their terms. By the Cramer formula and under the condition that $\det \Xi_n \neq 0$, we find

$$x_{n-k} = \det \Xi_n^{(n-k)} / \det \Xi_n, \ldots, x_n = \det \Xi_n^{(n)} / \det \Xi_n, \qquad (15.1.2)$$

where $\Xi_n^{(k)}$ is a matrix, obtained from the matrix Ξ_n by replacing the kth column vector by the vector $\vec{\eta}$.

Let us represent (15.1.2) in the form

$$x_{n-k} = \sum_{i=1}^{n} \eta_i \alpha_{in-k} \left[\sum_{i=1}^{n} \xi_{in-k} \alpha_{in-k} \right]^{-1}, \ldots, x_n$$

$$= \sum_{i=1}^{n} \eta_i \alpha_{in} \left[\sum_{i=1}^{n} \xi_{in} \alpha_{in} \right]^{-1},$$

where the variables α_{ik} are defined as in Theorem 15.1.1.

By using the method of proving the arctangent law, we obtain

$$\left\{ \sum_{i=1}^{n} \eta_i \alpha_{in-k}, \ldots, \sum_{i=1}^{n} \eta_i \alpha_{in} \right\} \sim \left\{ \sum_{i=1}^{n} \nu_i \alpha_{in-k}, \ldots, \sum_{i=1}^{n} \nu_i \alpha_{in} \right\},$$

where ν_i, $i = \overline{1, n}$ are independent random variables which do not depend on the matrix Ξ and are distributed by the normal law $N(0, 1)$.

Hence, $\mathbf{P}\{x_{n-k} < y_1, \ldots, x_n < y_k\} = \mathbf{P}\{\tilde{x}_{n-k} < y_1, \ldots, \tilde{x}_n < y_k\} + 0(1)$, where $\tilde{x}_s$, $s = n - k, n$ are components of the vector solution of the system $\Xi \tilde{\vec{x}} = \vec{\nu}$.

Decomposing the determinants of the matrices Ξ, $\widetilde{\Xi}^{(s)}$, $s = n - k, n$ by the nth column vector, where $\widetilde{\Xi}^{(s)}$ is the matrix obtained from the matrix Ξ_n by replacing the sth column vector by the vector $\vec{\nu}$, we have

$$\tilde{x}_{n-k} = \sum_{i=1}^{n} \xi_{in} \widetilde{\Xi}_{in}^{(n-k)} / \sum_{i=1}^{n} \xi_{in} \Xi_{in}, \ldots,$$

$$\tilde{x}_n = \sum_{i=1}^{n} \nu_i \Xi_{in}^{(n)} / \sum_{i=1}^{n} \xi_{in} \Xi_{in}.$$

We write

$$\beta_s = \left\{ \sum_{i=1}^{n} [\widetilde{\Xi}_{in}^{(s)}]^2 \right\}^{1/2}, \qquad \beta = \left[\sum_{i=1}^{n} \Xi_{in}^2 \right]^{1/2},$$

$$\alpha_{is} = \widetilde{\Xi}_{in}^{(s)} \beta_s^{-1}, \qquad \alpha_i = \Xi_{in} \beta^{-1}, \qquad s = \overline{n - k, n}.$$

If $\beta_s \neq 0$, $\beta \neq 0$, then

$$\tilde{x}_s = \sum_{i=1}^{n} \xi_{in} \alpha_{is} \left[\sum_{i=1}^{n} \xi_{in} \alpha_i \right]^{-1} (\beta_s / \beta), \qquad s = \overline{n - k, n}.$$

Analogously, as in the proof of the arctangent law,

$$\left\{ \sum_{i=1}^{n} \xi_{in}\alpha_{is}, \ s = \overline{n-k,n}, \sum_{i=1}^{n} \xi_{in}\alpha_i \right\}$$

$$\sim \left\{ \sum_{i=1}^{n} \nu_{in}\alpha_{is}, \ s = \overline{n-k,n}, \ \sum_{i=1}^{n} \nu_{in}\alpha_i \right\},$$

where ν_{in}, $i = \overline{1,n}$ are independent random variables not depending on the matrix Ξ_n and distributed by the normal law $N(0,1)$. It follows from the arctangent law, that $\lim_{n\to\infty} \mathbf{P}\{\tilde{x}_s < z\} = 1/2 + \pi^{-1} \operatorname{arctg} z$. Therefore, β_s/β cannot approach in probability to infinity. Hence,

$$\mathbf{P}\{x_{n-k} < y_1, \ldots, x_n < y_k\} = \mathbf{P}\{x'_{n-k} < y_1, \ldots, x'_n < y_k\} + 0(1),$$

where the x'_s are the components of the vector solution of the system $\tilde{\Xi}\vec{x}' = \vec{\nu}$, and the matrix $\tilde{\Xi}$ is obtained from the matrix Ξ by replacing the last column vector by the vector $(\nu_{1n}, \nu_{2n}, \ldots, \nu_{nn})$.

After k transformations, we obtain

$$\mathbf{P}\{x_{n-k} < y_1, \ldots, x_n < y_k\} = \mathbf{P}\{\hat{x}_{n-k} < y_1, \ldots, \hat{x}_n < y_k\} + 0(1),$$

where the $\hat{x}_s$ are the components of the vector solution of the system of equations $\hat{\Xi}\vec{x} = \vec{\nu}$, and the matrix $\hat{\Xi} = (\nu_{ij})$ is obtained from the Ξ_n by replacing the entries of the last k column vectors by independent random variables, which have a normal distribution $N(0,1)$

Let us multiply both sides of the equation $\hat{\Xi}\vec{x} = \vec{\nu}$ on the left by the orthogonal matrix T, the first row vector of which is collinear to the first column vector of matrix $\hat{\Xi}$ (since the matrix Ξ is asymptotically nondegenerate by virtue of (6.3.31), the probability that the matrix T exists, and tends to 1 as $n \to \infty$). After this transformation, the last k column vectors of the matrix $T\hat{\Xi}$ have the same distribution as that of the matrix $\hat{\Xi}$ and they do not depend on the rest of the column vectors of the matrix $T\hat{\Xi}$. The distribution of the vector $T\vec{\nu}$ does not change either.

The obtained system of equations has the following form,

$$\sum_{i=1}^{n} \nu_{i1}^2 \hat{x}_1 + \mu_{12}\hat{x}_2 + \cdots + \mu_{1n-k+1}\hat{x}_{n-k+1}$$

$$+ \nu_{1n-k}\hat{x}_{n-k} + \cdots + \nu_{1n}\hat{x}_n = \tilde{\nu}_1, \quad \Xi_{n-1}^{(1)}\hat{\vec{x}}_1 = \tilde{\vec{\nu}}_1,$$

where the matrix $\hat{\Xi}_{n-1}^{(1)}$ is obtained from the matrix $\hat{\Xi}_n$ by deleting the first column and the first row $\hat{\vec{x}}_1 = (\hat{x}_2, \ldots, \hat{x}_n)$, $\tilde{\vec{\nu}}_1 = (\tilde{\nu}_2, \ldots, \tilde{\nu}_n)$, $\mu_{12} = (\vec{t}_1, \vec{\nu}_2), \ldots,$ $\mu_{1n-k+1} = (\vec{t}_1, \nu_{n-k+1})$; $\vec{t}_1$ is the first row vector of the matrix T; $\vec{\nu}_i$ are the column vectors of the matrix $\hat{\Xi}_n$.

This system of equations is asymptotically nondegenerate, and $\hat{x}_1$ is expressed by the variables $\hat{x}_2, \ldots, \hat{x}_n$, which can be obtained from the system $\hat{\Xi}_{n-1}^{(1)} \hat{\vec{x}}_1 = \tilde{\vec{\nu}}_1$. If we perform such transformations consecutively $n - k + 1$ times, we obtain the following system of equations:

$$\nu_{n-k\,n-k}\hat{x}_{n-k} + \cdots + \nu_{n-k\,n}\hat{x}_n = \tilde{\nu}_{n-k},$$
$$\nu_{nn-k}\hat{x}_{n-k} + \cdots + \nu_{nn}\hat{x}_n = \tilde{\nu}_n. \tag{15.1.3}$$

The variables $\hat{x}_1, \ldots, \hat{x}_{n-k+1}$ are expressed in the form of some function by the variables $\hat{x}_{n-k}, \ldots, \hat{x}_n$. By virtue of (6.3.31), the system of equations for the variables $\hat{x}_1, \ldots, \hat{x}_n$ is asymptotically nondegenerate. The distribution of the solution of system (15.1.3) is obtained from Theorem 14.1.1. Theorem 15.1.2 is proved.

§2 Method of Integral Representations of the Solution of Systems of Linear Random Algebraic Equations

Let $\vec{x}_n$, $\vec{y}_n$, $n = 1, 2, \ldots$ be a sequence of random vectors of nth order. The expression $\vec{x}_n \sim \vec{y}_n$ denotes that the distribution functions of any arbitrary finite set of components of the vector $\vec{x}_n$ weakly converge to limit functions if and only if the distribution functions of the corresponding set of components of the vector $\vec{y}_n$ weakly converge to these limit functions. In what follows, we use the notations given in Chapter 8, in particular in Theorem 8.4.1.

Theorem 15.2.1.   *If for each n, the random vectors $(\xi_{ij}^{(n)}, \xi_{ji}^{(n)})$, $i \geq j$, $i, j = \overline{1, n}$ are independent and asymptotically constant,*

$$\lim_{h \to \infty} \overline{\lim_{n \to \infty}} \, \mathbf{P}\left\{ \left| \sum_{i=1}^n \nu_{ii}^{(n)} \right| + \sum_{i,j=1}^n \nu_{ij}^2 + \sum_{i=1}^n \eta_i^2 > h \right\} = 0; \tag{15.2.1}$$

$$\sup_n [|\mathrm{Tr}\, B_n| + \mathrm{Tr}\, B_n B_n'] < \infty; \tag{15.2.2}$$

$$\lim_{\varepsilon \downarrow 0} \overline{\lim_{n \to \infty}} \, \mathbf{P}\{|\det(I + \Xi_n)| < \varepsilon\} = 0, \tag{15.2.3}$$

then $\vec{x}_n \sim \vec{y}_n$, where $\vec{y}_n$ is the solution of the system of equations

$$\left(I - \mathrm{diag}\left\{ -\nu_{ii} + \sum_{p \in T_i \cup K_i} \nu_{pi}\nu_{ip}, \ i = \overline{1, n} \right\} + B_n \right) \vec{y}_n = \vec{\eta}_n.$$

Proof.   By the Cramer formula and under the condition that $\det(I + \Xi_n) \neq 0$, we have $x_k = \det \Xi_n^k \det(I + \Xi_n)^{-1}$, where Ξ_n^k is the matrix obtained from the matrix $(I + \Xi_n)$ by replacing the kth column vector by the vector $\vec{\eta}$.

Let us show that $x_k \sim y_k$. We transform the matrix Ξ_n^k to the form $I + C_n$, where all column vectors of the matrix C_n coincide with the column vectors of the matrix Ξ_n, except for the kth column vector, which is equal to $\vec{c} :=$ $(\eta_1, \eta_x, \ldots, \eta_{k-1}, \eta_k - 1, \eta_{k+1}, \ldots, \eta_n)$. According to Theorem 8.4.1,

$$\det(I + \Xi_n) \sim \prod_{i=1}^{n} (1 + \nu_{ii}) \det[\delta_{ij} - \delta_{ij} \sum_{p \in T_i \cup K_i} \nu_{pi} \nu_{ip} + b_{ij}]. \qquad (15.2.4)$$

By using (15.2.1) and (15.2.2) as in the proof of Theorem 8.4.1, we obtain the following relation:

$$\{\det(I + \Xi_n), \det(I + C_n)\}$$
$$\sim \{\det[\delta_{ij}(1 + \nu_{ii} - \sum_{p \in T_i \cup K_i} \nu_{pi}\nu_{ip}) + b_{ij}],$$
$$\det[\delta_{ij}(1 + \bar{\nu}_{ii} - \sum_{p \in T_i \cup K_i} \bar{\nu}_{pi}\bar{\nu}_{ip}) + \bar{b}_{ij}]\}, \qquad (15.2.5)$$

where $\bar{\nu}_{ij} = \nu_{ij}$ if $j \neq k$ and $\bar{\nu}_{ij} = 0$ if $j = k$; $\bar{b}_{ij} = b_{ij}$, $j \neq k$; $b_{ik} = \eta_i$, $i \neq k$; $b_{kk} = \eta_k - 1$.

It is obvious, that for $i \neq k$,

$$\left| \sum_{p \in T_i \cup K_i} \bar{\nu}_{pi}\bar{\nu}_{ip} - \sum_{p \in T_i \cup K_i} \nu_{pi}\nu_{ip} \right| \leq |\nu_{ki}\nu_{ik}|; \qquad (15.2.6)$$

moreover, by the construction of the sets T_i and K_i,

$$\operatorname*{plim}_{n \to \infty} \sum_{i \in T_k \cup K_k, i = \overline{1,n}} \nu_{ki}\nu_{ik} = 0. \qquad (15.2.7)$$

Lemma 15.2.1. *For each n, let the variables μ_{in}, $i = \overline{1,n}$ be independent, $\lim_{h \to \infty} \overline{\lim}_{n \to \infty} \mathbf{P}\left\{\sum_{i=1}^{n} \mu_{in}^2 + |\sum_{i=1}^{n} \mu_{in}| \geq h\right\} = 0$, let the matrix B_n satisfy the condition (15.2.2), and the random variables ε_{in}, $i = \overline{1,n}$ be independent and such that $\operatorname{plim}_{n \to \infty} \sum_{i=1}^{n} [\varepsilon_{in}^2 + \varepsilon_{in}] = 0$. Then $\det(\delta_{ij}(1 + \mu_{in} + \varepsilon_{in}) + b_{ij}) \sim \det(\delta_{ij}(1 + \mu_{in}) + b_{ij})$.*

The proof of Lemma 15.2.2 is analogous to that of Theorem 8.2.1. By using Lemma 15.2.1, (15.2.6), and (15.2.7), we obtain from (15.2.5)

$$\{\det(I + \Xi_n), \det(I + C_n)\} \sim \left\{ \det\left[\delta_{ij}\left(1 + \nu_{ii} - \sum_{p \in T_i \cup K_i} \nu_{pi}\nu_{ip}\right) + b_{ij} \right], \right.$$
$$\left. \det\left[\delta_{ij}\left\{1 + (1 - \delta_{ik})\left(\nu_{ii} - \sum_{p \in T_i \cup K_i} \nu_{pi}\nu_{ip}\right)\right\} + \bar{b}_{ij} \right] \right\}.$$

The assertion of Theorem 15.2.1 follows from this.

Corollary 15.2.1. *If under the conditions of Theorem 15.2.1, instead of (15.2.2),* $\lim_{n \to \infty}[|\operatorname{Tr} B_n| + \operatorname{Tr} B_n B_n'] = 0$ *holds, then*

$$x_k \sim \eta_k \left[1 + \nu_{kk} - \sum_{p \in T_k \cup K_k} \nu_{pk} \nu_{kp} \right]^{-1}. \qquad (15.2.8)$$

Note, that the random variables in the sum $\sum_{p \in t_k \cup K_k} \nu_{pk} \nu_{kp}$ are independent; therefore, all limit theorems for the sum of independent random variables can be applied for finding the limit distribution.

By passing to the limit as $n \to \infty$, instead of the system of equations $(I + \Xi_n)\vec{x}_n = \vec{\eta}_n$, we obtain some infinite-dimensional system whose solution is the infinite-dimensional vector x_∞, which is not obligatory of the finite length. To study the limit finite-dimensional distributions of vectors the $\vec{x}_n$ we consider the distributions of the scalar products $(\vec{x}_n, \vec{z}_n)$, where $(\vec{z}_n, \vec{z}_n) \leq 1$.

Corollary 15.2.2. *If the condition of Theorem 15.3.1 are fulfilled, then*

$$(\vec{x}_n, \vec{z}_n) \sim \det \left[\delta_{ij} \left(1 + \nu_{ii} - \sum_{p \in T_i \cup K_i} \nu_{pi} \nu_{ip} \right) + b_{ij} + \eta_i z_j \right]$$

$$\times \det \left[\delta_{ij} \left(1 + \nu_{ii} - \sum_{p \in T_i \cup K_i} \nu_{pi} \nu_{ip} \right) + b_{ij} \right]^{-1} - 1. \qquad (15.2.9)$$

Proof. By using the perturbation formulas for determinants (see §5, Chapter 2), we obtain

$$(\vec{x}_n, \vec{z}_n) = ((I + \Xi_n)^{-1} \vec{\eta}_n, \vec{z}_n) = \det(I + \Xi_n + \vec{\eta}_n \vec{z}_n) \det(I + \Xi_n)^{-1} - 1. \quad (15.2.10)$$

(in this formula, by $\vec{z}_n$ we mean a row vector). The rest of the proof coincides with that of Theorem 15.2.1.

§3 The Resolvent Method of Solutions of the Systems of Linear Random Algebraic Equations

Let $(I + \Xi_n)\vec{x} = \vec{\eta}$ be a system of linear random algebraic equations and $(I + \Xi_n)^{-1}$ exists with probability 1. Then $\vec{x} = R(1)\vec{\eta}$ with probability 1, where $R(z) = (\Xi + Iz)^{-1}$, and z is a complex number. Thus, to prove the limit theorem for distributions of components of vector $\vec{x}$, we need the limit theorems for resolvents of random matrices. Let us consider the system of equations of the form $(I + i\Xi_n)\vec{x} = \vec{\eta}$, where Ξ_n is a real random symmetric matrix of nth order, $\vec{\eta}$ is a random vector, systems $(I + \Xi_n \Xi_n')\vec{x} = \vec{\eta}$, $H_n \vec{x} = \vec{\eta}$, and H_n is a dominant matrix. $\vec{x} = R(1)\vec{\eta}$ is valid for $\vec{x}$, where $R(t) =$

$(I + it\Xi)^{-1}$, t is a real number. If the vector η has a normal distribution $N(0,1)$ and does not depend on the matrix Ξ_n, then

$$x_k \sim \eta_1[2^{-1}(d/dt)t^2 \operatorname{Re} r_{kk}(t)]^{1/2} + i\eta_2[-2^{-1}(d/dt)\operatorname{Re} r_{kk}(t)]\big|_{t=1}. \quad (15.3.1)$$

When the matrix Ξ is fixed, the random variables η_1 and η_2 have the joint normal distribution with mean zero and the coefficient of the correlation

$$\begin{aligned} \rho = &2^{-1}(d/dt)t \operatorname{Im} r_{kk}(t)[2^{-1}(d/dt)\operatorname{Re} r_{kk}(t)]^{-1/2} \\ &\times [-2^{-1}(d/dt)\operatorname{Re} r_{kk}(t)]^{-1/2}\big|_{t=1}, \end{aligned}$$

where $r_{kk}(t)$ are the entries of the matrix $(I + it\Xi_n)^{-1}$.

Formula (15.3.1) is obtained from the following relations

$$(\operatorname{Re} R(t))^2 = 0.5(d/dt)t^2 \operatorname{Re} R(t),$$
$$(\operatorname{Im} R(t))^2 = -0.5(d/dt)\operatorname{Re} R(t),$$
$$\operatorname{Re} R(t)\operatorname{Im} R(t) = \operatorname{Im} R^2(t)/2.$$

In the theorem proved below, we reduce the finding of the limit distribution of the random variables x_k to the solution of the following functional equation:

$$\begin{aligned} G_t(x, y, v) = \mathbf{P}\{&[1 + t^2\xi(G_t(\cdot, 1, \cdot), v)]^{-1} < x, \\ &- 2t\xi(G_t(\cdot, 1, \cdot), v) + t^2\xi(G_t(\cdot, 1, \cdot), v) \\ &\times [1 + t^2\xi(G_t(\cdot, 1, \cdot)]^{-2} < y\}, \end{aligned} \quad (15.3.2)$$

where $\xi(\cdot)$ is a random functional whose Laplace transformation of two-dimensional distributions is equal to

$$\begin{aligned} \mathbf{E}\exp\{&- s_1\xi(G_t(\cdot, 1, \cdot), v) - s_2\xi(G_t(1, \cdot, \cdot), v)\} \\ = \exp\Big\{&\int_0^\infty \Big[\int_0^\infty \exp\{-x[s_1y + s_2z]\} - 1)(1 + x^{-1}) \\ &\times dK(u, v, x)dG_t(y, z, u)\Big]du\Big\}, \end{aligned} \quad (15.3.3)$$

where $K(u, v, x)$ is a nondecreasing function of bounded variation with respect to x and continuous with respect to u, and v, $0 \le x \le \infty$, $0 \le u$, $v \le 1$.

Theorem 15.3.1. *For every n, let the random entries $\xi_{ij}^{(n)}$, $i \le j$, $i,j = \overline{1,n}$ of the symmetric matrix $\Xi_n = (\xi_{ij}^{(n)} - \alpha_{ij}^{(n)})$ be independent and infinitesimal $\alpha_{ij}^{(n)} = \int_{|x|<\tau} x d\mathbf{P}\{\xi_{ij}^{(n)} < x\}$, $\tau > 0$ be an arbitrary constant $K_n(u, v, z) \Rightarrow K(u, v, z)$, where $K_n(u, v, z) = n\int_0^z y(1+y)^{-1}d\mathbf{P}\{(\xi_{ij} - \alpha_{ij})^2 < y\}$ as $in^{-1} \le u < (i+1)n^{-1}$, $jn^{-1} \le v < (j+1)n^{-1}$, $K(u, v, z)$ be a nondecreasing function of bounded variation with respect to z and continuous with respect to the u and* *v functions,* $\lim_{n\to\infty} kn^{-1} = v$ *exists,* $0 \le v < 1$, *the vector* $\vec{\eta}$ *does not depend on* Ξ_n *and has normal distribution* $N(0,1)$, *the solution of equation (15.4.2) under a given* $K(u,v,z)$ *exist and it be uniquely in the class of functions* $G_t(x,y,v)$, *which are distribution function under all fixed t and* $v(0 \le v \le 1)$, *and such that the functions* $\int\int x^m y^l dG_t(x,y,v)$ *be analytical with respect to t for any positive integer m and l.*

Then $x_k \sim \eta_1(\xi_1 + \xi_2/2)^{1/2} + i\eta_2(-\xi_2/2)^{1/2}$ *is fulfilled for the components* x_k *of the vector solution of the system* $(I + i\Xi_n)\vec{x}_n = \vec{\eta}_n$, *where* η_1 *and* η_2 *are independent random variables, which do not depend on* ξ_1, ξ_2 *and are distributed normally* $N(0,1)$, *the distribution function of random variables* ξ_1, ξ_2 *is equal to the solution of the Eq. (15.3.2) as t = 1.*

Proof.   By the method of proving Theorem 9.3.2, we find $r_{kk}(t) = [1+itv_{kk}^{(n)}+ t^2(R_t^k\vec{\nu}_k, \vec{\nu}_k)]^{-1}$. Using the formulas

$$r_{kk}(t) = [1 + t^2 \sum_{i=1}^{n} r_{ii}(t)\nu_{ik}^2]^{-1} + \varepsilon_n',$$

$$(d/dt)r_{kk}(t) = -[1 + t^2 \sum_{i=1}^{n} r_{ii}(t)\nu_{ik}^2]^{-2}\left((d/dt)t^2 \sum_{i=1}^{n} r_{ii}(t)\nu_{ik}^2\right) + \varepsilon_n'',$$

$$\operatorname*{plim}_{n\to\infty}(|\varepsilon_n'| + |\varepsilon_n''|) = 0,$$

we obtain

$$\mathbf{P}\{\operatorname{Re} r_{kk}(t) < x, (d/dt)\operatorname{Re} r_{kk}(t) < y\}$$

$$= \mathbf{P}\{[1 + t^2 \operatorname{Re} r_{ii}(t)\nu_{ik}^2]^{-1} < x, (2t \sum_{i=1}^{n} \operatorname{Re} r_{ii}(t)$$

$$\times \nu_{ik}^2 + t^2 \sum_{i=1}^{n}(d/dt)\operatorname{Re} r_{ii}(t)\nu_{ik}^2)$$

$$\times [1 + t^2 \sum_{i=1}^{n} \operatorname{Re} r_{ii}(t)\nu_{ik}^2]^{-2} < y\} + 0(1),$$

$$\operatorname*{plim}_{n\to\infty} \operatorname{Im} r_{kk}(t) = 0.$$

Hence, we deduce Eq. (15.3.2) (see Theorem 9.3.2). Theorem 15.3.1 is proved.

Corollary 15.3.1.   *Let the system of random equations* $(I+i\Xi_n)\vec{x} = \vec{\eta}$ *be given, where the vector* $\vec{\eta}$ *does not depend on the random symmetric matrix* Ξ_n *and has normal distribution* $N(0,1)$, *for each n, the random entries* $\xi_{ij}^{(n)}$, $i \ge j$, $i,j = \overline{1,n}$ *of matrix* Ξ_n *be independent,* $\mathbf{E}\xi_{ij}^{(n)} = 0$, $n\mathbf{V}\xi_{ij}^{(n)} = \sigma^2$, *and the Lindeberg condition hold, for any* $\tau > 0$ *and* $k = \overline{1,n}$,

$$\lim_{n\to\infty} \sum_{i=1}^{n} \int_{|x|>\tau} x^2 d\mathbf{P}\{\xi_{ik} < x\} = 0.$$

Then for any k,

$$x_k \Rightarrow \eta_1 \{ 2[1 + (1 + 4\sigma^2)^{1/2}]^{-1}$$
$$- 8\sigma^2 (1 + 4\sigma^2)^{-1/2} (1 + (1 + 4\sigma^2)^{1/2})^{-2} \}^{1/2}$$
$$+ i\eta_2 [4\sigma^2 (1 + 4\sigma^2)^{-1/2} (1 + (1 + 4\sigma^2)^{1/2})^{-2}]^{1/2}.$$

According to the Wigner semicircle law when the conditions of Corollary 15.3.1 are fulfilled,

$$\operatorname*{plim}_{n \to \infty} r_{kk}(t) = [1 + (1 + 4t^2 \sigma^2)^{1/2}]^{-1} 2t^2 \sigma^2,$$

$$\operatorname*{plim}_{n \to \infty} (d/dt) r_{kk}(t) = (d/dt)([1 + (1 + 4t\sigma^2)^{1/2}]^{-1} 2t^2 \sigma^2).$$

These formulas imply the validity of the assertion of Corollary 15.3.1 . Let us study the solution of systems of the form $(I + B_n + \Xi_n)\vec{x} = \vec{\eta}$, where B_n is a positive-definite nonrandom matrix, and Ξ_n is a random Gramm matrix. For such equations $\vec{x} = R(1)\vec{\eta}$, where $R(t) = (I + tB_n + t\Xi_n)^{-1}$, $t \geq 0$ is a real number. If $\vec{\eta}$ is a random vector which has a normal distribution $N(0,1)$ and does not depend on matrix Ξ_n, then

$$x_k \approx \eta_k [(d/dt)(tr_{kk}(t))]_{t=1}^{1/2}. \tag{15.3.4}$$

To prove the theorem below, we need the following functional equation,

$$\begin{aligned} G_t(x,y,v) = \mathbf{P}\{ &[1 + t\xi_1(\{1 + t\xi_2 G_t(\cdot, 1, \cdot), \cdot)\}^{-1}, v)]^{-1} < x, \\ &- [1 + t\xi_1(\{1 + t\xi_2(G_t(\cdot, 1, \cdot), \cdot)\}^{-1}, v)]^{-2} \\ &\times [\xi_1(\{1 + t\xi_2(G_t(\cdot, 1, \cdot), \cdot)\}^{-1}, v) \\ &+ t\xi_1(-\{1 + t\xi_2(G_t(\cdot, 1, \cdot), \cdot)\}^{-2} \\ &\times \{\xi_2(G_t(\cdot, 1, \cdot), \cdot) + t\xi_2(G_t(1, \cdot, \cdot), \cdot)\}, v)] < y\}, \end{aligned} \tag{15.3.5}$$

where the random functional $\xi_1(\cdot, v)$ is given on the set L of the bounded positive random real functions $\eta(x, t, w)$ which do not depend on $\xi_1(\cdot, v)$, the random functional $\xi_2(\cdot, v)$ is given on the set of functions $G_t(x, y, v)$, which are the distribution functions with respect to x and y, and continuous with respect to v, $0 \leq x$, $y \leq 1$, $0 \leq t < \infty$, $0 \leq v \leq 1$.

The Laplace transformations of two-dimensional distributions of random functionals $\xi_1(\cdot, v)$ and $\xi_2(\cdot, v)$ have the form:

$$\mathbf{E}\exp\{-s_1\xi_2(G_t(\cdot, 1, \cdot), v) - s_2\xi_2(G_t, (1, \cdot, \cdot), v)\}$$

$$= \exp\left\{\int_0^1 \int_0^1 \int_0^1 \int_0^\infty (\exp\{-s_1 x + s_2 y)z\} - 1)(1 + z^{-1})\right.$$

$$\left. \times\, dK(v, u, z)dG(x, y, u)\, du\right\}, \tag{15.3.6}$$

$$\mathbf{E}\exp\{-s_1\xi_1(\eta_1(\cdot, t, \cdot), v) - s_2\xi_1(\eta_2(\cdot, t, \cdot), v)\}$$

$$= \exp\left\{\int_0^c \mathbf{E}\int_0^\infty (\exp\{-(s_1\eta_1(u, t, w) - s_n\eta_2(u, t, w))z\} - 1)\right.$$

$$\left. \times\, (1 + z^{-1})dK(v, u, z)\, du\}; \quad s_1, s_2 \geq 0;\right.$$

$$\eta_1(u, t, w), \eta_2(u, t, w) \in L,$$

$K(u, v, w)$ is a nondecreasing function of bounded variation with respect to x and continuous with respect to u and v, $0 \leq x < \infty$, $0 \leq u$, $v \leq 1$.

By the solutions of Eq. (15.3.5) we mean functions $G_t(x, y, v)$, which are the distribution functions with respect to x and y, $0 \leq x$, $y \leq 1$ continuous with respect to v, $0 \leq v \leq 1$ and such that for any integer $k > 0$, $l > 0$ the functions $\int_0^1 \int_0^1 x^k y^l dG_t(x, y, v)$ are analytical with respect to t, except for point zero.

Theorem 15.3.2. *For each n, let the random entries $\xi_{ij}^{(n)}$, $i = \overline{1,n}$, $j = \overline{1, m_n}$ of the matrix $\Xi = (\xi_{ij}^{(n)} - \alpha_{ij}^{(n)})$ be independent, infinitesimal, $\alpha_{ij}^{(n)} = \int_{|x|<\tau} x dP\{\xi_{ij}^{(n)} < x\}$, $\lim_{n\to\infty} m_n n^{-1} = c$, $0 < c < \infty$, $\tau > 0$ be an arbitrary constant, $K_n(u, v, z) \Rightarrow K(u, v, z)$, where $K_n(u, v, z) = n\int_0^z (1 + y)^{-1}y dP\{(\xi_{ij} - \alpha_{ij})^2 < y\}$ as $in^{-1} \leq u < (i+1)n^{-1}$, $jn^{-1} \leq v < (j+1)n^{-1}$. Let the function $K(u, v, z)$ be nondecreasing and have a bounded variation with respect to $z(0 \leq z < \infty)$ and continuous with respect to u and v in the domain $0 \leq u$, $v \leq 1$, $\lim kn^{-1} = v$, the solution of Eq. (15.3.5) exist and be unique. Then for the components of the vector solution of the system $(I+\Xi\Xi')\vec{x} = \vec{\eta}$,*

$$\lim_{n\to\infty} \mathbf{P}\{x_k^{(n)} < z\} = \int_0^1 \int_0^1 F(z(x + y)^{-1/2})dG_t(x, y, v)_{t=1}, \tag{15.3.7}$$

is valid, where $F(z) = (2\pi)^{-1}\int_{-\infty}^z \exp(-y^2/2)\, dy$.

Proof. By the method of proving Theorem 10.1.3, we obtain

$$(d/dt)r_{kk}(t) = (d/dt)\left[1 + t\sum_{l=1}^{m_n} |\nu_{kl}|^2\right.$$

$$\left. + \left\{1 + t\sum_{i=1}^n r_{ii}(t)|\nu_{il}|^2\right\}^{-1}\right]^{-1} + \varepsilon_n,$$

where $\lim_{n \to \infty} \mathbf{E} \, |\varepsilon_n| = 0$.

Equation (15.2.5) follows from this equation. Let us use formula (15.3.4) to complete the proof. From this formula, we obtain (15.3.7). Theorem 15.3.2 is proved.

Corollary 15.3.2. *If for the matrices Ξ_n the conditions of Corollary 10.2.1 are fulfilled and there exists the limit*

$$\lim_{n \to \infty} kn^{-1} = v, \ 0 \leq v \leq 1,$$

then

$$\lim_{n \to \infty} \mathbf{P}\{x_k^{(n)} < z\} = F(z(d/dt)tu(v,t))_{t=1},$$

where $u(v,t)$ is an analytical function with respect to t, which satisfies the equation $u(x,t) = [1 + t \int_0^c \sigma^2(x,y)[1 + t \int_0^1 u(z,t)\sigma^2(z,y)\,dz]^{-1}\,dy]^{-1}$.

The solution of this equation exists and is unique in the class of functions $u(x,t) \geq 0$, which are analytical with respect to t and continuous with respect to $x(0 \leq x \leq 1)$. This solution can be found by the method of successive approximations.

§4 Limit Theorems for Solutions of Difference Equations

We consider the integral representation method for solving typical differential equations of the second order. At first we approximate the solution of the differential equation by the solution of corresponding finite-difference boundary value problem and then, using the integral representation for random determinants (see Chapter 12), we obtain the exact solution as $n \to \infty$. Among the boundary value problems for typical differential equations, the main part belongs to the problem for equations of the second order. Let us consider one widespread boundary value problem. It is necessary to solve the equation

$$u''(x) - \xi(x)u(x) = \eta(x)$$

in the class of twice continually differentiable random functions defined on $[0,1]$ with the boundary conditions $u(0) = a, \ u(1) = b$, where ξ, η are continuous random processes defined on $[0,1]$, a,b are some random variables. Let us change the second derivative in this equation by the difference of the second order:

$$n^2(u_{k+1} - 2u_k + u_{k-1}), \quad u_k = u(k/n), \quad k = \overline{0, n-1}.$$

We obtain a system of differential equations

$$n^2(u_{j+1} - 2u_j + u_{j-1}) - \xi_j u_j = \eta_j, \quad j = \overline{1, n-1}, \tag{15.4.1}$$

where $\xi_j = \xi(j/n)$, $\eta_j = \eta(j/n)$, and the boundary conditions are replaced by $u_0 = a, \ u_n = b$.

Theorem 15.4.1. *If with probability 1,*

$$\xi(x) \geq 0, x \in [0,1], \quad \sup_{x \in [0,1]} \mathbf{E}[|\xi(x)| + |\eta(x)|] < \infty,$$

$$\lim_{h \to \infty} \sup_{|t'-t''|<h} [\mathbf{E}|\xi(t') - \xi(t'')| + \mathbf{E}|\eta(t') - \eta(t'')|] = 0,$$

then

$$u(z) = \left(\int_0^1 m(t)\, dt \right)^{-1} \left\{ b \int_0^z m(t)\, dt + a \int_z^1 m(t)\, dt \right.$$

$$+ \int_z^1 m(t)\, dt \int_0^x \left(\int_0^x m(t)\, dt \right) \eta(x)\, dx + \int_0^z m(t)\, dt$$

$$\times \left. \int_1^z \left(\int_x^1 m(t)\, dt \right) \eta(x)\, dx \right\}, \tag{15.4.2}$$

where $m(t) = \left[\mathbf{E} \exp \left\{ -\int_0^t \xi(y) w^2(y)\, dy/2 \right\} / \sigma \right]^{-2}$.

In this expression, by the stochastic integral, we understand the limit in mean of sums, constructed on random step functions, $w(y)$ is the Brownian motion process which does not depend on $\xi(y)$, σ is the minimal σ-algebra, with respect to which the random process $\xi(y)$ is measurable.

Proof. System (15.4.1) can be represented as follows, $A_n \vec{u}_n = \vec{\eta}_n$, where $A_n = [(-2 - n^{-2}\xi_i)\delta_{ij} + \delta_{ij-1} + \delta_{ij+1}]$; $\vec{u}_n = (u_1, \ldots, u_n)$, $\vec{\eta}_n = n^{-2}(\bar{\eta}_1, \ldots, \bar{\eta}_n)$, $\bar{\eta}_1 = \eta_1 - a$; $\bar{\eta}_k = \eta_k$, $k = \overline{2, n-1}$, $\bar{\eta}_n = \eta_n - b$. Then $u_k = \det A_n^{-1} \det A_n^{(k)}$, where (k) means that the kth row vector of matrix A_n is replaced by vector $\vec{\eta}_n$.

Let us write

$$B_k = \det[(-2 - n^{-2}\xi_j)\delta_{ij} + \delta_{ij-1} + \delta_{ij+1}]_{i,j=k+1}^n,$$

$$C_k = \det[(-2 - n^{-2}\xi_j)\delta_{ij} + \delta_{ij-1} + \delta_{ij+1}]_{i,j=1}^k.$$

Then

$$\det A_n^{(k)} = n^{-2} \sum_{i=1}^k (-1)^{i+k} \bar{\eta}_i C_{i-1} B_k + n^{-2} \sum_{i=k+1}^n (-1)^{i+k} \bar{\eta}_i C_{k-1} B_i,$$

and

$$u_k = \det A_n^{-1} \left(n^{-2} \sum_{i=1}^k \bar{C}_{i-1} B_k \eta_i \right.$$

$$\left. + n^{-2} \sum_{i=k+1}^n \eta_i \bar{C}_{k-1} B_i + a \bar{B}_k + b \bar{C}_{k-1}, \right. \tag{15.4.3}$$

where
$$\bar{C}_k = (-1)^k C_k, \quad \bar{B}_k = (-1)^{n-k} B_k.$$

Defining $u_n(z) = u_k$, $b_n(z) = n^{(-1)}\bar{B}_k$, $c_n(z) = n^{-1}\bar{C}_{k-1}$, $\eta_n(z) = \eta_k$, $k/n \leq z < (k+1)n$, and using the integral representation method for random determinants, we obtain

$$b_n(z) \Rightarrow \int_0^z m(t)\,dt,$$

$$c_n(z) \Rightarrow \int_z^1 m(t)\,dt$$

(see Chapter 12). But then, (15.4.3) implies that $u_n(z) \Rightarrow v(z)$, where $v(z)$ is the right part of (15.4.2). The solution of the equation

$$u''(x) - \xi(x)u(x) = \eta(x),$$
$$u(0) = a,$$
$$u(1) = b,$$

is unique with probability 1. It is easy to see that $v(x)$ satisfies this equation. Theorem 15.4.1 is proved.

CHAPTER 16

INTEGRAL EQUATIONS WITH
RANDOM DEGENERATE KERNELS

In this chapter, limit theorems for random determinants are applied to derive distributions of the solutions of some Fredholm integral equations of the second kind with random kernels $y(x,\omega) = \int_a^b K(x,z,\omega)y(z,\omega)dz + f(x,\omega)$, and of the first kind with random kernels $\int_a^b K_1(x,z,\omega)y(z,\omega)dz = f_1(x,\omega)$.

§1 Fredholm Integral Equations with Degenerate Random Kernels

Let us consider a Fredholm integral equation of the second kind,

$$y(x,\omega) = \lambda \int_0^1 K(x,z,\omega)y(z,\omega)dz + f(x,\omega), \qquad (16.1.1)$$

where the random functions $K(x,z,\omega)$ and $g(x,\omega)$, $0 \leq x \leq 1$, $0 \leq z \leq 1$ are measurable; defined on the probability space (Ω, σ, P), the integral $\int_0^1 \int_0^1 K^2(x,z,\omega)dx\,dz$ is finite with probability 1, and λ is a fixed number, which can also be a random variable.

The random kernel $K(x,z,\omega)$ of the integral equation is called degenerate, if it can be represented in the form of a finite sum of products of two random functions, one of which depends only on x, the other depends only on z:

$$K(x,z,\omega) = \sum_{i=1}^n \alpha_i(x,\omega)\beta_i(y,\omega), \quad 0 \leq x \leq 1, \qquad (16.1.2)$$

where $0 \leq y \leq 1$, $\alpha_i(x,\omega)$, $\beta_i(y,\omega)$ are measurable random processes, and with probability 1,

$$\int_0^1 \alpha_i^2(x,\omega)dx < \infty, \qquad \int_0^1 \beta_i^2(y,\omega)dy < \infty.$$

We need the following statement.

391

Lemma 16.1.1. *Let a system of linear algebraic equations* $(I-\lambda A(\omega))\vec{\gamma}(\omega) = \vec{b}(\omega)$, *where* $A(\omega) = (a_{ij}(\omega))$ *is a square matrix of the order* n, $a_{ij}(\omega) = \int_0^1 \alpha_i(x,\omega)\beta_i(x,\omega)dx$, $\vec{b}(\omega)$ *is a column vector with the components* $b_i(\omega) = \int_0^1 \beta_i(x,\omega)f(x,\omega)dx$, *have the unique strong solution* $\vec{\gamma}(\omega)$.

Then Equation (16.2.1), *with the kernel* (16.1.2), *has the unique strong solution*

$$y(x,\omega) = \lambda \sum_{i=1}^n \alpha_i(x,\omega)\gamma_i(\omega) + f(x,\omega). \qquad (16.1.3)$$

If the solution of the system $(I-\lambda A(\omega))\vec{\gamma}(\omega) = \vec{b}(\omega)$ is strong and unique, then $\vec{\gamma}(\omega) = (I - \lambda A(\omega))^{-1}b(\omega)$, with probability 1. By substituting $\vec{\gamma}(\omega)$ in (16.1.3), we find

$$y(x,\omega) = \lambda((I - \lambda A(\omega))^{-1}\vec{b}(\omega), \vec{\alpha}(x,\omega)) + f(x,\omega). \qquad (16.1.4)$$

According to §5, Chapter 2, if $\det(I - \lambda A(\omega)) \neq 0$ with probability 1, then

$$((I - \lambda A(\omega))^{-1}\vec{b}(\omega), \vec{\alpha}(x,\omega))$$
$$= \det(I - \lambda A(\omega) + \vec{b}'(\omega)\vec{\alpha}(x,\omega)) \det(I - \lambda A(\omega))^{-1} - 1.$$

By substituting this in (16.1.4), we get

$$y(x,\omega) = \lambda \det(I - \lambda A(\omega) + \vec{b}'(\omega)\vec{\alpha}(x,\omega)) \det(I - \lambda A(\omega))^{-1} - \lambda + f(x,\omega). \qquad (16.1.5)$$

Since $b_i(\omega) = \int_0^1 \beta_i(x,\omega)f(x,\omega)dx$, (16.1.5) is equivalent to the following formula,

$$y(x,\omega) = \int_0^1 \det[I - \lambda A(\omega) + \lambda \vec{\beta}'(y,\omega)\vec{\alpha}(x,\omega)] \det(I - \lambda A(\omega))^{-1} f(y,\omega)dy$$
$$- \int_0^1 f(y,\omega)dy - f(x,\omega). \qquad (16.1.6)$$

In formulas (16.1.5) (16.1.6), the matrices

$$A(\omega), \vec{b}'(\omega)\vec{\alpha}(x,\omega), \qquad \vec{\beta}'(y,\omega)\vec{\alpha}(x,\omega),$$

in general, are nonsymmetric, which impedes the investigation of the asymptotics of these formulas as $n \to \infty$. We use the integral representations for random determinants (see §4, Chapter 1):

$$\det(I + A) = [\mathbf{E}\exp\{-(A\vec{\eta}, \vec{\eta})\}]^{-2}; \qquad (16.1.7)$$
$$\det(I + \alpha_t C)^{-1} = \mathbf{E}\exp\{i\alpha_t((C - C')\vec{\xi}, \vec{\eta}) - \alpha_t(C\vec{\xi}, \vec{\xi}) - \alpha_t(C\vec{\eta}, \vec{\eta})\}, \qquad (16.1.8)$$

where A is nonnegative-definite, C is a square nonrandom real matrix of the order n, $\overrightarrow{\xi}$ and $\overrightarrow{\eta}$ are independent random vectors, distributed by the normal laws $N(0, 0.5I)$, $\alpha_t = t[q + |\operatorname{Tr} C| + \operatorname{Tr}(C + C')^2/4]^{-1}$, $0 \le t < 1$, $q \ge 1$.

We write $K_{n1} = \sum_{i=1}^{n} \alpha_i(x, \omega)\alpha_i(z, \omega)$,

$$K_{n2} = \sum_{i=1}^{n} \beta_i(x, \omega)\beta_i(z, \omega),$$

$$K_{n3} = \sum_{i=1}^{n} \alpha_i(x, \omega)\beta_i(z, \omega).$$

Theorem 16.1.1. *As $n \to \infty$, let*

$$\{K_{nl}(x, z, \omega), l = \overline{1,3}\} \Rightarrow \{p(x, z, \omega), q(x, z, \omega), r(x, z, \omega); x, z \in [0, 1]\},$$

$$\sup_n \mathbf{E} \sum_{i=1}^{n} \int_0^1 [\alpha_i^2(x, \omega) + \beta_i^2(x, \omega)]dx < \infty,$$

$$\sup_{0 \le x \le 1} \operatorname{Tr}(\overrightarrow{b}'(\omega)\overrightarrow{a}(x, \omega) + \overrightarrow{a}'(x, \omega)\overrightarrow{b}(\omega))^2 \le C,$$

$$\lim_{h \to 0} \lim_{n \to \infty} \sup_{\substack{|t_1 - t_2| \le h, \\ |s_1 - s_2| \le h}} \mathbf{E}|K_{nl}^2(t_1, s_1, \omega) - K_{nl}^2(t_2, s_2, \omega)| = 0,$$

where $p(x, y, \omega)$, $q(x, y, \omega)$, and $r(x, y, \omega)$ are measurable random functions, and the integrals

$$\int_0^1 \int_0^1 p(x, y, \omega)q(x, y, \omega)dx\,dy, \qquad \int_0^1 \int_0^1 r^2(x, y, \omega)dx\,dy$$

are finite with probability 1. Then,

$$y_n(x, \omega) \Rightarrow y(x, \omega), \quad x \in [0, 1],$$

where $y_n(x, \omega)$, $y(x, \omega)$ are strong solutions of the random equations

$$y_n(x, \omega) = \lambda_{nt}(\omega) \int_0^1 K_n(x, z, \omega)y_n(z, \omega)dz + f(x, \omega),$$

$$y(x, \omega) = \lambda_t(\omega) \int_0^1 r(x, z, \omega)y(z, \omega)dz + f(x, \omega),$$

$$\lambda_{nt}(\omega) = t[q + |\operatorname{Tr} A(\omega)| + 0.25 \operatorname{Tr}(A(\omega) + A'(\omega))^2 + C]^{-1},$$

$$\lambda_{nt}(\omega) = t[q + |\int_0^1 r(x, x)dx| + 0.5 \int_0^1 \int_0^1 p(x, y, \omega)q(x, y, \omega)dx\,dy$$

$$+ 0.5 \int_0^1 \int_0^1 r^2(x, y, \omega)dx\,dy + C]^{-1}.$$

Proof. By using formula (16.1.5), we find

$$y_n(x,\omega) = \eta_n(x,\omega) - 1 + f(x,\omega),$$

where

$$\eta(x,\omega) = \det[I - \lambda_{nt}(\omega)A(\omega) + \lambda_{nt}(\omega)\overrightarrow{b}'(\omega)\overrightarrow{a}(x,\omega)]\det(I - \lambda_{nt}(\omega)A(\omega))^{-1}.$$

We have to show that the finite-dimensional distributions of the functions $\eta_n(x,\omega)$ converge weakly. To show this, we consider the joint moments of the quantities $\Delta_{jn} = \det(I - \lambda_{nt}(\omega)A(\omega) + \lambda_{nt}(\omega)\overrightarrow{b}'(\omega)\overrightarrow{a}(x_j,\omega))$, $\gamma_n = \det(I - \lambda_{nt}(\omega)A(\omega))$, where $x_j, j = \overline{1,m}$ are arbitrary real numbers in $[0,1]$. Obviously, $\Delta_{jn}\gamma_n \geq \exp(t(1-t)^{-1})$. Consider the moments $\mathbf{E}\prod_{j=1}^m \Delta_{jn}^{-s_j}\gamma_n^{-s}$, where s_j, s are arbitrary positive integers. Without loss of generality, we find the moments $\mathbf{E}\Delta_{1n}^{-1}\gamma_n^{-1}$. By using (16.1.8), we obtain

$$\begin{aligned}
\mathbf{E}\Delta_{1n}^{-1}\gamma_n^{-1} = \; & \mathbf{E}\exp\{i\lambda_{nt}(\omega)((C_1 - C_1')\overrightarrow{\xi}_1, \overrightarrow{\eta}_1) - \lambda_{nt}(\omega)(C_1\overrightarrow{\xi}_1, \overrightarrow{\xi}_1) \\
& - \lambda_{nt}(\omega)(C_1\overrightarrow{\eta}_1, \overrightarrow{\eta}_1) + i\lambda_{nt}(\omega)((C_2 - C_2')\overrightarrow{\xi}_2, \overrightarrow{\eta}_2) \\
& - \lambda_{nt}(\omega)(C_2\overrightarrow{\xi}_2, \overrightarrow{\xi}_2) - \lambda_{nt}(\omega)(C_2\overrightarrow{\eta}_2, \overrightarrow{\eta}_2)\},
\end{aligned}$$

$$(16.1.9)$$

where $C_1 = \overrightarrow{b}'(\omega)\overrightarrow{a}(x,\omega) - A(\omega)$, $C_2 = -A(\omega)$, $\overrightarrow{\xi}_i, \overrightarrow{\eta}_i$ are independent random vectors with the distribution $N(0,0,5I)$.

It is obvious that $(C_2\overrightarrow{\eta}_2, \overrightarrow{\eta}_2) = \int_0^1 (\sum_{i=1}^n \alpha_i(x,\omega)\eta_i)(\sum_{i=1}^n \beta_i(x,\omega)\eta_i)dx$. Analogous formulas also hold for other bilinear and quadratic forms included in (16.1.9).

We need the following definition: the random process is called Gaussian process with zero mean and the random covariance function $r(x,y,\omega)$, if the characteristic function of its finite-dimensional distributions is given by the formula

$$\mathbf{E}\exp\{i\sum_{k=1}^m \xi(t_k)S_k\} = \mathbf{E}\exp\{-0.5(R_k(\omega)\overrightarrow{S}_m, \overrightarrow{S}_m)\},$$

where $R_k(\omega) = (r(t_i, t_j, \omega))_{i,j=\overline{1,k}}$.

The random processes $\nu(x) = \sum_{i=1}^n \alpha_i(x,\omega)\eta_i$, $\mu_n(x) = \sum_{i=1}^n \beta_i(x,\omega)\eta_i$ will be Gaussian with zero mean and random covariance functions

$$K_{n1}(x,y,\omega) = \sum_{i=1}^n \alpha_i(x,\omega)\alpha_i(y,\omega), \qquad K_{n2}(x,y,\omega) = \sum_{i=1}^n \beta_i(x,\omega)\beta_i(y,\omega).$$

According to the limit theorems for functionals of the integral type of random processes (see §1, Chapter 9), $(C_2\overrightarrow{\eta}_2, \overrightarrow{\eta}_2) \Rightarrow \int_0^1 \nu(x)\mu(x)dx$, where $\nu(x)$

and $\mu(x)$ are Gaussian processes with random covariance functions $R_{\nu(x)} = p(x, y, \omega)$, $R_{\mu(x)} = q(x, y, \omega)$, $R_{\nu(x),\mu(x)} = r(x, y, \omega)$.

Analogously, we can show that for any fixed $0 \le t \le 1$, $q \ge 1$, $\lambda_{nt}(\omega) \Rightarrow \lambda_t(\omega)$. By using this argument, we prove the convergence of the moments.

$$\lim_{n\to\infty} \mathbf{E} \prod_{j=1}^{m} \Delta_{jn}^{-s_j} \gamma_n^{-s} = \mathbf{E} \prod_{j=1}^{m} \Delta_j^{-s_j} \gamma^{-s}, \qquad (16.1.10)$$

where $\Delta_j(\omega)$ and $\gamma(\omega)$ are Fredholm determinants of the kernels $r(x, y, \omega) - \int_0^1 f(y, \omega) r(y, x, \omega) dy$, $r(x, y, \omega)$, respectively, with the parameters $\lambda_t(\omega)$.

We recall the definition of the Fredholm determinant of kernel $r(x, y)$, $0 \le x$, $y \le 1$ with parameter s. It is the expression $\prod_{i=1}^{\infty} (1 - s\lambda_i)$, where λ_i are eigenvalues of the kernel $r(x, y)$.

From (16.1.10) we obtain $\eta_n(x, \omega) \Rightarrow \Delta(x, \omega)/\gamma(\omega) - 1 + f(x, \omega)$. According to the theory of Fredholm integral equations, the solution of the equation $y(x, \omega) = \lambda_t(\omega) \int_0^1 r(x, z, \omega) y(z, \omega) dz + f(x, \omega)$ can be represented in the form $y(x, \omega) = f(x, \omega) + \lambda_t(\omega) \int_0^1 D(x, z, \lambda_i(\omega)) D^{-1}(\lambda_t(\omega)) f(z, \omega) dz$, where $D(\lambda_t(\omega))$ and $D(x, z, \lambda_t(\omega))$ are the Fredholm determinant and minor, respectively.

After simple transformations, we obtain the equality

$$\Delta(x, \omega)/\gamma(\omega) = \lambda_t(\omega) \int_0^1 [D(x, z, \lambda_t(\omega))/D(\lambda_t(\omega))] f(z, \omega) dz + 1.$$

Theorem 16.1.1 is proved.

By using the limit theorems for analytical random functions (see §1, Chapter 8), we derive the following corollaries from Theorem 16.1.1.

Corollary 16.1.1. *If in addition to the conditions of Theorem 16.1.1 such a λ exists, that for all n $\det(I - \lambda A(\omega)) \ne 0$ with probability 1 and the Fredholm determinant of the kernel $r(x, y, z)$ $D(\lambda) \ne 0$, then $y_n(x, \omega) \Rightarrow y(x, \omega)$, where $y_n(x, \omega)$, $y(x, \omega)$ are strong solutions of the random equations*

$$y_n(x, \omega) = \lambda \int_0^1 K_n(x, z, \omega) y_n(z, \omega) dz + f(x, \omega),$$

$$y_n(x, \omega) = \lambda \int_0^1 r(x, z, \omega) y(z, \omega) dz + f(x, \omega).$$

Corollary 16.1.2. *Let the random functions $\alpha_i(x, \omega)$, $\beta_i(x, \omega)$ be independent, identically distributed and measurable,*

$$r(x, y) := \mathbf{E} \alpha_i(x, \omega) \beta_i(y, \omega), \qquad \mathbf{E}[\alpha^2(x, \omega) + \beta_i^2(x, \omega)] < \infty,$$

the function $r(x,y)$ *be continuous in the square* $0 \le x \le 1$, $0 \le y \le 1$ *for all* $n \det(I - \lambda n^{-1} A(\omega)) \ne 0$ *with probability 1 and the Fredholm determinant* $D(\lambda)$ *of the kernel* $r(x,y)$, $0 \le x$, $y \le 1$ *not be equal to zero.*

Then $\operatorname{plim}_{n \to \infty} u_n(x,\omega) = u(x,\omega)$, $x \in [0,1]$, *where* $u_n(x,\omega)$, $u(x,\omega)$ *are strong solutions of the integral equations*

$$u_n(x,\omega) = \lambda n^{-1} \int_0^1 K_n(x,z,\omega) u_n(z,\omega) dz + f(x,\omega),$$

$$u(x,\omega) = \lambda \int_0^1 r(x,z) u(z,\omega) dz + f(x,\omega).$$

The proof follows form the fact that for the sums $K_n(x,z,\omega)$, under the conditions of this corollary, the law of large numbers holds:

$$\operatorname*{plim}_{n \to \infty} n^{-1} K_n(x,z,\omega) = r(x,z).$$

§2 Limit Theorem for Normalized Spectral Functions

The eigenvalues of the Fredholm integral random equation of the second kind,

$$y(x,\omega) = \lambda \int_0^1 K(x,z,\omega) y(z,\omega) dz + f(x,\omega), \quad x \in [0,1],$$

with the degenerate kernel $K(x,z,\omega) = \sum_{i=1}^n \alpha_i(x,\omega) \beta_i(z,\omega)$ are equal to eigenvalues of the random matrix

$$\left(\int_0^1 \alpha_i(x,\omega) \beta_j(x,\omega) dx \right)^{-1}_{i,j=\overline{1,n}}.$$

Let $\beta_i(x,\omega) = \alpha_i(x,\omega)$ with probability 1. Consider the spectral functions of random matrices

$$\Xi_n = \left(\int_0^1 \alpha_i(x,\omega) \alpha_j(x,\omega) dx \right).$$

We call a spectral function of the matrix Ξ_n the expression

$$\mu_n(x,\omega) = n^{-1} \sum_{i=1}^n F(x - \lambda_i(\omega)),$$

where

$$F(y) = \begin{cases} 1, & y > 0, \\ 0, & y \le 0, \end{cases}$$

and $\lambda_i(\omega)$ are the eigenvalues of the matrix Ξ_n. Let us find the conditions under which random spectral functions converge and the form of the limit spectral function as $n \to \infty$.

Theorem 16.2.1. *Let the random functions $\alpha_i(x,\omega)$ be independent, the limit*

$$\lim_{n\to\infty} n^{-1}\mathbf{E}\,\mathrm{Tr}(I + it\Xi_n)^{-1} = m(t)$$

exist, and $m(t)$ be continuous at zero.

Then at every point of the continuity of the function $\mu(x)$,

$$\lim_{n\to\infty} \mu_n(x,\omega) = \mu(x)$$

with probability 1, where $\mu(x)$ is a nonrandom distribution function, with the Stieltjes transform $\int(1+itx)^{-1}d\mu(x) = m(t)$.

Proof. Let

$$m_n(t,\omega) = \int (1+itx)^{-1}d\mu_n(x,\omega).$$

Let

$$m_n(t,\omega) - \mathbf{E}m_n(t,\omega) = \sum_{k=1}^{n}[\mathbf{E}\{m_n(t,\omega)/\sigma_{k-1}\} - \mathbf{E}\{m_n(t,\omega)/\sigma_k\}],$$

where σ_k is a minimal σ-algebra, with respect to which random functions are measurable,

$$\alpha_i(x,\omega), \quad i = \overline{k+1,n}.$$

The random variables $\gamma_k(\omega) = \mathbf{E}\{m_n(t,\omega)/\sigma_{k-1}\} - \mathbf{E}\{m_n(t,\omega)/\sigma_k\}$ are uncorrelated. By using inequalities for moments of the sums $\sum_k \gamma_k(\omega)$ (see §4, Chapter 5), we obtain

$$\mathbf{E}(\sum_{k=1}^{n}\gamma_k(\omega))^4 \leq C(\sum_{k=1}^{n}[\mathbf{E}\gamma_k^4(\omega)]^{1/2})^2,$$

where C is some constant.

We show that $|\gamma_k|^2 \leq C'$. To do this we consider the matrices Ξ_k, which are obtained from the matrices Ξ_n by substituting the entries of the kth column and the kth row of the matrix Ξ_n for zeros. Then

$$\gamma_k(\omega) = \mathbf{E}[m_n(t,\omega) - n^{-1}\,\mathrm{Tr}(I+it\Xi_k)^{-1}/\sigma_{k-1}]$$
$$- \mathbf{E}[m_n(t,\omega) - n^{-1}\,\mathrm{Tr}(I+it\Xi_k)^{-1}/\sigma_k].$$

According to §5, Chapter 2,

$$m_n(t,\omega) - n^{-1}\,\mathrm{Tr}(I+it\Xi_k)^{-1} = -n^{-1}t\frac{d}{dt}\ln\det[I + itR_t^k(\Xi - \Xi_k)],$$

where $R_t^k = (I+it\Xi_k)^{-1}$, r_{ei}^k are the entries of the matrix R_t^k, $R_t^k(\Xi - \Xi_k) = (\xi_{ij}\sigma_{ki} + b_{ij}\delta_{jk}(1-\delta_{kk}))$, $b_{ek} = \sum_{i\neq k}r_{ei}^k\xi_{ik}$, $\xi_{ik} = \int_0^1 \alpha_i(x,\omega)\alpha_k(x,\omega)dx$.

Hence, $|\gamma_k(\omega)| \leq n^{-1}C'$, where C' is a constant. Therefore, $\mathbf{E}|m_n(t,\omega) - \mathbf{E}m_n(t,\omega)| \leq n^{-2}C$. By using the limit theorems for random spectral functions (see Theorem 9.3.2), we obtain the assertion of Theorem 16.2.1.

In general, in the proved theorem, the degenerate kernels $K_n(x,y,\omega)$ can tend to infinity, and the normalized spectral functions can tend to the finite limit.

§3 Limit Theorems for Spectral Functions of Integral Equations with Random Kernels

Now we consider the spectral functions $\lambda_n(x,\omega) = \sum_{i=1}^n C_n^{-1}\lambda_i(\omega)F(x - C_n^{-1}\lambda_i(\omega))$, where C_n are constants, $\lambda_i(\omega)$ are the eigenvalues of the matrix $\Xi_n = (\int_0^1 \alpha_i(x,\omega)\alpha_j(x,\omega)dx)$.

Theorem 16.3.1. *As $n \to \infty$, let*

$$0.5 \sum_{i=1}^n C_n^{-2}\alpha_i(x,\omega)\alpha_i(y,\omega) \Rightarrow r(x,y,\omega), \quad x,y \in [0,1], \qquad (16.3.1)$$

where $r(x,y,\omega)$ is a measurable random function, the integral $\int_0^1 r(x,x)dx$ be finite with probability 1.

$$\lim_{h\to 0}\ \overline{\lim_{n\to\infty}}\ \sup_{\substack{|t_1-t_2|\le h \\ |s_1-s_2|\le h}} \mathbf{E}|\xi_n(t_1,s_1) - \xi_n(t_2,s_2)| = 0, \qquad (16.3.2)$$

where $\xi_n(t,s) = 0.5C_n^{-2}\sum_{i=1}^n \alpha_i(t,\omega)\alpha_i(s,\omega)$.

Then $\lambda_n(x,\omega) \Rightarrow \lambda(x,\omega)$, where $\lambda(x,\omega)$ is a random spectral function, with the Stieltjes transform

$$\int_0^1 (1+tx)^{-1}d\lambda(x,\omega) = \frac{d}{dt}\ln D(-t,\omega), \quad t \ge 0,$$

where $D(-t,\omega)$ is the Fredholm determinant of the kernel $r(x,y,\omega)$.

Proof. The Stieltjes transform is

$$\int_0^1 (1+tx)^{-1}d\lambda_n(x,\omega) = \frac{d}{dt}\ln\det(I + tC_n^{-1}\Xi_n).$$

By using the integral representation for random determinants, we obtain

$$\int_0^1 (1+tx)^{-1}d\lambda_n(x,\omega) = -2\frac{d}{dt}\ln\mathbf{E}[\exp\{-tC_n^{-1}(\Xi_n\overrightarrow{\eta},\overrightarrow{\eta})\}/\Xi_n]$$

$$= -2\frac{d}{dt}\ln\mathbf{E}[\exp\{-t\int_0^1 [\sum_{i=1}^n C_n^{-1}\alpha_i(x,\omega)\eta_i]^2dx\}/\Xi_n],$$

where $\overrightarrow{\eta}$ is a random vector, which does not depend on the matrix Ξ_n and has a normal distribution $N(0,0,5I)$.

We write $\eta_n(x,\omega) = C_n^{-1}\sum_{i=1}^n \alpha_i(x,\omega)\eta_i$. Then

$$\int_0^\infty (1+tx)^{-1}d\lambda_n(x,\omega) = -2\mathbf{E}[\exp\{-t\int_0^1 \eta_n^2(x,\omega)dx\}\int_0^1 \eta_n^2(x,\omega)dx/\Xi]$$

$$\times \mathbf{E}[\exp\{-t\int_0^1 \eta_n^2(x,\omega)dx\}/\Xi]\}^{-1}.$$

$$(16.3.3)$$

Evidently, $\eta_n(x,\omega)$ is a Gaussian process with zero mean and the random covariance function $0.5 C_n^{-2} \sum_{i=1}^{n} \alpha_i(x,\omega)\alpha_i(y,\omega) := r_n(x,y,\omega)$. We denote the numerator fraction (16.3.3.) by $\theta_{1n}(t)$ and the denominator by $\theta_{2n}(t)$. Obviously, $\theta_{1n}(t)$ and $\theta_{2n}(t)$ are bounded nonnegative random variables for $t \neq 0$. We consider the joint moments of finite-dimensional distributions $\theta_{1n}(t)$ and $\theta_{2n}(t)$:

$$\mathbf{E}\theta_{1n}^k(t)\theta_{2n}^l(t) = \mathbf{E}[\exp\{-t\sum_{i=1}^{k+l}\int_0^1 \eta_{ni}^2(x,\omega)dx\}\prod_{i=1}^{k}\eta_{ni}^2(x,\omega)dx], \quad (16.3.4)$$

where $\eta_{ni}^2(x,\omega)$ are Gaussian processes with zero mean and identical random covariance functions $r_n(x,y,\omega)$.

Under fixed minimal σ-algebra of events, generated by $r_n(x,y,\omega)$, the random processes $\eta_{ni}(x,\omega)$ are independent.

By using (16.3.1), (16.3.2), and limit theorems for the functionals of the integral type of random processes, we obtain

$$\lim_{n\to\infty}\mathbf{E}\theta_{1n}^k(t)\theta_{2n}^k(l) = \mathbf{E}\exp[-t\sum_{i=1}^{k+l}\int_0^1 \eta_i^2(x,\omega)dx]\prod_{i=1}^{k}\int_0^1 \eta_i^2(x,\omega)dx,$$
$$(16.3.5)$$

where $\eta_i(x,\omega)$ are Gaussian processes with zero mean and random covariance functions $r(x,y,\omega)$. Under fixed σ-algebra σ, processes $\eta_i(x,\omega)$ are independent. The formula (16.3.5) equals

$$\lim_{n\to\infty}\mathbf{E}\theta_{1n}^k(t)\theta_{2n}^l(t) = \mathbf{E}[\mathbf{E}[\exp\{-t\int_0^1 \eta^2(x,\omega)dx\}\int_0^1 \eta^2(x,\omega)dx/r(\cdot,\cdot,\omega)]]^k$$
$$\times[\mathbf{E}[\exp\{-t\int_0^1 \eta^2(x,\omega)dx\}/r(\cdot,\cdot,\omega)]]^l.$$

The analogous assertion holds for moments of the random variables $\theta_{1n}(ti)$, $\theta_{2n}(ts)$, $i = \overline{1,p_1}$, $s = \overline{1,p_2}$. Since the variables $\theta_{1n}(t)$ and $\theta_{2n}(t)$ are positive and bounded by 1, the convergence of their joint moments implies the convergence of their finite-dimensional distribution. Therefore,

$$\{\theta_{1n}(t),\theta_{2n}(t)\} \Rightarrow \{\mathbf{E}[\exp\{t\int_0^1 \eta^2(x,\omega)dx\}\int_0^1 \eta^2(x,\omega)dx/r(\cdot,\cdot,\omega)],$$

$$\mathbf{E}[\exp\{t\int_0^1 \eta^2(x,\omega)dx\}/r(\cdot,\cdot,\omega)\}.$$

By using (16.3.3), we obtain

$$\int_0^1 (1+tx)^{-1}d\lambda_n(x,\omega) \Rightarrow 2\mathbf{E}\{\exp[-t\int_0^1 \eta^2(x,\omega)dx]\int_0^1 \eta^2(x,\omega)dx/r(\cdot,\cdot,\omega)\}$$

$$\times [\mathbf{E}[\exp\{-t\int_0^1 \eta^2(x,\omega)dx\}/r(\cdot,\cdot,\omega)]]^{-1}.$$
$$(16.3.6)$$

Let $\lambda_k(\omega)$, $k = 1, 2, \ldots$ be the eigenvalues of the kernel $r(x, y, \omega)$, and $\varphi_k(\omega)$ be their corresponding eigenfunctions. The process $\eta(x, \omega)$ can be decomposed into the series $\eta(x, \omega) = \sum_{k=1}^{\infty} \eta_k \sqrt{\lambda_k(\omega)} \varphi_k(x, \omega)$, where η_k are independent random variables not depending on $\varphi(x, \omega)$ and $\lambda_k(\omega)$ and with the normal distribution $N(0, 1)$.

Evidently, $\int_0^1 \eta^2(x, \omega) dx = \sum_{k=1}^{\infty} \eta_k^2 \lambda_k(\omega)$. Substituting this into (16.3.6), we obtain

$$\int_0^1 (1 + tx)^{-1} d\lambda_n(x, \omega) \Rightarrow -2 \frac{d}{dt} \ln \prod_{k=1}^{\infty} (1 + t\lambda_k)^{1/2} = \frac{d}{dt} \ln D(-t).$$

Then we make use of the fact that the convergence of the Stieltjes transforms of random spectral functions $\lambda_n(x, \omega)$ to the Stieltjes transforms of a random spectral function $\lambda(x, \omega)$ implies $\lambda_n(x, \omega) \overset{\approx}{\to} \lambda(x, \omega)$. Theorem 6.3.1 is proved.

Corollary 16.3.1. *In addition to the conditions of Theorem 16.3.1, let the random functions $\alpha_i(x, \omega)$, $i = \overline{1, n}$ be independent for every n, $C_n \equiv \sqrt{n}$, $\sup_{i,x} \mathbf{E}\alpha_i^4(x, \omega) < \infty$, $(2n)^{-1} \sum_{i=1}^n \mathbf{E}\alpha_i(x, \omega)\alpha_i(y, \omega) \to r(x, y)$, $r(x, y)$ be continuous in the domain $0 \le x, y \le 1$. Then for any $\varepsilon > 0$,*

$$\lim_{n \to \infty} \mathbf{P}\{|\lambda_n(x, \omega) - \lambda(x)| > \varepsilon\} = 0,$$

at each point of continuity of the spectral nonrandom function $\lambda(x)$ whose Stieltjes transform is

$$\int_0^{\infty} (1 + tx)^{-1} d\lambda(x) = \frac{d}{dt} \ln D(-t), \quad t \ge 0,$$

where $D(t)$ is the Fredholm determinant of the kernel $r(x, y)$, $0 \le x, y \le 1$.

Proof. By the law of large numbers,

$$\plim_{n \to \infty} (2n)^{-1} \sum_{i=1}^n \alpha_i(x, \omega)\alpha_i(y, \omega) = r(x, y).$$

Therefore, by Theorem 16.3.1 for each $t \ge 0$,

$$\plim_{n \to \infty} \int_0^{\infty} (1 + tx)^{-1} d\lambda_n(x) = \frac{d}{dt} \ln D(-t), \quad t \ge 0.$$

Hence, the assertion of Corollary 16.3.1 follows.

CHAPTER 17

RANDOM DETERMINANTS IN
THE SPECTRAL THEORY OF
NON-SELF-ADJOINT RANDOM MATRICES

One of the unsolved problems of the spectral theory of non-self-adjoint random matrices is that of a general description of their limit normalized spectral functions under the assumptions that the entries of the random matrices are independent, infinitesimal, and the dimension of the matrices tends to infinity. Until recently, the problem was unsolved even for real random matrices whose entries are independent and normally distributed. For self-adjoint random matrices Ξ_n, the problem has been solved under some additional restrictions with the help of limit theorems for the Stieltjes transforms (see [50]),

$$\int (z-x)^{-1} d\mu_n(x) = n^{-1} \operatorname{Tr}(Iz - \Xi_n)^{-1} = n^{-1}(\partial/\partial z) \ln \det(Iz - \Xi_n),$$

where $Imz \neq 0$, and $\mu_n(x)$ is the normalized spectral function of Ξ_n.

For non-self-adjoint random matrices H_n, the limit theorems for the Stieltjes transforms cannot be applied, in general, since the integrals $\mathbf{E} \operatorname{Tr}(Iz - H_n)^{-1}$ as a rule do not exist for all n. Moreover, the formulas of perturbation theory for the resolvents $(Iz - H_n)^{-1}$ lose the validity. In this chapter we show that by means of the so-called $\mathbf{V}$-transform the study of the spectral functions of non-self-adjoint random matrices can be reduced to that of spectral functions of Hermitian random matrices.

§1 Limit Theorems for the Normalized Spectral Functions of Complex Gaussian Matrices

Let $H_n = (\xi_{pl})_{p,l=1}^n$ be a complex random matrix, real and imaginary parts of random elements of ξ_{pl} are distributed according to the normal law $N(0,1), \lambda_k$,
$k = \overline{1,n}$ random eigenvalues of matrix H_n arranged in increasing order. From

Chapter 3, we obtain that the distribution density of eigenvalues of H_n is equal to

$$(\pi^n \prod_{j=1}^n j! 2^{n(n+1)/2})^{-1} \exp\{-2^{-1} \sum_{k=1}^n |z_k|^2\}$$

$$\times \prod_{i>j} |z_i - z_j|^2, \quad \arg z_1 > \ldots \arg z_n. \tag{17.1.1}$$

Denote $\nu_n(B) = n^{-1} \sum_{k=1}^n x(\omega : \lambda_k n^{-1/2} \in B)$, where B is some Borel set on the plane.

Let us prove Mehta's theorem for spectral functions $\nu_n(B)$.

Theorem 17.1.1. *If the distribution density of the eigenvalues of complex random matrices H_n is defined by formula (17.1.1), then for any Borel set B on the plane R^2,*

$$\lim_{n \to \infty} \mathbf{E}\nu_n(B) = \nu(B),$$

where $\nu(B) = \pi^{-1} \int_{B \cap (z:|z|<1)} dx\, dy$.

We call the statement of Theorem 17.1.1 the circle law. In the following, we shall denote by the circle law any statement that $\operatorname{plim}_{n \to \infty} \nu_n(B) = \nu(B)$ or $\lim_{n \to \infty} \mathbf{E}\nu_n(B) = \nu(B)$.

Proof of Theorem 17.1.1. By using formula (17.1.1) for any bounded continuous function $f(z)$, $z \in R_2$, we have

$$n^{-1} \sum_{l=1}^n \mathbf{E}f(\lambda_l n^{-1/2}) = [\pi^n \prod_{j=1}^n j! 2^{n(n+1)/2}]^{-1}$$

$$\times \int \cdots \int f(z_1 n^{-1/2}) \exp\{-2^{-1} \sum_{i=1}^n |z_i|^2\} \det[\sum_{k=1}^n z_k^p$$

$$\times \bar{z}_k^l]_{p,l=0}^{n-1} \prod_{i=1}^n dx_i\, dy_i.$$

By using the formulas of perturbances for random determinants (Chapter 2, §5) from this equality, we have

$$n^{-1} \sum_{i=1}^n \mathbf{E}f(\lambda_i n^{-1/2}) = [\pi^n \prod_{j=1}^n j! 2^{n(n+1)/2}]^{-1}$$

$$\times \int f(z_1 n^{-1/2}) \exp\{-2^{-1} \sum_{i=1}^n |z_i|^2\} \det W[1$$

$$+ (W^{-1}\vec{z}, \vec{z})] \prod_{i=1}^n dx_i\, dy_i, \tag{17.1.2}$$

where

$$W = [\sum_{k=2}^{n} z_k^p \bar{z}_k^l]_{p,l=0}^{n-1}, \quad \vec{z} = (1, z_1, \ldots, z_1^{n-1}).$$

By (17.1.2) we can easily get that

$$\int \exp\{-2^{-1} \sum_{i=2}^{n} |z_i|^2\} \det W \prod_{i=2}^{n} dx_i dy_i \qquad (17.1.3)$$

$$= \prod_{j=2}^{n} j!(n-1)! \pi^n 2^{n(n+1)/2},$$

$$\int \exp\{-2^{-1} \sum_{i=2}^{n} |z_i|^2\} W^{-1} \det W \prod_{i=2}^{n} dx_i dy_i$$

$$= \pi^{n-1}(n-1)! \int \exp\{-2^{-1} \sum_{i=2}^{n} |z_i|^2\} Y^{-1} \det Y \prod_{i=2}^{n} dx_i dy_i, \qquad (17.1.4)$$

where

$$Y = [z_p^{p-2} \bar{z}_p^l]_{p=\overline{2,n}, l=\overline{1,n-1}}.$$

As

$$\int \exp(-|z|^2 2^{-1}) z^{-j} z^k dx dy = \pi \delta_{jk} j! 2^{j/2+2},$$

then from formula (17.1.4), we obtain

$$\int \exp\{-2^{-1} \sum_{i=2}^{n} |z_i|^2\} W^{-1} \det W \prod_{i} dx_i dy_i$$

$$= [\delta_{pl} \prod_{j \neq p} j! \pi^{n-1} 2^{p/2} + 2]_{p,l=1}^{n-1} (n-1)!$$

But then it follows from (17.1.2) and (17.1.3) that

$$n^{-1} \sum_{i=2}^{n} \mathbf{E} f(\lambda_i n^{-1/2}) = \int [1 + \sum_{k=1}^{n-1} |z_k|^{2k} (k!)^{-1}]$$

$$\times e^{-|z_1|^2} f(z_1 n^{-1/2}) dx dy \pi n^{-1} = \pi^{-1} \int [1 + \sum_{k=1}^{n-1} (k!)^{-1}$$

$$\times |z_1|^{2k} n^k] e^{-n|z_1|^2} f(z_1) dx_1 dy_1 \to \pi^{-1} \int \varphi(z_1)$$

$$\times f(z_1) dx_1 dy_1, \quad (n \to \infty),$$

where

$$\varphi(z) = \begin{cases} \pi^{-1}, & x^2 + y^2 < 1, \\ 0, & x^2 + y^2 \geq 1. \end{cases}$$

From this, the statement of Theorem 17.1.1 follows.

By applying formula (17.1.1), it is possible to obtain the limit distributions for averaged distances between the nearest eigenvalues. These limit distributions will be found in Chapter 27.

We make some remarks about studying the limit distributions of the eigenvalues of real symmetric Gaussian matrices. We use the notations and definitions introduced in Chapter 3, §5.

Under the condition that event H_s is realized, from Corollary 3.5.1 we obtain that the conditional distribution density of the eigenvalues of the matrix Ξ_s is equal to

$$c_s \exp\{-2^{-1}TrY_sY_s'\}\sqrt{J_s(Y_s)}\varphi(Y_s)\psi(Y_s),$$

where the coefficients c_s are defined by some system of equations mentioned in Corollary 3.5.1. This formula has a cumbersome form.

Making use of it, it is very difficult to prove the limit theorems, because there are no exact values for constant c_s. In connection with this, in the next paragraph we introduce the **V**-transform of spectral functions by means of which the proof of limit theorems for spectral functions of random matrices is considerably simplified.

§2 The **V**-Transform of Spectral Functions

We consider the normalized spectral functions

$$\nu_n(x,y) = n^{-1}\sum_{k=1}^{n}\chi(\omega : \operatorname{Re}\lambda_k < x, \operatorname{Im}\lambda_k < y),$$

where λ_k are the eigenvalues of a complex $n \times n$ random matrix H_n,

$$\mu_n(x,z) = n^{-1}\sum_{k=1}^{n}\chi(\omega : \lambda_k(z) < x),$$

where $\lambda_k(z)$ are the eigenvalues of the Hermitian matrix $(Iz - H_n)(Iz - H_n)^*$, $z = t + is$.

The **V**-transform of the spectral function $\nu_n(x,y)$ is given in the following expression:

$$m_n(p,q) := \iint e^{ipx+iqy}d\nu_n(x,y) = [(q^2+p^2)/4iq\pi]$$ (17.2.1)

$$\times \iint (\partial/\partial s)[\int_0^\infty \ln x\,d\mu_n(x,z)]e^{itp+isq}dt\,ds, q \neq 0.$$

It is easy to verify the validity of this equality by using the fact that

$$\int_0^\infty \ln x\,d\mu_n(x,z) = n^{-1}\sum_{k=1}^{n}\ln|z - \lambda_k|^2.$$

However, the transform (17.2.1) in this form is not yet suitable for the proof of limit theorems for spectral functions $\nu_n(x,y)$. The derivative $\partial/\partial s$ and infinite limits of integrations "hinder."

In the proof of limit theorems for $\nu_n(x,y)$, the **V**-transform is used in the following form:

$$m_n(p,q,c,d) := [(q^2+p^2)/4iq\pi] \iint [\int_{-(c-x)}^{c-x} [\int_{-(d-y)u^{-1}}^{(d-y)u^{-1}} \text{sign}\, u$$
$$\times (1+v^2)^{-1}e^{iqvu}dv]e^{ipu}du]e^{ipx+iqy}d\nu_n(x,y)$$
$$= [(q^2+p^2)/4iq\pi] \int_{-c}^{c} [\int_{0}^{\infty} \ln x d\mu_n(x,z)e^{itp}|_{t=-d}^{t=d} \qquad (17.2.2)$$
$$- \int_{-d}^{d} (\int_{0}^{\infty} \ln x d\mu_n(x,z))e^{itp}ipdt]e^{isq}ds,$$

where c,d are the positive constants.

Theorem 17.2.1. *If $q \neq 0$ and*

$$\lim_{h\to\infty}\lim_{n\to\infty} \mathbf{E} \iint_{|x|<h,|y|<h} d\nu_n(x,y) = 1, \qquad (17.2.3)$$

then

$$\lim_{c\to\infty}\lim_{d\to\infty}\sup_n \mathbf{E}|\varepsilon_n(p,q,c,d)| = 0, \qquad (17.2.4)$$

where $\varepsilon_n(p,q,c,d) = m_n(p,q) - m_n(p,q,c,d)$.

Proof. Obviously,

$$\varepsilon_n(p,q,c,d) = \sum_{i=1}^{4} J_i,$$

where

$$J_1 = \iint_{|x|<A,|y|<B} f(p,q,d,x,y)d\nu_n(x,y),$$

$$J_2 = \iint_{|x|<A,|y|\geq B} f(p,q,c,d,x,y)d\nu_n(x,y),$$

$$J_3 = \iint_{|x|\geq A,|y|<B} f(p,q,c,d,x,y)d\nu_n(x,y),$$

$$f(p,q,c,d,x,y)$$
$$= (p^2+q^2)(4\pi ip)^{-1} \int [\int \text{sign}\, u(1+v^2)^{-1}e^{iqvu}dv]e^{ipu}du$$
$$- \int_{-c-x}^{c-x} [\int_{-(d+y)u^{-1}}^{(d-y)u^{-1}} \text{sign}\, u(1+v^2)^{-1}e^{iqvu}dv]e^{ipu}du.$$

As

$$\int [\int \operatorname{sign} u (1+v^2)^{-1} e^{iqvu} dv] e^{ipu} du = 4ip\pi/(p^2+q^2),$$

then

$$|f(p,q,c,d,x,y)| \le 1 + 2c[(p^2+q^2)/4ip\pi] = c_1$$

Therefore,

$$|J_4| \le c_1 \int_{|y| \ge B} d\nu_n(x,y), \qquad (17.2.5)$$

$$|J_2| \le c_1 \int_{|y| \ge B} d\nu_n(x,y). \qquad (17.2.6)$$

Obviously, as $|y| \le B$,

$$\lim_{d \to \infty} \int_{-(d+y)u^{-1}}^{(d-y)u^{-1}} [e^{iqvu}/(1+v^2)] dv = e^{-|qu|}.$$

Therefore,

$$\int_{-c-x}^{c-x} [\int_{-(d+y)u^{-1}}^{(d-y)u^{-1}} \operatorname{sign} u (1+v^2)^{-1} e^{iqvu} dv]$$

$$\times e^{ipu} du = \int_{-c-x}^{c-x} [\operatorname{sign} u e^{-|qu|} + \varphi_n(B,d,u)]$$

$$\times e^{ipu} du,$$

where

$$\varphi_n(B,d,u) = \int_{-(d+y)u^{-1}}^{(d-y)u^{-1}} \operatorname{sign} u (1+v^2)^{-1} e^{iqvu} dv - \operatorname{sign} u e^{-|qu|},$$

It is easy to see that

$$\lim_{d \to \infty} \varphi_n(B,d,u) = 0.$$

Therefore, for J_1, we have

$$|J_1| \le c_1 e^{-c+A} + c_2 \psi(B,d) \qquad (17.2.7)$$

and

$$\lim_{d \to \infty} \psi(B,d) = 0.$$

Similarly, we obtain the expression

$$|J_3| \le [c_3 + c\psi(B,d)] \iint_{|x| \ge A} d\nu_n(x,y).$$

Thus, by virtue of the condition (17.2.3) for fixed c,

$$\lim_{B \to \infty} |J_4| = 0, \qquad (17.2.8)$$

$$\lim_{B \to \infty} |J_2| = 0; \qquad (17.2.9)$$

as A and B are fixed,

$$\lim_{c \to \infty} \lim_{d \to \infty} |J_1| = 0; \qquad (17.2.10)$$

as c and B are fixed,

$$\lim_{A \to \infty} \lim_{d \to \infty} |J_3| = 0. \qquad (17.2.11)$$

Thus, passing to the limit as $d \to \infty$ as the variables are fixed, and then by A, B, C as $B \to \infty$ are fixed, further by A, c as $c \to \infty$ is fixed, and finally by $A \to \infty$ by virtue of (17.2.3), and by making use of (17.2.5)–(17.2.11), we obtain (17.2.4). This proves Theorem 17.2.1.

§3 Limit Theorems like the Law of Large Numbers for Normalized Spectral Functions of Non-Self-Adjoint Random Matrices with Independent Entries

It was proved in §3 of Chapter 9 that the normalized spectral functions of symmetric random matrices $\Xi_n = (\xi_{ij})_{ij=1}^n$, for which the vectors $(\xi_{ii}, \xi_{ii+1}, \ldots, \xi_{in})i = \overline{1, n}$ are stochastically independent, are "self-averaged," that is, they approach with nonrandom functions in probability as the order of the matrices increases to infinity. Earlier, this statement on a physical level of rigor was repeatedly proved for different particular cases in the solutions of some problems of nuclear physics, irregular crystal structures. The proof of this statement was obtained in Chapter 9 with the help of the limit theorems for Stieltjes transforms. In this section, the analogous statement for non-self-adjoint random matrices with independent entries by means of limit theorems for $\mathbf{V}$–transforms of spectral functions of random matrices $(Iz - H_n)(IzH_n)^*$ is proved. We note that although these matrices are self-adjoint, their eigenvalues can possess considerably worst properties in comparison to the same properties of non-self-adjoint matrices H_n. For example, we consider the matrix

$$A = \begin{bmatrix} 1 & c \\ 0 & 1 \end{bmatrix}, \qquad c^2 = (1 - \varepsilon)^2 \varepsilon^{-1}, \quad 0 < \varepsilon < 1.$$

Its eigenvalues are all equal to 1. The eigenvalues of AA' are equal to $\varepsilon, \varepsilon^{-1}$. When ε are small, one of the eigenvalues of AA' is near zero. From this simple note we see that the proof of the limit theorems for $\mathbf{V}$-transforms of spectral functions of random matrices $(Iz - H_n)(Iz - H_n)^*$ is a matter of large analytical difficulties. In the present chapter, we used a regularized $\mathbf{V}$-transform which helps us avoid these difficulties.

Theorem 17.3.1. *For every value n, let the entries of the complex random matrices $H_n = (\xi_{pl}^{(n)} n^{-1/2}), p, l = 1 \div n$ be independent, $\mathbf{E}\xi_{pl}^{(n)} = 0, \mathbf{E}|\xi_{pl}^{(n)}|^2 = \sigma_{pl}^2, \sup_{p,l=1\div n} \sigma_{pl}^2 < \infty$, the real and imaginary parts of $\xi_{kl}^{(n)}$ have distribution densities $p_{kl}^{(n)}(x, y)$ satisfying the condition*

$$\sup_n \sup_{k,l=1\div n} \int_{-\infty}^{\infty} [p_{pl}^{(n)}(x)]^\beta dx < \infty, \quad \beta > 1$$

$$\nu_n(B) = n^{-1} \sum_{k=1}^{n} \chi(\omega : \lambda_k, \in B), \tag{17.3.1}$$

where the λ_k are the eigenvalues of H_n, B is any Borel set on a plane

$$\lim_{h \to \infty} \lim_{n \to \infty} \mathbf{E}\nu_n(x, y : |x| < h, |y| < h) = 1. \tag{17.3.2}$$

Then for almost every Borel set,

$$\text{plim}_{n \to \infty}[\nu_n(B) - \mathbf{E}\nu_n(B)] = 0 \tag{17.3.3}$$

Proof. For the quantities $m_n(p, q, c, d)$ defined by formula (17.2.2), we have for $q \neq 0$,

$$m_n(p, q, c, d) - \mathbf{E}m_n(p, q, c, d) = n^{-1} \sum_{k=1}^{n} \gamma_k,$$

where

$$\gamma_k = [(p^2 + q^2)/4iq\pi] \int_{-c}^{c} \int_{-d}^{d} (\partial/\partial s)\{\mathbf{E}_{k-1} \ln |\det(Iz - H_n)|^2 - \mathbf{E}_k \ln |\det(Iz - H_n)|^2\} \exp\{itp + isq\} dt ds,$$

and $\mathbf{E}_k$ is the conditional mathematical expectation for the fixed minimal σ-algebra with respect to which the components of the column $h_p, p = k+1 \div n$ are measurable.

We transform γ_k by expanding $\det(Iz - H_n)$ with respect to the kth row: $\det(Iz + H_n) = -n^{-1/2} \sum_{s=1}^{n} \xi_{ks}^{(n)} A_{ks} + z A_{kk}$, where the A_{ks} are the cofactors of matrix $(Iz + H_n)$,

$$\gamma_k = [(p^2 + q^2)/4iq\pi] \int_{-c}^{c} \int_{-d}^{d} (\partial/\partial s)\{\mathbf{E}_{k-1} \ln \theta_k(z) - \mathbf{E}_k \ln \theta_k(z)\} \exp(itp + isq) dt ds, \tag{17.3.4}$$

where

$$\beta_k = A_{k\bar{s}}, |A_{k\bar{s}}| = \max_{s=1\div n} |A_{ks}|, \theta_k(z)$$

$$= |z A_{kk} \beta_k^{-1} - \sum_{s=1}^{n} n^{-1/2} \xi_{ks}^{(n)} A_{ks} \beta_k^{-1}|.$$

It is easy to verify that the following simple inequalities hold,

$$\mathbf{E}\ln^2\theta_k(z)\chi(\theta_k(z) < 1) \le c_1\Big[\int_{|x|<1}|\ln|x||^{2\alpha}dx\Big]^{1/\alpha}$$

$$\times \sup_{k,l=1\div n}\Big[\int_0^1[p_{lk}^{(n)}(x)]^\beta dx\Big]^{1/\beta}n^{(1-\beta^{-1})/2} \le c_2 n^{1/2},$$

where $\alpha^{-1} + \beta^{-1} = 1, \alpha > 1, \beta > 1$,

$$\mathbf{E}\ln^2\theta_k(z)\chi(\theta_k(z) \ge 1) \le c_1(|z|^2 + 1).$$

From these two inequalities, we find that $\mathbf{E}|\gamma_k|^2 \le c_4 n^{1/2}$. Consequently, since $\mathbf{E}\gamma_k\bar{\gamma}_k = 0, k \ne l$, and by use of (17.3.1) and (17.3.4), we obtain

$$\lim_{n\to\infty}\mathbf{E}|m_n(p,q,c,d) - \mathbf{E}m_n(p,q,c,d)|^2 = 0.$$

But then, as (17.3.1) and (17.3.2) hold, by applying Theorem 17.2.1, we obtain (17.3.3). This proves Theorem 17.3.1.

We look for conditions under which with probability 1 for almost every Borel set,

$$\lim_{n\to\infty}[\nu_n(B) - \mathbf{E}\nu_n(B)] = 0.$$

Theorem 17.3.2. *If the conditions of Theorem 17.3.1 hold, then with probability 1 for almost every Borel set B on the plane,*

$$\lim_{n\to\infty}[\nu_B(B) - \mathbf{E}\nu_n(B)] = 0. \tag{17.3.5}$$

Proof. We examine the equality

$$\int_0^\infty [m_n(px, qx, c, d) - \mathbf{E}m_n(px, qx, c, d)]$$

$$\times e^{-x}dxpq = n^{-1}\sum_{k=1}^n \gamma_k,$$

where

$$\gamma_k = \int_0^\infty (p^2 + q^2)(4i\pi q)^{-1}xpq\int_{-c}^c\int_{-d}^d (\partial/\partial s)\{(\mathbf{E}_{k-1} - \mathbf{E}_k)$$

$$\times \ln|\det(Iz - H_n)|^2\}\exp\{itpx + isqx\}dtdse^{-x}dx.$$

As in the proof of Theorem 17.3.1, it is easy to verify that $\mathbf{E}|\gamma_k|^4 \le c_5 n^{1/2}$. Hence, using Theorem 9.3.1, we obtain that with the probability of 1 for every fixed p and q

$$\lim_{n\to\infty}\int_0^\infty [m_n(px, qx, c, d) - \mathbf{E}m_n(px, qx, c, d)]e^{-x}dxpq = 0, \tag{17.3.6}$$

and by using Theorem 17.2.1, we have

$$\lim_{c \to \infty} \lim_{d \to \infty} \sup_n \left| \iint pq(1 + ipx + iqy)^{-1} d\nu_n(x, y) \right.$$
$$\left. - pq \int_0^\infty m_n(px, qx, c, d) e^{-x} dx \right| = 0, \qquad (17.3.7)$$

where

$$\nu_n(x, y) = n^{-1} \sum_{k=1}^{n} \chi(\operatorname{Re} \lambda_k < x) \chi(\operatorname{Im} \lambda_k < y).$$

As the function $pq(1 + ipx + iqy)^{-1}$ is equicontinuous by the variables x and y, then by virtue of (17.3.6) and (17.3.7), we obtain that with the probability 1,

$$\lim_{n \to \infty} \sup_{p,q} \left| \iint [1 + ipx + iqy]^{-1} d\nu_n(x, y) - \mathbf{E}\nu_n(x, y)) \right| = 0$$

Hence, as (17.2.3) holds, we obtain that, with probability 1, for almost all x and y, $\lim_{n \to \infty} [\nu_n(x, y) - \mathbf{E}\nu_n(x, y)]$. This implies the statement of Theorem 17.3.2.

Similarly we prove the following assertion.

Theorem 17.3.3. *For every value of n, let the entries of the complex random matrices $H_n = (\xi_{pl}^{(n)})_{p,l=1 \div n}$ be independent, for some $\alpha > 0$,*

$$\sup_n \sup_{p=\overline{1,n}} \sum_{l=1}^{n} \mathbf{E}|\xi_{pl}^{(n)}|^\alpha < \infty, \qquad (17.3.8)$$

and the condition (17.3.1) is satisfied. Then (17.3.5) holds.

The proof of Theorem 17.3.3 almost entirely coincides with that of 17.3.1. Note that the inequality (17.3.8) is necessary so that for the finite z,

$$\mathbf{E} \ln^2 \theta_k^2(z) \chi(\theta_k(z) \geq 1) \leq c < \infty.$$

Theorem 17.3.4. *For every n, let the random vectors $(\xi_{pl}^{(n)}, \xi_{lp}^{(n)}), l \geq p, l, p = 1 \div n$ be stochastically independent, where $\xi_{pl}^{(n)}$ are the entries of the complex matrices $H_n = (\xi_{pl}^{(n)} n^{-1/2})$,*

$$l, p = 1 \div n, \quad \mathbf{E}\xi_{pl}^{(n)} = 0, \quad \mathbf{E}|\xi_{pl}^{(n)}| = \sigma_{pl}^2, \quad \sup_n \sup_{p,l=1 \div n} \sigma_{pl}^2 < \infty,$$

and the real and imaginary parts of random elements $\xi_{pl}^{(n)}, \xi_{lp}^{(n)}$ have the distribution densities $q_{pl}(x_1, x_2, y_1, y_2)$ satisfying the condition

$$\sup_n \sup_{p,l=\overline{1,n}} \sup_{y,x} q_{pl}(x, y) < \infty, \qquad (17.3.9)$$

where

$$q_{pl}(x, y) = \iint q(x, x_1, y, y_1) dx_1 dy_1.$$

Then with probability 1 for almost every Borel set B,

$$\lim_{n \to \infty} [\nu_n(B) - \mathbf{E}\nu_n(B)] = 0.$$

Proof. As in the proof of Theorem 17.3.2, we have

$$\int_0^\infty [m_n(px, qy, c, d) - \mathbf{E}m_n(px, qx, c, d)]e^{-x} dx pq$$

$$= n^{-1} \sum_{k=1}^n \gamma_k, \tag{17.3.10}$$

where

$$\gamma_k = \int_0^\infty (p^2 + q^2)(4\pi iq)^{-1} x pq \int_{-c}^c \int_{-q}^q (\partial/\partial s)\{\mathbf{E}_{k-1} \ln \kappa_k(z)$$

$$- \mathbf{E}_k \ln \kappa_k(z)\} \exp(itpx + isqx) ds dt e^{-x} dx,$$

$$\kappa_k(z) = |(z - \xi_{kk}^{(n)} n^{-1/2})A_{kk} + n^{-1} \sum_{\substack{p \neq k, l \neq k, p, l = 1}}^n \xi_{pl}$$

$$\times \xi_{lk} \alpha_{plk} |\beta_k^{-1}, \beta_k = \max_{k, p, l} [|A_{kk}|, |\alpha_{plk}|],$$

and α_{plk} are some complex random variables not depending on the random variables $\xi_{kp}, \xi_{lp}, l, p = 1 \div n$.

As in the proof of Theorem 17.3.2, it is easy to check that by virtue of (17.3.9), the inequalities

$$\mathbf{E} \ln^4 \kappa_k(z) \chi(\kappa_k(z) < 1) \leq \mathbf{E} \ln^4 \kappa_k(z)$$

$$\times \chi(\kappa_k(z) \geq 1) \leq c(|z| + |z|^2 + 1)$$

are satisfied. Further, the proof is similar to that of Theorem 17.3.2.

§4 The Regularized V-Transform for Spectral Functions

Let us find the conditions under which

$$\lim_{n \to \infty} n^{-1} |\mathbf{E} \ln \det[(Iz - H_n)(Iz - H_n)^* \gamma]$$

$$- \mathbf{E} \ln \det[I\alpha_n + (Iz + H_n)(Iz - H_n)^* \gamma]| = 0,$$

$$\tag{17.4.1}$$

where $\alpha_n > 0$ is a numerical sequence and γ is a random variable with distribution density $e^{-x}, x \geq 0$.

By virtue of (17.4.1),

$$m_n(p, q, c, d) := (p^2 + q^2)(4iq\pi)^{-1} \int_{-c}^{c} \int_{-d}^{d} (\partial/\partial s)\mathbf{E}$$
$$\times \ln \det[I\alpha_n + (Iz - H_n)(Iz - H_n)^*\gamma] \exp(itp + isq)dtds$$
$$= m_n(p, q, c, d) + 0(1).$$

The expression $m_n(p, q, c, d, \alpha_n)$ is called the regularized **V**-transform of the spectral function $\nu_n(B)$. With its help, it is considerably easier to prove limit theorems for the spectral functions $\nu_n(B)$.

Theorem 17.4.1. *For every n, let the entries of the random matrices $H_n = (\xi_{pl}^{(n)} n^{-1/2})_{p,l=1\div n}$, be independent, $\mathbf{E}\xi_{pl}^{(n)} = 0$, $\mathbf{E}|\xi_{pl}^{(n)}|^2 = \sigma_{pl}^2$, $\sigma_{pl}^2 < c < \infty$ and the real and imaginary parts of $\xi_{pl}^{(n)}$ have the distribution densities satisfying (17.3.1).*

Then (17.4.1) holds for finite t, s and α_n such that

$$\lim_{n \to \infty} \alpha_n^\varepsilon n = 0, \qquad 0 < \varepsilon < (\beta - 1)(2\beta)^{-1}.$$

Proof. We introduce the matrices Q_k whose first k diagonal entries are α_n and the other entries vanish, Q_o is the null matrix. We consider the equality

$$\ln \det[(Iz - H_n)(Iz - H_n)^*\gamma] - \ln \det[I\alpha_n + (Iz - H_n)$$
$$\times (Iz - H_n)^*\gamma] = \sum_{k=1}^{n} \kappa_k \qquad (17.4.2)$$

where

$$\kappa_k = \ln \det[Q_{k-1} + \gamma(Iz - H_n)(Iz - H_n)^*] - \ln \det[Q_k + \gamma(Iz - H_n)(Iz - H_n)^*].$$

By using formula (2.5.7), we have

$$\kappa_k = \ln \kappa_k - \ln(\alpha_n + x_k), \qquad (17.4.3)$$

where

$$x_k = (b_k, \bar{b}_k) - ((\tilde{Q}_k + B_k B_k^*)^{-1} B_k \bar{b}_k, \bar{B}_k, b_k),$$

$b_k = (b_{kl}, l = 1 \div n)$, b_{kl} are the entries of $B := (Iz - H_n)\sqrt{\gamma}$, B_k is the matrix obtained from B by deleting the kth row, and $\tilde{Q}_k$ is the matrix obtained from Q_k by deleting the kth row and the kth column.

Obviously,

$$x_k \geq (b_k, \bar{b}_k) - (B_k^*(B_k B_k^*)^{-1} B_k \bar{b}_k, b_k) \geq 0. \qquad (17.4.4)$$

The matrix B_k can be represented as $B_k = \sqrt{B_k B_k^*} H$, where H is an orthogonal $n \times (n-1)$ matrix. Then it follows from (17.4.4) that

$$x_k \geq (b_k, \bar{b}_k) - (HH^*\bar{b}_k, b_k). \tag{17.4.5}$$

Obviously, $H^*H = I - h_k h_k^*$, where h_k is a vector of dimension n and $(h_k, \bar{h}_k) = 1$. Therefore, the inequality (17.4.5) is equivalent to the following one:

$$x_k \geq |(h_k, \bar{b}_k)|^2.$$

Using this inequality and (17.4.3), we have

$$\mathbf{E}[\ln(x_k + \alpha_k) - \ln x_k]\chi(x_k < 1) < \mathbf{E}\ln[1 + \alpha_n |(h_k, \bar{b}_k)|^{-2}],$$
$$[\ln(x_k + \alpha_n) - \ln(x_k)]\chi(x_k \geq 1) \leq \alpha_n.$$

Obviously, among the components of vector h_k, at least one satisfies $|h_{kl}|^2 \leq cn^{-1}, c > 0$. Suppose, for certainty, that $(\operatorname{Re} h_{kl})^2 \geq cn^{-1}$. Therefore, as in the proof of Theorem 17.3.1, we have

$$\mathbf{E}\ln[1 + \alpha_n |(h_k, \bar{b}_k)|^{-2}]$$
$$\leq \alpha_n + c_1 \mathbf{E} \inf_{l=1\div n} \int_0^1 [1 + \alpha_n \gamma^{-1} x^{-2}]|\operatorname{Re} h_{kl}|^{-1} p_{lk}(x + \nu_l)(\operatorname{Re} h_{kl})^{-1} dx$$
$$\leq \alpha_n + c_1 [\int_0^1 \ln^s [1 + \alpha_n \gamma^{-1} x^{-2}] dx]^{s-1}$$
$$\times \mathbf{E} \inf_{l=1\div n} [\int_0^1 p_{lk}^\beta ((x + \nu_l)(\operatorname{Re} h_{kl})^{-1}) dx]^{\beta_n^{-1}} \leq c_2 \alpha_n^\varepsilon h,$$

where $0 < \varepsilon < (2s)^{-1}, s^{-1} = 1 - \beta^{-1}$, and ν_l are random variables. From this inequality, the assertion of Theorem 17.4.1 follows.

Theorem 17.4.1 implies that in the proof of limit theorems for the spectral functions $\nu_n(x, y)$, we can consider a functional which is called the regularized **V**-transform of the spectral function $\nu_n(x, y)$,

$$f(\nu_n(\cdot, \cdot), \alpha_n, z) := n^{-1}\operatorname{Tr}(I\alpha_n + (Iz - H_n)(Iz - H_n)^*)^{-1}. \tag{17.4.6}$$

If the conditions of Theorem 17.4.1 hold for matrices H_n, then

$$\iint e^{ipx+iqy} d\nu_n(x, y) = (q^2 + p^2)(4\pi iq)^{-1}$$
$$\times \int_{-c}^c \int_{-d}^d (\partial/\partial s) \int_{\alpha_n}^A f(\nu_n(\cdot, \cdot), z, y) dy e^{itp+isq} dt ds$$
$$+ \delta_n(\alpha_n, A, c, d, q, p) \tag{17.4.7}$$

and

$$\lim_{n \to \infty} \lim_{c \to \infty} \lim_{d \to \infty} \lim_{A \to \infty} \sup_n |\delta_n(\alpha_n, A, c, d, q, p)| = 0.$$

Thus, in the proof of the limit theorems for $\nu_n(x, y)$, with large enough n, we first find some formulas for $n^{-1} \operatorname{Tr}(I\alpha_n + (Iz - H_n)(Iz - H_n)^*)^{-1}$. By using them, we can take the limit as in the formula (17.4.7), then take the limit as $A \to \infty, C \to \infty, d \to \infty$.

We shall now study the random matrices with dependent random entries $\xi_{pl}^{(n)}, \xi_{lp}^{(n)}, l, p = 1 \div n$.

Theorem 17.4.2. *Let the conditions of Theorem 17.3.4 hold. Then(17.4.1) holds for finite t, s, and α_n such that $\lim_{n \to \infty} \alpha_n^\varepsilon n = o$, where $\alpha_n > 0, 0 < \varepsilon < 1$.*

Proof. We use the notations introduced in the proof of Theorem 17.4.1. As in the proof of 17.4.1, we obtain the inequality (17.4.5) and the following one:

$$\mathbf{E}[\ln(x_k + \alpha_k) - \ln x_k]\chi(x_k < 1) \leq \mathbf{E} \ln[1 + \alpha_n |(h_k, \bar{b}_k)|^{-2}]. \qquad (17.4.8)$$

However, in (17.4.8) the random vectors $h_k, \bar{b}_k$ are stochastically independent. It is easy to see that

$$\mathbf{E} \ln[1 + \alpha_n |(h_k, \bar{b}_k)|^{-2}] \leq \sum_{s=1}^n \mathbf{E}\chi(A_s) \int \ln[1 + \alpha_n$$

$$\times |\sum_{s=1}^n (x_s - iy_s)(q_{sk}(x_s, y_s) + ig_{sk}(x_s, y_s))|^{-2}]$$

$$\times p_{ks}(x_s, y_s, u_s, v_s)dx_s dy_s du_s dv_s, \qquad (17.4.9)$$

where

$$A_s = \{\omega : |q_{sk}(x_s, y_s, u_s, v_s)|^2 + |g_{sk}(x_k, y_s, u_s, v_s)|^2 \geq cn^{-1},$$
$$|q_{pk}|^2 + |g_{pk}|^2 < cn^{-1}, p \neq s\}, \quad c > 0,$$

and $h_{pk} = q_{pk} + ig_{pk}$ for fixed $\xi_{ks} = x_s + iy_s, \xi_{sk} = u_s + iv_s, q_{pk} = q_{pk}(x_s, y_s, u_s, v_s), q_{pk} = g_{pk}(x_s, y_s, u_s, v_s)$. Next, by changing the variables in the integral of (17.4.9), analogously to the proof of (17.4.1), we obtain Theorem 17.4.2. Theorem 17.4.2 is proved.

§5 An Estimate of the Rate of Convergence of the Stieltjes Transforms of Spectral Functions to the Limit Function

In this section, we present some auxiliary assertions that are needed in the proof of the circle and elliptic laws.

Theorem 17.5.1. *Suppose that for every $n = 1, 2, \ldots$, the random entries $\xi_{pl}^{(n)}, p, l = 1 \div n$ of the complex matrix $H_n = (\xi_{pl}^{(n)})$ are independent, that $\mathrm{E}\xi_{pl}^{(n)} = 0, \mathrm{E}|\xi_{pl}^{(n)}|^2 = \sigma^2 n^{-1}, 0 < \sigma^2 < \infty$, and the Lindeberg condition holds, for every $\tau > 0$,*

$$\lim_{n \to \infty} n^{-1} \sum_{p,l=1}^{n} \mathrm{E}|\xi_{pl}^{(n)}|^2 \chi(|\xi_{pl}^{(n)}| > \tau) = 0. \tag{17.5.1}$$

Then $\mathrm{plim}_{n \to \infty} \mu_n(x, z) = \mu(x, z)$ for almost all x and z, where $\mu(x, z)$ is a distribution function in x whose Stieltjes transform $m(\theta) := \int_0^\infty (1 + i\theta x)^{-1} d\mu$ (x, z) satisfies the equality

$$m(\theta) = [1 + i\theta \sigma^2 m(\theta) + i|z|^2 \theta (1 + i\theta \sigma^2 m(\theta))^{-1}]^{-1}. \tag{17.5.2}$$

where θ is a real parameter. If, in addition, instead of the Lindeberg condition (17.5.1) for some $0 < \delta \leq 2$,

$$\sup_{n} \sup_{p,l=1 \div n} \mathrm{E}|\xi_{p,l}^{(n)}|^{2+\delta} < \infty,$$

then

$$\left| \int_0^\infty (1 + i\theta x)^{-1} d\mu_n(x, z) - m(\theta) \right| \leq c(|\theta| + \theta^2) n^{-\delta(2+\delta)^{-1}}, \quad c > 0 \tag{17.5.3}$$

Proof. Let b_{ij} be the entries of the matrix $B := (Iz - H_n)$, r_{ij} the entries of the matrix $R := (I + i\theta BB^*)^{-1}$. For r_{kk} formula (2.5.7) holds.

$$r_{kk} = [1 + i\theta(b_k, \overline{b_k}) + \theta^2(B_k^* R_k B_k \overline{b_k}, b_k)]^{-1}, \tag{17.5.4}$$

where

$$R_k := (r_{ij}^k) = (I + i\theta B_k B_k^*)^{-1}.$$

We transform (17.5.4) to the following form:

$$r_{kk} = [1 + i\theta\sigma^2 + i\theta|z|^2 + \theta^2\sigma^2 n^{-1} \operatorname{Tr} R_k B_k B_k^* + \theta^2|z|^2 \operatorname{Tr} R_k T_k^k + \varepsilon_{1k}]^{-1}, \tag{17.5.5}$$

where

$$\varepsilon_{1k} = \theta \left\{ \sum_{l=1}^{n} |b_{kl}|^2 - \sigma^2 - |z|^2 \right\} + \theta^2 \{ (B_k^* R_k B_k \overline{b_k}, b_k)$$
$$- \sigma^2 n^{-1} \operatorname{Tr} R_k B_k B_k^* - |z|^2 \operatorname{Tr} R_k T_k^k \},$$

$T_p^k = (b_{ip} \overline{b_{jp}})_{i,j=1}^n$ are the square matrices with the kth column and kth row deleted.

By using Formula (2.5.4), we obtain that

$$\text{Tr } R_k T_k^k = (R_k^k b_k, \overline{b_k})[1 + i\theta(R_k^k b_k, \overline{b_k})]^{-1},$$

where

$$R_k^k = (I + i\theta \sum_{p \neq k} T_p^k)^{-1}.$$

On the basis of this equality, we transform (17.5.5) to the following form:

$$r_{kk} = [1 + i\theta\sigma^2 n^{-1} \text{Tr } R_k + i\theta|z|^2[1 + i\theta\sigma^2 n^{-1} \text{Tr } R_k^k]^{-1} + \varepsilon_{2k}]^{-1}, \quad (17.5.6)$$

where

$$\varepsilon_{2k} = \varepsilon_{1k} - i\theta|z|^2\{[1 + i\theta(R_k^k b_k \overline{b_k})]^{-1} - [1 + i\theta\sigma^2 n^{-1} \text{Tr } R_k^k]^{-1}\}.$$

Hence

$$m_n(\theta) = n^{-1} \sum_{k=1}^n [1 + i\theta\sigma^2 m_n(\theta) + i\theta|z|^2(1 + i\theta\sigma^2 m_n(\theta))^{-1} + \varepsilon_{3k}]^{-1}, \quad (17.5.7)$$

where

$$\varepsilon_{3k} = \varepsilon_{2k} + i\theta\sigma^2 n^{-1}[\text{Tr } R_k - \text{Tr } R] + i\theta|z|^2\{[1 + i\theta\sigma^2 n^{-1} \text{Tr } R_k^k]^{-1} - [1 + i\theta\sigma^2 n^{-1} \text{Tr } R]^{-1}\} + i\theta\sigma^2 n^{-1}[\text{Tr } R - \text{E Tr } R] + i\theta|z|^2[(1 + i\theta\sigma^2 n^{-1} \text{Tr } R)^{-1} - (1 + i\theta\sigma^2 n^{-1} \text{E Tr } R)^{-1}], \quad m_n(\theta) = \text{E} n^{-1} \sum_{k=1}^n r_{kk}.$$

Since $|m_n(\theta)| \leq 1, |1 + i\theta\sigma^2 m_n(\theta)|^2 \leq 1$, we find from (17.5.7) that

$$m_n(\theta) = [1 + i\theta\sigma^2 m_n(\theta) + i\theta|z|^2(1 + i\theta\sigma^2 m_n(\theta))^{-1}]^{-1} + \varepsilon_4, \quad (17.5.8)$$

where

$$|\varepsilon_4| \leq n^{-1} \sum_{k=1}^n \text{E}|\varepsilon_{3k}|.$$

From (17.5.8), we obtain that when $|z| \leq \sqrt{c^2 + d^2}$,

$$|m_n(\theta) - m(\theta)| \leq |\theta|\sigma^2|m_n(\theta) - m(\theta)| + \theta^2|z|^2\sigma^2|m_n(\theta) - m(\theta)| + |\varepsilon_4|.$$

So as $0 < \theta < c_1$,

$$|m_n(\theta) - m(\theta)| < c_2|c_4|, \quad (17.5.9)$$

where $c_1 > 0$ and c_2 are some constants.

From (17.5.8), we find that $m_n(\theta)$ is a root of the polynomial of the third degree $m_n^3(\theta) + a m_n^2(\theta) + b m_n(\theta) + c = 0$, where

$$a = -2i(\theta\sigma^2)^{-1} - \varepsilon_4,$$
$$b = -(1 + i\theta|z|^2)\theta^{-2}\sigma^{-4} + i\theta^{-1}\sigma^{-2} + 2i\varepsilon_4\theta^{-1}\sigma^{-2},$$
$$c = \theta^{-2}\sigma^{-4} + \varepsilon_4\theta^{-2}\sigma^{-4}.$$

For the roots of such a polynomial, the Cardano formulas hold:

$$m_i = y_i - a/3, \quad i = \overline{1,3}, \quad y_1 = A + B,$$
$$y_{2,3} = (-A + B)/2 \pm i(A - B)\sqrt{3}/2,$$
$$A = \sqrt[3]{-q/2 + \sqrt{Q}}, B = \sqrt[3]{-q/2 - \sqrt{Q}},$$
$$Q = (p/3)^3 + (q/2)^2, \quad p = -a^2/3 + b,$$
$$q = 2(a/3)^3 - ab/3 + c. \tag{17.5.10}$$

As A and B any of the values of cube roots satisfying the correlation $AB = -p/3$ are taken.

By using the Cardano formulas for the roots of polynomial of the third degree, we obtain for $\theta > c_1 > 0$ and $|\varepsilon_4| < \varepsilon$, where ε is a small value, the following expression,

$$|m_n(\theta) - m(\theta)| < c_3|\varepsilon_4|. \tag{17.5.11}$$

With regard to (17.5.9), (17.5.11) implies for $|\varepsilon_4| < \varepsilon$,

$$|m_n(\theta) - m(\theta)| \le c_4|\varepsilon_4|. \tag{17.5.12}$$

Let us estimate the ε_4.

Lemma 17.5.1. *If for every n the random entries $\xi_{pl}^{(n)}, p, l = 1 \div n$ are independent, $\mathbf{E}\xi_{pl}^{(n)} = 0, \mathbf{E}|\xi_{pl}^{(n)}|^2 = \sigma_{pl}^2, \le c < \infty$, and for some $0 < \delta \le 2$,*

$$\sup_n \sup_{p,l=1\div n} \mathbf{E}|\xi_{pl}^{(n)}|^{2+\delta} < \infty, \tag{17.5.13}$$

then

$$|\varepsilon_4| \le c_n^{-\delta(2+\delta)^{-1}}[|\theta| + \theta^2], \quad c < \infty. \tag{17.5.14}$$

Proof. By using (17.5.13), we have for $|z|^2 < c$,

$$\mathbf{E}\left|\sum_{l=1}^n |b_{kl}|^2 - \sigma^2 - |z|^2\right|^{1+p/2} \le c_5^{-p/2},$$

$$\mathbf{E}|\theta|\,|(B_k^* R_k B_k \overline{b_k}, b_k) - \sigma^2 n^{-1}\,\mathrm{Tr}\,R_k B_k B_k^*$$
$$- |z|^2\,\mathrm{Tr}\,R_k T_k^k| \le c_6 n^{-1/2},$$

where $0 \le p < \min(2, \delta)$.

Therefore,

$$\mathbf{E}|\varepsilon_{1k}| \leq c_6|\theta|n^{-\delta(2+\delta)^{-1}}. \tag{17.5.15}$$

Similarly, if we use (17.5.15), we obtain that

$$\mathbf{E}|\varepsilon_{2p}| \leq c_7|\theta|n^{-\delta(2+\delta)^{-1}}. \tag{17.5.16}$$

It follows from the proof of Theorem 3.6.1 that

$$|\operatorname{Tr} R_k - \operatorname{Tr} R| \leq c_8,$$
$$|\operatorname{Tr} R_k^t - \operatorname{Tr} R| < c_9,$$
$$\mathbf{E}|n^{-1}\operatorname{Tr} R - n^{-1}\mathbf{E}\operatorname{Tr} R| \leq c_{10}n^{-1/2}.$$

So, taking into account the inequalities (17.5.15) and (17.5.16), we have

$$\mathbf{E}|\varepsilon_{3k}| \leq c_{11}n^{-\delta(2+\delta)^{-1}}(|\theta| + \theta^2).$$

This inequality implies (17.5.14). Lemma 17.5.1 is proved. By using Lemma 17.5.1 and (17.5.12), we have (17.5.3).

Note that if the Lindeberg condition (17.5.1) holds, $\lim_{n\to\infty}|\varepsilon_4| = 0$. Therefore, $\lim_{n\to\infty} m_n(\theta) = m(\theta)$. But, from Theorem 9.3.1, $n \to \infty$, and from the uniqueness of the solution of Eq. (17.5.2) in the class of analytic functions, the assertion of Theorem (17.5.1) follows.

We generalize Theorem 17.5.1. Assume that the random entries $\xi_{pl}^{(n)}$ and $\xi_{lp}^{(n)}$ of the matrix $H_n = (\xi_{pl}^{(n)})$ are dependent.

Theorem 17.5.2. *Suppose that for every* $n = 1, 2, \ldots$ *the random vectors* $(\xi_{pl}^{(n)}, \xi_{pl}^{(n)}, p \geq l, p, l = \overline{1, n})$ *are stochastically independent, that* $\mathbf{E}|\xi_{pl}^{(n)}|^2 = n^{-1}, \mathbf{E}\xi_{pl}^{(n)}\xi_{lp}^{(n)} = \rho/n, 0 \leq |\rho| < 1, p \neq l,$ *and the Lindeberg condition (17.5.1) holds, then for almost all* x *and* z *with probability 1,* $\lim_{n\to\infty} \mu_n(x, z) = \mu(x, z)$, *where* $\mu(x, z)$ *is a distribution function on* x *for fixed* z *whose Stieltjes transform* $m(\theta) := \int_0^\infty (\theta + x)^{-1} d\mu(x, z), \theta > 0$ *satisfies the equation*

$$[|\rho|/[m(\theta)(1 + m(\theta))]] - (bt - as)^2(1 + m(\theta) - |\rho|m(\theta))^{-2}$$
$$- (at + sb)^2(1 - m(\theta) + |\rho|m(\theta))^2 - |\rho|\theta = 0, \tag{17.5.17}$$

where $a = \operatorname{Re}\sqrt{\rho}, b = \operatorname{Im}\sqrt{\rho}, z = t + is, \theta > 0.$

The solution of Eq. (17.5.17) exists and is unique in the class of analytic functions $m(\theta) > 0$ *for* $\theta > 0$.

If, in addition to the conditions listed instead of the Lindeberg condition (17.5.1), for some $0 < \delta, \sup_n \sup_{p,l=1\div n} \mathbf{E}|\xi_{pl}^{(n)}\sqrt{n}|^{2+\delta} < \infty,$ *then* $|\int_0^\infty (1 + i\theta x)^{-1} d\mu_n(x, z) - m(\theta)| \leq c(|\theta| + \theta + |\theta|^3)n^{-\gamma}, 0 < \gamma < 1.$

Proof. Suppose that the Lindeberg condition (17.5.1) is satisfied. We write $B := (Iz - H_n), R_k := (I\theta + B_k B_k^*)^{-1}$, where B is a matrix obtained from B

by replacing the kth vector row by the null vector, and that $\theta > 0$. By using formula (17.5.3), we have

$$n^{-1}\operatorname{Tr}(I\theta + BB^*)^{-1} = n^{-1}\sum_{k=1}^{n}[\theta + \|b_{1k}\|^2 - (R_k B_k \overline{b_k}, \overline{B_k}b_k)]^{-1}. \quad (17.5.19)$$

We consider in (17.5.19) that $\overrightarrow{b_k}$ are the column vectors. Obviously,

$$(R_k B_k \overline{b_k}, \overline{B_k}, b_k) = \operatorname{Tr} R_k c, \quad (17.5.20)$$

where

$$c = B_k \overline{b_k}(\overline{B_k}, b_k)' = B_k \overline{b_k} b_k' B_k^*.$$

We give the notation $\tilde{R}_k = (I\theta + \tilde{B}_k \tilde{B}_k^*)^{-1}$, where $\tilde{B}_k$ is the matrix obtained from B by replacing the kth column a_k and the kth row b_k by zeros.

From Chapter 2, §2, we obtain the following formula:

$$\operatorname{Tr} R_k c - \operatorname{Tr} \tilde{R}_k c = -(\tilde{R}_k c \tilde{R}_k a_k, a_k)[1 + (\tilde{R}_k a_k, a_k)]^{-1}, \quad (17.5.21)$$

where $\overrightarrow{a_k} = (b_{1k}, \ldots, b_{k-1k}, 0, b_{k+1k}, \ldots, b_{nk})$, b_{ij} are the entries of matrix B. By using Formula (17.5.21) and equality (17.5.20), from (17.5.19) we find

$$n^{-1}\operatorname{Tr}(I\theta + BB^*)^{-1} = n^{-1}\sum_{k=1}^{n}[\theta + \|b_k\|^2 + (\tilde{R}_k c \tilde{R}_k a_k, a_k)$$
$$\times [1 + (\tilde{R}_k a_k, a_k)]^{-1} - (\tilde{R}_k B_k \overline{b_k}, \overline{B_k}b_k)]^{-1}. \quad (17.5.22)$$

We remark that

$$b_k = (\xi_{1k}, \xi_{2k}, \ldots, \xi_{k-1k}, \xi_{kk} + z, \xi_{k+1k}, \ldots, \xi_{nk}),$$
$$B_k b_k = \tilde{B}_k \overline{b_k} + (\overline{z} + \overline{\xi_{kk}})a_k, \quad (17.5.23)$$

where $\widetilde{\overline{b_k}} = \overline{b_k}$ as $z = -\xi_{kk}$.

By using (17.5.23), we have

$$(\tilde{R}_k B_k \overline{b_k}, \overline{B_k}b_k) = (\tilde{R}_k(\tilde{B}_k \widetilde{\overline{b_k}} + (\overline{z} + \xi_{kk})a_k),$$
$$(\overline{B_k}\tilde{b_k} + (z + \xi_{kk})\overline{a_k})) = (\tilde{R}_k \tilde{B}_k \widetilde{\overline{b_k}}, \overline{\tilde{B}_k}b_k)$$
$$+ (\tilde{R}_k \tilde{B}_k \overline{b_k}a_k)(z + \xi_{kk}) + (\tilde{R}_k a_k \overline{\tilde{B}_k \tilde{b_k}})(\overline{z}$$
$$+ \overline{\xi_{kk}} + |z + \xi_{kk}|^2(\tilde{R}_k a_k, \overline{a_k}).$$

By using this equality and the conditions of this theorem, we obtain

$$
\begin{aligned}
(\tilde{R}_k B_k \overline{b_k}, \overline{B_k} b_k) &= n^{-1} \operatorname{Tr} \overline{B}_k^* \tilde{R}_k \tilde{B}_k + z \overline{\rho} n^{-1} \operatorname{Tr} \tilde{R}_k \tilde{B}_k \\
&+ \overline{z} \rho n^{-1} \operatorname{Tr} \tilde{R}_k \tilde{B}_k^* + |z|^2 n^{-1} \operatorname{Tr} \tilde{R}_k + \varepsilon_{kn} = 1 - n^{-1} \theta \operatorname{Tr} \tilde{R}_k \\
&+ 2 \operatorname{Re} \rho \overline{z} n^{-1} \operatorname{Tr} \tilde{R}_k \tilde{B}_k^* + |z|^2 n^{-1} \operatorname{Tr} R_k + \varepsilon_{kn}',
\end{aligned}
$$

where

$$
\operatorname{plim}_{n \to \infty}[|\varepsilon_{kn}| + |\varepsilon_{kn}'|] = 0. \tag{17.5.24}
$$

Obviously,

$$
\begin{aligned}
(\tilde{R}_k (B_k \overline{b_k}) (\overline{B_k} b_k)' \tilde{R}_k a_k, \tilde{a}_k) &= ((B_k \overline{b_k})(\overline{B_k} b_k)' \\
&\times \tilde{R}_k a_k, \tilde{R}_k \overline{a_k}) = (\tilde{R}_k a_k, \overline{B_k} b_k)(\tilde{R}_k \overline{a_k}, B_k \overline{b_k}) \\
&= |(\tilde{R}_k \tilde{a}_k, B_k \overline{b_k})|^2 = |(\tilde{R}_k \overline{a_k}, \tilde{B}_k \widetilde{\overline{b_k}}) + (\overline{z} + \overline{\xi_{kk}}) \\
&\times a_k)|^2 = |(\tilde{R}_k \overline{a_k}, \tilde{B}_k \widetilde{\overline{b_k}}) + (\tilde{R}_k \overline{a_k}, a_k)(\overline{z} + \overline{\xi_{kk}})|^2 \\
&= |\overline{\rho} n^{-1} \operatorname{Tr} \tilde{R}_k \tilde{B}_k + \overline{z} n^{-1} \operatorname{Tr} \tilde{R}_k|^2 + \varepsilon_n'',
\end{aligned} \tag{17.5.25}
$$

where $\operatorname{plim}_{n \to \infty} |\varepsilon_n''| = 0$.

By substituting (17.5.24) and (17.5.25) by (17.5.22), we have

$$
\begin{aligned}
n^{-1} \mathbf{E} \operatorname{Tr}(I\theta + BB^*)^{-1} &= n^{-1} \sum_{k=1}^{n} \mathbf{E}[\theta + 1 + |z|^2 + |\overline{\rho} n^{-1} \\
&\times \operatorname{Tr} \tilde{R}_k \tilde{B}_k + \overline{z} n^{-1} \operatorname{Tr} \tilde{R}_k|^2 (1 + n^{-1} \operatorname{Tr} \tilde{R}_k)^{-1} - 1 + n^{-1} \theta \\
&\times \operatorname{Tr} \tilde{R}_k - 2 \operatorname{Re} \rho z n^{-1} \operatorname{Tr} \tilde{R}_k \tilde{B}_k^* - |z|^2 n^{-1} \operatorname{Tr} \tilde{R}_k]^{-1} + 0(1).
\end{aligned} \tag{17.5.26}
$$

Since $\operatorname{Tr} \tilde{R}_k$ is a real number, from (17.5.26) we obtain

$$
\begin{aligned}
n^{-1} \mathbf{E} \operatorname{Tr}(I\theta + BB^*)^{-1} &= n^{-1} \sum_{k=1}^{n} \mathbf{E}[\theta(1 + n^{-1} \operatorname{Tr} \tilde{R}_k) \\
&+ |\rho n^{-1} \operatorname{Tr} \tilde{R}_k \tilde{B}_k - z|^2 (1 + n^{-1} \operatorname{Tr} R_k)^{-1}]^{-1} + 0(1).
\end{aligned} \tag{17.5.27}
$$

The equation (17.5.27) is cumbersome. By means of it we cannot find the functional equation for the limiting spectral function. We note that each addend in equation (17.5.27) contains no elements of the kth row vector and of the kth column vector of B. Therefore, the matrix B can be chosen in such a way that the analogous equation holds for it, too. But by choosing this matrix in a special manner, we can considerably simplify this equation.

Let us consider the matrices $n^{-1/2} \rho^{1/2} A + (1 - |\rho|)^{1/2} \Xi n^{1/2}$, where $A = (\nu_{ij})_{i,j=1}^{n}$ are the real symmetric matrices whose entries $\nu_{ij}, i \geq j$ are independent and distributed according to the normal law $N(0,1)$, the matrix Ξ

does not depend on the matrix A, and its entries are independent and distributed according to the normal law $N(0,1)$, too, and $\rho^{1/2}$ are the principal values of the root.

Let

$$Q = (I\theta + (Iz - n^{-1/2}\rho^{1/2}A - (1 - |\rho|)^{1/2}n^{-1/2}\Xi)$$
$$\times (Iz - \rho^{1/2}n^{-1/2}A - (1 - |\rho|)^{1/2}n^{-1/2}\Xi)^*)^{-1}.$$

$$(17.5.28)$$

Let us prove that

$$\lim_{n\to\infty} \mathbf{E}|n^{-1}\operatorname{Tr}(I\theta + BB^*)^{-1} - n^{-1}\operatorname{Tr}Q| = 0.$$

To do this, we introduce the matrices T_k whose entries of the first row vectors and column vectors are the entries of the matrix $I\theta - \rho^{1/2}n^{-1/2}A - (1 - |\rho|)^{1/2}n^{1/2}\Xi$, and the rest of the entries are equal to the corresponding entries of B.

Let us consider the equality

$$n^{-1}\operatorname{Tr}(I\theta + BB^*)^{-1} - n^{-1}\operatorname{Tr}Q = n^{-1}\sum_{k=1}^{n}[\operatorname{Tr}(I\theta$$
$$+ T_{k-1}T_{k-1}^*)^{-1} - \operatorname{Tr}(I\theta + T_kT_k^*)^{-1}] = \sum_{k=1}^{n}n^{-1}[\operatorname{Tr}(I\theta$$
$$+ T_{k-1}T_{k-1}^*)^{-1} - \operatorname{Tr}(I\theta + \tilde{T}_{k-1}\tilde{T}_{k-1})^{-1} - \{\operatorname{Tr}(I\theta + T_kT_k^*)^{-1}$$
$$- \operatorname{Tr}(I\theta + \tilde{T}_{k-1}T_{k-1}^*)^{-1}],$$

$$(17.5.29)$$

where $\tilde{T}_k$ is a matrix whose kth row vector $(17.5.29)$ has zero entries.

From Chapter 2, §5, we obtain that

$$\operatorname{Tr}(I\theta + T_{k-1}T_{k-1}^*)^{-1} - \operatorname{Tr}(I\theta + \tilde{T}_{k-1}\tilde{T}_{k-1}^*)^{-1}$$
$$= (\partial/\partial\theta)\ln[\theta + \|b_k\|^2 - (\hat{R}_k\tilde{T}_{k-1}\overline{b_k}, \tilde{T}_{k-1}b_k)],$$

$$(17.5.30)$$

where $\hat{R}_k = (I\theta + \tilde{T}_{k-1}\tilde{T}_{k-1}^*)^{-1}$,

$$\operatorname{Tr}(I\theta + T_kT_k^*)^{-1} - \operatorname{Tr}(I\theta + \tilde{T}_{k-1}\tilde{T}_{k-1}^*)^{-1}$$
$$= (\partial/\partial s)\ln[\theta + \|\nu_k\|^2 - (\hat{R}_k\tilde{T}_{k-1}\overline{\nu_k}, \tilde{T}_{k-1}\nu_k)],$$

$$(17.5.31)$$

where ν_k is the kth row vector of matrix T_k.

Since

$$\|b_k\|^2 - (\hat{R}_k\tilde{T}_{k-1}\overline{b_k}, \tilde{T}_{k-1}b_k) = ((I\theta + \tilde{T}_{k-1}^*\tilde{T}_{k-1})^{-1}\overline{b_k}, b_k),$$

then for every $\varepsilon > 0$,

$$\sup_{\theta > \epsilon > 0} \{|(\partial/\partial\theta)\ln[\theta + \|b_k\|^2 - (\hat{R}_k\tilde{T}_{k-1}\overline{b_k}, \tilde{T}_{k-1}b_k)]|$$

$$+ |\partial/\partial\theta)\ln[\theta + \|\nu_k\|^2 - (\hat{R}_k\tilde{T}_{k-1}\nu_k, \tilde{T}_{k-1}\nu_k)|\} < \infty.$$

$$(17.5.32)$$

By using (17.5.30)–(17.5.32), as in the proof of (17.5.26), we obtain that for all $\theta > 0$,

$$\varlimsup_{n\to\infty} \mathbf{E}|\operatorname{Tr}(I\theta + T_{k-1}T_{k-1}^*)^{-1} - \operatorname{Tr}(I\theta + T_kT_k^*)^{-1}| = 0.$$

Hence, (17.5.28) holds.

The matrix A can be represented in the form of $A = H\Lambda H'$ where H is an orthogonal matrix of the eigenvectors and $\Lambda = (\lambda_i\delta_{ij})$ is a diagonal matrix of the eigenvalues. Since the distributions of the matrices $H\Xi H'$ and Ξ coincide,

$$\mathbf{E}n^{-1}\operatorname{Tr}Q = n^{-1}\mathbf{E}\operatorname{Tr}(I\theta + (Iz - \rho^{1/2}n^{-1/2}\Lambda)$$
$$- (1 - |\rho|)^{1/2}n^{-1/2}\Xi)(Iz - \rho^{1/2}n^{-1/2}\Lambda - (1$$
$$- |\rho|^{1/2}n^{-1/2}\Xi)^*)^{-1}.$$

For such an expression with the matrix Λ fixed, we use the proof of Theorem 17.5.1. Repeating the proof of Theorem 17.5.1 almost literally, we have

$$m_n(\theta) = n^{-1}\mathbf{E}\sum_{k=1}^{n}\{\theta[1 + (1 - |\rho|)m_n(\theta)]$$
$$+ |z - \rho^{1/2}n^{-1/2}\lambda_k|^2(1 + (1 - |\rho|)m_n(\theta))^{-1}\}^{-1}$$
$$+ 0(1), \quad \theta > 0,$$

where

$$m_n(\theta) = n^{-1}\mathbf{E}\operatorname{Tr}(I\theta + BB^*)^{-1}.$$

From this equation, we obtain

$$m_n(\theta) = \int \{\theta[1 + (1 - |\rho|)m_n(\theta)] + |z - \sqrt{\rho}x|^2(1$$
$$+ (1 - |\rho|)m_n(\theta)^{-1}\}^{-1}d\mathbf{E}\mu_n(x) + 0(1),$$

$$(17.5.33)$$

where $\mu_n(x)$ is a normalized spectral function of $An^{-1/2}$.

The convergent subsequence $m'_n(\theta) \to m(\theta)$, as $n \to \infty$, $\theta > 0$, is chosen. Since the functions $m_n(\theta)$ are analytical for $\theta > 0$ and are the Stieltjes transform, $m(\theta)$ is an analytic function. Then, by using the semicircle Wigner law (Chapter 3, §4), we obtain from (17.5.33)

$$m(\theta) = (2\pi)^{-1}\int_{-2}^{2}\{(4 - x^2)^{1/2}\{\theta[1 + (1 - |\rho|)m(\theta)]$$
$$+ |z - \rho^{1/2}x|^2[1 + (1 - |\rho|)m(\theta)^{-1}\}^{-1}\}dx.$$

$$(17.5.34)$$

It is not hard to check that this equation has the unique solution in the class of analytic functions. Therefore, $\lim_{n\to\infty} m_n(\theta) = m(\theta)$, where $m(\theta)$ is the solution of Eq. (17.5.34).

Let $\rho^{1/2} = a + ib$. Transform (17.5.34) to the following form:

$$m(\theta) = (2\pi)^{-1} \int_{-2}^{2} (4 - x^2)^{1/2} [c(x - k)^2 + p]^{-1} dx$$
$$\times [1 + (1 - |\rho|)m],\tag{17.5.35}$$

where

$$c = a^2 + b^2,$$
$$k = (at + sb)(a^2 + b^2)^{-1},$$
$$p = (bt - as)^2 (a^2 + b^2)^{-1} + \theta[1 + (1 - |\rho|)m]^2.$$

Taking advantage of

$$(2\pi)^{-1} \int_{-2}^{2} (4 - x^2)^{1/2} [c(x - k)^2 + p]^{-1} dx$$
$$= \{[1 + 4cp^{-1} - cp^{-1}k^2 + [1 + 4cp^{-1} - cp^{-1}k^2)^2$$
$$+ 4cp^{-1}k^2]^{1/2}]/2 - 1\}c/2,$$

we obtain from (17.5.35) that $m(\theta)$ satisfies Eq. (17.5.17). Further, as in the proof of Theorem 17.4.2, we obtain that

$$\lim_{n\to\infty} \mathbf{E}|m_n(\theta) - \mathbf{E}m_n(\theta)| = 0.$$

Thus, by using Theorem 3.6.1, we obtain the assertion of Theorem 17.5.2.

Let us find the estimate of a rate of convergence of the Stieltjes transform to the limit functions. As in the proof of Theorem 17.5.1, we find that the function

$$m_n(\theta) = \int_0^\infty (1 + i\theta x)^{-1} d\mu_n(x, z)$$

satisfies the equation

$$m_n(\theta) = (2\pi)^{-1} \int_{-2}^{2} [(4 - x^2)^{1/2} [1 + i\theta(1 - |\rho|)m_n(\theta)$$
$$+ i\theta|z - \rho^{1/2} x|^2 [1 + i\theta(1 - |\rho|)m_n(\theta)]^{-1}]^{-1} dx + \varepsilon_n,\tag{17.5.36}$$

where

$$|\varepsilon_n| \le cn^{-\gamma}[|\theta| + \theta^2], \quad 0 < \gamma < 1.\tag{17.5.37}$$

By using (17.5.35), we obtain

$$\Delta_n(1 + \Delta_n)[(bt + as)^2|\rho|^{-1}(1 + \Delta_n + |\rho|\Delta_n)^2$$
$$+ (at + sb)^2|\rho|^{-1}(1 + \Delta_n - |\rho|\Delta_n)^2 - [(1 + \Delta_n)^2$$
$$- |\rho|^2\Delta_n^2]^2 + (i\theta)^{-1}\Delta_n(1 + \Delta_n)[(1 + \Delta^2)$$
$$- |\rho|^2\Delta_n^2]^2 + \varepsilon_{n1} = 0, \tag{17.5.38}$$

where ε_{n1} satisfies the inequality (17.5.37) $\Delta_n = i\theta m_n(\theta)$.

The analogous equation is satisfied by the function $m(\theta) = \int_0^\infty (1+i\theta x)^{-1}d\mu$ (x, z). As in the proof of Theorem 17.5.1, we find for $0 \le |\theta| \le c_1$,

$$|m_n(\theta) - m(\theta)| \le c|\varepsilon_{n1}|, \tag{17.5.39}$$

where $c_1 > 0$ is a constant.

We write the equation (17.5.38) in the following form,

$$\sum_{i=0}^{6} \Delta^i(\theta)f_i(\theta) + \varepsilon_{n1} = 0, \tag{17.5.40}$$

where $f_i(\theta)$ are the polynomial coefficients.

Obviously, the function $\Delta(\theta) = i\theta m(\theta)$ satisfies the equation

$$\sum_{i=0}^{6} \Delta^i(\theta)f_i(\theta) = 0. \tag{17.5.41}$$

By subtracting Eq. (17.5.40) and (17.5.41) and by taking into account $\Delta_n = \Delta + 0(1)$ and (17.5.38), we obtain

$$(1 - m_n(\theta)m^{-1}(\theta))(\sum_{i=1}^{6} i\Delta^i(\theta)f_i(\theta) + 0(1)) = -\varepsilon_{n1}. \tag{17.5.42}$$

We notice that

$$\sum_{i=1}^{6} i\Delta^i(\theta)f_i(\theta) = (d/d\theta)\sum_{i=0}^{6}\Delta^i(\theta)f_i(\theta)\Delta(\theta)[(d/dt)\Delta(\theta)]^{-1}$$
$$- \sum_{i=1}^{6}\Delta^i(\theta)(d/d\theta)f_i(\theta)\Delta(d/d\theta))\Delta(\theta)$$
$$= -\sum_{i=0}^{6}\Delta^i(\theta)(d/d\theta)f_i(\theta)$$
$$= (i\theta^2)^{-1}\Delta(1 + \Delta)[(1 + \Delta)^2 - |\rho|^2\Delta^2]^2.$$

By using this equality and (17.5.42), we find

$$|m(\theta) - m_n(\theta)| \le |\varepsilon_{n1}| \, |\theta| \{(1 + i\theta m(\theta))[(1 + i\theta m(\theta))^2 + |\rho|^2 \theta^2 m^2(\theta)]^2 + 0(1)\}. \qquad (17.5.43)$$

Taking into account that

$$|1 + i\theta m(\theta)| \ge (1 + \int_0^\infty (1 + \theta^2 x)^{-1} \theta^2 x d\mu(x, z)) \ge 1,$$

$$|(1 + \Delta)^2 - |\rho|^2 \Delta^2| \ge |1 - i\theta m(\theta)|$$

$$= |1 - i\theta \int_0^\infty (1 + \theta^2 x^2)^{-1} d\mu(x, z)$$

$$- \theta^2 \int_0^\infty (1 + \theta^2 x^2)^{-1} x d\mu(x, z)| \ge \theta^2$$

$$\times \int_0^\infty (1 + \theta^2 x^2)^{-1} d\mu(x, z)$$

$$\ge c_2 \theta^2 (1 + \theta^2 c_1)^{-1}, \quad c_1 > 0, c_2 > 0.$$

From (17.5.43), we have

$$|m(\theta) - m_n(\theta)| \le c_3 |\varepsilon_{n1}| \theta^{-1} (1 + \theta^2); \qquad (17.5.44)$$

and (17.5.37), (17.5.39), and (17.5.44) imply that for all θ,

$$|m(\theta) - m_n(\theta)| \le c_4 n^{-\gamma} [|\theta| + \theta^2 + |\theta|^3]. \qquad (17.5.45)$$

Theorem 17.5.2 is proved.

§6 The Estimates of the Deviations of Spectral Functions from the Limit Functions

Theorem 17.6.1. *Let $\mu_n(x)$ be the normalized spectral function of a random matrix $(Iz - H_n)(Iz - H_n)^*$ satisfying the hypotheses of Theorem 17.5.1, and let $G(x)$ be a spectral function whose Stieltjes transform satisfies (17.5.2). Then for $|z|^2 < c$,*

$$\sup_x |\mathbf{E}\mu_n(x\gamma^{-1}) - \mathbf{E}G(x\gamma^{-1})| \le c_1 n^{-\beta}, \qquad (17.6.1)$$

where $0 < \beta < 1$, and γ is a random variable with distribution density e^{-x}, $x \ge 0$.

Proof. The inverse formula for $G(x)$ is (3.2.4),

$$G(x_2) - G(x_1) = \pi^{-1} \lim_{\varepsilon \downarrow 0} \int_{x_1}^{x_2} \operatorname{Im} m(y + i\varepsilon) \, dy, \qquad (17.6.2)$$

where

$$m(z) = \int_0^\infty (z - x)^{-1} dG(x), \quad \mathrm{Im}\, z \neq 0.$$

From Theorem 17.5.1, we obtain that $m(y + i\varepsilon)$ satisfies

$$m^3 + am^2 + bm + c = 0, \tag{17.6.3}$$

where

$$a = 2, \quad b = [1 + (|z|^2 - 1)(y + i\varepsilon)^{-1}], \quad c = -[y + i\varepsilon]^{-1}.$$

Using Cardano formulas, (17.5.10) and (17.6.2), we obtain

$$G'(y) = \begin{cases} \sqrt{3}/2[\sqrt[3]{q + \sqrt{Q}} - \sqrt[3]{q - \sqrt{Q}}], & Q > 0 \\ 0, & Q < 0, \end{cases}$$

where $y > 0$,

$$q = 1/27 + 3^{-1}(|z|^2 - 1)y^{-1} + (2y)^{-1},$$
$$Q = (-1/9 + 3^{-1}(|z|^2 - 1)y^{-1})^3 + q^2.$$

Obviously,

$$G'(y) = \frac{\sqrt{3}}{2} \frac{2\sqrt{Q}}{\left(\sqrt[3]{q + \sqrt{q^2 + b^3}}\right)^2 - b + \left(\sqrt[3]{q - \sqrt{q^2 + b^3}}\right)^2},$$

$$q^2 + b^3 > 0, \qquad b = -1/9 + 1/3(|z|^2 - 1)y^{-1}, \quad y > 0, \quad q > 1/27.$$

If $b < 0$, then $G'(y) \leq c\sqrt{Q}/\sqrt[3]{Q} = c\sqrt[6]{Q}$,

$$Q = (c_1 y^{-3} + c_2 y^{-2} + c_3 y^{-1}).$$

Now, let $b > 0$. We note that for $0 < y < A$, where $A > 0$ is some number,

$$G'(y) \leq c[\sqrt[3]{q} + \sqrt[6]{Q}]\chi(|y| < A) \leq c_1 \chi(0 < y < A)$$
$$+ c_2 y^{-1/3} + y^{-1/2} \leq c_1' y^{-1/2} + c_2 y^{-1/3}.$$

If A is chosen large enough, then for $y > A$, the quantity b is negative and we return to the first case. Thus, for $G'(y)$, we have the inequality

$$G'(y) \leq c_2(y^{-1/6} + y^{-1/3} + y^{-1/2}), \quad y > 0.$$

But then

$$|\mathbf{E}G((x + y)\gamma^{-1} - \mathbf{E}G(x\gamma^{-1})| \leq c_3[y^{1-1/6} + y^{1-1/3} + y^{1-1/2}].$$

By using this inequality, we have for $T > 1$,

$$T \sup_{x} \int_{|y| \le T^{-1}} |\mathbf{E}G((x+y)\gamma^{-1}) - \mathbf{E}G(x\gamma^{-1})| dy \le c_4 T^{-1/2}. \qquad (17.6.4)$$

It follows from [142, p. 131] that

$$\sup_{x} |F(x) - G(x)| \le b \int_{-T}^{T} t^{-1} |f(t) - g(t)| dt$$
$$+ bT \sup_{x} \int_{|y| \le c(b)T^{-1}} |G(x+y) - G(x)| dy, \qquad (17.6.5)$$

where $c(b)$ is a positive constant depending only on b, and $F(x)$ is a nondecreasing bounded function, $G(x)$ is the function of bounded variation on a real straight line. $F(-\infty) = G(-\infty), f(t) = \int e^{itx} dF(x), g(t) = \int e^{itx} dG(x), T$ is an arbitrary positive number, $b > (2\pi)^{-1}$.

By the inequalities (17.6.4), (17.6.5), and (17.5.3), and since the characteristic functions of the distributions $\mathbf{E}G(x\gamma^{-1})$ and $\mathbf{E}\mu_n(x\gamma^{-1})$ are equal to $m_n(t)$ and $m(t)$, respectively,

$$\sup_{x} |\mathbf{E}\mu_n(x\gamma^{-1}) - \mathbf{E}G(x\gamma^{-1})| \le c_5[T + T^2]n^{-\delta(2+\delta)^{-1}} + c_5 T^{-1/2}.$$

This implies (17.6.1). Theorem 17.6.1 is proved.

Theorem 17.6.2. *Let* $\tilde{\mu}_n(x)$ *be the normalized spectral function whose Stieltjes transform satisfies Eq. (17.5.17). Then for* $|z^2| < c$,

$$\sup_{x} |\mathbf{E}\tilde{\mu}_n(x\gamma^{-1}) - \mathbf{E}\tilde{\mu}(x\gamma^{-1}) \le c_1 n^{-\beta_1},$$
$$0 < \beta_1 < 1. \qquad (17.6.6)$$

The proof of Theorem 17.6.2 is almost the same as that of Theorem 17.6.1. The difference lies in finding an estimate for $|\mathbf{E}\tilde{\mu}_n(x\gamma^{-1}) - \mathbf{E}\tilde{\mu}(x\gamma^{-1})|$. We shall discuss this in more detail. Since $m(\theta) = \int_0^{\infty} (\theta + x^{-1}) d\mu(x)$ satisfies (17.5.34), we obtain from it

$$m(\theta) = (2\pi)^{-1} \sqrt{4 - f(\theta)} \{|\theta[1 + (1 - |\rho|)m(\theta)]$$
$$+ |z - \sqrt{\rho} f(\theta)|^2 [1 + (1 - |\rho| m(\theta)]^{-1} \}^{-1}, \qquad (17.6.7)$$

where $f(\theta)$ is an unknown real function satisfying the condition $0 < f(\theta) \le 4$.

From the equation (17.6.7), we have

$$m^3(\theta) + am^2(\theta) + bm(\theta) + c = 0, \qquad a = 2(1 - |\rho|)^{-1},$$
$$b = [\theta + |z - \sqrt{\rho} f(\theta)|^2 - (2\pi)^{-1} \sqrt{4 - f(\theta)}](1 - |\rho|)[\theta(1 - |\rho|)^2]^{-1}$$
$$c = -(2\pi)^{-1} \sqrt{4 - f(\theta)} [\theta(1 - |\rho|)^2]^{-1}.$$

By using the inverse formula (17.6.2), we obtain that

$$G'(\theta) = \begin{cases} \sqrt{3}/2[\sqrt[3]{q+\sqrt{Q}} - \sqrt[3]{q-\sqrt{Q}}], & Q > 0 \\ 0, & Q < 0, \end{cases}$$

where

$$\theta > 0, \qquad p = -a^2/3 + b,$$
$$q = 2(a/3)^3 - ab/3 + c, \qquad Q = (p/3)^3 + (q/2)^2.$$

Obviously, from this formula it follows,

$$G'(\theta) = (\sqrt{3}/2)2\sqrt{Q}\left[\left(\sqrt[3]{q+\sqrt{q^2+b^3}}\right)^2 - \beta\right.$$
$$\left. + \left(\sqrt[3]{q-\sqrt{q^2+b^3}}\right)^2\right]^{-1}, \quad q^2 + b^3 > 0$$

where $\beta = p/3$. If $\beta < 0$, then

$$G'(\theta) \le c\sqrt[6]{Q} \le c_1\theta^{-1/2} + c_2\theta^{-1/3} + c_3\theta^{-1/6}, \quad \theta > 0. \tag{17.6.8}$$

If $\beta > 0$, then for $0 < \theta < A$,

$$G'(\theta) \le c\sqrt[3]{q + \sqrt{Q}}\chi(|\theta| < A) \le c_1'\theta^{-1/2} + c_2\theta^{-1/3}, \quad \theta > 0.$$

We note that for $y > A$ and A being large enough, the quantity β becomes negative. Consequently, for all $\theta > 0$, (17.6.8) holds. Further, as in the proof of Theorem (17.6.1), we obtain (17.6.6). Theorem 17.6.2 is proved.

§7 The Circle Law

By using the statements of the six previous sections, we prove the basic statement of the chapter.

Theorem 17.7.1. *For every n, let the random entries $\xi_{pl}^{(n)}, l, p = 1 \div n$ of a complex matrix $H_n = (\xi_{pl}^{(n)} n^{-1/2})$ be independent, $E\xi_{pl}^{(n)} = 0, E|\xi_{pl}^{(n)}|^2 = \sigma^2, 0 < \sigma < \infty$ and the quantities $\operatorname{Re}\xi_{kl}^{(n)}, \operatorname{Im}\xi_{kl}^{(n)}$ have the distribution densities $p_{kl}(x)$ and $p_{lk}(x)$ satisfying the condition: for some $\beta > 1$,*

$$\sup_n \sup_{k,l=1\div n} \int [p_{kl}^\beta(x) + q_{kl}^\beta(x)]dx < \infty,$$

for some $\delta > 0$,

$$\sup_{n} \sup_{k,l=1 \div n} E|\xi_{pl}^{(n)}|^{2+\delta} < \infty.$$

Then for any x and y, $\text{plim}_{n \to \infty} \nu_n(x, y) = \nu(x, y)$, where $[\partial^2 \nu(x, y)/\partial x \partial y] = \sigma^{-2} \pi^{-1}$ for $x^2 + y^2 < \sigma^2, = 0, x^2 + y^2 \geq \sigma^2, \nu_n(x, y) = n^{-1} \sum_{k=1}^{n} \chi(\text{Re } \lambda_k < x) \chi(\text{Im } \lambda_k < y)$, λ_k are the eigenvalues of the matrix H_n.

Proof. By using Theorem 17.4.1, we have for $q \neq 0$,

$$m_n(p, q, c, d, \alpha_n) = (p^2 + q^2)(4i\pi q)^{-1}$$
$$\times \int_{-c}^{c} \int_{-d}^{d} (\partial/\partial s) E \ln \det[I\alpha_n + \gamma(Iz - H_n)(Iz - H_n)^*]$$
$$\times \exp(itp + isq) dt ds + 0(1), \quad 0 < \varepsilon < 1,$$
$$\lim_{n \to \infty} \alpha_n^{\varepsilon(\beta-1)(2\beta)^{-1}} n = 0. \tag{17.7.1}$$

We represent (17.7.1) in the following form,

$$m_n(\cdot) = (p^2 + q^2)(4\pi i q)^{-1} E \int_{-c}^{c} \int_{-d}^{d} (\partial/\partial s) \int_{0}^{\infty} \ln(\alpha_n$$
$$+ x) d\mu_n(x\gamma^{-1}, z) e^{itp+isq} dt ds + 0(1).$$

By using this equality and Theorem 17.6.1, we obtain

$$m_n(\cdot) = (p^2 + q^2)(4\pi i q)^{-1} \int_{-c}^{c} \int_{-d}^{d} (\partial/\partial s) \int_{0}^{\infty} \ln(\alpha_n$$
$$+ \gamma x) dG(x) e^{itp+isq} dt ds + 0(1). \tag{17.7.2}$$

Note that

$$E \int_{0}^{\infty} \ln(\alpha_n \gamma^{-1} + x) dG(x) = E \lim_{A \to \infty} \left[-\int_{\alpha_n \gamma^{-1}}^{A} m(\theta, z) d\theta + \ln A \right], \tag{17.7.3}$$

where $m(\theta, z)$ is a positive solution of the equation (17.6.3) for $\varepsilon \equiv 0$. By using (17.7.3), it is easy to make sure that

$$E(\partial/\partial s) \int_{0}^{\infty} \ln(\alpha_n \gamma^{-1} + x) dG(x) = -E \int_{\alpha \gamma^{-1}}^{\infty} (\partial/\partial s) m(\theta, z) d\theta. \tag{17.7.4}$$

The Cardano formula for the solution of Eq. (17.6.3) is cumbersome and finding with its help of the integral (17.7.4) requires complex calculations. We make use of the following method of calculation of the integral (17.7.4), of which the author was informed by Litvin. Without loss of generality, we can consider that $\sigma^2 = 1$.

From (17.6.3), we find

$$(\partial/\partial s)m(\theta, z) = 2s(x+1)x^2[2|z|^2x^2 - (2x+1)(x+1)]^{-1},$$
$$\theta = x^{-1}(1-x)^{-1} - |z|^2(1+x)^{-2}, \qquad (17.7.5)$$

where for simplification of the formulas, we consider

$$x = m(\theta, z).$$

We make the change of variables in the integral (17.7.4).

$$\theta = x^{-1}(1+x)^{-1} - |z|^2(1+x)^{-2}, \qquad x \geq 0.$$

After such a change of variables on the basis of (17.7.5), the integral (17.7.4) takes the form:

$$\mathbf{E}(\partial/\partial s) \int_0^\infty \ln(x + \alpha_n \gamma^{-1})dG(x) = -2s\mathbf{E} \int_{m(\alpha_n\gamma^{-1})}^{m(\infty)} (1+x)^{-2}dx. \quad (17.7.6)$$

It is easy to obtain from Eq. (17.6.3) that

$$m(\infty) = 0$$

$$\lim_{\alpha\downarrow 0} m(\alpha) = \begin{cases} (|z|^2 - 1)^{-1}, & |z|^2 > 1 \\ \infty, & |z|^2 \leq 1. \end{cases}$$

Thus, for the integral (17.7.6), we have

$$\lim_{\alpha\downarrow 0} \mathbf{E}(\partial/\partial s) \int_0^\infty \ln(\alpha\gamma^{-1} + x)dG(x)$$
$$= \begin{cases} 2s|z|^{-2}, & |z|^2 > 1 \\ 2s, & |z|^2 \leq 1. \end{cases} \qquad (17.7.7)$$

If we put this expression into the V-transform, we can find a limit spectral function. However, the calculations become cumbersome. By the uniqueness of the inverse formula for the Fourier transform, the validity of Theorem 17.7.1 can be established in a different way. Obviously, if

$$\pi^{-1} \iint_{x^2+y^2\leq 1} 2(s-x)[(s-x)^2 + (t-y)^2]^{-1}dxdy$$
$$= \begin{cases} 2s|z|^{-2}, & |z|^2 > 1 \\ -2s, & |z|^2 \leq 1, \end{cases} \qquad (17.7.8)$$

then by (17.7.7), Theorem 17.7.1 will be proved. With the help of the polar change, this integral is reduced to the following:

$$\pi^{-1}[2(t^2 + s^2)^{-1}] \int_0^1 \int_0^\pi [(sz - z^2 \cos u \cos \theta)$$
$$\times (1 - 2a \cos u + a^2)^{-1}] du\, dz, \tag{17.7.9}$$

where
$$a = z(t^2 + s^2)^{-1/2}, \qquad \cos \theta = t(t^2 + s^2)^{-1/2}.$$

Using the fact that

$$\int_0^\pi \cos(nx)[1 - a2 \cos x + a^2]^{-1} dx$$
$$= \begin{cases} \pi a^n (1 - a^2)^{-1}, & a^2 < 1 \\ \pi[(a^2 - 1)a^n]^{-1}, & a^2 > 1 \end{cases}$$

we get (17.7.8) from (17.7.9).

It is easy to see that

$$|1 - \mathbf{E} \iint_{|x|<h,|y|<h} d\nu_n(x,y)| \le 2n^{-1}h^{-2}\mathbf{E}\,\mathrm{Tr}\, H_n$$
$$\times H_n^* \le 2\sigma^2 h^{-2}.$$

Thus, (17.2.3) holds, and (17.2.4) is satisfied.

But then, taking the limit as $c \to \infty$ and $d \to \infty$, and by using Theorems 17.3.1 and 17.5.1, we obtain from (17.7.2) and (17.7.8) the statement of Theorem 17.7.1.

§8 The Elliptic Law

In this section, we use the notations introduced in the proof of Theorem 17.5.2.

Theorem 17.8.1. *If for the random matrices $H_n = (\xi_{pl}^{(n)})$ the conditions of Theorem 17.5.2 and Theorem 17.3.4 hold, then for almost all x and y, we have*

$$\mathrm{plim}_{n\to\infty} \nu_n(x,y) = \lambda(x,y),$$

where

$$(\partial^2/\partial x \partial y)\lambda(x,y) = \pi^{-1}[1 - (a^2 + b^2)]^{-1}$$
$$\times \chi[(bx - ay)^2(1 - a^2 - b^2)^{-2}(a^2 + b^2)^{-1} + (ax + by)^2$$
$$\times (1 + a^2 + b^2)^2(a^2 + b^2)^{-1} < 1],$$
$$a = \mathrm{Re}\, \rho^{1/2}, \qquad b = \mathrm{Im}\, \rho^{1/2}. \tag{17.8.1}$$

Proof. By using the assertions of Theorem 17.3.4, 17.5.2, and 17.6.2, as in the proof of the circle law, we have

$$(\partial/\partial s) \int_0^\infty \ln(\alpha + x)d\mu(x, z) = -\int_\alpha^\infty (\partial/\partial s)m(\theta, z)d\theta \qquad (17.8.2)$$

where $m(\theta, z)$ is the positive solution of Eq. (17.5.17) in the class of analytic functions in θ for $\theta > 0$, and $\mu(x, z)$ is defined in Theorem 17.5.2.

The explicit form of the solution of (17.5.17) is also cumbersome, therefore in order to calculate the integral (17.8.2), we proceed as in the proof of the circle law.

From equation (17.5.17), we find for $0 < |\rho| < 1$,

$$
\begin{aligned}
(\partial m(\theta, z)/\partial s) = &[-(2bat + 2a^2 s)|\rho|^{-1}(1 + m(\theta, z) \\
&- |\rho|m(\theta, z))^{-2} + (2bat + 2b^2 s)|\rho|^{-1} \\
&\times (1 + m(\theta, z) + |\rho|m(\theta, z))^{-2}][-m(\theta, z))^{-2} \\
&+ |\rho|(1 + m(\theta, z))^{-2} + (bt - as)^2|\rho|^{-1}(1 - |\rho|) \\
&\times (1 + m(\theta, z) - |\rho|m(\theta, z))^{-2} + (at + sb)^2|\rho|^{-1}(1 + |\rho|) \\
&\times (1 + m(\theta, z) + |\rho|m(\theta, z))^{-2} + (at + sb)^2[2(1 + m(\theta, z) \\
&+ |\rho|m(\theta, z))^{-2} - 4m(\theta, z)(1 + |\rho|)(1 + m(\theta, z) + |\rho|m(\theta, z))^{-3}]].
\end{aligned}
$$
$$(17.8.3)$$

In the integral (17.8.2), we change the variables

$$
\begin{aligned}
\theta = &x^{-1}(1 + x)^{-1} - (bt - as)^2|\rho|^{-1}(1 + x - |\rho|x)^{-2} \\
&- (at + sb)^2|\rho|^{-1}(1 - x - |\rho|x)^{-2}, \quad x > 0.
\end{aligned}
$$
$$(17.8.4)$$

We note that the variable x of the function $f(x)$ in the right part of this equality is always negative on the set of those x whose $f^{-1}(x)$ is the Stieltjes transform $\int_0^\infty (x + y)^{-1}d\mu(y)$ of a distribution function. This is clear from the following argument:

$$f'(x) = [(f^{-1}(y))'_{y=f(x)}]^{-1} < 0.$$

After the change of variables (17.8.3) and by virtue of the equality (17.8.1), the integral (17.8.2) takes the form, for $\alpha > 0$,

$$
\begin{aligned}
(\partial/\partial s) \int_0^\infty &\ln(\alpha + x)d\mu(x, z) \\
= \int_0^{m(\alpha)} &[-(2bat + 2a^2 s)|\rho|^{-1}(1 + x - |\rho|x)^{-2} \\
&+ (2bat + 2b^2 s)|\rho|^{-1}(1 + x + |\rho|x)^{-2}]dx.
\end{aligned}
$$
$$(17.8.5)$$

We find $m(\theta, z)$. If the function $m(0, z) < \infty$ for some values of z, then we have the equation for it:

$$m(0, z) = (2\pi)^{-1} \int_{-2}^{2} (4 - x^2)^{1/2}[(t - ax)^2 + (s - bx)^2]^{-1}$$
$$\times [1 + (1 - |\rho|)m(0, z).$$

Hence,

$$m(0, z) = \Delta(1 - \Delta(1 - |\rho|))^{-1};$$

$$\Delta = (2\pi)^{-1} \int_{-2}^{2} (4 - x^2)^{1/2}[(t - ax)^2 + (s - bx)^2]^{-1} dx$$

$$(17.8.6)$$

If $m(0, z) > 0$, then this equality holds for $\Delta \leq (1 - |\rho|)^{-1}$. If $\Delta > (1 - |\rho|)^{-1}$, then $m(0, z) = \infty$.

Therefore, as $\Delta > (1 - |\rho|)^{-1}$, for the density $p(x, y)$ of the limit spectral functions, we obtain from (17.8.5),

$$\iint (z - x - iy)^{-1} p(x, y) dx dy = -i\{(a^2 s + bat)$$
$$\times [|\rho|(1 - |\rho|)]^{-1} + (b^2 s + bat)[|\rho|(1 + |\rho|)]^{-1}\} + (b^2 t - abs)$$
$$\times [|\rho|(1 - |\rho|)]^{-1} + (a^2 t + bas)[|\rho|(1 + |\rho|)]^{-1}.$$

But then, by using the inverse formula, we find (17.8.1). Theorem 17.8.1 is proved.

§9 Limit Theorems for the Spectral Functions of Non-Self-Adjoint Random Jacobi Matrices

In the previous sections of this chapter, a regularized V-transform was used to prove the limit theorems for the normalized spectral functions of non-self-adjoint random matrices. However, the functional equations for the auxiliary limit singular spectral functions obtained with the help of a regularized V-transform in some cases proved to be cumbersome. For random Jacobi matrices with independent entries, we fail to obtain even such a cumbersome functional equation. If we take advantage of the V-transform and put some conditions on the entries of the Jacobi matrix, then we can find some transform for the limiting spectral function.

Let the real random entries $\xi_i, \eta_i, \zeta_i, i = \overline{1, n}$ of the matrix $H_n = \{\xi_i \delta_{ij} + \eta_j \delta_{ij+1} + \zeta_i \delta_{ij-1}\}_{i,j=1}^{n}$ be independent and identically distributed. It follows from Chapter 12 that

$$\det(Iz + H_{n,k}) = (z - \xi_k) \det(Iz - H_{n,k+1})$$
$$+ \eta_k \xi_k \det(Iz - H_{n,k+2}), \qquad (17.9.1)$$

where

$$H_{n,k} = \{\xi_i \delta_{ij} + \eta_j \delta_{ij+1} + \zeta_i \delta_{ij-1}\}_{i,j=k}^n, \quad z = t + is.$$

Let

$$\nu_n(B) = n^{-1} \sum_{k=1}^n \chi(\omega : \lambda_k \in B),$$

where λ_k are the eigenvalues of the matrix H_n, B is a Borel set on the plane.

Theorem 17.9.1. *Let the random entries $\xi_i, \eta_i, \zeta_i, i = 1, 2, \ldots$ of the matrices H_n be independent and have the same distribution densities $p(u), q(u), r(u)$; let the solution of the equation*

$$\theta_z(x_1, x_2, z) = \int \cdots \int_{L_z} du_1 du_2 \theta_z(u_1, u_2)$$
$$\times \, dP\{\xi_1 < y_1, \eta_1 < y_2, \zeta_1 < y_3\},$$

where

$$L_z = \{y_i, i = \overline{1,3}, u_1, u_2 : \operatorname{Re}(z - y_1 - y_2 y_3(u_1 + iu_2)^{-1}) < x, \operatorname{Im}(z - y_1 - y_2 y_3(u_1 + iu_2)^{-1}) < x_2\}$$

be unique in the class of nondegenerate distribution functions $\theta_z(x_1, x_2)$ dependent on the parameter z, $|z| < \infty$. For some $\beta > 1$, $\int_{|y|<1}[p^\beta(u) + q^\beta(u) + r^\beta(u)]du < \infty$, and for some $0 < \varepsilon < 1 - \beta^{-1}$,

$$\mathbf{E}[|\xi_1|^\varepsilon + |\eta_1|^\varepsilon + |\zeta|^\varepsilon] < \infty. \tag{17.9.2}$$

Then, with probability 1, for almost every Borel set B with respect to the Lebesgue measure on the plane, $\lim_{n\to\infty}(\nu_n(B) - F(B)) = 0$, the characteristic function of the probabilistic measure $F(B)$ is equal to

$$\iint e^{ixp+iqy} F(dB) = (p^2 + q^2)(4ip\pi)^{-1}$$
$$\times \lim_{c\to\infty} \lim_{d\to\infty} \int \left\{ \int_{-c}^c \left[\int_{-d-u}^{d+u} iq \int \ln(x_1^2 + x_2^2) \right. \right.$$
$$\times \, d\theta_z(x_1, x_2) e^{isq} ds - \int \ln(x_1^2 + x_2^2) d\theta_z(x_1, y_2)$$
$$\left. \left. \times e^{isq} \Big|_{s=-d-u}^{s=d+u} \right] e^{itp} dt \right\} \pi^{-1/2} e^{-u^2} du, \quad p \ne 0.$$

Proof.
Obviously,

$$\ln|\det(Iz - \tilde{H}_n)|^2 = \ln \prod_{k=1}^n |r_k|^2, \tag{17.9.3}$$

where $r_k := \det(Iz - H_{n,k}) \det^{-1}(Iz - H_{n,k-1})$ satisfy the equation

$$r_k = z - \xi_k - \eta_k \zeta_k r_{k+1}^{-1}, r_{n+1}^{-1} \equiv 0. \tag{17.9.4}$$

By using (17.9.3), we represent the **V**-transform of the spectral function $\nu_n(B)$ in the form

$$
\mathbf{E} \iint e^{ipx+iqy} d\nu_n(B)
$$

$$
= \mathbf{E} \int \left\{ \int_{-c}^{c} \left[\int_{-d-u}^{d+u} I_1 iq\, e^{isq} ds - I_1 e^{isq} \big|_{s=-d-u}^{s=u+d} \right] e^{itp} dt \right\}
$$

$$
\times \pi^{-1/2} e^{-u^2} du + f_n(c,d),
$$

where

$$
I_1 = n^{-1} \sum_{k=1}^{n} \int \ln(x_1^2 + x_2^2) dG_z^{(k)}(x_1, x_2),
$$

$$
\lim_{c \to \infty} \lim_{d \to \infty} \sup_{n} |f_n(c,d)| = 0,
$$

$$
G_z^{(k)}(x_1, x_2) = \mathbf{P}\{\operatorname{Re} r_k, < x_1, \operatorname{Im} r_k < x_2\}.
$$

We show that

$$
\sup_{n} n^{-1} \sum_{k=1}^{n} \int \ln^2(x_1^2 + x_2^2) dG_z^{(k)}(x_1, x_2) < \infty.
$$

Consider the inequality

$$
I := \left| n^{-1} \sum_{k=1}^{n} \int_{\sqrt{x_1^2+x_2^2}<1} \ln^2(x_1^2 + x_2^2) dG_z^{(k)}(x_1, x_2) \right|
$$

$$
\leq n^{-1} \sum_{k=1}^{n} \int \cdots \int_{|z-y+u_1 u_2 (x_1+ix_2)^{-1}|<\epsilon} \ln^2 |z - y + u_1 u_2 (x_1 + ix_2)^{-1}|
$$

$$
\times P_1(y) dy\, P_2(u_1) du_1\, P_3(u_2) du_2\, dG_z^{(k)}(x_1, x_2). \tag{17.9.5}
$$

By taking into account that $z = t + is$, we make the change of variables $y = t - \tilde{y} + u_1 u_2 \operatorname{Re}(x_1 + ix_2)^{-1}$ for fixed t, u_1, u_2, x_1, x_2. Then, by using the

Hölder inequality, we have

$$
|I| \leq n^{-1} \sum_{k=1}^{n} c_1 \int \cdots \int_{|\tilde{y}| < c} \ln^2 \tilde{y}^2 p_1(t - \tilde{y}
$$
$$
+ u_1 u_2 \operatorname{Re}(x_1 + ix_2)^{-1}) d\tilde{y} p_2(u) du_1 p_3(u_2) du_2
$$
$$
\times \, dG_z^{(k)}(x_1, x_2) \leq c_2 \left[\int_{|\tilde{y}| < \epsilon} (\ln^2 \tilde{y}^2)^{\alpha} d\tilde{y} \right]^{1/\alpha}
$$
$$
\times \left[\int \int p_1^{\beta}(t - \tilde{y} + u_1 u_2 \operatorname{Re}(x_1 + ix_2)^{-1}) d\tilde{y} \right]^{1/\beta} p_2(u_1)
$$
$$
\times \, p_3(u_2) dG_z^{(k)}(x_1, x_2) du_1 du_2 = c_2 \left[\int_{|y| < 1} (\ln^2 y^2)^{\alpha} dy \right]^{1/\alpha}
$$
$$
\times \left[p_1^{\beta}(y) dy \right]^{1/\beta}.
$$

Thus,

$$
\sup_n |I| < \infty.
$$

Now, we estimate

$$
I := |n^{-1} \sum_{k=1}^{n} \int_{x_1^2 + x_2^2 < 1} \ln^2(x_1^2 + x_2^2)
$$
$$
\times \, dG_z^{(k)}(x_1, x_2)| \leq cn^{-1} \sum_{k=1}^{n} \mathbf{E}|z - \xi_k - \eta_k \zeta_k|
$$
$$
\times \, r_{k+1}^{-1}|^{\epsilon} \leq c_2 n^{-1} \sum_{k=1}^{n} [|z|^{\epsilon} + \mathbf{E}|\xi_k|^{\epsilon}
$$
$$
+ \mathbf{E}|\eta_k \xi_k|^{\epsilon} \mathbf{E}|(z - \xi_{k+1} - \eta_{k+1} \zeta_{k+1} r_{k+2}^{-1})^{-1}|^{\epsilon}.
$$

By using the condition (17.9.2), we obtain for some $1 > \varepsilon > 0$,

$$
I_n \leq c_3 n^{-1} \sum_{k=1}^{n} \mathbf{E}|(z - \xi_{k+1} - \eta_{k+1} \zeta_{k+1} r_{k+2}^{-1})^{-1}|^{\epsilon}
$$
$$
+ \chi(|z - \xi_{k+1} - \eta_{k+1} r_{k+2}^{-1}| < 1) + c_4
$$
$$
\leq c_3 n^{-1} \sum_{k=1}^{n} \int_{|\tilde{y}| \leq 1} |\tilde{y}|^{-\epsilon} p_1(t - \tilde{y} + u_1 + u_2
$$
$$
\times \operatorname{Re}(x_1 + ix_2)^{-1}) p_2(u_1) p_3(u_2) du_1 du_2 \leq c_5 + c_6
$$
$$
\times \left[\int_{|\tilde{y}| < 1} |\tilde{y}|^{-\epsilon \alpha} dy \right]^{1/\alpha} \left[\int_{|y| < 1} p_1^{\beta}(y) dy \right]^{1/\beta}.
$$

If we choose ε such that $0 < \varepsilon\alpha < 1$, we obtain

$$\sup_n I_n < \infty.$$

Thus, (17.9.5) holds.
We write

$$\theta_z^{(n)}(x_1, x_2) = n^{-1} \sum_{k=1}^{n} G_z^{(k)}(x_1, x_2).$$

Obviously,

$$n^{-1} \sum_{k=1}^{n} G_z^{(k)}(x_1, x_2) = n^{-1} \sum_{k=1}^{n} G_z^{(k+1)}(x_1, x_2) + 0(1).$$

Then, by using (17.9.4), we have

$$\theta_z^n(x_1, x_2) = \int \cdots \int_{L_z} dP\{\xi_1 < y_1, \eta_1 < y_2, \zeta_1 < y_3\} d\theta_z^{(n)}(u_1, u_2) + 0(1).$$

By uniqueness of the solution of Eq. (17.9.2) in the class of distribution functions $\theta_z(x_1, x_2)$, we obtain that $\theta_z^{(n)}(x_1, x_2) \Rightarrow \theta_z(x_1, x_2)$. Hence, by virtue of (17.9.5), $\lim_{n\to\infty}[E\nu_n(B) - F(B)] = 0$. But then, by using Theorem 17.3.3, we obtain the statement of Theorem 17.9.1.

Corollary 17.9.1. *If, in addition to the conditions of Theorem 17.9.1, the nonproper integral $\int_0^\infty \ln(x_1^2 + x_2^2)d\theta_z(x_1, x_2)$ exists for all $|z| < \infty$ and the integral $\int_0^\infty \ln(x_1^2 + x_2^2)d\theta_z(x_1, x_2)$ is differentiable by t, then*

$$\iint e^{ipx + iqy} F(dB) = (p^2 + q^2)(4\pi ip)^{-1}$$

$$\times \iint (\partial/\partial t) \int_0^\infty \ln(x_1^2 + x_2^2)d\theta_z(x_1, x_2)$$

$$\times e^{itp + isq} dt ds, \quad p \neq 0.$$

§10 The Unimodal Law

In this section, the limit spectral function for a random matrix AB^{-1} is found when the entries of the random matrices A and B are independent and satisfying certain conditions.

Theorem 17.10.1. *(Unimodal Law). Suppose that for every n the entries of the random matrices* $A_n = (\xi_{ij}^{(n)})_{i,j=1}^n, B_n = (\eta_{ij}^{(n)})_{i,j=1}^n$ *are independent, and* $\mathbf{E}\xi_{ij}^{(n)} = \mathbf{E}\eta_{ij}^{(n)} = 0,$

$$\mathbf{E}[\xi_{ij}^{(n)}]^2 = \delta_{in}^2, \qquad \mathbf{E}[\eta_{ij}^{(n)}]^2 = \sigma_{in}^2,$$
$$0 < c_1 \le \sigma_{in}^2 \le c_2 < \infty,$$
$$0 < c_1 \le \delta_{in}^2 \le c_2 < \infty, \tag{17.10.1}$$

and for random entries $\xi_{pl}^{(n)}, \eta_{ij}^{(n)}$, *the condition*

$$\sup_n \sup_{i,j=\overline{1,n}} \mathbf{E}[|\xi_{ij}^{(n)}|^{4+\delta} + |\eta_{ij}^{(n)}|^{4+\delta} < \infty \text{ is satisfied}, \delta > 0. \tag{17.10.2}$$

Then for almost all x, y,

$$\text{plim}_{n\to\infty}[\mu_n(x,y) - \int_{-\infty}^x \int_{-\infty}^x p_n(u,v)dudv] = 0, \tag{17.10.3}$$

where

$$p_n(u,v) = n^{-1}\sum_{k=1}^n [\sigma_{kn}^2 + (u^2 + v^2)\delta_{kn}^2]^{-2}\alpha,$$

$$\alpha = (\pi n^{-1}\sum_{k=1}^n \sigma_k^{-2}\delta_k^{-2})^{-1},$$

$$\mu_n(x,y) = n^{-1}\sum_{k=1}^n \chi(\text{Re }\lambda_{kn} < x)\chi(\text{Im }\lambda_k < y),$$

and λ_{kn} *are the eigenvalues of the matrix* $A_n^{-1}B_n$.

Proof. By using the **V**-transform, we have for $q \ne 0$,

$$\iint e^{ipx+iqy}d\mu_n(x,y) = (p^2 + q^2)(4i\pi q)^{-1}\int_{-c}^c [f_n(z)\alpha_n e^{isq}|_{s=-d}^{s=d}$$
$$- \int_{-d}^d f_n(z)\alpha_n iqe^{isq}ds]e^{itp}\,dt$$
$$+ \varepsilon_n(p,q,c,d) + \Delta(p,q,c,d,\alpha), \tag{17.10.4}$$

where

$$\Delta(p,q,c,d,\alpha)$$
$$= (p^2 + q^2)(4i\pi q)^{-1}\int_{-c}^c \int_{-d}^d (\partial/\partial s)[f_n(z)(1 - \alpha_n)]e^{itp+isq}dtds,$$
$$\alpha_n = [1 + \alpha f_n^2(z)]^{-1},$$
$$f_n(z) = n^{-1}\ln|\det((A_n z - B_n)n^{-1/2})|^2, \qquad \alpha > 0,$$
$$\lim_{d\to\infty}\lim_{c\to\infty}\sup_n |\varepsilon_n(p,q,c,d)| = 0$$

Obviously,

$$\Delta(p,q,c,d,\alpha) = (p^2+q^2)(4i\pi q)^{-1} \int_{-c}^{c} \int_{-d}^{d} [3\alpha f^2(z) + \alpha^2 f^4(z)]$$
$$\times (1+\alpha f^2(z))^{-2}(\partial/\partial s)f(z)e^{itp+iqs} dtds.$$

$$(17.10.5)$$

From (17.10.5), we obtain for fixed p, q, c, d,

$$\mathbf{E}|\Delta(p,q,c,d,\alpha)|$$
$$\leq c_3 n^{-1} \sum_{k=1}^{n} [\mathbf{E} \int_{-d}^{d} \int_{-c}^{c} (|s - \operatorname{Im}\lambda_{kn}||z - \lambda_{kn}|^{-2})^\varepsilon dtds]^{\varepsilon^{-1}}$$
$$\times [\mathbf{E} \int_{-d}^{d} \int_{-c}^{c} |3\alpha f^2(z) + \alpha^2 f^4(z)|(1+\alpha f^2(z))^{-2}|^\beta dtds]^{\beta^{-1}},$$

$$(17.10.6)$$

where $2 > \varepsilon > 1, \beta > 1, \varepsilon^{-1} + \beta^{-1} = 1$. Note that

$$\ln|\det \Xi_n(z)|^2 = \sum_{k=1}^{n} \ln[\sigma_{kn}^2 + (t^2+s^2)\delta_{kn}^2] + \ln|\det H_n|^2,$$

where

$$H_n = ((\xi_{pl}^{(n)}z - \eta_{pl}^{(n)})(\delta_{pn}^2|z|^2 + \sigma_{pn}^2)^{-1/2})_{p,l=1}^{n} n^{-1/2},$$
$$\Xi_n(z) = (A_n z - B_n)n^{-1/2}.$$

For the entries H_n, Theorem 6.4.1 can be applied. Therefore,

$$\operatorname{plim}_{n\to\infty} |n^{-1}\ln|\det \Xi_n(z)|^2 + 1 - n^{-1}\sum_{k=1}^{n} \ln[\sigma_{kn}^2 + (t^2+s^2)\delta_{kn}^2]| = 0.$$

Using this relation, (17.10.4), (17.10.6), and the fact that the quantities $|n^{-1}\ln|\det \Xi_n(z)|^2|\alpha_n$ are bounded, we obtain

$$\lim_{\alpha \downarrow 0} \overline{\lim_{n\to\infty}} \mathbf{E}|\Delta(p,q,c,d,\alpha)| = o,$$

$$\lim_{n\to\infty} \mathbf{E}| \int |e^{ipx+iqy} d\mu_n(x,y) - (p^2+q^2)(4i\pi q)^{-1}$$
$$\times \iint (\partial/\partial s)[n^{-1}\sum_{k=1}^{n} \ln[\sigma_{kn}^2 + (t^2+s^2)\delta_{kn}^2]]e^{itp+isq} dtds| = 0.$$

$$(17.10.7)$$

Applying the inverse formula for an integral equation

$$F(x,y) = \pi^{-1} \iint (z - x - iy)^{-1} p(x,y) dx dy,$$

it is easy to verify that

$$(p^2 + q^2)(4i\pi q)^{-1} \iint (\partial/\partial s)[n^{-1} \sum_{k=1}^{n} \ln[\sigma_{kn}^2 + (t^2 + s^2)\delta_{kn}^2$$

$$\times \, e^{itp+isq} dt ds$$

$$= \iint e^{ipx+iqy} \alpha[n^{-1} \sum_{k=1}^{n} [\sigma_{kn}^2 + (x^2 + y^2)\delta_{kn}^2]^{-2}] dx dy.$$

This equality and (17.10.7) imply the statement of Theorem 17.10.1.

If, in addition to the conditions of Theorem 17.10.1,

$$\lim_{n \to \infty} \lambda_n(x) = \lambda(x),$$

$$\lambda_n(x) = n^{-1} \sum_{k=1}^{n} \delta_{kn}^{-4} \chi(\sigma_{kn}^2 \delta_{kn}^{-2} < x),$$

$\lambda(x)$ is a nondecreasing function of bounded variation, and $\int_0^\infty x^{-1-\epsilon} d\lambda(x) < \infty, \epsilon > 0$, then

$$\text{plim}_{n \to \infty} \mu_n(x, y) = c_2 \int_{-\infty}^{x} \int_{-\infty}^{y} [\int_0^{-\infty} (x + u^2 + v^2)^{-2} d\lambda(x)] du dv,$$

$$c_2 = (\pi \int_0^\infty x^{-1} d\lambda(x))^{-1}.$$

If, in addition to the conditions of Theorem 17.10.1, $\delta_{in}^2 \equiv \sigma_{in}^2 \equiv 1$, then

$$\text{plim}_{n \to \infty} \mu_n(x, y) = \pi^{-1} \int_{-\infty}^{x} \int_{-\infty}^{y} (1 + u^2 + v^2)^{-2} du dv.$$

Theorem 17.10.1 can be generalized as follows. We can require that under the conditions of this theorem the random vectors $(\xi_{ij}^{(n)}, \xi_{ji}^{(n)}), (\eta_{ij}^{(n)}, \eta_{ji}^{(n)})$ be independent and such that $\mathbf{E}\xi_{ij}^{(n)}\xi_{ji}^{(n)} = \rho_1, \mathbf{E}\eta_{ij}^{(n)}\eta_{ji}^{(n)} = \rho_2$. Then, if in addition to the conditions of Theorem 17.10.1 $\delta_{in}^2 = \sigma_{in}^2 = 1$ and the random vectors $(\xi_{ij}^{(n)}, \xi_{ji}^{(n)}), (\eta_{ij}^{(n)}, \eta_{ji}^{(n)})$ have the distribution densities $p_{ij}(x, y), q_{ij}(x, y)$, satisfying the condition

$$\sup_{i,j} \int [p_{ij}^{\beta}(x, y) + q_{ij}^{\beta}(x, y)] dx dy < \infty, \quad \beta > 1,$$

then

$$\text{plim}_{n \to \infty} \mu_n(x, y) = \pi^{-1} \int_{-\infty}^{x} \int_{-\infty}^{y} p(u, v) du dv,$$

$$p(t, s) = (4\pi)^{-1}(\partial^2/\partial t^2) f(t, s) + (4\pi)^{-1}(\partial^2/\partial s^2) f(t, s),$$

where

$$f(t,s) = \ln[1 + t^2 + s^2] + \pi^{-1} \iint_{g(x,y,z)\leq 1} \ln(x^2 + y^2)dxdy,$$

$$g(x,y,z) = (bx - ay)^2(1 - |\rho|^2)^{-2}|\rho|^{-2} + (ax + by)^2(1 + |\rho|^2)^{-2}|\rho|^{-2},$$

$$a = \operatorname{Re}\rho^{1/2}, \quad b = \operatorname{Im}\rho^{1/2}, \quad \rho = \rho_1 + z^2\rho_2.$$

To prove this, we need to use the equality

$$\ln|\det n^{-1/2}(A_n + zB_n)|^2 = n\ln[1 + |z|^2] + \ln|\det c_n|^2,$$

where

$$c_n = \{(\xi_{ij}^{(n)} + z\eta_{ij}^{(n)})(1 + |z|^2)^{-1/2}n^{-1/2}\}_{i,j=1}^n.$$

Further, by using the elliptic law, we prove that

$$\operatorname{plim}_{n\to\infty} n^{-1}\ln|\det c_n|^2 = \pi^{-1} \iint_{q(x,y,z)\leq 1} \ln(x^2 + y^2)dxdy.$$

CHAPTER 18

THE DISTRIBUTION OF EIGENVALUES AND EIGENVECTORS OF ADDITIVE RANDOM MATRIX-VALUED PROCESSES

In this chapter, the explicit formulas for the distributions of the eigenvalues and eigenvectors of random matrix-valued processes and stochastic differential equations for finding the eigenvalues and eigenvectors are derived.

§1 Distribution of Eigenvalues and Eigenvectors of Random Symmetric Matrix-Valued Processes

Let $\Xi_n(t)$ be a random symmetric square matrix of order n, with real random processes $\xi_{ij}(t), t \geq 0$ being its elements. Let $\lambda_1(t), \ldots, \lambda_n(t), \overrightarrow{\theta}_1(t), \cdots, \overrightarrow{\theta}_n(t)$, respectively be the eigenvalues and eigenvectors of the matrix $\Xi_n(t)$. We arrange the eigenvalues $\lambda_1(t) \geq \cdots \geq \lambda_n(t)$ for every t and choose the eigenvectors $\overrightarrow{\theta}_i(t), i = \overline{1, n}$ in such a way that their first nonzero component be positive. Assume that the finite-dimensional kth dimensional distributions of the random process $\Xi_n(t)$ have the densities $p(t_1, \ldots, t_k, x_1, \ldots, x_k)$, where $t_1, \ldots, t_k$ are some values of time parameter, $x_i, i = \overline{1, k}$ are real symmetric matrices of order n. Using the proof of Theorem 3.2.1, we obtain that the density of finite-dimensional k-dimensional distributions of the eigenvector $\{\lambda_1(t), \ldots, \lambda_n(t)\}$ is equal to

$$q(t_1, \ldots, t_k, Y_1, \ldots, Y_k) := c^k \int y_{1s} > \cdots > y_{ns}, \quad h_{1n}^{(s)} > 0, \quad s = \overline{1, k} \int$$

$$\times p(t_1, \ldots, t_k, H_1 Y_1 H_1', \ldots, H_k Y_k H_k')$$

$$\times \prod_{s=1}^{k} \prod_{i>j}(y_{is} - y_{js}) \prod_{s=1}^{k} \mu(dH_s),$$

where $Y_s = (\delta_{ij} y_{is})$, $H_s = (h_{ij}^{(s)})$ are real orthogonal matrices of order $n, c = \pi^{n(n+1)/4} \prod_{i=1}^{n} \{\Gamma((n-i+1)/2)\}^{-1}$. Similarly, we get the joint distributions for the eigenvalues and eigenvectors of the matrix $\Xi_n(t)$ as well as formulas for the distribution of the eigenvalues and eigenvectors of random matrix-valued processes $\Xi_n(t)$, where $\Xi_n(t)$ for every t is one of the matrices examined in §§1–8 of Chapter 3.

Let $\Lambda(t)$ be the diagonal matrix of random eigenvalues of the matrix $H_n(t)$ and $\theta(t)$ the matrix of random eigenvectors of the matrix $H_n(t)$. Suppose that $H_n(t)$ is the random process with independent increments, and (Γ_1, B_1) is the measurable set of matrices of order n. Considering the equation

$$\mathbf{P}\{\Lambda(t) \in \mathbf{E}_1, \qquad \theta(t) \in \mathbf{E}_2/\Lambda(S) = Y_1, \qquad \Lambda(u) = Y_2, \qquad \theta(s) = x_1,$$
$$\theta(u) = x_2\} = \mathbf{P}\{\Lambda(H(t) - H(s) + H(s)) \in \mathbf{E}_1, \quad \theta(H(t)$$
$$- H(s) + H(s)) \in \mathbf{E}_2/\Lambda(S) = Y_1, \quad \Lambda(u) = Y_2, \quad \theta(s) = x_1,$$
$$\theta(u) = x_2\} = \mathbf{P}\{\Lambda(H(t) - H(s) + X_1 Y_1 X_1^{-1}) \in \mathbf{E}_1, \quad \theta(H(t)$$
$$- H(s) + X_1 Y_1 X_1^{-1}) \in \mathbf{E}_2/\Lambda(S) = Y_1, \quad \theta(s) = X_1\},$$

where $\mathbf{E}_1 \in B_1, \mathbf{E}_2 \in B_1, t > s > u$, we obtain that the process $\{\Lambda(t), \theta(t)\}$ will be a Markov one.

We obtain an analogous statement for the eigenvalues and eigenvectors of the random matrix processes $H_n(t)$ with multiplicative right-hand independent increments. We consider the process $H_n(t)$ to be a random one with multiplicative right (left)-hand independent increments if for $t > 0$ with probability 1 $H_n^{-1}(t)$ exists and for any $0 \le t_1 < t_2 < \cdots < t_n$ the random matrices $H_n(t_k) H_n^{-1}(t_{k-1}), k = \overline{1, n}, \{H_n^{-1}(t_{k-1}) H_n(t_k), k = \overline{1, n}\}$ are independent.

Consider a two-parameter right (left)-hand set of random matrices $H_n(s, t)$, $0 \le s \le t < \infty$ having the following property: for any

$$s \le t_1 < t_2 < \cdots < t_n, H_n(s, t) = H_n(s, t_1) \ldots H_n(t_{n-1}, t_n),$$

and the random matrices $H_n(t_i, t_{i+1}), i = \overline{1, n}$ are independent. In the same manner as above, we obtain that the random vector of eigenvalues $\lambda(H_n(s, t))$ and the random matrix of eigenvectors $\theta(H_n(s, t))$ with fixed s are a multidimensional Markov process as $t \ge s$.

For this multidimensional Markov process, the Chapman–Kolmogorov equation can be written. In this chapter, we shall find sufficient conditions for this Markov process to be a diffusion one. We need some auxiliary formulas which will be presented in the next section.

§2 Perturbation Formulas

If we assume the existence of the distribution density of a random matrix, its eigenvalues with probability 1 will be different. Therefore, we need formulas for the perturbations of matrices with different eigenvalues. First, we consider the case of symmetric matrices. Let A and B be the symmetric square matrices of order n, let $\lambda_1 > \cdots > \lambda_n$ be the eigenvalues of the matrix

A, with $\lambda_i \neq \lambda_j, i \neq j, l_i, i = \overline{1,n}$ being the eigenvectors of the matrix A corresponding to these eigenvalues.

Let $\lambda_i(A + \varepsilon B)$ and $l_j(A + \varepsilon B)$ denote the eigenvalues and eigenvectors of the matrix A, where ε is some arbitrary real parameter, with the eigenvalues $\lambda_j(A+\varepsilon B)$, satisfying the relation $\lim_{\varepsilon \to 0} \lambda_j(A+\varepsilon B) = \lambda_j$. The coefficients of the characteristic equation for the matrix $A+\varepsilon B$ are analytical functions of ε. Therefore, the eigenvalues of such matrix are analytical functions of ε having only algebraic singularities. Then for ε sufficiently small, the expansions hold:

$$\lambda_j(A + \varepsilon B) = \sum_{m=0}^{\infty} \lambda_j^{(m)} \varepsilon^{(m)}, \quad \lambda_j^{(0)} = \lambda_j,$$

$$l_j(A + \varepsilon B) = \sum_{m=0}^{\infty} l_j^{(m)} \varepsilon^m, \quad l_j^{(0)} = l_j.$$

We prove the following theorem.

Theorem 18.2.1. *[116] If $\lambda_i \neq \lambda_j, i \neq j$, then*

$$\lambda_j^{(m)} = \sum_{s_1+s_2+s_3=m-2, s_i \geq 0} \mathrm{Tr}(S_j B)^{s_1} E_j B (S_j B)^{s_2} E_j B (S_j B)^{s_3}$$

$$+ \, \mathrm{Tr} \, E_j B (S_j B)^{m-1}, m \geq 2,$$

$$\lambda_j^{(1)} = \mathrm{Tr} \, E_j B,$$

$$E_j^{(m)} = \sum_{s_1+s_2=m-1, s \geq 0} (S_j B)^{s_1} E_j B (S_j B)^{s_2} S_j + (S_j B)^{m-1} E_j,$$

where

$$S_j = \sum_{k \neq j, k = \overline{1,n}} E_k (\lambda_k - \lambda_j)^{-1}.$$

We find the matrices $E_j^{(m)}, E_i$ from the expansion

$$E_j(\varepsilon) = \sum_{m=0}^{\infty} \varepsilon^m E_j^{(m)},$$

$$A = \sum_{i=1}^{n} \lambda_i E_i,$$

$$A + \varepsilon B = \sum_{i=1}^{n} \lambda_i(\varepsilon) E_i(\varepsilon).$$

Obviously, $E_i = l_i l_i', E_j(\varepsilon) = l_j(A + \varepsilon B) l_j'(A + \varepsilon B)$.

Proof. Let $R_z(\varepsilon) = (Iz - A - \varepsilon B)^{-1}, R_z = (Iz - A)^{-1}$ be the resolvents of the matrices $A + \varepsilon B$ and A.

It is evident that for ε sufficiently small,

$$R_z(\varepsilon) = \sum_{k=0}^{\infty} (\varepsilon R_z B)^k R_z = \sum_{k=0}^{\infty} \varepsilon^k (R_z B R_z B \ldots R_z B R_z).$$

Let the matrices $R_z, R_z(\varepsilon)$ be represented as

$$R_z = \sum_{k=1}^{n} E_k (z - \lambda_k)^{-1}, \qquad R_z(\varepsilon) = \sum_{k=1}^{n} E_k(\varepsilon)(z - \lambda_k(\varepsilon))^{-1}.$$

Let Γ be the positively oriented circumference of a sufficiently small radius with the center in λ_j containing the point $\lambda_j(A+\varepsilon B)$ and containing no other points $\lambda_0, \lambda_k(A+\varepsilon B)$. It is obvious that such a domain Γ can always be chosen by the proper choice of ε.

We use the Cauchy integral formula

$$f^{(n)}(a)(n!)^{-1} = (2\pi i)^{-1} \int_{\Gamma} f(z)(z-a)^{-n-1}dz, \quad n = 0, 1, 2, \ldots$$

where $f(z)$ is the analytic function and a is inside the circumference Γ, $f^{\circ}(z) = f(z)$, $f^{(n)}(z), n \geq 1$ is the nth derivative with respect to z, the integration is performed around the circumference counterclockwise.

$$(2\pi i)^{-1} \int_{\Gamma} R_z(\varepsilon)dz = E_j(\varepsilon), \qquad (2\pi i)^{-1} \int_{\Gamma} R_z dz = E_j,$$

$$(2\pi i)^{-1} \int_{\Gamma} (R_z B)^m R_z dz = (2\pi i)^{-1} \int_{\Gamma} [(S_j B + E_j B(z - \lambda_j)^{-1})$$
$$\times (S_j B + E_j B(z - \lambda_j)^{-1}) \ldots (S_j B + E_j B(z - \lambda_j)^{-1})]$$
$$\times [S_j + E_j(z - \lambda_j)^{-1}]dz = E_j^{(m)}.$$

Similarly, we obtain

$$\lambda_j(A + \varepsilon B) - \lambda_j(A) = (2\pi i)^{-1} \operatorname{Tr} \int_{\Gamma} R_z(\varepsilon)(z - \lambda_j)dz$$

$$= \sum_{k=1}^{\infty} \varepsilon^k \lambda_j^{(k)}.$$

Theorem 18.2.1 is proved.

In particular, from Theorem 18.2.1, we get the formulas

$$\lambda_k^{(1)} = (Bl_k, l_k), \quad \lambda_k^{(2)} = \sum_{m \neq k} (Bl_k, l_m)^2 (\lambda_k - \lambda_m)^{-1}, \tag{18.2.1}$$

$$l_k^{(1)} = \sum_{m \neq k} (Bl_k, l_m)(\lambda_k - \lambda_m)^{-1} l_m,$$

$$l_k^{(2)} = \sum_{m \neq k} \sum_{s \neq k} (Bl_m, l_s)(Bl_s, l_k)(\lambda_k - \lambda_s)^{-1}(\lambda_k - \lambda_m)^{-1}$$

$$\times\, l_m - \sum_{m \neq k} (Bl_k, l_k)(Bl_m, l_k)(\lambda_k - \lambda_m)^{-2}$$

$$- 0.5 l_k \sum_{m \neq k} (Bl_k, l_m)^2 (\lambda_k - \lambda_m)^{-2}. \tag{18.2.2}$$

Similarly, the formulas for the nonsymmetric square matrices A of order n can be obtained. If the eigenvalues of the matrix A are different, it can be represented as $A = Z\Lambda Z^{-1} = \sum_{k=1}^{n} \lambda_k T_k$, where $T_k = (z_{ik} z_{jk}^{(-1)})_{i,j=1}^{n}$, z_{ik} are the elements of the matrix Z, $\Lambda = (\lambda_i, \delta_{ij})$. Theorem 18.2.1 is valid for such matrices on condition that the matrix E_j must be replaced by T_j. Theorem 18.2.1 is also valid for such a case when the matrix A has only one eigenvalue λ_k isolated from all other eigenvalues.

Suppose now that the symmetric matrix A is random and has distribution density. The eigenvalues of such a matrix do not coincide with probability 1. Thus, the formulas of Theorem 18.2.1 are valid with probability 1. But here the problem of ordering the eigenvalues arises. It is evident that the eigenvalues cannot be arranged in increasing order in this case, since in formulas (18.2.1) and (18.2.2) the right-hand members of the equations must be arranged in increasing order. Therefore, we arrange in increasing order only the eigenvalues of the matrix A, and assign the number of eigenvalues of the matrix A to the eigenvalues $\lambda_k(\varepsilon)$ of the matrix $A + \varepsilon B$, with the eigenvalues of the matrix $A + \varepsilon B$ being expressed in terms of these numbers according to formulas (18.2.1) and (18.2.2). It is obvious that the eigenvalues of the matrix $A + \varepsilon B$ determined in such a way will be chosen in a unique manner. The eigenvectors of the matrix A are chosen such that their first nonzero components be positive, then the eigenvalues of the matrix $A + \varepsilon B$ determined by formulas (18.2.2) will also be chosen uniquely.

§3 Continuity and Nondegeneration of Eigenvalues of Random Matrix-Valued Processes with Independent Increments

In this section, a method for studying the spectrum of random matrices based on their random perturbations is examined. Stochastic (singular) differential equations for the eigenvalues and eigenvectors of random matrices were found using their random perturbations.

The scheme of application of random perturbations of matrices for finding the distribution of eigenvalues and eigenvectors of the random matrix Ξ can be briefly summarized in the following way: instead of the matrix Ξ, we consider some random matrix process $\Xi(t)$ with independent increments. (We assume that the matrix Ξ is such that $\Xi(1) = \Xi$.) Using the formulas for perturbations of the matrix $\Xi(t)$, we find the stochastic differential equation which the eigenvalues and eigenvectors of the matrix satisfy. They are measurable with respect to the underlying probability space on which the process $\Xi(t)$ is defined. Then the distribution of eigenvalues of the matrix Ξ will be equal to that of the solution of this stochastic differential equation where $t = 1$.

Let the random symmetric matrix-valued process $\Xi(t)$, of order n with independent increments $\Xi(t_{k+1}) - \Xi(t_k), 0 = t_0 < t_1 < \cdots < t_s = T$ be defined on the probability space (Ω, F, P). We assume that for any subdivision of the segment $[O, T]$, the distribution densities $p(x, t_{k+1}, t_k)$ of the random matrices $\Xi(t_{k+1}) - \Xi(t_k)$ exist. The process $\Xi(t)$ will be called a random matrix process of Brownian motion and will be denoted by $w(t)$ if $\Xi(0) = A$, where A is some nonrandom matrix, and

$$p(x, t_{k+1}, t_k) = (\pi \Delta t_k)^{-n(n+1)/2} 2^{-n/2} \exp\{-(2\Delta t_k)^{-1} \operatorname{Tr} x^2\},$$
$$\Delta t_k = t_{k+1} - t_k.$$

We assume that $\Xi(0)$ is the nonrandom matrix whose eigenvalues a_i are different and arranged in increasing order. Let us assume that the eigenvectors of the matrix A are fixed. We choose as eigenvalues of the matrix $\Xi(t)$ the eigenvalues $\lambda_i(t), i = \overline{1, n}$ determined by formulas (18.2.1). It can be seen from the formulas of perturbations of the random matrix spectrum that $\lambda_i(t)$ will be random variables defined on the probability space (Ω, F, P). The eigenvectors $\overrightarrow{\varphi}_i(t)$ of the matrix $w(t)$ will be chosen in a single possible way, since $\overrightarrow{\varphi}_i(t) \to \overrightarrow{\varphi}_i$ for $t \to 0$.

We calculate the joint distribution density of the eigenvalues $\lambda_i(t)$ of the matrix $w(t)$, given the matrix $w(s), t > s \geq 0$. From Chapter 3 we obtain that the density of the eigenvalues $\lambda_i(t)$ arranged in increasing order is equal to

$$p(Y/\theta_n) = c(\pi(t - s))^{-n(n+1)/2} 2^{-n/2}) \int_{h_{1i} \geq 0} \exp\{-(2(t - s))^{-1}$$

$$\times \operatorname{Tr}(\theta_n - HYH')^2\} \prod_{i>j} |y_i - y_j| \mu(dH), \quad y_1 > \cdots > y_n,$$
$$\tag{18.3.1}$$

where $\theta_n = (\delta_{ij}\theta_i)$ is the diagonal matrix of eigenvalues of the matrix $w(s)$, $Y = (\delta_{ij}y_i)$ is the diagonal matrix, μ is the Haar probability measure on the group G of real orthogonal matrices H,

$$c = 2^n \pi^{n(n+1)/4} \prod_{i=1}^{n} \{\Gamma[(n - i - 1)/2]\}^{-1}.$$

We derive the proof from Theorem 3.7.1 by using the fact that the density of the matrix $w(t)$, given the matrix $w(s)$, equals

$$(\pi(t - s))^{-n(n+1)/2} 2^{-n/2} \exp\{-(2(t - s))^{-1} \operatorname{Tr}(x - w(s))^2\}.$$

Then Theorem 3.7.1 and the invariance of the measure μ should be used.

It follows from formula (18.3.1) that the process $\Lambda(t) = (\delta_{ij}\lambda_i(t))^n_{i,j=1}$ will be a Markov one. Subsequently, we shall consider that this process is separable. Let us prove that it is continuous with probability 1.

Theorem 18.3.1. *The separable random process $\Lambda(t)$ is continuous with probability 1.*

Proof. We should verify the following condition [40],

$$\mathbf{E}|\lambda_i(t) - \lambda_i(s)|^p \le L|t - s|^{1+r}, \tag{18.3.2}$$

where $p > 0, r > 0, L > 0$. Using (18.3.1), we have for $t > s > 0$,

$$\sum_{i=1}^{n} \mathbf{E}|\lambda_i(t) - \lambda_i(s)|^p = c[(2\pi s)2\pi(t - s)]^{-n(n+1)/2}$$

$$\times \int \exp\{-2^{-1}(t - s)^{-1} \operatorname{Tr}(Y_1 + TXT')^2 - (2s)^{-1}X^2\} \sum_{i=1}^{n} |y_i - x_i|^p$$

$$\times \prod_{i>j}[(y_i - y_j)(x_i - x_j)]\mu(dT)dY\,dX,$$

where $Y = (\delta_{ij}y_i), X = (\delta_{ij}x_i), T$ is the real orthogonal matrix, μ is the Haar measure on the group G of the matrices T. We get from this formula,

$$\sum_{i=1}^{n} \mathbf{E}|\lambda_i(t) - \lambda_i(s)|^p$$

$$\le \mathbf{E}[\sum_{i=1}^{n} |\lambda_i(t) - \lambda_i(s)|^p / |\lambda_i(s) - \lambda_j(s)| > \varepsilon, \quad i \ne j]$$

$$\times \mathbf{P}\{|\lambda_i(s) - \lambda_j(s)| > \varepsilon, \quad i \ne j\} + \varepsilon^{n(n-1)/2}c, \tag{18.3.3}$$

where $\varepsilon > 0$ is an arbitrary real value which we shall choose later and c is some constant.

We write $\|w(t) - w(s)\|(\Delta t)^{-1} = \alpha(\Delta t), \Delta t = \sqrt{t - s}$, where by the norm of the matrix w, we shall mean the expression $\|w\| = \sqrt{\operatorname{Tr} ww}$. From Theorem 18.2.1, we obtain that $|\lambda_j^{(m)}| \le m^{-1}\alpha^m(\Delta t)(2/\varepsilon)^{m-1}$, provided that $|\lambda_i(s) - \lambda_j(s)| \ge \varepsilon, i \ne j$. By using this estimate and the perturbation formulas for the eigenvalues, we get

$$|\lambda_i(t) - \lambda_i(s)| \le \Delta t\alpha(\Delta t)(2/\varepsilon)[1 - \Delta t\alpha(\Delta t)^2/\varepsilon]^{-1}, \tag{18.3.4}$$

where $\Delta t \alpha(\Delta t) 2/\varepsilon < 1$. Therefore, by choosing $\varepsilon = \sqrt{\Delta t}/4$, we obtain

$$\mathbf{E} \sum_{i=1}^{n} [|\lambda_i(t) - \lambda_i(s)|^p / |\lambda_i(s) - \lambda_j(s)| > \varepsilon, \quad i \neq j]$$

$$\times \mathbf{P}\{|\lambda_i(s) - \lambda_j(s)| > \varepsilon, \, i \neq j\}$$

$$\leq \sum_{i=1}^{n} \int_{|x| < \Delta t^{-1/2}} |\Delta t x (1 - \varepsilon^{-1} \Delta t x^2)^{-1}|^p$$

$$\times d\mathbf{P}\{\alpha(\Delta t) < x\} + c_1 \Delta t^p \leq c_2 \Delta t^p,$$

where c_1, c_2, p are some positive constants.

It follows from this inequality and (18.3.3) that

$$\sum_{i=1}^{n} \mathbf{E}|\lambda_i(t) - \lambda_i(s)|^4 \leq L|t - s|^m,$$

where $m > 1$ is some integer.

Theorem 18.3.1 is proved.

A crucial point in the investigation of perturbation formulas for random matrices is the proof that for any $T > 0$,

$$\mathbf{P}\{\inf_{t \in [0,T]} |\lambda_i(t) - \lambda_j(t)| > 0, i \neq j, i, j = \overline{1,n}\} = 1. \tag{18.3.5}$$

Theorem 18.3.2. *For any real $T > 0$, the separable continuous random process $\Lambda(t)$ satisfies the condition (18.3.5).*

Proof. For any $\varepsilon > 0$,

$$\mathbf{P}\{\inf_{t \in [0,T]} |\lambda_i(t) - \lambda_j(t)| > \varepsilon, i \neq j, i, j = \overline{1,n}\}$$

$$\geq \mathbf{P}\{\inf_{t \in (0,T]} |\lambda_i(t) - \lambda_j(t)| > \varepsilon, i \neq j, i, j = \overline{1,n}/|\lambda_i(t_k)$$

$$- \lambda_j(t_k)| > \delta_1, i \neq j, k = \overline{1,m}\}\mathbf{P}\{|\lambda_i(t_k) - \lambda_j(t_k)| > \delta_1,$$

$$i \neq j, k = \overline{1,m}, \quad \sup_{t \in [t_{k-1}, t_k]} |\alpha_k(\Delta t_k)| < \delta_2\},$$

where $\alpha_k(\Delta t_k) = \|\Xi(t_k) - \Xi(t_{k-1})\|, \delta_1 > 0, \delta_2 > 0, 0 \leq t_1 \leq \cdots \leq t_m$ is an arbitrary subdivision of the half-segment $(0,T]$.

By using the estimate (18.3.4), we get for $|\lambda_i(t_k) - \lambda_j(t_k)| > \delta_1$,

$$i \neq j, k = \overline{1,m}, \quad \sup_{t \in [t_{k-1}, t_k]} |\alpha_k(\Delta t_k)| < \delta_2:$$

$$\inf_{t \in [t_{k-1}, t_k]} |\lambda_i(t) - \lambda_j(t)| \geq |\lambda_i(t_{k-1}) - \lambda_j(t_{k-1})|$$

$$- \sup_{t \in [t_{k-1}, t_k]} |\lambda_i(t) - \lambda_j(t) - \lambda_i(t_{k-1}) - \lambda_j(t_{k-1})| \geq \delta_1$$

$$- \sup_{t \in [t_{k-1}, t_k]} \alpha_k(\Delta t_k) 2\delta_1^{-1} [1 - \sup_{t \in [t_{k-1}, t_k]} \alpha_k(\Delta t_k) 2\delta_1^{-1}]$$

$$\geq \delta_1 - \delta_2 \delta_1^{-1} 2(1 - \delta_2 \delta_1^{-1} 2)^{-1}.$$

Now we choose δ_1 and δ_2 such that

$$\delta_1 - \delta_2\delta_1^{-1}2(1 - \delta_2\delta_1^{-1}2) = \varepsilon.$$

Then

$$\mathbf{P}\{\inf_{t\in[0,T]}|\lambda_i(t) - \lambda_j(t)|, i \neq j, i, j = \overline{1,n}\}$$
$$\geq \mathbf{P}\{|\lambda_i(t_k) - \lambda_j(t_k)| > \delta_1, i \neq j, \sup_{t\in[t_{k-1},t_k]}|\alpha_k(\Delta t_k)| < \delta_2\}$$
$$\geq 1 - \delta_1^{3/2}\sum_{i\neq j,k=\overline{1,m}}\mathbf{E}|\lambda_i(t_k) - \lambda_j(t_k)|^{-3/2}$$
$$- \delta_2\sum_{k=1}^{m}\mathbf{E}\{\sup_{t\in[t_{k-1},t_k]}|\alpha_k(\Delta t_k)|\}^2,$$

where $s \geq 0$. From the inequalities for martingales, we have (Ref. [40, p. 137])

$$\mathbf{E}[\sup_{t\in[t_{k-1},t_k]}|\alpha_k(\Delta t_k)|]^2$$
$$\leq (s/(s-1))^s\mathbf{E}|\alpha_k(\Delta t_k)|^s \leq c_s\Delta t_k^{s/2}, \qquad (18.3.6)$$

where c_s is some constant. It follows from formula (18.3.1) that

$$\mathbf{E}|\lambda_i(t_k) - \lambda_j(t_k)|^{-3/2} \leq L < \infty. \qquad (18.3.7)$$

Using (18.3.6) and (18.3.7), we have

$$\mathbf{P}\{\inf_{t\in(0,T]}|\lambda_i(t) - \lambda_j(t)| > \varepsilon, i \neq j, i, j = \overline{1,n}\}$$
$$\geq 1 - L\delta_1^{3/2}m - m\Delta t^{s/2}\delta_2^{-s}.$$

It is easy to observe that s, δ_1, δ_2 and Δt_k can be chosen in such a way that for $\varepsilon > 0$,

$$\mathbf{P}\{\inf_{t\in(0,T]}|\lambda_i(t) - \lambda_j(t)| > \varepsilon, i \neq j, i, j = \overline{1,n}\} = 1.$$

But then (18.3.6) follows from $a_i \neq a_j$, $i \neq j$. Theorem 18.3.2 is proved.

§4 Straight and Back Spectral Kolmogorov Equations for Distribution Densities of Eigenvalues of Random Matrix Processes with Independent Increments

Suppose that $\Xi_n(t)$ is the random symmetric matrix process of order n with independent increments. If finite-dimensional distribution densities of such a process exist, then the eigenvalues $\lambda_i(t), i = \overline{1,n}$ of such a matrix will be distinct with probability 1 when t is fixed.

Let $\overline{\lim}_{\Delta t \downarrow 0} \mathbf{E}(\Delta t)^{-1}\|\Xi_n(t + \Delta t) - \Xi_n(t)\| < \infty$ exist. Then for Δt sufficiently small, the statement of Theorem 18.2.1 is valid.

$$\lambda_j[\Xi_n(t) + \Delta t(\Xi_n(t + \Delta t) - \Xi_n(t))(\Delta t)^{-1}]$$
$$= \sum_{m=0}^{\infty} \lambda_j^{(m)}(\Delta t)^m, \qquad \lambda_j^{(0)} = \lambda_j(\Xi_n(t)),$$
$$l_j[\Xi_n(t) + \Delta t(\Xi_n(t + \Delta t) - \Xi_n(t))(\Delta t)^{-1}]$$
$$= \sum_{m=0}^{\infty} l_j^{(m)}(\Delta t)^m, \qquad l_j^{(0)} = l_j(\Xi_n(t)). \tag{18.4.1}$$

Let the eigenvalues λ_j of the matrix $\Xi_n(0)$ be arranged, as in the previous section.

Consider a particular form of the matrix $\Xi_n(t)$.

Let $w_n(t)$ be the symmetric matrix process of Brownian motion of order n, i.e., the elements of the matrix $w_n(t)$ are random processes of the form $\delta_{ij}\mu_i + w_{ij}(t)(1 + \delta_{ij})/2$, where $w_{ij}(t), i \geq j$ are independent processes of Brownian motion, $\mu_1 > \mu_2 > \cdots > \mu_n$ are arbitrary real nonrandom values. It was demonstrated in the previous section that for the Markov process $\Lambda(t)$ determined by the perturbation formulas, the transition probability density exists,
$$p(s, \overrightarrow{x}, t, \overrightarrow{y}), \ \overrightarrow{x} = (x_1, \ldots, x_n), \ \overrightarrow{y} = (y_1, \ldots, y_n).$$

We prove the following theorem.

Theorem 18.4.1 (Back Kolmogorov Equation). *Let $f(\overrightarrow{x})$ be continuous and bounded,*

$$u(s, \overrightarrow{x}) := \int_{R^n} f(\overrightarrow{y})p(s, \overrightarrow{x}, t, \overrightarrow{y})d\overrightarrow{y},$$
$$M^n := \{\overrightarrow{x} : x_i \neq x_j, i \neq j\}.$$

Then $u(s, \overrightarrow{x})$ for $\overrightarrow{x} \in M^n, s \in (0, t)$ satisfies the equation

$$-\frac{\partial u(s, \overrightarrow{x})}{\partial s} = \sum_{i=1}^{n} a_i(\overrightarrow{x})/2\frac{\partial u(s, \overrightarrow{x})}{\partial x_i} + \frac{1}{2}\sum_{i=1}^{n}\frac{\partial^2 u(s, \overrightarrow{x})}{\partial x_i^2} \tag{18.4.2}$$

where

$$a_i(\overrightarrow{x}) = \sum_{k \neq i}[1/(x_i - x_k)]$$

and the boundary condition $\lim_{s \uparrow t} u(s, \overrightarrow{x}) = f(\overrightarrow{x})$.

Proof. We introduce $M_\varepsilon^n = \{\overrightarrow{x} : |x_i - x_j| > \varepsilon, i \neq j\}, \varepsilon > 0$. By using estimates (18.3.4) for residual sums of series (18.4.1), it is easy to verify that for some $\delta > 0$ for all the values of the vectors $\overrightarrow{x} \in M_\varepsilon^n$,

$$\int \|\overrightarrow{y} - \overrightarrow{x}\|^{2+\delta} p(s, \overrightarrow{x}, t, \overrightarrow{y}) d\overrightarrow{y} = 0(s - t),$$

$$\int (y_i - x_i) p(s, \overrightarrow{x}, t, \overrightarrow{y}) d\overrightarrow{y} = a_i(\overrightarrow{x})(s - t) + 0(s - t),$$

$$\int (\overrightarrow{z}, \overrightarrow{y} - \overrightarrow{x})^2 p(s, \overrightarrow{x}, t, \overrightarrow{y}) d\overrightarrow{y} = (\overrightarrow{z}, \overrightarrow{z})(s - t) + 0(s - t),$$

$$\|\overrightarrow{x}\|^2 = (\overrightarrow{x}, \overrightarrow{x}), \qquad d\overrightarrow{y} = \prod_{i=1}^{n} dy_i. \tag{18.4.3}$$

We prove, for example, the second relation

$$\int (y_i - x_i) p(s, \overrightarrow{x}, t, \overrightarrow{y}) d\overrightarrow{y} = a_i(\overrightarrow{x})\Delta t + \beta(s, t), \overrightarrow{x} \in M_\varepsilon^n,$$

where $\Delta t = (s - t), \beta(s, t)$ is estimated as follows:

$$|\beta(s, t)| \leq \int_{|x| \leq (\Delta t)^{-1/2}\varepsilon/2} |1 - (2/\varepsilon)\Delta t(x)|^{-1}(\Delta t)^2 x^2(2/\varepsilon)$$
$$\times dP\{\alpha(\Delta t) < x\} + \mathbf{E}|\lambda_i(w(t)) - \lambda_i(w(s)) - a_i(\overrightarrow{x})\Delta t|$$
$$\times \lambda_i(w(s)) = x_i, \quad |\alpha(\Delta t)| \geq (\Delta t)^{-1/2}\varepsilon/2|P\{|\alpha(\Delta t)|$$
$$\geq (\Delta t)^{-1/2}\varepsilon/2\} \leq \Delta t/(1 - \sqrt{\Delta t}) + \Delta t c_\varepsilon,$$

where c_ε is some constant. Thus, the second relation in formula (18.4.3) holds.

We obtain the boundary condition for (18.4.2) from the following relation,

$$u(s, \overrightarrow{x}) - f(\overrightarrow{x}) = \int_{\|\overrightarrow{x} - \overrightarrow{y}\| \leq \varepsilon} [f(\overrightarrow{y}) - f(\overrightarrow{x})]p(s, \overrightarrow{x}, t, \overrightarrow{y}) d\overrightarrow{y} + 0_{\varepsilon_1}(t - s),$$

where $\varepsilon_1 > 0, 0_{\varepsilon_1}(t - s)(t - s)^{-1} \to 0$ as $t - s \to 0$ and any $\varepsilon_1 > 0$. One can see from the formula for the density of eigenvalues that the continuous derivatives $\frac{\partial u(s, \overrightarrow{x})}{\partial x_i}, \frac{\partial^2 u(s, \overrightarrow{x})}{\partial x_i^2}, x \in M_\varepsilon^k$ exist. Therefore, using the Taylor expansion of the function $u(s, \overrightarrow{x})$, we have

$$u(s, \overrightarrow{y}) - u(s, \overrightarrow{x}) = \sum_{i=1}^{n}(y_i - x_i)\frac{\partial u(s, \overrightarrow{x})}{\partial x_i}$$
$$+ \frac{1}{2}\sum_{i,j=1}^{n}(y_i - x_i)(y_i - x_j)\frac{\partial^2 u(s, \overrightarrow{x})}{\partial x_i \partial x_j} + r(\overrightarrow{x}, \overrightarrow{y}, s),$$

and for $\overrightarrow{y} \in s_{\epsilon_2}(\overrightarrow{x})|r(\overrightarrow{x}, \overrightarrow{y}, s)| \leq \|\overrightarrow{y} - \overrightarrow{x}\|^2 \delta_{\epsilon_1}$,

$$\delta_\epsilon = \sup_{i,j,s,\overrightarrow{y} \in s_{\epsilon_2}(\overrightarrow{x})} \left| \frac{\partial^2 u(s, \overrightarrow{x} + \theta(\overrightarrow{y} - \overrightarrow{x}))}{\partial x_i \partial x_j} - \frac{\partial^2 u(s, \overrightarrow{x})}{\partial x_i \partial x_j} \right| \to 0,$$

as $\epsilon_2 \to 0, x \in M_\epsilon^n, \|\theta\| < 1, s_{\epsilon_2}(\overrightarrow{x})$ is the sphere in R^n of the radius ϵ_2 with the center in the point $\overrightarrow{x}$.

Obviously,

$$u(s, \overrightarrow{y}) = \int p(s, \overrightarrow{y}, t, \overrightarrow{z}) u(t, \overrightarrow{z}) d\overrightarrow{z}, s < t.$$

Therefore, using the Taylor expansion for $u(s, \overrightarrow{y}) - u(s, \overrightarrow{x})$, we have as $\overrightarrow{x} \in M_\epsilon^n$,

$$u(t, \overrightarrow{x}) - u(s, \overrightarrow{x}) = \int [u(s, \overrightarrow{y}) - u(s, \overrightarrow{x})] p(t, \overrightarrow{x}, s, \overrightarrow{y}) d\overrightarrow{y}$$

$$= \int_{\|\overrightarrow{y} - \overrightarrow{x}\| < \epsilon} [u(s, \overrightarrow{y}) - u(s, \overrightarrow{x})] p(t, \overrightarrow{x}, s, \overrightarrow{y}) d\overrightarrow{y} + 0(t - s)$$

$$= \sum_{i=1}^n \frac{\partial u(s, \overrightarrow{x})}{\partial x_i} \int_{\|\overrightarrow{y} - \overrightarrow{x}\| < \epsilon} (y_i - x_i) p(t, \overrightarrow{x}, s, \overrightarrow{y}) d\overrightarrow{y}$$

$$+ 2^{-1} \sum_{i,j=1}^n \frac{\partial^2 u(s, \overrightarrow{x})}{\partial x_i \partial x_j} \int_{\|\overrightarrow{y} - \overrightarrow{x}\| \leq \epsilon} (y_i - x_i)(y_j - x_j)$$

$$\times p(t, \overrightarrow{x}, s, \overrightarrow{y}), d\overrightarrow{y} + 0(t - s).$$

Using formulas (18.4.3), we obtain for $x \in M_\epsilon^n$,

$$u(t, \overrightarrow{x}) - u(s, \overrightarrow{x})/(s - t) = \sum_{i=1}^n \frac{\partial u(s, \overrightarrow{x})}{\partial x_i} a_i(\overrightarrow{x})$$

$$+ 2^{-1} \sum_{i=1}^n \frac{\partial^2 u(s, \overrightarrow{x})}{\partial x_i^2} + \delta_{\epsilon_1}(\Delta t)] + 0(t - s)/(s - t),$$

where $\lim_{\epsilon_1 \downarrow 0} \lim_{\Delta t \downarrow 0} \delta_{\epsilon_1}(\Delta t) = 0$. Passing to the limit as $\Delta t \to 0$ and $\epsilon_1 \to 0$, and then $\epsilon \to 0$ we get (18.4.1). Theorem 18.4.1 is proved.

Now we find the straight Kolmogorov equation for $p(s, \overrightarrow{x}, t, \overrightarrow{y})$ which is also called the Fokker–Planck or Einstein–Smoluchowski equation.

Theorem 18.4.2. *The function* $p(s, \overrightarrow{x}, t, \overrightarrow{y})$ *in the domain* $s \in (t, T), \overrightarrow{x} \in R_n, \overrightarrow{y} \in M^n$ *satisfies the equation*

$$\frac{\partial p(s, \overrightarrow{x}, t, \overrightarrow{y})}{\partial t} = -2^{-1} \sum_{i=1}^n \frac{\partial}{\partial y_i} [a_i(\overrightarrow{y}) p(s, \overrightarrow{x}, t, \overrightarrow{y})]$$

$$+ 2^{-1} \sum_{i=1}^n \frac{\partial}{\partial y_i^2} p(s, \overrightarrow{x}, t, \overrightarrow{y}). \tag{18.4.4}$$

For any fixed s, the solution of this equation exists and is unique for all initial functions $p(s, \overrightarrow{x}, t, \overrightarrow{y})$ belonging to the class of everywhere compact in the metric of uniform convergence on the space of all continuous and differentiable functions $p(s, \overrightarrow{x}, t, \overrightarrow{y})$, once with respect to s and twice with respect to $y_i, i = \overline{1, n}$.

Proof. Let $g(\overrightarrow{x})$ be an arbitrary twice continuously differentiable function which is equal to zero outside some finite measurable set in R^n. Similarly to the proof of Theorem 18.4.1, we obtain for $\overrightarrow{x} \in M_\varepsilon^n, \varepsilon > 0$,

$$\lim_{h \downarrow 0} h^{-1}[\int g(\overrightarrow{y})p(s, \overrightarrow{x}, s+h, \overrightarrow{y})d\overrightarrow{y} - g(\overrightarrow{x})]$$

$$= \sum_{i=1}^{n} a_i(\overrightarrow{x})(\partial/\partial x_i)g(\overrightarrow{x}) + 0.5 \sum_{i=1}^{n}(\partial^2/\partial x_i^2)g(\overrightarrow{x}).$$

Using the Chapman–Kolmogorov equation and this equation, we have

$$(\frac{\partial}{\partial t}) \int p(s, \overrightarrow{x}, t, \overrightarrow{y})g(\overrightarrow{y})d\overrightarrow{y} = \lim_{h \downarrow 0} h^{-1} \int [p(s, \overrightarrow{x}, t+h, \overrightarrow{y})$$

$$- p(s, \overrightarrow{x}, t, \overrightarrow{y})]g(\overrightarrow{y})d\overrightarrow{y} = \lim_{h \downarrow 0} \int p(s, \overrightarrow{x}, , t, \overrightarrow{y})h^{-1}$$

$$\times [\int p(t, \overrightarrow{x}, t+h, \overrightarrow{z})g(\overrightarrow{z})d\overrightarrow{z} - g(\overrightarrow{y})]d\overrightarrow{y} = \int_{\overrightarrow{y} \in M_{\varepsilon_1}^n} p(s, \overrightarrow{x}, t, \overrightarrow{y})$$

$$\times \left[\sum_{i=1}^{n} a_i(\overrightarrow{y})(\frac{\partial}{\partial y_i})g(\overrightarrow{y}) + 0.5 \sum_{i=1}^{n}(\frac{\partial^2}{\partial y_i^2})g(\overrightarrow{y}) \right] d\overrightarrow{y}$$

$$+ \lim_{h \downarrow 0} \int_{\overrightarrow{y} \in M_{\varepsilon_1}^n} p(s, \overrightarrow{x}, t, \overrightarrow{y})[h^{-1} \int p(t, \overrightarrow{y}, t+h, \overrightarrow{z})g(\overrightarrow{z})d\overrightarrow{z}$$

$$- g(\overrightarrow{y})]d\overrightarrow{y} + \frac{\partial}{\partial t} \int_{\overrightarrow{y} \in M_{\varepsilon_1}^n} p(s, \overrightarrow{x}, t, \overrightarrow{y})g(\overrightarrow{y})d\overrightarrow{y}.$$

By integrating this expression by parts ($g(\overrightarrow{y}) = 0$ outside some finite set) and by using it for $\varepsilon_1 \rightarrow 0$

$$(\frac{\partial}{\partial t}) \int_{\overrightarrow{y} \in M_{\varepsilon_1}^n} p(s, \overrightarrow{x}, t, \overrightarrow{y})g(\overrightarrow{y})d\overrightarrow{y} \rightarrow 0,$$

$$\int_{\overrightarrow{y} \in M_{\varepsilon_1}^n} p(s, \overrightarrow{x}, t, \overrightarrow{y})[\sum_{i=1}^{n} a_i(\overrightarrow{y})(\frac{\partial}{\partial y_i})g(\overrightarrow{y})$$

$$+ \sum_{i=1}^{n}(\frac{\partial^2}{\partial y_i^2})g(\overrightarrow{y})]d\overrightarrow{y} \rightarrow 0,$$

we get as $x \in M_\varepsilon^n$,

$$\int (\frac{\partial}{\partial t}) p(s, \overrightarrow{x}, t, \overrightarrow{y}) g(\overrightarrow{y}) d\overrightarrow{y} = - \int \{\sum_{i=1}^{n} (\frac{\partial}{\partial y_i})[a_i(\overrightarrow{y})$$

$$\times p(s, \overrightarrow{x}, t, \overrightarrow{y})] + 0.5 \sum_{i=1}^{n} (\frac{\partial^2}{\partial y_i^2}) p(s, \overrightarrow{x}, t, \overrightarrow{y})\} g(\overrightarrow{y}) d\overrightarrow{y}.$$

With the function $g(\overrightarrow{y})$ and ε being arbitrary, we come to the statement of Theorem 18.4.2.

By writing Eq. (18.4.4) for $p(0, \overrightarrow{x}, t, \overrightarrow{y})$ and by integrating it over $\overrightarrow{x}$, we get the following equation for the density $p(t, \overrightarrow{y})$ of the vector $(\lambda_1(t), \ldots, \lambda_n(t))$,

$$(\frac{\partial}{\partial t}) p(t, \overrightarrow{y}) = \sum_{i=1}^{n} (\frac{\partial}{\partial y_i})[a_i(\overrightarrow{y}) p(t, \overrightarrow{y})]$$

$$+ 0.5 \sum_{i=1}^{n} (\frac{\partial^2}{\partial y_i^2}) p(t, \overrightarrow{y}),$$

with the initial condition

$$p(0, \overrightarrow{y}) = \prod_{i=1}^{n} \delta(y_i - a_i),$$

where δ is the delta function.

An interesting peculiarity of the theorems in this section is that the distributions of the eigenvectors of the matrix $w_n(t)$ were not used for finding the distribution densities of the eigenvalues of the matrix $w_n(t)$. It happened because in finding relations (18.4.2), the expression $\mathbf{E}[((w_n(t) - w_n(s)) \overrightarrow{l}_k, \overrightarrow{l}_k)/\overrightarrow{l}_k]$, where $\overrightarrow{l}_k$ is the eigenvector of the matrix $w_n(s)$, will be independent of $\overrightarrow{l}_k$. It is not valid for a general case, and when the Kolmogorov equations for distribution densities of eigenvalues are derived, it is necessary to use the distributions of eigenvectors. But the distribution densities of eigenvectors in R_{n^2} do not exist in a general case. Therefore, when Kolmogorov-type equations are derived, analytical difficulties arise; besides, these equations become cumbersome. It was mentioned in Chapter 1 that the components of eigenvectors can be expressed as certain differentiable functions of the Euler angles. But these functions are complex. Thus, in this case, stochastic integro-differential equations which are compact and can contribute to thorough knowledge of the problems of existence and uniqueness of equation solutions are more convenient.

§5 Spectral Stochastic Differential Equations for Random Symmetric Matrix Processes with Independent Increments

Consider the matrix process $w(t)$, defined in the previous section. Let us prove the following statement for the eigenvalues $\lambda_i(t)$ of this matrix process determined by the perturbation formulas.

Theorem 18.5.1. *The eigenvalues $\lambda_k(t)$ satisfy the system of spectral stochastic differential equations*

$$d\lambda_k(t) = \sum_{m \neq k} (\lambda_k(t) - \lambda_m(t))^{-1} dt + dw_k(t),$$

$$\lambda_k(0) = a_k, \quad k = \overline{1,k} \tag{18.5.1}$$

$w_k(t)$ *are independent random processes of Brownian motion, a weak solution of the system of equations (18.5.1) exists and is unique in a strong sense.*

Proof. For the eigenvalues $\lambda_i(t)$ of the matrix process $w(t)$ determined by perturbation formulas, we have the following equations under the condition that

$$\inf_{t \in [0,T]} |\lambda_i(t) - \lambda_j(t)| > \varepsilon, \quad i \neq j, \quad i,j = \overline{1,n}, \quad \varepsilon > 0. \tag{18.5.2}$$

$$\lambda_s(t) - \lambda_s(0) = \sum_{k=1}^{m} (w(\Delta t_k)\overrightarrow{\varphi}_s(t_{k-1}), \varphi_s(t_{k-1}))\sqrt{\Delta t_k}$$

$$+ \sum_{k=1}^{m} \sum_{m \neq s} \Delta t_k (w(\Delta t_k)\overrightarrow{\varphi}_s(t_{k-1}), \overrightarrow{\varphi}_m(t_{k-1}))^2$$

$$\times (\lambda_s(t_{k-1}) - \lambda_m(t_{k-1}))^{-1} + \delta_s(\varepsilon, \max_k \Delta t_k), \tag{18.5.3}$$

where

$$w(\Delta t_k) = [w(t_k) - w(t_{k-1})](\Delta t_k)^{-1/2},$$

$$0 \leq t_0 < t_1 < \cdots < t_m = t, \quad \Delta t_k = t_k - t_{k-1},$$

and $\overrightarrow{\varphi}_s(t_k)$ are the eigenfunctions of the process $w(t_k)$ corresponding to the eigenvalues $\lambda_k(t_k)$. For $\delta(\varepsilon_1 \max_k \Delta t_k)$ because of estimates for the residual series of perturbation formulas, we have the following relation under the condition (18.5.2),

$$\text{plim}_{\max_k \Delta t_k \downarrow 0} \delta(\varepsilon_1 \max_k \Delta t_k) = 0. \tag{18.5.4}$$

We rewrite formulas (18.5.3) in the following form,

$$\lambda_s(t) - \lambda_s(0) = \sum_{k=1}^{m}(w(\Delta t_k)\overrightarrow{\varphi}_s(t_{k-1}), \overrightarrow{\varphi}_s(t_{k-1}))\sqrt{\Delta t_k}$$

$$+ \sum_{k=1}^{m}\sum_{m\neq s}\Delta t_k(w(\Delta t_k)\overrightarrow{\varphi}_s(t_{k-1}), \overrightarrow{\varphi}_m(t_{k-1}))^2(\lambda_s(t_{k-1})$$

$$- \lambda_m(t_{k-1}))^{-1}\chi(|\lambda_s(t_{k-1}) - \lambda_m(t_{k-1})|^{-1} < c)$$

$$+ \delta_1(\varepsilon, \max_k \Delta t_k, c), \tag{18.5.5}$$

where under the condition (18.5.2),

$$\text{plim}_{\max_k \Delta t_k \downarrow 0, c\to\infty} \delta_1(\cdot) = 0.$$

Since

$$\mathbf{E}[(w(\Delta t_k)\overrightarrow{\varphi}_s(t_{k-1}), \overrightarrow{\varphi}_m(t_{k-1}))/w(t_{k-1})] = 0,$$
$$\mathbf{E}[(w(\Delta t_k)\overrightarrow{\varphi}_s(t_{k-1}), \overrightarrow{\varphi}_m(t_{k-1}))^2/w(t_{k-1})]^2 = 1,$$

and the functions $\lambda_i(t)$ are continuous

$$\text{plim}_{\max_k \Delta t_k \downarrow 0} \sum_{k=1}^{m}\sum_{m\neq s}\Delta t_k(w(\Delta t_k)\overrightarrow{\varphi}_s(t_{k-1}), \overrightarrow{\varphi}_m(t_{k-1}))^2$$

$$\times (\lambda_s(t_{k-1}) - \lambda_m(t_{k-1}))^{-1}\chi(|\lambda_s(t_{k-1}) - \lambda_m(t_{k-1})|^{-1} < c)$$

$$= \int_0^T \sum_{m\neq s}(\lambda_s(t) - \lambda_m(t))^{-1}\chi(|\lambda_s(t) - \lambda_m(t)|^{-1} < c)dt,$$

$$\{\sum_{k=1}^{m}(w(\Delta t_k)\overrightarrow{\varphi}_s(t_{k-1}), \overrightarrow{\varphi}_s(t_{k-1})), s = \overline{1, n}\}$$

$$\approx \{w_1(t), \ldots, w_n(t)\}, \tag{18.5.6}$$

where the random vector process $\{w_1(t), \ldots, w_n(t)\}$ is nonanticipative with respect to the flow F_t of σ-algebras, the random matrix-valued processes $w(s), s < t$ being measurable with respect to them.

By using (18.5.2), (18.5.5), and (18.5.6), we find

$$\{\lambda_s(t) - \lambda_s(0) - \int_0^T [\sum_{m\neq s}(\lambda_s(t) - \lambda_m(t))^{-1}$$

$$+ \delta_s(\varepsilon, \max_k \Delta t_k)]dt, s = \overline{1, n}\} \approx \{w_1(t), \ldots, w_n(t)\}.$$

From this relation and (18.5.4), passing to the limit as we derive equation (18.5.1).

Let us demonstrate that Eq. (18.5.1) has a weak solution which is unique in a strong sense. The existence of the weak solution $\Lambda(t)$ of Eq. (18.5.1) has just been proved. Suppose that two weak solutions $\Lambda_1(t)$ and $\Lambda_2(t)$ of Eq. (18.5.1) exist which are set on one probability space. For these solutions, we have for any $\varepsilon > 0$,

$$\mathbf{E}\{ \sup_{0 \leq t \leq T} \|\overrightarrow{\Lambda}_1(t) - \overrightarrow{\Lambda}(t)\| / \inf_{t \in [0,T]}[|\lambda_{1i}(t) - \lambda_{1j}(t)|$$
$$+ |\lambda_i(t) - \lambda_j(t)|] > \varepsilon, \quad i \neq j \} = 0.$$

It follows from

$$\mathbf{P}\{ \inf_{t \in [0,T]} |\lambda_i(t) - \lambda_j(t)| > 0, i \neq j, i, j = \overline{1,n} \} = 1,$$

that $\overrightarrow{\Lambda}_1(t) = \overrightarrow{\Lambda}(t)$. Theorem 18.5.1 is proved.

Similarly, using the perturbation formulas for eigenvectors, we obtain the following statement.

Theorem 18.5.2.   *The eigenvectors $\overrightarrow{\varphi}_k(t), k = \overline{1,n}$ of the matrix process $w(t)$ satisfy the system of stochastic differential equations*

$$d\overrightarrow{\varphi}(t) = -0.5C(t)\overrightarrow{\varphi}(t)dt + R^{1/2}d\overrightarrow{w}(t),$$
$$\overrightarrow{\varphi}'(t) = \{\overrightarrow{\varphi}_1(t), \dots, \overrightarrow{\varphi}_n(t)\},$$
$$C(t) = \mathrm{diag}\{I_n \sum_{m \neq k}(\lambda_k(t) - \lambda_m(t))^{-2}, \quad k = \overline{1,n}\},$$
$$R = [B_{ij}]_{i,j=1}^n,$$
$$B_{kk} = \sum_{m \neq k} \overrightarrow{\varphi}_m(t)\overrightarrow{\varphi}'_m(t)(\lambda_k(t) - \lambda_m(t))^{-2},$$
$$B_{ij} = -\overrightarrow{\varphi}_i(t)\overrightarrow{\varphi}'_j(t)(\lambda_i(t) - \lambda_j(t))^{-2}, i \neq j;$$
$$\overrightarrow{w}'(t) = \{w_i(t), \quad i = \overline{1,n^2}\}, \quad \overrightarrow{\varphi}_k(0) = \overrightarrow{a}_k,$$

where $\overrightarrow{a}_k$ are the eigenvectors of the matrix A corresponding to the eigenvalues μ_k. For Eq. (18.5.7), a weak solution exists being unique in a strong sense.

Analogously, the equations for the eigenvalues and eigenvectors of random symmetric matrix processes with Gaussian independent increments as well as of random nonsymmetric matrix processes with Gaussian independent increments can be found. But in this case, a joint system of stochastic differential equations for eigenvalues and eigenvectors should be found. Since the latter is cumbersome, it is not given here.

It should be noted that in [130] the other derivation of Eq. (18.5.1) was proposed as follows. Let L be the set of real symmetric matrices x of order n,

$$L = \{x : \lambda_i(x) \neq \lambda_j(x), i \neq j\},$$

where $\lambda_i(x)$ are the eigenvalues of the matrix x arranged in increasing order.

Consider the function

$$p(X) = \int_L [\operatorname{Tr}(X - Y)^2]^d m(dY), \tag{18.5.8}$$

where $d = -n(n + 1)/2$, $m(Y)$ is the probability measure on the set $\bar{L}$ absolutely continuous relative to the Lebesgue measure on each of the sets $L_k = \{Y : \lambda_1 = \cdots = \lambda_k < \cdots < \lambda_n\}$ with the distribution density $c \exp^{-\operatorname{Tr}(Y)^2}$, $Y \in L_k$.

We introduce the matrix $A(\theta) = X + \theta(Y - X)$, and $\lambda_k(\theta), \varphi_k(\theta)$ are its eigenvalues and eigenvectors. Differentiating the equality $(A(\theta) - \lambda_k(\theta)I)\varphi_k (\theta) = 0$ and multiplying the obtained vector by $\varphi_k(\theta)$, we have $\lambda'_k(\theta) = (A'(\theta)\varphi_k(\theta), \varphi_k(\theta))$. From this equality for almost all values of the matrices X and Y with respect to the Lebesgue measure on the set L, it follows that

$$\lambda_k(Y) - \lambda_k(X) = \int_0^1 ((Y - X)\varphi_k(\theta), \varphi_k(\theta))d\theta.$$

Hence,

$$|\lambda_k(X) - \lambda_k(Y)|^2 \leq \operatorname{Tr}(X - Y)^2, \quad k = \overline{1, n}.$$

By using this inequality, we get from (18.5.8) for almost all matrices X,

$$p(x) \leq [\min_{i \neq j} |\lambda_i(x) - \lambda_j(x)|^2/2]^{-d/2+1}. \tag{18.5.9}$$

If $X \in L_k$, then

$$\int_{L_k} [\operatorname{Tr}(X - Y)^2]^{-d/2+1} m(dY) = \infty. \tag{18.5.10}$$

Let us prove (18.5.10). It is obvious that as $X \in L_k$ for any $\varepsilon > 0$,

$$\int_{L_k} [\operatorname{Tr}(X - Y)^2]^{-d/2+1} m(dY) \geq c \int [\operatorname{Tr} Y^2]^{-d/2+1} dY$$

$$L_k \cap \{Y : \operatorname{Tr} Y^2 \leq \varepsilon\}, \tag{18.5.11}$$

where $c > 0$ is some constant.

We change the variables $Y = H\Lambda H'$ in the integral from the right-hand side of this inequality, where H is the orthogonal matrix $h_{1i} \geq 0, i = \overline{1, n}, \Lambda$

is the diagonal matrix and its diagonal elements satisfy the condition $\lambda_1 = \cdots = \lambda_k < \lambda_{k+1} < \cdots < \lambda_n$. On the set L_k this transformation will be one-to-one and differentiable with respect to the variables $\lambda_k, \ldots, \lambda_n$, where θ_p are the Euler angles of the matrix (Chapter 3). The Jacobian $I(H, \Lambda)$ of such a transformation is equal to the module of some homogeneous polynomial function of the variables $\lambda_k, \ldots, \lambda_n$ of power $(n - k + 1)(n - k)/2$.

Thus, because of (18.5.11),

$$\int_{L_k} [\mathrm{Tr}(X - Y)^2]^{-d/2+1} m(dY) \geq c_1 \int [k\lambda_k^2 + \lambda_{k+1}^2 + \cdots + \lambda_n^2]^{-d/2+1}$$
$$\times \varphi(\Lambda) d\Lambda, \quad k\lambda_k^2 + \cdots + \lambda_n^2 \leq \varepsilon,$$

where $\varphi(\Lambda) = \int_{h_{1i} \geq 0, i=\overline{1,n}} I(H, \Lambda)\mu(dH)$, μ is the normalized Haar measure on the group of real orthogonal matrices of order n; then using the polar replacement of variables, it is quite obvious that (18.5.10) is valid.

Let us denote ν as

$$\nu = \inf(t \geq 0 : w(t) \in \bar{L}).$$

Since ν is the Markov moment and $w(0) \in L$, then as $0 \leq t < \nu$, using the Ito formula, we have

$$dp(w(t)) = \sum_{i \geq j}(\partial p(w(t))/\partial x_{ij})dw_{ij}(t)$$
$$+ [0.5 \sum_{i=1}(\partial^2 p(w(t))/\partial x_{ii}^2$$
$$+ 4^{-1} \sum_{i<j}(\partial^2 p(w(t))/\partial x_{ij}^2)]dt. \qquad (18.5.12)$$

It can be easily seen that the drift coefficient of this equation is identically equal to zero. Therefore, as $t < \nu$ because of (18.5.9),

$$f(t) := \int_0^t [\sum_{i \geq j}(\partial p(w(t))/\partial x_{ij})]^2 dt < \infty,$$

performing a random change of time in Eq. (18.5.12),

$$f^{-1}(t) = \min(s : f(s) = t),$$

we obtain that the random process $\xi(t) := p(w(f^{(-1)}(t))) - p(w(0))$ is a Brownian one until the moment $f(\nu)$.

If $\nu < \infty$, then $\xi(t) \to \infty$ as $t \uparrow f(\nu)$ since (18.5.10) is valid. But it is impossible, since $|\xi(t)| < \infty$, if $f(\nu) < \infty$. Otherwise, if $f(\nu) = \infty$, then it is possible to choose such a subsequence t that $\xi(t) \leq 0$. Thus, $\nu = \infty$. But then, by using the perturbation formulas for the eigenvalues and eigenvectors of random matrices, we derive Eq. (18.5.1).

Similarly, we prove that the moment of the hit of the process $w(t)$ in the set $(x : f(\lambda_i(x), i = \overline{1,n}) = 0)$ is equal to the infinity, where $f(x_1, \ldots, x_n)$ is any polynomial function which is not identically equal to zero.

§6 Spectral Stochastic Differential Equations for Random Matrix-Valued Processes with Multiplicative Independent Increments

Let w_0^s be the random matrix-valued process of dimension $m \times m$ satisfying the following conditions; for any $0 < t_1 < t_2 < \cdots < t_k < s$,

$$w_0^s = w_0^{t_1} w_{t_1}^{t_2} \ldots w_{t_k}^s, \quad w_0^0 = A, \quad w_t^t = I,$$

the random matrices $w_{t_i}^{t_{i+1}}, i = 1, 2, \ldots$ are independent, and their distributions depend only on the difference $t_{i+1} - t_i$, A is a real deterministic matrix, the eigenvalues α_i of the matrix AA' are different, $\alpha_1 > \alpha_2 > \cdots > \alpha_n$, $\lim_{\Delta t \downarrow 0} \mathbf{E} w_t^{t+\Delta t} = A$ for any vectors $\overrightarrow{x}$ and $\overrightarrow{y}$ of dimension m,

$$\lim_{\Delta t \to 0} (\Delta t)^{-1} \mathbf{E}[((w_0^{t+\Delta t} - w_0^t)\overrightarrow{x}, \overrightarrow{y})^2 / w_0^t]$$
$$= (\overrightarrow{x}, \overrightarrow{x})(\overrightarrow{y}, w_0^t (w_0^t)^* \overrightarrow{y}). \tag{18.6.1}$$

In the following, this process will be called the Wiener random matrix process with independent multiplicative increments. To simplify the formulas, we find equations for the eigenvalues of the process $w_0^t (w_0^t)^*$.

Let $\lambda_k(t)$ and $\overrightarrow{l}_k(t)$, respectively, be the eigenvalues and eigenvectors of the process $\xi(t) := w_0^t (w_0^t)$ defined by formulas (18.2.1).

Let $L = \{X := \lambda_i(X) \neq \lambda_j(X), i \neq j\}$, where X are nonnegatively defined matrices of dimension $m \times m$,

$$\nu = \inf\{t \geq 0 : w(t) \in \bar{L}\}.$$

By using the perturbation formulas for the eigenvalues and (18.6.1), we obtain that the eigenvalues $\lambda_k(t)$ satisfy the following system of stochastic differential equations as $t < \nu$,

$$d\lambda_c(t) = \sum_{s \neq k} (\lambda_s(t) + \lambda_k(t))(\lambda_k(t) - \lambda_s(t))^{-1} dt + n \, dt$$
$$+ 2\sqrt{\lambda_k(t)} dw_k(t), \quad \lambda_k(0) = \alpha_k;$$

$w_k(t)$ are independent processes of Brownian motion.

As in §§3–6 of this chapter, it is possible to prove that $\nu = \infty$. We must use the fact that the process $\xi(t)$ satisfies the stochastic diffusion equation, the entries of the diffusion matrix being linear functions of the matrix $\xi(t)$ and the coefficients of the drift vector being equal to 1. For such a process, we can find that it has finite-dimensional distribution densities which satisfy some inequalities. Using these inequalities, the perturbation formulas for eigenvalues as well as the inequality,

$$\mathbf{E}[\sup_{t\in[t_{k-1},t_k]} \|\xi(t_k) - \xi(t_{k-1})\|]^2 \leq \mathbf{E}[(\|w(t_{k-1})\|$$
$$+ \|w(t_k)\|) \sup_{t\in[t_{k-1},t_k]} \|w(t_k) - w(t_{k-1})\|]^2 \leq c_s \Delta t_k,$$

where c_s is some constant, we obtain that (18.3.5) for the process $\Lambda(t)$ holds, consequently, $\nu = \infty$.

Let us find a stochastic differential equation for the unitary random matrix-valued processes U_0^t of dimension $m \times m$ with multiplicative independent increments. Assume that $U_0^0 = A$ where A is a unitary matrix, $\exp(i\alpha_k)$ are its eigenvalues, and $0 \leq \alpha_1 < \alpha_2 < \cdots < \alpha_m < 2\pi$ for any t as $\Delta t \downarrow 0$,

$$U_0^{t+\Delta t} - U_0^t = iw(\Delta t)U_0^t, (\Delta t)^{-1/2}w(\Delta t) \Rightarrow w,$$

where w is the Hermitian matrix with the distribution density

$$c\exp\{-\frac{1}{2}\operatorname{Tr} X^2\}.$$

Define $L = \{H : \lambda_k(H) \neq \lambda_p(H), k \neq p\}$, where H are the unitary matrices, $\lambda_k(H)$ are the arguments of eigenvalues of the matrix H,

$$\mu = \inf\{t \geq 0 : U(t)\bar{\in}L\}.$$

Let $e^{i\lambda_k(t)}$ and $\varphi_k(t)$, respectively, be the eigenvalues and eigenvectors of the random process $U(t)$ found from formulas (18.2.1). Using the perturbation formulas for eigenvalues, we have as $\Delta t \geq 0, t \leq \mu$,

$$\exp(i\lambda_k(t))i[\lambda_k(t + \Delta t) - \lambda_k(t)] = i(w(\Delta t)$$
$$\times U_0^t\varphi_k(t), \bar{\varphi}_k(t)) - \sum_{s\neq k}(w(\Delta t)U_0^t\varphi_k(t), \bar{\varphi}_s(t))^2$$
$$\times (\exp\{i\lambda_k(t)\} - \exp\{i\lambda_s(t)\})^{-1} + \varepsilon(\Delta t),$$

where $\operatorname{plim}_{\Delta t\downarrow 0} \varepsilon(\Delta t) = 0$.

From this equation, we obtain the stochastic diffusion equation for the process $\Lambda(t)$ as $t < \mu$,

$$d\lambda_k(t) = -0.5\sum_{s\neq k} ctq(\lambda_k(t) - \lambda_s(t))/2dt$$
$$+ 0.5dw_k(t), \quad \lambda_k(0) = \lambda_k, \tag{18.6.2}$$

where $w_k(t)$ are independent processes of Brownian motion.

We can see from Eq. (18.6.1) and (18.6.2) that the distributions of eigen-values in these two cases do not depend on the distributions of eigenvectors. The eigenvectors corresponding to the eigenvalues also satisfy some stochas-tic diffusion equations. They can easily be written using proper perturbation formulas.

Similarly, we can find stochastic differential equations for eigenvalues of the stochastic matrizant which can be found in the following way: let $\{\Omega, F, P\}$ be the probability space on which the Wiener random matrix process $w(t)$ of $m \times m$ dimension is given, $A(t)$ and $B(t)$ measurable matrix functions of the same dimension, $F_t \in F$ a sequence of expanding σ-algebras, such that the random matrices $w(t)F_t$ are measurable, and the increments $w(t+s) - w(t)$ do not depend on F_t for any $t \geq 0, s > 0$. Suppose that $A(t)$ and $B(t)$ are measurable for every $t \in [0, T]$ and with probability 1,

$$\int_0^T [\|A(t)\|^2 + \|B(t)\|^2]dt < \infty.$$

Let $A(t) = A(t_k), B(t) = B(t_k), t \in [t_k, t_{k+1}]$, where $0 = t_0 < t_1 < \cdots < t_n = T$ is an arbitrary division of a segment $[0, T]$, and consider the products (the product of the factors should be taken in increasing order of index k),

$$J_0^n(n) = \prod_{k=0}^{n} [I + A(t_k)\Delta t_k + B(t_k)(w(t_{k+1}) - w(t_k))].$$

Using the proof of limit existence for the Ito integral sums, one can easily verify that there is a limit with respect to the probability of the products $J_0^T(n)$ as $\max(t_k - t_n) \to 0$, and this limit does not depend on the way of dividing the segment $[0, T]$ with points t_k.

The random matrix is a limit for $J_0^T(n)$, which we define as

$$J_0^T := \prod_0^T [I + A(t)dt + B(t)dw(t)].$$

§7 Stochastic Differential Equations for Differences of Eigenvalues of Random Matrix-Valued Processes

Let $w(t)$ be a symmetric matrix-valued process of Brownian motion (see §5). By using the perturbation formulas for the eigenvalues of the matrix $w(t)$, we have

$$\lambda_k(w(t + \Delta t)) - \lambda_{k-1}(w(t + \Delta t)) = \lambda_k w(t))$$

$$- \lambda_{k-1}(w(t)) + (c(\Delta t)\vec{e}_k, \vec{e}_k) - (c(\Delta t)\vec{e}_{k-1}, \vec{e}_{k-1})$$

$$+ \sum_{m \neq k} (c(\Delta t)\vec{e}_k, \vec{e}_k)^2 (\lambda_k(w(t)) - \lambda_m(w(t)))^{-1}$$

$$- \sum_{m \neq k-1} (c(\Delta t)\vec{e}_{k-1}, \vec{e}_{k-1})(\lambda_{k-1}(w(t)) - \lambda_m(w(t)))^{-1} + \varepsilon(\Delta t, t),$$

where λ_k and $\overrightarrow{e}_k$ are respectively the eigenvalues and eigenvectors of the matrix $w(t)$ defined by formulas (18.2.1),

$$\text{plim}_{\Delta t \downarrow 0} |\varepsilon(\Delta t, t)| \chi(t < \nu) = 0,$$

where

$$\nu = \inf\{t : \lambda_i(t) \in \bar{L}, \quad i = \overline{1, n}\},$$
$$L = \{x : \lambda_i(x) \neq \lambda_j(x), \quad i \neq j\}.$$

Let

$$\overrightarrow{\eta}(t) = \{\lambda_k(w(t)) - \lambda_{k-1}(w(t)), k = \overline{2, n}\}$$
$$=: \{\eta_1(t), \ldots, \eta_{m-1}(t)\}.$$

It is obvious that

$$\sum_{m \neq k} [\lambda_k(w(t)) - \lambda_m(w(t))]^{-1}$$

$$= \sum_{m=1}^{k-1} [\sum_{s=m}^{k-1} \eta_s(t)]^{-1} - \sum_{m=k+1}^{n} [\sum_{s=k-1}^{m} \eta_s(t)]^{-1} \qquad (18.7.1)$$

With the account of (18.7.1), we find drift and diffusion coefficients for the process $\overrightarrow{\eta}(t)$.

It is obvious that

$$\mathbf{E}[(c(\Delta t)\overrightarrow{e}_k, \overrightarrow{e}_k) - (c(\Delta t)\overrightarrow{e}_{k-1}, \overrightarrow{e}_{k-1})]$$
$$\times [(c(\Delta t)\overrightarrow{e}_m, \overrightarrow{e}_m - (c(\Delta t)\overrightarrow{e}_{m-1}, \overrightarrow{e}_{m-1})]$$

is equal to 0 if $k \neq m$, is equal to $2\Delta t$ if $k = m$, and is equal to $-\Delta t$ if $k - 1 = m$ or $k = m - 1$.

By using this relation and (18.7.1), we obtain the equation

$$d\overrightarrow{\eta}(t) = \{\sum_{m=1}^{k-1} [\sum_{s=m}^{k-1} \eta_s(t)]^{-1} - \sum_{m=k+1}^{n} [\sum_{s=k+1}^{m} \eta_s(t)]^{-1}$$
$$- \sum_{m=1}^{k-2} [\sum_{s=m}^{k} \eta_s(t)]^{-1} + \sum_{m=k}^{n} [\sum_{s=k}^{m} \eta_s(t)]^{-1},$$
$$k = \overline{1, n-1}\}dt + Ld\overrightarrow{w}(t),$$

where $L = (2\delta_{ij} - \delta_{ij-1} - \delta_{i-1j})$ is the matrix of order $n - 1, w(t)$ is the $n - 1$-dimensional process of Brownian motion,

$$\overrightarrow{\eta}(0) = \{\alpha_s - \alpha_{s-1}, s = \overline{2, n}\}.$$

It follows from Theorem 18.5.1 that the solution of this equation exists and is unique.

§8 Resolvent Stochastic Differential Equation for Self-Adjoint Random Matrix-Valued Processes

When solving some problems of the spectral theory of random matrices, it is necessary to be able to calculate the integrals $\mathbf{E}f(\xi(t))$, where $\xi(t)$ is a diffusion matrix process, f is an analytical function. For example, of great interest is the problem of calculating the integral $\mathrm{Tr}\,\mathbf{E}\exp(w+A)$, where w is the Hermitian matrix order n with the distribution density $\pi^{-n^2}\exp\{-\mathrm{Tr}\,XX^*\}$, A is the deterministic Hermitian matrix of order n. This problem seemed to be solved in the following way. Instead of the integral $\mathrm{Tr}\,\mathbf{E}\exp(w+A)$, consider the expression $\mathrm{Tr}\,\mathbf{E}\exp w(t)$ where $w(t)$ is the Wiener matrix-valued process, its increments $w(t+\Delta t) - w(t), \Delta t \geq 0$ being distributed just as the matrix $\Delta t^{1/2}w, w(0) = A$ and then for the random process $\exp w(t)$, we find a stochastic diffusion equation. However, with the help of simple calculations we obtain that

$$\lim_{\Delta t \downarrow 0} (\Delta t)^{-1}\mathbf{E}[\mathrm{Tr}\exp w(t + \Delta t) - \mathrm{Tr}\exp w(t)]$$

$$= \sum_{k=1}^{\infty}(2k!)^{-1} \sum_{\substack{s+p=2k-2 \\ s,p\geq 0}} 0.5\mathbf{E}\,\mathrm{Tr}\,w^s(t)\,\mathrm{Tr}\,w^p(t).$$

It can be seen from this formula that a stochastic diffusion equation for $\exp w(t)$ will be rather complicated.

Let us consider another method. Let $\mu(x) = \sum_{k=1}^{n}\chi(\lambda_k < x)$, where λ_k are the eigenvalues of the matrix w. Then

$$\mathrm{Tr}\exp w = \int e^x d\mu(x).$$

Consider the Stieltjes transform for $\mu(x)$:

$$\int (z+x)^{-1}d\mu(x) = \mathrm{Tr}(zI + w)^{-1}, z = t + is, s \neq 0.$$

It appears that for the traces of resolvents $R(z,t) := \mathrm{Tr}(zI + w(t))^{-1}$ the stochastic diffusion equation has a simple form:

$$d_t\begin{pmatrix} \mathrm{Tr}\,\mathrm{Re}\,R(z,t) \\ \mathrm{Tr}\,\mathrm{Im}\,R(z,t) \end{pmatrix} = 0.5\begin{pmatrix} \mathrm{Tr}\,\mathrm{Re}\,R^2(z,t)\,\mathrm{Tr}\,R(z,t) \\ \mathrm{Tr}\,\mathrm{Im}\,R^2(z,t)\,\mathrm{Tr}\,R(z,t) \end{pmatrix} dt$$

$$+ \begin{pmatrix} \mathrm{Tr}(\mathrm{Re}\,R^2(z,t))^2 & \mathrm{Tr}\,\mathrm{Re}\,R^2(z,t)\,\mathrm{Im}\,R^2(z,t) \\ \mathrm{Tr}\,\mathrm{Re}\,R^2(z,t)\,\mathrm{Im}\,R^2(z,t) & \mathrm{Tr}(\mathrm{Im}\,R^2(z,t))^2 \end{pmatrix}^{1/2}$$

$$\times \begin{pmatrix} dw_1(t) \\ dw_2(t) \end{pmatrix}, \tag{18.8.1}$$

where $w_i(t)$ are independent processes of Brownian motion

$$R(z,0) = \text{Tr}(Iz + A)^{-1}.$$

To derive the equation (18.8.1), we must use the equality

$$R(z, t + \Delta t) - R(z, t) = [\sum_{k=1}^{\infty}(-R(z,t)B)^k]R(z,t),$$

$$B = w(t + \Delta t) - w(t)$$

valid for sufficiently small Δt.

§9 Resolvent Stochastic Differential Equation for Non-Self-Adjoint Random Matrix-Valued Processes

Let $\xi(t) = (w_{ij}(t))_{i,j=1}^n$ be the random matrix-valued process, where $A := (a_{ij})_{i,j=1}^n$ is the matrix with the elements being constant, $w_{ij}(t)$ are the independent processes of Brownian motion. For the resolvents of such random processes, the stochastic diffusion equation can also be derived. (Ref. 8) but is not suitable for finding an equation for $\mathbf{E}\,\text{Tr}\,R_z^{(v)}$, since the integrals $\mathbf{E}\,\text{Tr}\,R_z^k$ diverge as $k = 2, 3, \ldots$. Consider the integrated resolvents

$$u(z, A, v) := \mathbf{E}\int_0^t \ln\det(I(u + is) + A + \xi(v))du,$$

$$s \neq 0, z = t + is,$$

which can be used for the finding of an equation for $\mathbf{E}\,\text{Tr}\,Rz(v)$.

It is obvious that

$$\lim_{\Delta v \downarrow 0} \mathbf{E}(\Delta v)^{-1}[u(z, A, v + \Delta v) - u(z, A, v)]$$

$$= \lim_{\Delta v \downarrow 0}(\Delta v)^{-1}\mathbf{E}\int_0^t \ln\det\{I + [I(u + is) + A$$

$$+ \xi(v)]^{-1}(\xi(v + \Delta v) - \xi(v))\}du$$

$$= -0.5\mathbf{E}\int_0^t \text{Tr}[I(u + is) + A + \xi(v)]^{-1}[I(u + is) + A$$

$$+ \xi(v)]^{-1\prime}du.$$

It is easy to verify that the integral on the right-hand side of this equality exists.

From this equality, we obtain the equation

$$(\partial/\partial v)u(z, A, v) = 0.5\sum_{i,j=1}^n (\partial^2/\partial a_{ij}^2)u(z, A, v),$$

$$u(z, A, 0) = \int_0^t \ln\det(I(u + is) + A)du. \qquad (18.9.1)$$

It is obvious that

$$\mathbf{E}\operatorname{Tr} R_z(v) = \iint (z - x - iy)^{-1} d\mu_n(x, y, v)$$
$$= (\partial^2/\partial t^2)u(t, A, v),$$

where

$$\mu_n(x, y, v) = \sum_{k=1}^{n} \mathbf{P}\{\operatorname{Re}\lambda_k(v) < x, \operatorname{Im}\lambda_k(v) < y\};$$

$\lambda_k(v)$ are the eigenvalues of the process $\xi(t)$.

To find the function $\mu_n(x, y, v)$, we can use the inverse formula for the Stieltjes transforms.

Similarly, we obtain the stochastic diffusion equation for the process

$$\eta(z, A, v) = \int_0^t \ln \det(I(u + is) + A + \xi(v)) du.$$

CHAPTER 19

THE STOCHASTIC LJAPUNOV PROBLEM
FOR SYSTEMS OF STATIONARY
LINEAR DIFFERENTIAL EQUATIONS

Let $\vec{x}'(t) = A\vec{x}(t)$, $\vec{x}(0) = \vec{x}_0$, $\vec{x}_0 \neq \vec{0}$ be a system of linear differential equations with a random matrix of coefficients A. The stochastic Ljapunov problem for such systems is that of finding a probability of the event

$$\{w : \vec{x}(t) \to \vec{0}, t \to \infty\}.$$

A lot of works have been devoted to the deterministic theory of stability founded by Ljapunov. The beginning of the stochastic theory of stability was laid in 1933. However, the first result of solving the problem of stability in our statement was obtained only in 1979 [75]. This theory has been further developed mainly due to efforts of Girko and Litvin [89]. To examine the stochastic Ljapanov problem, some results of the theory of random determinants have been used in this chapter.

§1 The Stochastic Ljapunov Problem for Systems of Linear Differential Equations with the Symmetric Matrix of Coefficients

Let us consider a system $\vec{x}'(t) = A\vec{x}(t)$, $\vec{x}(0) = \vec{c}$ of linear differential equations with constant real coefficients, where A is a square matrix of order n, and $\vec{x}$ and $\vec{c}$ are n vectors. The solution of such an equation converges to the null vector, as $t \to \infty$, for any vector $\vec{c} \neq \vec{0}$ if and only if $\operatorname{Re} \lambda_i < 0$, where λ_i are the eigenvalues of A. A matrix A for which $\operatorname{Re} \lambda_i < 0$ will be said to be *stable*. To prove the stability of A we can use Ljapunov's theorem: A is stable if and only if the matrix Y determined by the equation $A'Y + YA = -I$ is positive-definite. However, if A is a random matrix, this stability criterion is inefficient.

468

In the previous section, we determine the probability of the event that the solution of a system $\overrightarrow{x}''(t) = \Xi\overrightarrow{x}, \overrightarrow{x}(0) = \overrightarrow{c}$ of stochastic linear differential equations (where Ξ is a random symmetric matrix of order n whose entries on and above the diagonal have probability densities) is asymptotically stable.

Theorem 19.1.1. *Let Ξ be a random symmetric matrix of order n with probability density $p(x)$ and let λ_i be its eigenvalues. Then,*

$$\mathbf{P}\{\lambda_i < 0, i = \overline{1,n}\} = c \int p(-Z_{n\times(n+1)}Z'_{n\times(n+1)})dZ_{n\times(n+1)}, \qquad (19.1.1)$$

where $Z_{n\times(n+1)}$ is an $n \times (n+1)$ real matrix, and

$$c = \pi^{-n(n+3)/4} \prod_{i=1}^{n} \Gamma[(i+1)/2], dZ_{n\times(n+1)} = \prod_{\substack{i=\overline{1,n} \\ j=\overline{1,n+1}}} dZ_{ij}.$$

Proof. It is obvious that

$$\mathbf{P}\{\lambda_i < 0, i = \overline{1,n}\} = \int p(-Q)dQ, \qquad (19.1.2)$$

where $Q = (q_{ij})$ is a positive-semidefinite matrix and dQ is the element of the Lebesgue measure on the set of positive-semidefinite matrices.

Instead of (19.1.2), we consider the integral

$$J_\epsilon = c_{n,m} \int p(-Q)\exp\{-0.5\epsilon \operatorname{Tr} Q\}\det Q^{(m-n+1)/2}dQ$$
$$\times \epsilon^{[n(n+1)+n(m-n+1)]/2},$$

where $m = n+1, \epsilon > 0$, and

$$c_{n,m} = [2^{nm/2}\pi^{n(n-1)/4}\prod_{i=1}^{n}\Gamma[(m+1-i)/2]]^{-1}(2\pi)^{nm/2}.$$

As $\epsilon \to 0$, we obtain

$$J_\epsilon\epsilon^{-0.5[n(n+1)+n(m-n+1)]} \to \mathbf{P}\{\lambda_i < 0, i = \overline{1,n}\}. \qquad (19.1.3)$$

Let $H = (h_{ij})$ be an $m \times n$ matrix whose entries are independent and have $N(0,1)$ distribution. The probability density of HH' is called the *Wishart density*. It is well known that this density equals $c_{nm}e^{-\operatorname{Tr} Q}\det Q^{(m-n+1)/2}$. By using this, we write (19.1.3) in the following form:

$$J_\epsilon = c_{n,m}\int p(-Z_{m\times n}Z'_{m\times n})\exp\{-\epsilon\operatorname{Tr} Z_{m\times n}Z'_{m\times n}/2\}$$
$$\times dZ_{m\times n}\epsilon^{[n(n+1)+n(m-n+1)]/2}.$$

Then,

$$\lim_{\epsilon \to 0} I_\epsilon \epsilon^{-[n(n+1)+n(m-n+1)]/2} = c \int p(-Z_{n+1 \times n} Z'_{n+1 \times n}) dZ_{n+1 \times n}.$$

Theorem 19.1.1 is proved.

Corollary 19.1.1. *If the elements $\xi_{ij}, i \geq j$ of the symmetric matrix Ξ are independent and have $N(a_{ij}, \sigma_{ij}^2)$ distributions $(\sigma_{ij}^2 \neq 0)$, then*

$$\mathbf{P}\{\lambda_i < 0, i = \overline{1,n}\}$$

$$= (2\pi)^{-n(n+1)/2} \prod_{i \geq j} \sigma_{ij}^{-1} c \int \cdots \int \exp\{-\frac{1}{2} \sum_{i > j} \sigma_{ij}^{-2} (a_{ij} - \sum_{k=1}^{n+1} z_{ik} z_{jk})^2$$

$$-\frac{1}{2} \sum_{i=1}^{n} \sigma_{ii}^{-2} (a_{ii} - \sum_{k=1}^{n+1} z_{ik}^2)^2\} \prod_{i=\overline{1,n}, j=\overline{1,n+1}} dz_{ij}.$$

§2 Hyperdeterminants

Let the elements $\xi_{ij}, i \geq j, i, j = \overline{1,n}$ of the symmetric matrix A be independent random values distributed by the normal law $N(0, \sigma_{ij}^2), \sigma_{ij}^2 \neq 0$. According to Corollary 19.1.1, the probability of the fact that a system $\overrightarrow{x}'(t) = A \overrightarrow{x}(t), \overrightarrow{x}(0) = \overrightarrow{c} \neq 0$ is asymptotically stable is equal to

$$\mathbf{P}\{\alpha_i < 0, i = \overline{1,n}\} = (2\pi)^{-n(n+1)/2} c \prod_{1 \leq i \leq j \leq n} \sigma_{ij}^{-1}$$

$$\times \int \cdots \int \exp\{-(D\overrightarrow{z}, \overrightarrow{z}, \overrightarrow{z}, \overrightarrow{z})\} \prod_{i=\overline{1,n}, k=\overline{1,n+1}} dz_{ik}, \qquad (19.2.1)$$

where $\overrightarrow{z}$ is an $n(n+1)$-dimensional vector:

$$\overrightarrow{z} = (z_{11}, \ldots, z_{n1}, z_{12}, \ldots, z_{n2}, \ldots, z_{1n-1}, \ldots, z_{nn+1})$$

$$= (\hat{z}_1, \ldots, \hat{z}_m), \quad m = n(n+1),$$

$$(D\overrightarrow{z}, \overrightarrow{z}, \overrightarrow{z}, \overrightarrow{z}) = \sum_{p,q,r,s=1}^{m} d_{pqrs} \hat{z}_p \hat{z}_q \hat{z}_r \hat{z}_s.$$

We shall call the functional $(D\overrightarrow{z}, \overrightarrow{z}, \overrightarrow{z}, \overrightarrow{z})$ a hyperform.

$$D = \{d_{pqrs}, p, q, r, s = \overline{1, n(n+1)}\}$$

is a symmetric hypermatrix (or multidimensional matrix [156]) of the dimension $[n(n+1)]^4$, which can be defined in the following way:

$$d_{pppp} = (2\sigma_{ii}^2)^{-1}; \quad p = (k-1)n+i, \quad i = \overline{1,n}, \quad k = \overline{1,n+1},$$

$$d_{ppqq} = (6\sigma_{ii}^2)^{-1}; \quad p = (u-1)n+i, \quad q = (v-1)n+i; \quad i = \overline{1,n};$$
$$u = \overline{1,n}; \quad v = \overline{1,n+1}$$

$$d_{ppqq} = (12\sigma_{ij}^2)^{-1}; \quad p = (k-1)n+i; \quad q = (k-1)n+j; \quad i = \overline{1,n};$$
$$j = \overline{1,n}; \quad k = \overline{1,n+1}$$

$$d_{pqrs} = (24\sigma_{ij}^2)^{-1}; \quad p = (u-1)n+i; \quad q = (u-1)n+j; \quad r = (v-1)n+i;$$
$$s = (v-1)n+j, \quad i = \overline{1,n}; \quad j = \overline{1,n}; u = \overline{1,n+1}; \quad v = \overline{1,n+1}$$

All the other elements of the hypermatrix D are equal to zero. Thus, the problem to be solved may be reduced to the evaluation of the integral of the form

$$\int_{R^m} \exp\{-(D\overrightarrow{z}, \overrightarrow{z}, \overrightarrow{z}, \overrightarrow{z})\} \prod_{i=1}^{m} d\hat{z}_i.$$

Definition. The hyperdeterminant of the hypermatrix $D = \{d_{i_1 i_2 i_3 i_4}\}$ is an expression

$$g \det D = \sum [(-1)^{\sum_{\mu=1}^{4} I_\mu} d_{i_1^{(1)} \dots i_4^{(1)}} \dots d_{i_1^{(m)} \dots i_4^{(m)}}],$$

where $i_\mu^{(1)} \dots i_\mu^{(m)}$, $\mu = \overline{1,4}$ is a permutation of numbers $1, 2, \dots, m$; I_μ is the number of inversions in this permutation, and the sum is over all such permutations of all four indices.

The following assertion is proved in [89].

Theorem 19.2.1. *The expression (19.2.1) is equal to*

$$\mathbf{P}\{d_i < 0, i = \overline{1,n}\} = (2\pi)^{-n(n+1)/2} c \prod_{1 \le i \le j \le n} \sigma_{ij}^{-1}$$
$$\times \Gamma^m(4^{-1}) 2^{-m} (g \det D)^{-1/4},$$

$$(19.2.2)$$

where Γ is a gamma-function.

§3 The Stochastic Ljapunov Problem for Systems of Linear Differential Equations with a Nonsymmetric Matrix of Coefficients

Let a system of equations

$$\overrightarrow{x}'(t) = A\overrightarrow{x}(t), \overrightarrow{x}(0) = \overrightarrow{x}_0, \overrightarrow{x}_0 \ne \overrightarrow{0} \tag{19.3.1}$$

be given.

Theorem 19.3.1 [89]. *Let A be a random matrix, $p(x)$ be a joint density of distribution of the elements in this matrix. Then the system (19.3.1) is asymptotically stable with probability*

$$\mathbf{P}\{\operatorname{Re}\alpha_i < 0, i = \overline{1,n}\} = 2^n \int_0^{+\infty} \cdots \int_0^{+\infty} \prod_{i=1}^n dy_{ii} \int \cdots \int \prod_{i<j} dy_{ij}\, dh_{ij}$$

$$\times \{\prod_{i=1}^n y_{ii}^i |I(YY^*, H)| p(YY^*(H - I2^{-1}))\},$$

$$\tag{19.3.2}$$

where $H = -H^* = \{h_{ij}\}$ is a skew-symmetric matrix, Y is a triangular matrix, $y_{ii} > 0, i = \overline{1,n}, J(X,H)$ is the Jacobian of the matrix transform $A = X(H - 2^{-1}I)$, and X is a symmetric matrix. (All the matrices are $n \times n$-dimensional.)

Proof. Let us prove that the matrix A with property $\operatorname{Re}\alpha_i, i = \overline{1,n}$ if and only if it can be uniquely represented in the form $A = X(H - 2^{-1}I)$. In fact, let $\operatorname{Re}\alpha_i < 0, i = \overline{1,n}$. Then, there is such a symmetric-positive definite matrix V, satisfying the Ljapunov's equation $A^*V + VA = -I$ ([12], p.284). But then $(VA)^* + VA = -I$. We write the matrix VA in the form $VA = H + S$, where H is a skew-symmetric matrix, and S is a symmetric matrix. It is clear that $S = 2^{-1}I$. Consequently, $VA = H - 2^{-1}I, A = V^{-1}(H - 2^{-1}I)$, where V^{-1} is a symmetric positive-definite matrix. Now let $A = X(H - 2^{-1}I)$. But then X^{-1} will satisfy the Ljapunov equation. In fact,

$$(H^* - 2^{-1}I)XX^{-1} + X^{-1}X(H - 2^{-1}I) = H^* + H - I = -I.$$

Since X^{-1} is a symmetric positive-definite matrix, the eigenvalues of matrix A will satisfy the condition $\operatorname{Re}\alpha_i < 0, i = \overline{1,n}$ ([12], p. 284).

The representation of $A = X(H - 2^{-1}I)$ for stable matrices is unique due to the uniqueness of the Ljapunov's solution in relation to V.

Using this assertion, we prove Theorem 19.3.1 like Theorem 19.1.1.

Methods of calculating the Jacobian $J(X, H)$ of the transform are described in Ref. [89].

§4 The Spectral Method of Calculating a Probability of Stationary Stochastic Systems Stability

Let us find some formulas for a probability of the event $\{\operatorname{Re}\alpha_i < 0, i = \overline{1,n}\}$, where α_i are eigenvalues of the random matrix A, with the help of some formulas for the density of distribution of eigenvalues of random matrices found in Chapter 3.

For example, let $A = (\xi_{ij})$ be a symmetric random matrix with the density of distribution $p(x)$ of its elements $\xi_{ij}, i \geq j, i, j = \overline{1,n}, X = (x_{ij})$. Then

$$
\begin{aligned}
& \mathbf{P}\{\alpha_i < 0, i = \overline{1,n}\} \\
& = c_{1n} \int_{h_{1i}>0,0>y_1>\cdots>y_n} p(H_n Y_n H_n') \prod_{i>j}(y_i - y_j)\mu(dH_n)dY_n,
\end{aligned}
\tag{19.4.1}
$$

where μ is the normalized Haar measure indicated for a group of orthogonal matrices H_n, $Y_n = \{\delta_{ij}y_i\}$

$$
c_{1n} = 2^n \pi^{n(n+1)/4} \prod_{i=1}^{n}\{\Gamma(i/2)\}^{-1}.
$$

The formula (19.4.1) is more complicated than formulas (19.1.1) and (19.2.2), however in some cases it can be used to obtain simpler formulas than with the help of formulas (19.1.1) and (19.2.2). For example, let

$$
p(x) = (2\pi)^{-n(n+1)/4} \exp\{-2^{-1}\operatorname{Tr} X^2\}.
$$

Then

$$
\begin{aligned}
\mathbf{P}\{\alpha_i < 0, i = \overline{1,n}\} = {} & (2\pi)^{-n(n+1)/4}2^{-n}c_{1n}(n!)^{-1} \\
& \times \int_{y_i<0,i=\overline{1,n}} \exp\left\{-2^{-1}\sum_{i=1}^{n}y_i^2\right\}\prod_{i>j}|y_i - y_j|dY_n.
\end{aligned}
$$

This integral may be evaluated with the help of the Mehta theorem (see Chapter 2).

Similarly, using Theorem 3.5.1, we obtain the following assertion (see the notation used in this theorem): if the elements of matrix A are independent and are distributed according to a standard normal law, then

$$
\begin{aligned}
\mathbf{P}\{\operatorname{Re}\alpha_i < 0, i = \overline{1,n}\} = {} & \sum_{s=0}^{[n/2]}c_s\int_{\substack{\operatorname{Re}q_i<0,i=\overline{1,n}\\|y_1|>\cdots>|y_n|}}\exp\{-2^{-1} \\
& \times \operatorname{Tr}Y_s Y_s'\}\sqrt{I_s(Y_s)}\psi(Y_s)dY_s.
\end{aligned}
$$

§5 The Resolvent Method of Proving the Stability of the Solutions of Stochastic Systems

We shall call the solution of the system of equations $\vec{x}'(t) = A\vec{x}(t), \vec{x}(0) = \vec{c} \neq 0$, where A is a random matrix asymptotically stable in the mean if

$\lim_{t\to\infty} \mathbf{E}\|\overrightarrow{x}(t)\|^2 = 0$. When proving the stability in the mean of solutions of such equations, the integrals $\mathbf{E}f(\xi(t))$ should be found, where $\xi(t)$ is a matrix diffusion process, and f is some analytic function. For example, the problem of evaluating the integral $\text{Tr}\,\mathbf{E}\exp[t(W+A)]$, where W is a Hermitian random matrix with the density of distribution $c\exp\{-\text{Tr}\,XX^*\}$, A is a nonrandom Hermitian matrix of the nth order, is of interest.

Indeed, let a system of equations $\overrightarrow{x}'(t) = (W+A)\overrightarrow{x}(t)$, $\overrightarrow{x}(0) \neq \overrightarrow{0}$ be given. Then $\overrightarrow{x}(t) = \exp[t(W+A)]\overrightarrow{x}_0$. As $\mathbf{E}\|\overrightarrow{x}(t)\|^2 = (\mathbf{E}\exp[2t(W+A)]\overrightarrow{x}_0, \overrightarrow{x}_0) < \infty$ then, in order that the solution $\overrightarrow{x}(t)$ be asymptotically stable in the mean, it is necessary and sufficient that $\lim_{t\to\infty}\mathbf{E}\exp[2t(W+A)] = 0$, and this relation will be equivalent to the following:

$$\lim_{t\to\infty}\text{Tr}\,\mathbf{E}e^{t(W+A)} = 0.$$

Let $\mu(t,x) = \sum_{k=1}^{n}\chi(\lambda_k^{(t)} < x)$, where $\lambda_k^{(t)}$ are eigenvalues of the matrix $W(t)$, $W(s)$ is the matrix Wiener process whose increments $W(s+\Delta s) - W(s)$, $\Delta s \geq 0$ are independent and distributed similarly to those of the matrix $\sqrt{\Delta s}W$, $W(0) = A$. For $\mu(t,x)$, we consider the Stieltjes transforms

$$\int(z+ix)^{-1}d\mu(t,x) = \text{Tr}(Iz+iW(t))^{-1}, \qquad z = q+is, \quad s \neq 0.$$

For resolvent traces $R(z,t) := (Iz+W(t))^{-1}$, the stochastic diffusion equation has a simple form,

$$d_t\begin{pmatrix} \text{Tr}\,\text{Re}\,R(z,t) \\ \text{Tr}\,\text{Im}\,R(z,t) \end{pmatrix} = -\begin{pmatrix} \text{Tr}\,\text{Re}\,R^2(z,t)\,\text{Tr}\,R(z,t) \\ \text{Tr}\,\text{Im}\,R^2(z,t)\,\text{Tr}\,R(z,t) \end{pmatrix}dt$$
$$+\begin{pmatrix} \text{Tr}(\text{Re}\,R^2(z,t))^2 & \text{Tr}\,\text{Re}\,R^2(z,t)\,\text{Im}\,R^2(z,t) \\ \text{Tr}\,\text{Re}\,R^2(z,t)\,\text{Im}\,R^2(z,t) & \text{Tr}(\text{Im}\,R^2(z,t))^2 \end{pmatrix}^{1/2}\begin{pmatrix} dw_1(t) \\ dw_2(t) \end{pmatrix},$$
$$(19.5.1)$$

where $\omega_i(t)$ are independent processes of the Brownian motion

$$R(z,0) = (Iz+A)^{-1}.$$

To derive Eq. (19.5.1), the equality

$$R(z,t+\Delta t) - R(z,t) = \left[\sum_{k=0}^{\infty}(-R(z,t)B)^k\right]R(z,t),$$
$$B = W(t+\Delta t) - W(t)$$

is valid for sufficiently small Δt.

Let

$$u(z,t) = \text{Tr } \mathbf{E}R(z,t); \quad u(z,0) = \text{Tr}(Iz + A)^{-1}, \quad \text{Im } z = 0.$$

If the solution of this equation $u(z,t)$ is known, then the function $\mu(1,x)$ may be found with the help of the inverse formula $\mu(x_1) - \mu(x_2) = \pi^{-1} \lim_{\epsilon \to 0} \int_{x_1}^{x_2} \text{Im } u(y + i\epsilon, 1)dy$, where x_1 and x_2 are points of continuity of the function $\mu(1,x)$. Hence we can obtain the expression

$$\text{Tr } \mathbf{E}f(W) = \int_{-\infty}^{+\infty} f(x)d\mu(x),$$

where $f(x)$ is an analytic function.

In particular, assuming $f(x) = \exp(x)$, we obtain the expression for $\text{Tr } \mathbf{E} \exp(t(W + A))$. If the function $\mu(x)$ has a derivative $p(x) := \mu'(x)$, then from the inverse formula it follows that $p(x) = \pi^{-1} \text{Im } u(x, 1)$. Therefore,

$$\text{Tr } \mathbf{E}f(W) = \pi^{-1} \int_{-\infty}^{+\infty} f(x) \text{Im } u(x, 1)dx.$$

Let us show how the resolvent method can be applied to nonsymmetric matrices.

Let $\Xi(t) = (a_{ij} + w_{ij}(t))_{i,j=1}^n$ be an arbitrary matrix process, where $A := (a_{ij})_{i,j=1}^n$ is a matrix whose elements have nonrandom values $w_{ij}(t)$ are independent processes of the Brownian motion. For resolvents of such random processes $R(z,v) := (Iz - \Xi(v))^{-1}$, one can also find a stochastic diffusion equation, however it is not suitable for finding an equation for $\mathbf{E} \text{Tr } R(z,v)$, since the integrals $\mathbf{E} \text{Tr } R^k(z,v)$ become divergent at $k = 2, 3, \ldots$. Consider the integrated resolvents

$$u(z, A, v) := \mathbf{E} \int_0^t \ln \det(I(u + is + A) + \Xi(v))du,$$

with the help of which we shall find the equation for $\mathbf{E} \text{Tr } R(z, v)$. It is obvious that

$$\lim_{\Delta v \to 0} (\Delta v)^{-1}[u(z, \alpha, v + \Delta v) - u(z, \alpha, v)] = \lim_{\Delta v \to 0} (\Delta v)^{-1}$$

$$\times \mathbf{E} \int_0^t \ln \det\{I + [I(u + is + \alpha) + \Xi(v)]^{-1}(\Xi(v + \Delta v) - \Xi(v))\}du.$$

It is easy to check (see Chapter 14) that the integral exists o the right-hand side of this equality.

From the same equality, we obtain the equation (18.9.1). It is obvious that

$$\mathbf{E} \text{Tr } R(z, v) = \iint (z - x - iy)^{-1}d\mu_n(x, y, v),$$

where

$$\mu_n(x, y, v) = \sum_{k=1}^n \mathbf{P}\{\text{Re } \lambda_k(v) < x, \text{Im } X_k(v) < y\},$$

and $\lambda_k(v)$ are eigenvalues of the process $\Xi(v)$. See Chapter 17 (§6,§7) for the inverse formula of this transform.

§6 The Spectral Method of Calculating Mathematical Expectations of Exponents of Random Matrices

Let A be a random Hermitian matrix of the nth order with the density of distribution

$$p(x) = 2^{(n-1)n/2}\pi^{-n^2/2}\exp\{-\operatorname{Tr} XX^*\},$$

where X is a Hermitian matrix of the nth order. In this section, we shall compute the integral $\mathbf{E}\exp(SA)$, where S is a complex number. From Chapter 3 it follows that the matrix A is distributed like the matrix $U\Lambda U^*$, where U is a random unitary matrix, $\arg u_{1i} = c_i,$, $0 \le c_i \le 2\pi$ are some constants, $\Lambda = (\delta_{ij}\alpha_i)$ is a diagonal matrix $\alpha_1 > \cdots > \alpha_n$; the matrices U and Λ are stochastically independent, and the density of distribution of random values α_i is equal to

$$\mathbf{P}(\alpha_1, \ldots, \alpha_n) = [2^{-n(n-1)/2}n!\pi^{n/2}\prod_{j=i}^{n-1} j!]^{-1}$$

$$\times \exp\{-\frac{1}{2}\sum_{i=1}^{n}\alpha_i^2\}\prod_{i>j}(\alpha_i - \alpha_j)^2, \quad \alpha_1 > \alpha_2 > \cdots > \alpha_n.$$

As c_i are arbitrary constants, we set all of them equal to zero. From Chapter 3 it follows that the matrix U is distributed by the Haar conditional probability measure, given that $\arg u_{1i} = 0, i = \overline{1,n}$. Then

$$\mathbf{E}\exp(sA))_{pl} = \begin{cases} 0, & p \ne l \\ n^{-1}\mathbf{E}\sum_{k=1}^{n}\exp(s\alpha_k), & p = l. \end{cases}$$

Since $p(\alpha_1, \ldots, \alpha_n)$ is a symmetric function,

$$\alpha(s) := n^{-1}\mathbf{E}\sum_{k=1}^{n}e^{s\alpha_k} = (n!)^{-1}\int\cdots\int e^{s\alpha_1}p(\alpha_1, \ldots, \alpha_n)\prod_{i=1}^{n} d\alpha_i.$$

Thus, $\mathbf{E}\exp(sA) = \alpha(S)I$.

We shall find the function $\alpha(s)$ with the help of some methods of computing multidimensional integrals proposed by Mehta (see Chapter 2).

Let us represent $\Pi_{i>j}(\alpha_i - \alpha_j)$ in the form of the Vandermonde determinant,

$$\det\begin{bmatrix} 1 & 1 & \cdots & 1 \\ \alpha_1 & \alpha_2 & \cdots & \alpha_n \\ \alpha_1^{n-1} & \alpha_2^{n-1} & \cdots & \alpha_n^{n-1} \end{bmatrix}.$$

Multiplying the jth row by 2^{j-1} and adding it to the corresponding linear combination of other rows with degrees of variables which are less then j, replace the jth row by $H_{j-1}(\alpha_1), \ldots, H_{j-1}(\alpha_k)$, where $H_j(\alpha)$ is the Hermitian polynomial of the jth order.

After such transformations, the density p takes the form

$$p(\alpha_1,\ldots,\alpha_n) = [2^{-n(n-1)/2}n!\pi^{n/2}\prod_{j=0}^{n-1}j!]^{-1}\prod_{j=0}^{n-1}[2^{-j}(2^j j!\sqrt{\pi})^{1/2}]^2$$

$$\times\{\det[\varphi_k(\alpha_j)]_{\substack{k=0,n-1\\j=1,n}}\} = (n!)^{-1}\{\det[\varphi_{k-1}(\alpha_j)]_{k,j=\overline{1,n}}\}^2,$$

where

$$\varphi_j(\alpha) = (2^j j!\sqrt{\pi})^{-1/2}e^{\alpha^2/2}(-d/d\alpha)^j e^{-\alpha^2}.$$

By using this expression, we obtain

$$\alpha(s) = (n-1)(n!)^{-2}\sum_{j=0}^{n-1}\int e^{sx}\varphi_j^2(\alpha)d\alpha.$$

From the formula by Kristoffel–Darboux

$$\sum_{j=0}^{n-1}\varphi_j^2(\alpha) = n\varphi_n^2(\alpha) - [n(n+1)]^{1/2}\varphi_{n-1}(\alpha)\varphi_{n+1}(\alpha).$$

Consequently,

$$\alpha(s) = (n!n)^{-1}\int e^{s\alpha}[n\varphi_n^2(\alpha) - [(n+1)n]^{1/2}$$

$$\times\varphi_{n-1}(\alpha)\varphi_{n+1}(\alpha)]d\alpha. \tag{19.6.1}$$

It follows from Chapter 2 that

$$\varphi_n(\alpha) = \sum_{j=0}^{[n/2]}h_{jn}\alpha^{n-2j}e^{-\alpha^2/2} = (2^n n!\sqrt{\pi})^{-1/2}e^{-\alpha^2/2}H_n(\alpha),$$

where

$$h_{jn} = [n!2^{-n}\pi^{-1/2}]^{1/2}\frac{(-1)^j 2^{n-2j}}{j!(n-2j)!};$$

$$H_n(\alpha) = n!\sum_{\nu=0}^{[n/2]}\frac{(-1)^\nu}{\nu!}\cdot\frac{(2\alpha)^{n-2\nu}}{(n-2\nu)!}.$$

Substituting this expression into the formula (19.6.1), we have

$$\alpha(s) = (n!)^{-1}\int e^{s\alpha}\sum_{p,l=0}^{[n/2]}h_{pn}h_{ln}\alpha^{2(n-p-l)}e^{-\alpha^2}d\alpha$$

$$-\frac{\sqrt{n(n+1)}}{n!n}\int e^{s\alpha}\sum_{p,l=0}^{[n/2]}h_{pn}h_{ln}\alpha^{2(n-p-l)}e^{-\alpha^2}d\alpha.$$

It is clear that for any integer $m \geq 1$,

$$
\int e^{s\alpha} \alpha^m e^{-\alpha^2} d\alpha = \frac{\partial^m}{\partial s^m} \int e^{s\alpha} e^{-\alpha^2} d\alpha
$$

$$
= \sqrt{\pi} \frac{\partial^m}{\partial s^m} e^{s^2/4}
$$

$$
= H_m(s/2) e^{s^2/4} \sqrt{\pi}.
$$

Thus, the formula to be found has the form

$$
\mathbf{E} \exp\{sA\} = I\sqrt{\pi}\{(n!)^{-1} \sum_{p,l=0}^{[n/2]} h_{pn} h_{ln} H_{2(n-p-l)}(s/2)
$$

$$
- \sqrt{n(n+1)}(n!n)^{-1} \sum_{p,l=0}^{[(n-1)/2]} h_{pn} h_{ln} H_{2(n-p-l)}(s/2)\} e^{s^2/4}.
$$

§7 Method of Stochastic Diffusion Equations

In this section, we consider a problem of computing a probability of the event $\{\operatorname{Re}\alpha_i(A+W) < 0, \quad i = \overline{1,n}\}$, where A is a nonrandom matrix of the nth order, W is a random matrix of the same order, whose elements are distributed by the general normal law. The method of stochastic equations of diffusion for finding a probability of the asymptotic stability of solutions to systems of differential equations of the first order with constant Gaussian random coefficients is essentially the following: instead of the random vector $\{\operatorname{Re}\alpha_i(A+W), i = \overline{1,n}\}$, we consider a random vector process $\{\lambda_i(W(t)), i = \overline{1,n}\}$ where $W(t)$ is a homogeneous Gaussian random matrix process with independent additive increments $W(0) = A, W(1) \approx W$. We find a stochastic equation of diffusion for it. With the help of this equation for densities of distribution of the vector $\{\lambda_i(W(t)), i = \overline{1,n}\}$, we find the Fokker–Planck–Kolmogorov equations which can be solved with the help of the variables separation method. We apply this method to symmetric matrices A, W.

Let A be a symmetric real matrix of nth order with different eigenvalues α_i, W be a Gaussian matrix with the density of distribution $2^{-n/2}\pi^{-n(n+1)/4} \times \exp\{-2^{-1}\operatorname{Tr}x^2\}$, $W(t)$ be a random matrix process with additive independent increments $W(0) = A, [W(t+\Delta t) - W(t)](\Delta t)^{-1/2} \approx W$. It follows from Chapter 18 that for all $t, (0 \leq t < \tau)$,

$$
d\overrightarrow{\lambda}(W(t)) = \{\sum_{s \neq i}(\lambda_i(W(t)) - \lambda_s(W(t)))^{-1}, i = \overline{1,n}\}dt + d\overrightarrow{w}(t),
$$

where

$$
\overrightarrow{\lambda}(W(t)) = \{\lambda_i(W(t)), i = \overline{1,n}\}', \lambda_i(0) = \alpha_i(A),
$$

and $\vec{w}(t)$ is a vector process of the Brownian motion

$$\tau = \inf_{0 \le t}\{t := \vec{\lambda}(W(t)) \in D\},$$

where

$$\overline{D} = \{\vec{x} \in R^n : x_1 \ne x_j, i \ne j, i, j = \overline{1, n}\}.$$

Using this equation, we obtain from Chapter 18 that the density of distribution $q(t, \vec{y})$ of eigenvalues of the process $A + W(t)$ satisfies the Fokker–Planck–Kolmogorov equation

$$\frac{\partial}{\partial t}q(t\,\vec{y}) = 2^{-1}\left\{\sum_{i=1}^{n}\left(\frac{\partial}{\partial y_i}\right)[a_i(\vec{y})q(t, \vec{y})] + \sum_{i=1}^{n}\left(\frac{\partial^2}{\partial y_i^2}\right)q(t, \vec{y})\right\}$$

(19.7.1)

with the initial condition $q(0, \vec{y}) = \prod_{i=1}^{n}\delta(y_i - a_i)$, where

$$a_i(\vec{y}) = \sum_{j \ne i}(y_i - y_j)^{-1}.$$

Equation (19.7.1) can be solved with the help of the method of variables separation. We shall look for a solution in the form of the product of two functions $q(t, \vec{y}) = f(t)\varphi(\vec{y})$. Then

$$\varphi(\vec{y})\frac{\partial}{\partial t}f(t) = 2f(t)\sum_{i=1}^{n}\frac{\partial}{\partial y_i}[a_i(\vec{y})\varphi(\vec{y})] + f(t)\sum_{i=1}^{n}\left(\frac{\partial^2}{\partial y_i^2}\right)\varphi(\vec{y}).$$

Hence,

$$[f(t)]^{-1}\left(\frac{\partial}{\partial t}\right)f(t) = 2[\varphi(\vec{y})]^{-1}\sum_{i=1}^{n}\left(\frac{\partial}{\partial y_i}\right)[a_i(\vec{y})\varphi(\vec{y})] + [\varphi(\vec{y})]^{-1}$$

$$\times \sum_{i=1}^{n}\left(\frac{\partial^2}{\partial y_i^2}\right)\varphi(\vec{y}).$$

From this equation, we obtain two differential equations,

$$\left(\frac{\partial}{\partial t}\right)f(t) = \gamma f(t), \quad t \ge 0,$$

$$2\sum_{i=1}^{n}\left(\frac{\partial}{\partial y_i}\right)[a_i(\vec{y})\varphi(\vec{y})] + \sum_{i=1}^{n}\left(\frac{\partial^2}{\partial y_i^2}\right)\varphi(\vec{y}) = \gamma\varphi(\vec{y}), \quad \vec{y} \in R^n,$$

(19.7.2)

where γ is an arbitrary constant.

From this system of equations with the indicated initial condition, we obtain that

$$q(t, \overrightarrow{y}) = \sum_{k=0}^{\infty} \varphi_k(\overrightarrow{a})\varphi_k(\overrightarrow{y})e^{\gamma_k t},$$

where γ_k and φ_k are eigenvalues and normalized eigenvalues of the equation (19.7.2), $\overrightarrow{a} = (\alpha_i, i = \overline{1,n})'$, respectively. Obviously,

$$\mathbf{P}\{\alpha_i(A + W) < 0, i = \overline{1,n}\}$$

$$= \int \cdots \int_{y_i < 0, i = \overline{1,n}} \left\{ \sum_{k=0}^{\infty} \varphi_k(\overrightarrow{a})\varphi_k(\overrightarrow{y})e^{\gamma_k} \right\} \prod_{i=1}^{n} dy_i.$$

CHAPTER 20

RANDOM DETERMINANTS IN THE THEORY OF ESTIMATION OF PARAMETERS OF SOME SYSTEMS

This chapter deals with the problems of estimation of parameters of some equations that are solved using different functions of random matrices which can be expressed by the determinant of the matrix. Some problems of this type where studied in Chapter 14.

§1 The Estimation of Solutions of Equation Systems with Multiplicative Errors in the Series of Observation

The method for estimation of the solution proposed in this section is based on the perturbation formulas for linear operator eigenvalues.

Suppose a system of linear equations

$$A\overrightarrow{x} = \overrightarrow{h} + \overrightarrow{\xi_1}. \tag{20.1.1}$$

is given, where A is a square nonsingular matrix of nth order; $\overrightarrow{x}, \overrightarrow{h}, \overrightarrow{\xi_1}$ are n-dimensional vectors; the matrix A and the vector $\overrightarrow{h}$ are known, the value of the vector $\overrightarrow{\xi_1}$ is not defined; it is only known if it takes values in some set.

Let us look at a vector y of m dimension connected with the vector $\overrightarrow{x}$ by the equation

$$\overrightarrow{y} = \Xi\overrightarrow{x} + \overrightarrow{\xi_2}, \tag{20.1.2}$$

where Ξ is a random matrix, $\overrightarrow{\xi_2}$ is an m-dimensional vector whose value is unknown and satisfies the inequality

$$||\overrightarrow{\xi_1}||^2 + ||\overrightarrow{\xi_2}||^2 \leq 1, \quad \text{where} \quad ||\overrightarrow{\xi_1}||^2 = (\overrightarrow{\xi_1}, \overrightarrow{\xi_1}).$$

The problem of the estimation of the vector $\overrightarrow{x}$ is to find by some linear transformations of the vector $\overrightarrow{y}$ the estimate (optimal is some sense) of the

481

vector $\vec{x}$. To be exact, it is necessary to find the matrix K^* of dimension $n \times m$ and vector $\vec{l^*}$ of dimension n such that the expression

$$\mathbf{E} \max_{\|\vec{\xi_1}\|^2 + \|\vec{\xi_2}\|^2 \leq 1} \|\vec{x} - K\vec{y} - \vec{l}\|^2 \qquad (20.1.3)$$

would take the minimal value. The vector $\hat{\vec{x}} = K^*\vec{y} + \vec{l^*}$ will be called the minimax estimate of the vector $\vec{x}$. Without loss of generality the vector $\vec{h}$ can be chosen to be zero.

Let L_1 be a set of real matrices of dimension $n \times m$, L_2 a set of real vectors of dimension m.

Theorem 20.1.1.

$$\min_{\substack{K \in L_1 \\ \vec{l} \in L_2}} \mathbf{E} \max_{\|\vec{\xi_1}\|^2 + \|\vec{\xi_2}\|^2 \leq 1} \|\vec{x} - K\vec{y} - \vec{l}\|^2$$

$$= \mathbf{E}\lambda_1[(I - K^*\Xi)A^{-1}(I - K^*\Xi)'A^{-1'} + K^*K'^*],$$

where the matrix K^* satisfies the equation

$$\mathbf{E}(\Xi A^{-1}(I - K^*\Xi)A^{-1} + K^*)\vec{\varphi_1}(D)\vec{\varphi_1'}(D) = 0, \qquad \vec{l^*} = 0,$$

where λ_1 is a maximal eigenvalue of the matrix, φ_1 is the eigenvector corresponding to the maximal eigenvalue of the matrix D. We suppose that there exists an expectation in this formula,

$$D = (I - K^*\Xi)A^{-1}(I - K^*\Xi)'A^{-1'} + K^*K'^*.$$

Proof. By using the Rayleigh formula, we get

$$\max_{\|\vec{\xi_1}\|^2 + \|\vec{\xi_2}\|^2 \leq 1} \|\vec{x} - K\vec{y}\|^2 = \lambda_1(B'(K)B(K)), \qquad (20.1.4)$$

where

$$B(K) = \{(I - K\Xi)A^{-1}, K\}. \qquad (20.1.5)$$

Since λ_1 is a continuous function of the elements of the matrix $B'(K)B(K)$ and $\lambda_1(B'(K^* + t\theta)B(K^* + t\theta)) \geq \lambda_1(B'(K^*)B(K^*))$ (where K^* is a point of minimum) for any real values of t and matrices θ, the matrix K satisfies the equation

$$\frac{d}{dt}\mathbf{E}\lambda_{\max}(B'(K + t\theta)B(K + t\theta))|_{t=0} = 0, \qquad (20.1.6)$$

or $\lim_{t\to 0} t^{-1}\mathbf{E}\{\lambda_{\max}(B'(K + t\theta)B(K + t\theta) - \lambda_{\max}(B'(K)B(K))\} = 0$. We use the linear operators of perturbation theory to find this derivative.

Let $B'(K+t\theta)B(K+t\theta) = D(t)$, and suppose that the maximal eigenvalue is simple. Then by using the perturbation theory formulas (see Chapter 18), we obtain

$$\mathbf{E}[\lambda_1(D(t)) - \lambda_1(D(0))] = t\mathbf{E}\left(\left\{\frac{D(t) - D(0)}{t}\right\}\right.$$
$$\left. \times \overrightarrow{\varphi_1}(D(0)), \overrightarrow{\varphi_1}(D(0))\right) + t^2(\ldots) + \ldots \qquad (20.1.7)$$

where $\overrightarrow{\varphi_1}(D(0))$ is the eigenvector corresponding to the eigenvalue $\lambda_1(D(0))$. Then Eq. (20.1.6) can be written in the following form,

$$\lim_{t\to 0}\mathbf{E}\frac{\lambda_1(D(t)) - \lambda_1(D(0))}{t} = \mathbf{E}\left(\frac{dD(0)}{dt}\varphi_1(D(0)), \varphi_1(D(0))\right).$$

Find

$$\mathbf{E}\frac{dD(0)}{dt} = \mathbf{E}\frac{d}{dt}[B'(K + t\theta)B(K + t\theta)]_{t=0}$$
$$= \mathbf{E}(-\theta\Xi A^{-1}A'^{-1}(I - K\Xi)' + \theta K'),$$

then the matrix K^* satisfies the equation

$$\mathbf{E}((-\theta\Xi A^{-1}A'^{-1}(I - K\Xi) + \theta K)\overrightarrow{\varphi_1}(D), \overrightarrow{\varphi_1}(D)) = 0. \qquad (20.1.8)$$

From Eq. (20.1.8) and by virtue of the arbitrariness of the matrix θ, we get the equation for the matrix K,

$$\mathbf{E}(\Xi A^{-1}(I - K\Xi)A^{-1} + K)\overrightarrow{\varphi_1}(D)\overrightarrow{\varphi_1}(D) = 0. \qquad (20.1.9)$$

We note that formula (20.1.7) is valid if we assume that the maximal eigenvalue is simple. Otherwise if it is not so, we write $D(0) + \varepsilon\mathbf{P}$ instead of the matrix $D(0)$, where ε is some small number and the matrix $\mathbf{P}$ is chosen so that the maximal eigenvalue of the matrix $D(0) + \varepsilon\mathbf{P}$ should be simple. For such a matrix, we carry out all the previous constructions and let ε tend to zero in the latter formula. Thus we get the same result for the general case. Theorem (20.1.1) is proved.

Equation (20.1.9) can be solved on a computer with the help of various methods of numerically solving functional equations, for example, with the help of the principle of contractile mappings.

In Eq. (20.1.9), a sum of independent realizations of the random matrices can be taken instead of the expectations, and the Monte-Carlo method can be used.

§2 Spectral Equations for Minimax Estimations of Parameters of Linear Systems

Suppose that the unknown n-dimensional vector c satisfies the system of equations

$$y = Xc + \varepsilon, \tag{20.2.1}$$

where y is an n-dimensional observation vector, $X = (x_{ij}), j = \overline{1,m}, i = \overline{1,n}$ is a random matrix, and ε is an n-dimensional random vector which does not depend on the matrix X,

$$\mathbf{E}\varepsilon = 0, \qquad \mathbf{E}\varepsilon\varepsilon' = R.$$

Besides, let the vector c satisfy the inequality

$$(Dc, c) \le a, \tag{20.2.2}$$

where D is a nonnegative-definite matrix of $m \times m$ dimension, $0 < a < \infty$.

The problem of estimation of the vector c is the fact that with the help of some linear transformation on the vector y: $Ty + t$, where T is a matrix of $n \times m$ dimension, t is an m-dimensional vector, matrix $\widehat{T}$ and the vector $\hat{t}$ should be found such that the expression

$$\max_{c:(Dc,c)\le a} \mathbf{E}\|Ty + t - c\|^2 \tag{20.2.3}$$

takes the minimal value. The vector $\hat{c} = \widehat{T}y + \hat{t}$ is called the minimax estimation of the vector c. Denote the set of real $n \times m$-dimensional matrices by L_1, the set of real m-dimensional vectors by L_2.

Theorem 20.2.1. *If R and D are nondegenerate matrices, there exists $\mathbf{E}XX'$, $a > 0$, then*

$$\min_{T\in L_1, T\in L_2} \max_{c:(Dc,c)\le a} \mathbf{E}\|Ty + t - c\|^2 = a\lambda_1\{ED^{-1/2}(\widehat{T}X - I)'$$
$$\times (\widehat{T}X - I)D^{-1/2}\} + \operatorname{Tr}\widehat{T}R\widehat{T}', \quad \hat{t} = 0, \tag{20.2.4}$$

where λ_1 is the maximal eigenvalue of the matrix

$$\mathbf{E}D^{-1/2}(\widehat{T}X - I)'(\widehat{T}X - I)D^{-1/2}.$$

The matrix $\widehat{T}$ is the solution of the equation

$$[\mathbf{E}aXD^{-1}(\widehat{T}X - I)']e_1e_1' + R\widehat{T}' = 0, \tag{20.2.5}$$

where e_1 is an eigenvector corresponding to the eigenvalue λ_1.

Proof. It is evident that

$$\mathbf{E}\|Ty + t - c\|^2 = \mathbf{E}\|TXc + t - c + T\varepsilon\|^2$$
$$= \mathbf{E}\|(TX - I)c + t\|^2 + \operatorname{Tr} TRT'.$$
(20.2.6)

Since

$$\max_{c:(Dc,c)\leq a} \mathbf{E}\|(TX - I)c + t\|^2 \geq \max_{c:(Dc,c)\leq a} \mathbf{E}\|(TX - I)c\|^2,$$

in the expression (3.2.6) we will minimize on t when $\hat{t} = 0$.

From the Rayleigh formula it follows that

$$\max_{c:(Dc,c)\leq a} \mathbf{E}\|(TX - I)c\|^2 = a\lambda_1\{D^{-1/2}\mathbf{E}(TX - I)'(TX - I)D^{-1/2}\}.$$

Let us find the equation for the unknown matrix T. Since $\lambda_1(B(T))$ is a continuous and almost everywhere differentiable function of the elements of the matrix $B(T) = \mathbf{E}D^{-1/2}(TX - I)'(TX - I)D^{-1/2}$, the unknown matrix $\hat{T}$ satisfies the equation

$$\frac{d}{d\gamma}[a\lambda_1\{B(T + \gamma\theta)\} + \operatorname{Tr}(T + \gamma\theta)R(T + \gamma\theta)']_{\gamma=0} = 0,$$
(20.2.7)

where θ is an arbitrary $n \times m$-dimensional matrix, and γ is a real variable. Using the perturbation formula for eigenvalues (see Chapter 18) from (20.2.7), we get

$$a(\{\mathbf{E}D^{-1/2}(\theta X)'(TX - I)D^{-1/2} + \mathbf{E}D^{-1/2}(TX - I)'$$
$$\times (\theta X)D^{-1/2}\}e_1, e_1) + \operatorname{Tr}[\theta RT' + TR\theta'] = 0.$$

By the arbitrariness of the matrix θ, this equation implies Eq. (20.2.5). Theorem (20.2.1) is proved.

In general, Eq. (20.2.5) is difficult to solve with respect to the unknown matrix $\hat{T}$. If the problem of the minimax estimation is formulated as follows, find $\hat{T}$ and $\hat{t}$ such that the expression

$$\max_{c:(Dc,c)\leq a} \mathbf{E}((Ty + t)'b - c'b)$$

will take the minimax value, where b is an arbitrary m-dimensional vector, then an explicit expression can be found for the unknown matrix $\hat{T}$.

§3 The Estimation of the Parameters of Stable Discrete Control Systems

Mathematical models of some linear discrete control systems are given with the help of recurrent equations

$$\overrightarrow{y_k} = \theta \overrightarrow{y}_{k-1} = \overrightarrow{b}_{k-1} + \overrightarrow{\varepsilon_k}, \tag{20.3.1}$$

where $\theta = (\theta_{ij})_{i,j=1}^m$ is an unknown matrix, $\overrightarrow{y_0}, \overrightarrow{b_k}, k = 1, 2, \ldots$ are unknown m-dimensional control vectors, $\overrightarrow{y_k}$ are the observed m-dimensional vectors, and $\overrightarrow{\varepsilon_k}$ are random m-dimensional vectors.

The problem is to estimate the matrix θ under the condition that we observe the vectors $\overrightarrow{y_1}, \ldots, \overrightarrow{y_n}$.

The estimation θ_n of the matrix θ can be found from the following equation,

$$\hat{\theta}_n \sum_{k=1}^n \overrightarrow{y}_{k-1} \overrightarrow{y}_{k-1} = \sum_{k=1}^n \overrightarrow{z_k} \overrightarrow{y}_{k-1},$$

where $\overrightarrow{z_k} = \overrightarrow{y_k} - \overrightarrow{b}_{k-1}$.

From this equation, substituting $\overrightarrow{z_k} = \theta \overrightarrow{y}_{k-1} + \overrightarrow{\varepsilon_k}$, we have

$$(\hat{\theta}_n - \theta) \sum_{k=1}^n \overrightarrow{y}_{k-1} \overrightarrow{y}_{k-1} = \sum_{k=1}^n \overrightarrow{\varepsilon_k} \overrightarrow{y}_{k-1}.$$

If $\mathbf{E}(\overrightarrow{\varepsilon_k}|\overrightarrow{\varepsilon_1}, \ldots, \overrightarrow{\varepsilon}_{k-1}) = 0$, then under the condition that there exists an inverse matrix $(\sum_{k=1}^n \overrightarrow{y}_{k-1} \overrightarrow{y}_{k-1})^{-1}$, the estimator $\hat{\theta}_n$ will be biased, i.e., $\mathbf{E}\hat{\theta}_n \neq \theta$. We prove that under some additional conditions it will be consistent, i.e., $\text{plim}_{n\to\infty} \hat{\theta}_n = \theta$, and find the rate of convergence $\hat{\theta} - \theta$ to 0 as $n \to \infty$, in the sense of a weak convergence of distributions.

We call the vectors $\overrightarrow{\varepsilon_k}$ G-asymptotically independent, if for almost all values of the vectors $\overrightarrow{x}_{i_p}$ and $\overrightarrow{y}_{j_p}$, where i_p, j_p are integer nonnegative number variables,

$$\varlimsup_{\substack{h\to\infty \\ |i_p - j_p| \geq h, \\ p=\overline{1,3}}} \sup |\mathbf{P}\{\varepsilon_{i_p} < x_{i_p}, \varepsilon_{j_p} < y_{j_p}, p = \overline{1,3}\}$$

$$- \mathbf{P}\{\varepsilon_{i_p} < x_{i_p}, p = \overline{1,3}\}\mathbf{P}\{\overrightarrow{\varepsilon}_{j_p} < \overrightarrow{y}_{j_p}, p = \overline{1,3}\}| = 0. \tag{20.3.2}$$

Theorem 20.3.1. *Let* $\mathbf{E}(\overrightarrow{\varepsilon}_k|\overrightarrow{\varepsilon}_1, \ldots, \overrightarrow{\varepsilon}_{k-1}) = 0, k = 2, 3, \ldots$ *and the random vectors* $\overrightarrow{\varepsilon}_k$ *be G-asymptotically independent,*

$$\|\theta\| < 1, \quad \sup_i \|\overrightarrow{b}_i\| < c < \infty, \tag{20.3.3}$$

$$\lim_{h\to\infty} \sup_k \mathbf{E}\|\overrightarrow{\varepsilon}_k\|^2 \chi(\|\overrightarrow{\varepsilon}_k\| \geq h) = 0. \tag{20.3.4}$$

Then

$$\mathbf{E} \exp\{i\sqrt{c_n}\,\mathrm{Tr}\,Q[\hat{\theta}_n - \theta](\sum_{k=1}^{n} \mathbf{E}\overrightarrow{y}_{k-1}\overrightarrow{y}'_{k-1}c_n^{-1})\}$$

$$= \exp\{-\frac{1}{2}\sum_{k=1}^{n}\mathrm{Sp}\,\mathbf{E}\overrightarrow{y}'_{k-1}Q\overrightarrow{\varepsilon}_k\overrightarrow{\varepsilon}'_kQ'\overrightarrow{y}_{k-1}c_n^{-1}\} + 0(1), \quad (20.3.5)$$

where $R_k^{(n)} = \mathbf{E}\overrightarrow{\varepsilon}_k\overrightarrow{\varepsilon}'_k, Q = (q_{ij})_{i,j=1}^{m}$ *is a matrix of parameters, and* $c_n \geq n$ *is some sequence of constants.*

Proof. Let us prove that

$$\mathrm{plim}_{n\to\infty}\,c_n^{-1}\sum_{k=1}^{n}(\overrightarrow{y}_k\overrightarrow{y}'_k - \mathbf{E}\overrightarrow{y}_k\overrightarrow{y}'_k) = 0. \quad (20.3.6)$$

Consider the following transformation,

$$\mathbf{E}e^{-c_n^{-1}\,\mathrm{Tr}\,S\sum_{k=1}^{n}\overrightarrow{y}_k\overrightarrow{y}'_k} = \mathbf{E}e^{-c_n^{-1}\sum_{k=1}^{n}\overrightarrow{y}'_kS\overrightarrow{y}_k}, \quad (20.3.7)$$

where $S = (S_{ij})_{i,j=1}^{m}$ is a positive-definite matrix. It is evident that (20.3.7) equals

$$\mathbf{E}\exp\{-C_n^{-1}\sum_{k=1}^{n}(S\overrightarrow{y}_k, \overrightarrow{y}_k)\} = \mathbf{E}\exp\{iC_n^{-1/2}\sum_{k=1}^{n}(\overrightarrow{y}_k, \overrightarrow{\eta}_k)\}, \quad (20.3.8)$$

where $\overrightarrow{\eta}_k, k = 1, 2, \ldots$ are independent random vectors not depending on the vectors $\overrightarrow{\varepsilon}_k$ and are distributed by the normal law $N(0, 2S)$.

From Eq. (20.3.1), we get

$$\overrightarrow{y}_k = \theta^k\overrightarrow{y}_0 + \sum_{i=0}^{k-1}\theta^i(\overrightarrow{\varepsilon}_{k-i} + \overrightarrow{b}_{k-i-1}). \quad (20.3.9)$$

Using this equality, we get

$$\sum_{p=1}^{n}(\overrightarrow{y}_p, \overrightarrow{\eta}_p) = \sum_{p=1}^{n}(\theta^p(\overrightarrow{y}_0, \overrightarrow{\eta}_p) + \sum_{p=1}^{n}\sum_{i=0}^{p-1}(\theta^i b_{p-i-1}, \overrightarrow{\eta}_p)$$

$$+ \sum_{k=1}^{n-1}(\overrightarrow{\varepsilon}_k, \sum_{p=1}^{n-k}\theta'^p\overrightarrow{\eta}_{p+1}). \quad (20.3.10)$$

Let

$$\gamma_k = (\overrightarrow{\varepsilon}_k, \sum_{s=k}^{n}\theta'^{s-k}\overrightarrow{\eta}_s).$$

The variables γ_k for the fixed $\overrightarrow{\eta}_p$ are martingale differences. Therefore, we apply the central limit theorem for these variables (see Chapter 5): if for some sequence of the constants C_n,

$$\text{plim}_{n\to\infty} \sum_{k=1}^{n} C_n^{-1}(\widetilde{\mathbf{E}}_k \gamma_k^2 - \widetilde{\mathbf{E}}\gamma_k^2) = 0, \tag{20.3.11}$$

$$\varlimsup_{n\to\infty} C_n^{-1} \sum_{k=1}^{n} \mathbf{E}\gamma_k^2 < \infty \tag{20.3.12}$$

and the Lindeberg condition holds for any $\tau > 0$,

$$\lim_{n\to\infty} \sum_{k=1}^{n} C_n^{-1}\mathbf{E}\gamma_k^2 \chi(|\gamma_k|^2 C_n^{-1/2} > \tau) = 0, \tag{20.3.13}$$

then for every real t,

$$\mathbf{E}e^{it\sum_{k=1}^{n} \gamma_k C_n^{-1/2}} = \mathbf{E}e^{-\frac{t^2}{2C_n}\sum_{k=1}^{n} \widetilde{\mathbf{E}}\gamma_k^2} + 0(1), \tag{20.3.14}$$

where the wave under the mathematical expectation means that the variables η_k are fixed, and $\mathbf{E}_k$ is an arbitrary mathematical expectation under the fixed minimal σ-algebra σ_{nk}, with respect to which the random vectors $\overrightarrow{\varepsilon}_1, \ldots, \overrightarrow{\varepsilon}_k$ are measurable.

Let us check the validity of the conditions (20.3.11)–(20.3.13).

It is evident that condition (20.3.12) is valid if $C_n \geq n$, since by force of conditions (20.3.3) and (20.3.4),

$$\mathbf{E}\gamma_k^2 = \mathbf{E}(R_k^{(n)} \sum_{p=k}^{n} \theta'^{p-k}\overrightarrow{\eta}_p \sum_{p=1}^{n-k} \theta'^p \overrightarrow{\eta}_{p+1}$$

$$= \sum_{p=1}^{n-1} \mathbf{E}(R_k^{(n)}\theta'^p \overrightarrow{\eta}_{p+1}, \theta'^p \overrightarrow{\eta}_{p+1})$$

$$= \sum_{p=1}^{n-k} \frac{1}{2} \text{Tr } S\theta^p R_k^{(n)}\theta'^p$$

$$\leq \|s\| \sup_k \|R_k^{(n)}\|(1 - \|\theta\|)^{-1} < \infty.$$

We prove that (20.3.11) is valid. Analogously to the proof of (20.3.12), we have

$$\sum_{k=1}^{n} C_n^{-1}(\widetilde{\mathbf{E}}_k \gamma_k^2 - \widetilde{\mathbf{E}}\gamma_k^2)$$

$$= \text{Tr} \sum_{k=1}^{n} C_n^{-1}([\widetilde{\mathbf{E}}\overrightarrow{\varepsilon}_k \overrightarrow{\varepsilon}_k' - \mathbf{E}\overrightarrow{\varepsilon}_k \overrightarrow{\varepsilon}_k'] \sum_{p=1}^{n-k} \theta'^p \overrightarrow{\eta}_{p+1}, \sum_{p=1}^{n-k} \theta'^p \overrightarrow{\eta}_{p+1}).$$

But by force of (20.3.4) and the asymptotic independence of the random vectors $\vec{\varepsilon}_k$, the probability of this difference approaches zero for $n \to \infty$.

By using (20.7.4), we make sure of the validity of (20.3.13).

Consequently, (20.3.14) holds.

Note that

$$1/C_n \sum_{k=1}^{n} \tilde{E}\gamma_k^2 = C_n^{-1} \sum_{k=1}^{n} \sum_{p=1}^{n-k} (R_k^{(n)} \theta'^p \vec{\eta}_{p+1}, \theta'^p \vec{\eta}_{p+1})$$

$$= C_n^{-1} \sum_{p=1}^{n} \sum_{i=1}^{p-1} (\theta^i R_{p-1} \theta'^i \vec{\eta}_p, \vec{\eta}_p).$$

Since the vectors $\vec{\eta}_p, p = 1, 2, \ldots$ are independent and by force of the conditions (20.3.3) and (20.3.4),

$$\left\| \sum_{i=1}^{p-1} \theta^i R_{p-i} \theta'^i \right\| \le C < \infty,$$

by the law of large numbers,

$$\text{plim}_{n \to \infty} C_n^{-1} \sum_{k=1}^{n} (\tilde{E}\gamma_k^2 - E\gamma_k^2) = 0.$$

Therefore, instead of (20.3.14), we get

$$E e^{it \sum_{k=1}^{n} \gamma_k C_n^{-1/2}}$$

$$= E e^{-t^2 (2C_n)^{-1} \sum_{p=1}^{n} \sum_{i=1}^{p-1} \text{Tr } S\theta^i R_{p-i}\theta'^i}, \quad \left(\sum_{i=1}^{0} \equiv 0 \right). \tag{20.3.15}$$

Analogously, using condition (20.3.3) and the central limit theorem for sums of independent random vectors, we get

$$E \exp[itC_n^{-1/2} \{ \sum_{p=1}^{n} (\theta^p (\vec{y}_0 + \vec{b}_0), \vec{\eta}_p)$$

$$+ \sum_{p=1}^{n} \sum_{i=1}^{p-1} (\theta^i \vec{b}_{p-i}, \vec{\eta}_p) \}] = \exp[-\frac{t^2}{2C_n} \sum_{p=1}^{n} (S(\theta^p(\vec{y}_0 + \vec{b}_0)$$

$$+ \sum_{i=1}^{p-1} \theta^i \vec{b}_{p-i}), (\theta^p(\vec{y}_0 + \vec{b}_0) + \sum_{i=1}^{p-1} \theta^i \vec{b}_{p-i}))] + 0(1). \tag{20.3.16}$$

From (20.3.15) and (20.3.16) and on the basis of (20.3.8), it follows that

$$\mathbf{E} \exp\{-C_n^{-1} \operatorname{Tr} S \sum_{k=1}^{n} \vec{y}_k \vec{y}_k'\}$$

$$= \exp\{-\frac{C_n^{-1}}{2} \operatorname{Tr} S[\sum_{p=1}^{n}\sum_{i=1}^{p-1} \theta^i R_{p-i}\theta'^i + \sum_{p=1}^{n}(\theta^p(\vec{y}_0 + \vec{b}_0) + \sum_{i=1}^{p-1} \theta^i \vec{b}_{p-i})$$

$$\times (\theta^p(\vec{y}_0 + \vec{b}_0) + \sum_{i=1}^{p-1} \theta^i \vec{b}_{p-i})')]\} + 0(1).$$

But then, from this relation and due to the limit theorems for Laplace transformations, we have

$$\operatorname{plim}_{n\to\infty}(1/C_n)\|\sum_{k=1}^{n} \vec{y}_k \vec{y}_k' - \sum_{p=1}^{n}\sum_{i=1}^{p-1} \theta^i R_{p-i}\theta'^i - \sum_{p=1}^{n}(\theta^p(\vec{y}_0 + \vec{b}_0)$$

$$+ \sum_{i=1}^{p-1} \theta_i \vec{b}_{p-i})(\theta^p(\vec{y}_0 + \vec{b}_0) + \sum_{i=1}^{p-1} \theta^i \vec{b}'_{p-i})\| = 0. \qquad (20.3.17)$$

Let us consider the sums $\frac{1}{\sqrt{C_n}} \sum_{k=1}^{n} \vec{\epsilon}_k \vec{y}'_{k-1}$. The matrices $\vec{\epsilon}_k \vec{y}'_{k-1}$ are martingale differences.

The multidimensional control limit theorem is valid for them if the following conditions (see Chapter 5) hold:

$$\operatorname{plim}_{n\to\infty} \frac{1}{C_n} \sum_{k=1}^{n}\{y_{k-1}\mathbf{E}_k \vec{\epsilon}'_k \vec{\epsilon}_k \vec{y}'_{k-1} - \mathbf{E}\vec{y}_{k-1} \vec{\epsilon}_k \vec{\epsilon}_k \vec{y}'_{k-1}\} = 0. \qquad (20.3.18)$$

$$\overline{\lim_{n\to\infty}} \frac{1}{C_n} \sum_{k=1}^{n} \mathbf{E}\|\vec{\epsilon}_k \vec{y}'_{k-1}\|^2 < \infty, \qquad \text{for any} \quad \tau > 0 \qquad (20.3.19)$$

$$\lim_{n\to\infty} \sum_{k=1}^{n} C_n^{-1}\mathbf{E}\|\vec{\epsilon}_k \vec{y}'_{k-1}\|^2 \chi(\|\vec{\epsilon}_k \vec{y}'_{k-1}\|^2 C_n^{-1/2} > \tau) = 0. \qquad (20.3.20)$$

By using (20.3.9) and conditions (20.3.4), we make sure that conditions (20.3.19) and (20.3.20) hold.

Indeed, from the limit theorems for the sums of martingale differences (Chapter 5), it follows that condition (20.3.18) is valid if the condition

$$\operatorname{plim}_{n\to\infty} C_n^{-1} \sum_{k=1}^{n}\{\vec{y}_{k-1} \vec{\epsilon}'_k \vec{\epsilon}_k \vec{y}'_{k-1} - \mathbf{E}\vec{y}_{k-1} \vec{\epsilon}'_k \vec{\epsilon}_k \vec{y}'_{k-1}\} = 0.$$

$$(20.3.21)$$

holds.

It is possible to use (20.3.21) for the proof of relation (20.3.17). All calculations for this case hold, except some trivial changes.

Let us point out these changes. By using conditions (20.3.3) and (20.3.4), it is easy to make sure of the fact that all the vectors in condition (20.3.21) can be substituted for the vectors

$$\tilde{\vec{\varepsilon}}_k = \begin{cases} \vec{\varepsilon}_k, & \|\vec{\varepsilon}_k\| \le h, \\ 0, & \|\vec{\varepsilon}_k\| > h, \end{cases}$$

where $h > 0$ is some arbitrary number.

Note that if the vectors $\vec{\varepsilon}_k$ are asymptotically independent and their norms are bounded, then

$$\operatorname{plim}_{n\to\infty} \|C_n^{-1} \sum_{k=1}^{n} (\vec{\varepsilon}_k \vec{\varepsilon}'_{k+s} A_k - \mathbf{E}\,\vec{\varepsilon}_k \vec{\varepsilon}'_{k+s} A_k)\| = 0, \qquad (20.3.22)$$

where $s > 0$ is any small number not depending on n, and A_k are some nonrandom matrices with $\|A_k\| \le c < \infty$.

From formula (20.3.9) it follows that the vector $\vec{y}_k$ for $k \to \infty$ is stochastically independent from the random vectors $\vec{\varepsilon}_1, \ldots, \vec{\varepsilon}_{k-m}$, where $m > 0$ is any integer such that $k - m \to \infty$ if $k \to \infty$.

Consequently, we can approximately consider that in the sum (20.3.21) the vectors $\vec{y}_{k-1}$ depend only on the random vectors $\vec{\varepsilon}_1, \ldots, \vec{\varepsilon}_{k-m}$, where $k \to \infty, k - m \to \infty, k > m$.

Then, by using the conditions of the asymptotic independence and condition (20.3.22), conditions (20.3.18)–(20.3.20) are valid. Therefore, taking into account (20.3.17), we come to the assertion of Theorem (20.3.1).

Corollary 20.3.1. *If in addition to the conditions of Theorem (20.3.1) there exist inverse matrices* $\sup_n \|(\sum_{k=1}^{n} \mathbf{E}\,\vec{y}_{k-1} \vec{y}'_{k-1} C_n^{-1})^{-1}\| < \infty$, *then*

$$\mathbf{E} \exp\{i\sqrt{C_n}\,\mathrm{Tr}\,Q[\hat{\theta}_n - \theta]\}$$

$$= \exp\{-\frac{1}{2}\,\mathrm{Tr} \sum_{k=1}^{n} \mathbf{E}\,\vec{\varepsilon}_k \vec{\varepsilon}'_k Q' (\sum_{k=1}^{n} \mathbf{E}\,\vec{y}_{k-1} \vec{y}'_{k-1} C_n^{-1})^{-1'}$$

$$\times \vec{y}_{k-1} \vec{y}'_{k-1} Q (\sum_{k=1}^{n} \mathbf{E}\,\vec{y}_{k-1} \vec{y}'_{k-1} C_n^{-1})^{-1} C_n^{-1}\} + 0(1).$$

§4 The Parameter Estimation of Nonlinear Control Systems

Let us consider discrete control systems which are described by the equations

$$\vec{x}_{k+1} = \vec{g}_k(\theta, \vec{x}_k) + \vec{b}_k + \vec{\varepsilon}_k,$$

where the vector function $\vec{g}_k(\theta, \vec{x}_k)$ is some nonlinear measurable function of the matrix of parameters $\theta = (\theta_{ij})_{i,j=1}^m$ and the components of vector $\vec{x}_k$.

In some cases, this equation can be linearized by the matrix θ and approximately substituted for the following equation

$$\vec{x}_{k+1} = \theta \vec{f}_k(\vec{x}_k) + \vec{b}_k + \vec{\varepsilon}_k, \tag{20.4.1}$$

where $\theta = (\theta_{ij})_{i,j=1}^m$ is an unknown matrix, b_k are known control vectors of the dimension m, $\vec{x}_k$ are observed vectors, $\vec{\varepsilon}_k$ are random vectors, and the vector $\vec{x}_0$ is given. The estimation $\hat{\theta}_n$ of the matrix θ can be found from the following equation. (This equation for normal independent random variables can be obtained with the help of the method of maximum likelihood or by direct solving of the system of equations (20.4.1):

$$\hat{\theta}_n \sum_{k=1}^n \vec{f}_k(\vec{x}_k) \vec{f}'_k(\vec{x}_k) = \sum_{k=1}^n \vec{z}_{k+1} \vec{f}'_k(\vec{x}_k), \tag{20.4.2}$$

where $\vec{z}_{k+1} = \vec{x}_{k+1} - \vec{b}_k$.

If the random vectors $\vec{\varepsilon}_k$ are independent, $\mathbf{E}\vec{\varepsilon}_k = 0$, then under the condition that with probability 1 there exist an inverse matrix

$$(\sum_{k=1}^n \vec{f}_k(\vec{x}_k) \vec{f}'_k(\vec{x}_k))^{-1},$$

the estimate $\hat{\theta}_n$ in the general case will be biased, i.e.,

$$\mathbf{E}\hat{\theta}_n \neq \theta.$$

Let us prove that under some additional conditions it will be consistent, i.e., $\text{plim}_{n\to\infty} \hat{\theta}_n = \theta$, and find the asymptotic distribution of the vector $\hat{\theta}_n - \theta$ as $n \to \infty$.

Obviously,

$$(\hat{\theta}_n - \theta) \sum_{k=1}^n \vec{f}_k(\vec{x}_k) \vec{f}'_k(\vec{x}_k) = \sum_{k=1}^n \vec{\varepsilon}_k \vec{f}'_k(\vec{x}_k). \tag{20.4.3}$$

If $\mathbf{E}(\vec{\varepsilon}_k | \vec{\varepsilon}_1, \ldots, \vec{\varepsilon}_{k-1}) = 0, k = 2, 3, \ldots$ then in expression (20.4.3), the sum $\sum_{k=1}^n \vec{\varepsilon}_k \vec{f}'_k(\vec{x}_k)$ is the sum of vector martingale differences. In order to use limit theorems for the sums of martingale differences, we prove the asymptotic stochastic independence of the random vectors $\vec{x}_k, \vec{x}_s$, as $|k - s| \to \infty$. This is the main moment of the theorem proved below.

Theorem 20.4.1. *Let* $\mathbf{E}(\overrightarrow{\varepsilon}_k | \overrightarrow{\varepsilon}_1, \ldots, \overrightarrow{\varepsilon}_{k-1}) = 0, k = 2, 3, \ldots,$ *and the random vectors* $\overrightarrow{\varepsilon}_k, k = 1, 2, \ldots$ *be G-asymptotically stochastically indepedent,*

$$\sup_k \mathbf{E}\|\overrightarrow{\varepsilon}_k\|^4 < \infty, \tag{20.4.4}$$

and let the function $\overrightarrow{f}_k(\overrightarrow{x}_k)$ *satisfy the Lipschitz condition*

$$\|\overrightarrow{f}_s(\overrightarrow{x}) - \overrightarrow{f}_s(\overrightarrow{y})\| \leq L\|\overrightarrow{x} - \overrightarrow{y}\|, \quad L > 0, \quad s = 1, 2, \ldots \tag{20.4.5}$$

$$\|\theta\| < L^{-1}. \tag{20.4.6}$$

And for all values $n > m, \|R_{m^2}\| < c < \infty,$ *where*

$$R_{m^2} = \left[\sum_{k=1}^{n} n^{-1} \mathbf{E} \overrightarrow{z}_k \overrightarrow{z}_k' \right]^{-1/2},$$

$$\overrightarrow{z}_k = \{ \overrightarrow{\varepsilon}_k \overrightarrow{f}_k'(\overrightarrow{x}_k)_{ij}, \quad i, j = \overline{1, m} \}.$$

Then for any Borel set $B \in R^{m^2},$

$$\lim_{n \to \infty} \mathbf{P}\{R_{m^2} \overrightarrow{q} \in B\} = \frac{1}{(2\pi)}^{m^2/2} \int \cdots \int_B e^{-1/2 \sum_{i=1}^{m^2} x_i^2} \prod_{i=1}^{m^2} dx_i.$$

$$\overrightarrow{q} = ([\hat{\theta}_n - \theta]\sqrt{n}C_n)_{i,j}, \quad i, j = \overline{1, m}), \tag{20.4.7}$$

where

$$C_n = \sum_{k=1}^{n} \frac{\mathbf{E}\overrightarrow{f}_k(\overrightarrow{x}_k)\overrightarrow{f}_k'(\overrightarrow{x}_k)}{n}.$$

Proof. We shall prove that if the conditions of Theorem (20.4.1) hold, the vectors $\overrightarrow{x}_k$ and $\overrightarrow{x}_s$ are stochastically asymptotically independent as $|k - s| \to \infty.$

Let $L_k(\overrightarrow{x})$ be nonlinear random operators acting by the formula

$$L_k(\overrightarrow{x}) = \theta \overrightarrow{f}_k(\overrightarrow{x}_k) + \overrightarrow{b}_k + \overrightarrow{\varepsilon}_k.$$

It is obvious that

$$\overrightarrow{x}_{k+1} = \prod_{p=2}^{k} L_p L_1(\overrightarrow{x}_0).$$

In this case, by the product of random operators, we mean their multiplication from right to left in order of increasing the index p.

In particular, for $k > s$,

$$\vec{x}_{k+1} = \prod_{p=s+1}^{k} L_p(\vec{x}_p).$$

By using condition (20.4.5), we have

$$\left\| \vec{x}_{k+1} - \prod_{p=s+1}^{k} L_p(\vec{0}) \right\| \le L\|\theta\| \left\| \vec{x}_k - \prod_{p=s+1}^{k-1} L_p(\vec{0}) \right\| \le (L\|\theta\|)^{k-s} \quad (20.4.8)$$

for $k > s$.

Since condition (20.4.6) holds and the random vectors $\vec{\varepsilon}_k$ are asymptotically independent, (20.4.8) implies that the random vectors $\vec{x}_k$ and $\vec{x}_s$ are asymptotically stochastically independent as $k - s \to \infty$.

The distribution of the sums $\sum_{k=1}^{n} \vec{\varepsilon}_k \vec{f}'_k(x_k) n^{-1/2}$ will be asymptotically normal, if the following conditions (see Chapter 5) hold:

$$\mathrm{plim}_{n\to\infty} [n^{-1} \sum_{k=1}^{n} \mathbf{E}\vec{\varepsilon}_k \vec{f}'_k(\vec{x}_k) \vec{f}_k(\vec{x}_k) \vec{\varepsilon}'_k]^{-1}$$

$$\times n^{-1} \sum_{k=1}^{n} \mathbf{E}\{\vec{\varepsilon}_k \vec{f}'_k(\vec{x}_k) \vec{f}_k(\vec{x}_k) \vec{\varepsilon}'_k / \vec{x}_k\} = I,$$

$$(20.4.9)$$

for any $\tau > 0$,

$$\lim_{n\to\infty} \sum_{k=1}^{n} \mathbf{E}\| R_n \vec{\varepsilon}_k \vec{f}'_k(\vec{x}_k) n^{-1/2} \|^2 \chi(\| R_n \vec{\varepsilon}_k \vec{f}_k(\vec{x}_k) n^{-1/2} \| > \tau\|) = 0.$$

$$(20.4.10)$$

Since the vectors $\vec{x}_k$ and $\vec{x}_s$ are asymptotically stochastically independent, by condition (20.4.4), the conditions (20.4.9) and (20.4.10) hold.

We consider the following transformation

$$\mathbf{E}e^{-\mathrm{Tr}\, Q \sum_{k=1}^{n} \vec{f}_k(\vec{x}_k) \vec{f}'_k(\vec{x}_k) n^{-1}}, \qquad (20.4.11)$$

where Q is a nonnegative-definite symmetric matrix of mth order.

We represent the expression (20.4.11) in the following form,

$$\mathbf{E}e^{-\mathrm{Tr}\, Q \sum_{k=1}^{n} \vec{f}_k(\vec{x}_k) \vec{f}'_k(\vec{x}_k) n^{-1}} = \mathbf{E}e^{i \sum_{k=1}^{n} (\vec{\eta}_k, \sqrt{Q} \vec{f}_k(\vec{x}_k)) n^{-1/2}},$$

$$(20.4.12)$$

where $\eta_k, k = 1, 2, \ldots$ are independent random vectors, distributed by the multidimensional normal law $N(0, I/2)$. The sums

$$\sum_{k=1}^{n} (\eta_k, \sqrt{Q}\,\vec{f}_k(\vec{x}_k))$$

are the sums of martingale differences. The following theorem (see Chapter 5) is valid for them if

$$\text{plim}_{n\to\infty} \frac{1}{n}\sum_{k=1}^{n}\{\mathbf{E}[\vec{\eta}_k, \sqrt{Q}\,\vec{f}_k(\vec{x}_k))^2/x_k] - \mathbf{E}[(\eta_k, \sqrt{Q}\,\vec{f}_k(\vec{x}_k))^2]\} = 0,$$

$$(20.4.13)$$

for any $\tau > 0$,

$$\lim_{n\to\infty} \frac{1}{n}\sum_{k=1}^{n}\mathbf{E}(\vec{\eta}_k, \sqrt{Q}\,\vec{f}_k(\vec{x}_k))^2\chi((\vec{\eta}_k, \sqrt{Q}\,\vec{f}_k(\vec{x}_k))n^{-1/2} > \tau) = 0,$$

$$(20.4.14)$$

then

$$\lim_{n\to\infty}[\mathbf{E}e^{i\sum_{k=1}^{n}(\vec{\eta}_k, \sqrt{Q}\,\vec{f}_k(\vec{x}_k))} - \mathbf{E}e^{-\text{Tr}\,Q\sum_{k=1}^{n}\vec{f}_k(\vec{x}_k)\vec{f}'_k(\vec{x}_k)n^{-1}}].$$

Conditions (20.4.13) and (20.4.14) follow from (20.4.4)–(20.4.6).
     Theorem 20.4.1 is proved.

§5 Limit Theorems of the General Form for the Parameter Estimation of Discrete Control Systems

This paragraph again examines discrete systems of control (see notations of §3 of this chapter). The estimate $\hat{\theta}_n$ is found from the following equation,

$$(\hat{\theta}_n - \theta)\sum_{k=1}^{n}\vec{x}_{k-1}\vec{x}'_{k-1} = \sum_{k=1}^{n}\vec{\varepsilon}_k\vec{x}'_{k-1}. \qquad (20.5.1)$$

We find the general form of limiting distributions for the matrices $(\hat{\theta}_n - \theta)\sqrt{n}$. We shall prove the following statement.

Theorem 20.5.1.   *Let*

$$\mathbf{E}(\vec{\varepsilon}_k|\vec{\varepsilon}_1, \ldots, \vec{\varepsilon}_{k-1}) = 0, \quad k = 2, 3, \ldots, \qquad (20.5.2)$$

for any $\delta > 0$,

$$\text{plim}_{n\to\infty} \sup_{k=\overline{1,n}} \mathbf{P}\{\|\vec{\varepsilon}_k\vec{y}_{k-1}\|n^{-1} > \delta/\sigma_k\} = 0, \qquad (20.5.3)$$

for any $\delta > 0$,

$$\mathrm{plim}_{n \to \infty} \sup_{k=\overline{1,n}} \mathbf{P}\{\|\overrightarrow{\eta}_k \overrightarrow{y}'_{k-1}\| > \delta/\sigma_k\} = 0, \qquad (20.5.4)$$

where $\overrightarrow{\eta}_k, k = \overline{1,n}$ are independent random variables not depending on $\overrightarrow{y}_k$, $k = \overline{1,n}$, and distributed according to the standard normal law $N(0, I)$; $\mathbf{E}_k$ is the conditional expectation given for the σ-algebra of events σ_k with respect to which the random vectors $\overrightarrow{\varepsilon}_1, \ldots, \overrightarrow{\varepsilon}_k$ are measurable

$$\mathrm{plim}_{n \to \infty} \sum_{k=1}^{n} [\mathbf{E}_k \exp\{i\, \mathrm{Tr}\, Q_1 \overrightarrow{\varepsilon}_k \overrightarrow{y}'_{k-1} n^{-1/2} + i\, \mathrm{Tr}\, Q_2 \overrightarrow{\eta}_k \overrightarrow{y}'_{k-1} n^{-1/2}\}$$
$$- \mathbf{E} \exp\{i\, \mathrm{Tr}\, Q_1 \overrightarrow{\varepsilon}_k \overrightarrow{y}'_{k-1} n^{-1/2} + i\, \mathrm{Tr}\, Q_2 \overrightarrow{\eta}_k \overrightarrow{y}'_{k-1} n^{-1/2}\}] = 0, \qquad (20.5.5)$$

$$\mathrm{plim}_{n \to \infty} \sum_{k=1}^{n} \mathbf{E}_k [\|\overrightarrow{\varepsilon}_k \overrightarrow{y}'_{k-1}\|^2 + \|\overrightarrow{\eta}_1 \overrightarrow{y}'_{k-1}\|^2] n^{-1} = c < \infty. \qquad (20.5.6)$$

Then for any real square matrix Q of order m,

$$\mathbf{E} \exp\{i\, \mathrm{Tr}(\hat{\theta} - \theta) Q \sqrt{n}\} = \mathbf{E} exp\{i\, \mathrm{Tr}\, Q \Gamma_1^{(n)} (\Gamma_2^{(n)})^{-1}\} + 0(1), \qquad (20.5.7)$$

where $\Gamma_1^{(n)}$ and $\Gamma_2^{(n)}$ are random matrices of order m with the characteristic function of the elements

$$\mathbf{E} \exp\{i\, \mathrm{Tr}\, Q_1 \Gamma_1^{(n)} - \frac{1}{2} \mathrm{Tr}\, Q_2 \Gamma_2^{(n)}\}$$
$$= \exp\{\sum_{k=1}^{n} [\mathbf{E} \exp\{i\, \mathrm{Tr}\, Q_1 \overrightarrow{\varepsilon}_k \overrightarrow{y}'_{k-1} n^{-1/2}$$
$$+ i\, \mathrm{Tr}\, Q_2^{1/2} \overrightarrow{\eta}_1 \overrightarrow{y}'_{k-1} n^{-1/2}\} - 1]\} + 0(1),$$

where Q_1 is a square matrix of order m, and Q_2 is a nonnegative-definite matrix of order m.

Proof. Analogously to §3 of this chapter, we have

$$(\hat{\theta}_n - \theta) \sum_{k=1}^{n} \overrightarrow{y}_{k-1} \overrightarrow{y}'_{k-1} = \sum_{k=1}^{n} \overrightarrow{\varepsilon}_k \overrightarrow{y}'_{k-1}.$$

Let us consider the joint characteristic function and the Laplace transform of the random matrices $n^{-1/2} \sum_{k=1}^{n} \overrightarrow{\varepsilon}_k \overrightarrow{y}'_{k-1}, n^{-1} \sum_{k=1}^{n} \overrightarrow{y}_{k-1} \overrightarrow{y}'_{k-1}$:

$$\mathbf{E} \exp\{i\, \mathrm{Tr}\, Q_1 \sum_{k=1}^{n} \overrightarrow{\varepsilon}_k \overrightarrow{y}'_{k-1} n^{-1/2} - \frac{1}{2} \mathrm{Tr}\, Q_2 \sum_{k=1}^{n} \overrightarrow{y}_{k-1} \overrightarrow{y}'_{k-1} n^{-1}\}$$

$$= \mathbf{E} \exp\{i\, \mathrm{Tr}\, Q_1 \sum_{k=1}^{n} \overrightarrow{\varepsilon}_k \overrightarrow{y}'_{k-1} n^{-1/2} + i\, \mathrm{Tr}\, Q_2^{1/2} \sum_{k=1}^{n} \overrightarrow{\eta}_k \overrightarrow{y}'_{k-1} n^{-1/2}\}. \qquad (20.5.8)$$

By applying the theorems of §3 of this chapter to expression (20.5.8) and by using conditions (20.5.1)–(20.5.6), we get (20.5.7). Theorem 20.5.1 is proved.

Let us simplify the conditions of Theorem 20.5.1.

Theorem 20.5.2. *Let for any value of n the random vectors* $\vec{\varepsilon}_k^{(n)}, k = \overline{1,n}$ *be independent,* $\mathbf{E}\varepsilon_k^{(n)} = 0, \vec{b}_k = 0,$

$$\sup_{k,n} \mathbf{E}\|\varepsilon_k\|^2 < \infty, \quad \varlimsup_{\substack{n\to\infty \\ m\to\infty \\ \frac{m}{n}\to 0}} \mathbf{E}\|\vec{\varepsilon}_k \vec{\beta}'_k(m)\|^2 < \infty, \tag{20.5.9}$$

$$\|\theta\| < 1, \quad \lim_{\varepsilon\downarrow 0}\varliminf_{n\to\infty} \mathbf{P}\{\det \Gamma_2^{(n)} > \varepsilon\} = 1, \tag{20.5.10}$$

$$\beta_k(m) = \theta^{m-1}\varepsilon_{k-m} + \cdots + \varepsilon_k.$$

Then (20.5.7) holds.

Proof. We consider the sums

$$S_n = \sum_{k=1}^{n} \vec{\varepsilon}_k \vec{y}'_{k-1}, \qquad P_n = \sum_{k=1}^{n} \vec{\eta}_k \vec{y}'_{k-1}. \tag{20.5.11}$$

Obviously,

$$\vec{y}_k = \theta^k \vec{y}_0 + \sum_{i=0}^{k-1} \theta^i \vec{\varepsilon}_{k-i}.$$

Consider that $\vec{y}_k = \alpha_k(m) + \beta_k(m), \quad \alpha_k(m) = \theta^k y_0 + \theta^{k-1}\varepsilon_1 + \cdots + \theta^m \varepsilon_{k-m}, \beta_k(m) = \theta^{m-1}\varepsilon_{k-m} + \cdots + \varepsilon_k,$ where $0 < m < k$ are arbitrary integers.

We represent the sums (20.5.11) in the following form,

$$S_n = \sum_{k=m}^{n} \vec{\varepsilon}_k \alpha'_k(m) + \sum_{k=m}^{n} \varepsilon_k \beta'_k(m) + \sum_{k=1}^{m} \vec{\varepsilon}_k \vec{y}'_{k-1},$$

$$P_n = \sum_{k=m}^{n} \vec{\eta}_k \alpha'_k(m) + \sum_{k=m}^{n} \vec{\eta}_k \beta'_k(m) + \sum_{k=1}^{m} \vec{\eta}_k \vec{y}'_{k-1}.$$

Due to conditions (20.5.9) and (20.5.10) for $\frac{m}{n} \to 0, m \to \infty,$

$$\lim_{m\to\infty} \text{plim}_{n\to\infty} \left[\sum_{k=m}^{n} \vec{\varepsilon}_k \alpha'_k(m) + \sum_{k=1}^{m} \vec{\varepsilon}_k \vec{y}'_{k-1}\right] n^{-1/2} = 0, \tag{20.5.12}$$

$$\lim_{m\to\infty} \text{plim}_{n\to\infty} \left[\sum_{k=m}^{n} \vec{\eta}_k \alpha'_k(m) + \sum_{k=1}^{m} \vec{\eta}_k \vec{y}'_{k-}\right] n^{-1/2} = 0. \tag{20.5.13}$$

Note that the vectors $\beta_k(m), \beta_s(m)$ are independent as $k - s > m$. Therefore, the previous theorem can be applied to the sums

$$S' = \sum_{k=m}^{n} \vec{\varepsilon}_k \beta'_k(m) n^{-1/2}, \qquad \mathbf{P}'_n = \sum_{k=m}^{n} \vec{\eta}_k \beta'_k(m) n^{-1/2}.$$

Let us verify the conditions of this theorem. Since the vectors $\beta_k(m), \beta_s(m)$ are independent for $k - s > m$, by the law of large numbers, (20.5.5) holds, since

$$S_n = \sum_{s=1}^{n} \gamma_s, \qquad P_n = \sum_{s=1}^{n} \tilde{\gamma}_s,$$

$$\gamma_s = \mathbf{E}(S_n/\overrightarrow{\varepsilon}_k \overrightarrow{y}'_{k-1}, k = \overline{1,s}) - \mathbf{E}(S_n/\overrightarrow{\varepsilon}_k \overrightarrow{y}'_{k-1}, k = \overline{1,s-1}),$$

$$\tilde{\gamma}_s = \mathbf{E}(P_n/\overrightarrow{\eta}_k \overrightarrow{\beta}'_k, k = \overline{1,s}) - \mathbf{E}(P_n/\overrightarrow{\eta}_k \overrightarrow{\beta}'_k, k = \overline{1,s-1}).$$

The other conditions of Theorem 20.5.1 are verified analogously. Then we use again limit theorems of the type of law of large numbers and the independence of the random vectors $\overrightarrow{y}_k(m)$ and $\overrightarrow{y}_s(m)$ as $|k - s| > m$. Theorem (20.5.2) is proved.

§6 Limit Theorem for Estimating Parameters of Discrete Control Systems with Multiplicative Noises

Consider the discrete control systems described by the equations

$$\overrightarrow{x}_{k+1} = (\theta + \Xi_k)\overrightarrow{x}_k + \overrightarrow{b}_k + \overrightarrow{\varepsilon}_k,$$

where $\theta = (\theta_{ij})_{i,j=1}^m$ is an unknown matrix, $\overrightarrow{b}_k$ are known control vectors, the vector $\overrightarrow{x}_0$ is given, $\overrightarrow{\varepsilon}_k$ and Ξ_k are independent random vectors and matrices: the matrix elements are called *inner noises* or *multiplicative noises*. The estimate of the matrix $\hat{\theta}_n$ can be found from the equation

$$\hat{\theta}_n \sum_{k=1}^{n} \overrightarrow{x}_{k-1} \overrightarrow{x}'_{k-1} = \sum_{k=1}^{n} \overrightarrow{z}_k \overrightarrow{x}'_{k-1},$$

where $z_k = \overrightarrow{x}_k - \overrightarrow{b}_{k-1}$. Hence, by substituting the value of the vector $\overrightarrow{z}_k$ in this expression, we get

$$(\hat{\theta}_n - \theta) \sum_{k=1}^{n} \overrightarrow{x}_{k-1} \overrightarrow{x}'_{k-1} = \sum_{k=1}^{n} \Xi_{k-1} \overrightarrow{x}_{k-1} \overrightarrow{x}'_{k-1} + \sum_{k=1}^{n} \overrightarrow{\varepsilon}_{k-1} \overrightarrow{x}'_{k-1}.$$

Theorem 20.6.1. *Let the random vectors $\overrightarrow{\varepsilon}_k$, $k = 1, 2, \ldots$ and the matrices Ξ_k, $k = 1, 2, \ldots$ be independent,*

$$\mathbf{E}\overrightarrow{\varepsilon}_k = 0, \qquad \mathbf{E}\Xi_k = 0,$$

$$\sup_k \|\theta + \Xi_k\| < 1, \qquad \sup_k \|\overrightarrow{b}_k\| < c < \infty, \qquad (20.6.1)$$

$$\|R_{m^2}\| < \infty, \qquad \|C_n^{-1}\| < c < \infty, \qquad (20.6.2)$$

$$\sup_k \mathbf{E}\{||\varepsilon_k||^{2+\delta} + ||\Xi_k||^{2+\delta}\} < \infty, \quad \delta > 0, \tag{20.6.3}$$

$$R_{m^2} = [n^{-1} \sum_{k=1}^{n} \mathbf{E}\,\vec{\beta}_k\,\vec{\beta}'_k]^{-1/2},$$

$$\vec{\beta}_k = \{((\Xi_k x_{k-1} + \varepsilon_k)x'_{k-1})_{ij}, \quad i,j = \overline{1,m}\}.$$

Then for any Borel set $B \in R^{m^2}$,

$$\lim_{n\to\infty} \mathbf{P}\{R_{m^2}\vec{q}\} = (2\pi)^{-m^2/2} \int_B \cdots \int \exp(-\frac{1}{2}\sum_{i=1}^{m^2} x_i^2) \prod_{i=1}^{m^2} dx_i,$$

where

$$\vec{q} = \{([\hat{\theta}_n - \theta]\sqrt{n}C_n)_{ij}, \quad i,j = \overline{1,m}\},$$

$$C_n = n^{-1} \sum_{k=1}^{n} \mathbf{E}\,\vec{x}_{k-1}\,\vec{x}'_{k-1}.$$

Proof. Obviously,

$$\sum_{k=1}^{n} \Xi_k\,\vec{x}_{k-1}\,\vec{x}'_{k-1} + \sum_{k=1}^{n} \vec{\varepsilon}_k\,\vec{x}'_{k-1}$$

are the sums of the martingale differences, and by condition (20.6.1) the random vectors $\vec{x}_s$, $\vec{x}_k$ are asymptotically independent for $|k - s| \to \infty$ (see Theorem 20.3.1). Therefore, by using conditions (20.6.1)–(20.6.3) and the theorems of §3 of this chapter, we get the assertion of the theorem.

§7 Estimating Spectra of Stochastic Linear Control Systems and Spectral Equations in the Theory of the Parameters Estimation

Let us study the systems of equations

$$\vec{y}_k = \theta\vec{y}_{k-1} + \vec{b}_{k-1} + \vec{\varepsilon}_{k-1} \tag{20.7.1}$$

(see the notations of the previous sections).

As the estimations of the eigenvalues λ_k of the matrix θ, we take the eigenvalues λ_k of the matrix $\hat{\theta}_n$ defined by the following equation,

$$\hat{\theta}_n \sum_{k=1}^{n} \vec{y}_{k-1}\vec{y}'_{k-1} = \sum_{k=1}^{n} (\vec{y}_k - \vec{b}_{k-1})\vec{y}'_{k-1}. \tag{20.7.2}$$

Prove the asymptotic normality of distributions of these estimations of the spectrum of the matrix θ.

Theorem 20.7.1. *Let* $\mathbf{E}(\overrightarrow{\varepsilon}_k / \overrightarrow{\varepsilon}_1, \ldots, \overrightarrow{\varepsilon}_{k-1}) = 0, k = 2, 3, \ldots$ *and the random vectors* $\overrightarrow{\varepsilon}_k$ *be G-asymptotically independent*

$$\|\theta\| < 1 \qquad \sup_i \|b_i\| < c < \infty; \tag{20.7.3}$$

let the eigenvalues of the matrix θ *be distinct,*

$$\lim_{h \to \infty} \sup_k \mathbf{E}\|\overrightarrow{\varepsilon}_k\|^2 \chi(\|\overrightarrow{\varepsilon}_k\| > h) = 0, \tag{20.7.4}$$

and let there exist inverse matrices for all $n \leq m$,

$$\lim_{n \to \infty} \|[\sum_{k=1}^{n} \mathbf{E}\overrightarrow{y}_{k-1} \overrightarrow{y}'_{k-1} c_n^{-1}]\| < c < \infty, \tag{20.7.5}$$

where $c_n \geq n$ *are some sequence of constants. Then*

$$\mathbf{E}\exp\{is\,\mathrm{Re}[\lambda_k(\hat{\theta}_n) - \lambda_k(\theta)]\sqrt{c_n}\}$$

$$= \mathbf{E}\exp\{-\frac{s^2}{2c_n}\sum_{p=1}^{n}\mathbf{E}[\mathrm{Re}(\overrightarrow{\varepsilon}_p, \overrightarrow{\psi}_k)$$

$$\times \overrightarrow{y}'_p(\sum_{s=1}^{n}\mathbf{E}\overrightarrow{y}_{s-1}\overrightarrow{y}'_{s-1}c_n^{-1})^{-1}\overrightarrow{\varphi}_k)]^2\} + 0(1), \tag{20.7.6}$$

where $\overrightarrow{\varphi}_k$ *is the eigenvector of the matrix* θ *corresponding to the eigenvalue* λ_k, *and* ψ_k *is the eigenvector of the matrix* θ' *corresponding to the eigenvalue* λ_k.

Proof. The perturbation formulas introduced in Chapter 18 are not suitable in this case, as in general, the matrix θ can be nonsymmetric. But since under the conditions of Theorem 20.7.1, the eigenvalues of the matrix θ are distinct, we can substitute the formulas of Chapter 18 for the following perturbation formulas for nonsymmetric matrices,

$$\lambda_k(A + \varepsilon B) = \lambda_k(A) + \varepsilon(B\varphi_k, \psi_k) + \varepsilon^2 \sum_{l \neq k} \frac{(B\varphi_k, \psi_l)^2}{\lambda_k(A) - \lambda_l(A)} + \delta_n(\varepsilon),$$

$$\varphi_k(A + \varepsilon B) = \varphi_k(A) + \varepsilon \sum_{l \neq k} \frac{(B\varphi_k, \psi_l)\varphi_l}{\lambda_k(A) - \lambda_l(A)} + \gamma_n(\varepsilon), \tag{20.7.7}$$

where $\lambda_k(A + \varepsilon B), \varphi_k(A + \varepsilon B)$ are eigenvalues and eigenvectors of the $n \times n$ matrices $a + \varepsilon B$, ψ_k are eigenvalues of the matrix A', and $\delta_n(\varepsilon)$ and $\gamma_n(\varepsilon)$ are residual terms of decomposition (20.7.7). The estimation for the values $\delta_n(\varepsilon)$ and $\gamma_n(\varepsilon)$ can be also found. By $\lambda_i \neq \lambda_j, i \neq j$, we find

$$|\delta_n(\varepsilon)\| \leq \frac{|\varepsilon|^3\|B\|^3 C_1}{1 - |\varepsilon|\|B\|C_2}, \qquad |\gamma_n(\varepsilon)\| \leq \frac{|\varepsilon|^3\|B\|^3 C_3}{1 - |\varepsilon|\|B\|C_4}, \tag{20.7.8}$$

where ε is chosen so small that $|\varepsilon|\|B\|C_2 < 1, |\varepsilon|\|B\|C_4 < 1$.

If follows from the proof of Theorem 20.3.1 that

$$\text{plim}_{n\to\infty}[(\sum_{k=1}^{n} \vec{y}_{k-1}\vec{y}'_{k-1} - \sum_{k=1}^{n} \mathbf{E}\vec{y}_{k-1}\vec{y}'_{k-1})\frac{1}{C_n}] = 0.$$

Therefore, as in Theorem 20.3.1 as $n \to \infty$ with probability close to 1, we get

$$(\hat{\theta}_n - \theta)\sqrt{C_n} = \sum_{k=1}^{n} \vec{\varepsilon}_k \vec{y}'_{k-1}\frac{1}{\sqrt{C_n}}(\sum_{k=1}^{n} \mathbf{E}\vec{y}_{k-1}\vec{y}'_{k-1}\frac{1}{C_n})^{-1}. \qquad (20.7.9)$$

By using the perturbation formulas (20.7.7) and the inequalities (20.7.8), we have

$$[\lambda_k(\hat{\theta}_n) - \lambda_k(\theta)]\sqrt{C_n} = [\lambda_k(\theta + \sqrt{C_n}(\hat{\theta}_n - \theta)\frac{1}{C_n}) - \lambda_k(\theta)]\sqrt{C_n}$$

$$= \sqrt{C_n}((\hat{\theta}_n - \theta)\varphi_k(\theta), \psi_k(\theta)) + \frac{1}{\sqrt{C_n}}\varepsilon_n \qquad (20.7.10)$$

for sufficiently large n under the condition that $\|\hat{\theta}_n - \theta\|\sqrt{C_n} < 1$, where ε_n is some sequence of random bounded variables.

From (20.7.9) and (20.7.10), we obtain

$$\mathbf{E}\exp\{is\,\text{Re}[\lambda_k(\hat{\theta}) - \lambda_k(\theta)]\sqrt{C_n}\}$$

$$= \mathbf{E}\exp\{is\,\text{Re}\,\sqrt{C_n}((\hat{\theta}_n - \theta)\vec{\varphi}_k(\theta), \vec{\psi}_k(\theta))\} + 0(1)$$

$$= \mathbf{E}\exp\{is\,\text{Re}(\sum_{p=1}^{n} \vec{\varepsilon}_p \vec{y}'_{p-1}\frac{1}{\sqrt{C_n}}(\sum_{p=1}^{n} \mathbf{E}\vec{y}_{p-1}\vec{y}'_{p-1}\frac{1}{C_n})^{-1}$$

$$\times \vec{\varphi}_k(\theta), \vec{\psi}_k(\theta))\} + 0(1) = \mathbf{E}\exp\{is \sum_{p=1}^{n} \frac{\text{Re}(\vec{\varepsilon}_p, \vec{\psi}_k(\theta))}{\sqrt{C_n}}$$

$$\times (\vec{y}'_{p-1}(\sum_{p=1}^{n} C_n\mathbf{E}\vec{y}_{p-1}\vec{y}'_{p-1})^{-1}\vec{\varphi}_k(\theta))\} + 0(1).$$

As in the proof of Theorem 20.3.1 and by taking into account that

$$\|\psi_k(\theta)\| < c < \infty, \qquad \|\varphi_k(\theta)\| < c < \infty,$$

and

$$\sup_{n\geq m} \|(\sum_{p=1}^{n} C_n^{-1}\mathbf{E}\vec{y}_{p-1}\vec{y}'_{p-1})^{-1}\| < c < \infty,$$

we get (20.7.6).

The analogous assertion can be proved for the eigenvectors of the matrix $\hat{\theta}_n$.

Theorem 20.7.2. *If the conditions of Theorem 20.7.1 hold, then*

$$\mathbf{E}\exp\{is'\operatorname{Re}[\varphi_k(\hat{\theta}_n) - \varphi_k(\theta)]\sqrt{C_n}\}$$

$$= \mathbf{E}\exp\{\frac{1}{2}\frac{1}{C_n}\sum_{p=1}^{n}\mathbf{E}[\operatorname{Re}(\vec{\varepsilon}_p, \sum_{l\neq k}\frac{\psi_l(\theta)s'\varphi_1}{\lambda_k(\theta) - \lambda_l(\theta)}$$

$$\times (\vec{y}_p'(\sum_{s=1}^{n}\mathbf{E}\,\vec{y}_{s-1}\,\vec{y}_{s-1}'C_n^{-1})^{-1}, \vec{\varphi}_k)]^2\} + 0(1).$$

Analogous statements are also valid for the random variables

$$\operatorname{Im}[\lambda_k(\hat{\theta}_n) - \lambda_k(\theta))]\sqrt{C_n}, \quad \operatorname{Im}[\vec{\varphi}_k(\hat{\theta}_n) - \vec{\varphi}_k(\theta)]\sqrt{C_n},$$

and for their joint distributions.

   Besides, analogous assertions can be proved for the joint distributions of the random variables

$$\operatorname{Re}[\lambda_k(\hat{\theta}_n) - \lambda_k(\theta)]\sqrt{C_n}, \quad \operatorname{Re}[\vec{\varphi}_k(\hat{\theta}_n) - \vec{\varphi}_k(\theta)]\sqrt{C_n}.$$

From Theorem 20.7.1 and on the basis of the limit theorems for characteristic functions, we get the following statement.

Corollary 20.7.1. *If the conditions of Theorem (20.7.1) hold, then for any real α and β,*

$$\lim_{n\to\infty}\mathbf{P}\left\{\alpha < \frac{\operatorname{Re}[\lambda_k(\hat{\theta}_n) - \lambda_k(\theta)]\sqrt{C_n}}{[L(\theta)]} < \beta\right\} = \frac{1}{\sqrt{2\pi}}\int_{\alpha}^{\beta}e^{-x^2/2}dx.$$

$$(20.7.11)$$

   From formula (20.7.11), we get the spectral equation for $\operatorname{Re}\lambda_k(\theta)$,

$$\operatorname{Re}\lambda_k(\theta) = \operatorname{Re}\lambda_k(\hat{\theta}_n) + \gamma L(\theta), \qquad (20.7.12)$$

where

$$L(\theta) = [\frac{1}{C_n}\sum_{p=1}^{n}\mathbf{E}[\operatorname{Re}(\vec{\varepsilon}_p, \vec{\psi}_k(\theta))(\vec{y}_p'$$

$$\times (\sum_{s=1}^{n}\mathbf{E}\,\vec{y}_{s-1}\,\vec{y}_{s-1}'C_n^{-1})^{-1}, \vec{\varphi}_k(\theta))]^2]^{1/2},$$

and γ is any value from the segment $[-\frac{\alpha}{\sqrt{C_n}}, \frac{\alpha}{\sqrt{C_n}}]$. The value α can be found from the equation

$$\frac{1}{\sqrt{2\pi}}\int_{-\alpha}^{\alpha}e^{-x^2/2}dx = p,$$

where $0 < p < 1$ is a given value.
   The analogous equations can be obtained for $\operatorname{Im}\lambda_k(\theta)$.
   Note that Theorems 20.7.1 and 20.7.2 are the examples for applying Theorem 20.3.11 to solving Eq. (20.7.12). Analogous assertions can be proved by virtue of §§4–6 of this chapter.

CHAPTER 21

RANDOM DETERMINANTS IN SOME PROBLEMS
OF CONTROL THEORY OF STOCHASTIC SYSTEMS

This chapter deals with some problems of control theory for some dynamic systems. In order to solve these problems, we use random determinants.

§1 The Kalman Stochastic Condition

The basis of the theory of linear stationary controllable systems is the following mathematical model,

$$\dot{\vec{x}} = A\vec{x} + B\vec{u}; \qquad \vec{z} = G\vec{x}, \tag{21.1.1}$$

where $\vec{x}(t) = (x_1(t), \ldots, x_n(t))^*$ is the vector of a state; $\vec{u}(t) = (u_1(t), \ldots, u_r(t))^*$ is the entry of the system, or control; $\vec{z}(t) = (z_1(t), \ldots, z_m(t))^*$ is the exit of the system, or observation; $A = \{a_{ij}\}$ is an $n \times n$ matrix; $B = \{b_{ij}\}$ is an $n \times r$ matrix; and $G = (g_{ij})$ is an $m \times n$ matrix.

System (21.1.1) satisfies the condition of generality of the position if and only if

$$\det(\vec{b_i} A\vec{b_i} \ldots A^{n-1}b_i) \neq 0, \quad i = 1, 2, \tag{21.1.2}$$

where b_i is the ith column of the matrix B.

If $B = \vec{b}$ is a vector column, then the condition

$$\det(\vec{b} A\vec{b} \ldots A^{n-1}\vec{b}) \neq 0, \tag{21.1.3}$$

is a criterion of the system's complete controllability. The homogeneous system corresponding to the system (21.1.1),

$$\dot{\vec{x}} = A\vec{x}, \tag{21.1.4}$$

will be observable by the ith, $i = \overline{1, m}$ component of the vector z if and only if

$$\det(\vec{g_i} A^* \vec{g_i} \ldots A^{*n-1}\vec{g_i}) \neq 0, \quad i = \overline{1, m}, \tag{21.1.5}$$

503

where $\vec{g}_i^*$ is the ith row of the matrix G.

System (21.1.4) can be identified by a state $\vec{x}(t_1) = \vec{x}_1$ if and only if $\det(\vec{x}_1 A \vec{x}_1 \ldots A^{n-1}\vec{x}_1) \neq 0$. Note also that conditions (21.1.3) and (21.1.5) are the same for controllable systems with discrete time:

$$\vec{x}(k) = A\vec{x}(k-1) + B\vec{u}(k-1); \vec{z}(k) = G\vec{x}(k). \qquad (21.1.6)$$

Thus the value $\det(\vec{b}A\vec{b} \ldots A^{n-1}\vec{b}) = D_n$ takes one of the central places in the theory of stationary controllable systems.

However, in real systems with respect to a number of always existing factors (obstacles, noises, inaccuracies in measurings, wrong information), the elements of the matrix A and of the vector $\vec{b}$ cannot be regarded as deterministic quantities. The purpose of this paragraph is to investigate D_n as a random variable.

Obviously, if elements of the matrix A and the vector b are continuous random variables, then $p\{D_n \neq 0\} = 1$. A much more interesting (but rather complicated) problem is to find the probability

$$\mathbf{P}\{|D_n| > \varepsilon\}, \quad \varepsilon > 0, \qquad (21.1.7)$$

This problem, in general, is far from being solved. We shall consider only one particular important case.

Let A be a random symmetric matrix which does not depend on the vector $\vec{b}$, and its entries, arranged on the main diagonal and above it, are independent and distributed according to the normal law $N(0,1)$; components of the vector b are also independent and distributed according to the standard normal law. Then, according to Corollary 3.1.2, the eigenvalues of matrix A arranged by increasing $(\lambda_1 \geq \lambda_2 \geq \cdots \geq \lambda_n)$ have the following distribution density:

$$\mathbf{P}(y_1, \ldots y_n) = 2^{-3n/2}\pi^{-n(n+1)2}c_{1n} \exp\left(-\sum_{i=1}^{n}\frac{y_i}{2}\right)^2$$

$$\times \prod_{i<j}(y_i - y_j), \quad y_1 \geq y_2 \geq \cdots \geq y_n, \qquad (21.1.8)$$

where c_{1n} is a normalizing constant

$$c_{1n} = 2^n \pi^{n(n+1)/4}\left\{\prod_{j=1}^{n}\Gamma(j/2)\right\}^{-1}.$$

The matrix of normalized eigenvectors is stochastically independent of the eigenvalues and is distributed according to the conditional Haar measure.

Since the eigenvalues are distinct with probability 1,

$$|D_n| = \prod_{i=1}^{n} |(\vec{b}, \vec{h}_i^*)| \prod_{i<j} (\lambda_i - \lambda_j), \qquad (2.1.9)$$

where $\vec{h}_i^*$ are the rows of the matrix of eigenvectors.

This formula is easy to verify. It also can be obtained from a more complicated formula by taking into account the multiplicity of the eigenvalues.

Since the vectors $\vec{h}_i^*$ are orthogonal and the vector $\vec{b}$ is distributed normally $N(0, 1)$, then the values $(\vec{b}, \vec{h}_i^*), i = \overline{1, n}$ are independent, distributed normally $N(0, 1)$, and do not depend on the eigenvalues $\lambda_i, i = \overline{1, n}$.

For the analysis of probability (21.1.7), we use the method of moments. Since in formula (21.1.9) the first product does not depend on the second one,

$$\mathbf{E}|D_n|^k = \mathbf{E}\left(\prod_{i=1}^{n} |(\vec{b}, \vec{h}_i^*)|^k\right) \mathbf{E}\left(\prod_{i<j} (\lambda_i - \lambda_j)^k\right). \qquad (2.1.10)$$

We calculate the first factor of the expression (21.1.10)

$$\mathbf{E}\prod_{i=1}^{n} |(\vec{b}, \vec{h}_i^*)|^k = \left[\frac{1}{\sqrt{2\pi}} \int_{-\infty}^{\infty} |x|^k e^{-\frac{x^2}{2}} dx\right]^n$$

$$= 2^{n/2} \pi^{-n/2} \left[\int_{0}^{\infty} x^k e^{-\frac{x^2}{2}} dx\right]^n$$

$$= 2^{nk/2} \pi^{-n/2} [\Gamma\left(\frac{k+1}{2}\right)]^n. \qquad (21.1.11)$$

Let

$$c = c_{1n} 2^{-3n/2} \pi^{-n(n+1)/2} = 2^{-n/2} \pi^{-n(n+1)/4} \left[\prod_{i=1}^{n} \Gamma(j/2)\right]^{-1}. \qquad (21.1.12)$$

Then the second factor of the expression (21.1.10) is

$$\mathbf{E}\prod_{i>j} (\lambda_i - \lambda_j)^k = c \int \cdots \int_{y_1 \ge \cdots > y_n} \exp\left(-\sum_{i=1}^{n} \frac{y_i^2}{2}\right) \prod_{i>j} (y_i - y_j)^{k+1} dy_1 \ldots dy_n$$

$$= \frac{c}{n!} \int_{-\infty}^{\infty} \cdots \int_{-\infty}^{\infty} \exp\left(-\sum_{i=1}^{n} \frac{y_i^2}{2}\right) \prod_{i>j} |y_i - y_j|^{k+1} dy_1 \ldots dy_n$$

$$= \frac{c}{n!} (2\pi)^{n/2} \mathbf{E}|V_n|^{k+1}, \qquad (21.1.13)$$

where $V_n = \{\beta_i^{j-1}\}_{i,j=1}^n$ is a Vandermonde determinant; β_i are independent and identically distributed according to the normal law variables $N(0,1)$.

The value $\mathbf{E}|V_n|$ was calculated in §2, Chapter 2.

$$\mathbf{E}|V_n|^m = [\Gamma(1 + \frac{m}{2})]^{-n} \prod_{i=1}^n \Gamma(1 + \frac{mi}{2}). \qquad (21.1.14)$$

Hence,

$$\mathbf{E} \prod_{i>j} (\lambda_i - \lambda_j)^k = \frac{c}{n!}(2\pi)^{n/2}[\Gamma\left(1 + \frac{k+1}{2}\right)^{-n} \prod_{i=1}^n \Gamma\left(1 + \frac{(k+1)j}{2}\right), \qquad (21.1.15)$$

and (21.1.12) yields

$$\mathbf{E} \prod_{i<j} (\lambda_i - \lambda_j)^k = \frac{\pi^{-n(n-1)/4} \prod_{j=1}^n \Gamma(1 + \frac{(k+1)j}{2})}{n! \prod_{j=1}^n \Gamma(j/2)[\Gamma(1 + \frac{k+1}{2})]^n}. \qquad (21.1.16)$$

By multiplying (21.1.11) by (21.1.16), we find

$$\mathbf{E}|D_n|^k = \frac{2^{n(k+2)/2} \pi^{-n(n+1)/4}}{n!(k+1)^n} \prod_{j=1}^n \frac{\Gamma(1 + \frac{(k+1)j}{2})}{\Gamma(j/2)}.$$

In conclusion, we recall that (under some weak assumptions) the characteristic function can be constructed with the help of moments of the random variable

$$\varphi(s) = 1 + \sum_{k=1}^\infty \frac{i^k \mathbf{E}|D_n|^k}{k!} s^k.$$

The probability (21.1.7) can be estimated with the help of Chebyshev-type inequalities. For example,

$$\mathbf{P}\{|D_n| > \varepsilon\} \le \frac{\mathbf{E}|D_n|^h}{\varepsilon^k}.$$

Solving some control problems involves proving limit theorems for

$$\mathbf{P}\{|\det(\vec{b}(\omega), A(\omega)\vec{b}(\omega), \ldots, A^{n-1}(\omega)\vec{b}(\omega)| > \varepsilon\}.$$

Let us prove the following statement.

Theorem 21.1.1. *Let $\mu_n(x) \Rightarrow \mu(x)$ on some everywhere dense set c on the straight line R_1, where $\mu_n(x)$ are the normalized spectral functions of the symmetric matrices $A_n(\omega)$, $\mu(x)$ is a nondecreasing random function of bounded variation, for some $\delta > 0$,*

$$\sup_n n^{-2} \sum_{i>j} \mathbf{E}\left|\ln |\lambda_i - \lambda_j|\right|^{1+\delta} < \infty, \qquad (21.1.17)$$

and let the integral

$$\iint \ln |x - y| d\mu(x) d\mu(y) < \infty \qquad (21.1.18)$$

be finite with probability 1.
    Then

$$n^{-2} \sum_{i>j} \ln |\lambda_i - \lambda_j| \Rightarrow 0.5 \iint \ln |x - y| d\mu(x) d\mu(y). \qquad (21.1.19)$$

Proof.   Consider the sums

$$n^{-2} \sum_{i>j} \ln |\lambda_i - \lambda_j| \chi(|\ln |\lambda_i - \lambda_j|| < A), \qquad (21.1.20)$$

where $A > 0$ is an arbitrary constant.
    Evidently, for some sufficiently small ε,

$$\left| 0.5 n^{-2} \sum_{i,j=1}^{n} \ln \left[|\lambda_i - \lambda_j| + \varepsilon \right] \chi(|\ln |\lambda_i - \lambda_j|| < A) \right.$$

$$\left. - n^{-2} \sum_{i>j} \ln |\lambda_i - \lambda_j| \chi(|\ln |\lambda_i - \lambda_j|| < A) \right.$$

$$\leq 0.5 n^{-1} \ln \varepsilon + n^{-2} \sum_{i>j} \ln |1 + \varepsilon |\lambda_i - \lambda_j|^{-1}| \chi(|\ln |\lambda_i - \lambda_j|| < A)$$

$$\leq 0.5 n^{-1} \ln \varepsilon + \ln |1 + \varepsilon e^A|.$$

From this inequality and by using (21.1.17) and (21.1.18), we obtain

$$n^{-2} \sum_{i>j} \ln |\lambda_i - \lambda_j| \chi(|\ln |\lambda_i - \lambda_j|| < A)$$

$$\Rightarrow 0.5 \iint_{|\ln |x-y|| < A} \ln |x - y| d\mu(x) d\mu(y).$$

Hence, by using (21.1.17), we arrive at the statement of Theorem 21.1.1.

Corollary 21.1.1.   *If in addition to the conditions of Theorem 21.1.1, a) with probability 1, $\iint \ln |x-y| d\mu(x) d\mu(y) < 0$ and $\lim_{\delta \to \infty} \underline{\lim}_{n \to \infty} \mathbf{P}\{| \prod_{k=1}^{n} (b(\omega), \eta_k(\omega))| < \delta\} = 1$, then $\operatorname{plim}_{n \to \infty} \det(\vec{b}(\omega), A(\omega)\vec{b}(\omega), \ldots, A^{n-1}(\omega), \vec{b}(w)) = 0$; b) with probability 1, $\iint \ln |x - y| d\mu(x) d\mu(y) > 0$ and $\lim_{\delta \to 0} \underline{\lim}_{n \to \infty} \mathbf{P}\{| \prod_{k=1}^{n} (\vec{b}(\omega), \vec{\eta}_k(\omega))| > \delta\} = 1$, then for any $c > 0$,*

$$\lim_{n \to \infty} \mathbf{P}\{| \det(\vec{b}(\omega), A(\omega)b(\omega), \ldots A^{n-1}(\omega)\vec{b}(\omega))| > c\} = 1.$$

Note that the limit theorems for normalized spectral functions $\mu_n(x)$ involved in the proof of Theorem 21.1.1 were proved in Chapter 9.

§2 Spectrum Control in Systems Described by Linear Equations in Hilbert Spaces

In this paragraph, the spectrum control problem for dynamic systems is formulated in abstract mathematical language and its physical and technical applications are examined. A new method of investigation of this problem is considered which enables us to transform the whole discrete spectrum of a given operator into the desired one. The possibility of controlling the electron energetic levels is discussed.

The methods of the theory of modal control (spectrum control) are being vigorously developed at the present time. This is due to the fact that in terms of modes it is possible to formulate and solve many problems of investigation of linear systems such as controllability, observability, sensitivity, invariance, etc. Morever, the development of these methods was also prompted by the requirements of engineering practice. By assigning the spectra during the design of systems, it is possible to achieve their stabilization and the decoupling of subsystems, and to ensure the desired dynamic performance of the transient processes. By transformation of spectra, it is possible to construct the performance functionals in the form of moments, correlation functions, or other statistical averages.

Methods of control of the spectrum of operators must be developed also for the purpose of solving various applied problems. Among them let us mention the construction of spectra of quantum mechanical systems for the purpose of varying the scattering matrix, the preparation of crystals, and the design of control arrays possessing the required physical characterstics that strongly depend on the energy spectrum, the design of stable mechanical systems with assigned oscillation spectra, and the design of systems of automatic stabilization for a plasma.

Let the equations of motion of a lumped-parameter system be

$$\frac{dx}{dt} = Ax(t) + bu(t), \qquad (21.2.1)$$

where $x(t) = \{x_1(t), \ldots, x_n(t)\}$ is a vector that specifies the state of the system: $A = \{a_{ij}\}$ is a matrix of parameters a_{ij} of the system with n distinct eigenvalues: $b = (b_1, \ldots b_n)$ is a preassigned vector; and $u(t)$ is a control function. The control $u(t)$ is realized in the form of feedback with respect to the state, i.e.,

$$u(t) = -c'x(t) = -\sum_{i=1}^{n} c_i x_i(t),$$

where $c = \{c_1, \ldots c_n\}$ is an undertermined vector. Then the closed-loop system will be

$$\frac{dx}{dt} = (A - bc')x(t), \qquad (21.2.2)$$

and the eigenvalues of the matrix $A - bc' = [a_{ij} - b_i c_j]$ will form the spectrum of this system. It is required to find a vector c that provides the desired spectrum for the closed-loop dynamic system (21.2.2).

It is well known that if system (21.1.1) is completely controllable, then there exists a control function $u(t) = -c'x(t)$ such that the closed-loop system (21.2.2) will have a preassigned spectrum $\{\alpha_k\}$. The components of the vector c_s can be expressed in terms of the eigenvalues $\{\lambda_i\}$ of the matrix A by the formula

$$c_s = \sum_{k=1}^n t_{sk}^{-1} \left(\sum_{s=1}^n t_{sk}^{-1} b_s \right)^{-1} \frac{\prod_{s=1}(\lambda_k - \alpha_s)}{\prod_{s \neq k}(\lambda_k - \lambda_s)}, \qquad (21.2.3)$$

where the t_{sk}^{-1} are elements of the matrix T^{-1}, which is the inverse of the matrix of orthonormalized eigenvectors. Equation (21.2.3) can also be obtained by using the perturbation determinant

$$\det[(A - bc - \lambda I)(A - \lambda I)^{-1}] = \det[I - (A - \lambda I)^{-1} bc'] = 1 - ((A - \lambda I)^{-1}b, c),$$

and by representing the matrix A in the form $A = T \Lambda T$, where

$$\Lambda = \mathrm{diag}\{\lambda_1, \ldots \lambda_n\}.$$

Such a method of investigation of spectrum control problems can be easily extended to the case of operators acting in infinite-dimensional Hilbert spaces, for which it is possible to introduce the concept of characteristic determinants. Indeed, let A, B, and C be linear operators acting in a Hilbert space H, where the operators A and B as well as the spectrum $\sigma(A)$ of the operator A are known. It is required to find an operator C such that the operator $A + BC$ has a preassigned spectrum $\sigma(A + Bc)$.

In many particular cases, the operator A has a discrete spectrum, and the action of the operator Bc on an element x of H can be defined as follows: $Bcx = b(c, x)$, where b and c are elements of the space H. For controlling the discrete spectrum $\{\lambda_i\}_{i=1}^\infty$ of the operator A, it is therefore necessary to find a vector c such that the operator $A + b(c, \cdot)$ has the desired spectrum $\{\alpha_i\}_{i=1}^\infty$.

Thus, if the operators A and $A - b(c, \cdot)$ are nuclear and self-adjoint, then the vector c will be determined from the relation

$$\det[I - z(A - b(c, \cdot))] = \lim_{n \to \infty} \det[\delta_{ij} - z((A - b(c, \cdot))\varphi_i, \varphi_j]_{i,j=1}^n$$

$$= \prod_{i=1}^\infty (1 - z\alpha_i),$$

where $\{\varphi_i\}_{i=1}^\infty$ are eigenvectors of the operator A taken as a basis in H.

The operator $A - b(c_i)$, under the condition that the series

$$\sum_{k=1}^\infty |\alpha_k|, \sum_{k=1}^\infty [(b, \varphi_k)^{-1} \prod_{s \neq k} (\lambda_k - \alpha_s)(\lambda_k - \lambda_s)^{-1}(\lambda_k - \alpha_k)]^2 \qquad (21.2.4)$$

is convergent, will have the desired spectrum if and only if

$$c = \sum_{k=1}^{\infty} \varphi_k (b, \varphi_k)^{-1} \prod_{s \neq k} (\lambda_k - \alpha_s)(\lambda_k - \lambda_s)^{-1}(\lambda_k - \alpha_k). \qquad (21.2.5)$$

The numbers $\alpha_k, k = 1, 2 \ldots$ satisfying the condition (21.2.4) exist. Indeed, let the spectrum $\{\lambda_k\}_{k=1}^{\infty}$ of the nuclear operator A be positive and decreasing monotonously. Let us take $\{\alpha_k\}_{k=1}$ in such a way that $\sum_{k=1}^{\infty} |\alpha_k| < \infty$ and $\lambda_{k+1} < \alpha_k < \lambda_k, k = 1, 2 \ldots$.

If the inequality

$$(b, \varphi_k)^2 \geq \lambda_k \prod_{i \neq k} (1 - \alpha_i \lambda_k^{-1})(1 - \lambda_i \lambda_k^{-1})^{-1}(1 - \alpha_k \lambda_k^{-1}) \qquad (21.2.6)$$

holds, then the vector c will have a finite squared norm. In fact, $\mathrm{Tr}(A - b(c, \cdot)) = \sum_{i=1}^{\infty} \alpha_i$, and therefore, $\sum_{k=1}^{\infty} (b, \varphi_k)(c, \varphi_k) = \sum_{k=1}^{\infty} \lambda_k - \sum_{k=1}^{\infty} \alpha_k$. It follows from (21.2.5) that

$$\sum_{k=1}^{\infty} (b, \varphi_k)(c, \varphi_k) = \lambda_k \prod_{i \neq k} (1 - \alpha_i \lambda_k^{-1})(1 - \lambda_i \lambda_k^{-1})^{-1}(1 - \alpha_k \lambda_k^{-1}).$$

Hence, it follows by virtue of the conditions of selection of the numbers $\{\alpha_k\}_{k=1}^{\infty}$ and $\{\lambda_k\}_{k=1}^{\infty}$ that $(b, \varphi_k)(c, \varphi_k) > 0$, i.e., (b, φ_k) and (c, φ_k) have the same sign. By virtue of inequality (21.2.6), we have $|(b, \varphi_k)| \geq |(c, \varphi_k)|$. Therefore,

$$\|c\|^2 = \sum_{k=1}^{\infty} (c, \varphi_k)^2 \leq \sum_{k=1}^{\infty} (c, \varphi_k)(b, \varphi_k) = \sum_{k=1}^{\infty} \lambda_k - \sum_{k=1}^{\infty} \alpha_k < \infty,$$

and hence the series (21.2.4) will be convergent.

The characteristic determinant of a nuclear non-self-adjoint operator A can be specified in a biorthogonal basis constructed from the eigenvectors $\{\varphi_k\}_{k=1}^{\infty}$ of the operator A and the eigenvectors $\{\psi_k\}_{k=1}^{\infty}$ of the adjoint operator A^* if the series $\sum_{k=1}^{\infty} \|A\varphi_k\|\|\psi_k\|$ is convergent [103]. Then the formula for the vector takes the form

$$c = \sum_{k=1}^{\infty} \varphi_k (b, \psi_k)^{-1} \prod_{s \neq k} (\lambda_k - \alpha_s)(\lambda_k - \lambda_s)^{-1}(\lambda_k - \alpha_k). \qquad (21.2.7)$$

Another important class of completely continuous operators are the Hilbert–Schmidt operators. If the operators A and $A - b(c, \cdot)$ belong to this class, then the vector c can be obtained from the equation

$$\det[I - z(A - b(c, \cdot))]e^{z \operatorname{Tr}(A - b(c, \cdot))}$$

$$= \lim \det[\delta_{ij} - z(A\varphi_i, \varphi_j)$$

$$+ z(b, \varphi_i)(c, \varphi_j)]_{i,j=1}^n \exp\{z \sum_{i=1}^{\infty} [(A\varphi_i, \varphi_i) - (b, \varphi_i)(c, \varphi_i)]\}$$

$$= \prod_{i=1}^{\infty} (1 - z\alpha_i)e^{z\alpha_i}.$$

Here the characteristic determinant is used in the regularized form [103].

The operator $A - b(c, \cdot)$ has the desired spectrum $\{\alpha_i\}_{i=1}^{\infty}$ if and only if

$$c = \sum_{k=1}^{\infty} \varphi_k(b, \varphi_k)^{-1} e^{-\Delta \lambda_k^{-1}} P_k, \qquad (21.2.8)$$

where

$$P_k = \prod_{i \neq k} (\lambda_k - \alpha_i)e^{\alpha_i \lambda_k^{-1}} (\lambda_k - \lambda_i)^{-1} e^{-\lambda_i \lambda_k^{-1}} (\lambda_k - \alpha_k)e^{\alpha_k \lambda_k^{-1}},$$

Δ satisfies the equation $\Delta = \sum_{k=1}^{\infty} \lambda_k e^{-\Delta \lambda_k^{-1}} P_k$, and the series $\sum_{k=1}^{\infty} \alpha_k^2$ and

$$\sum_{k=1}^{\infty} [(b, \varphi_k)^{-1} \exp\{-\Delta \lambda_k^{-1}\} P_k]^2$$

are convergent.

Nuclear operators and Hilbert–Schmidt operators form a class of completely continuous and bounded operators. However, the behavior of systems with a distributed type of action between the plant and the controller can be described also by operators that do not have these properties. For example, the differentiation operator is defined on an everywhere dense (in $L_2[0, a]$) set of square-summable functions, and it is unbounded on this set. The method developed here can also be extended to the case of unbounded operators whose inverse operators are nuclear.

Indeed, let us assume that the series $\sum_{k=1}^{\infty} |\alpha_k^{-1}|$ is convergent. Then the operator $(A + b(c, \cdot))^{-1}$ will be nuclear, and for it we can consider a characteristic determinant in the form

$$\det[I - z(A + b(c, \cdot))^{-1}]$$

$$= \lim_{n \to \infty} \det[\delta_{ij} - z((A - b(c, \cdot))^{-1}\varphi_i, \varphi_j]_{i,j=1}^n$$

$$= \prod_{i=1}^{\infty} (1 - z\alpha_i^{-1}). \qquad (21.2.9)$$

The vector c in (21.2.9) can be obtained by a method described in the theorem. It is based on perturbation theory.

Theorem 21.2.1.   *Let:*

a) A be an unbounded self-adjoint operator such that A^{-1} is nuclear;

b) $\{\lambda_k\}_{k=1}^\infty$ and $\{\varphi_k\}_{k=1}^\infty$ are the eigenvalues and the eigenfunctions of the operator A, and the $\{\varphi_k\}_{k=1}^\infty$ form a basis in H;

c) there exist numbers $\{\alpha_k\}_{k=1}^\infty$ such that the series

$$\sum_{k=1}^\infty |\alpha_k^{-1}|, \sum_{k=1}^\infty [\lambda_k b^{-1}(\varphi_k) \prod_{i \neq k}(1 - \lambda_k \alpha_i^{-1})(1 - \lambda_k \lambda_i^{-1})^{-1}(1 - \theta)^{-1}]^2, \quad (21.2.10)$$

are convergent, with

$$\theta = \sum_{k=1}^\infty \prod_{i \neq k}(1 - \lambda_k \alpha_i^{-1})(1 - \lambda_k \lambda_i^{-1})(1 - \lambda_k \alpha_k^{-1}).$$

Then the spectrum of the operator $A + b(c, \cdot)$ is equal to $\{\alpha_k\}_{k=1}^\infty$, if and only if

$$c = (1-\theta)^{-1} \sum_{k=1}^\infty \lambda_k \varphi_k(b, \varphi_k) \prod_{i \neq k}(1 - \lambda_k a_i^{-1})(1 - \lambda_k \lambda_i^{-1})(1 - \lambda_k \alpha_k^{-1}). \quad (21.2.11)$$

Proof.   Suppose that the operator $A + b(c, \cdot)$ has a spectrum $\{\alpha_k\}_{k=1}^\infty$. Since the series $\sum_{k=1}^\infty |\alpha_k^{-1}|$ is convergent, it follows that the operator $(A + b(c, \cdot))^{-1}$ is nuclear. Its characteristic determinant

$$
\begin{aligned}
\det[I &- z(A + b(c, \cdot))^{-1}] \\
&= \det[(I - zA^{-1}) + A^{-1}b(c, \cdot)][I + A^{-1}b(c, \cdot)]^{-1} \\
&= \det[(I - zA^{-1})(I + A^{-1}(I - zA^{-1})^{-1}b(c, \cdot)][I + A^{-1}b(c, \cdot)] \\
&= \prod_{i=1}^\infty (1 - z\alpha_i^{-1}).
\end{aligned}
\quad (21.2.12)
$$

As an orthonormalized basis in H, let us take the eigenfunctions $\{\varphi_k\}_{k=1}^\infty$ of the operator A. By the Cauchy–Bunyakovskii inequality,

$$\sum_{k=1}^\infty (A^{-1}b(c, \cdot)\varphi_k, \varphi_k) \leq \sum_{k=1}^\infty |(A^{-1}\varphi_k, \varphi_k) \sum_{k=1}^\infty |(c, \varphi_k)(b, \varphi_k)| < \infty,$$

$$\sum_{k=1}^\infty ((A^{-1}(I - zA^{-1})^{-1}b(c, \cdot)\varphi_k, \varphi_k)$$

$$\leq \sum_{k=1}^\infty (A^{-1}(I - zA^{-1})^{-1}\varphi_k, \varphi_k)| \sum_{k=1}^\infty |(c, \varphi_k)(b, \varphi_k)| < \infty,$$

it follows that the matrix trace of the operators $A^{-1}b(c,\cdot)$ and $A^{-1}(I - zA^{-1})^{-1}$ is finite, and hence the operators are nuclear. Therefore, we can write (21.2.12) in the form

$$\prod_{i=1}^{\infty}(1 - z\alpha_i^{-1})(1 - z\lambda_i^{-1})^{-1} = \det[I - z(A + b(c,\cdot))]^{-1}\det[I - zA^{-1}]$$

$$= \det[I + A^{-1}(I - zA^{-1})b(c,\cdot)][\det[I + A^{-1}b(c,\cdot)]]^{-1}.$$

By using the perturbation determinant, we obtain

$$\det[I + A^{-1}b(c,\cdot)] = \det[I + b(c,\cdot)A^{-1}]$$

$$= \lim_{n\to\infty}\det[\delta_{ij} + \lambda_i^{-1}(b,\varphi_i)(c,\varphi_j)]_{i,j=1}^n = 1 + \sum_{k=1}^{\infty}\lambda_k^{-1}(b,\varphi_k)(c,\varphi_k);$$

$$\det[I + A^{-1}(I - zA^{-1})b(c,\cdot)]$$

$$= \lim_{n\to\infty}\det[\delta_{ij} + \lambda_i^{-1}(1 - \lambda_i^{-1})^{-1}(b,\varphi_i)(c,\varphi_j)]_{i,j=1}^n$$

$$= 1 + \sum_{k=1}^{n}(1 - z\lambda_k^{-1})^{-1}\lambda_k^{-1}(b,\varphi_k)(c,\varphi_k).$$

Therefore,

$$\frac{\prod_{i=1}^{\infty}(1 - z\alpha_i^{-1})}{\prod_{i=1}^{\infty}(1 - z\lambda_i^{-1})} = \frac{1 + \sum_{k=1}^{\infty}\lambda_k^{-1}(b,\varphi_k)(c,\varphi_k)(1 - z\lambda_k^{-1})^{-1}}{1 + \sum_{k=1}^{\infty}\lambda_k^{-1}(b,\varphi_k)(c,\varphi_k)}. \qquad (21.2.13)$$

By multiplying (21.2.13) by $(1 - z\lambda_k^{-1})$ and by taking the limit as $z \to \lambda_k$, we obtain

$$\frac{\lambda_k^{-1}(b,\varphi_k)(c,\varphi_k)}{1 + \sum_{k=1}^{\infty}\lambda_k^{-1}(b,\varphi_k)} = \prod_{l\neq k}\frac{(1 - \lambda_k\alpha_i^{-1})}{1 - \lambda_k\lambda_i^{-1}}(1 - \lambda_k\alpha_k^{-1}). \qquad (21.2.14)$$

Let $\sum_{k=1}^{\infty}\lambda_k^{-1}(b,\varphi_k)(c,\varphi_k) = \Delta$. Then it is easy to see that (21.2.14) takes the form

$$\frac{\Delta}{1 + \Delta}\sum_{k=1}^{\infty}\prod_{i\neq k}\frac{1 - \lambda_k\alpha_i^{-1}}{1 - \lambda_k\lambda_i^{-1}}(1 - \lambda_k\alpha_k^{-1});$$

hence, it follows that

$$\Delta = \frac{\sum_{k=1}^{\infty}\{\prod_{i\neq k}(1 - \lambda_k\alpha_i^{-1})(1 - \lambda_k\lambda_i^{-1})^{-1}\}(1 - \lambda_k\alpha_k^{-1})}{1 - \sum_{k=1}^{\infty}\{\prod_{i\neq k}(1 - \lambda_k\alpha_i^{-1})(1 - \lambda_k\lambda_i^{-1})^{-1}\}(1 - \lambda_k\alpha_k^{-1})}. \qquad (21.2.15)$$

On the other hand, (21.2.14) implies that

$$c_k = (c,\varphi_k)$$

$$= \lambda_k(1 + \Delta)(b,\varphi_k)^{-1}\{\prod_{i\neq k}(1 - \lambda_k\alpha_i^{-1})(1 - \lambda_k\lambda_i^{-1})^{-1}\}(1 - \lambda_k\alpha_k^{-1}).$$

Since the vector $c \in H$, it follows that $c = \sum_{k=1}^{\infty} c_k \varphi_k$. Hence,

$$c = \frac{\sum_{k=1}^{\infty} \frac{\lambda_k \varphi_k}{(b, \varphi_k)} \prod_{i \neq k} \frac{(1 - \lambda_k \alpha_i^{-1})}{(1 - \lambda_k \lambda_i^{-1})} (1 + \lambda_k \alpha_k^{-1})}{1 - \sum_{k=1}^{\infty} \prod_{i \neq k} \frac{1 - \lambda_k \alpha_i^{-1}}{1 - \lambda_k \lambda_i^{-1}} (1 - \lambda_k \alpha_k^{-1})}, \tag{21.2.16}$$

and we have proved the necessity. The sufficiency of the conditions is obvious.

If we must vary only the first s eigenvalues of the unbounded operator A, then we must take only the first s terms in (21.2.11). Then the vector c will satisfy the equation

$$\det[I - z(\tilde{A} + b(c, \cdot))^{-1}] = \prod_{k=1}^{s} (1 - z\alpha_k^{-1}),$$

where $\tilde{A}$ is a finite-dimensional operator with a finite number of eigenvalues $\lambda_k, k = \overline{1, s}$ and corresponding eigenfunctions $\varphi_k, k = \overline{1, s}$.

Since $\tilde{A}$ is a finite-dimensional operator, it follows that the previous expression is equivalent to the equation

$$\det[I - z(\tilde{A} + b(c, \cdot))] = \prod_{k=1}^{s} (1 - z\alpha_k).$$

As in the case of nuclear operators, a vector c that satisfies (21.2.13) will then be specified as

$$c = \sum_{k=1}^{s} \varphi_k (b, \varphi_k)^{-1} \prod_{i \neq k} (\lambda_k - \alpha_i)(\lambda_k - \lambda_i)^{-1}(\lambda_k - \alpha_k).$$

In this form, the expression for the vector c, with the help of which it is possible to transform only partially the spectrum of an unbounded operator, was obtained for the first time in [171] and [164]. Note that the problem being solved is a particular case of the well-known inverse spectral problem.

Example. Let us consider the motion of an electron that is in a rectangular potential well of width a and with infinitely high walls [23] and whose state is obtained in the form of the solution of the Schrödinger equation

$$ih\frac{d\psi(t)}{dt} = H\psi(t) + xu(t),$$

where $H = \frac{h^2}{2m_l} \frac{\partial^2}{\partial x^2} + \mathbf{V}(x)$ is the Hamiltonian of the electron, h is the Planck constant, $m(l)$ is the mass of the electron, $\mathbf{V}(x)$ is the potential function of the well, x is a coordinate, and $u(t)$ is the control.

This form is assumed by the Schrödinger equation when it is interpreted as a function of the variable t that takes its values in the space $L_2[0, a]$. Suppose that the potential function $V(x)$ is defined as follows:

$$V(x) = \begin{cases} \infty, & x > a, \quad x < 0, \\ 0, & 0 < x < a, \end{cases}$$

where a is the width of the potential well. Then the Hamiltonian of the electron will have the eigenvalues $\{\lambda_n\}_{n=1}^{\infty} = \{\pi^2 h^2 n^2 (2m_l a^2)^{-1}\}_{n=1}^{\infty}$ and the eigenfunctions $\{\varphi_n = \sqrt{2a^{-1}}\sin(n\pi x a^{-1})\}_{n=1}^{\infty}$. The values $\{\lambda_n\}_{n=1}^{\infty}$ specify the energetic levels of the electron. It is required to find a control in the form of a feedback $u(t) = (c, \psi(t))$ with the help of which it is possible to change the value of the electron energy $\{\lambda_n\}_{n=1}^{\infty}$ by replacing $\{\lambda_n\}_{n=1}^{\infty}$ by $\{\alpha_n\}_{n=1}^{\infty}$. For this purpose, we must find the element c in $L_{2[0,a]}$ such that the operator $H + x(c, \cdot)$ will have a spectrum $\{\alpha_n\}_{n=1}^{\infty}$. Then, in a certain state, the electron will have an energy equal to $\{\alpha_n\}_{n=1}^{\infty}$. The Hamiltonian operator satisfies all the conditions of the theorem. Therefore, its spectrum can be tranformed into the desired one by using formula (21.2.11), and $\{\alpha_k\}_{k=1}^{\infty}$ can be selected in such a way that the series (21.2.10) are convergent. Indeed, let $\{\alpha_k = \gamma_k \lambda_k\}_{k=1}^{\infty}$, where $\lambda_k = \lambda_0 k^2, \gamma_k = (1 - \exp\{-\gamma\lambda_k\})^{-1}, \lambda_0 = \pi^2 h^2 |2m_l a^2$, and γ is a positive constant. Let us consider

$$f(\lambda_k) = (1 - \lambda_k \alpha_k^{-1}) \prod_{i \neq k} (1 - \lambda_k a_i^{-1})(1 - \lambda_k \lambda_i^{-1})^{-1}$$

$$= (1 - \lambda_k \alpha_k^{-1}) \prod_{i \neq k} (1 - \lambda_k \alpha_i^{-1})(\lambda_i - \lambda_k)^{-1}\lambda_i$$

$$= (1 - \lambda_k \alpha_k^{-1}) \prod_{i \neq k} [1 + (1 - \lambda_i \alpha_i^{-1})(\lambda_i - \lambda_k)^{-1}\lambda_k].$$

Let

$$\Delta_i = \min\{|\lambda_i - \lambda_{i-1}|, |\lambda_i - \lambda_{i+1}|\}, \qquad d_i = e^{-\gamma\lambda_i}, \qquad c_i = d_i/\Delta_i.$$

Since $\alpha_k = \gamma_k \lambda_k$, it follows that

$$f(\lambda_k) \leq d_k \exp(c\lambda_k), \qquad c = \sum_{i=1}^{\infty} c_i$$

It follows therefore that for $\gamma > c$, the series (21.2.10) will be convergent.

If the potential function $V(x)$ has the form

$$V(x) = \begin{cases} \infty, & x > a, \quad x < 0, \\ p\cos(2\pi x a^{-1}), & 0 < x < a, \end{cases}$$

where p is the depth of the potential well, then the eigenvalues and eigenfunctions of the Hamiltonian will be

$$\lambda_1 = \lambda_0 - p/2 = \lambda_0(1 - q), \quad q = p/2\lambda_0,$$
$$\lambda_n = \lambda_0 n^2, \quad n = 2, 3, \ldots,$$
$$\varphi_1 = p(2a)^{-1/2} \sin(\pi x a^{-1})\lambda_0(n^2 - 1 + q)^{-1},$$
$$\varphi_n = \sqrt{2}a^{-1/2} \sin(n\pi x a^{-1}), \quad n = 2, 3, \ldots$$

In this case, the series (21.2.10) will also be convergent, provided that $\gamma > e^{-\gamma\lambda_0(1-q)} + \sum_{m=2}^{\infty} e^{-\gamma\lambda_0 m^2}$. But the depth of the potential well changes by a factor $(\gamma_1 p - 2\lambda_0(\gamma_1 - 1))p^{-1}$. Thus a change in the parameters of the potential well does not affect the convergence of the series (21.2.10).

Let $\mathfrak{S}_p$ be the class of all complete continuous random operators in a separable Hilbert space H for which $\sum_{i=1}^{n} s_i^p(A(\omega)) < \infty$, where $s_i(A(\omega))$ are singular random numbers of operator $A(\omega)$, and p is a real number. One of the tasks concerning the control of the spectrum of differential and integral operators with random parameters is formulated as follows. Let $A(\omega), B(\omega), c(\omega)$ be random linear operators, which act in H, and $\sigma(A(\omega))$ the spectrum of the operator $A(\omega)$. It is necessary to find an operator $c(\omega)$ such that the spectrum $\sigma(A(\omega) + B(\omega)c(\omega))$ is a given set.

Consider the particular case of this problem when the operator $B(\omega)c(\omega)$ is one-dimensional: for every $x \in H$, $B(\omega)c(\omega)x = (b(\omega), x)c(\omega)$, where $b(\omega)$ and $c(\omega)$ are measurable random functions with the paths in the space L_2, $A(\omega)$ is a random linear operator, which acts from $L_2(\Omega, B, p)$ to $L_2(\Omega, B, p)$, $L_2(\Omega, B, p)$ is a Hilbert space of random functions $\xi(\omega, x)$, measurable and square integrable.

Let $A(\omega)$ be a self-adjoint random linear operator of the class $\mathfrak{S}_1$ (i.e. nuclear kernel), $\lambda_i(\omega), i = 1, 2, \ldots$ be its random eigenvalues, and $\varphi_i(\omega), i = 1, 2, \ldots$ be the uniquely chosen orthonormal random eigenfunctions.

Theorem 21.2.2. *With probability 1, let*

$$\sum_{k=1}^{\infty} |\alpha_k(\omega)| < \infty,$$

$$\sum_{k=1}^{\infty} \lambda_k^{-2}(b(\omega), \varphi_k(\omega))^{-2} \prod_{s \neq k} \{(1 - \lambda_k^{-1}(\omega)\alpha_s(\omega))$$
$$\times (1 - \lambda_k^{-1}(\omega)\lambda_s(\omega)^{-2}\}(1 - \lambda_k^{-1}(\omega)\alpha_k(\omega))^2 < \infty,$$

where $\alpha_k(\omega), k = 1, 2 \ldots$ are random variables, among which there are complex conjugate pairs.

Then $\sigma(A(\omega) + (c(\omega), \cdot)b(\omega)) = \{\alpha_k(\omega), k = 1, 2 \ldots\}$, where

$$
\begin{aligned}
c(\omega) = \sum_{k=1}^{\infty} & \lambda_k^{-1}(\omega)(b(\omega), \varphi_k(\omega))^{-1}\varphi_k(\omega) \prod_{s \neq k} \\
& \times \{(1 - \lambda_k^{-1}(\omega)\alpha_s(\omega))(1 - \lambda_k^{-1}(\omega)\lambda_s(\omega))^{-1}\} \\
& \times (1 - \lambda_k^{-1}(\omega)\alpha_k(\omega)). \quad (21.2.17)
\end{aligned}
$$

Proof. Suppose that the operator $A(\omega) + (c(\omega), \cdot)b(\omega)$ has a spectrum $\{\alpha_i(\omega), i = 1, 2, \ldots\}$.

Then, with probability 1,

$$
\begin{aligned}
\det[I - z(A(\omega) + (c(\omega), \cdot)b(\omega)] = \lim_{n \to \infty} & \det[\delta_{jk} - z(A(\omega) \\
& \times \varphi_j(\omega), \varphi_k(\omega)) - z(b(\omega), \varphi_j(\omega)) \\
& \times (c(\omega), \varphi_k(\omega)], \quad j, k = \overline{1, n}.
\end{aligned}
$$

Under $z \neq \lambda_k^{-1}(\omega)), k = 1, 2, \ldots,$

$$
\begin{aligned}
\det[I - z(A(\omega) &+ ((c(\omega), \cdot)b(\omega))] \det[I - zA(\omega)]^{-1} \\
&= \prod_{i=1}^{\infty} \{(1 - z\alpha_i)(1 - z\lambda_i)^{-1}\}.
\end{aligned}
$$

By multiplying both parts of this equation by $1 - z\lambda_k$ and by taking the limit as $z \to \lambda_k^{-1}$, we obtain (21.2.17). Theorem 21.2.2 is proved.

We consider the following problem in the same manner. There exists a sequence of symmetric real random matrices $\Xi_n, n = 1, 2, \ldots$ and the random vectors $\vec{b}_n, n = 1, 2 \ldots$. It is necessary to choose a sequence of random vectors $\vec{c}_n$ such that the eigenvalues of the matrix $\Xi_n + \vec{b}_n\vec{c}_n$ tend (in the sense of convergence of distributions) to some definite random values.

Theorem 21.2.3. *For random matrices $\Xi_n, n = 1, 2, \ldots$, let the conditions of Theorem 8.4.3 hold, let the vector rows of the matrix Ξ_n be asymptotically constant:* $\lim_{n \to \infty} \operatorname{Tr} B_n B_n' = 0$ *for all finite numbers $i_1, \ldots i_s$,*

$$
\operatorname{plim}_{n \to \infty} \sum_{i=1}^{n} \nu_{ij}^2 = \theta_j, \quad j = i_1, \ldots, i_s, \quad (21.2.18)
$$

θ_j being nonrandom numbers, with probability 1,

$$
\sum_{k=1}^{\infty} |\alpha_k(\omega)| < \infty,
$$

$$
\sum_{k=1}^{\infty} \theta_k^{-2}b_k^{-2}(\omega) \prod_{s \neq k} \{(1 - \theta_n^{-1}\alpha_s)^2(1 - \theta_k^{-1}\alpha_s)^{-2}\}(1 - \theta_k^{-1}\alpha_k)^2, \sum_{k=1}^{\infty} b_k^2 < \infty,
$$

where $\alpha_k(\omega)$ are random variables, among which there are complexly conjugate pairs and which do not depend on Ξ_n.

Then the eigenvalues λ_k of the matrix $\Xi_n\Xi' + \vec{b}_n\vec{c}_n$ converge (in the sense of convergence of distributions) to random variables,

$$c_k = \sum_{k=1}^{\infty} \theta_k^{-1} b_k^{-1} \ldots,$$
$$\vec{b}_n = (b_1, \ldots, b_n),$$
$$\vec{c}_n = (c_1, \ldots, c_n),$$

the eigenvalues $\lambda_k, k = \overline{1, n}$ are ordered by increasing their moduli (if the moduli are equal, then we order eigenvalues by increasing their argument).

Proof. By the same argument as in Theorem 8.4.3 (see also Corollary 8.4.3), we deduce that we can choose a sequence n' such that

$$\det(I + z\Xi_{n'}\Xi'_{n'} - z\vec{b}_{n'}\vec{c}_{n'})$$
$$\sim \det(1 - z \, \text{diag}\{\sum_{i=1}^{n'} \nu_{ij}^2, j = \overline{1, n'}\} - z\vec{b}_{n'}\vec{c}_{n'}).$$

According to (21.2.18),

$$\det(I - z\Xi_n\Xi'_n - z\vec{b}_n\vec{c}_{n'}) \Rightarrow \det(I - z \, \text{diag}(\theta_j, j = 1, \infty) - zb_\infty\vec{c}_\infty),$$
$$\vec{b}_\infty = (b_1, b_2, \ldots), \qquad \vec{c}_\infty = (c_1, c_2, \ldots).$$

Now, by considering the Hilbert space of infinite-dimensional vectors and random linear nuclear operators $A(\omega)$, which are equal to infinite-dimensional diagonal matrices, on the basis of Theorem 13.4.1, we obtain

$$\det(I - z\Xi_n\Xi'_n - z\vec{b}_n\vec{c}_n) \Rightarrow \prod_{i=1}^{\infty}(1 - z\alpha_i).$$

From this and from Theorem 13.2.1, we obtain the assertion of Theorem 21.2.3. Analogous statements are true for other classes of matrices considered in Chapter 8.

§3 Adaptive Approach to the Control of Manipulator Motion

Increasing demands for the quality of the operation of industrial robots lead to the necessity of creating more perfected methods of control that take into account dynamic characteristics of manipulators. In order to construct such control systems, it is necessary, as a rule, to have rather full knowledge of a mathematical model of the manipulator. The dynamic model of the manipulator is a system of nonlinear differential equations. Coefficients of these equations are connected in a rather complicated fashion via trigonometric functions with generalized coordinates of the manipulator. Such a system is complicated for practical application due to essential nonlinearity and mutual influence of links, which makes practical use of such a model difficult. Therefore, the use of a simplified mathematical model with adaptive adjustment of the parameters in the control process proves to be expedient. Control of the manipulator with the use of a standard model has been studied in various papers [119]. The standard model was given by linear differential equations of the second order in which the desired characteristics of motion were pointed out. An adaptive regulator in accordance with the exit of the standard model "adjusts" control of the manipulator according to the desired motion.

Linearized with respect to nominal motion, the mathematical model was used in a procedure of control synthesis on the basis of asymptotic linear regulators as well as for constructing autoregressive models, representing displacements in separate links [119]. Parameters of the model are estimated in the process of motion, proceeding from the optimization of some quality criterion. Another approach to the adaptive control of the manipulator was offered in paper [96].

The dynamics are described by a Lagrange equation of the second kind, which depends on unknown parameters of the manipulator. Locally optimal finitely convergent methods of solving purpose inequalities were used as algorithms of adaptation.

In this section, a method of adaptive control of the manipulator without full knowledge of the mathematical model is proposed and its characteristics are studied. The estimation of the parameters of the model is made by observations of the manipulator in the block of adaptation. With the aid of these estimates, a linear regulator optimizing generalized energy is constructed. The estimate of the parameters and the controls is made recurrently. The algorithm offered is locally optimal. The dynamic model of the manipulator representing the open kinematic chain with m links can be written in the form

$$u(t) = H(\theta)\ddot{\theta} + \mathbf{E}(\theta, \dot{\theta}) + G(\theta), \qquad (21.3.1)$$

where $\theta = (\theta_1, \ldots \theta_m)'$ is the vector of the generalized coordinates; $H(\theta)$ is the symmetric nonnegative-definite matrix of inertia; $\mathbf{E}(\theta, \dot{\theta})$ is the vector, representing centrifugal and Coriolis moments; $G(\theta)$ is the vector of gravitation moments; $u(t)$ is the vector of control moments developed by drives in the joints.

The motion of the manipulator is uniquely defined via generalized coordinates and their first derivatives. Therefore, in the following, we shall use the model described in terms of these variables.

Let $x_i = \theta_i$, $x_{i+m} = \dot{\theta}_i$, $i = \overline{1, m}$. Then equation (21.3.1) may be represented in the form

$$\dot{x}(t) = A(x(t))x(t) + B(x(t))u(t), \tag{21.3.2}$$

where the matrices $A(\cdot), B(\cdot) \in R^{2m \times 2m}$ are nonlinear functions of the variables of states $x(\cdot) \in R^{2m}$. The discrete analog of model (21.3.2) can be represented in the form

$$x_{n+1} = A(x_n)x_n + B(x_n)u_n. \tag{21.3.3}$$

We define the trajectory of motion of the manipulator in the form of a sequence of points $a_i \in R^{2m}$, $i = 1, 2, \ldots$, through which the manipulator has to pass.

We approximate the dynamic model of the manipulator by a linear model

$$x_{n+1} = A_n x_n + B_n u_n + \varepsilon_{n+1}, \tag{21.3.4}$$

where A_n, B_n are unknown matrices, and ε_{n+1} are the errors of modelling.

Assume that the matrices $A(x_n), B(x_n)$ in (21.3.3) are constant but unknown. Such assumption will be true with local displacements of the manipulator. Then (21.3.3) can be written in the form $x_{n+1} = Ax_n + Bu_n$.

We make $n > m$ observations of the manipulator under some fixed controls. From the observations, we construct estimates of the matrices $\hat{A}_n, \hat{B}_n$. By these estimates, we can find the extrapolated position of the manipulator

$$x_{n+1}^e = \hat{A}_n x_n + \hat{B}_n U_n.$$

We choose the control to minimize the functional

$$I_n(\tilde{u}) = \min_{u_n}\{\|a_{n+1} - x_{n+1}^e\|^2 + \delta\|u_n\|^2\}, \quad \delta > 0. \tag{21.3.5}$$

The observed position of the manipulator under this control will be

$$x_{n+1} = \hat{A}_n x_n + \hat{B}_n \tilde{U}_n + \varepsilon_{n+1}. \tag{21.3.6}$$

Without loss of generality, we consider that B is a known square matrix which has an inverse one. The matrix A will be estimated by the least squares method

$$\hat{A}_n = \sum_{s=1}^{n}(x_s - B\tilde{u}_{s-1})x_{s-1}' \left(\sum x_{s-1}x_{s-1}'\right)^{-1}. \tag{21.3.7}$$

Controls from (21.3.5) will be given in the form

$$\tilde{u}_s = (\delta I + BB')^{-1}B'(a_{s+1} - \tilde{A}_s x_s), \tag{21.3.8}$$

where

$$\tilde{A}_s = \hat{A}_s \chi(\|\hat{A}_s\| < \|A\|) + \hat{A}_{s-1}\chi(\|\hat{A}_s\| \geq \|A\|),$$

and $\chi(\cdot)$ is the indicator of an event.

Given $\tilde{u}_n$, we observe the vector x_{n+1} again, find $\tilde{u}_{n+1}$, and continue such calculations up to the moment of time s when $\|a_s - \tilde{x}_s\|^2 < \varepsilon$, where $\varepsilon > 0$ is a given number.

By using some results for random determinants, we prove convergence of the method proposed above and asymptotic normality of estimates of parameters of the matrix A.

Theorem 21.3.1. *Let the following conditions hold:*

1. $\mathbf{E}(\varepsilon_{n+1}/\varepsilon_1, \ldots, \varepsilon_n) = 0, \quad n = 1, 2, \ldots,$

2. $\sup\limits_{n} \|a_n\| < \infty,$

3. $\|A\|(1 + \|BB'(I\delta + BB')^{-1}\|) < 1,$

4. $\sup\limits_{n} \mathbf{E}\|\varepsilon_n\|^4 < \infty,$

5. $\sup\limits_{n} \|(n^{-1}\sum\limits_{s=1}^{n}\mathbf{E}\varepsilon_{s-1}\varepsilon'_{s-1})^{-1}\| < \infty,$

6. $\overline{\lim\limits_{h\to\infty}} \sup\limits_{|i_p - j_p| \geq h, p=\overline{1,3}} |\mathbf{P}\{\varepsilon_{i_p} < x_{i_p}, \varepsilon_{j_p} < x_{j_p}, p = \overline{1,3}\}$
$$- \mathbf{P}\{\varepsilon_{i_p} < x_{i_p}, p = \overline{1,3}\}\mathbf{P}\{\varepsilon_{j_p} < x_{j_p}, p = \overline{1,3}\}| = 0.$$

Then

$$\lim_{n\to\infty} \mathbf{E}\|a_n - \tilde{x}_n\|^2 \leq c\delta\|(I\delta + BB')^{-1}\|^2,$$

$$\lim_{n\to\infty} \mathbf{P}\{R_{m^2}\vec{q} \in B\} = (2\pi)^{-m^2/2}\int_B \cdots \int \exp(-\frac{1}{2}\sum_{i=1}^{m^2}x_i^2)\prod_{i=1}^{m^2} dx_i,$$

where

$$R_{m^2} = [n^{-1}\sum_{k=1}^{n}\mathbf{E}\vec{\beta}_k\vec{\beta}'_k]^{-1/2},$$

$$\vec{\beta}_k = \{(\vec{\varepsilon}_k\vec{x}'_{k-1})_{ij,i,j=\overline{1,m}}\},$$

$$\vec{q} = \{((\hat{A}_n - A)\sqrt{n}C_n)_{ij,i,j=\overline{1,m}}\},$$

$$C_n = n^{-1}\sum_{k=1}^{n}\mathbf{E}\vec{x}_{k-1}\vec{x}_{k-1},$$

and B is an arbitrary Borel subset from the set of R^{m^2}.

Proof. By substituting the values of control from (21.3.8) in (21.3.6), we obtain

$$x_{s+1} = [A - B(\delta I + BB')^{-1}B'\tilde{A}_s]x_s + B(\delta I + BB')^{-1}B'a_{s+1} + \varepsilon_{s+1}. \quad (21.3.9)$$

Estimate the first summand

$$\|A - B(\delta I + BB')^{-1}B'\tilde{A}_s\| = \|A - C_\delta \hat{A}_s \chi(\|\tilde{A}_s\| < \|A\|)\|$$
$$\leq \|A\| + \|C_\delta\| \|A\| = \|A\|(1 + \|C_\delta\|) < 1,$$

where $C_\delta = B(\delta I + BB')^{-1}B'$.

Let $C_s = A - C_\delta \hat{A}_s, \|C_s\| < 1$.

Then we obtain from (21.3.8) that

$$x_{s+1} = C_s x_s + C_\delta A_{s+1} + \varepsilon_{s+1}. \qquad (21.3.10)$$

Show that the values $\mathbf{E}\|x_s\|$ are bounded,

$$\mathbf{E}\|x_{s+1}\| \leq \|c_s\| \mathbf{E}\|x_s\| + \|c_\delta\| \|a_{s+1}\| + \mathbf{E}\|\varepsilon_{s+1}\| < c < \infty. \qquad (21.3.11)$$

We can easily see that from (21.3.7) by the least squares method it follows that the estimate $\hat{A}_n$ of the matrix A is

$$n^{1/2}(\hat{A}_n - A) = n^{-1/2} \sum_{s=1}^{n} \varepsilon_s x'_{s-1} \left(n^{-1} \sum_{s=1}^{n} x_{s-1} x'_{s-1}\right)^{-1}. \qquad (21.3.12)$$

Estimate the first factor in (21.3.12). The matrices $\varepsilon_s x'_{s-1}$ are martingale differences. By using Condition 4 of the theorem and (21.3.11), we obtain that

$$\mathbf{E}\|x_s\|^2 < \infty.$$

Then it is easy to show that

$$\mathbf{E}\left\| \sum_{s=1}^{n} \varepsilon_s x'_{s-1} n^{-1/2} \right\| \leq n^{-1} \sum_{s=1}^{n} \mathbf{E}\|\varepsilon_s x'_{s-1}\|^2 \leq c,$$

where $c > 0$ is some constant.

Estimate the second factor in (21.3.12). Introduce in (21.3.10) the following denotations:

$$x_s = d_{s-1} + \varepsilon_s,$$
$$d_{s-1} = c_{s-1} x_{s-1} + c_\delta a_s,$$

and we obtain

$$n^{-1} \sum_{s=1}^{n} x_s x'_s = n^{-1} \sum_{s=1}^{n} d_{s-1} d'_{s-1} + n^{-1} \sum_{s=1}^{n} (d_{s-1} \varepsilon'_s$$
$$+ \varepsilon_s d'_{s-1}) + n^{-1} \sum_{s=1}^{n} \varepsilon_s \varepsilon'_s. \qquad (21.3.13)$$

The variables d_s are bounded in probability owing to the boundness of $\mathbf{E}\|x_s\|$. By virtue of Condition 1 of the theorem and the boundedness in probability of the variables d_s, we obtain that the mathematical expectation of the second summand in (21.3.13) tends to zero in probability.

Due to conditions 1, 4, and 6 of the theorem and the law of large numbers, the sum $n^{-1}\sum_{s=1}^{n}\varepsilon_{s-1}\varepsilon'_{s-1}$ can be approximately replaced by $n^{-1}\sum_{s=1}^{n}\mathbf{E}\varepsilon_{s-1}\varepsilon'_{s-1}$. The matrix $n^{-1}\sum_{s=1}^{n}\mathbf{E}\varepsilon_{s-1}\varepsilon'_{s-1}$ is nonsingular. Hence, we obtain that

$$\left\|n^{-1}\sum_{s=1}^{n}X_x X'_s - n^{-1}\sum_{s=1}^{n}d_{s-1}d'_{s-1} - n^{-1}\sum_{s=1}^{n}\mathbf{E}\varepsilon_s\varepsilon'_s\right\| \xrightarrow{P} 0, \quad n\to\infty.$$

By making use of the familiar inequality,

$$\|(Q+R)^{-1}\| \le \|R^{-1}\|,$$

where Q is a nonnegative-definite matrix, R is a nonsingular matrix, and from (21.3.12), we obtain

$$\lim_{h\to\infty}\lim_{n\to\infty} p\{n^{1/2}\|\hat{A}_n - A\| > h\} = 0.$$

Hence,

$$\|\hat{A}_n - A\| \xrightarrow{P} 0, \quad n\to\infty.$$

Thus it is proved that the estimate $\hat{A}_n$ is consistent. Since

$$\hat{A}_n \xrightarrow{P} A, \|\hat{A}_n\| \to \|A\|, \quad n\to\infty,$$

if follows that

$$\chi(\|\hat{A}_n\| < \|A\|) \xrightarrow{P} 1,$$

and consequently,

$$\|\tilde{A}_n - A\| \xrightarrow{P} 0, \quad n\to\infty.$$

By using the fact that the estimate $\hat{A}_n$ is consistent, we obtain that the first assertion of the theorem follows from the inequality

$$\mathbf{E}\|x_{s+1}^e - a_{s+1}\|^2 = \mathbf{E}\|c_s x_x + c_\delta a_{s+1} - a_{s+1}\|$$
$$\le \mathbf{E}\|c_s\|^2\mathbf{E}\|x_s\|^2 + \|a_{s+1}\|^2\|(c_\delta - I)\|^2$$
$$\le c_\delta\|(\delta I + BB')^{-1}\|^2.$$

Let us prove the second assertion of the theorem. By using Condition 6 of the asymptotic independence of the variables ε_s, we can easily show the asymptotic independence of the x_s for the system

$$x_{s+1} = c_s x_s + c_\delta a_{s+1} + \varepsilon_{s+1}.$$

For the system described above, we build an accompanying system

$$z_{s+1} = cz_s + c_\delta a_{s+1} + \varepsilon_{s+1},$$

where $c = (A - c_\delta A)$. Since

$$\mathbf{E}\|\tilde{A}_s - A\|^2 \to 0, \quad s \to \infty,$$

it is easy to show that

$$\lim_{s \to \infty} \mathbf{E}\|x_s - z_s\| = 0.$$

But then

$$\|n^{1/2}(\hat{A}_n - A) - n^{-1/2} \sum_{s=1}^{n} \varepsilon_s z'_{s-1} (n^{-1} \sum_{s=1}^{n} z_{s-1} z'_{s-1})^{-1}\| \overset{p}{\to} 0, \quad n \to \infty.$$

By using this expression, we prove the asymptotic normality of the estimates of the parameters of matrix $\hat{A}_n$ in the same way as in Chapter 20. Theorem 21.3.1 is proved.

§4 The Perturbation Method of Linear Operators in the Theory of Optimal Control of Stochastic Systems

Consider a system of equations

$$A(\omega)\vec{x} = B(\omega)\vec{u} + \vec{\xi}, \tag{21.4.1}$$

where $A(\omega)$ is a random square matrix of order n, $B(\omega)$ is a nonsingular random matrix of dimension $(n \times m)$, n is a number of rows, m is a number of columns of the matrix $B(\omega)$, $\vec{\xi}$ is a vector of interferences of dimension n which satisfies the condition $\|\vec{\xi}\| \le 1$, and $\vec{u}$ is a control vector of dimension m. In such a form, we find $\vec{u} = K\vec{x}$, where K is a matrix of dimension $(m \times n)$. Let L be a set of real matrices of dimension $m \times n$.

We are interested in

$$\min_{K \in L} \mathbf{E} \max_{\|\vec{\xi}\| \le 1} J(\vec{u}), \tag{21.4.2}$$

where $J(\vec{u})$ is a criterion of quality of the control system, which is equal to

$$J(\vec{u}) = (Q_1\vec{x}, \vec{x}) + (Q_2\vec{u}, \vec{u}),$$

and Q_1 and Q_2 are positive-definite matrices of dimension $(n \times n)$ and $(m \times m)$, respectively (matrix Q_2 is nonsingular).

Theorem 21.4.1. *If the elements of the random matrices $A(\omega)$ and $B(\omega)$ have joint probability density, then the solutions $K_0 \in L$ of equation*

$$\min_{K \in L} \mathbf{E} \max_{\|\vec{\xi}\| \le 1} J(\vec{u}) = \mathbf{E}\lambda_1 \{ (A(\omega) - B(\omega)K_0)^{-1}$$
$$\times (Q_1 + K_0'Q_2K_0)(A(\omega) - B(\omega)K_0)^{-1} \},$$

where λ_1 is the maximum eigenvalue of the matrix

$$(A(\omega) - B(\omega)K_0)^{-1'}(Q_1 + K_0'Q_2K_0)(A(\omega) - B(\omega)K_0)^{-1},$$

satisfy the equation

$$\mathbf{E}\{ B'(\omega)(A(\omega) - B(\omega)K_0)^{-1}(Q_1 + K_0'Q_2K_0) + Q_2K_0 \}$$
$$\times (A(\omega) - B(\omega)K_0)^{-1} e_1 e_1' = 0, \tag{21.4.3}$$

where $\vec{e}_1$ is an eigenvector corresponding to the eigenvalue λ_1.

We see in Equation (21.4.3) that the matrices $A(\omega)$ and $B(\omega)$ have such properties that mathematical expectations exist in each part of this equation.

Proof. Substitute in (21.4.3) the values $\vec{u}$ and $\vec{x}$ from (21.4.1). Then,

$$J(\vec{u}) = ((Q_1 + K'Q_2K)\vec{x}, \vec{x}) = ((Q_1 + K'Q_2K)$$
$$\times (A(\omega) - B(\omega)K)^{-1}\vec{\xi}, \quad (A(\omega) - B(\omega)K)^{-1}\vec{\xi}).$$

By using the Rayleigh formula, we have

$$\max_{\|\vec{\xi}\| \le 1} J(\vec{u}) = \lambda_1 \{ (A(\omega) - B(\omega)K)^{-1'}(Q_1 + K'Q_2K)(A(\omega) - B(\omega)K)^{-1} \}. \tag{21.4.4}$$

Let

$$D(K) = (A(\omega) - B(\omega)K)^{-1'}(Q_1 + K'Q_2K)(A(\omega) - B(\omega)K)^{-1}.$$

Find $\min_k \mathbf{E}\lambda_1(D(K))$. Since λ_1 is a continuous function of the elements of matrix $D(K)$ and

$$\lambda_1(D(K_0 + t\theta)) \ge \lambda_1(D(K_0)),$$

(where K_0 is a point of minimum) for all real numbers t and matrices θ, the unknown matrix K satisfies the equation

$$\frac{\partial}{\partial t} \mathbf{E}\lambda_1(D(K + t\theta))_{t=0} = 0,$$

or

$$\lim_{t \to 0} \mathbf{E}t^{-1}\{\lambda_1(D(K + t\theta)) - \lambda_1(D(K))\} = 0.$$

We use the perturbation theory of linear operators for finding this derivative (Chapter 18). Since the elements of the random matrices $A(\omega)$ and $B(\omega)$ have joint probability density, the maximal eigenvalue λ_1 is simple. Let $D(K+t\theta) = D(t)$, $D(K) = D(0)$. By applying perturbation formulas of linear operators, we obtain

$$\lambda_1\left(D(t) - \lambda_1(D(0))\right) = \lambda_1\left(D(0) + t\left\{\frac{D(t) - D(0)}{t}\right\}\right) - \lambda_1(D(0))$$

$$= t\left(\left\{\frac{D(t) - D(0)}{t}\right\}\vec{e}_1(D(0)), \vec{e}_1(D(0))\right) + t^2(\dots) + \dots,$$

$$(21.4.5)$$

where $\vec{e}_1(D(0))$ is the eigenvector corresponding to the eigenvalue $\lambda_1(D(0))$. Then from (21.4.5), we obtain

$$\lim_{t\to 0} t^{-1}\mathbf{E}\{\lambda_1(D(t)) - \lambda_1(D(0))\} = \mathbf{E}\left(\frac{dD(0)}{dt}\vec{e}_1(D(0)), \vec{e}_1(D(0))\right) = 0.$$

$$(21.4.6)$$

One of the solutions of equation (21.4.6) is equal to

$$\mathbf{E}\frac{\partial}{\partial t}\left\{[A^{(\omega)} - B^{(\omega)}(K + t\theta)]^{-1}[Q_1 + (K + t\theta)'Q_2\right.$$

$$\left.\times (K + t\theta)][A^{(\omega)} - B^{(\omega)}(K + t\theta)]^{-1}\right\}_{t=0} = 0.$$

Hence,

$$\mathbf{E}(A(\omega) - B(\omega)K)^{-1'}\theta'B'(\omega)(A(\omega) - B(\omega)K)^{-1'}[Q_1 + K'Q_2K][A(\omega)$$

$$- B(\omega)K]^{-1} + (A(\omega) - B(\omega)K)^{-1'}(\theta'Q_2K + K'Q_2\theta)(A(\omega)$$

$$- B(\omega)K)^{-1} + (A(\omega) - B(\omega)K)^{-1'}(Q_1 + K'Q_2K)(A(\omega)$$

$$- B(\omega)K)^{-1}B(\omega)\theta(A(\omega) - B(\omega)K)^{-1} = 0.$$

Since the matrix θ is arbitrary, we obtain the equation (21.4.3) from this equation. Theorem 21.4.1 is proved.

CHAPTER 22

RANDOM DETERMINANTS IN SOME LINEAR STOCHASTIC PROGRAMMING PROBLEMS

The solving of linear stochastic programming problems often proves to be difficult because of large dimensions. The real meaning of the term "large-dimension problem," usually depends upon computing algorithm capabilities, computation speed, internal storage capacity, and other characteristics of the electronic computer being used. In practice, problems often arise which are extremely difficult to solve, even when using the most advanced algorithms and modern computers. The present state of applications of economical and mathematical methods for planning, design, and control is characterized by a transition to more complex large-scale problems of the development of optimal systems of the economy. All sorts of "perturbances" and "noise" reduce accuracy of the results and greatly distort the solution to the problems, containing hundreds and thousands of restrictions. Control, planning, and design methods under conditions of incomplete information were studied by Yudin, and effective methods for solving a broad class of stochastic programming problems were proposed by Ermol'ev.

In this chapter, on the basis of the integral representation for determinants, the limit theorem is proved under general assumptions about the solution $\vec{x}_n^*$ of the equation

$$\min_{\vec{x}_n : A_n(\omega)\vec{x}_n \leq \vec{b}_m(\omega), \vec{x}_n \geq 0,} \mathbf{E}f((\vec{c}_n(\omega), \vec{x}_n)) = \mathbf{E}f((\vec{c}_n(\omega), \vec{x}_n^*)),$$

where $A_n(\omega)$ is a random matrix of dimension $n \times m$, $\vec{c}_n(\omega)$, $\vec{x}_n$ and $\vec{b}_m(\omega)$ are random vectors, $(\vec{c}_n, \vec{x}_n)$ is the scalar product of vectors $\vec{c}_n$ and $\vec{x}_n$, and f is a measurable function. By the notation $\vec{a}_n \leq \vec{b}_n$, where $\vec{a}_n$ and $\vec{b}_n$ are vectors of the same dimension, we mean componentwise inequality.

The main result of the chapter is that under certain conditions the matrix $A_n(\omega)$ can be replaced by the approximate matrix, with diagonal entries equal to the sums of entries of the matrix $A_n(\omega)$. Provided that the law of large numbers holds for such sums, these diagonal entries can be replaced by deterministic variables. The obtained result enables us to simplify the calculation of the solution $\vec{x}^*$ considerably. The original stochastic problem can be reduced to a deterministic one.

§1 Formulation of the Linear Stochastic Programming Problem

The classical linear stochastic programming problem tries to find a vector satisfying the restrictions

$$\sum_{j=1}^{n} a_{ij} x_j \le b_j, \quad i = \overline{1, n}, \tag{22.1.1}$$

$$x_j \ge 0, \quad j = \overline{1, n} \tag{22.1.2}$$

and to minimize the purpose function

$$\sum_{j=1}^{n} c_j x_j. \tag{22.1.3}$$

The notation, which is entirely definite in the case of deterministic values of the problem's parameters, loses its definiteness and needs further interpretation in the case of random values of input information parameters. Meanwhile, in many applied problems, the purpose function coefficients, entries of condition matrix or restriction components of vectors are random variables. Replacing random parameters by their mean values is not always justified, for it can violate adequacy of the studied model and the object. Yet, planning estimates must be stochastic in nature, i.e., the matrix A and the vectors $\vec{x}, \vec{c}, \vec{b}$ are random, so the formulation must be defined more exactly, and it is necessary to state a few formulations of the linear stochastic programming problem.

1. Find

$$\inf(\vec{c}(\omega), \vec{x}(\omega))$$

on the set of random solutions of the system of inequalities

$$A(\omega)\vec{x}(\omega) \le \vec{b}(\omega), \quad \vec{x}(\omega) \ge 0.$$

This problem can be solved with fixed ω, and, in fact, it does not differ from the linear programming problem.

2. Find

$$\inf \mathbf{E}(\vec{c}(\omega), \vec{x}(\omega))$$

on the set of distributions

$$G(\vec{u}_1, \vec{u}_2) = P\{\vec{c}(\omega) < \vec{u}_1, \vec{x}(\omega) < \vec{u}_2, \vec{x}(\omega) \ge 0\}.$$

3. Find
$$\inf \mathbf{E}(\vec{c}(\omega), \vec{y})$$

on the nonrandom set of values of vectors

$$\vec{y} : \mathbf{P}\{A(\omega)\vec{y} \leq \vec{b}(\omega)\} \geq \mathbf{P}.$$

We study a particular case, when there exists an inverse matrix A_n^{-1}. In this case, the second problem can be formulated as follows: find

$$\inf \mathbf{E}(\vec{c}(\omega), A_m^{-1}(\vec{b}_m + \hat{A}\vec{x}_{n-m}))$$

on the set of all joint distributions

$$G(\vec{u}_1, \vec{u}_2, \vec{x}) = \mathbf{P}\{\vec{b}(\omega) < \vec{u}_1, \vec{x}_{n-m}$$
$$< \vec{u}_2, \vec{x}_{n-m} \leq 0, A_m^{-1}(\vec{b}_m + \hat{A}\vec{x}_{n-m}) \geq 0\},$$

where

$$\vec{x}_{n-m} = (x_{m+1}, \ldots, x_n), \hat{A} = (a_{ij})_{i=m+1,n}^{j=\overline{1,n}}, n > m.$$

§2 Systems of Inequalities with Random Coefficients in Linear Stochastic Programming

Many economic characteristics have stable distributions. Great attention was paid to the application of stable laws in economy in the works of Mandelbrot [126]. He pointed out the possibility of describing the income distribution process by means of the stable Pareto–Lévy law with finite mean value and gave motivation for the choice of the Pareto–Lévy distribution.

We shall explain beforehand what we mean by solutions of the system of inequalities with random coefficients.

Consider the system of inequalities

$$A_n(\omega)\vec{x}_n \leq \vec{b}_m, \qquad (22.2.1)$$

where $A_n(\omega)$ is a random matrix of dimension $n \times m$ (n is the number of columns, m is the number of rows), $\vec{x}_n = (x_1, \ldots, x_n)$, $\vec{b}_m = (b_1, b_2, \ldots, b_m)$ is a random vector.

By a set of solutions of the system of inequalities (22.2.1), we mean a set of random vectors satisfying the system with probability 1. Note that beside random solutions (if its number is greater than 1), there is an infinite number of nonmeasurable solutions of the system (22.2.1).

Indeed, let $\vec{x}_1(\omega)$, $\vec{x}_2(\omega)$ be any two random solutions of system (22.2.1). Then the vectors $\vec{y}(\omega) = \vec{x}_1(\omega)$ if $\omega = c$, and $\vec{y}(\omega) = \vec{x}_2(\omega)$ if $\omega \bar{\in} c$, where c is a nonmeasurable set, will be a solution of the system (22.2.1). It is obvious that set c can be chosen in such a way that vectors $\vec{y}(\omega)$ are nonmeasurable.

Instead of the inequality (22.2.1), it is more convenient to deal with the system of algebraic equations

$$\tilde{A}\vec{x}_{n+m} = \vec{b}_m, \qquad (22.2.2)$$

where $\tilde{A} = (A \oplus I)$ is a matrix of dimension $(n \times m) \times n$, $\vec{x}_{n+m} = (x_1, \ldots, x_{n+m})$, I is a unit matrix of order n. By the notation $A \oplus B$, where A and B are matrices with the same number of rows, we mean the matrix $(a_1, \ldots, a_n, b_1, \ldots, b_m)$, where a_i are vector columns of matrix A, b_i are vector columns of matrix B.

By a set of solutions of system (22.2.2) we mean a set of random vectors $\vec{x}_{n+m}$, which satisfies the system with probability 1 and has nonnegative components which are measurable with respect to the minimal σ-algebra generated by the entries of matrix A and the components of vector b.

Obviously, every random solution of the system of inequalities (22.2.1) corresponds to the solution of the system (22.2.2), and vice versa.

We shall consider only those solution of system (22.2.2) for which components $x_{n+1}, \ldots, x_{n+m}$ are deterministic and nonnegative. The generalized case will be considered by the author in subsequent publications on the subject.

Consider the system of algebraic inequalities

$$A\vec{x} \leq \vec{b}, \qquad (22.2.3)$$

where $A = (\xi_{ij})_{i,j=1}^n$ is a random matrix of specific production expenditures for the ith ingredient by the jth technology, b_i is the number of ingredients i available within the economic system: x_i is an intensity of the jth technology.

Reduce system (22.2.3) to the canonical form

$$A\vec{x} = \vec{b} + \vec{u}, \quad \vec{u} \leq 0, \qquad (22.2.4)$$

where $\vec{u}$ is a random vector, generally stochastically independent of the entries of the matrix A and of the entries of the vector $\vec{x}(u)$.

By the solution of such a system of equations, we mean the set of random vectors $\vec{x}(u)$ which depend on the random vector $\vec{u}$, where $u_i \leq 0 (i = \overline{1, n})$, and which satisfy condition (22.1.2).

Theorem 22.2.1. *If the entries of the matrix $A = (\xi_{ij})_{i,j=1}^n$ are independent and equally distributed by the symmetric stable law, with the characteristic index $0 < \alpha \leq 2$, then the ratio of components of the solution $x_k(u)$ and $x_s(u)$, $k = s$ at nonrandom vector $\vec{u}$ of system (22.2.4) will be distributed as the ratio of two independent random variables, distributed by the symmetric stable law with the characteristic index α.*

Proof. It is obvious, that

$$\mathbf{P}\{\det A_n = 0\}, \qquad \mathbf{P}\left\{ \left[\sum_{p=1}^n |A_{ps}|^\alpha \right]^{1/\alpha} = 0 \right\} = 0, \qquad (22.2.5)$$

where A_{ps} is a cofactor of the entry ξ_{ps} of matrix A_n. Then, with probability 1, and by using the Cramer formula, we have $x_k(u) = \det A_k(u) \det A^{-1}$, where $A_k(u)$ means that the kth column of the matrix A is replaced by $\vec{b} + \vec{u}$, $x_s(u) = \det A_s(u) \det A^{-1}$, where $A_s(u)$ means that the sth column of matrix A is replaced by $\vec{b} + \vec{u}$.

Then

$$x = \frac{x_k(u)}{x_s(u)} = \frac{\det A_k(u)}{\det A_s(u)} = \frac{\begin{pmatrix} \xi_{11} & \cdots & \xi_{1s} & \cdots & b_1 + u_1 & \cdots & \xi_{1n} \\ \xi_{21} & \cdots & \xi_{2s} & \cdots & b_2 + u_2 & \cdots & \xi_{2n} \\ \cdots & & \cdots & & \cdots & & \cdots \\ \xi_{n1} & \cdots & \xi_{ns} & \cdots & b_n + u_n & \cdots & \xi_{nn} \end{pmatrix}}{\begin{pmatrix} \xi_{11} & \cdots & b_1 + u_1 & \cdots & \xi_{1k} & \cdots & \xi_{1n} \\ \xi_{21} & \cdots & b_2 + u_2 & \cdots & \xi_{2k} & \cdots & \xi_{2n} \\ \cdots & \cdots & \cdots & & \cdots & \cdots & \cdots \\ \xi_{n1} & \cdots & b_n + u_n & \cdots & \xi_{nk} & \cdots & \xi_{nn} \end{pmatrix}}.$$

We change the sth and jth columns in the denominator, so that the determinant changes its sign. We obtain

$$x = -\frac{\sum_{p=1}^{n} \xi_{ps} A_{ps}(u)}{\sum_{p=1}^{n} \xi_{pk} A_{ps}(u)}, \tag{22.2.6}$$

and $(A_{ps}(u)$ is a cofactor of the entry $A_k(u)$ of matrix A_n).

When taking into account condition (22.2.5), the expression (22.2.6) would not change if numerator and denominator are divided by the same expression

$$x = \eta_1 \eta_2^{-1},$$

where

$$\eta_1 = \frac{\sum_{p=1}^{n} \xi_{ps} A_{ps}(u)}{[\sum_{p=1}^{n} |A_{ps}(u)|^{\alpha}]^{1/\alpha}},$$

$$\eta_2 = -\frac{\sum_{p=1}^{n} \xi_{pk} A_{ps}(u)}{[\sum_{p=1}^{n} |A_{ps}(u)|^{\alpha}]^{1/\alpha}}.$$

Consider the characteristic function η_1:

$$\mathbf{E} \exp\{is\eta_1\}.$$

Since $A_{ps}(u)[\sum_{p=1}^{n} |A_{ps}(u)|^{\alpha}]^{-1/\alpha}$ is independent of ξ_{ps},

$$\mathbf{E} \exp\{is\eta_1\} = \mathbf{E} \prod_{i=1}^{n} [\mathbf{E} \exp\left\{ -\frac{is A_{ps}(u)}{[\sum_{p=1}^{n} |A_{ps}|^{\alpha}]^{1/\alpha}} \right.$$

$$\times \xi_{ps}\} / A_{ps}(u), \quad p = \overline{1,n}]. \tag{22.2.7}$$

But ξ_{ps} is a random variable distributed by the symmetric stable law with the characteristic index α. Consequently, the equality (22.2.7) can be written in the form

$$\mathbf{E}\exp\{is\eta_1\} = \mathbf{E}\prod_{p=1}^{n}\exp\left\{-\frac{c|s|^{\alpha}|A_{ps}(u)|^{\alpha}}{\sum_{p=1}^{n}|A_{ps}(u)|^{\alpha}}\right\} = e^{-c|s|^{\alpha}},$$

i.e., the variable η_1 is symmetrically stably distributed.

Similarly we can prove that η_2 is also stably distributed and that η_1 is independent of η_2.

By taking into account that ξ_{pk} and ξ_{ps} are independent, we obtain

$$\mathbf{P}\{x_k(u)x_s^{-1}(u) < z\} = \mathbf{P}\{\eta_1\eta_2^{-1} < z\}.$$

The theorem is proved.

Note that finding the distributions of the variables x_k/x_s, in general, is a very difficult task. The expressions for distribution functions are extremely complicated. Therefore, we are interested in limit theorems, where the order of the system increases up to infinity, since in various practical problems, we have to solve systems of inequalities of large dimensions. The analysis of many economical problems prompts the restrictions which are imposed upon coefficients of systems of inequalities, namely:

a) the random coefficients ξ_{ij} are independent,

$$\mathbf{E}\xi_{ij} = 0, \quad \text{Var}\,\xi_{ij} = \sigma^2;$$

b) the order of the system is high (the quantity $\text{Var}\,\xi_{ij}/n$ is considered to be equivalent to the accepted error of the solution.

c) the coefficients are infinitesimal.

Consider each of the conditions in detail. Assumptions of independence of values ξ_{ij} prove to be natural, due to the fact that the noises in real economic models, influencing specific expenditure coefficients, are independent. The assumption of $\mathbf{E}\xi_{ij} = 0$, $\text{Var}\,\xi_{ij} = \sigma^2$ is accepted for simplification of the problem. Condition (b) needs an explanation, because it is not clear what is meant by the high order of the system. Among various concepts, we choose the following: if n is the order of the system, then we consider it to be high when n^{-1} is equivalent to the accepted error of the solution. Point (c) also proves to be natural, because the noise acting on every specific expenditure coefficient influences the behaviour of the whole system somewhat. This concept also needs interpretation: we consider, that

$$\text{Var}\,\xi_{ij} \equiv n^{-1}.$$

Under these rather common assumptions, we prove the following statement.

Theorem 22.2.2. *For every* $n = 1, 2, \ldots$, *let the random variables* $\xi_{ij}^{(n)}$, $i, j = \overline{1, n}$, $b_j^{(n)}$, $j = \overline{1, n}$ *be independent,* $\mathbf{E}\xi_{ij}^{(n)} = \mathbf{E}b_{ij}^{(n)} = 0$, $\operatorname{Var}\xi_{ij}^{(n)} = \operatorname{Var}b_j^{(n)} = 1$, *for some* $\delta > 0$:

$$\sup_n \sup_{i,j\overline{1,n}} [\mathbf{E}|\xi_{ij}^{(n)}|^{4+\delta} + \mathbf{E}|b_j|^{4+\delta}] < \infty. \tag{22.2.8}$$

Then for any nonrandom $\vec{u}$ *such that* $\sup_n \|\vec{u}\| < \infty$,

$$\lim_{n\to\infty} \mathbf{P}\{x_k(u)x_s^{-1}(u) < z\} = \tfrac{1}{2} + \pi^{-1}\arctan z.$$

Proof. Note that the solution of system (22.2.4) cannot exist. We consider that $x = \infty$, if $\det A = 0$. When the conditions of the theorem hold, it follows from formula (6.3.31) that $\operatorname{plim}_{n\to\infty} |\det A| = \infty$. Therefore, with the probability tending to one, as $n \to \infty$, $x \neq \infty$.

Without loss of generality, we assume that $\sigma^2 = 1$, since in the expression $x_k(u)/x_s(u)$ all entries ξ_{ij} entering the numerator and denominator can be divided by σ.

It follows from Theorem 6.3.2, that for every $\varepsilon > 0$,

$$\lim_{n\to\infty} \mathbf{P}\{A_{1k}^2(u) \geq (n-1)! \exp(-\varepsilon C_n)\} = 1,$$

where C_n is an arbitrary sequence of positive numbers, satisfying condition $\lim_{n\to\infty} C_n/\ln n = \infty$. Therefore,

$$\lim_{n\to\infty} \mathbf{P}\left\{\frac{x_k(u)}{x_s(u)} < z\right\} = \lim_{n\to\infty} \mathbf{P}\{x_k(u)x_s^{-1}(u) < z/A_{1k}^2(u) > \varepsilon\}.$$

We shall prove the lemma.

Lemma 22.2.1. *Under the conditions of Theorem 22.2.2,*

$$\sum_{k=1}^n \xi_{1k}\alpha_{1k} \sim \sum_{k=1}^n \eta_{1k}\alpha_{1k}, \tag{22.2.9}$$

where

$$\alpha_{1k} = \begin{cases} A_{1k}(u)\left[\sum_{k=1}^n A_{1k}^2(u)\right], & \text{if } \sum_{k=1}^n A_{1k}^2(u) \neq 0, \\ n^{-1/2}, & \text{if } \sum_{k=1}^n A_{1k}^2(u) = 0; \end{cases}$$

and η_{1k} *are independent of one another and of the random variables* ξ_{ij}; *random variables, normally distributed according to* $N(0,1)$; $A_{1k}(u)$ *is a cofactor of the entry* ξ_{pk} *of the matrix* $A_k(u)$.

Proof. From Chapter 5, it follows that condition (22.2.9) will be realized, if we have

$$\lim_{n\to\infty} \sum_{k=1}^n |\alpha_{1k}|^{2+\delta} = 0.$$

Note that

$$\sum_{k=1}^{n} A_{1k}^2(u) = \mathbf{E}\{\det \Xi_n^2 / A_k^2(u)\},$$

where the matrix Ξ_n is obtained from the matrix $A_k(u)$ by replacing the first column vector by the vector η, which is independent of the matrix A and is normally distributed according to $N(0,1)$. Hence,

$$\alpha_{11} = \begin{cases} [1 + (B_{11}(u)\xi_1, \xi_1)], & \text{if } \sum_{k=1}^{n} A_{1k}^2(u) \neq 0, \\ n^{-1/2}, & \text{if } \sum_{k=1}^{n} A_{1k}^2(u) = 0, \end{cases} \tag{22.2.10}$$

where $B_{kk}(u) = \Xi_{kk}\Xi_{kk}'$ (the matrix Ξ_{kk} is obtained from matrix A_n by deleting the kth column and the kth row);

$$\xi_k = (\xi_{1k}, \ldots, \xi_{k-1k}, \xi_{k+1k}, \ldots, \xi_{nk}).$$

We assume that the random variables ξ_{ij} have variances n^{-1}, therefore all entries of matrix A_n entering α_{ks} can be divided by n^{-1}.

We obtain similar formulas for the variables α_{1k}, $k \neq 1$. For this purpose it is necessary to interchange the kth column and the first one in the matrix Ξ_n, and also to insert the first column after the $(k-1)th$ one in the matrix A_{1k}. Then,

$$\alpha_{1k} = \begin{cases} [1 + (\widetilde{B}_{kk}^{-1}(u)\zeta_k, \zeta_k)]^{-1/2}, & \text{if } \sum_{k=1}^{n} A_{1k}^2(u) \neq 0, \\ n^{-1/2}, & \text{if } \sum_{k=1}^{n} A_{1k}^2 = 0; \end{cases}$$

the vector ζ_k equals the vector $(\xi_{2k}, \ldots, \xi_{nk})$, $\widetilde{B}_{kk} = \widetilde{\Xi}_{kk}\widetilde{\Xi}_{kk}'$, the matrix $\widetilde{\Xi}_{kk}$ is obtained from the matrix Ξ_{kk} by displacement of the first column after the $(k-1)$th one. This formula is identical to formula (22.2.10) without loss of generality, and we consider that

$$\alpha_{1k} = \begin{cases} [1 + (B_{kk}(u)\xi_k, \xi_k)]^{-1/2}, & \text{if } \sum_{k=1}^{n} A_{1k}^2(u) \neq 0, \\ n^{-1/2}, & \text{if } \sum_{k=1}^{n} A_{1k}^2(u) = 0. \end{cases}$$

It is obvious, that $(B_{kk}^{-1}\vec{\xi_k}, \vec{\xi_k}) \geq (R_k(t)\vec{\xi_k}, \vec{\xi_k})$, where $R_k(t) = (It + B_{kk})^{-1}$, $t > 0$ is a real variable.

We note that $B_{kk} = A_{kk}(u)A_{kk}'(u) = \sum_{k=1}^{n} \nu_k \nu_k'$, where ν_k is the column vector of the matrix $A_{kk}(u)$, with one of those being equal to $\vec{b} + \vec{u}$. Then, by using formula (2.5.4), we have

$$(R_k(t)\vec{\xi_k}, \vec{\xi_k}) \sim \frac{1}{n} \operatorname{Tr} R_k(t)$$

and

$$\frac{1}{n}R_k(t) - \frac{1}{n}\operatorname{Tr}(It + B_{kk} - (\vec{u} + \vec{b})(\vec{u} + \vec{b})')^{-1} \leq ct^{-1}n^{-1}\|\vec{u}\|^2.$$

Thus, in the expressions for the variables α_{1k}, we can neglect the vector $\vec{u}$. Therefore, all evaluations given for Theorem 6.3.1 are valid. The lemma is proved.

By employing the lemma, provided that $A_{1k}^2(u) > \varepsilon$, we obtain

$$\frac{x_k(u)}{x_s(u)} \sim \frac{\sum_{p=1}^n \eta_{ps}\alpha_{ps}}{\sum_{p=1}^n \eta_{pk}\alpha_{pk}},$$

where

$$\alpha_{pk} = A_{pk}(u)\left[\sum_{p=1}^n A_{pk}^2(u)\right]^{-1/2}, \eta_{ps}, \eta_{ps}, k \neq s$$

are independent random variables, not being dependent of A_n, and $(\vec{b} + \vec{u})$ and are distributed normally according to $N(0, 1)$. Hence,

$$\lim_{n \to \infty} \mathbf{P}\{x_k(u)x_s^{-1}(u) < z\} = 2^{-1} + \pi^{-1}\arctan z.$$

The theorem is proved.

§3 Integral Representation Method for Solving Linear Stochastic Programming Problems

In this section, under rather common assumptions and by means of integral representations for determinants, the limit theorem is proved for the solution x^* of the equation

$$\min_{\vec{x}_n, A_n(\omega)\vec{x}_n \leq \vec{b}_m(\omega), \vec{x}_n \geq 0} \mathbf{E}f((\vec{c}_n(\omega), \vec{x}_n)) = \mathbf{E}f((\vec{c}_n(\omega), \vec{x}_n^*)),$$

where $A_n(\omega)$ is a random matrix of dimension $n \times m$, $\vec{c}_n(\omega)$ and $\vec{x}_n$, $\vec{b}_m(\omega)$ are random vectors, $(\vec{c}_n, \vec{x}_n)$, is a scalar vector product of $\vec{c}_n$ and $\vec{x}_n$, and f is a measurable function.

The basic result is that under certain conditions, the matrix $A_n(\omega)$ can be replaced by the approximate matrix which has only diagonal random entries being equal to the sums of entries of the matrix $A_n(\omega)$. If the law of large numbers holds for these sums, then diagonal entries can be replaced by deterministic values. The obtained result makes it possible to considerably simplify the calculation of the solution x^*, as well as to reduce the original stochastic problem to a determinate one under certain conditions.

Now we formulate a linear stochastic problem to be solved here. Suppose we have to solve the following linear stochastic problem, then find

$$\inf \mathbf{E}f((\vec{\xi}_n(\omega), \vec{x}_n(\omega))) = \inf_{\vec{v} \leq \vec{\delta}} \int f((\vec{u}_1, \vec{u}_2))dG(\vec{u}_1, \vec{u}_2, \vec{v})$$

on the distribution function set

$$G(\vec{u}_1, \vec{u}_2, \vec{v}) = \mathbf{P}\{\vec{x}_n(\omega) < \vec{u}_1, \vec{\xi}_n(\omega) < \vec{u}_2,$$
$$\|\vec{x}_n(\omega)\| \leq 1, \vec{x}_n(\omega) \geq 0\}, \vec{v} \leq \vec{0},$$

where $\vec{x}(\omega)$ is a solution of the system of equations

$$(I + A_n(\omega))\vec{x}_n(\omega) = \eta_n(\omega) + \vec{v},$$

$A_n(\omega)$ is a random matrix of order n, $\vec{\xi}_n$ and $\vec{\eta}_n$ are random vectors, $\vec{v} = (v_1, \ldots, \vec{v}_n)$ is a nonrandom vector, $f((\vec{\xi}(\omega), \vec{x}_n(\omega))$ is a certain measurable function chosen in such a way that there exists the integral

$$\mathbf{E}f((\vec{\xi}, \vec{x})).$$

Theorem 22.3.1. *Let the vectors $\vec{\xi}_n$ and $\vec{\eta}_n$ be independent of the matrices $A_n = (\xi_{ij}^{(n)})_{i,j=1}^n$, for all values n, let the vectors $(\xi_{ij}^{(n)}, \xi_{ji}^{(n)})$, $i \geq j$, $i,j = \overline{1,n}$ be independent and asymptotically constant,*

$$\lim_{h \to \infty} \lim_{n \to \infty} \mathbf{P}\left\{|\sum_{i=1}^n \nu_{ii}^{(n)}| + \sum_{i,j=1}^n (\nu_{ij}^{(n)})^2 \geq h\right\} = 0,$$
$$\sup_n [|\operatorname{Tr} B_n| + \operatorname{Tr} B_n B_n'] < \infty,$$

where

$$\nu_{ij}^{(n)} = \xi_{ij}^{(n)} - a_{ij}^{(n)} - \rho_{ij}^{(n)}, \qquad \rho_{ij}^{(n)} = \int_{|x| < \tau} x dP\{\xi_{ij}^{(n)} - a_{ij}^{(n)} < x\},$$
$$b_{ij} = \rho_{ij} + a_{ij}, \qquad B_n = (b_{ij}^{(n)}),$$

$\tau > 0$ is an arbitrary constant, for all values n, there exist the inverse matrices $(I + A_n)^{-1}$,

$$\lim_{h \to \infty} \overline{\lim_{n \to \infty}} \mathbf{P}\{\|\vec{\xi}_n\| + \|\vec{\eta}_n\| \geq h\} = 0,$$

and the function $|f|$ is bounded by a nonrandom constant. Then,

$$\inf_G \mathbf{E}f((\vec{\xi}(\omega), \vec{x}(\omega))) = \inf_F \mathbf{E}f((\vec{\xi}(\omega), \vec{y}(\omega))) + 0(1),$$

where F is a set of distribution functions

$$F(\vec{u}_1, \vec{u}_2, \vec{v}_n) = \mathbf{P}\{\vec{y}(\omega) < \vec{u}_1, \vec{\xi}(\omega) < \vec{u}_2,$$
$$\|\vec{y}_n(\omega)\| \leq 1, \vec{y}(\omega) \geq \vec{0}\}, \vec{v}_n \leq \vec{0},$$

and $\vec{y}_n(\omega)$ is a solution of the system of equations

$$(I + \text{diag}\{\nu_{ii} - \sum_{p \in T_i \cup K_i} \nu_{pi}\nu_{ip}, i = \overline{1,n}\} + B_n)\vec{y}_n(\omega) = \vec{\eta}_n(\omega) + \vec{v}_n.$$

Here the sets T_i and K_i are arranged as follows (see Chapter 15): the set of vectors $\{(\nu_{ij}, \nu_{ji}), i > j, i, j = \overline{1,n}\}$, is divided onto $2n$ nonintersecting sets R'_{in}, R''_{in}, $i = \overline{1,n}$, so that the vectors μ_i, composed of the entries from each set, are infinitesimal, and also that the set R'_{in} contains only entries of the ith vector row of the matrix $(A_n + A'_n)$ and the set R''_{in} contains only those entries of the ith vector column of matrix $(A_n + A'_n)$; T_{jn} is a set of index values i of $(\nu_{ij}, \nu_{ji}) \in \bigcup_{p=1}^n R'_{pn}$, K_{in} is a set of index values j of $(\nu_{ij}, \nu_{ji}) \in \bigcup_{p=1}^n R'_{pn}$. Assume that $\sum_{i \in \emptyset} = 0$.

Proof. Using the formulas for perturbations of the determinant, we have:

$$\mathbf{E}f((\vec{\xi}_n, \vec{x}_n)) = \mathbf{E}f(\{\det(I + A_n + (\vec{\eta} + \vec{v})\vec{\xi}') \det(I + A_n)^{-1} - 1\}) \times X((I + A_n)^{-1}(\vec{\eta} + \vec{v})),$$

where

$$X(x) = \begin{cases} 1, & \text{if } ||\vec{x}|| \le 1, x_i \ge 0, i = \overline{1,n}, \\ 0, & \text{otherwise}. \end{cases}$$

Since $||\vec{\eta}_n + \vec{v}_n||^2 \le ||I + A_n||^2$, by using the conditions of the theorem, we have: $\overline{\lim}_{n \to \infty} ||\vec{v}_n|| < \infty$.

But in this case, for random variables, $\det(I + A_n + (\vec{\eta}_n + \vec{v}_n)\vec{\xi}'_n) \det(I + A_n)^{-1}$, the method of integral representation can by applied (see the proof of Theorem 15.2.1), based on the formula

$$\det(I + \alpha_t A)^{-1} = \mathbf{E} \exp\{i\alpha_t((A - A')\vec{\xi}, \vec{\eta}) - \alpha_t(A\vec{\xi}, \vec{\xi}) - \alpha_t(A\vec{\eta}, \vec{\eta})\},$$

where

$$\alpha_t = t[1 + 1/2|\text{Tr}(A + A')| + 1/4\text{Tr}(A + A')^2]^{-1};$$

$\vec{\xi}$ and $\vec{\eta}$ are independent random normally distributed $N(0, \frac{1}{2}I)$ vectors.

Consider the joint characteristic function for the vector $(I + A_n)^{-1}(\vec{\eta}_n + \vec{v})$ and the random variable $\det(I + A_n + (\vec{\eta} + \vec{v})\vec{\xi}') \det(I + A_n)^{-1}$:

$$\mathbf{E} \exp\{i((I + A_n)^{-1}(\vec{\eta}_n + \vec{v}_n), \vec{s}_n) \\ + i\theta \det(I + A_n + (\vec{\eta}_n + \vec{v}_n)\vec{\xi}') \det(I + A_n)^{-1}\} \\ = \mathbf{E} \exp\{i \det(I + A_n + (\vec{\eta}_n + \vec{v}_n)s') \det(I + A_n)^{-1} \\ - 1 + i\theta \det(I + A_n + (\vec{\eta} + \vec{v})\vec{\xi}') \det(I + A_n)^{-1} - 1\},$$

where θ is a parameter, and $\vec{s}$ is a parameter vector.

By using the proof of the above-mentioned theorem, we have

$$\mathbf{E} \exp\{i(\vec{x}_n, \vec{s}_n) + i\theta \det(I + A_n + (\vec{\eta}_n + \vec{v}_n)\vec{\xi}) \det(I + A_n)^{-1} - 1\}$$
$$= \mathbf{E} \exp\{i(\vec{y}_n, \vec{s}_n) + i\theta(\vec{y}_n, \xi_n)\} + 0(1),$$

where $\sup \|\vec{s}_n\| < \infty$. Since $\|\vec{x}_n\| < 1$, then for each $\vec{v}_n$ (such that $\|\vec{v}_n\| < \infty$)

$$\mathbf{E} f((\vec{\xi}_n, \vec{x}_n)) = \mathbf{E} f((\vec{\xi}_n, \vec{y}_n)) + 0(1).$$

The theorem is proved.

Note that the diagonal entries $\nu_{ii} - \sum_{p \in T_j \cup K_i} \nu_{pi}\nu_{ip}$, $i = \overline{1, n}$ are the sums of independent random variables, and all known limit theorems for sums of independent random variables can be applied to them. In particular, the conditions for stochastic convergence of such sums to nonrandom constants could be given.

CHAPTER 23

RANDOM DETERMINANTS IN
GENERAL STATISTICAL ANALYSIS

The general statistical analysis of observations (G-analysis) is a mathematical theory studying some complex systems S such that the number m_n of parameters of their mathematical models can increase together with the growth of the number n of observations over the system S. The purpose of this theory consists of finding by the observations of the system S such mathematical models (G-estimates) that would approach the system S, in a certain sense, with a given rate at the minimal number of observations and under the general assumptions on the observations: The existence of the distribution densities of observed random vectors and matrices is not necessary. Only the existence of several first moments of their components is required, the numbers m_n and n satisfy the G-condition $\overline{\lim}_{n\to\infty} f(m_n, n) < \infty$, where $f(x, y)$ is some positive function increasing along y and decreasing along x. In most cases, the function $f(x, y)$ is equal to yx^{-1}. In this case, the G-condition is also called the *Kolmogorov condition*.

In the general statistical analysis, two conditions (postulates) are assumed:
(1) The dimension (a number of parameters) of estimated characteristics of this system does not change with the increase of the number m_n of parameters of the mathematical models of the system S.
(2) The dimension m_n of mathematical models can increase with the growth of the number n of observations over the system S and, on the contrary, depends on m_n and cannot grow arbitrarily fast with the increase of m_n.

As a rule, in solving the limit theorems of the G-analysis problem for sums of martingales difference, perturbations formula for resolvents of covariance matrices, random matrices theory are applied, which forms the theoretical part of G-analysis. In G-analysis, complicated mathematical models of the observed systems are substituted by the simplified ones constructed by the minimal number of observations providing the required accuracy of a solution. It enables us to avoid tremendous difficulties in matching mathematical models and in computations.

The criticism of the multivariate statistical analysis of large-dimensional observations was due to the fact that the error of the scores of estimates is equivalent to $mn^{-1/2}$, where m is the number of parameters to be estimated, and n is the number of observations. It is evident that the number of observations needed for estimation with given accuracy increases sharply with the growth of m. In this connection, publications appeared that pointed out the inconsistency of multivariate statistical analysis for solving practical problems involving observations over large-dimensional vectors.

In 1972–1980 [6], corrections for multivariate statistical analysis estimates were found. The method was based on proving some theorems of the type concerning the law of large numbers for some functions of random matrix entries. It should be noted that analogous theorems were proved by Wigner, Mehta, and Pastur earlier in the fifties [162, 132, 141], although for other purposes.

After many years of investigations, it was thought that if the G-condition $\overline{\lim}_{n\to\infty} f(m_n, n) < \infty$ holds, where f is some function, then there exist no consistent and asymptotically normal estimations of functions $\varphi(R)$, where φ is some function of the matrix R entries.

However, the advanced theory of random matrices leads to the conclusion that under the Kolmogorov condition, $\lim_{n\to\infty} mn^{-1} = c$, $0 < c < \infty$ for some functions φ,

$$\text{plim}_{n\to\infty}[\varphi(\widehat{R}_{m_n}) - \psi(R_{m_n})] = 0,$$

where ψ is a certain measurable function of the entries of the matrices R_{m_n}. This is the principal statement that the G- analysis is based on.

By using this equation, we can find a measurable function $G(R_{m_n})$ (the G-estimate) such that

$$\text{plim}_{n\to\infty}[G(\widehat{R}_{m_n}) - \varphi(R_{m_n})] = 0,$$

or

$$[G(\widehat{R}_{m_n}) - \varphi(\widehat{R}_{m_n})]c_n^{-1/2} \Rightarrow N(0,1),$$

where c_n is some sequence of numbers.

We show that the error of G-estimators of some functions $\varphi(R_{m_n})$ is equivalent to $(m_n n)^{-1/2}$, whereas for estimators of the $\varphi(\widehat{R}_{m_n})$, it is equivalent to $m_n n^{-1/2}$.

§1 The Equation for Estimation of Parameters of Fixed Functions

Suppose that in R^{m_n} an absolutely integrable R^{m_n}-valued Borel function $f(x)$, having partial derivatives of the second order, and observations $x_1, \ldots, x_n$ of an m_n-dimensional random vector η, distributed according to the normal law $N(a, R_{m_n})$, are given and that we need a consistent estimate of the value $f(a)$. Many problems of the control of mechanical and radiotechnical systems and of multivariate statistical analysis can be formulated in these terms. We note that in some problems the function is given, and changing it entails the reconstruction of the system, which involves large financial expenditures. If we take $\hat{a} = n^{-1} \sum_{i=1}^{n} x_i$ as the estimations of a, then, obviously, we have for fixed m, $\text{plim}_{n \to \infty} f(\hat{a}) = f(a)$, provided f is continuous. But the application of this relation in solving practical problems is unsatisfactory due to the fact that the number of observations n necessary to solve the problem with given accuracy increases sharply for large m.

It is possible to reduce significantly the number of observations n by making use of the fact that under some conditions, including

$$\lim_{n \to \infty} mn^{-1} = c, \quad 0 < c < \infty,$$

the relation

$$\text{plim}_{n \to \infty}[f(\hat{a}) - \mathbf{E}f(\hat{a})] = 0 \qquad (23.1.1)$$

holds.

We call Eq. (23.1.1) and similar ones the basic equations of the G-analysis of the large dimensional observations, in which the methods of estimating functions of some characteristics of random vectors are studied.

Hence, we have the equation for the G-estimate $\tilde{a}$,

$$(2\pi)^{-m/2} \det R_m^{-1/2} \int f(\tilde{a} + n^{-1/2}\vec{y}\,) \exp\left\{-\frac{1}{2}(R_m^{-1}y, y)\right\} prod_{i=1}^{m} dy_i = f(\hat{a}). \qquad (23.1.2)$$

If the function f has partial derivatives of the third order,

$$\overline{\lim_{n \to \infty}} \sup_z \int_0^1 \int \left[\sum_{k=1}^{m} x_k \left(\frac{\partial}{\partial z_k}\right)\right]^3 f(\vec{z} + t\,\vec{x}\,n^{-1/2})$$
$$\times (2\pi)^{-m/2} \exp\left\{-\frac{1}{2}(R_m^{-1}x, x)\right\} \det R_m^{-1/2} dx\,dt\,n^{-3/2} = 0,$$

then under the conditions

$$\lim_{n \to \infty} mn^{-1} = c,$$

Equation (23.1.2) can approximately be replaced by the following one,

$$f(\tilde{a}) + \frac{1}{2} \sum_{i,j=1}^{m} \frac{\partial^2 f(\tilde{a})}{\partial y_i \partial y_j} r_{ij} n^{-1} = f(\hat{a}) + 0(1),$$

where r_{ij} are the elements of the matrix R_m.

§2 The Equations for Estimation of Twice-Differentiable Functions of Unknown Parameters

Consider the functions

$$u(t,z) = \mathbf{E}f(z + a + \nu t^{1/2}n^{-1/2}), \qquad (23.2.1)$$

where $t > 0$ is a real parameter, $z \in R^{m_n}$, and ν is a random vector distributed by the normal law $N(0, R_{m_n})$.

These functions satisfy the equation

$$\frac{\partial}{\partial t}u(t,z) = Au(t,z), \qquad u(1,z) = \mathbf{E}f(z + \hat{a}),$$

$$A = (2n)^{-1}\sum_{i,j=1}^{m_n} r_{ij}\frac{\partial^2}{\partial z_i \partial z_j}, \qquad u(0,z) = f(z + a),$$

$$(23.2.2)$$

where r_{ij} are the entries of the matrix R_{m_n}.

Suppose that the random vector ξ has arbitrary distribution and that there exists $\mathbf{E}\xi\xi'$. Let

$$\alpha_n(kn^{-1}, z) = \mathbf{E}f(z + a + \sum_{p=1}^{k}(x_p - \mathbf{E}x_p)n^{-1}),$$

$$\alpha_n(t,z) = \alpha_n(kn^{-1}, z), \qquad kn^{-1} \le t < (k+1)n^{-1}, \quad k = 1, \ldots, n,$$

and

$$\lim_{n\to\infty} n\mathbf{E}\int_0^1 (1-t)^2 2^{-1}(\sum_{i=1}^{m_n} n^{-1}(x_{ik} - a_i)(\frac{\partial}{\partial z_i}))^3$$

$$\times f(z + a + n^{-1}\sum_{i=1}^{k-1}(x_i - a_i) + tn^{-1}(x_k - a))dt = 0.$$

Then, by using the expansion of the function f in Taylor series, we obtain

$$n[\alpha_n(kn^{-1}, z) - \alpha_n((k-1)n^{-1}, z)]$$

$$= (2n)^{-1}\sum_{i,j=1}^{m_n} r_{ij}\left(\frac{\partial^2}{\partial z_i \partial z_j}\right)\alpha_n((k-1)n^{-1}, z) + \varepsilon_n,$$

$$(23.2.3)$$

where

$$\lim_{n\to\infty} \varepsilon_n = 0.$$

From Eq. (23.2.3), we have

$$\alpha_n(t,z) = \alpha_n(0,z) + (2n)^{-1}\int_0^t \sum_{i,j=1}^{m_n} r_{ij}\left(\frac{\partial^2}{\partial z_i \partial z_j}\right)\alpha_n(u,z)du + 0(1). \quad (23.2.4)$$

We write Eq. (23.2.4) in the form

$$\alpha_n(t, z) = \alpha_n(0, z) + \int_0^t A\alpha_n(u, z)du + 0(1),$$

where A is a linear differential operator of the second order,

$$Af(z) = (2n)^{-1} \sum_{i,j=1}^{m_n} r_{ij} \left(\frac{\partial^2}{\partial z_i \partial z_j} \right) f(z).$$

§3 The Quasiinversion Method for Solving G_1-Equations

We deduce the finding of G-estimations of the functions $f(a)$ to the solution of inverse problem for Eq. (23.2.4). The latter consists of finding $\alpha_n(0, z)$ by the function $\alpha_n(1, z)$, which is replaced by the function $f(z + \hat{a})$ obtained from observations of the random vector ξ. Of course, the solution of the inverse problem with such a replacement cannot exist in the class of functions $W_2^{(0,2)}$. Therefore, it appears expedient to find some generalized solution of the problem.

Let $\psi(x) \in L_2$ and the functional

$$I(\varphi) = \int_D |\alpha_n(1, x, \varphi) - \varphi(x)|dx \qquad (23.3.1)$$

be determined by the functions $\varphi \in W_2^{(0,2)}$. Here D is a domain of the m-dimensional Euclid space of points $x = (x_1, \ldots, x_m)$, which is bounded by the piecewise smooth surface S, and $\alpha_n(1, x, \varphi)$ are solutions of equation

$$\alpha_n(t, x, \varphi) = \varphi(x) + \int_0^1 (2n)^{-1} \sum_{i,j=1}^m r_{ij} \left(\frac{\partial^2}{\partial x_i \partial x_j} \right) \alpha_n(u, x, \varphi)du + 0(1),$$

at the point $t = 1$.

The function $\hat{\varphi}(x)$ is the solution of the inverse problem if

$$\inf_{\varphi \in W_2^{(0,2)}} I(\varphi) = I(\hat{\varphi}).$$

To solve this problem, we do the following. First, we solve the direct problem

$$\alpha_n(t, x, \varphi) = \varphi(x) + \int_0^1 A\alpha_n(u, x, \varphi)du + 0(1),$$

$$\alpha_n(u, x, \varphi) = 0, \quad x \in S;$$

here S is the piecewise smooth boundary of some connected domain D, where $\alpha_n(1, x, \varphi) = \psi(x)$ is a given function.

Then we have the approximate value $\alpha_n(0, x, \varphi)$ for the initial condition of the function $\varphi(x)$. But it is quite possible that, in general, such a problem has no solution for the given function. Therefore, it is appropriate to solve the inverse problem approximately with the help of the so-called quasiinversion method. Namely, we consider the following equation,

$$\left(\frac{\partial}{\partial t}\right) u(t, z) = A_\delta u(t, z), \qquad u(1, z) = \alpha_n(1, z), \qquad (23.3.2)$$

instead of Eq. (23.2.4); here A_δ is some operator similar, in some sense, to the operator A and such that the solution to Eq. (23.3.2) is stable. We can let $A_\delta = A + \delta A^2$, $\delta > 0$.

By obtaining the solution of Eq. (23.3.2), we can apply the spectral theory of the operator A_δ. Its spectrum is, however, continuous. Therefore, it would be appropriate to substitute operator A by an operator A_ε, such that its spectrum is discrete and the eigenfunctions would form the complete orthonormal basis in the Hilbert space L_2. For example, instead of such an operator A_ε, we can choose

$$A_\varepsilon = A + \varepsilon q(z) + \delta[A + \varepsilon q(z)]^2; \quad \varepsilon, \delta > 0,$$

where $q(z)$ is any measurable function such that the operator $A + \varepsilon q(z)$, $z \in R^{mn}$ satisfies the above-mentioned condition. From the operator spectral theory, it follows that instead of function $q(z)$, we can choose any measurable function in such a way that

$$\lim_{\|z\| \to \infty} q(z) = \infty.$$

Let $\lambda_k(\varepsilon)$ and $\varphi_{k\varepsilon}(z)$, $k = 1, 2, \ldots$ denote the eigenvalues and eigenfunctions of the operator $A + \varepsilon q(z)$, respectively. As the standard estimate of the function $f(a)$, we take

$$G = \sum_{k=1}^{\infty} \exp\{-\lambda_k(\varepsilon) - \delta\lambda_k^2(\varepsilon)\}\varphi_{k\varepsilon}(0) \int f(\hat{a} + z)\varphi_{k\varepsilon}(z)dz.$$

Now we obtain $\mathbf{E}G$. Suppose that $f \in L_2$ and $\mathbf{E}(\partial^2/\partial z_i \partial z_j)f(z + \hat{a}) < \infty$. Since the function $\mathbf{E}f(\hat{a} + z)$ satisfies Eq. (23.2.2),

$$\left(\frac{\partial}{\partial t}\right) u(t, z) = [A + \varepsilon q(z)]u(t, z) - \varepsilon q(z)u(t, z),$$

$$u(t, 0) = f(a + z).$$

The solution of this equation is

$$\mathbf{E}f(\hat{a} + z) = \sum_{k=1}^{\infty} e^{\lambda_k(\varepsilon)} \varphi_{k\varepsilon}(z) \int f(a + x)\varphi_{k\varepsilon}(x)dx$$

$$- \varepsilon \int_0^1 \{\sum_{k=1}^{\infty} e^{(1-\tau)\lambda_k(\varepsilon)} \varphi_{k\varepsilon}(z) \int q(x)u(\tau, x)\varphi_{k\varepsilon}(x)dx\}d\tau.$$

By substituting the value of this integral for $\mathbf{E}G$, we obtain

$$\mathbf{E}G = \sum_{k=1}^{\infty} e^{-\delta\lambda_k^2(\varepsilon)} \varphi_{k\varepsilon}(0) \int f(a + x)\varphi_{k\varepsilon}(x)dx$$

$$- \varepsilon \int_0^1 \{\sum_{k=1}^{\infty} (e^{-\tau\lambda_k(\varepsilon) - \delta\lambda_k^2(\varepsilon)} \varphi_{k\varepsilon}(0) \int q(x)u(\tau, x)\varphi_{k\varepsilon}(x)dx\}d\tau.$$

Note that $\lambda_k(\varepsilon) \to -\infty$ as $k \to \infty$, $q(x) \to \infty$, $\|x\| \to \infty$. Therefore, if $q(x)u(\tau, x) \in L_2$, $\tau \in (0, 1)$; $f(x) \in L_2$,

$$\lim_{\varepsilon \downarrow 0} \lim_{\delta \downarrow 0} \mathbf{E}G = f(a).$$

If the function $u(\tau, x)$ does not tend to zero as $\|x\| \to \infty$, we introduce the function $f(a + x)\theta(x)$ instead of $f(a + x)$, where the function θ is chosen in such a way that $q(x)u(\tau, x) \in L_2$ and $\theta(0) \neq 0$. Then, as the G-estimate of function $f(a)$, we can take the expression $G_1\theta^{-1}(0)$, where $G_1(z, 0)$ satisfies the equation

$$\frac{\partial u(t, z)}{\partial t} = (2n)^{-1} \sum_{i,j=1}^{m} r_{ij} \frac{\partial^2}{\partial z_i \partial z_j} [\theta^{-1}(z)u(t, z)],$$

$$u(1, z) = f(\hat{a} + z)\theta(z), \quad U(0, z) = G(z, 0).$$

§4 The Fourier Transformation Method

Let us apply the Fourier transform to both parts of Eq. (23.2.2). Then for the function

$$\varphi(t, s) = \int e^{i(s, z)} u(t, z)dz, \quad s \in R^{m_n}, \quad dz = \prod_{i=1}^{m_n} dz_i,$$

we have

$$\left(\frac{\partial}{\partial t}\right) \varphi(t, s) = -(2n)^{-1}(R_{m_n}s, s)\varphi(t, s),$$

and we find from this equation,

$$\varphi(0,s) = \exp\{(2n)^{-1}(R_{m_n}s,s)\}\varphi(1,s).$$

The G-estimate of the $\varphi(0,s)$ is chosen as follows:

$$G_n(s) = \exp\{(2n)^{-1}(R_{m_n}s,s)\}\hat{\varphi}(1,s),$$

where $\hat{\varphi}(1,s) = \int e^{i(s,z)} f(z + \hat{a})dz$.

By using the estimate $G_n(s)$ for the Fourier transforms of the function $f(a + z)$, we embed the regularized estimate of the $f(a + z)$:

$$G_n(\varepsilon) = (2\pi)^{-m_n/2} \int e^{-\frac{1}{2}(z,z)} f(\varepsilon(I - \varepsilon^{-2}n^{-1}R_{m_n})^{1/2}z + \hat{a})dz,$$

or

$$G_{n1}(\varepsilon) = c_\delta \int_{((R_{m_n} - \varepsilon I)s,s)\le\delta} G_n(s)e^{-\frac{\varepsilon}{2}(s,s)}ds,$$

where $\varepsilon > 0$, $\delta > 0$.

In proving the consistency of such estimates, the following assertions are often used.

Theorem 23.4.1. *Let*

$$\lim_{n\to\infty} n^{-1} \sum_{i,j=1}^{m_n} r_{ij}\mathbf{E}\left(\frac{\partial}{\partial z}\right) f(z + \hat{a} - x_1 n^{-1})(\partial/\partial z_j)f(z + \hat{a} - x_1 n^{-1}) = 0, \tag{23.4.1}$$

$$\lim_{n\to\infty} \iint [\int_0^1 |\left(\sum_{k=1}^{m_n}(R_{m_n}^{1/2}x)_k\left(\frac{\partial}{\partial z_k}\right)\right)^2 f(z + a + R_{m_n}^{-1/2}$$
$$\times [y(n-1)^{1/2}n^{-1} + txn^{-1}])|dt]^2 \exp\{-(y,y)/2 - (x,x)/2\}$$
$$\times (2\pi)^{-m} dy\, dx\, n^{-3} = 0. \tag{23.4.2}$$

Then for any z,

$$\lim_{n\to\infty} \mathbf{E}[f(\hat{a} + z) - \mathbf{E}f(\hat{a} + z)]^2 = 0. \tag{23.4.3}$$

Proof. We represent the difference $f(\hat{a} + z) - \mathbf{E}f(\hat{a} + z)$ in the form

$$f(\hat{a} + z) - \mathbf{E}f(\hat{a} + z) = \sum_{k=1}^n \gamma_k(z),$$

where $\gamma_k(z) = \mathbf{E}_{k-1}f(z + \hat{a}) - \mathbf{E}_k f(z + \hat{a})$, and $\mathbf{E}_k$ is the conditional mathematical expectation under the fixed minimal σ-algebra. We assume that the random vectors $x_{k+1}, \ldots, x_n$ are measurable with respect to this algebra.

Obviously,

$$\mathbf{E}[f(\hat{a}+z)-\mathbf{E}f(\hat{a}+z)]^2 = \sum_{k=1}^{n} \mathbf{E}\gamma_k^2(z),$$

$$\gamma_k(z) = \mathbf{E}_{k-1}\theta(z) - \mathbf{E}_k\theta(z),$$

where

$$\theta(z) = f(z + a + R_{m_n}^{1/2}\nu_k \frac{\sqrt{n-1}}{n} + R_{m_n}^{1/2}\mu_k n^{-1})$$

$$- f(z + a + R_{m_n}^{1/2}\nu_k \frac{\sqrt{n-1}}{n}),$$

$$\nu_k = (n-1)^{-1/2}\sum_{i\neq k}(x_i - a), \qquad \mu_k = x_k - a.$$

For the function f, the Taylor formula holds:

$$f(z+h) - f(z) = \sum_{k=1}^{p-1}(k!)^{-1}\left(\sum_{i=1}^{m_n} h_i \frac{\partial}{\partial z_i}\right)^k f(z) + \alpha_p(z,h),$$

where

$$\alpha_p(z,h) = \int_0^1 \frac{(1-t)^{p-1}}{(p-1)!}\left(\sum_{i=1}^{m_n} h_i \frac{\partial}{\partial z_i}\right)^p f(z+th)dt, \ h \in R^{m_n}.$$

By using this formula, we derive (23.4.3) from (23.4.1) and (23.4.2). Theorem 23.4.1 is proved.

§5 Equations for Estimations of Functions of Unknown Parameters

Let us find the equations for G-estimates of the differentiable Borel functions $\varphi(a, R_{m_n})$ of the vector of expectations and the covariance matrix. Let $x_k, k = 1, 2, \ldots, n$ be observations over the normally distributed $N(a, R_{m_n})$ random vector ξ. Consider the functions

$$u(X_{m_n}, z, kn^{-1}) = \mathbf{E}\varphi(a + z + R_{m_n}^{1/2}\eta_n n^{-1/2}, R_{m_n}$$

$$+ X_{m_n} + R_{m_n}^{1/2}\sum_{s=1}^{k}(\eta_s\eta_s'(n-1)^{-1} - I)R_{m_n}^{1/2}),$$

where η_s are independent m_n-dimensional random vectors distributed according to the normal law $N(0, I)$, and $X_{m_n} = (x_{ij})$ is a matrix of the parameters of the same order as the matrix $R_{m_n}, z \in R^{m_n}$.

To simplify the formula index, m_n is omitted.

If the functions $u(X, z, kn^{-1})$ can be represented as

$$u(X, z, kn^{-1}) - u(X, z, (k-1)n^{-1}) = Au(X, z, (k-1)n^{-1}) + \varepsilon_n n^{-1},$$

where

$$A = \frac{1}{2} \sum_{i,j,p,l=1}^{m_n} \mathbf{E} \left(R^{1/2} \frac{\eta_1 \eta_1' - I}{n-1} R^{1/2} \right)_{ij} \left(R^{1/2} \frac{\eta_1 \eta_1' - I}{n-1} R^{1/2} \right)_{pl}$$

$$\times \frac{\partial^2}{\partial x_{ij} \partial x_{pl}} + \frac{1}{2} \sum_{i,j=1}^{m_n} r_{ij} \frac{\partial^2}{\partial z_i \partial z_j}; \quad \lim_{n \to \infty} \varepsilon_n = 0,$$

then we obtain the equations

$$\psi_n(X, z, t) = \varphi(z + a, X + R) + \int_0^t A\psi_n(X, z, y)dy + \varepsilon_n,$$

$$\psi_n(X, z, 1) = \mathbf{E}\varphi(z, \hat{a}, X + \hat{R}),$$

for the functions

$$\psi_n(X, z, t) = u(X, z, kn^{-1}), \qquad kn^{-1} \le t < (k+1)n^{-1},$$

where

$$\hat{R} = \frac{1}{n-1} \sum_{k=1}^{n} (x_k - \hat{a})(x_k - \hat{a})'.$$

We can also use the quasiinversion and the Fourier method of finding G-estimates of the functions $\varphi(a, R)$ to this equation.

Following are some notes concerning the search for of G-estimates if functions are not differentiable:

Let us give the Borel functions $\varphi(x)$, $x \in R^{m_n}$ and the independent observation x_k, $k = 1, \ldots, n$ over the random m_n-dimensional vector ξ, $\mathbf{E}\xi = a$.

Let

$$u(z, kn^{-1}) = \mathbf{E}\varphi(z + a + n^{-1} \sum_{s=1}^{k} (x_s - a)),$$

$$\psi(z, t) = u(z, kn^{-1}), \quad kn^{-1} \le t < (k+1)n^{-1}.$$

If the limit exists,

$$\lim_{n \to \infty} \{n[u\left(z, \frac{k}{n}\right) - u\left(z, \frac{k-1}{n}\right)] - \theta(u\left(z, \frac{k}{n}\right))\} = 0,$$

where $\theta(y)$ is a certain function continuous on $[0,1]$, then for the functions $\psi_n(z,t)$, we have

$$\psi_n(z,t) = \varphi(z+a) + \int_0^t \theta(\psi_n(z,y))dy + \varepsilon_n.$$

By solving this functional equation under the condition that $\psi_n(z,1) = \psi(z+\hat{a})$, we can find the G-estimate of the quantity $\varphi(a)$.

From the equations obtained, it follows that the deduction of equations for G-estimates of differentiable functions is considerably simpler than solving these equations with the help of the Fourier transform or quasiinversion method.

§6 *G*-Equations of Higher Orders

Let the Borel functions $f(x)$, $x \in R^{m_n}$ be given, having mixed particular derivatives to order p inclusively; let ξ be a certain m_n-dimensional random vector, $E\xi = a$, $x_1, \ldots, x_n$ independent observations over the vector ξ.

If the condition holds for every $z \in R^{m_n}$, and $k = 1, \ldots, n$,

$$\lim_{n\to\infty} nE \int_0^1 \frac{(1-t)_{p-1}}{(p-1)!} \left(\sum_{i=1}^{m_n} \frac{(x_{ik} - a_i)}{n} \frac{\partial}{\partial z_i} \right)^p$$

$$\times f(z+a+ \sum_{i=1}^{k-1} \frac{x_i - a}{n} + t\frac{x_k - a}{n})dt = 0,$$

$$\sup_{z\in R^{m_n}} E|f(z+a+ \sum_{i=1}^{k-1} \frac{x_i - a}{n})| < \infty,$$

then

$$Ef(z+a+\nu_k) = f(z+a) + n^{-1} \sum_{s=0}^{k} BEf(z+a+\nu_{s-1}) + \varepsilon_n, \tag{23.6.1}$$

$$\nu_k = n^{-1} \sum_{i=1}^{k} (x_i - a),$$

where

$$B = \sum_{l=1}^{p-1} (l!)^{-1} E \left(\sum_{i=1}^{m_n} \frac{x_{is} - a_i}{n} \frac{\partial}{\partial z_i} \right)^l, \quad \lim_{n\to\infty} \varepsilon_n = 0.$$

Let

$$u_n(t,z) = Ef(z+a+\nu_k), \quad kn^{-1} \le t < (k+1)n^{-1}, \quad k = 1, \ldots, n-1.$$

Then Eq. (23.6.1) takes the form

$$u_n(t, z) = f(z + a) + \int_0^t Bu_n(y, z)dy + \varepsilon_n,$$

$$u_n(1, z) = \mathbf{E}f(z + \hat{a}).$$

In order to find the G-estimates of the quantities $f(z+a)$, it is also possible to apply the Fourier transform and quasiinversion method to this equation.

In the next sections, we shall pursue another way. We shall consider certain functions of covariance matrices and find some functional equations with the help of the analytical methods of probability theory without finding differential equations for these functions. The advantage of the functional equations is that we do not use regularization methods to find the G-estimates. This causes a considerable gain in the rate of their convergence to the estimated functions.

§7 G-Equation for the Resolvent of Empirical Covariance Matrices if the Lindeberg Condition Holds

Let R be a covariance matrix of the m_n-dimensional random vector ξ, $\mathbf{E}\xi = a$. The expression

$$\mu_{m_n}(x, R) = m_n^{-1} \sum_{k=1}^{m_n} F(x - \lambda_k)$$

is called the normalized spectral function of the matrix R, where $F(x-\lambda_k) = 1$ if $\lambda_k < x$ and $F(x - \lambda_k) = 0$ if $\lambda_k \geq x$, and λ_k are roots of the characteristic equation

$$\det(Iz - R) = 0.$$

Theorem 23.7.1. *Let $x_1, x_2, \ldots, x_n$ be the observations over the m-dimensional random vector ξ. Suppose this vector has the covariance matrix R,*

$$\xi_i = (\xi_{ij}, j = \overline{1,m})' = HR^{-1/2}(x_i - a), \quad a = \mathbf{E}\xi,$$

where $H = (h_1, \ldots, h_m)$, h_p is an eigenvector corresponding to the eigenvalue λ_p, and the random variables ξ_{ij} are independent,

$$0 < c_1 \leq \lambda_k \leq c_2 < \infty, \tag{23.7.1}$$

$$\lim_{n \to \infty} m_n n^{-1} = c, \quad 0 < c < 1. \tag{23.7.2}$$

Then, in order that for every $t > 0$,

$$\mathrm{plim}_{n \to \infty} \left[\int_0^\infty (t + x)^{-1} d\mu(x, \widehat{R}_{m_n}) - a_{m_n}(t) \right] = 0, \tag{23.7.3}$$

where

$$\hat{R}_m = (n-1)^{-1}\sum_{k=1}^{n}(x_k - \hat{x})(x_k - \hat{x})', \quad \hat{x} = n^{-1}\sum_{k=1}^{n}x_k,$$

and $a_{m_n}(t)$ *are nonnegative analytical functions satisfying the equation*

$$a_{m_n}(t) = \int_0^\infty \{t + [(1 - m_n n^{-1}) + m_n n^{-1} t a_{m_n}(t)]x\}^{-1}d\mu_{m_n}(x, R_{m_n}), \quad t \geq 0,$$
$$(23.7.4)$$

it is sufficient, and in the case of symmetric variables ξ_{ij}, *it is also necessary, that the Lindeberg condition holds, i.e., for every* $\tau > 0$,

$$\lim_{n\to\infty} m_n^{-1}\sum_{i=1}^{m_n}\sum_{j=1}^{n}\mathbf{E}|\xi_{ij}(n-1)^{-1/2}|^2\chi(|\xi_{ij}|(n-1)^{-1/2} > \tau) = 0. \quad (23.7.5)$$

The solution of Eq. (23.7.4) exists and is unique in the class of analytic non-negative functions and can be obtained by means of the method of successive approximations.

Proof. *Sufficiency.* Using the formula

$$\mathrm{Tr}(tI + A + xx')^{-1} - \mathrm{Tr}(It + A)^{-1} = (d/dt)\ln[1 + ((It + A)^{-1}x, x)], \quad (23.7.6)$$

where A is a nonnegative-definite matrix of order m_n, x is an m_n-dimensional vector, and the equality

$$\hat{R} = (n-1)^{-1}\sum_{k=1}^{n}(x_k - a)(x_k - a)' - \frac{n}{n-1}(\hat{x} - a)(\hat{x} - a)', \quad (23.7.7)$$

we have

$$|m_n^{-1}\mathrm{Tr}(It + \hat{R})^{-1} - m_n^{-1}\mathrm{Tr}(It + Q)^{-1}| \leq m_n^{-1}|((It + \hat{R})^{-2}\eta, \eta)|$$
$$\times n(n-1)^{-1}[1 + ((It + \hat{R})^{-1}\eta, \eta)n(n-1)^{-1}]^{-1} \leq m_n^{-1}t^{-1},$$
$$(23.7.8)$$

where $\eta = \hat{x} - a$, $Q = \frac{1}{n-1}\sum_{k=1}^{n}(x_k - a)(x_k - a)'$.
Consequently, for every $t > 0$,

$$\lim_{m\to\infty}\mathbf{E}|\int_0^\infty (t + x)^{-1}d\mu_{m_n}(x, \hat{R}_{m_n}) - m_n^{-1}\mathrm{Tr}(It + Q)^{-1}| = 0.$$

Let us consider the equality

$$m_n^{-1}\mathrm{Tr}(It + Q)^{-1} - m_n^{-1}\mathbf{E}\,\mathrm{Tr}(It + Q)^{-1} = m_n^{-1}\sum_{k=1}^{n}\gamma_k,$$

where

$$\gamma_k = \mathbf{E}_{k-1} \operatorname{Tr}(It+Q)^{-1} - \mathbf{E}_k \operatorname{Tr}(It+Q)^{-1},$$

$\mathbf{E}_k$ is a conditional mathematical expectation for the fixed minimal σ-algebra with respect to which the random vectors x_s, $s = \overline{k+1, n}$ are measurable.

The random variables γ_k are the martingale differences, and therefore, we have for them

$$\mathbf{E}\left(\sum_{k=1}^{n} \gamma_k\right)^2 = \sum_{k=1}^{n} \mathbf{E}\gamma_k^2. \tag{23.7.10}$$

Let

$$Q_k = Q - (x_k - a)(x_k - a)'(n-1)^{-1}.$$

On the basis of (23.7.6), we get the estimation for γ_k,

$$|\gamma_k| \leq |\mathbf{E}_{k-1}\operatorname{Tr}(It+Q)^{-1} - \mathbf{E}_{k-1}\operatorname{Tr}(It+Q_k)^{-1}| + |\mathbf{E}_k \operatorname{Tr}(It+Q)^{-1} - \mathbf{E}_k$$
$$\times \operatorname{Tr}(It+Q_k)^{-1}| \leq (\mathbf{E}_{k-1} + \mathbf{E}_k)|((It+Q_k)^{-1}\eta_k, \eta_k)(n-1)^{-1}$$
$$\times [1 + ((It+Q_k)^{-1}\eta_k, \eta_k)(n-1)^{-1}]^{-1} \leq 2t^{-1},$$

where $\eta_k = x_k - a$. Using this inequality and equality (23.7.10), we have

$$\operatorname{plim}_{n \to \infty} m^{-1}[\operatorname{Tr}(It+Q)^{-1} - \mathbf{E}\operatorname{Tr}(It+Q)^{-1}] = 0. \tag{23.7.11}$$

Let us derive the equation for the function $m_n^{-1}\mathbf{E}\operatorname{Tr}(It+Q)^{-1}$. For this, we introduce the following notations: Let B_p be the matrix obtained by deleting the pth row of the matrix B,

$$B = (\nu_1, \nu_2, \ldots, \nu_n)(n-1)^{-1/2}, \quad \nu_i = (\Lambda m_n)^{1/2}\xi_i,$$
$$\xi_k = H R_{m_n}^{-1/2}\eta_k, \quad \Lambda_{m_n} = \operatorname{diag}(\lambda_i, i = \overline{1, m_n}),$$

where H is a matrix whose column vectors are equal to the eigenvectors of the matrix R_{m_n}, $\Gamma_p = (It + B_p B_p')^{-1}$, and b_p is the row vector of matrix B.

Obviously,

$$\operatorname{Tr}(It+Q)^{-1} = \operatorname{Tr}(It+BB')^{-1}.$$

Therefore, by using the formulas of the matrix perturbations, we have

$$m_n^{-1}\mathbf{E}\operatorname{Tr}(It+Q)^{-1} = m_n^{-1}\sum_{k=1}^{m_n} \mathbf{E}[t + (b_k, b_k) - (B_k'\Gamma_k B_k b_k, b_k)]^{-1}$$

$$= m_n^{-1}\sum_{p=1}^{m_n} \mathbf{E}[t + \lambda_p - \lambda_p(n-1)^{-1}\operatorname{Tr}\Gamma_p B_p B_p' + \varepsilon_{1p}]^{-1},$$

$$\tag{23.7.12}$$

where

$$\varepsilon_{1p} = \sum_{l=1}^{n} b_{pl}^2 - \lambda_p - \{(B_p' \Gamma_p B_p b_p, b_p) - \lambda_p (n-1)^{-1} \operatorname{Tr} \Gamma_p B_p B_p'\}.$$

It follows from Eq. (23.7.12) that

$$m_n^{-1} \mathbf{E} \operatorname{Tr}(It+Q)^{-1} = m_n^{-1} \sum_{p=1}^{m_n} \mathbf{E}[t + \lambda_p(1 - k_n + tn^{-1}\mathbf{E}\operatorname{Tr}(It+Q)^{-1}] + \varepsilon_{2p}]^{-1},$$

$$(23.7.13)$$

where

$$\varepsilon_{2p} = \varepsilon_{1p} + \lambda_p(n-1)^{-1}(\operatorname{Tr}\Gamma_p - (n-1)^{-1}\mathbf{E}\operatorname{Tr}(tI + Q)^{-1}),$$
$$k_n = m_n(n-1)^{-1}.$$

Using the perturbation formulas for the random matrices, we get

$$\delta_n := (n-1)^{-1}[\operatorname{Tr}\Gamma_p - \operatorname{Tr}(tI + Q)^{-1}]$$
$$= (n-1)^{-1}(d/dt)\ln[t + (b_p, b_p) - (B_p'\Gamma_p B_p b_p, b_p)]$$
$$= (n-1)^{-1}(d/dt)\ln[t + ((It + B_p'B_p)^{-1}b_p, b_p)].$$

It is easy to derive

$$|\delta_n| \le (n-1)^{-1}ct^{-1} \qquad (23.7.14)$$

from this equality. Obviously,

$$\left|\frac{\mathbf{E}\operatorname{Tr}(It+Q)^{-1}}{m_n} - a_{m_n}(t)\right|$$

$$\le t^{-2} m_n^{-1} \sum_{p=1}^{m_n} \{t k_n | \frac{\mathbf{E}\operatorname{Tr}(It+Q)^{-1}}{m_n} - a_{m_n}(t)| + \mathbf{E}|\varepsilon_{2p}|.$$

Using the Lindeberg condition (23.7.5), the inequality (23.7.14), and the limit relation (23.7.11), we get for $t > 0$,

$$\lim_{n\to\infty} m_n^{-1} \sum_{p=1}^{m_n} \mathbf{E}|\varepsilon_{2p}| = 0.$$

On the basis of this expression, we have

$$\alpha_{m_n}(t) \le t^{-1}\alpha_{m_n}(t) + 0(1),$$

where

$$\alpha_{m_n}(t) = |m_n^{-1}\mathbf{E}\operatorname{Tr}(It+Q)^{-1} - a_{m_n}(t)|.$$

It follows from this inequality that for sufficiently large t, $\lim_{n \to \infty} \alpha_{m_n}(t) = 0$. But since the functions $\frac{1}{m_n} \mathbf{E} \operatorname{Tr}(It + Q)^{-1} - a_m(t)$ are analytical for $t > 0$, then $\lim_{n \to \infty} \alpha_m(t) = 0$ for all $t > 0$. Consequently, on the basis of (23.7.11), we have that (23.7.4) is valid. The sufficiency of the conditions of Theorem (23.7.1) is proved.

Let us prove the necessity. Since the conditions (23.7.3) and (23.7.4) hold,

$$\lim_{n \to \infty} [\mathbf{E} m_n^{-1} \operatorname{Tr}(It+Q)^{-1} - m_n^{-1} \sum_{p=1}^{m_n} [t + \lambda_p [1 - m_n n^{-1} + t n^{-1} \operatorname{Tr}(it+Q)^{-1}] = 0.$$

(23.7.15)

By using the equality (23.7.12), we have

$$m_n^{-1} \mathbf{E} \operatorname{Tr}(It + Q)^{-1} = m_n^{-1} \sum_{k=1}^{m_n} \mathbf{E}[t + (\{I - B_p' \Gamma_p B_p\} b_p, b_p)]^{-1}$$

$$= m_n^{-1} \sum_{k=1}^{m_n} \mathbf{E}[t + \lambda_p \sum_{i=1}^{n} \theta_{ii} \nu_{pi}^2 n^{-1}]^{-1} + 0(1),$$

(23.7.16)

where θ_{ij} are the entries of the matrix $I - B_p' \Gamma_p B_p$, and ν_{pi} are components of vector $\vec{\nu}_p$.

Since the inequality (23.7.14) is valid, it is possible to change (23.7.15) to the following form

$$\mathbf{E} m_n^{-1} \operatorname{Tr}(It + Q)^{-1} = m_n^{-1} \sum_{p=1}^{m_n} \mathbf{E}[t + \lambda_p n^{-1} \sum_{i=1}^{n} \theta_{ii}]^{-1} 0(1).$$

(23.7.17)

If the matrix $B_p B_p'$ is nonsingular, then the matrix B_p can be represented in the following form,

$$B_p = (B_p B_p')^{1/2} H, \qquad H = (B_p B_p')^{-1/2} B_p.$$

Since the matrix H is orthogonal,

$$I - B_p' \Gamma_p B_p = H'(I - (I + t(B_p B_p')^{-1})^{-1})H = t(It + B_p' B_p)^{-1}.$$

If the matrix $B_p B_p'$ is degenerate, then we carry out the proof with the help of the condition of continuity. We replace the matrix $B_p B_p'$ by the matrix $B_p B_p' + \varepsilon I$, $\varepsilon > 0$ and pass to the limit as $\varepsilon \downarrow 0$ in the final form. Consequently, the variables θ_{ii} are nonnegative, since the matrix $B_p B_p'$ is nonnegative-definite.

By using (23.7.16) and (23.7.17), we have for $t > 0$,

$$\lim_{n \to \infty} m_n^{-1} \sum_{p=1}^{m_n} \mathbf{E}\{[1 + t^{-1} n^{-1} \sum_{i=1}^{n} \theta_{ii} \nu_{pi}^2]^{-1} - [1 + \lambda_p n^{-1} t^{-1} \sum_{i=1}^{n} \theta_{ii}]^{-1}\} = 0.$$

(23.7.18)

Obviously,

$$\mathbf{E}[1 + n^{-1}t^{-1}\sum_{i=1}^{n}\theta_{ii}\nu_{pi}^2]^{-1} = \mathbf{E}\exp\{-\gamma n^{-1}t^{-1}\sum_{s=1}^{n}\theta_{ss}\nu_{ps}^2\},$$

where γ is the random variable with the distribution density $\exp\{-x\}$, $x \geq 0$. We get from this equality,

$$\mathbf{E}[1 + t^{-1}\sum_{i=1}^{n}\theta_{ii}n^{-1}\nu_{pi}^2]^{-1} = \mathbf{E}\exp\left\{i\sqrt{\gamma\lambda_p t^{-1}n^{-1}}\sum_{i=1}^{n}\xi_{pi}\beta_i\theta_{ii}^{1/2}\right\},$$

where β_s are random variables distributed according to the normal law and independent of the random variables γ, ξ_{ps}. Let

$$\alpha_{pi} = \mathbf{E}[\exp\{i\sqrt{\gamma\lambda_p t^{-1}}\xi_{pi}\beta_i\theta_{ii}^{1/2}n^{-1/2}\} - 1/\gamma].$$

Let us prove that

$$\lim_{n\to\infty}\mathbf{E}\sum_{i=1}^{n}\sum_{p=1}^{m_n}|\alpha_{pi}|^2 = 0. \qquad (23.7.19)$$

To do this, we consider the inequalities

$$|\alpha_{pi}| \leq \sqrt{\gamma\lambda_p}t^{-1/2}\varepsilon + 2\mathbf{P}\{\xi_{pi}^2 n^{-1} > \varepsilon^2\},$$
$$|\alpha_{pi}| \leq \gamma\lambda_p t^{-1}n^{-1}\theta_{ii}\beta_i^2.$$

By using these inequalities, we get

$$\lim_{n\to\infty}\mathbf{E}\sum_{p=1}^{n}\sum_{i=1}^{n}|\alpha_{pi}|^2 = 0.$$

Consequently,

$$m_n^{-1}\sum_{p=1}^{m_n}\mathbf{E}[1 + t^{-1}\sum_{i=1}^{n}\theta_{ii}n^{-1}\nu_{pi}^2]^{-1} = m_n^{-1}\sum_{p=1}^{m_n}\exp\left\{\sum_{i=1}^{n}\alpha_{pi}\right\} + 0(1).$$
$$(23.7.20)$$

By using (23.7.18) and (23.7.20), we find

$$\lim_{n\to\infty}m_n^{-1}\sum_{p=1}^{m_n}\left[\mathbf{E}\exp\left\{\sum_{i=1}^{n}\alpha_{pi}\right\} - \mathbf{E}\exp\left\{-\gamma t^{-1}\lambda_p\sum_{i=1}^{n}\beta_i^2\theta_{ii}n^{-1}\right\}\right] = 0.$$

Since the expression in square brackets is nonnegative,

$$\lim_{n\to\infty}\mathbf{E}c_m m_n^{-1}\sum_{p=\overline{1,m_n},i=\overline{1,n}}\{\alpha_{pi} + \gamma t^{-1}\lambda_p\beta_i^2\theta_{ii}n^{-1}\} = 0, \qquad (23.7.21)$$

where

$$c_n = \exp\{-\gamma t^{-1}\lambda_p \sum_{i=1}^{n} \beta_i^2 \theta_{ii} n^{-1}\}.$$

It is obvious, if $t > 0$, that

$$\underline{\text{plim}}_{n \to \infty} c_n > 0.$$

Therefore, from the equality (23.7.21), we get

$$\lim_{n \to \infty} m_n^{-1} \mathbf{E} \sum_{p=\overline{1,m_n}, i=\overline{1,n}} \{\alpha_{pi} + \gamma t^{-1}\theta_{ii}n^{-1}\} = \lim_{n \to \infty} m_n^{-1}$$

$$\times \mathbf{E} \sum_{p=\overline{1,m_n}, i=\overline{1,n}} [(1 + t^{-1}\lambda_p \xi_{pi}^2 n^{-1}\theta_{ii})^{-1} - 1 + t^{-1}\lambda_p \theta_{ii}n^{-1}] = 0.$$

From this equality, we have

$$\lim_{n \to \infty} m_n^{-1} \mathbf{E} \sum_{p,i} (t^{-1}\lambda_p \xi_{pi}^2 \theta_{ii}n^{-1})^2 (1 + t^{-1}\lambda_p \xi_{pi}^2 \theta_{ii}n^{-1})^{-1} = 0.$$

In this case, it follows from this expression that

$$\lim_{n \to \infty} m_n^{-1} \sum_{p,i} \mathbf{E}\xi_{pi}^2 n^{-1} \chi(\xi_{pi}^2 n^{-1} > \varepsilon^2) t^{-2} \lambda_p^2 \theta_{ii}^2 (\varepsilon^{-2} + t^{-1}\lambda_p \theta_{ii}) = 0.$$

Since the quantities λ_p satisfy the inequality (23.7.2) and the vectors $\vec{\xi}_p$ are uniformly distributed, we have

$$\lim_{n \to \infty} \sum_{p=1}^{m_n} n^{-1} \mathbf{E}\xi_{p1}^2 \chi(\xi_{p1}^2 n^{-1} > \varepsilon^2) m_n^{-1} \sum_{i=1}^{n} t^{-2} c_2^2 \theta_{ii}^2 (\varepsilon^{-2} + t^{-1}c_2\theta_{ii})^{-1} = 0.$$

Note that θ_{pp} are the entries of the matrix $t(It + B_p' B_p)^{-1}$. Hence, the variables θ_{pp} have satisfied the inequality $\theta_{pp} \leq 1$. Therefore, from (23.7.22), we get

$$\lim_{n \to \infty} \sum_{p=1}^{m_n} \mathbf{E}\xi_{p1}^2 n^{-1} \chi(\xi_{p1}^2 n^{-1} > \varepsilon^2) m_n^{-1} \mathbf{E} \operatorname{Tr}(It + BB')^{-1} = 0.$$

Since (23.7.3) is fulfilled and the function $a_n(t) > 0$ as $t > 0$, this implies the condition of (23.7.5). Theorem 23.7.1 is proved.

We note that under the conditions of the theorem proved, it is possible to get rid of the independence of the vector components $HR^{-1/2}(x_k - a)$. For this purpose, for any nonnegative m-dimensional matrix,

$$\text{plim}_{n \to \infty} [((tI + A)^{-1}\tilde{\xi}, \tilde{\xi}) - \operatorname{Tr} R(It + A)^{-1}]n^{-1} = 0 \qquad (23.7.23)$$

will be necessary as $t > 0$, $\tilde{\xi} = \xi - a$.

Under this condition, we make the following supplementary transformations in the proof of Theorem 23.7.1. Let us introduce the matrices

$$Q_k = (n-1)^{-1} \sum_{s=1}^{k} R_{m_n} Q_s Q_s' + (n-1)^{-1} \sum_{s=k+1}^{n} (x_s - a)(x_s - a)',$$

where Q_s, $s = 1, 2, \ldots$ are random independent vectors distributed according to the normal law $N(0, I)$ and independent of the random vectors x_s, $Q_0 = Q$, $p_k = Q_k - (n-1)^{-1} R_{m_n} \theta_k \theta_k'$. Further, by using the expression (23.7.6), we have,

$$\mathrm{Tr}(It + Q)^{-1} - \mathrm{Tr}(It + Q_n)^{-1}$$

$$= \sum_{k=1}^{n} \mathrm{Tr}(It + Q_{k-1})^{-1} - \mathrm{Tr}(It + Q_k)^{-1}$$

$$= \sum_{k=1}^{n} \{ [\mathrm{Tr}(It + Q_{k-1})^{-1} - \mathrm{Tr}(It + p_k)^{-1}]$$

$$\quad - [\mathrm{Tr}(It + Q_k)^{-1} - \mathrm{Tr}(It + p_k)^{-1}]$$

$$= \sum_{k=1}^{n} \{ (d/dt) \ln[1 + (n-1)^{-1}((It + p_k)^{-1}(x_k - a), (x_k - a))]$$

$$\quad - (d/dt) \ln[1 + (n-1)^{-1}((It + p_k)^{-1} R_m Q_k, Q_k)] \}.$$

Therefore, we are using (23.7.23) and the expression

$$\mathrm{plim}_{n \to \infty}(n-1)^{-1}[((It + p_k)^{-1} R_{m_n} Q_k, Q_k) - \mathrm{Tr}\, R_{m_n}(It + p_k)^{-1}] = 0.$$

From this equality, we have,

$$\lim_{n \to \infty} \mathbf{E} m_n^{-1}[\mathrm{Tr}(It + Q)^{-1} - \mathrm{Tr}(It + Q_n)^{-1}] = 0, \qquad (23.7.29)$$

and for the expressions of $m_n^{-1}\, \mathrm{Tr}(It + Q_n)^{-1}$, Theorem 23.7.1 holds.

§8 G-Equation for the Stieltjes Transformation of Normal Spectral Functions of the Empirical Covariance Matrices Beam

Let R_1 and R_2 be nonsingular covariance matrices of the independent m-dimensional random vectors ξ_1 and ξ_2, $a_1 = \mathbf{E}\xi_1$, $a_2 = \mathbf{E}\xi_2$.

The expression

$$\mu_n(x_1, R_1, R_2) = m^{-1} \sum_{k=1}^{m} F(x - \lambda_k)$$

is called the normalized spectral function of the covariance R_1 and R_2 beam, where $F(x - \lambda_k) = 1$ if $\lambda_k < x$, and $F(x - \lambda_k) = 0$ if $\lambda_k \geq x$; λ_k are roots of the characteristic equation

$$\det(R_1 z - R_2) = 0, \quad 0 < d_1 \leq \lambda_k \leq d_2 < \infty.$$

Theorem 23.8.1.   *Let $x_1, \ldots, x_{n_1}$, $y_1, \ldots, y_{n_2}$ be observations of the random vectors ξ_1 and ξ_2,*

$$\xi_i = (\xi_{ij}, j = \overline{1, m})' = R_1^{-1/2}(x_i - a_1),$$
$$\eta_i = (\eta_{ij}, j = \overline{1, m})' = R_2^{-1/2}(y_i - a_2),$$

let the random variables ξ_{ij}, η_{ij}, $i, j = 1, 2, \ldots$ be independent,

$$\lim_{m \to \infty} \frac{m}{n_1} = c_1, \quad \lim_{n \to \infty} \frac{m}{n_2} = c_2, \quad c_1^{-1} + c_2^{-1} < 1, \quad c_1 \neq 1, \qquad (23.8.1)$$

and let the Lindeberg condition be fulfilled, i.e., we have

$$\lim_{n \to \infty} \left[\sum_{i=1}^{m} \mathbf{E}|\xi_{i1}(n_1 - 1)^{-1/2}|^2 \chi(|\xi_{i1}|(n_1 - 1)^{-1/2} > \tau) \right.$$
$$\left. + m_n^{-1} \sum_{i=1}^{m} \mathbf{E}|\eta_{i1}(n_2 - 1)^{-1/2}|^2 \chi(|\eta_{i1}|(n_2 - 1)^{-1/2} > \tau) \right] = 0,$$
$$(23.8.2)$$

for every $\tau > 0$. Then

$$\text{plim}_{m \to \infty} \left[\int_0^\infty (t + x)^{-1} d\mu_m(x, \hat{R}_1, \hat{R}_2) - a_m(t) \right] = 0, \quad t > 0, \qquad (23.8.3)$$

where the function $a_m(t)$, $t > 0$ is equal to

$$a_m(t) = \int_\infty^0 (\frac{\partial}{\partial t}) b_m(t, x) dx,$$

and the function $b_m(t, x)$ satisfies the equation

$$b_m(t, \alpha) = \int_0^\infty [\alpha + t(1 + tc_1 b_m(t, \alpha))^{-1} + x[1 - c_2 + \alpha c_2$$
$$\times b_m(t, \alpha)(\alpha + tc_1 b_m(t, \alpha))^{-1}]]^{-1} d\mu_m(x, R_1, R_2).$$

The solution of the equation for the function $b_m(t, \alpha) > 0$ exists and is unique in the class of function analytic on $t, t > 0$.

Proof. Let us consider the Stieltjes transformation

$$\int_0^\infty (t+x)^{-1} d\mu_m(x, \widehat{R}_1, \widehat{R}_2) = m^{-1}\,\mathrm{Tr}\,\widehat{R}_1 R_1^{-1}[R_1^{-1/2}\widehat{R}_1$$
$$\times R_1^{-1/2}t + R_1^{-1/2}\widehat{R}_2 R_1^{-1/2}]^{-1}, \quad t > 0. \tag{23.8.4}$$

It is evident that for any $\alpha > 0$,

$$|m^{-1}\,\mathrm{Tr}\,\widehat{R}_1[\widehat{R}_1 t + \widehat{R}_2]^{-1} - m^{-1}\,\mathrm{Tr}\,R_1^{-1}\widehat{R}_1[R_1^{-1/2}\widehat{R}_1 R_1^{-1/2}t$$
$$+ R_1^{-1/2}\widehat{R}_2 R_1^{-1/2} + I_\alpha]^{-1}| \le t^{-1}\alpha m^{-1}\,\mathrm{Tr}(I_\alpha$$
$$+ R_1^{-1/2}\widehat{R}_1 R_1^{-1/2}t + R_1^{-1/2}\widehat{R}_2 R_1^{-1/2})^{-1}, \quad \left(\frac{0}{0}=0\right). \tag{23.8.5}$$

We need the following auxiliary statements.

Lemma 23.8.1. *If the conditions of Theorem 23.8.1 are fulfilled, then for any* $\alpha > 0$,

$$\mathrm{plim}_{n\to\infty}\, m^{-1}\{\mathrm{Tr}[I_\alpha + R_1^{-1/2}\widehat{R}_1 R_1^{-1/2}t + R_1^{-1/2}\widehat{R}_2 R_1^{-1/2}]^{-1}$$
$$- \mathbf{E}\,\mathrm{Tr}[I_\alpha + t\sum_{k=1}^{n_1}\xi_k\xi_k'(n_1-1)^{-1} + \sum_{k=1}^{n_2} R_3\eta_k\eta_k'$$
$$\times R_3'(n_2-1)^{-1}]^{-1}\} = 0, \quad R_3 = R_1^{-1/2} R_2^{1/2}. \tag{23.8.6}$$

Proof. Using formula (23.7.6), we have

$$|\,\mathrm{Tr}(I\alpha + R_1^{-1/2}\widehat{R}_1 R_1^{-1/2}t + R_1^{-1/2}\widehat{R}_2 R_1^{-1/2})^{-1} - \mathrm{Tr}\,Q^{-1}|$$
$$\le (Q^{-2}(\hat{x}_1-a_1),(\hat{x}_1-a_1))(n_1-1)^{-1}n_1[1+(n_1-1)^{-1}n_1$$
$$\times (Q^{-1}(\hat{x}_1-a_1),(\hat{x}_1-a_1))]^{-1},$$

where

$$Q = I\alpha + t\sum_{k=1}^{n_1}\xi_k\xi_k'(n_1-1)^{-1} + R_1^{-1/2}\widehat{R}_2 R_1^{-1/2},$$
$$|\,\mathrm{Tr}\,Q^{-1} - \mathrm{Tr}\,U^{-1}| \le (R_3 U^{-2} R_3(\hat{x}_2-a_2),(\hat{x}_2-a_2))(n_2-1)^{-1}n_2$$
$$\times [1 + (R_3 U^{-1} R_3(\hat{x}_2-a_2),(\hat{x}_2-a_2))(n_2-1)^{-1}n_2]^{-1},$$
$$U = I\alpha + t\sum_{k=1}^{n_1}\xi_k\xi_k'(n_1-1)^{-1} + \sum_{k=1}^{n_2} R_3\eta_k\eta_k'(n_2-1)^{-1}R_3.$$

Using these two inequalities, we get (23.8.6). Lemma 23.8.1 is proved.

Let
$$b_m(t, \alpha) = m^{-1} \mathbf{E} \operatorname{Tr} U^{-1}.$$

Lemma 23.8.2. *The function $b_m(t, \alpha)$ satisfies the equation*

$$b_m(t, \alpha) = m^{-1} \sum_{k=1}^{m} [\alpha + t(1 + tc_1 b_m(t, \alpha))^{-1} + \lambda_k$$
$$\times [1 - c_2 + \alpha b_m(t, \alpha) c_2 (\alpha + t(1 + tc_1 b_m(t, \alpha))^{-1})]]^{-1} + 0(1).$$
$$(23.8.7)$$

Proof. It follows from the matrix theory that $R_3 = T_1 \Lambda^{1/2} T_2'$, where T_1, T_2 is the orthogonal matrix, and Λ is a diagonal matrix of eigenvalues of the matrix $(R_3 R_3')^{1/2}$. Then [see (23.7.24)],

$$b_m(t, \alpha) = \mathbf{E} m^{-1} \operatorname{Tr}(I\alpha + t \sum_{k=1}^{n_1} \xi_k \xi_k' (n_1 - 1)^{-1}$$
$$+ \sum_{k=1}^{n_2} \sqrt{\Lambda} \eta_k \eta_k' (n_2 - 1)^{-1} \sqrt{\Lambda})^{-1} + 0(1).$$

Let us introduce the notations

$$\nu_{ij} = \sqrt{t} \xi_{ij} (n_1 - 1)^{-1/2}, \quad i = \overline{1, m}, \quad j = \overline{1, n_1};$$
$$\nu_{ij} = \lambda_i^{1/2} \eta_{ij} (n_2 - 1)^{-1/2}, \quad i = \overline{1, m}, \quad j = \overline{n_1 + 1, n_1 + n_2};$$
$$c = (\nu_{ij}), \quad i = \overline{1, m}, \quad j = \overline{1, n_1 + n_2}.$$

Using the formula

$$\det T_n = \det T_{n-1}(a_{11} - (T_{n-1}^{-1} a_1, b_1)),$$

where $T_n = (a_{ij})_{i,j=1}^{n}$ are nondegenerate matrices, $a_1 = (a_{11}, \dots, a_{1n})$, $b_1 = (a_{21}, \dots, a_{2n})$, we have

$$m^{-1} \mathbf{E} \operatorname{Tr}(I\alpha + CC')^{-1} = m^{-1} \sum_{k=1}^{m} \mathbf{E}[\alpha + (d_k, d_k) - (C_k' R_k C_k d_k, d_k)]^{-1}$$
$$= m^{-1} \sum_{k=1}^{m} \mathbf{E}[\alpha + \sum_{l=1}^{n_1+n_2} \mathbf{E}\nu_{kl}^2 - \sum_{l=1}^{n_1+n_2} \nu_{kl}^2 (\sum_{i,j=1}^{m} r_{ij}^k \nu_{il} \nu_{jl}) + \varepsilon_k^{(n)}]^{-1},$$

where $d_k = (\nu_{k1}, \dots, \nu_{k(n_1+n_2)})$ is the kth row vector of the matrix C; C_k is a matrix obtained from matrix C by deleting the kth row,

$$R_k = (I\alpha + C_k C_k')^{-1} = (r_{ij}^k)_{i,j=1}^{m},$$

$$\varepsilon_k^{(n)} = \sum_{l=1}^{n_1+n_2} (\nu_{kl}^2 - \mathbf{E}\nu_{kl}^2) - (C_k R_k C_k' d_k, d_k) + \sum_{l=1}^{n_1+n_2} \nu_{kl}^2 \left(\sum_{i,j \neq k} r_{ij}^k \nu_{il} \nu_{jl} \right).$$

Further, by using the proof of Theorem 23.7.1, we have

$$\mathbf{Er}_{kk} = \left[\alpha + \sum_{l=1}^{n_1+n_2} \mathbf{E}\nu_{kl}^2(1 + \sum_{i=1}^{m} \mathbf{Er}_{ii}\mathbf{E}\nu_{il}^2)^{-1}\right]^{-1} + 0(1), \qquad (23.8.8)$$

where r_{ii} are the entries of the matrix $(I\alpha + CC')^{-1}$. From the expression (23.8.8), we get

$$\mathbf{Er}_{kk} = [\alpha + \sum_{l=1}^{n_1} tn^{-1}(1 + tn_1^{-1}\sum_{i=1}^{m}\mathbf{Er}_{ii})^{-1} + \sum_{l=n_1+1}^{n_1+n_2} \lambda_k n_2^{-1}$$

$$\times (1 + \sum_{i=1}^{m}\lambda_i n_2^{-1}\mathbf{Er}_{ii})^{-1}]^{-1} + 0(1) = [\alpha + t(1 + tmn_1^{-1}$$

$$\times b_m(t,\alpha))^{-1} + \lambda_k(1 + \sum_{i=1}^{m}\lambda_i n_2^{-1}\mathbf{Er}_{ii})^{-1}]^{-1} + 0(1).$$

Let

$$a = m^{-1}\sum_{k=1}^{m}\lambda_k \mathbf{Er}_{kk}$$

$$= m^{-1}\sum_{k=1}^{m}\lambda_k[\alpha + t(1 + tmn_1^{-1}b_m(t,\alpha))^{-1} + \lambda_k(1 + \sum_{i=1}^{m}\lambda_i n_2^{-1}\mathbf{Er}_{ii})^{-1}]^{-1}.$$

Then

$$a = m^{-1}\sum_{k=1}^{m}\lambda_k[\alpha + t(1 + tmn_1^{-1}b_m(t,\alpha))^{-1} + \lambda_k(1 + mn_2^{-1}a)^{-1}]^{-1};$$

besides,

$$b_m = m^{-1}\sum_{k=1}^{m}[\alpha + t(1 + mn_1^{-1}b_m(t,\alpha))^{-1} + \lambda_k(1 + mn_2^{-1}a)^{-1}]^{-1}.$$

Then for a, we have the equation

$$(1 + c_2 a)^{-1} = 1 - c_2 + bc_2[\alpha + t(1 + tc_1 b)^{-1}].$$

Then

$$b = m^{-1}\sum_{k=1}^{m}[\alpha + t(1 + tc_1 b)^{-1} + \lambda_k[1 - c_2 + bc_2(\alpha + t(1 + tc_1 b)^{-1})]^{-1}.$$

As in the previous section, we have

$$\operatorname*{plim}_{m \to \infty} [m_n^{-1} \operatorname{Tr} Q - m_n^{-1} \mathbf{E} \operatorname{Tr} Q] = 0. \tag{23.8.9}$$

Let us show that, as $t > 0$ and $c_1 \neq 1$,

$$\lim_{\alpha \downarrow 0} \lim_{m \to \infty} \alpha b_m(t, \alpha) = 0. \tag{23.8.10}$$

From Lemma (23.8.2), we have that

$$0 \leq b_m(t, \alpha) \leq [\alpha + t[1 + tc_1 b_m(t, \alpha)]^{-1}]^{-1} + 0(1).$$

By solving this inequality with respect to $b_m(t, \alpha)$, we get

$$0 \leq b_m(t, \alpha) \leq 2[\sqrt{(t(1 - c_1) + \alpha)^2 + 4tc_1\alpha} + t(1 - c_1) + \alpha]^{-1} + 0(1).$$

Therefore, as $c_1 \neq 1$, we have

$$\lim_{\alpha \downarrow 0} \lim_{n \to \infty} \alpha b_m(t, \alpha) = 0.$$

It is evident that

$$m^{-1} \operatorname{Tr} R_1^{-1} \widehat{R}_1 [R_1^{-1/2} \widehat{R}_1 R_1^{-1/2} t + R_1^{-1/2} \widehat{R}_2 R_1^{-1/2} + I\alpha]^{-1}$$
$$= \lim_{\beta \to \infty} \int_\beta^\alpha \left(\frac{\partial}{\partial t}\right) b_m(t, x) dx. \tag{23.8.11}$$

Therefore, by using this equality, Lemma 23.8.2, and (23.8.9)–(23.8.11), we have the statement of Theorem 23.8.1.

Theorem 23.8.2. *Let the entries of the random matrices* $A = (\xi_{ij}^{(n)})$, $i = \overline{1, m_n}$, $j = \overline{1, p_n}$, $B = (\eta_{ij}^{(k)})$, $i = \overline{1, m_n}$, $j = \overline{1, q_n}$ *be independent for each* n, *given on the same probability space,* $\mathbf{E}\xi_{ij}^{(n)} = \mathbf{E}\eta_{ij}^{(n)} = 0$, $\operatorname{Var} \xi_{ij}^{(n)} = \operatorname{Var} \eta_{ij}^{(n)} = 1$;

$$\lim_{n \to \infty} p_n m_n^{-1} = c_1, \quad \lim_{n \to \infty} q_n m_n^{-1} = c_2, \quad c_1 + c_2 > 1, \tag{23.8.12}$$

let $\mu_n(x)$ *be the normalized spectral function of the matrices* AA' *and* BB' *beam, and let the Lindeberg condition hold for any* $\tau > 0$,

$$\lim_{n \to \infty} m_n^{-2} \sum_{i=1}^{m_n} \left\{ \sum_{j=1}^{p_n} \mathbf{E} \left[\xi_{ij}^{(n)}\right]^2 \chi(|\xi_{ij}^{(n)}| m_n^{-1/2} > \tau) \right.$$
$$\left. + \sum_{j=1}^{q_n} \mathbf{E} \left[\eta_{jj}^{(n)}\right]^2 \chi(|\eta_{ij}^{(n)}| m_n^{-1/2} > \tau) \right\} = 0.$$

Then with probability 1,

$$\lim_{n \to \infty} \mu_{m_n}(x) = \mu_1(x) + \mu_2(x),$$

$$\mu_1'(x) = \delta(x)(1 - c_2)\chi(c_2 < 1), \tag{23.8.13}$$

where

$$\mu_2'(x) = \begin{cases} [4x(c_1 - 1 + c_2) - (x(1 - c_1) + c_2 - 1)^2]^{1/2} & \\ \quad \times (2\pi x(1 + x))^{-1}, & \gamma_1 \le x \le \gamma_2; \\ 0, & x \in (\gamma_1, \gamma_2); \end{cases}$$

$$\gamma_{1,2} = (c_1 - 1)^{-2}[c_1 c_2 + c_1 + c_2 - 1 \pm 2[(c_1 + c_2 - 1)c_1 c_2]^{1/2}],$$

if $c_1 \ne 1$,

$$\mu_2'(x) = \begin{cases} [4xc_2 - (c_2 - 1)^2]^{1/2}(2\pi x(1 + x))^{-1}, & x > (4c_2)^{-1}(c_2 - 1)^2; \\ 0, & 0 \le x < (c_2 - 1)^2(4c_2)^{-1}; \end{cases}$$

if $c_1 = 1$; $\mu_2'(x) = 0$, if $c_1 = 0$.

Proof. Let us consider the Stieltjes transformation

$$\int_0^\infty (t + x)^{-1} d\mu_{m_n}(x) = m_n^{-1} \operatorname{Tr} m_n^{-1} AA'[AA'm_n^{-1}t + m_n^{-1}BB']^{-1}, \quad t > 0.$$

By using Theorem (23.8.1), we obtain

$$\lim_{n \to \infty} m_n^{-1} \mathbf{E} \operatorname{Tr} AA'm_n^{-1}(m_n^{-1}tAA' + m_n^{-1}BB' + I\alpha)^{-1} = \int_\infty^\alpha \left(\frac{\partial}{\partial t}\right) a(t, x)dx, \tag{23.8.14}$$

where the function $a(t, \alpha)$ satisfies the equation

$$a(t, \alpha) = [\alpha + c_1 t(1 + ta(t, \alpha))^{-1} + c_2(1 + a(t, \alpha))^{-1}].$$

Lemma 23.8.3.

$$\lim_{\alpha \downarrow 0} \lim_{\beta \to \infty} \left(\frac{\partial}{\partial t}\right) \int_\beta^\alpha a(t, x)dx = -c_1 a(t, 0)(1 + ta(t, 0))^{-1}, \tag{23.8.15}$$

where

$$a(t, 0) = 2[t(c_1 - 1) + c_2 - 1 + ((t(c_1 - 1) + c_2 - 1)^2 + 4t(c_1 + c_2 - 1))^{1/2}]^{-1}.$$

Proof. In the integral

$$f(\alpha, \beta) := \left(\frac{\partial}{\partial t}\right) \int_\alpha^\beta a(t, x)dx,$$

we make the change of variables

$$x = y^{-1} - c_1 t(1 + ty)^{-1} - c_2(1 + y)^{-1}, \quad y > 0.$$

Then

$$f(\alpha, \beta) = \left(\frac{\partial}{\partial t}\right) \int_{a(t,\beta)}^{a(t,\beta)} y\left(\frac{d}{dy}\right)[y^{-1} - c_1 t(1 + ty)^{-1} - c_2(1 + y)^{-1}] dy$$

$$= \left(\frac{\partial}{\partial t}\right) y[y^{-1} - c_2 t(1 + ty)^{-1} - c_2(1 + y)^{-1}]_{a(t,\beta)}^{a(t,\alpha)}$$

$$- \left(\frac{\partial}{\partial t}\right) \int_{a(t,\beta)}^{a(t,\alpha)} [y^{-1} - c_1 t(1 + ty)^{-1} - c_2(1 + y)^{-1}] dy.$$

$$(23.8.16)$$

Using the equality

$$a(t,\alpha)^{-1} - c_2 t(1 + ta(t,\alpha))^{-1} - c_2(1 + a(t,\alpha))^{-1} = \alpha$$

and (23.8.16), we obtain

$$f(\alpha, \beta) = c_1 t^{-1}[(1 + ta(t,\alpha))^{-1} - (1 + ta(t,\beta))^{-1}].$$

By passing to the limit as $\alpha \to 0$, $\beta \to \infty$, we obtain (23.8.15).

Lemma 23.8.3 is proved. Since the function $a(t,0)$ is analytic as $t > 0$ and the functions $\mu_{m_n}(x)$ converge with probability 1 to μ, we obtain the equation for the limiting deterministic function $\mu(x)$ continuing analytically the function $-c_1 a(t,0) \times [1 + a(t,0)]^{-1}$ on the whole analytical plane

$$\int_0^\infty (x - z)^{-1} d\mu(x) = -c_1 a(-z, 0)(1 - za(-z, 0))^{-1}, \quad \text{Im } z \neq 0. \quad (23.8.17)$$

By the inverse formula for the Stieltjes transform, we find from (23.8.17),

$$\mu(x_1) - \mu(x_2) = -\pi^{-1} \lim_{\varepsilon \downarrow 0} \int_{x_2}^{x_1} \text{Im}[c_1 a(-x - i\varepsilon, 0)(1 - (x + i\varepsilon)a(-x - i\varepsilon, 0))^{-1}] dx.$$

$$(23.8.18)$$

Let $a(-z, 0) = m(z)$. Then

$$m(z) = [-c_1 z(1 - zm(z))^{-1} + c_2(1 + m(z))^{-1}]^{-1},$$

$$m(z) = 2[-z(c_1 - 1) + c_2 - 1$$

$$\pm \{-z(c_1 - 1) + c_2 - 1\}^2 - 4z[c_1 - 1 + c_2]\}^{1/2}]^{-1},$$

where the '+' or '−' sign is chosen in such a way that the function $m(z)$ is the Stieltjes transform.

Therefore,

$$\int_0^\infty (z - x)^{-1} d\mu(x) = -2c_1[-z(c_1 + 1) + c_2 - 1$$

$$+ \{[-z(c_1 - 1) + c_2 - 1]^2 - 4z[c_1 - 1 + c_2]\}^{1/2}]^{-1}.$$

By using formula (23.8.18), we get (23.8.13). Theorem 23.8.2 is proved.

§9 G_1-Estimate of Generalized Variance

Let the independent observations $x_1, \ldots, x_m$ over the m_n-dimensional random vector ξ, $n > m_n$ be given,

$$\widehat{R} := (n-1)^{-1} \sum_{k=1}^{n} (x_k - \hat{x})(x_k - \hat{x})', \qquad \hat{x} = n^{-1} \sum_{k=1}^{n} x_k.$$

The expression $\det R$ is called a generalized variance. If the vectors x_i, $i = \overline{1, n}$ are independent and distributed according to the multidimensional normal law $N(a, R)$, then (see Chapter 2, §1),

$$\det \widehat{R} \approx \det R(n-1)^{-m} \prod_{i=n-m}^{n-1} \chi_i^2,$$

where χ_i^2 are independent random variables distributed according to the χ^2-law with i degrees of freedom. In the general case, the distribution of $\det \widehat{R}$ is inconvenient, and therefore finding the G-estimates for $\det R$ is a very complicated problem. Let us prove that under certain conditions the G-estimates for the variables $c_n^{-1} \ln \det R$, where c_n is such a sequence of constants that $\lim_{n \to \infty} c_n^{-2} \ln n(n - m_n)^{-1} = 0$, can be represented in the form

$$G_1(\widehat{R}) := c_n^{-1} \{ \ln \det \widehat{R} + \ln[(n-1)^m (A_{n-1}^m)^{-1} n(n - m_n)^{-1}] \},$$

where $A_{n-1}^m = (n-1) \ldots (n-m)$.

Theorem 23.9.1. *For every value $n > m_n$, let the random m_n-dimensional vectors $x_1^{(n)}, \ldots, x_n^{(n)}$ be independent and identically distributed with a mean vector $\vec{a}$ and nondegenerate covariance matrices R_{m_n}, for certain $\delta > 0$,*

$$\sup_n \sup_{i = \overline{1,n}, j = \overline{1,m_n}} \mathbf{E} |\tilde{x}_{ij}^{(n)}|^{4+\delta} < \infty, \qquad (23.9.1)$$

where $\tilde{x}_{ij}^{(n)}$ are vector components $\tilde{x}_i = R_{m_n}^{-1/2}(x_i^{(n)} - a)$,

$$\lim_{n \to \infty} n - m_n = \infty, \quad \lim_{n \to \infty} nm_n^{-1} = 1; \qquad (23.9.2)$$

and for each value of $n > m_n$, the random variables $x_{ij}^{(n)}$, $i = \overline{1, n}$, $j = \overline{1, m_n}$ are independent.
    Then

$$\mathrm{plim}_{n \to \infty}[G_1(\widehat{R}_{m_n}) - c_n^{-1} \ln \det R_{m_n}] = 0. \qquad (23.9.3)$$

If in addition to the conditions of Theorem 23.9.1,

$$\mathbf{E}(\tilde{x}_{ij}^{(n)})^4 = 3, \quad i = \overline{1, n}, \quad j = \overline{1, m_n}, \qquad (23.9.4)$$

then

$$\lim_{n\to\infty} \mathbf{P}\{(c_n G_1(\widehat{R}_{m_n}) - \ln\det R_{m_n})(-2\ln(1 - m_n n^{-1}))^{-1/2}$$

$$< x\} = (2\pi)^{-1/2} \int_{-\infty}^{x} e^{-y^2/2} dy. \tag{23.9.5}$$

Proof. Since

$$\widehat{R} = (n-1)^{-1}\left[\sum_{k=1}^{n} x_k x_k' - n\hat{x}\hat{x}'\right],$$

we represent the matrix $\widehat{R}$ in the following form,

$$\widehat{R} = \sqrt{R}\sum_{k=1}^{n-1} z_k z_k'(n-1)^{-1}\sqrt{R},$$

where $z_k = \sum_{i=1}^{n} h_{ik}\tilde{x}_i$, h_{ik} are the entries of a real orthogonal matrix, $h_{in} = n^{-1/2}$.

Then

$$G_1(\widehat{R}) - c_n^{-1}\ln\det R_{m_n}$$

$$= \left[\ln\det\sum_{k=1}^{n-1} z_k z_k'(A_{n-1}^m)^{-1}\right]c_n^{-1} + c_n^{-1}\ln[n(n - m_n)^{-1}]. \tag{23.9.6}$$

We write the expression (23.9.6) in the following form (see Chapter 4, §1),

$$\det\left(\sum_{k=1}^{n-1} z_k z_k'\right)(A_{n-1}^{m_n})^{-1} = \prod_{k=1}^{m_n} \gamma_{n-k}, \tag{23.9.7}$$

where

$$\gamma_{n-1} = (n-1)^{-1}\sum_{i=1}^{n-1} z_{1i}^2,$$

$$\gamma_{n-k} = \sum_{j=k}^{n-1}(n-k)^{-1}\left[\sum_{L_k} z_{kp_1} t_{p_1 p_2}^{(1)} t_{p_2 p_3}^{(2)}\cdots t_{p_{k-1}j}^{(k-1)}\right]^2, \quad k = \overline{2,m},$$

$$L_k = \{p_1 = \overline{1, n-1}, p_2 = \overline{2, n-1}, \ldots, p_{k-1} = \overline{k-1, n-1}\};$$

$t_{ij}^{(k)}(i, j = \overline{k, n-1})$ are entries of a real orthogonal matrix T_k, measurable for the fixed smallest σ-algebra with respect to which the random variables $\tilde{x}_{pi}$,

$i = \overline{1, n-1}$, $p = \overline{1, k}$, are measurable. The first vector column of the matrix $T_k (k = \overline{2, n+1})$ is equal to

$$\left\{ \sum_{L_k} z_{kp_1} t^{(1)}_{p_1 p_2} t^{(2)}_{p_2 p_3} \cdots t^{(k-1)}_{p_{k-1} j} (n-k+1)^{-1/2} \gamma_{n-k}^{-1/2}, \quad j = \overline{k, n-1} \right\}$$

if $\gamma_{n-k} \neq 0$, and the arbitrary nonrandom real vector of unit length, if $\gamma_{n-k} = 0$; the first vector column of the matrix T_1 is equal to the vector

$$(n-1)^{-1/2} \left\{ z^2_{1j} \gamma_{n-1}^{-1/2}, \quad j = \overline{1, n-1} \right\}$$

if $\gamma_{n-1} \neq 0$, and the arbitrary nonrandom real vector of unit length, if $\gamma_{n-1} = 0$.

Explain the formula (23.9.7). Let $\gamma_{i-1} \neq 0$. Then

$$\det \left(\sum_{k=1}^{n-1} z_k z'_k \right) (A^{m_n}_{n-1})^{-1} = \gamma_{n-1} (A^{m_n}_{n-2})^{-1} \det B T_1 T'_1 B',$$

where

$$B = \begin{bmatrix} z_{11}[\gamma_{n-1}(n-1)]^{-1/2} & z_{12}[\gamma_{n-2}(n-1)]^{-1/2} & \cdots & z_{1n-1}[\gamma_{n-1}(n-1)]^{-1/2} \\ z_{21} & z_{22} & \cdots & z_{2n-1} \\ \cdots & \cdots & \cdots & \cdots \\ z_{m_n 1} & z_{m_n 2} & \cdots & z_{m_n n-1} \end{bmatrix}.$$

By multiplying the matrices B and T, we have

$$\det \left(\sum_{k=1}^{n} z_k z'_k \right) (A^{m_n}_{n-1})^{-1} = \gamma_{n-1} (A^{m_n}_{n-2})^{-1} \det CC', \qquad (23.9.8)$$

where

$$C = \begin{bmatrix} 1 & 0 & \cdots & 0 \\ y_{21} & y_{22} & \cdots & y_{2n-1} \\ y_{m_n 1} & y_{m_n 2} & \cdots & y_{m_n n-1} \end{bmatrix}, \qquad y_{ij} = \sum_{k=1}^{n-1} z_{ik} t^{(1)}_{kj}.$$

We complete the matrix C by a certain random matrix D so that the new matrix K will have the dimension of $(n-1) \times (n-1)$, where the entries of the random matrix D must satisfy the following conditions: Its first vector column consists of zero elements, and the row vectors are orthogonal to the matrix row vector of $Y = (y_{ij})$, $i = \overline{2, m_n}$, $j = \overline{1, n-1}$.

It is evident then that such a matrix always exists. Owing to the properties of matrix D,

$$\det K^2 = \det CC' \det DD' = \det YY' \det DD'.$$

By using this equation, from (23.9.8), we have

$$\det \left(\sum_{k=1}^{n} z_k z_k' \right) (A_{n-1}^{m_n})^{-1} = \gamma_{n-1}(A_{n-2}^{m_n})^{-1} \det \sum_{k=2}^{n-1} y_k y_k',$$

$$y_k' = (y_{ik}, i = \overline{2, m_n}).$$

By continuing this process further, we obtain formula (23.9.7). It is evident that for any $0 < \varepsilon < 1$ and for a certain $\delta > 0$,

$$\mathbf{P}\{|\gamma_{n-k} - 1| < \varepsilon, k = \overline{1, m_n}\} \geq \mathbf{P}\left\{ \sum_{k=1}^{m_n} |\gamma_{n-1} - 1|^{2+\delta} < \varepsilon^{2+\delta} \right\}$$

$$\geq 1 - \sum_{k=1}^{m_n} \mathbf{E}|\gamma_{n-k} - 1|^{2+\delta} \varepsilon^{-2-\delta}.$$

$$(23.9.9)$$

We write γ_{n-k} in the following form,

$$\gamma_{n-k} = \sum_{j=k}^{n-1} (n-k)^{-1} \eta_j^2, \qquad \eta_j = \sum_{p=1}^{n-1} z_{kp} \theta_{pj},$$

where

$$\theta_{pj} = \sum_{p_2 = \overline{2,n-1}, \dots, p_{k-1} = \overline{k-1, n-1}} t_{pp_2}^{(1)} t_{p_2 p_3}^{(2)} \cdots t_{p_{k-1} j}^{(k-1)}.$$

It is evident that $\theta_j = (\theta_{pj}, p = \overline{1, n-1})$, $j = \overline{k, n-1}$ is the orthonormalized random vector not depending on random values z_{kp}, $p = \overline{1, n-1}$,

$$\gamma_{n-k} = \sum_{p,l=1}^{n-1} a_{pl} z_{kp} z_{kl}, \qquad a_{pl} = \sum_{j=k}^{n-1} \theta_{pj} \theta_{lj} (n-k)^{-1}.$$

But then, by using Theorems 6.3.2 and 6.4.1, we obtain (23.9.3) and (23.9.9). Theorem 23.9.1 is proved.

§10 G_2-Estimate of the Stieltjes Transform of the Normalized Spectral Function of Covariance Matrices

Consider the main problem of G-analysis—the estimation of Stieltjes transforms of normalized spectral functions $\mu_{m_n}(x) = m_n^{-1} \sum_{k=1}^{m_n} \chi(\lambda_k < x)$ of the covariance matrices R_{m_n} by the observations over the random vector ξ with covariance matrix R_{m_n}, where λ_k are eigenvalues of the R_{m_n} matrix. Note that many analytic functions of the covariance matrices that are used in multivariate statistical analysis can be expressed through the spectral functions $\mu_{m_n}(x)$. For example, $m_n^{-1} \operatorname{Tr} f(R_{m_n}) = \int_0^\infty f(x) d\mu_{m_n}(x)$, where f is an analytical function.

The expression $\varphi(t, R_{m_n}) = \int_0^\infty (1+tx)^{-1} d\mu_{m_n}(x) = m_n^{-1} \operatorname{Tr}(I + tR_{m_n})^{-1}$ is called the Stieltjes transform of the function $\mu_{m_n}(x)$. The G-estimate of Stieltjes' transform $\varphi(t, R_{m_n})$ is by definition the following expression: $G_2(t, \widehat{R}_{m_n}) = \varphi(\hat{\theta}_n(t), \widehat{R}_{m_n})$, where $\hat{\theta}_n(t)$ is the solution of the equation

$$\theta(1 - m_n(n-1)^{-1} + m_n(n-1)^{-1}\varphi(\theta, \widehat{R}_{m_n})) = t, \quad t \geq 0. \tag{23.10.1}$$

It is obvious that the positive solution of Eq. (23.10.1) exists and is unique as $t \geq 0, m_n(n-1)^{-1} < 1$.

Theorem 23.10.1. *Let the independent observations $x_1, \ldots, x_n$ over the m_n-dimensional random vector ξ be given, let the G-condition be fulfilled:*

$$\varlimsup_{n\to\infty} m_n n^{-1} < 1, \quad 0 < c_1 \leq \lambda_i \leq c_2 < \infty, \quad i = \overline{1, m_n},$$

let the components of the vector $\eta_k := (\eta_{1k}, \ldots, \eta_{m_n k}) = H R_{m_n}^{-1/2}(\xi - \mathbf{E}\xi)$ be independent, and

$$\sup_n \sup_{k=\overline{1,n}} \sup_{i=\overline{1,m_n}} \mathbf{E}|\eta_{ik}|^{4+\delta} < \infty, \quad \delta > 0.$$

Then

$$\lim_{n\to\infty} \mathbf{P}\{[G_2 - \varphi(t, R_{m_n})]\sqrt{(n-1)m_n} a_n(t) + c_n(t)$$

$$< x\} = (2\pi)^{-1/2} \int_{-\infty}^x e^{-y^2/2} dy,$$

as $t > 0$, where $a_n(t) = q_1^{-1}(t)q_2^{-1}(t)$,

$$q_1(t) = 1 - \frac{\mathbf{E}\hat{\varphi}'(\theta_n)k_n\theta_n}{1 - k_n + k_n\theta_n\mathbf{E}\hat{\varphi}'(\theta_n) + k_n\mathbf{E}\hat{\varphi}(\theta_n)},$$

$$q_2(t) = [\mathbf{E}2m_n^{-1}SpA^2 + \mathbf{E}m_n^{-1}\sum_{i=1}^{m_n} d_i a_{ii}^2]^{-1/2}, \quad d_i = \mathbf{E}[(\eta_{i1})^2 - 1]^2 - 2,$$

$$A := (a_{ij}) = [1 + \theta_n(n-1)^{-1}\mathbf{E}SpSR_{m_n}]^{-1}\theta_n S^2 R_{m_n} + \theta^2 SR_{m_n}$$
$$\times \mathbf{E}(n-1)^{-1}SpS^2 R_{m_n}[1 + \theta_n(n-1)^{-1}\mathbf{E}SpSR_{m_n}]^{-2},$$
$$S = (I + \theta_n Q)^{-1}, \quad k_n = m_n(n-1)^{-1};$$

θ_n *is the solution of the equation* $\theta_n(1 - k_n + k_n \mathbf{E}\hat{\varphi}(\theta_n)) = t$;

$$\hat{\varphi}(\theta_n) = m_n^{-1} SpS, \quad Q = (n-1)^{-1} \sum_{k=1}^{n} (x_k - \mathbf{E}x_k)(x_k - \mathbf{E}x_k)',$$

$$c_n(t) = -a_n(t)\theta_n(n-1)^{-1/2}m_n^{-1/2}\mathbf{E}(S\eta, \eta)[1 - \theta_n(n-1)^{-1}\mathbf{E}(S\eta, \eta)]^{-1}q_1$$

$$\times (t) + (n-1)^{1/2}m_n^{-1/2}\sum_{p=1}^{m_n}\{-[1 + \lambda_p t]^{-3}\mathbf{E}\varepsilon_{2p}^2 + [1 + \lambda_p t]^{-2}\mathbf{E}\varepsilon_{2p}\},$$

$$\eta = \sum_{i=1}^{n} n^{-1/2}(x_i - \mathbf{E}x_i),$$

$$\varepsilon_{2p} = \theta_n \sum_{l=1}^{n}(b_{pl}^2 - \lambda_p) - \theta_n^2\{(b_p' B_p \Gamma_p B_p b_p, b_p)$$

$$- \lambda_p(n-1)^{-1}SpR_p B_p B_p'\} + \theta_n \lambda_p[m_n^{-1}S_p\Gamma_p - \mathbf{E}\hat{\varphi}(\theta_n)];$$

B_p *is a matrix obtained by deleting the pth row of the matrix*

$$B = (\xi_1, \xi_2, \ldots, \xi_n)(n-1)^{-1/2}, \quad \xi_i = \Lambda_{m_n}^{1/2}\eta_i,$$

$$\eta_k = HR_{m_n}^{-1/2}(\xi_k - \mathbf{E}\xi_k), \quad \Lambda_{m_n} = \operatorname{diag}(\lambda_i, i = \overline{1, m_n});$$

H *is a matrix of eigenvectors of* R_{m_n}, $\Gamma_p = (I + \theta_n B_p B_p')^{-1}$; b_p *is the row-vector of the matrix* B; *and the functions* $a_n(t)$ *and* $c_n(t)$ *satisfy the inequality*

$$\sup[|a_n(t)| + |c_n(t)|] < \infty \tag{23.10.2}$$

as $0 < t < c < \infty$.

Proof. We divide Theorem 23.10.1 into a series of intermediate statements.

Lemma 23.10.1.

$$G_2(t, \widehat{R}_{m_n}) - \varphi(t, R_{m_n}) = [\hat{\varphi}(\theta_n) - \mathbf{E}\hat{\varphi}(\theta_n)]c_n + [\mathbf{E}\hat{\varphi}(\theta_n) - \varphi(t, R_{m_n})] + \varepsilon_n,$$

$$c_n = 1 - \hat{\varphi}'(\theta_n)k_n\theta_n[1 - k_n + k_n\hat{\theta}_n\hat{\varphi}'(\theta_n) + k_n\hat{\varphi}(\theta_n)]^{-1},$$

$$\varepsilon_n = \hat{\varphi}'(\theta_n)k_n\hat{\theta}_n\hat{\varphi}''(\theta_n + \alpha(\hat{\theta}_n - \theta_n))(\hat{\theta}_n - \theta_n)^2[1 - k_n + k_n\hat{\theta}_n\hat{\varphi}'(\theta_n)$$

$$+ k_n\hat{\varphi}(\theta_n)]^{-1}/2 + \varphi''(\theta_n + \alpha(\hat{\theta}_n - \theta_n))(\hat{\theta}_n - \theta_n)^2/2; \quad 0 \le \alpha \le 1. \tag{23.10.3}$$

Proof. By expanding the function $\hat{\varphi}$ into the Taylor series and by using the equation $\theta_n(1 - k_n + k_n\mathbf{E}\hat{\varphi}(\theta_n)) = t$, $\hat{\theta}_n(1 - k_n + k_n\hat{\varphi}(\hat{\theta}_n)) = t$, we have

$$\hat{\theta}_n(1 - k_n) + \hat{\theta}_n k_n[\hat{\varphi}(\theta_n) + \hat{\varphi}'(\theta_n)(\hat{\theta}_n - \theta_n) + \hat{\varphi}''$$

$$\times (\theta_n + \alpha(\hat{\theta}_n - \theta_n))/2(\hat{\theta}_n - \theta_n)^2] = t,$$

$$\theta_n(1 - k_n) + \theta_n k_n\mathbf{E}\hat{\varphi}(\theta_n) = t.$$

By subtracting these two equations, we obtain

$$\hat{\theta}_n - \theta_n = -\{\theta_n k_n[\hat{\varphi}(\theta_n) - \mathbf{E}\hat{\varphi}(\theta_n)] + \hat{\theta}_n k_n \hat{\varphi}''(\theta_n + \alpha(\hat{\theta}_n - \theta_n))$$
$$\times (\theta_n - \hat{\theta}_n)^2/2\}[1 - k_n + k_n \hat{\varphi}(\theta_n) + \hat{\theta}_n k_n \hat{\varphi}'(\theta_n)]^{-1}.$$
$$(23.10.4)$$

Obviously,

$$G_2(t, \widehat{R}_{m_n}) - \varphi(t, R_{m_n}) = \hat{\varphi}'(\theta_n)(\hat{\theta}_n - \theta_n) + \hat{\varphi}''(\theta_n + \alpha(\hat{\theta}_n - \theta_n))$$
$$\times (\hat{\theta}_n - \theta_n)^2/2 + \hat{\varphi}(\theta_n) - \varphi(t, R_{m_n}).$$
$$(23.10.5)$$

The equalities (23.10.4) and (23.10.5) imply formula (23.10.3). Lemma 23.10.1 is proved.

Lemma 23.10.2. *Let the nondegenerate matrices $A_p^n = (a_{ij})_{i,j=p}^n$ be given. Then [see (2.5.7)],*

$$\det A_1^n = \det A_2^n(a_{11} - ((A_2^n)^{-1}a_1, b_1)),$$
$$a_1 = (a_{12}, \ldots, a_{1n}), \quad b_1 = (a_{21}, \ldots, a_{n1}); \qquad (23.10.6)$$
$$\mathrm{Tr}(I + itA_1^n)^{-1} - \mathrm{Tr}(I + itA_2^n)^{-1}$$
$$= -t(\frac{d}{dt})\ln[1 + ita_{11} + t^2((I + itA_2^n)^{-1}a_1, b_1)].$$

Let x_k be the m_n-dimensional column vectors, and let the matrix $R_{m_n} = \sum_{k=1}^{n-1} x_k$ x_k', $n > m_n$ be nondegenerate. Then [see (2.5.3)],

$$(I + R_{m_n} + x_n x_n')^{-1} - (I + R_{m_n})^{-1} = -(I + R_{m_n})^{-1}x_n x_n'(I + R_{m_n})^{-1}$$
$$\times [1 + ((I + R_{m_n})^{-1}x_n, x_n)]^{-1}.$$
$$(23.10.7)$$

Lemma 23.10.3.

$$[\mathbf{E}\hat{\varphi}(\theta_n) - \varphi(t, R_{m_n})]\sqrt{(n-1)m_n}$$
$$= (n-1)^{1/2}m_n^{-1/2}\sum_{p=1}^{m_n}\{-[1 + \lambda_p t]^{-3}\mathbf{E}\varepsilon_{2p}^2 + [1 + \lambda_p t]^{-2}\mathbf{E}\varepsilon_{2p}\} + 0(1).$$
$$(23.10.8)$$

Besides,

$$\mathbf{E}\varepsilon_{2p}^2 \le c_1(n-1)^{-1}, \qquad |\mathbf{E}\varepsilon_{2p}| \le c_2(n-1)^{-1}. \qquad (23.10.9)$$

Proof. Since $t = \theta_n(1 - k_n + k_n \mathbf{E}\hat{\varphi}(\theta_n))$,

$$\varphi(t, R_{m_n}) = m_n^{-1} \sum_{p=1}^{m_n} [1 + \lambda_p \theta_n (1 - k_n + k_n \mathbf{E}\hat{\varphi}(\theta_n))]^{-1}.$$

Evidently, $\mathrm{Tr}(I + \theta_n Q)^{-1} = \mathrm{Tr}(I + \theta_n BB')^{-1}$. Therefore, by using formula (23.10.6), we have

$$\mathbf{E}\hat{\varphi}(\theta_n) = m_n^{-1} \sum_{k=1}^{m_n} \mathbf{E}[1 + \theta_n(b_k, b_k) - \theta_n^2 (B_k' \Gamma_k B_k b_k, b_k)]^{-1}$$

$$= m_n^{-1} \sum_{p=1}^{m_n} \mathbf{E}[1 + \theta_n \lambda_p - \theta_n^2 \lambda_p (n-1)^{-1} \mathrm{Tr}\,\Gamma_p B_p B_p' + \varepsilon_{1p}]^{-1}, \tag{23.10.10}$$

where

$$\varepsilon_{1p} = \theta_n \left(\sum_{l=1}^{n} b_{pl}^2 - \lambda_p \right) - \theta_n^2 \{ (B_p' \Gamma_p B_p b_p, b_p) - \lambda_p (n-1)^{-1} S_p \Gamma_p B_p B_p' \}.$$

From equation (23.10.10), it follows that

$$\mathbf{E}\hat{\varphi}(\theta_n) = m_n^{-1} \sum_{p=1}^{m_n} \mathbf{E}[1 + \theta_n \lambda_p [1 - k_n + k_n \mathbf{E}\hat{\varphi}(\theta_n)] + \varepsilon_{2p}]^{-1},$$

where

$$\varepsilon_{2p} = \theta_n \left(\sum_{l=1}^{n} b_{pl}^2 - \lambda_p \right) - \theta_n^2 \{ (B_p' \Gamma_p B_p b_p, b_p) - \lambda_p (n-1)^{-1} \mathrm{Tr}\,\Gamma_p B_p B_p' \}$$
$$+ \theta_n \lambda_p [(n-1)^{-1} \mathrm{Tr}\,\Gamma_p - m_n(n-1)^{-1} \mathbf{E}\hat{\varphi}(\theta_n)].$$

It is easy to verify that

$$\mathbf{E}\varepsilon_{1k} = 0, \qquad \mathbf{E}\varepsilon_{1k}^2 \leq c(n-1)^{-1}(\theta_n^2 + \theta_n^4). \tag{23.10.11}$$

By using Eq. (23.10.6), we obtain

$$\delta_n := (n-1)^{-1} \mathrm{Tr}\,\Gamma_p - m_n(n-1)^{-1}\hat{\varphi}(\theta_n) = -(n-1)^{-1}\theta_n \left(\frac{d}{d\theta_n} \right)$$
$$\times \ln[1 + \theta_n(b_p, b_p) - \theta_n^2 (B_p' \Gamma_p B_p b_p, b_p)] = -(n-1)^{-1}\theta_n \left(\frac{d}{d\theta_n} \right)$$
$$\times \ln[1 + \theta_n((I + \theta_n B_p' B_p)^{-1} b_p, b_p)].$$

It is easy to derive from this equation that

$$|\delta_n| \leq (n-1)^{-1}c. \tag{23.10.12}$$

Obviously, $\hat{\varphi}(\theta_n) - \mathbf{E}\hat{\varphi}(\theta_n) = \sum_{k=1}^{n} \gamma_k$, where $\gamma_k = \mathbf{E}_{k-1}\hat{\varphi}(\theta_n) - \mathbf{E}_k\varphi(\theta_n)$, $\mathbf{E}_k$ is a conditional expectation with respect to the minimal σ-algebra generated by $\xi_{k+1}, \ldots, \xi_n$. The variables γ_k are martingale differences. Hence, by using (23.10.12) and (23.10.13), we obtain

$$\mathbf{E}[\hat{\varphi}(\theta_n) - \mathbf{E}\hat{\varphi}(\theta_n)]^2 = \sum_{k=1}^{n} \mathbf{E}\gamma_k^2 \leq nm_n^{-2}c, \tag{23.10.13}$$

$$\mathbf{E}|\varepsilon_{2p}|^{2+\delta}(n-1)^{2+\delta/2} \leq c. \tag{23.10.14}$$

By using (23.10.11)–(23.10.14) from this equation and by taking into account that $\mathbf{E}\varepsilon_{2k} = \theta_n\lambda_k\mathbf{E}\delta_n$, we obtain (23.10.8) and (23.10.9).

Lemma (23.10.3) is proved.

We shall need the following statement.

Lemma 23.10.4. *[See Theorem 5.4.3]. If for some $\delta > 0$,*

$$\lim_{n \to \infty} \sum_{k=1}^{n} \mathbf{E}|\gamma_k \sqrt{(n-1)m_n}a_n|^{2+\delta} = 0, \tag{23.10.15}$$

$$\lim_{n \to \infty} \sum_{k=1}^{n} \mathbf{E}|\mathbf{E}_k\gamma_k^2 - \mathbf{E}\gamma_k^2|(n-1)m_na_n^2 = 0, \tag{23.10.16}$$

where

$$a_n^{-2} = \sum_{k=1}^{n} \mathbf{E}\gamma_k^2(n-1)m_n,$$

then

$$\lim_{n \to \infty}[a_n^{-2} - m_n^{-1}\mathbf{E}\operatorname{Tr}A^2 - \mathbf{E}m_n^{-1}\sum_{i=1}^{m_n}a_{ii}^2d_i] = 0,$$

$$\lim_{n \to \infty}\mathbf{P}\{[\hat{\varphi}(\theta_n) - \mathbf{E}\hat{\varphi}(\theta_n)]\sqrt{(n-1)m_n}a_n < x\}$$

$$= (2\pi)^{-1/2}\int_{-\infty}^{x} \exp(-y^2/2)dy. \tag{23.10.17}$$

We verify (23.10.15) and (23.10.16). By virtue of (23.10.7), we have

$$[\hat{\varphi}(\theta_n) - \mathbf{E}\hat{\varphi}(\theta_n)]\sqrt{(n-1)m_n}$$

$$= -\sum_{k=1}^{n-1}\{\mathbf{E}_{k-1}\theta_n(S_k^2\xi_k,\xi_k)(n-1)^{-1}$$

$$\times [1+\theta_n(S_k\xi_k,\xi_k)(n-1)^{-1}]^{-1} - \mathbf{E}_k\theta_n(S_k^2\xi_k,\xi_k)(n-1)^{-1}$$

$$\times [1+\theta_n(S_k\xi_k,\xi_k)(n-1)^{-1}]^{-1}\}(n-1)^{1/2}m_n^{-1/2}$$

$$= -n^{-1/2}\sum_{k=1}^{n}(\mathbf{E}_{k-1}\delta_k - \mathbf{E}_k\delta_k), \qquad (23.10.18)$$

where

$$S_k = (I + \theta_n(n-1)^{-1}\sum_{p\neq k}\xi_p\xi_p')^{-1},$$

$$\delta_k = \frac{[\theta_n(S_k^2\xi_k,\xi_k) - \theta_n\,\text{Tr}\,S_k^2\Lambda_{m_n}]m_n^{-1/2}}{1+\theta_n(S_k\xi_k,\xi_k)(n-1)^{-1}} + \theta_n\frac{\text{Tr}\,S_k^2\Lambda_{m_n}}{n-1}$$

$$\times \frac{[\theta_n(S_k\xi_k,\xi_k) - \theta_n\,\text{Tr}\,S_k\Lambda_{m_n}]m_n^{-1/2}}{[1+\theta_n(n-1)^{-1}(S_k\xi_k,\xi_k)][1+\theta_n(n-1)^{-1}\,\text{Tr}\,S_k\Lambda_{m_n}]}.$$

The equality (23.10.18) shows that (23.10.15) holds. It is easy to see, that

$$\text{plim}_{n\to\infty}[(S_k\xi_k,\xi_k) - \mathbf{E}\,\text{Tr}\,S\Lambda_{m_n}](n-1)^{-1} = 0,$$

$$\text{plim}_{n\to\infty}[\text{Tr}\,S_k^2\Lambda_{m_n} - \mathbf{E}\,\text{Tr}\,S^2\Lambda_{m_n}](n-1)^{-1} = 0,$$

and the condition (23.10.16) can be replaced by the following one,

$$\lim_{n\to\infty} m_n(n-1)a_n^2\sum_{k=1}^{n}\mathbf{E}|\mathbf{E}_k(\mathbf{E}_{k-1}\delta_k)^2 - \mathbf{E}(\mathbf{E}_{k-1}\delta_k)^2| = 0. \qquad (23.10.19)$$

We represent $\mathbf{E}_{k-1}\delta_k$ in the form

$$\mathbf{E}_{k-1}\delta_k = m_n^{-1/2}[(B_k\xi_k,\xi_k) - \text{Tr}\,B_k\Lambda_{m_n}],$$

where

$$B_k = \mathbf{E}_{k-1}\frac{\theta_n S_k^2}{1+\theta_n(S_k\xi_k,\xi_k)(n-1)^{-1}}$$

$$+ \mathbf{E}_{k-1}\frac{\theta_n^2 S_k\,\text{Tr}\,S_k^2\Lambda_{m_n}(n-1)^{-1}}{[1+\theta_n(n-1)^{-1}(S_k\xi_k,\xi_k)][1+\theta_n(n-1)^{-1}\,\text{Tr}\,S_k R_m]}.$$

Obviously,

$$\mathbf{E}_k(\mathbf{E}_{k-1}\delta_k)^2 = 2m_n^{-1}\,\text{Tr}(B\Lambda_{m_n}) + m_n^{-1}\sum_{i=1}^{m_n}[(B\Lambda_{m_n})_{ii}]^2 d_i.$$

We note that due to formula (23.10.7),

$$S - \mathbf{E}S = \sum_{p=1}^{n}(\mathbf{E}_{p-1}S - \mathbf{E}_p S) = -(n-1)^{-1}$$

$$\times \sum_{p=1}^{n}\{\mathbf{E}_{p-1}\theta_n S_p \xi_p \xi_p' S_p[1 + \theta_n(n-1)^{-1}(S_p \xi_p, \xi_p)]^{-1}$$

$$- \mathbf{E}_p \theta_n S_p \xi_p \xi_p' S_p[1 + \theta_n(n-1)^{-1}(S_p \xi_p, \xi_p)]^{-1}\}.$$

$$(23.10.19)$$

Using this relation, it is easy to prove that the condition (23.10.19) holds and also that

$$\mathbf{E}(\mathbf{E}_{k-1}\delta_k)^2 = \mathbf{E}\delta_k^2 + 0(1) = a_n^{-2} + 0(1).$$

Hence (23.10.17) is fulfilled.

Therefore, Lemma (23.10.4) is valid.

Lemma 23.10.5. *Under the conditions of Theorem (23.10.1),*

$$\text{plim}_{n\to\infty}\{[G(t, \widehat{R}_{m_n}) - \widehat{\varphi}(\theta_n)]\sqrt{(n-1)m_n} - \theta_n(n-1)^{-1/2}m_n^{-1/2}$$
$$\times \mathbf{E}(S^2\eta, \eta)[1 - \theta_n(n-1)^{-1}\mathbf{E}(S\eta, \eta)]^{-1}\} = 0. \qquad (23.10.20)$$

Proof. According to (23.10.7), we obtain

$$[m_n^{-1} \text{Tr}(I + \theta_n \widehat{R}_{m_n})^{-1} - m_n^{-1} \text{Tr } S]\sqrt{(n-1)m_n}$$
$$= \theta_n(n-1)^{-1/2}m_n^{-1/2}(S^2\eta, \eta)(n-1)n^{-1}[1 - \theta_n(S\eta, \eta)n^{-1}]^{-1}. \qquad (23.10.21)$$

We again make use of the sums of the martingale difference,

$$(n-1)^{-1}(S\eta, \eta) - (n-1)^{-1}\mathbf{E}(S\eta, \eta) = (n-1)^{-1}\sum_{k=1}^{n}\gamma_k, \qquad (23.10.22)$$

where $\gamma_k = \mathbf{E}_{k-1}(S\eta, \eta) - \mathbf{E}_k(S\eta, \eta)$; $\mathbf{E}_k$ is a conditional expectation with respect to $\xi_{k+1}, \ldots, \xi_n$.

Let $\eta_k = \sum_{i\neq k}\xi_i n^{-1/2}$. It is evident that

$$\gamma_k = \mathbf{E}_{k-1}\Delta_k - \mathbf{E}_k\Delta_k, \quad \Delta_k = (S\eta_k, \eta_k) - (S_k\eta_k, \eta_k) + 2n^{-1/2}$$
$$\times ((S - S_k)\xi_k, \eta_k) + 2n^{-1/2}(S_k\xi_k, \eta_k) + n^{-1}(S\xi_k, \xi_k).$$

Due to the formula (23.10.7), we have for γ_k,

$$\gamma_k = \mathbf{E}_{k-1}\kappa_k - \mathbf{E}_k\kappa_k,$$

where

$$\kappa_k = [\theta_n(n-1)^{-1}(S_k\xi_k,\eta_k)^2 + 2\theta_n n^{-1/2}(n-1)^{-1}(S_k\xi_k,\xi_k)(S_k\xi_k,\eta_k)]$$
$$\times [1 + \theta_n(n-1)^{-1}(S_k\xi_k,\xi_k)]^{-1} + 2n^{-1/2}(S_k\xi_k,\eta_k) + n^{-1}(S\xi_k,\xi_k).$$

Then,

$$\mathbf{E}\left((n-1)^{-1}\sum_{k=1}^{n}\gamma_k\right)^2 = (n-1)^{-2}\sum_{k=1}^{n}\mathbf{E}\gamma_k^2 \le cn(n-1)^{-2}\sup_k[\theta_n(n-1)^{-2}$$
$$\times \mathbf{E}(S_k\xi_k,\eta_k)^4 + n^{-1}\mathbf{E}(S_k\xi_k,\eta_k)^2 + n^{-2}\mathbf{E}(S\xi_k,\xi_k)^2].$$

Due to the conditions of Theorem 23.10.1, it is easy to verify that

$$\mathbf{E}(S_k\xi_k,\eta_k)^4 \le c\mathbf{E}(\eta_k,\eta_k)^2 \le c_1 m_n^2, \qquad \mathbf{E}(S_k\xi_k,\eta_k)^2 \le c_2 m_n,$$
$$\mathbf{E}(S\xi_k,\xi_k)^2 \le c\mathbf{E}(\xi_k,\xi_k)^2 \le c_3 m_n^2.$$

Using these inequalities, we obtain $\mathbf{E}((n-1)^{-1}\sum_{k=1}^{n}\gamma_k)^2 \le cn^{-1} + 0(1)$. Hence, due to (23.10.22),

$$\operatorname{plim}_{n\to\infty}[(S\eta,\eta) - \mathbf{E}(S\eta,\eta)](n-1)^{-1} = 0. \tag{23.10.23}$$

Similarly,

$$(n-1)^{-1/2}m_n^{-1/2}[(S^2\eta,\eta) - \mathbf{E}(S^2\eta,\eta)]$$
$$= \sum_{k=1}^{n}(\theta_n(\frac{\partial}{\partial\theta_n})\gamma_k + \gamma_k)(n-1)^{-1/2}m_n^{-1/2}.$$

After the same simple transformations, we obtain that this difference also tends to zero in probability.

Therefore, by using (23.10.23) and (23.10.21), we obtain (23.10.20).

We note in general that

$$\lim_{n\to\infty}(n-1)^{-1}|\mathbf{E}(S\eta,\eta) - \mathbf{E}\operatorname{Tr}SR_{m_n}| > 0.$$

We show that the expression $\rho_n = (1 - \theta_n(n-1)^{-1}\mathbf{E}(S\eta,\eta))^{-1}$ is bounded. Due to formula (23.10.1), it is evident that

$$1 - \theta_n(n-1)^{-1}(S\eta,\eta) = \det[I + \theta_n(Q - (n-1)^{-1}\eta\eta')]\det[I + \theta_n Q]^{-1}$$
$$= [1 + \theta_n(n-1)^{-1}((I + \theta_n(Q - \eta\eta'(n-1)^{-1}))^{-1}\eta,\eta,)]^{-1}.$$

From this equation, we obtain

$$|\rho_n| \le 1 + c\theta_n + 0(1). \tag{23.10.24}$$

Lemma 23.10.6 is proved.

By using Lemma (23.10.5) and (23.10.6), it is easy to establish that (23.10.2) is valid, and that $\operatorname{plim}_{n\to\infty}|\theta_n - \hat\theta_n| = 0$, $\operatorname{plim}_{n\to\infty}|\hat\varphi(\theta_n) - \mathbf{E}\hat\varphi(\theta_n)| = 0$, $\operatorname{plim}_{n\to\infty}|\hat\varphi'(\theta_n) - \mathbf{E}\hat\varphi'(\theta_n)| = 0$, $\operatorname{plim}_{n\to\infty}|\varepsilon_n\sqrt{m_n(n-1)}| = 0$. Therefore, by using Lemma (23.10.1)–(23.10.5), and formulas (23.10.3), (23.10.5), and (23.10.24), we obtain the statement of Theorem (23.10.1).

§11 G_3-Estimation of the Inverse Covariance Matrix

In the previous two sections, the existence of asymptotically normal estimations of generalized variance and trace of resolvent of the growing dimensional covariance matrix was established. From these estimations, we can obtain the estimation of the inverse covariance matrix. The aim of the present section is to prove the asymptotical normality of this estimation.

Let $x_1, \ldots, x_n$ be observations on the m_n-dimensional random vector ξ with the nondegenerate matrix R. Since

$$m_n^{-1} \operatorname{Tr} R^{-1} = \lim_{t \to \infty} tm_n^{-1} \operatorname{Tr}(I + tR)^{-1},$$

by using the estimation G_2, we obtain that the estimation $G_3 = \widehat{R}^{-1}(1 - m(n-1)^{-1})$ has to be taken as the G-estimation of the matrix R^{-1}; here $\widehat{R}$ is the empirical covariance matrix. We prove that the entries of the matrix G_3 are consistent if the G-condition is fulfilled. We need the following statement.

Lemma 23.11.1. *Let $H = (h_{ij})$, $i = \overline{1,m}$, $j = \overline{1, n-1}$, $n > m$ be random matrices, and let the inverse matrix $B = (b_{ij}) = (HH')^{-1}$ exist with probability 1. Then with probability 1, we have, if $i \neq j$,*

$$b_{ij} = \left(\sum_{L_{m-1}} \tilde{h}_{jp_1} t^{(1)}_{p_1 p_2} \ldots t^{(m-2)}_{p_{m-2} m-1} \right) \gamma^{-1/2}_{n-m+1}(i,j) \gamma^{-1}_{n-m}(i,j),$$

$$b_{ii} = \gamma^{-1}_{m-n}(i,j), \tag{23.11.1}$$

where

$$\gamma_{n-1}(i,j) = \sum_{i=1}^{n-1} \tilde{h}^2_{1i},$$

$$\gamma_{n-k}(i,j) = \sum_{j=k}^{n-1} \left[\sum_{L_k} \tilde{h}_{kp_1} t^{(1)}_{p_1 p_2} t^{(2)}_{p_2 p_3} \ldots t^{(k-1)}_{p_{k-1} j} \right]^2, \quad k = \overline{2, m_n},$$

$$L_k = \{ p_1 = \overline{1, n-1}, p_2 = \overline{2, n-1}, \ldots, p_{k-1} = \overline{k-1, n-1};$$

$t^{(k)}_{ij} (i, j = \overline{k, n-1})$ *are elements of the orthogonal real matrix T_k that are measurable with respect to the smallest σ-algebra provided by the random variables $\tilde{h}_{pi}$, $i = \overline{1, n-1}$, $p = \overline{1, k}$; the first vector column of the matrix $T_k(k = 2, n+1)$ is equal to the vector*

$$\left\{ \sum_{L_k} \tilde{h}_{kp_1} \gamma^{-1/2}_{n-1}(i,j) t^{(1)}_{p_1 p_2} t^{(2)}_{p_2 p_3} \ldots t^{(k-1)}_{p_{k-1} j}, j = \overline{k, n-1} \right\}$$

if $\gamma_{n-k}(i,j) \neq 0$, *and to the arbitrary nonrandom real vector of unity length if* $\gamma_{n-k}(i,j) = 0$; *the first vector column of matrix* T_1 *is equal to the vector*

$$\{\tilde{h}_{1s}\gamma_{n-1}^{-1/2}(i,j), \quad s = \overline{1, n-1}\}$$

if $\gamma_{n-1}(i,j) \neq 0$, *and to the arbitrary nonrandom real vector of unity length if* $\gamma_{n-1}(i,j) = 0$; $\tilde{h}_{m-1} = h_i$, $\tilde{h}_m = h_j$, *the vectors* $\tilde{h}_p$, $p = \overline{1, m-2}$ *are equal to the vectors* $h_p = (h_{pi}, i = 1, n-1)$, $p \neq i, j$.

Proof. Obviously,

$$b_{ij} = \det H_i H_j [\det H H']^{-1},$$

where H_i is the matrix obtained from the matrix H by deleting the ith row. It is easy to verify that

$$b_{ij} = \det \tilde{H}_{m-1} \tilde{H}'_m [\det \tilde{H} \tilde{H}']^{-1},$$

where $\tilde{H} = (\tilde{h}_1, \ldots, \tilde{h}_m)$. It is easy to get from this expression the formula (23.11.1) (see §1, Chapter 4). Lemma 23.11.1 is proved.

We note that if the random variables h_{ij} are independent and distributed according to the normal law $N(0, I)$,

$$b_{ij} \approx y_{m-1} \left(\sum_{s=m-1}^{n-1} x_s^2 \right)^{-1/2} \left(\sum_{s=m}^{n-1} y_s^2 \right)^{-1}, \quad i \neq j; \qquad b_{ii} \approx \left(\sum_{s=m}^{n-1} y_s^2 \right)^{-1}$$
$$(23.11.2)$$

and the quantities b_{ii}, b_{jj} are independent for any $i \neq j$, $i, j = \overline{1, m}$, where the random quantities x_s and y_s, $s = 1, 2, \ldots$ are independent and distributed according to the normal law $N(0, I)$.

Let x_k, $k = 1, 2, \ldots$ be independent observations on the random m-dimensional vector ξ, which is distributed according to the normal law $N(a, R)$, and the matrix R is nondegenerate. It follows from [2] that

$$\tilde{R} \approx \sqrt{R} H H' (n-1)^{-1} \sqrt{R}. \tag{23.11.3}$$

Theorem 23.11.1. *If*

$$\varlimsup_{m_n \to \infty} mn^{-1} < 1, \quad \sup_n \sup_{p=1,m} |(R^{-1})_{pp}| < \infty,$$

then for any $p, l = \overline{1, m}$,

$$\operatorname{plim}_{n \to \infty} \mathbf{P}\{[(\hat{R}^{-1})_{pl}(1 - \frac{m_n}{n}) - (R^{-1})_{pl}]\sqrt{2n}(r_{pl}^2(1 - m_n n^{-1})^{-1}$$
$$+ r_{pp} r_{ll}(1 - m_n n^{-1})^{-3})^{-1/2} < x\} = (2\pi)^{-1/2} \int_\infty^x \exp\{-0.5y^2\}dy,$$
$$(23.11.4)$$

where r_{pl} are elements of the matrix R^{-1}.

Proof. Obviously, $(\widehat{R}^{-1})_{pl} = (\widehat{R}^{-1}a, b)$, $a = (\delta_{ip}, i = \overline{1,m})$, $b = (\delta_{il}, i = \overline{1,m})$. By using this equality and formula (23.11.3), we have

$$(\widehat{R}^{-1})_{pl} \approx \left(\left(\frac{HH'}{n-1}\right)^{-1} c, d\right) = \frac{1}{2} \operatorname{Tr}\left(\frac{HH'}{n-1}\right) K, \qquad (23.11.5)$$

where $c = R^{-1/2}a$, $d = R^{-1/2}b$, $K = cd' + cd'$. Since $HH' \approx THH'T$ for any orthogonal matrix T of mth order, we get from (23.11.5),

$$(\widehat{R}^{-1})_{pl} \approx (n-1)[(HH')_{11}^{-1}\lambda_1 + (HH')_{22}^{-1}\lambda_2]/2, \qquad (23.11.6)$$

where λ_j, $i = \overline{1,2}$ are nonzero eigenvalues of the matrix K, which are equal to

$$\lambda_{1,2} = (c,d) \pm \sqrt{(c,c)(d,d)}.$$

Hence (23.11.6) implies

$$(\widehat{R}^{-1})_{pl} \approx (n-1)2^{-1}\{[(HH')_{11}^{-1} + (HH')_{22}^{-1}](c,d)$$
$$+ [(HH')_{11}^{-1} - (HH)_{22}^{-1}]\sqrt{(c,c)(d,d)}\}. \qquad (23.11.7)$$

By (23.11.7), under the conditions of the theorem, (23.11.2) implies (23.11.4). Theorem 23.11.1 is proved.

Let us consider one more remarkable property of the G_3 estimation. If x_k are independent observations on the random m-dimensional vector ξ distributed according to the normal law, condition (23.7.2) is satisfied, and $\lim_{n\to\infty} n^{-1} \operatorname{Tr} \times R^{-2} = 0$, then

$$\lim_{n\to\infty} n^{-1} \mathbf{E} \operatorname{Tr}(G_3 - R^{-1})^2 = 0. \qquad (23.11.8)$$

It is easy to verify the validity of (23.11.8) by considering the equality

$$\operatorname{Tr}(G_3 - R^{-1})^2 \approx \sum_{j,i=1}^{m} \lambda_i^{-2} a_{ij}^2 (1-c)^2 - 2\sum_{i=1}^{m} \lambda_i^{-2}(1-c)a_{ii} + \sum_{i=1}^{m} \lambda_i^{-2},$$

where λ_i are eigenvalues of the matrix R, and formula (23.11.2), where a_{ij} are elements of the matrix $(HH')^{-1}(n-1)$, $c = mn^{-1}$.

We now consider the general case. Suppose the components of vector $R^{-1/2}(\xi - \mathbf{E}\xi)$ are independent.

Theorem 23.11.2. *Let* $\lim_{n\to\infty} m_n n^{-1} < 1$, *and let the distribution density of components of vector* ξ *exist,*

$$\sup_n \sup_{p=\overline{1,m}} r_{pp} < \infty, \qquad \mathbf{E}\tilde{x}_{i1}^4 = 3, \quad i = \overline{1,m},$$

for some $\delta > 0$,

$$\sup_n \sup_{i=\overline{1,m}} \mathbf{E}|x_{i1}|^{4+\delta} \le c < \infty.$$

Then for any $p, l = \overline{1,m}$, *the formula (23.11.4) holds.*

Proof. Since $\widehat{R} = (n-1)^{-1}[\sum_{k=1}^n x_k x_k' - n\hat{x}\hat{x}']$, we can represent the matrix $\widehat{R}$ in the following way,

$$\widehat{R} = \sqrt{R} \sum_{k=1}^{n-1} z_k z_k' (n-1)^{-1} \sqrt{R},$$

here $z_k = \sum_{i=1}^n h_{ik}\tilde{x}_i$, h_{ik} are elements of the real orthogonal matrix, $h_{in} = n^{-1/2}$, $\tilde{x} = R^{-1/2}(x - \mathbf{E}x)$; then by using (23.11.7), we have

$$(\widehat{R}^{-1})_{pl} = (n-1)2^{-1}\{[(BB')_{11}^{-1} + (BB')_{22}^{-1}](c,d)$$
$$+ [(BB')_{11}^{-1} - (BB')_{11}^{-1}]\sqrt{(c,c)(d,d)}\},$$

where $B = TZ$, $z = (z_1, \ldots, z_{n-1})$, and T is an orthogonal matrix of eigenvectors of the matrix $cd' + dc'$.

By using formula (23.11.1), we obtain if $\mathbf{E}\tilde{x}_{i1}^4 = 3$, $i = \overline{1,m}$, for some $\delta > 0$, $\mathbf{E}|\tilde{x}_{i1}|^{4+\delta} \le c < \infty$, then the variables

$$[(BB')_{11}^{-1}(1-c) - 1]\sqrt{n(1-c)}, [(BB')_{22}^{-1}(1-c) - 1]\sqrt{n(1-c)}$$

are independent asymptotically and distributed according to the standard normal law (see Chapter 6). Consequently, the statement of Theorem 23.11.2 is valid.

§12 G_4-Estimates of Traces of Covariance Matrix Powers

The G_2-estimate is the basic estimate in G-analysis. With the help of this estimate, we can obtain the G-estimates of traces of analytical functions of covariance matrices. Let us demonstrate how it is possible to obtain the G-estimates of traces of covariance matrix powers with the help of the G_2-estimate. We recall that the expression

$$m_n^{-1} \operatorname{Tr}(I + \theta(t)\widehat{R})^{-1}, \quad t > 0$$

is called the G_2-estimate of the quantity $m_n^{-1} \operatorname{Tr}(I + tR)^{-1}$; here $\theta(t)$ is the positive solution of the equation $\theta(t)(1 - m_n(n-1)^{-1} + (n-1)^{-1} \operatorname{Tr}(I + \theta(t)\widehat{R})^{-1}) = t$, $t \geq 0$, and $\widehat{R}$ is an empirical covariance matrix.

It is evident that for $k = 1, 2, \ldots,$

$$m^{-1} \operatorname{Tr} R^k = \frac{(-1)^k}{k!} \left(\frac{\partial^k}{\partial t^k} \right) m^{-1} \operatorname{Tr}(I + tR)^{-1}|_{t=0}. \qquad (23.12.2)$$

By using the formula (23.12.2) and Eq. (23.10.1), we obtain that the $G_4^{(1)}$ estimate of the quantity $m_n^{-1} \operatorname{Tr} R$ is equal to $m_n^{-1} \operatorname{Tr} \widehat{R}$. In order to find the $G_4^{(2)}$-estimate of the quantity $m_n^{-1} R^2$, it is necessary to do some calculations. It is easy to verify that this estimation is equal to

$$G_4^{(2)} = 2^{-1}[2m_n^{-1} \operatorname{Tr}(I + \theta(t)\widehat{R})^{-3}[\theta'(t)]^2$$
$$\times \widehat{R}^2 - m_n^{-1} \operatorname{Tr}(I + \theta(t)\widehat{R})^{-2}\widehat{R}\theta''(t)]_{t=0}.$$
$$(23.12.3)$$

It follows from the equation for the function $\theta(t)$ that $\theta(0) = 0$ and

$$\theta'(t)(1 - m_n(n-1)^{-1} + m_n(n-1)^{-1}m_n^{-1} \operatorname{Tr}(I + \theta(t)\widehat{R})^{-1})$$
$$+ \theta(t)[-(n-1)^{-1} \operatorname{Tr}(I + \theta(t)\widehat{R})^{-2}\widehat{R}\theta'(t)] = 1,$$
$$\theta''(t)[1 - m_n(n-1)^{-1} + m_n(n-1)^{-1}m_n^{-1} \operatorname{Tr}(I + \theta(t)\widehat{R})^{-1}]$$
$$+ 2\theta'(t)[-(n-1)^{-1} \operatorname{Tr}(I + \theta(t)\widehat{R})^{-2}\widehat{R}\theta'(t)] + \theta(t)[2(n-1)^{-1}$$
$$\times \operatorname{Tr}(I + \theta(t)\widehat{R})^{-3}\widehat{R}^2[\theta'(t)]^2 - (n-1)^{-1} \operatorname{Tr}(I + \theta(t)\widehat{R})^{-2}\widehat{R}\theta''(t)] = 0.$$

By taking into account that $\theta(0) = 0$, we get from these equations, as $t = 0$,

$$\theta'(0) = 1, \qquad \theta''(0) + 2[-(n-1)^{-1} \operatorname{Tr} \widehat{R}] = 0.$$

In view of these equalities, we find from (23.12.3) that

$$G_4^{(2)} = m_n^{-1} \operatorname{Tr} \widehat{R}^2 - (n-1)^{-1}m_n^{-1}(\operatorname{Tr} \widehat{R})^2.$$

The estimates $G_4^{(k)}$, $k = 3, 4, \ldots$ are similarly determined.

§13 G_5-Estimates of Smoothed Normalized Spectral Functions of Empirical Covariance Matrices

Let $\mu_{m_n}(x)$ be the normalized spectral function of the covariance matrix R_{m_n}. The G_2-estimation of the Stieltjes transformation $\int_0^\infty (1+tx)^{-1} d\mu_m(x)$ of the spectral function $\mu_m(x)$ is equal to $m_n^{-1} \operatorname{Tr}(I + \theta(t)\widehat{R})^{-1}$; here $\widehat{R}$ is an empirical covariance matrix and $\theta(t)$ is a positive solution of the equation

$$\theta(t)(1 - m_n(n-1)^{-1} + (n-1)^{-1} \operatorname{Tr}(I + \theta(t)\widehat{R})^{-1}) = t, \quad t \geq 0. \quad (23.13.1)$$

By using the G_2-estimation, we try to find the G_5-estimate of the function $\mu_m(x)$. But there are two unsolved problems. The first one: Is the G-estimate of the Stieltjes transformation of some spectral function that under certain conditions and by virtue of the properties of the G_2-estimate will be the consistent estimation of the function $\mu_m(x)$? The second one: Is the G_5-estimate of the function $\mu_m(x)$ normal asymptotically, if the G_2-estimate is normal asymptotically? The solution to these problems faces great analytical difficulties. In order to overcome them, it is possible to consider instead of the functions $\mu_{m_n}(x)$ their different regularized analogies, i.e., the function $\mu_{m_n}(x)$ is replaced by a function in some sense close to it, whose G-estimate will be normal asymptotically. Let us consider the so-called smoothed normalized spectral functions

$$\mu_{m_n}^{(\varepsilon)}(x) = \pi^{-1} \int_{-\infty}^{\infty} \mu_{m_n}(z + \varepsilon y)(1 + y^2)^{-1} dy;$$

here $\varepsilon > 0$ is an arbitrary real number that is as small as desired. Smoothed normalized spectral functions satisfy the condition

$$\mu_{m_n}^{(\varepsilon)}(x) = t^{-1} \int_{-\infty}^{x} \operatorname{Im} m_n^{-1} \operatorname{Tr}[I(u + i\varepsilon) + R_{m_n}]^{-1} du. \qquad (23.13.2)$$

With the help of this formula, it is possible to determine the asymptotic normality of these functions. In order to find the G-estimates of the functions $\tilde{\mu}_{m_n}(x)$, it is necessary to obtain the G_2-estimate of the Stieltjes transformation

$$m_n^{-1} \operatorname{Tr}[z - R_{m_n}]^{-1}, \quad z = t + is, \quad s \neq 0.$$

By repeating the proof of Theorem (23.10.1), we get the following statement.

Theorem 23.13.1.   *If the conditions of Theorem (23.10.1) hold, then for any t and $s \neq 0$,*

$$\lim_{n \to \infty} \mathbf{P}\{\operatorname{Im}[\tilde{G}_2(z, \tilde{R}_{m_n}) - m_n^{-1} \operatorname{Tr}[Iz - R_{m_n}]^{-1}]\sqrt{(n-1)m_n}$$

$$\times b_n(z) + d_n(z) < x\} = (2\pi)^{-1/2} \int_{-\infty}^{x} \exp\{-y^2 0.5\} dy; \qquad (23.13.3)$$

here $\tilde{G}_2(z, \tilde{R}_{m_n}) = \tilde{\theta}(z)z^{-1}m_n^{-1} \operatorname{Tr}(I\tilde{\theta}(z) - \tilde{R}_{m_n})^{-1}$, $\tilde{\theta}(z)$ is the analytical solution of the equation

$$(1 - k_n + k_n \tilde{\theta}(z)m_n^{-1} \operatorname{Tr}(\theta(z)I - \hat{R}_{m_n})^{-1}) = \tilde{\theta}(z)z^{-1},$$

and $b_n(z)$ and $d_n(z)$ are some functions that satisfy the condition

$$\sup_n [|b_n(z)| + |d_n(z)|] < \infty.$$

The same statement holds also for the joint distribution of random quantities

$$\mathrm{Im}\{\widetilde{G}_2(z, R_{m_n}) - m_n^{-1}\,\mathrm{Tr}[Iz - R_{m_n}]^{-1}\},$$
$$\mathrm{Re}\{\widetilde{G}_2(z, \widetilde{R}_{m_n}) - m_n^{-1}\,\mathrm{Tr}[Iz - R_{m_n}]^{-1}\}.$$

The expression

$$G_5(x) = t^{-1}\int_{-\infty}^{x}\mathrm{Im}\,\widetilde{\theta}(u+i\varepsilon)(u+i\varepsilon)^{-1}m_n^{-1}\,\mathrm{Tr}(I\widetilde{\theta}(u+i\varepsilon) - \widehat{R}_{m_n})^{-1}$$

is called the G_5-estimate of the function $\mu_{m_n}^{(\varepsilon)}(x)$.

Using formulas (23.13.2) and (23.13.5), we obtain, in view of the conditions of Theorem (23.10.1), that the following statement

$$\lim_{\varepsilon\downarrow 0}\mathrm{plim}_{n\to\infty}[G_5(x) - \mu_{m_n}(x)] = 0$$

holds at the point x of the continuity of the function μ_{m_n}, and the random quantity $[G_5(x) - \mu_{m_n}^\varepsilon(x)]\sqrt{m(n-1)}$ is normal asymptotically. To prove the last statement, it is necessary to use the central limit theorem for the sum of martingale differences and also the proof of Theorem (23.10.1).

§14 Parameter Estimation of Stable Discrete Control Systems under G-Conditions

This section considers the parameter estimation of the systems of linear recurrent equations, with the latter having m_n unknown parameters (the number of rows or columns of the matrix Θ to be comparable with the number n of observations). Let us study the statistical qualities of the estimation of such systems under the G-conditions:

$$\lim_{n\to\infty} m_n n^{-1} = c, \quad 0 < c < \infty.$$

Let us consider the system of recurrent equations

$$\vec{y}_k = \Theta\vec{y}_{k-1} + \vec{b}_{k-1} + \vec{\varepsilon}\,'_{k-1}, \tag{23.14.1}$$

where $\Theta = (\theta_{ij})_{i,j=1}^{m_n}$ is an unknown matrix, $\vec{y}_0, \vec{b}_k$, $k = 1, 2, \ldots$ are known control vectors of dimension m_n, $\vec{y}_k$ are the observed vectors of dimension m_n, and $\vec{\varepsilon}_k$ are random vectors of dimension m_n.

Note that, in general, the matrix $\sum_{k=1}^{n}\vec{y}_{k-1}\vec{y}\,'_{k-1}$ can be degenerate. We find the estimation $\hat{\theta}_n$ of the matrix Θ for degenerate matrices in the regularized form

$$\hat{\theta} = \sum_{k=1}^{n} c_n^{-1}(\vec{y}_k - \vec{b}_{k-1})\vec{y}\,'_{k-1}(I\alpha + c_n^{-1}\sum_{k=1}^{n}\vec{y}_{k-1}\vec{y}\,'_{k-1})^{-1}.$$

From this, we have

$$\hat{\theta}_n - \theta c_n^{-1} \sum_{k=1}^{n} y_{k-1} y'_{k-1} (I\alpha + c_n^{-1} \sum_{k=1}^{n} y_{k-1} y'_{k-1})^{-1}$$

$$= \sum_{k=1}^{n} c_n^{-1} \varepsilon_k \vec{y}'_{k-1} (I\alpha + c_n^{-1} \sum_{k=1}^{n} \vec{y}_{k-1} \vec{y}'_{k-1})^{-1}, \tag{23.14.2}$$

where $\alpha > 0$, c_n is a sequence of constants.

We represent the expression (23.14.2) in the following form,

$$n^{-1} \operatorname{Tr} Q(\hat{\theta}_n - \theta c_n^{-1} \sum_{k=1}^{n} y_{k-1} y'_{k-1} (I\alpha + c_n^{-1} \sum_{k=1}^{n} y_{k-1} y'_{k-1})^{-1})$$

$$= \operatorname{Tr} Q \sum_{k=1}^{n} \vec{\varepsilon}_k \vec{y}'_{k-1} c_n^{-1} [n^{-1} (I\alpha + c_n^{-1} \sum_{k=1}^{n} \vec{y}_{k-1} \vec{y}'_{k-1})^{-1}$$

$$- n^{-1} \mathbf{E}(I\alpha + c_n^{-1} \sum_{k=1}^{n} \vec{y}_{k-1} \vec{y}'_{k-1})^{-1}] + \operatorname{Tr} Q \sum_{k=1}^{n}$$

$$\times \vec{\varepsilon}_k \vec{y}'_{k-1} c_n^{-1} [n^{-1} \mathbf{E}(I\alpha + c_n^{-1} \sum_{k=1}^{n} \vec{y}_{k-1} \vec{y}'_{k-1})^{-1}], \tag{23.14.3}$$

where $Q = (q_{ij})_{i,j=1}^{m,n}$ is a matrix of real parameters.

Let us find the conditions, with the estimation $\hat{\theta}_n$ being carried out consistently. We need the following additional statements.

Lemma 23.14.1.   *If the random vectors ε_k, $k = 1, 2, \ldots$ are independent,*

$$\|\theta\| < 1, \quad \sup_n \sup_{p=\overline{1,n}} \mathbf{E} \|\varepsilon_p\|^2 c_n^{-1} < \infty,$$

then

$$\operatorname{plim}_{n \to \infty} [n^{-1} \operatorname{Tr}(I\alpha + c_n^{-1} \sum_{k=1}^{n} \vec{y}_{k-1} \vec{y}'_{k-1} - n^{-1}$$

$$\times \mathbf{E} \operatorname{Tr}(I\alpha + c_n^{-1} \sum_{k=1}^{n} \vec{y}_{k-1} \vec{y}'_{k-1})^{-1}] = 0. \tag{23.14.4}$$

Proof.   Consider the expression

$$n^{-1} \operatorname{Tr} C(I\alpha + c_n^{-1} \sum_{k=1}^{n} \vec{y}_{k-1} \vec{y}'_{k-1})^{-1} - n^{-1} \mathbf{E} \operatorname{Tr} C(I\alpha + c_n^{-1}$$

$$\times \sum_{k=1}^{n} \vec{y}_{k-1} \vec{y}'_{k-1})^{-1} = n^{-1} \sum_{p=1}^{n} \gamma_p, \tag{23.14.5}$$

where $C = (c_{ij})_{i,j=1}^{m_n}$ is a nonnegative-definite symmetric real matrix of order m_n,

$$\gamma_p = \mathbf{E}_{p-1} \operatorname{Tr} C (I\alpha + c_n^{-1} \sum_{k=1}^{n} \vec{y}_{k-1} \vec{y}_{k-1}')^{-1} - \mathbf{E} \operatorname{Tr} C$$

$$\times (I\alpha + c_n^{-1} \sum_{k=1}^{n} \vec{y}_{k-1} \vec{y}_{k-1}')^{-1};$$

$\mathbf{E}_k$ is a conditional (mathematical) expectation for the fixed minimal σ-algebra with respect to which the random variables $\vec{\varepsilon}_{k=1}, \ldots, \vec{\varepsilon}_n$ are measurable.

The variables γ_p are martingale differences. The theorem (see Chapter 5) is valid for them if

$$\lim_{n \to \infty} n^{-2} \sum_{p=1}^{n} \mathbf{E}\gamma_p^2 = 0. \tag{23.14.6}$$

Then

$$\operatorname{plim}_{n \to \infty} n^{-1} \sum_{p=1}^{n} \gamma_p = 0. \tag{23.14.7}$$

With the help of the perturbation formula for random determinants, we prove that the variables γ_p are bounded; for this, we represent the sum $\sum_{k=1}^{n} \vec{y}_{k-1} \vec{y}_{k-1}'$ in the following form:

$$\sum_{k=1}^{n} \vec{y}_{k-1} \vec{y}_{k-1}'$$

$$= \sum_{k=1}^{n} \{\theta^{k-1}(\vec{y}_0 + \vec{b}_0) + \sum_{i=1}^{k-1} \theta^{i-1}(\vec{\varepsilon}_{k-i} + \vec{b}_{k-i})\}$$

$$\times \{\theta^{k-1}(\vec{y}_0 + \vec{b}_0) + \sum_{j=1}^{k-1} \theta^{j-1}(\vec{\varepsilon}_{k-j} + \vec{b}_{k-j})\}'$$

$$= \sum_{k=p}^{n} \theta^{k-p-1} \varepsilon_p \varepsilon_p' \theta^{k-p-1} + \sum_{k=p}^{n} \theta^{k-p-1} \varepsilon_p \{\theta^{k-1}(\vec{y}_0 + \vec{b}_0)$$

$$+ \sum_{\substack{j \neq k-p}}^{k-1} \theta^{j-1}(\vec{\varepsilon}_{k-j} + \vec{b}_{k-j})\}' + \sum_{k=p}^{n} \{\theta^{k-1}(\vec{y}_0 + \vec{b}_0)$$

$$+ \sum_{\substack{i \neq k-p}}^{k-1} \theta^{i-1}(\vec{\varepsilon}_{k-i} + \vec{b}_{k-i})\} \varepsilon_p' \theta^{k-p-1} + A_{np}$$

$$= \sum_{k=p}^{n} \theta^{k-p-1} \varepsilon_p \varepsilon_p' \theta^{k-p-1} + \sum_{k=p}^{n} \theta^{k-p-1} \varepsilon_p c_k' + \sum_{k=p}^{n} c_k \varepsilon_p' \theta^{k-p-1} + A_{np}, \tag{23.14.8}$$

where $A_n = (a_{ij})_{i,j=1}^n$ is some nonnegative-definite matrix stochastically independent of the vector ε_p,

$$\vec{c}_k = \theta^{k-1}(\vec{y}_0 + \vec{b}_0) + \sum_{\substack{i \neq k-p}}^{k-1} \theta^{i-1}(\vec{\varepsilon}_{k-i} + \vec{b}_{k-i}).$$

We write the variable γ_p in the following form:

$$\gamma_p = \mathbf{E}_{p-1}\{\operatorname{Tr} C(I\alpha + c_n^{-1}\sum_{k=1}^n \vec{y}_{k-1}\vec{y}_{k-1}')^{-1} - \operatorname{Tr} C(I\alpha + c_n^{-1}A_{np})^{-1}\}$$

$$- \mathbf{E}_p\{\operatorname{Tr} C(I\alpha + c_n^{-1}\sum_{k=1}^n \vec{y}_{k-1}\vec{y}_{k-1}')^{-1} - \operatorname{Tr} C(I\alpha + c_n^{-1}A_{np})^{-1}\}. \tag{23.14.9}$$

Let us make the following transformations:

$$\operatorname{Tr} C(I\alpha + c_n^{-1}\sum_{k=1}^n \vec{y}_{k-1}\vec{y}_{k-1}')^{-1} - \operatorname{Tr} C(I\alpha + c_n^{-1}A_{np})^{-1}$$

$$= \left(\frac{\partial}{\partial\theta}\right)\ln\det(I\alpha + c\theta + c_n^{-1}\sum_{k=1}^n \vec{y}_{k-1}\vec{y}_{k-1}')_{\theta=0}$$

$$- \left(\frac{\partial}{\partial\theta}\right)\ln\det(I\alpha + c\theta + c_n^{-1}(A_{np}))_{\theta=0}. \tag{23.14.10}$$

Using the perturbation formulas, we get

$$\det(I + ab') = 1 + ab', \tag{23.14.11}$$

where a and b are arbitrary vectors of dimension m_n, $\{\ ,\ \}$ is a scalar product of the vectors. From (23.14.11) follows the following important formula:

$$\left(\frac{\partial}{\partial\theta}\right)\ln[1 + \{(I\alpha + c\theta + c_n^{-1}A_{np})^{-1}a, b\}]_{\theta=0} = -\{(I\alpha + c_n^{-1}A_{np})^{-1}$$

$$\times c(I\alpha + c_n^{-1}A_{np})^{-1}a, b\}[1 + \{(I\alpha + c_n^{-1}A_{np})^{-1}a, b\}]^{-1}.$$

From this formula, we get

$$\left|\left(\frac{\partial}{\partial\theta}\right)\ln[1 + \{(I\alpha + c\theta + c_n^{-1}A_{np})^{-1}a, b\}]_{\theta=0}\right| \le \alpha^{-2}\|c\|\,\|a\|\,\|b\|. \tag{23.14.12}$$

By applying the formulas (23.14.11) and (23.14.12) to the equality (23.14.10) as many times as there exist components in the last row of the equality (23.14.8), we have

$$|\operatorname{Tr} C(I\alpha + c_n^{-1}\sum_{k=1}^n \vec{y}_{k-1}\vec{y}_{k-1}')^{-1} - \operatorname{Tr} C(I\alpha + c_n^{-1}A_{np})^{-1}|$$

$$\le \alpha^{-2}\|c\|c_n\left[\sum_{k=p}^n \|\theta^{k-p}\varepsilon_p\|^2 + 2\sum_{k=p}^n \|\theta^{k-p}\varepsilon_p\|\,\|c_k\|\right].$$

From this, using the fact that $\|\theta\| < 1$ and $\sup_n \sup_{p=\overline{1,n}} c_n^{-1}\mathbf{E}\|\varepsilon_p\|^2 < \infty$, we get $|\gamma_p| < c < \infty$ from (23.14.9), where $c > 0$ is some constant. But then (23.14.7) holds. Consequently, (23.14.4) is valid. Lemma 23.14.1 is proved.

If the conditions of Lemma 23.14.1 hold and the random variables $\|\sum_{k=1}^n \vec{\varepsilon}_k \vec{y}'_{k-1} c_n^{-1}\|$ are bounded in probability, then due to (23.14.4),

$$n^{-1}\operatorname{Tr} Q(\hat{\theta}_n - \theta c_n^{-1}\sum_{k=1}^n y_{k-1}y'_{k-1}(I\alpha + c_n^{-1}\sum_{k=1}^n y_{k-1}y'_{k-1})^{-1})$$

$$\sim \operatorname{Tr} Q \sum_{k=1}^n \vec{\varepsilon}_k \vec{y}'_{k-1} c_n^{-1}[n^{-1}\mathbf{E}(I\alpha + c_n^{-1}\sum_{k=1}^n \vec{y}_{k-1}\vec{y}'_{k-1})^{-1}].$$

This expression is quite convenient for proving limit theorems, since there exists an expectation instead of the matrix $(I\alpha + c_n^{-1}\sum_{k=1}^n \vec{y}_{k-1}\vec{y}'_{k-1})^{-1}$, and limit theorems for the sums of martingale differences can be applied to the sums

$$\sum_{k=1}^n \vec{\varepsilon}_k \vec{y}_{k-1} c_n^{-1}.$$

Lemma 23.14.2. *If the conditions of Lemma 23.14.1 hold, then*

$$\operatorname*{plim}_{n\to\infty} n^{-1}[\operatorname{Tr} Q\theta c_n^{-1}\sum_{k=1}^n y_{k-1}y'_{k-1}(I\alpha + c_n^{-1}\sum_{k=1}^n y_{k-1}y'_{k-1})^{-1}$$

$$- \mathbf{E}\operatorname{Tr} Q\theta c_n^{-1}\sum_{k=1}^n y_{k-1}y'_{k-1}(I\alpha + c_n^{-1}\sum_{k=1}^n y_{k-1}y'_{k-1})^{-1}] = 0.$$

To prove Lemma (23.14.2), it is necessary to use the formula

$$\operatorname{Tr} Q\theta c_n^{-1}\sum_{k=1}^n y_{k-1}y'_{k-1}(I\alpha + c_n^{-1}\sum_{k=1}^n y_{k-1}y'_{k-1})^{-1}$$

$$= \operatorname{Tr} Q\theta - \alpha\left(\frac{\partial}{\partial\gamma}\right)\ln\det[I\alpha + \gamma Q\theta + c_n^{-1}\sum_{k=1}^n y_{k-1}y'_{k-1}]_{\gamma=0},$$

and the proof of Lemma (23.14.1).

Therefore, if the conditions of Lemma (23.14.1) hold and the random variables $\|\sum_{k=1}^n \vec{\varepsilon}_k \vec{y}'_{k-1} c_n^{-1}\|$ are bounded in probability, then

$$n^{-1}\operatorname{Tr} Q(\hat{\theta}_n - \theta\mathbf{E}c_n^{-1}\sum_{k=1}^n y_k y'_{k-1}(I\alpha + c_n^{-1}\sum_{k=1}^n y_{k-1}y'_{k-1})^{-1})$$

$$\sim \operatorname{Tr} Q \sum_{k=1}^n \vec{\varepsilon}_k \vec{y}'_{k-1} c_n^{-1}[n^{-1}\mathbf{E}(I\alpha + c_n^{-1}\sum_{k=1}^n y_{k-1}y'_{k-1})^{-1}].$$

CHAPTER 24

ESTIMATE OF THE SOLUTION OF THE
KOLMOGOROV–WIENER FILTER

The problem of signal filtration is reduced to solving the Kolmogorov–Wiener integral equation

$$Q(t,x) = \int_0^1 \varphi(t,y)R(x,y)dy, \qquad (24.0.1)$$

where $Q(t,x)$, $R(x,y)$ are covariance functions of a random process, $\varphi(t,y)$ is the solution to Eq. (24.0.1). The covariance functions $Q(t,x)$, $R(x,y)$ are found from the observations $\alpha_k(x)$, $\beta_k(y)$, $k = 1,2,\ldots$ over the random processes $\xi(x)$, $\eta(y)$:

$$\widehat{R}(x,y) = (n_1 - 1)^{-1}\sum_{k=1}^{n_1}(\alpha_k(x) - \hat{\alpha}(x))(\alpha_k(y) - \hat{\alpha}(y)),$$

$$\widehat{Q}(x,y) = (n_2 - 1)^{-1}\sum_{k=1}^{n_2}(\alpha_k(x) - \hat{\alpha}(x))(\beta_k(y) - \hat{\beta}_k(y)),$$

where n_1 and n_2 are the numbers of observations

$$\hat{\alpha}(x) = n_1^{-1}\sum_{k=1}^{n_1}\alpha_k(x), \qquad \hat{\beta}(y) = n_2^{-1}\sum_{k=1}^{n_2}\beta_k(y).$$

Note, that in many practical applications, these empirical covariation functions are found in the discrete points $x_i, y_1, i = 1,2,\ldots$.

Equation (24.0.1) is called the Fredholm integral equation of the first kind and belongs to a class of ill-posed equations.

In order to simplify the formulas, we shall assume that the function $Q(t,x)$ is known. Obviously, $R(x,y) = R_1(\dot{x},y) + a(x)a(y)$, where $a(x) = M\xi(x)$.

588

As a rule, the covariance function $R(x,y)$ is unknown in different applied problems; so instead of it, the empirical covariance function is used,

$$\widehat{R}(u,v) = (n-1)^{-1} \sum_{k=1}^{n} (x_k(u) - \hat{a}(u))(x_k(v) - \hat{a}(v)) + \hat{a}(u)\hat{a}(v),$$

$$\hat{a}(u) = n^{-1} \sum_{k=1}^{n} x_k(u),$$

where $x_k(u), k = \overline{1,n}$ are the observations over the random process $\xi(x)$.

Note that often the random process $\xi(x)$ can be observed only in some discrete set of points.

The explicit solution of Eq. (24.0.1), as a rule, is unavailable. Therefore, we solve it numerically by substituting the integral by the integral sum. Thus, we have the system of equations instead of Eq. (24.0.1) (for the sake of simplification of the formulas, we chose the constant step of partition of the interval $(0,1)$),

$$\vec{b}(t) = R\vec{\varphi}(t), \tag{24.0.2}$$

where $\vec{\varphi}(t) = (m^{-1}\varphi(t, km^{-1}), k = \overline{1,m})'$,

$$R = \{(R_1(sm^{-1}, km^{-1}) + a(sm^{-1})a(km^{-1}))\}_{k,s=1}^{m},$$
$$\vec{b}(t) = (Q(t, sm^{-1}), s = \overline{1,m})'.$$

§1 The G_9-Estimate of the Solution of the Kolmogorov–Wiener Filter

In this section, we consider solving the system (24.0.2) with the empirical covariance matrix $\widehat{R}$ rather than with the precision of the approximation of Eq. (24.0.1) by the system (24.0.2).

Note that in many problems we need some functional of the vector $\vec{\varphi}$ rather than the vector itself. Here we deal with the linear functionals $(\vec{\varphi}, \vec{c})$, where $\vec{c}$ is a given vector. Thus, the problem is as follows: to find the estimate $\tilde{\vec{\varphi}}$ of $\vec{\varphi}$, such that the expression

$$(\vec{\varphi}, \vec{c}) - \tilde{\vec{\varphi}}, \vec{c})$$

is tending to zero in measure as $n \to \infty$, $m \to \infty$, where $\tilde{\vec{\varphi}} = \widehat{R}\vec{b}$,

$$\widehat{R} = (\widehat{R}_{ij} + \hat{a}(i)\hat{a}(j))_{i,j=1}^{m_n},$$

$$\widehat{R}_{ij} = n^{-1} \sum_{k=1}^{n} [x_k(im_n^{-1}) - \hat{a}(im^{-1})][x_k(jm_n^{-1}) - \hat{a}(jm_n^{-1})],$$

$$\hat{a}(i) = n^{-1} \sum_{k=1}^{n} x_k(im_n^{-1}), \vec{x}_k = (x_k(i), i = \overline{1,m_n})', \qquad \hat{\vec{a}} = (\hat{a}(i), i = 1, m_n)'.$$

Theorem 24.1.1. *If the random process $\xi(x)$ is Gaussian,*

$$\varlimsup_{n \to \infty} m_n n_m^{-1} < 1, \qquad \varlimsup_{n \to \infty} n^{-1} \operatorname{Tr} R^{-1} < \infty,$$

$$\varlimsup_{n \to \infty} [(R^{-1}\vec{b}, \vec{b}) + |(R^{-1}\vec{a}, \vec{a})| + |(R^{-1}\vec{c}, \vec{c})|] < \infty, \tag{24.1.1}$$

$$\tilde{\varphi} = K\vec{b} - (K\hat{a}, b)(K\hat{a})[1 + (K\hat{a}, \hat{a}) - \frac{1}{n} \operatorname{Tr} K]^{-1}$$

$$K^{-1} = n^{-1} \sum_{k=1}^{n} (\vec{\hat{x}}_k - \hat{a})(\vec{\hat{x}}_k - \hat{a})'(1 - m_n n^{-1})^{-1}, \tag{24.1.2}$$

then

$$\operatorname{plim}_{n \to \infty}[(\tilde{\varphi} - \varphi), c] = 0. \tag{24.1.3}$$

The estimate $\tilde{\varphi}$ will be called a G_9-estimate of the solution to the Kolmogorov-Wiener filter.

Proof. Obviously,

$$\hat{R} = n^{-1} \sum_{k=1}^{n} (\vec{\hat{x}}_k - \hat{a})(\vec{\hat{x}}_k - \hat{a})' + \hat{a}\hat{a}',$$

where $\vec{\hat{x}}_k = (x_k(i), i = \overline{1, m_n})'$, $\hat{a} = (\hat{a}(i), i = \overline{1, m_n})'$.

It follows from the multivariate statistical analysis that the random matrix $n^{-1} \sum_{k=1}^{n} (\vec{\hat{x}}_k - \hat{a})(\vec{\hat{x}}_k - \hat{a})$ and the vector $\hat{a}$ are stochastically independent and have the distributions $R^{1/2}QR^{1/2}$, $\vec{a} + n^{-1/2}R^{1/2}\vec{h}_n$, respectively, where $\vec{h}_i$, $i = 1, 2, \ldots$ are independent random vectors, distributed by the normal law $N(0, 1)$,

$$\vec{a} = (a(im_n^{-1}), i = \overline{1, m_n})', \qquad Q = n^{-1} \sum_{k=1}^{n-1} \vec{h}_k \vec{h}_k'.$$

We use the following formula (Chapter 2, §5),

$$[Q + \vec{a}\vec{a}']^{-1} - Q^{-1} = -\frac{Q^{-1}\vec{a}\vec{a}'Q^{-1}}{1 + (Q^{-1}\vec{a}, \vec{a})}. \tag{24.1.4}$$

By using this relation, we obtain

$$((\hat{\tilde{\varphi}} - \varphi), \vec{c}) = ((Q(1 - m_n n^{-1})^{-1} + \vec{\beta}\vec{\beta}')^{-1}R^{-1/2}\vec{b}, R^{-1/2}\vec{c}) - (R^{-1}\vec{b}, \vec{c})$$

$$= -(Q^{-1}(1 - m_n n^{-1})\vec{\beta}, R^{-1/2}\vec{b})(Q^{-1}(1 - m_n n^{-1})\vec{\beta}, R^{-1/2}c)$$

$$\times [1 + (Q^{-1}(1 - m_n n^{-1})\vec{\beta}, \vec{\beta})]^{-1} + (Q^{-1}(1 - m_n n^{-1})$$

$$\times R^{-1/2}b, R^{-1/2}c) - (R^{-1}\vec{b}, \vec{c}), \tag{24.1.5}$$

where $\vec{\beta} = R^{-1/2}\hat{a}$.

Since the distribution of the matrix Q coincides with that of the matrix HQH', where H is an arbitrary real orthogonal $m \times m$ matrix, then by appropriately choosing the orthogonal matrices H, we obtain in the same manner as in Chapter 23 that

$$(Q^{-1}(1 - m_n n^{-1})\vec{\beta}, \vec{\beta}) \approx (\vec{\beta}, \vec{\beta}) \left(\sum_{s=1}^{n-m_n} \nu_s^2 (n - m_n)^{-1} \right)^{-1},$$

$$(Q^{-1}(1 - m_n n^{-1})\vec{\beta}, R^{-1/2}\vec{b})$$

$$\approx \frac{1}{2} \left[(R^{-1/2}\vec{b}, \vec{\beta}) + \sqrt{(R^{-1}\vec{b}, \vec{b})(\vec{\beta}, \vec{\beta})} \right] \times \left(\sum_{s=1}^{n-m_n} \nu_s^2 (n - m_n)^{-1} \right)^{-1}$$

$$+ \frac{1}{2} \left[(R^{-1/2}\vec{b}, \vec{\beta}) - \sqrt{(R^{-1}\vec{b}, \vec{b})(\vec{\beta}, \vec{\beta})} \right] \times \left(\sum_{s=1}^{n-m_n} \mu_s^2 (n - m_n)^{-1} \right)^{-1},$$

$$(Q^{-1}(1 - m_n n^{-1})\vec{\beta}, R^{-1/2}\vec{c})$$

$$\approx \frac{1}{2} \left[(R^{-1/2}\vec{c}, \vec{\beta}) + \sqrt{(R^{-1}\vec{c}, \vec{c})(\vec{\beta}, \vec{\beta})} \right] \left(\sum_{s=1}^{n-m_n} \nu_s^2 (n - m_n)^{-1} \right)^{-1}$$

$$+ \frac{1}{2} \left[(R^{-1/2}\vec{c}, \vec{\beta}) - \sqrt{(R^{-1}\vec{c}, \vec{c})(\vec{\beta}, \vec{\beta})} \right] \times \left(\sum_{s=1}^{n-m_n} \mu_s^2 (n - m_n)^{-1} \right)^{-1},$$

where ν_s, μ_s, $s = 1, 2, \ldots$ are independent random variables, distributed by the normal law $N(0, 1)$.

This formula, conditions (24.1.1), (24.1.2), and the law of larger numbers imply

$$\text{plim}_{n \to \infty}((\tilde{\vec{\varphi}} - \vec{\varphi}), \vec{c}) = \text{plim}_{n \to \infty}[-(P^{-1}\vec{a}, \vec{b})(P^{-1}\vec{a}, \vec{c})[1 + (P^{-1}\vec{a}, \vec{a})]^{-1}$$
$$+ (P^{-1}\vec{b}, \vec{c}) - (R^{-1}\vec{b}, \vec{c})],$$

where $P = [R_1(im_n^{-1}, jm_n^{-1})]_{i,j=1}^{m_n}$.

Again by using formula (24.1.4), we obtain (24.1.3). Theorem (24.1.1) is proved.

§2 Asymptotic Normality of the G_9-Estimate of the Solution of the Kolmogorov–Wiener Equation

We use the notations of the previous section.

Theorem 24.2.1. *If the conditions of Theorem 24.3.1 hold and*

$$\lim_{n \to \infty} |(R^{-1}b, a)| > 0, \tag{24.2.1}$$

then

$$\lim_{n \to \infty} \mathbf{P}\{[(G_9 - \varphi), c] n^{1/2} a_n^{-1} < x\} = (2\pi)^{-1/2} \int_{-\infty}^{x} e^{-y^2/2} dy, \qquad (24.2.2)$$

where

$$\begin{aligned}
a_n^2 &= 4^{-1}(f_+^2(b,c) + f_-^2(b,c))k_n + [(4^{-1}f_+^2(b,a) + 4^{-1}f_-^2(b,a))k_n \\
&\quad + (R^{-1}b, b))(R^{-1}c, a)^2 + (4^{-1}f_+^2(c,a) + 4^{-1}f_-^2(c,a))k_n \\
&\quad + (R^{-1}c, c))(R^{-1}b, a)^2][1 + (R^{-1}a, a)]^2 + (R^{-1}b, a)^2(R^{-1}c, a)^2 \\
&\quad \times [(R^{-1}a, a)^2 k_n + 4(R^{-1}a, a)][1 + (R^{-1}a, a)]^{-4}, \\
f_\pm(x, y) &= (R^{-1}x, y) \pm \sqrt{(R^{-1}x, x)(R^{-1}y, y)}, \qquad k_n = (1 - m_n n^{-1})^{-1}.
\end{aligned}$$

Proof.   As in the previous paragraph, we have

$$\begin{aligned}
((G_9 - \varphi), c) &= (Q^{-1}k_n^{-1}R^{-1/2}b, R^{-1/2}c) - (Q^{-1}k_n^{-1}R^{-1/2}b, R^{-1/2}\hat{a}) \\
&\quad \times (Q^{-1}k_n^{-1}R^{-1/2}c, R^{-1/2}\hat{a})[1 + (Q^{-1}k_n^{-1}R^{-1/2}\hat{a}, R^{-1/2}\hat{a})]^{-1} \\
&\quad - ((R + aa')^{-1}b, c). \qquad (24.2.3)
\end{aligned}$$

Since $Th_k \approx h_k$, where T is an arbitrary orthogonal real matrix of the mth order, then by appropriately choosing orthogonal matrices from the latter formula, we obtain as in the previous paragraph,

$$\begin{aligned}
((G_9 - \varphi), c) &\approx \frac{1}{2}[L_{11}f_+(b,c) + L_{22}f_-(b,c)] - \frac{1}{4}[L_{33}f_+(b,\hat{a}) \\
&\quad + L_{44}f_-(b,\hat{a})][L_{55}f_+(c,\hat{a}) + L_{66}f_-(c,\hat{a})] \\
&\quad \times [1 + L_{77}(R^{-1}\hat{a}, \hat{a})]^{-1} - ((R + aa')^{-1}b, c), \qquad (24.2.4)
\end{aligned}$$

where $L = [HH'n^{-1}(1 - m_n n^{-1})^{-1}]^{-1}$, H is a random $m \times n$ matrix, whose entries are independent and distributed by the normal law $N(0, 1)$.

It follows from Lemma (23.11.1) that the random variables L_{ii}, $i = 1, 2, \ldots$ are independent and identically distributed as well as the random variable

$$\left[\sum_{s=1}^{n-m_n} y_s^2 (n - m_n)^{-1} \right]^{-1},$$

where the random variables y_s are independent and distributed according to the normal law $N(0, 1)$.

It is obvious that

$$L_{ii} - 1 \approx \sum_{s=1}^{n-m_n} (y_s^2 - 1)(n - m_n)^{-1} \left[\sum_{s=1}^{n-m_n} y_s^2 (n - m_n)^{-1} \right]^{-1} \sim y_i (n - m_n)^{-1/2}.$$

Thus by using formula (24.4.4), we obtain

$$((G_9 - \varphi), c) \sim \frac{1}{2}(y_1 f_+(b, c) + y_2 f_-(b, c))(n - m_n)^{-1/2}$$
$$+ (R^{-1}b, c) - [\frac{1}{2}(y_3 f_+(b, \hat{a}) + y_4 f_-(b, \hat{a}))(n - m_n)^{-1/2}$$
$$+ (R^{-1}b, \hat{a})][\frac{1}{2}(y_5 f_+(c, \hat{a}) + y_6 f_-(c, \hat{a}))(n - m_n)^{-1/2}$$
$$+ (R^{-1}c, \hat{a})][1 + y_7(R^{-1}\hat{a}, \hat{a})(n - m_n)^{-1/2} + (R^{-1}\hat{a}, \hat{a})]^{-1}$$
$$- ((R + aa')^{-1}b, c). \tag{24.2.5}$$

In view of the conditions of Theorem (24.2.1),

$$(R^{-1}b, \hat{a}) \sim (R^{-1}b, a) + n^{-1/2}y_8(R^{-1}b, b)^{1/2},$$
$$(R^{-1}c, \hat{a}) \sim (R^{-1}c, a) + n^{-1/2}y_9(R^{-1}c, c)^{1/2}, \tag{24.2.6}$$
$$(R^{-1}\hat{a}, \hat{a}) \sim (R^{-1}a, a) + n^{-1/2}2y_{10}(R^{-1}a, a)^{1/2} + \frac{1}{n} \operatorname{Tr} R^{-1}.$$

By using (24.2.5) and (24.2.6), we have

$$((G_9 - \varphi), c) \sim \frac{1}{2}y_1[f_+^2(b, c) + f_-^2(b, c)]^{1/2}(n - m_n)^{-1/2} + (R^{-1}b, c)$$
$$- [y_3\{(4^{-1}f_+^2(b, a) + 4^{-1}f_-^2(b, a))k_n + (R^{-1}b, b)\}^{1/2}n^{-1/2}$$
$$+ (R^{-1}b, a)][y_5\{(4^{-1}f_+^2(c, a) + 4^{-1}f_-^2(c, a))k_n + (R^{-1}c, c)\}^{1/2}$$
$$+ (R^{-1}c, a)][1 + y_7\{(R^{-1}a, a)^2k_n + 4(R^{-1}a, a)\}^{1/2}n^{-1/2}$$
$$+ (R^{-1}a, a)]^{-1} - ((R + aa')^{-1}b, c) + \varepsilon_n, \tag{24.2.7}$$

where $\operatorname{plim}_{n \to \infty} \varepsilon_n \sqrt{n} = 0$.

Since $((R + aa')^{-1}b, c) = (R^{-1}b, c) - \frac{(R^{-1}b, a)(R^{-1}c, a)}{1 + (R^{-1}a, a)}$, we have from formula (24.2.7),

$$\sqrt{n}((G_9 - \varphi), c) \sim y_1 a_n.$$

Hence by using (24.2.1), we obtain (24.2.2).

Theorem (24.2.1) is proved.

§3 The G_{10}-Estimate of the Solution of the Regularized Kolmogorov–Wiener Filter

The discrete analog of a regularized Kolmogorov–Wiener filter has the form

$$\vec{b} = (I\varepsilon + R_m)\vec{\varphi}, \tag{24.3.1}$$

where $\varepsilon > 0$ is a parameter (see the notations in §1 of this chapter). As mentioned in previous sections of this chapter, the large order of system (24.3.1) requires a large number of observations over stochastic processes, so that the estimate $\tilde{\varphi} = (I\varepsilon + \widehat{R})^{-1}\vec{b}$ converges in measure to $\vec{\varphi}$. Therefore, it is of interest to obtain more accurate estimates. Employing the G-analysis technique, which is described in Chapter 23, we can obtain an estimate $\vec{\varphi}$, such that it would approach in measure $\vec{\varphi}$, provided $\lim_{n\to\infty} mn^{-1} = c$, $0 < c < 1$. This estimate will be referred to as the G_{10}-estimate. It is

$$G_{10} = (I\varepsilon + \varepsilon\hat{\theta}^{-1}\widehat{R})^{-1}\vec{b}, \qquad (24.3.2)$$

where $\hat{\theta}$ is a nonnegative solution of the equation

$$1 - k_m + k_m\theta m^{-1}\mathrm{Tr}(I\theta + \widehat{R})^{-1} = \theta\varepsilon^{-1}, \quad \varepsilon > 0, \qquad k_m = mn^{-1},$$
$$R_{ij} = M(\xi(im^{-1}) - M\xi(im^{-1}))(\xi(jm^{-1}) - M\xi(jm^{-1})).$$

Theorem 24.3.1. *If the stochastic process $\xi(x)$ is Gaussian,*

$$\varlimsup_{n\to\infty} mn^{-1} < \infty,$$
$$\sup_m[(\vec{b},\vec{b}) + (\vec{c},\vec{c})] < \infty, \quad \lambda_i \le c < \infty, \qquad (24.3.3)$$

where $\vec{c} \in R^m$ is a vector, λ_i are eigenvalues of matrix R_m; then as $\varepsilon > 0$,

$$\mathrm{plim}_{n\to\infty}[(G_{10},\vec{c}) - (\vec{\varphi},\vec{c})] = 0. \qquad (24.3.4)$$

Proof. By using formula (24.3.2) and equality (24.3.3), we have

$$(G_{10},\vec{c}) \approx \left(\left(I\varepsilon + (n-1)^{-1}R_m^{1/2}\sum_{k=1}^{n-1}\vec{h}_k\vec{h}_k' R_m^{1/2}\right)^{-1}\vec{b}, \vec{c}\right), \qquad (24.3.5)$$

where the random vectors $\vec{h}_k$ are mutually independent and distributed by the normal law $N(0, I)$.

By employing Theorem 23.10.1, we obtain that the expression (24.3.5) satisfies the relation (24.3.4). Theorem (24.3.1) is proved.

CHAPTER 25

RANDOM DETERMINANTS
IN PATTERN RECOGNITION

In this chapter we prove basic assertions concerning the classification observations over normally distributed vectors with known vectors of mean values and covariance matrices. We note that in solving various classification problems, we have to substitute these vectors and covariance matrices by empirical ones. Such substitution is sometimes referred to in the literature as "classification rule." As was indicated in Chapter 23, this substitution can lead to large errors. The errors can be easily avoided if we take advantage of the G-estimates for certain variables used in observation classification.

§1 The Bayes Method for Classification of Two Populations

Assume that we can observe certain parameters $x_1, \ldots, x_n$ of a given system S (of an individual), and on the basis of these observations, the problem is to determine to which of the classes chosen in advance the system belongs. Let R^m be an m-dimensional Euclidean space, R_1 and R_2 are the states of system S, referred to as classes, and system S can belong only to one of these states. In other words, the problem can be stated as follows: Determine some measurable set $D \in R^m$ such that, if $\vec{x} \in D$, then the systems belongs to the class K_1, otherwise it belongs to the class K_2. This constitutes the basic problem of deterministic observation classification (pattern recognition). If the observations are those of random vector ξ realizations, then while observing system S, events $B_1(B_2)$ may occur, when system S belongs to the class $K_1(K_2)$, yet nevertheless, $x \in \overline{D}(x \in D)$. The probability of misclassification will be equal to $p(B_1) + p(B_2)$. Sometimes, instead of this expression, the following one is considered: $c_1 p(B_1) + c_2 p(B_2)$, where c_1, c_2 are losses (costs) associated with relating the object to the class $K_1(K_2)$, while in fact, $x \in \overline{D}(x \in \overline{D})$. This expression is referred to as mean loss value. For observation under the random vector of system S, the classification problem is formulated in the following way: Let B be the σ-algebra of Borel subsets of set R^m. Determine

$$\text{ess} \inf_{D \subset B}[c_1 p(B_1) + c_2 p(B_2)]. \tag{25.1.1}$$

Assume that there exists a probability density of the vector ξ being equal to $p_1(x)$, $x \in R^m$, if $S \subset K_1$, and $p_2(x)$, $x \in R^m$, if $S \subset K_2$. Then by using conditional probabilities, we obtain that expression (25.1.1) equals

$$\text{ess} \inf_{D \subset B}[c_1 \int_D p_1(x)dx q_1 + c_2 \int_{\bar{D}} p_2(x)dx q_2], \tag{25.1.2}$$

where

$$q_i = \mathbf{P}\{S \in K_i\}, \quad i = 1,2, \quad dx = \prod_{i=1}^{m} dx_i.$$

The probabilities q_i are referred to as a priori probabilities that observations are performed over the population with probability density $p_i(x)$.

Transform (25.1.2) to the form

$$\text{ess} \inf_{d \subset B}[\int_D (c_1 q_1 p_1(x) - c_2 q_2 p_2(x))dx + c_2 q_2].$$

Hence, we get the set $\tilde{D}$ to be determined,

$$\tilde{D} = \{x : c_1 q_1 p_1(x) - c_2 q_2 p_2(x) < 0\}. \tag{25.1.3}$$

Thus, if the vector ξ has realization, then objects are classified according to the rule: if $y \in \tilde{D}$, then $S \in K_1$. The probability of such classification error is minimal and equals

$$c_1 q_1 \int_{\tilde{D}} p_1(x)dx + c_2 q_2 \int_{\tilde{D}} p_2(x)dx.$$

This classification rule is referred to as Bayesian. All previous calculations can be easily applied to the case when S can belong to L states (classes) K_l, $l = \overline{1, L}$.

Let $D = U_{l=1}^{L} D_l$, where $D_l \subset B$ are the sets separating the observations into classes. If $x \subset D_l$, then we consider $S \in K_l$. Let the probability density of the vector ξ be equal to $p_l(x)$, $x \in R^m$, with $S \subset K_l$. Then, the mean loss value is

$$\text{ess inf} \sum_{l=1}^{L} \sum_{k=1}^{L} c_{lk} q_k \int_D p_k(x)dx, \tag{25.1.4}$$

where c_{lp} are losses associated with relating the system S to the class K_l, when $S \in K_p$, $c_{ll} = 0$, q_p is the a priori probability of the fact, that the observation is performed over the system $S \in K_p$. From expression (25.1.4) with $c_{lp} = 1$, $l \neq p$, we get that $D = U_{k=1}^{L} D_k$,

$$D_k = \{x : q_k p_k(x) > q_l p_l(x), \quad l \neq k\}. \tag{25.1.5}$$

§2 Observation Classifications in the Case of Two Populations Having Known Multivariate Normal Distributions with Identical Covariance Matrices

In this section, we consider a special case of observation classifications, namely when the densities $p_i(x)$ are normal:

$$p_i(x) = (2\pi)^{-m/2} \det R^{-1/2} \exp\{-1/2(x - \mu_i)'R^{-1}(x - \mu_i)\}, \quad i = 1, 2,$$

where μ_i are vectors of the mean values of dimension m, and R is the nonsingular covariance matrix of order m.

Inequality (25.1.5) is equivalent to the following one:

$$\ln \frac{p_1(x)}{p_2(x)} < \ln k, \qquad k = c_2 q_2 / c_1 q_1.$$

Hence, we have

$$-1/2 \sum_{i=1}^{2} [(x - \mu_i)'R^{-1}(x - \mu_i)] < \ln k.$$

After simple transformations, this inequality takes the form

$$(x - 1/2(\mu_1 + \mu_2))'R^{-1}(\mu_1 - \mu_2) < \ln k. \qquad (25.2.1)$$

The expression $U(x)$ on the left-hand side of the inequality is referred to as the Anderson–Fisher discriminant function.

Determine the minimal probability of misclassification. Let X be an observation on the random vector ξ. If the vector X is normally distributed according to the normal law $N(\mu_1, R)$, then the variable $U(x)$ is normally distributed with the mean value

$$EU(x) = (\mu_1 - 1/2(\mu_1 + \mu_2))'R^{-1}(\mu_1 - \mu_2) = 1/2(\mu_1 - \mu_2)'R^{-1}(\mu_1 - \mu_2),$$

and the variance

$$\begin{aligned} \operatorname{Var} U(x) &= E[(x - \mu_1)'R^{-1}(\mu_1 - \mu_2)]^2 = E[(\mu_1 - \mu_2)'R^{-1}(x - \mu_1) \\ &\times (x - \mu_1)'R^{-1}(\mu_1 - \mu_2)] = (\mu_1 - \mu_2)'R^{-1}(\mu_1 - \mu_2). \end{aligned}$$

The variable $\alpha := (\mu_1 - \mu_2)'R^{-1}(\mu_1 - \mu_2)$ is referred to as the Mahalanobis "distance."

Thus, if the vector X is normally distributed $N(\mu_1, R)$, then the random variable $U(x)$ is normally distributed $N(1/2\alpha, \alpha)$. But, if X is distributed $N(\mu_2, R)$, then

$$EU(x) = 1/2(\mu_2 - \mu_1)'R^{-1}(\mu_1 - \mu_2) = -1/2\alpha, \quad \operatorname{Var} U(x) = \alpha.$$

The probability of classification is

$$c_1 q_1 \int_{x < \ln k} dN(x, 1/2\alpha, \alpha) + c_2 q_2 \int_{x > \ln k} dN(x, -1/2\alpha, \alpha).$$

In the course of simple transformations, we obtain that this expression is equal to

$$c_1 q_1 \int_{-\infty}^{(\ln k - \alpha/2)\alpha^{-1/2}} 1/\sqrt{2\pi} e^{-y^2/2} dy + c_2 q_2 \int_{(c+\alpha/2)\alpha^{1/2}}^{\infty} 1/\sqrt{2\pi} e^{-y^2/2} dy.$$

$$(25.2.2)$$

If $c_1 q_1 = c_2 q_2 = c$, then expression (25.2.2) is equal to

$$2c \int_{\sqrt{\alpha/2}} 1/\sqrt{2\pi} e^{-y^2/2} dy.$$

If all distribution densities $p_l(x)$, $l = \overline{1, L}$ are normal with different covariance matrices, then a separating function can be constructed that takes a simple form and minimizes errors of classification. By employing the inequality (25.1.5), we obtain

$$D_k = \{x : -1/2(x - \mu_k)' R_k^{-1}(x - \mu_k) + 1/2(x - \mu_l)' R_l^{-1}(x - \mu_l) - \ln(q_l/q_k)$$
$$- 1/2 \, \ln(\det R_k \det R_l^{-1}) > 0, l \neq k\}.$$

§3 G_{11}-Estimate of the Mahalanobis Distance

Let $x_1, \ldots, x_{n_1}, y_1, \ldots, y_{n_2}$ be observations over m-dimensional random vectors ξ and η, respectively, the vectors ξ and η are independent and distributions $N(a_1, R)$, $N(a_2, R)$. As the empirical mean value vectors and the covariance matrix R, we take:

$$\hat{a}_1 = n_1^{-1} \sum_{i=1}^{n_1} x_i, \qquad \hat{a}_2 = n_2^{-1} \sum_{i=1}^{n_2} y_i,$$

$$\widehat{R} = (n_1 + n_2 - 2)^{-1} \left\{ \sum_{k=1}^{n_1} (x_k - \hat{a}_1)(x_k - \hat{a}_1)' + \sum_{p=1}^{n_2} (y_p - \hat{a}_2)(y_p - \hat{a}_2)' \right\}.$$

Lemma 25.3.1. *If R is a nonsingular matrix, then*

$$(\hat{a}_1 - \hat{a}_2)' \widehat{R}^{-1}(\hat{a}_1 - \hat{a}_2)$$

$$\approx (a_1 - a_2)' R^{-1}(a_1 - a_2) \left(\frac{H H'}{n_1 + n_2 - 2} \right)_{11}^{-1}$$

$$+ \left\{ 2(a_1 - a_2)' R^{-1/2} \nu \sqrt{\frac{1}{n_1} + \frac{1}{n_2}} \right.$$

$$\left. + \nu' \nu \left(\frac{1}{n_1} + \frac{1}{n_2} \right) \right\} \left(\frac{H H'}{n_1 + n_2 - 2} \right)_{11}^{-1},$$

$$(25.3.1)$$

where ν is the $N(0, I)$ m_n-dimensional vector stochastically independent of the random vectors x_i, y_i, $H = (\nu_{ij})$, $i = \overline{1, m_n}$, $j = \overline{1, n_1 + n_2 - 2}$ is a random matrix whose elements are independent of the random vectors x_i, y_i, ν and normally distributed according to the law $N(0, 1)$; $(HH')_{ij}^{-1}$ are elements of the matrix $(HH')^{-1}$.

Proof. It follows from Chapter 2, that $\hat{a}_1$, $\hat{a}_2$, R are stochastically independent and

$$\hat{a}_1 \approx a_1 + \sqrt{R}\nu n_1^{-1/2}, \qquad \hat{a}_2 \approx a_2 + \sqrt{R}\nu n_2^{-1/2},$$
$$(25.3.2)$$

$$\hat{R} \approx \sqrt{R}HH'(n_1 + n_2 - 2)^{-1}\sqrt{R}. \tag{25.3.3}$$

From the expressions (25.3.2) and (25.3.3), formula (25.3.1) follows. Lemma 25.3.1 is proved.

By using the consistency property of estimate G_3 and Lemma 25.3.1, we obtain the following assertion.

Theorem 25.3.1. *Let* $n_1 + n_2 - 2 > m$, R *be nonsingular matrices,*

$$\lim_{n_1 \to \infty} \frac{m}{n_1} = c_1, \quad \lim_{n_2 \to \infty} \frac{m}{n_2} = c_2,$$
$$0 \leq c_1, c_2 < \infty, c_1 + c_2 \neq c_1 c_2, \tag{25.3.4}$$
$$\lim_{n_i \to \infty} (a_1 - a_2)'R^{-1}(a_1 - a_2)\left[\frac{1}{n_1} + \frac{1}{n_2}\right] = 0. \tag{25.3.5}$$

Then

$$\text{plim}_{n_i \to \infty} \{[(\hat{a}_1 - \hat{a}_2)'\hat{R}^{-1}(\hat{a}_1 - \hat{a}_2)][(n_1 + n_2 - 2 - m)/(n_1 + n_2 - 2)]$$
$$- (mn_1^{-1} + mn_2^{-1}) - (a_1 - a_2)'R^{-1}(a_1 - a_2)\} = 0 \tag{25.3.6}$$

Under the conditions of Theorem (25.3.1), we shall refer to the expressions

$$[(\hat{a}_1 - \hat{a}_2)'\hat{R}^{-1}(\hat{a}_1 - \hat{a}_2)]\frac{n_1 + n_2 - 2 - m}{n_1 + n_2 - 2} - \left(\frac{m}{n_1} + \frac{m}{n_2}\right)$$

as the G_{11}-Mahalanobis distance estimate.

Proof. Using the proof of consistency for estimate G_3 and condition (25.3.4), we get

$$\text{plim}_{n_i \to \infty} \left(\frac{HH'}{n_1 + n_2 - 2}\right)^{-1}_{11} = [1 - [c_1^{-1} + c_2^{-1}]^{-1}]^{-1}.$$

Since $\text{plim}_{n_1 \to \infty} m^{-1}(\nu, \nu) = 1$ and condition (25.3.5) holds, by virtue of Lemma 25.3.1, the assertion of Theorem 25.3.1 follows.

§4 Asymptotic Normality of Estimate G_{11}

Let $x_1,\ldots,x_n$, $y_1,\ldots,y_n$ be observations under m_n-dimensional random independent vectors ξ and η that have the distribution $N(a_1,R)$, $N(a_2,R)$, $\hat{a}_1$, $\hat{a}_2$; $\hat{R}$ are empirical mean vectors and the covariance matrix, respectively.

Theorem 25.4.1. *Let R be a nonsingular matrix, condition (25.3.4) hold, and $n_1 + n_2 - 2 > m$. Then*

$$\lim_{n_i \to \infty} \mathbf{P}\left\{ \frac{(G_{11} - \alpha_m)\sqrt{n_1 + n_2 - m - 2}}{\sqrt{D_m}} < x \right\} = (2\pi)^{-1/2} \int_{-\infty}^{x} e^{-y^2/2}\,dy, \tag{25.4.1}$$

where

$$D_m = 2\left(\alpha_m + \frac{m}{n_1} + \frac{m}{n_2}\right)^2 + 2\left[\frac{m}{n_1} + \frac{m}{n_2}\right]^2 \frac{n_1 + n_2 - m - 2}{m}$$
$$+ 4\alpha_m(n_1 + n_2 - m - 2)(n_1^{-1} + n_2^{-1}).$$

Proof. By using formulas (23.11.2) and (25.3.1), we have

$$(G_{11} - \alpha_m)\sqrt{n_1 + n_2 - m - 2}$$

$$\approx \left(\alpha_m + \frac{m}{n_1} + \frac{m}{n_2}\right)\left[-\sum_{i=m}^{n_1+n_2-2} \times \frac{(\xi_i^2 - 1)}{\sqrt{n_1 + n_2 - 2 - m}}\right]$$

$$\times \left[\sum_{i=m}^{n_1+n_2-2} \frac{\xi_i^2}{n_1 + n_2 - m - 2}\right]^{-1} + \sum_{i=1}^{m} \frac{(\nu_i^2 - 1)}{\sqrt{m}}\left[\frac{m}{n_1} + \frac{m}{n_2}\right]$$

$$\times \frac{\sqrt{n_1 + n_2 - m - 2}}{m}\left[\sum_{i=m}^{n_1+n_2-2} \frac{\xi_i^2}{n_1 + n_2 - m - 2}\right]^{-1}$$

$$+ 2(a_1 - a_2)'R^{-1/2}\nu\sqrt{\frac{1}{n_1} + \frac{1}{n_2}}\sqrt{n_1 + n_2 - m - 2}$$

$$\times \left[\sum_{i=m}^{n_1+n_2-2} \frac{\xi_i^2}{n_1 + n_2 - m - 2}\right]^{-1}, \tag{25.4.2}$$

where ξ_i, ν_i, $i = 1, 2, \ldots$ are independent random variables normally distributed according to the law $N(0,1)$, and ν_i are components of vector ν.

Obviously,

$$\operatorname{plim}_{n \to \infty} \sum_{i=m}^{n_1+n_2-2} \frac{\xi_i^2}{n_1 + n_2 - m - 2} = 1. \tag{25.4.3}$$

Let

$$2\sqrt{\frac{1}{n_1} + \frac{1}{n_2}}\sqrt{n_1 + n_2 - m - 2}(a_1 - a_2)'R^{-1/2} = c = (c_1, \ldots, c_m),$$

$$d = \left[\frac{m}{n_1} + \frac{m}{n_2}\right]\sqrt{\frac{n_1 + n_2 + m - 2}{m}}.$$

Consider the characteristic function

$$\varphi(s) = \mathbf{E}\exp\left\{is\frac{\alpha_m + \frac{m}{n_1} + \frac{m}{n_2}}{\sqrt{D_m}}\sum_{k=m}^{n_1+n_2-2}\frac{(\xi_k^2 - 1)}{\sqrt{n_1 + n_2 - m - 2}}\right.$$

$$\left. + isD_m^{-1/2}\sum_{k=1}^{m}\nu_k c_k + is\frac{d}{\sqrt{D_m}}\sum_{k=1}^{m}\frac{\nu_k^2 - 1}{\sqrt{m}}\right\}.$$

The first sum in the exponent index is stochastically independent of the other sums, and for it the central limit theorem is valid. Therefore, by using the central limit theorem for the first sum and calculating the mean value for function $\varphi(s)$, we have

$$\varphi(s) = e^{-s^2\frac{(\alpha_m + mn_1^{-1} + mn_2^{-1})^2}{D_m}}\prod_{k=1}^{m}\exp\left\{-\frac{isd}{\sqrt{mD_n}} - \frac{s^2}{2}\frac{c_k^2}{D_m}[1 - \frac{2isd}{\sqrt{mD_m}}]^{-1}\right\}$$

$$\times\left[1 - \frac{2isd}{\sqrt{mD_m}}\right]^{-1/2} + 0(1).$$

By passing to the limit as $n_i \to \infty$ in this expression and by taking into account that $\alpha_m^2 D_m^{-1} \leq 1$, $d(D_m)^{-1/2} \leq 1$, $c_k^2 D_m^{-1} \leq 1$, we have

$$\varphi(s) = e^{-s^2/2} + 0(1).$$

Therefore, by using the limit (25.4.3) and formula (25.4.2), we obtain formula (25.4.1). Theorem (25.4.1) is proved.

§5 The G_{12}-Estimate of the Regularized Mahalanobis Distance

If matrix R is singular or ill-conditioned, then instead of the distance α, its regularized analog is considered,

$$\alpha_\varepsilon = (a_1 - a_2)'(I\varepsilon + R)^{-1}(a_1 - a_2),$$

where $\varepsilon > 0$ is a certain number.

The regularized distance has better properties than the distance α. While proving asymptotic normality of its G-estimates, it is not necessary that the random vectors x_i and y_i be normally distributed.

As was mentioned in Chapter 23, the estimate $(\hat{x}_1 - \hat{x}_2)'(I\varepsilon + \hat{R})^{-1}(\hat{x}_1 - \hat{x}_2)$, where $\hat{x}_1$, $\hat{x}_2$, $\hat{R}$ are empirical mean values and the covariance matrix, is inappropriate for solving the multivariate classification problems, since with the increase of the number m of the components of the vectors ξ and η, the number of observations needed for obtaining a given accuracy in the Mahanalobis distance estimation rapidly grows. In this section, the assertion is proved, that under some conditions, the asymptotically normal G-estimate for the regularized Mahalanobis distance exists, provided that $\overline{\lim}_{n \to \infty} mn^{-1} < \infty$.

Let $x_1, \ldots, x_{n_1}$, $y_1, \ldots, y_{n_2}$ be independent observations under the m-dimensional random vectors ξ and η, respectively, with the vectors being normally distributed with the laws $N(a_1, R), N(a_2, R)$;

$$\hat{x}_1 = n_1^{-1} \sum_{i=1}^{n_1} x_i, \quad \hat{x}_2 = n_2^{-1} \sum_{i=1}^{n_2} y_i,$$

$$\widehat{R} = (n_1 + n_2 - 2)^{-1} \left[\sum_{i=1}^{n_1} (x_i - \hat{x}_1)(x_i - \hat{x}_1)' + \sum_{i=1}^{n_2} (y_i - \hat{x}_2)(y_i - \hat{x}_2)' \right].$$

We consider the numbers n_1 and n_2 depending on m tending to infinity as $m \to \infty$.

We call the G_{12}-regularized Mahalanobis distance estimate the expression

$$G_{12} = \left(\left(I\varepsilon + \frac{\varepsilon}{\hat{\theta}_m} \widehat{R} \right)^{-1} (\hat{x}_1 - \hat{x}_2), (\hat{x}_1 - \hat{x}_2) \right) - (n_1^{-1} + n_2^{-1})$$

$$\times \operatorname{Tr} \varepsilon \hat{\theta}_m^{-1} \widehat{R} \left(I\varepsilon + \frac{\varepsilon}{\hat{\theta}_m} \widehat{R} \right)^{-1}, \tag{25.5.1}$$

where $\hat{\theta}_m$ are nonnegative solutions of the equation

$$1 - k_m + k_m \theta_m^{-1} \operatorname{Tr}(I\theta + \widehat{R})^{-1} = \theta \varepsilon^{-1}, \quad \varepsilon > 0, \quad k_m = \frac{m}{n_1 + n_2 - 2}. \tag{25.5.2}$$

It is obvious that a nonnegative solution of this equation exists and is unique.

Theorem 25.5.1. *Let the G-condition be satisfied:*

$$\lim_{m \to \infty} \frac{m}{n_1 + n_2 - 2} = c < \infty, \quad \overline{\lim}_{m \to \infty} \left[\frac{n_1}{n_2} + \frac{n_2}{n_1} \right] < \infty,$$
$$\tag{25.5.3}$$

$$\lambda_i \le c < \infty, \quad i = \overline{1, m}, \tag{25.5.4}$$

where λ_i are eigenvalues of the matrix R_m,

$$\sup_m (b, b) < \infty, \quad b = a_1 - a_2. \tag{25.5.5}$$

Then, as $\varepsilon > 0$,

$$\lim_{m \to \infty} \mathbf{P}\{[G_{12} - b'(I\varepsilon + R)^{-1}b]\sqrt{n_1 + n_2 - 2} a_m(\varepsilon) < x\}$$
$$= (2\pi)^{-1/2} \int_{-\infty}^{x} e^{-y^2/2} dy,$$

where

$$a_m^{-2}(\varepsilon) = (n_1 + n_2 - 2)^{-1} \sum_{k=1}^{n_1+n_2-2} \mathbf{E}\delta_k^2 + 4\mathbf{E}(SR_m Sb, b)(n_1 + n_2 - 2)$$

$$\times (n_1^{-1} + n_2^{-1}) + 2\mathbf{E}\,\mathrm{Tr}(RS^2(n_1 + n_2 - 2)(n_1^{-1} + n_2^{-1})^2,$$

$$\mathbf{E}\delta_k^2 = \frac{\varepsilon^2}{\theta_m^2} \frac{\mathbf{E}[\mathbf{E}_{k-1}(S_k b, \xi_k)^2 - \mathbf{E}_k(S_k b, \xi_k)^2]^2}{1 + \varepsilon\theta_m^{-1}(n_1 + n_2 - 2)^{-1}\mathbf{E}\,\mathrm{Tr}\,RS_k},$$

$$S_k = \left(I\varepsilon + \varepsilon\theta_m^{-1}(n_1 + n_2 - 2)^{-1} \sum_{p=1, p\neq k}^{n_1+n_2-2} \xi_p \xi_p' \right)^{-1},$$

$$S = (I\varepsilon + \varepsilon\theta_m^{-1}\widehat{R})^{-1};$$

ξ_p *are m-dimensional random independent vectors, normally distributed according to the law $N(0, R)$, and θ_m is a nonnegative solution of the equation*

$$1 - k_m + k_m\theta_m^{-1}\mathbf{E}\,\mathrm{Tr}(I\theta + \widehat{R})^{-1} = \theta\varepsilon^{-1}, \quad \varepsilon > 0. \qquad (25.5.6)$$

Proof. We consider that in the formulations of the lemmas stated below, the conditions of the theorem are satisfied. We shall omit the indexes m of some variables. By letter the c we denote the constants which can be distinct in different formulas.

Lemma 25.5.1.

$$(I\varepsilon + \varepsilon\hat{\theta}^{-1}\widehat{R})^{-1} - S = c_m[m^{-1}\,\mathrm{Tr}(I\theta + \widehat{R})^{-1} - m^{-1}\mathbf{E}\,\mathrm{Tr}(I\theta + \widehat{R})^{-1}] + \varepsilon_m,$$
$$(25.5.7)$$

where c_m and ε_m are square matrices of order m.

$$c_m = -S^2\varepsilon^2 k_m \hat{\theta}^{-1}[1 - \varepsilon k_m m^{-1}\,\mathrm{Tr}(I\theta + \widehat{R})^{-1} + \varepsilon k_m \hat{\theta}m^{-1}\,\mathrm{Tr}(I\theta + \widehat{R})^{-2}]^{-1},$$

$$\varepsilon_m = [-S^2\varepsilon^2\hat{\theta}^{-1}k_m m^{-1}\,\mathrm{Tr}(I(\theta + \alpha(\hat{\theta} - \theta)) + \widehat{R})^{-3}\{1 - \varepsilon k_m m^{-1}$$

$$\times \,\mathrm{Tr}(I\theta + \widehat{R})^{-1} + \varepsilon k_m \hat{\theta}m^{-1}\,\mathrm{Tr}(I\theta + \widehat{R})^{-2}\}^{-1} + (I\varepsilon + \varepsilon\hat{\theta}^{-1}\widehat{R})^{-1}$$

$$\times (I\varepsilon + \varepsilon\theta^{-1}\widehat{R})^{-2}\varepsilon^2\hat{\theta}^{-2}\theta^{-2}](\hat{\theta} - \theta)^2, \quad 0 \leq \alpha \leq 1.$$

Proof. By expanding the function $\mathrm{Tr}(I\hat{\theta} + \widehat{R})^{-1}$ in the Taylor series and using equations (25.5.2) and (25.5.6), we have

$$\hat{\theta} - \theta = \varepsilon k_m\{(\hat{\theta} - \theta)m^{-1}\,\mathrm{Tr}(I\theta + \widehat{R})^{-1} + \theta[m^{-1}\,\mathrm{Tr}(I\theta + \widehat{R})^{-1}$$

$$- m^{-1}\mathbf{E}\,\mathrm{Tr}(I\theta + \widehat{R})^{-1}] - \hat{\theta}m^{-1}\,\mathrm{Tr}(I\theta + \widehat{R})^{-2}(\hat{\theta} - \theta)$$

$$+ \hat{\theta}m^{-1}\,\mathrm{Tr}(I(\theta + \alpha(\hat{\theta} - \theta)) + \widehat{R})^{-3}(\hat{\theta} - \theta)^2.$$

From this equation, we obtain

$$\hat{\theta} - \theta = \varepsilon k_m \frac{\theta[m^{-1}\operatorname{Tr}(I\theta + \widehat{R})^{-1} - m^{-1}\mathbf{E}\operatorname{Tr}(I\theta + \widehat{R})^{-1}]}{1 - \varepsilon k_m m^{-1}\operatorname{Tr}(I\theta + \widehat{R})^{-1} + \varepsilon k_m \hat{\theta} m^{-1}\operatorname{Tr}(I\theta + \widehat{R})^{-2}}.$$

$$+ \frac{\theta m^{-1}\operatorname{Tr}(I(\theta + \alpha(\hat{\theta} - \theta)) + \widehat{R})^{-3}(\hat{\theta} - \theta)^2}{1 - \varepsilon k_m m^{-1}\operatorname{Tr}(I\theta + \widehat{R})^{-1} + \varepsilon k_m \hat{\theta} m^{-1}\operatorname{Tr}(I\theta + \widehat{R})^{-2}} \quad (25.5.8)$$

It is obvious that

$$(I\varepsilon + \varepsilon\hat{\theta}^{-1}\widehat{R})^{-1} - S = -S^2(\varepsilon\hat{\theta}^{-1} - \varepsilon\theta^{-1}) + (I\varepsilon + \varepsilon\hat{\theta}^{-1}\widehat{R})^{-1}$$

$$\times (I\varepsilon + \varepsilon\theta^{-1}\widehat{R})^{-2}(\varepsilon\hat{\theta}^{-1} - \varepsilon\theta^{-1})^2. \quad (25.5.9)$$

By substituting the value $\hat{\theta} - \theta$ in Eq. (25.5.9) by the value from equation (25.5.8), we obtain (25.5.7). Lemma (25.5.1) is proved.

Using Lemma (25.5.1), we rewrite the estimate G_{12} in the form

$$G_{12} = \beta_m + (C_m(\hat{x}_1 - \hat{x}_2), (\hat{x}_1 - \hat{x}_2))[m^{-1}\operatorname{Tr}(I\theta + \widehat{R})^{-1} - m^{-1}$$

$$\times \mathbf{E}\operatorname{Tr}(I\theta + \widehat{R})^{-1}] + (\varepsilon_m(\hat{x}_1 - \hat{x}_2), (\hat{x}_1 - \hat{x}_2)) - (n_1^{-1} + n_2^{-1})\operatorname{Tr} C_m$$

$$\times [m^{-1}\operatorname{Tr}(I\theta + \widehat{R})^{-1} - m^{-1}\operatorname{Tr}(I\theta + \widehat{R})^{-1}] - (n_1^{-1} + n_2^{-1})\operatorname{Tr}\varepsilon_m,$$

$$(25.5.10)$$

where $\beta_m = (Sb, b) - 2(S\nu, b) + (S\nu, \nu) - (n_1^{-1} + n_2^{-1})\operatorname{Tr} S$.

It is obvious that

$$\beta_m = (S(\hat{x}_1 - \hat{x}_2), (\hat{x}_1 - \hat{x}_2)) - (n_1^{-1} + n_2^{-1})\operatorname{Tr} S, \quad (25.5.11)$$

where $\nu = \hat{x}_1 - \hat{x}_2 - b$.

Lemma 25.5.2.

$$\det A_n = \det A_{pl}(a_{pl} - (A_{pl}^{-1}a_p, b_l)),$$

where A_{pl} is a nonsingular matrix obtained from A by deleting the pth row and the lth column;

$$A_n = (a_{ij})_{i,j=1}^n, \qquad a_p = (a_{pi}, i \neq l), \qquad b_l = (a_{il}, i \neq p).$$

Lemma 25.5.3. *Let x_k be the m-dimensional column vectors and $R = \sum_{k=1}^n x_k x_k'$. Then*

$$(I + R + x_{n+1}x_{n+1}')^{-1} - (I + R)^{-1} = -\frac{(I + R)^{-1}x_{n+1}x_{n+1}'(I + R)^{-1}}{1 + ((I + R)^{-1}x_{n+1}, x_{n+1})}.$$

It follows from [2] that the matrix $\widehat{R}$ and the vectors $\hat{x}_1, \hat{x}_2$ are stochastically independent, and

$$(Sb, b) \approx \left(\left(I\varepsilon + \frac{\varepsilon}{\theta}\sum_{k=1}^{n_1+n_2-2} h_k h_k'(n_1 + n_2 - 2)^{-1}\right)^{-1} c, c\right),$$

where $h_k = H'\xi_k$, $c = H'b$, H is an orthogonal matrix of eigenvectors of the matrix R, and the ξ_k are defined in the conditions of the theorem.

Lemma 25.5.4.

$$\mathbf{E}\left(I\varepsilon + \varepsilon\theta^{-1}(n_1 + n_2 - 2)^{-1}\sum_{p=1}^{n_1+n_2-2} h_p h_p'\right)_{kk}^{-1} - (\varepsilon + \lambda_k)^{-1}$$
$$= \mathbf{E}\varepsilon_{3k}^2(\varepsilon + \lambda_k + \varepsilon_{3k})^{-1} - \mathbf{E}\varepsilon_{3k}(\varepsilon + \lambda_k)^{-2}, \qquad (25.5.13)$$

where

$$\varepsilon_{3k} = \frac{\varepsilon}{\theta}\left\{\sum_{l=1}^{n_1+n_2-2} b_{kl}^2 - \lambda_k - [(B_k'\Gamma_k B_k b_k, b_k) - \lambda_k(n_1 + n_2 - 2)^{-1}\right.$$
$$\left.\times \operatorname{Tr}\Gamma_k B_k B_k' + \theta_m\lambda_k(n_1 + n_2 - 2)^{-1}[\operatorname{Tr}\Gamma_k - \mathbf{E}(I\theta_m + \widehat{R})^{-1}],\right.$$
$$\Gamma_p = (I\theta + B_p B_p')^{-1}, \quad B = (\eta_1, \ldots, \eta_{n_1+n_2-2})(n_1 + n_2 - 2)^{-1/2},$$

b_p *are row-vectors of the matrix B, and B_p is a matrix obtained from the matrix B by deleting its pth row.*

$$\mathbf{E}\varepsilon_{3k}^2 \leq c(n_1+n_2-2)^{-1}, \quad |\mathbf{E}\varepsilon_{3k}| \leq c(n_1+n_2-2)^{-1}, \quad 0 \leq \varepsilon+\lambda_k+\varepsilon_{3k} \leq \varepsilon^{-1}.$$

Proof. By using Lemma 25.5.2, we have

$$\mathbf{E}(I\theta + BB')_{kk}^{-1} = \mathbf{E}[\theta + (b_k, b_k) - (B_k'\Gamma_k B_k b_k, b_k)]^{-1}$$
$$= [\theta + \lambda_k - \lambda_k(n_1 + n_2 - 2)^{-1}\operatorname{Tr}\Gamma_k B_k B_k' + \varepsilon_{1k}]^{-1}, \qquad (25.5.14)$$

where

$$\varepsilon_{1k} = \sum_{l=1}^{n_1+n_2-2} b_{kl}^2 - \lambda_k - \{(B_k'\Gamma_k B_k b_k, b_k) - \lambda_k(n_1 + n_2 - 2)^{-1}\operatorname{Tr}\Gamma_k B_k B_k'\}.$$

It follows from equality (25.5.14) that

$$\mathbf{E}(I\theta + BB')_{kk}^{-1} = [\theta + \lambda_k - \lambda_k m(n_1 + n_2 - 2)^{-1}$$
$$+ \theta m(n_1 + n_2 - 2)^{-1}\lambda_k m^{-1}\mathbf{E}\operatorname{Tr}(I\theta + \widehat{R})^{-1} + \varepsilon_{2k}]^{-1}, \qquad (25.5.15)$$

where

$$\varepsilon_{2k} = \varepsilon_{1k} + \theta\lambda_k[\operatorname{Tr}\Gamma_p - \mathbf{E}\operatorname{Tr}(I\theta + \widehat{R})^{-1}](n_1 + n_2 - 2)^{-1}.$$

Since θ_m satisfies the equation

$$(1 - k_m) + k_m\theta m^{-1}\mathbf{E}\operatorname{Tr}(I\theta + \widehat{R})^{-1} = \theta\varepsilon^{-1},$$

(25.5.15) implies

$$\theta \varepsilon^{-1} \mathbf{E}(I\theta + BB')^{-1}_{kk} = [\varepsilon + \lambda_k + \varepsilon_{2k} \varepsilon \theta^{-1}]^{-1}. \tag{25.5.16}$$

It is easy to see that

$$\mathbf{E}\varepsilon_{1k} = 0, \qquad \mathbf{E}\varepsilon^2_{1k} \le c(n_1 + n_2 - 2)^{-1}.$$

By using Lemma (25.5.2), we obtain

$$\delta_n := (n_1 + n_2 - 2)^{-1}[\operatorname{Tr} \Gamma_p - \operatorname{Tr}(I\theta + \widehat{R})^{-1}]$$
$$= -(n_1 + n_2 - 2)^{-1} \frac{d}{d\theta} \ln[\theta + (b_p, b_p) - (B'_p \Gamma_p B_p b_p, b_p)].$$

Then matrix B_p with probability 1 can be presented in the form $B_p = \sqrt{B_p B'_p} H$, $H = (B_p B'_p)^{1/2} B_p$. Using this representation and the formula for δ_n, we have

$$\delta_n = -(n_1 + n_2 - 2)^{-1} \frac{d}{d\theta} \ln[\theta + (H'(I - [I + \theta(B_p B'_p)^{-1}]^{-1})Hb_p, b_p)]$$
$$= -(n_1 + n_2 - 2)^{-1} \frac{d}{d\theta} \ln[\theta + \theta((I\theta + B'_p B_p)^{-1} b_p, b_p)]$$
$$= -(n_1 + n_2 - 2)^{-1} \left[\theta^{-1} - \frac{((I\theta + B'_p B_p)^{-2} b_p, b_p)}{1 + ((I\theta + B'_p B_p)^{-1} b_p, b_p)} \right]. \tag{25.5.17}$$

It follows from this equality that

$$|\delta_n| \le 2(n_1 + n_2 - 2)^{-1} \theta^{-1}. \tag{25.5.18}$$

Equation (25.5.6) implies

$$\theta^{-1} = \varepsilon^{-1}(1 - k_m + k_m \theta m^{-1} \mathbf{E} \operatorname{Tr}(I\theta + \widehat{R})^{-1})^{-1} \le \varepsilon^{-1}(1 - k_m)^{-1}. \tag{25.5.19}$$

Hence from (25.5.18),

$$|\delta_n| \le c(n_1 + n_2 - 2)^{-1}. \tag{25.5.20}$$

It is obvious that

$$\varphi(\theta) - \mathbf{E}\varphi(\theta) = \sum_{k=1}^{n_1+n_2-2} \gamma_k,$$

where

$$\varphi(\theta) = (n_1 + n_2 - 2)^{-1} \operatorname{Tr}(I\theta + \widehat{R})^{-1},$$
$$\gamma_k = \mathbf{E}_{k-1}\varphi(\theta) - \mathbf{E}_k\varphi(\theta),$$

E_k is the mean value, given the minimal σ-algebra of events, with respect to which the random vectors $\xi_{k+1}, \ldots, \xi_n$ are measurable, γ_k are martingale differences, and by using Lemma 25.5.3 and inequality (25.5.19), we get

$$E[\varphi(\theta) - E\varphi(\theta)]^2 = \sum_{k=1}^{n_1+n_2-2} E\gamma_k^2$$

$$\leq (n_1 + n_2 - 2)^{-2} \sum_{k=1}^{n_1+n_2-2} E[((I\theta + \sum_{s \neq k} \xi_s \xi_s')^{-1} \xi_k, \xi_k)^2$$

$$\times \{1 + ((I\theta + \sum_{s \neq k} \xi_s \xi_s')^{-1} \xi_k, \xi_k)\}^{-1}]^2$$

$$\leq (n_1 + n_2 - 2)^{-1} \theta^{-1}.$$

Obviously,

$$(\varepsilon + \lambda_k + \varepsilon_{3k})^{-1} - (\varepsilon + \lambda_k) = \varepsilon_{3k}^2 (\varepsilon + \lambda_k + \varepsilon_{3k})^{-1} - \varepsilon_{3k}(\varepsilon + \lambda_k)^{-2},$$

where $\varepsilon_{3k} = \varepsilon_{2k} \varepsilon \theta^{-1}$. By taking into account that $|E\varepsilon_{3k}| \leq c(n_1 + n_2 - 2)^{-1}$ and that the inequalities (25.5.20) and (25.5.21) imply $E\varepsilon_{3k}^2 \leq c(n_1 + n_2 - 2)^{-1}$ and also

$$0 \leq (\varepsilon + \lambda_k + \varepsilon_{3k})^{-1} = \varepsilon^{-1} \theta [I\theta + BB']_{kk}^{-1} \leq \varepsilon^{-1},$$

we have

$$|E(\varepsilon + \lambda_k + \varepsilon_{3k})^{-1} - (\varepsilon + \lambda_k)| \leq c(n_1 + n_2 - 2)^{-1}.$$

Therefore,

$$[\theta \varepsilon^{-1} E(I\theta + BB')_{kk}^{-1} - (\varepsilon + \lambda_k)^{-1}]$$
$$= E\varepsilon_{3k}^2 (\varepsilon + \lambda_k + \varepsilon_{3k})^{-1} - E\varepsilon_{3k}(\varepsilon + \lambda_k)^{-2} + 0(1).$$

Hence (25.5.13) is valid. Lemma (25.5.4) is proved.

Lemma 25.5.5.   *When $p \neq l$,*

$$E(I\varepsilon + \varepsilon \theta^{-1} BB')_{pl}^{-1} = 0.$$

Proof.   Let $a_{ij} = (b_i, b_j)$, $A = (a_{ij})_{i,j=1}^m$ $A_{j_1, \ldots, j_k}^{i_1, \ldots, i_m}$ be a matrix obtained from A by deleting the rows with numbers $i_1, \ldots, i_m$ and the columns with numbers $j_1, \ldots, j_k$. By using Lemma 25.5.2, we have

$$(I\varepsilon + \varepsilon \theta^{-1} BB')_{pl}^{-1} = (-1)^{p+l} \det(I\varepsilon + \varepsilon \theta^{-1} BB')_l^p \det(I\varepsilon + \frac{\varepsilon}{\theta} BB')^{-1}$$

$$= (-1)^{p+l} \times$$

$$\frac{\varepsilon \theta^{-1} a_{pl} - ([(I\varepsilon + \varepsilon \theta^{-1} BB')_{lp}^{lp}]^{-1} a_p^{(l)}, a_l^{(p)})}{[\varepsilon + \varepsilon \theta^{-1} a_{pp} - ([(I\varepsilon + \varepsilon \theta^{-1} BB')_p^p]^{-1} a_p, a_p)][\varepsilon + \varepsilon \theta^{-1} a_{ll} - ([(I\varepsilon + \varepsilon \theta^{-1} BB')_{pl}^{pl}]^{-1} a_p^{(p)}, a_p^{(P)}],}$$

where

$$a_p^{(l)} = \{(b_p, b_i), i \neq l, p\}, \qquad a_p = \{(b_p, b_i), i \neq p\}.$$

Since $b_p \approx -b_p$ and the vectors b_p are stochastically independent, $a_p^{(l)} \approx -a_p^{(l)}$ if vector $a_l^{(p)}$ is fixed. Thus, the distribution of the value $(I\varepsilon + \varepsilon\theta^{-1}BB')_{pl}^{-1}$, $p \neq l$ is symmetric. Consequently, Lemma (25.5.5) holds.

By using Lemmas 25.5.4 and 25.5.5, we obtain

$$[\mathbf{E}(Sb, b) - ((I\varepsilon + R)^{-1}b, b)] = \sum_{k=1}^{m}[\mathbf{E}\varepsilon_{3k}^2(\varepsilon + \lambda_k + \varepsilon_{3k})^{-1} - \mathbf{E}\varepsilon_{3k}(\varepsilon + \lambda_k)^{-2}]c_k^2,$$

$$(25.5.22)$$

where $c = (c_1, \ldots, c_m) = Hb$. By uniting formulas (25.5.10), (25.5.13), and (25.5.22), we get

$$\begin{aligned}
[G_{12} - ((I\varepsilon + R)^{-1}b, b)] &= (Sb, b) - \mathbf{E}(Sb, b) \\
&+ \sum_{k=1}^{m}[\mathbf{E}\varepsilon_{3k}^2(\varepsilon + \lambda_k + \varepsilon_{3k})^{-1} - \mathbf{E}\varepsilon_{3k}(\varepsilon + \lambda_k)^{-2}]c_k^2 \\
&- 2((S\nu, b) + (S\nu, \nu) - (n_1^{-1} + n_2^{-1})\operatorname{Tr} SR \\
&+ \{(c_m(\hat{x}_1 - \hat{x}_2), (\hat{x}_1 - \hat{x}_2)) + \varepsilon(n_1^{-1} + n_2^{-1})\operatorname{Tr} c_n\} \\
&\times m^{-1}\operatorname{Tr}(I\theta + \widehat{R})^{-1} - m^{-1}\mathbf{E}\operatorname{Tr}(I\theta + \widehat{R})^{-1}] \\
&+ (\varepsilon_m(\hat{x}_1 - \hat{x}_2), (\hat{x}_1 - \hat{x}_2)) - (n_1^{-1} + n_2^{-1}) \\
&\times \operatorname{Tr}\varepsilon_m + (n_1^{-1} + n_2^{-1})\operatorname{Tr} R[S - (I\varepsilon + R)^{-1}] \\
&+ \varepsilon(n_1^{-1} + n_2^{-1})\operatorname{Tr}[S - (I\varepsilon + R)^{-1}].
\end{aligned} \qquad (25.5.23)$$

Lemma 25.5.6. *Let γ_k, $k = \overline{1, n}$, $n = 1, 2, \ldots$ be random variables. $\mathbf{E}_k\gamma_k = 0$, where $\mathbf{E}_k\gamma_k$ is the conditional expectation with fixed random variables $\gamma_{k=1}$, $\ldots, \gamma_n$. If for any $\delta > 0$,*

$$\lim_{n \to \infty} \sum_{k=1}^{n} \mathbf{E}|\gamma_k a_n|^{2+\delta} = 0, \qquad (25.5.24)$$

$$\lim_{n \to \infty} \sum_{k=1}^{n} \mathbf{E}|\mathbf{E}_k\gamma_k^2 - \mathbf{E}\gamma_k^2|a_n^2 = 0, \qquad (25.5.25)$$

where

$$a_n^{-2} = \sum_{k=1}^{n} \mathbf{E}\gamma_k^2,$$

then

$$\lim_{n \to \infty} \mathbf{P}\left\{\sum_{k=1}^{n} \gamma_k a_k < x\right\} = \frac{1}{\sqrt{2\pi}} \int_{-\infty}^{x} e^{-y^2/2}dy. \qquad (25.5.26)$$

Lemma 25.5.7.

$$\mathrm{plim}_{n\to\infty}(n_1 + n_2 - 2)^{-1}[\mathrm{Tr}\, S - \mathbf{E}\,\mathrm{Tr}\, S] = 0,$$

$$(25.5.27)$$

$$\mathrm{plim}_{n\to\infty}(n_1 + n_2 - 2)^{-1}[\mathrm{Tr}\, RS - \mathbf{E}\,\mathrm{Tr}\, RS] = 0,$$

$$(25.5.28)$$

provided that $\underline{\lim}_{n\to\infty} \kappa_m > 0$,

$$\lim_{n\to\infty} \mathbf{P}\{[(Sb,b) - \mathbf{E}(Sb,b)]\kappa_m \sqrt{n_1 + n_2 - 2} < x\} = \frac{1}{\sqrt{2\pi}} \int_{-\infty}^{x} e^{-y^2/2} dy,$$

$$(25.5.29)$$

where

$$\kappa_m^{-2} = (n_1 + n_2 - 2)^{-1} \sum_{k=1}^{n_1+n_2-2} \mathbf{E}\delta_k^2.$$

Proof. We prove (25.5.27). It is obvious, that

$$\mathrm{Tr}\, S - \mathbf{E}\,\mathrm{Tr}\, S = \sum_{k=1}^{n_1+n_2-2} \gamma_k, \quad \gamma_k = \mathbf{E}_{k-1}\,\mathrm{Tr}\, S - \mathbf{E}_k\,\mathrm{Tr}\, S,$$

$\mathbf{E}_k\,\mathrm{Tr}\, S$ is the conditional expectation given the minimal σ-algebra of events, with respect to which the random vectors $\xi_{k+1}, \ldots, \xi_n$ are measurable.

By using Lemma 25.5.3, we have

$$\begin{aligned}
\gamma_k &= \mathbf{E}_{k-1}\varepsilon\theta^{-1}(S_k^2\xi_k, \xi_k)(n_1 + n_2 - 2)^{-1}[1 + \varepsilon\theta^{-1}(S_k\xi_k, \xi_k) \\
&\quad \times (n_1 + n_2 - 2)^{-1}]^{-1} - \mathbf{E}_k\varepsilon\theta^{-1}(S_k^2\xi_k, \xi_k)(n_1 + n_2 - 2)^{-1} \\
&\quad \times [1 + \varepsilon\theta^{-1}(S_k\xi_k, \xi_k)(n_1 + n_2 - 2)^{-1}]^{-1}.
\end{aligned}$$

From this equality it is clear that $|\gamma_k| \le 2\varepsilon^{-1}$. Therefore, by virtue of the law of large numbers, (25.5.27) is valid. Similarly, we prove (25.5.28). Now prove (25.5.29). It is obvious, that

$$(Sb,b) - \mathbf{E}(Sb,b) = \sum_{k=1}^{n_1+n_2-2} \rho_k,$$

where

$$\begin{aligned}
-\rho_k &= \mathbf{E}_{k-1}\varepsilon\theta^{-1}(S_k b, \xi_k)^2(n_1 + n_2 - 2)^{-1}[1 + \varepsilon\theta^{-1}(S_k\xi_k, \xi_k) \\
&\quad \times (n_1 + n_2 - 2)^{-1}]^{-1} - \mathbf{E}_k\varepsilon\theta^{-1}(S_k b, \xi_k)^2(n_1 + n_2 - 2)^{-1} \\
&\quad \times [1 + \varepsilon\theta^{-1}(S_k\xi_k, \xi_k)(n_1 + n_2 - 2)^{-1}]^{-1}.
\end{aligned}$$

$$(25.5.30)$$

Note, that by virtue of Lemma 25.3.3,

$$S_k - \mathbf{E}S_k = \sum_{p \neq k}^{n_1+n_2-2} (\mathbf{E}_{p-1}S - \mathbf{E}_p S) = -(n_1 + n_2 - 2)^{-1}\varepsilon\theta^{-1}$$

$$\times \sum_{p \neq k} \{\mathbf{E}_p S_{pk}\}\xi_p \xi_p' S_{pk}[1 + \varepsilon\theta^{-1}(n_1 + n_2 - 2)^{-1}(S_{pk}\xi_p, \xi_p)]^{-1}$$

$$- \mathbf{E}_p S_{pk}\xi_p \xi_p' S_{pk}[1 + \varepsilon\theta^{-1}(n_1 + n_2 - 2)^{-1}(S_{pk}\xi_p, \xi_p)]^{-1}\},$$

where

$$S_{pk} = (\varepsilon + \varepsilon\theta^{-1}\sum_{l \neq p,k} \xi_l\xi_l')^{-1}. \tag{25.5.31}$$

It is obvious, that

$$(S_k b, \xi_k)^2 - \mathbf{E}(S_k b, \xi_k)^2 = \sum_{p \neq k} q_p,$$

where

$$q_p = \mathbf{E}_{p-1}[((S_k - S_{kp})b, \xi_k)^2 + 2((S_k - S_{kp})b, \xi_k)(S_{kp}b, \xi_k)]$$
$$- \mathbf{E}_p[((S_k - S_{kp})b, \xi_k)^2 + 2((S_k - S_{kp})b, \xi_k)(S_{kp}b, \xi_k)].$$

By using formula (25.5.31), we obtain

$$\mathbf{E}q_p^2 \leq \mathbf{E}[(n_1 + n_2 - 2)^{-2}(S_{kp}b, \xi_p)^2(S_{kp}\xi_k, \xi_p)^2 + 2(n_1 + n_2 - 2)^{-1}$$
$$\times |(S_{kp}b, \xi_p)(S_{kp}\xi_k, \xi_p)(S_{kp}b, \xi_k)|]^2 \leq c(n_1 + n_2 - 2)^{-2}$$
$$\times \mathbf{E}(S_{kp}b, \xi_p)^2(S_{kp}\xi_k, \xi_p)^2(S_{kp}b, \xi_k)^2 \leq c_1(n_1 + n_2 - 2)^{-2}.$$

Therefore,

$$\mathrm{plim}_{m \to \infty}[(S_k b, \xi) - \mathbf{E}((S_k b, \xi_k)/\xi_k)] = 0. \tag{25.5.32}$$

Condition (25.5.24) for random variables ρ_k are obviously satisfied. Prove, that condition (25.5.25) holds for it.

Taking into account (25.5.28), (25.5.30), and (25.5.32), we have

$$\sum_k \mathbf{E}|\mathbf{E}_k\rho_k^2 - \mathbf{E}\rho_k^2|$$

$$= (n_1 + n_2 - 2)^{-1}\sum_{k=1}^{n_1+n_2-2} \mathbf{E}|\mathbf{E}_k[\mathbf{E}_{k-1}\varepsilon\theta^{-1}(S_k b, \xi_k)^2$$

$$\times [1 + \varepsilon\theta^{-1}(n_1 + n_2 - 2)^{-1}\mathbf{E}\,\mathrm{Tr}\,S_k R]^{-1}$$

$$- \varepsilon\theta^{-1}\mathbf{E}(S_k R S_k b, b)[1 + \varepsilon\theta^{-1}(n_1 + n_2 - 2)^{-1}\mathbf{E}\,\mathrm{Tr}\,S_k R]^{-1}]^{-2}$$

$$- \mathbf{E}[\mathbf{E}_{k-1}\varepsilon\theta^{-1}(S_k b, \xi_k)^2[1 + \varepsilon\theta^{-1}(n_1 + n_2 - 2)^{-1}\mathbf{E}\,\mathrm{Tr}\,S_k R]^{-1}$$

$$- \varepsilon\theta^{-1}\mathbf{E}(S_k R S_k b, b)[1 + \varepsilon\theta^{-1}(n_1 + n_2 - 2)^{-1}\mathbf{E}\,\mathrm{Tr}\,S_k R]^{-1}]^2|$$

$$= (n_1 + n_2 - 2)^{-1}\sum_{k=1}^{n_1+n_2-2} \mathbf{E}|\mathbf{E}_k[\mathbf{E}\varepsilon\theta^{-1}(S_k b, \xi_k)^2/\xi_k]^2$$

$$- \mathbf{E}[\mathbf{E}\varepsilon\theta^{-1}(S_k b, \xi_k)^2|\xi_k]^2|[1 + \varepsilon\theta^{-1}(n_1 + n_2 - 2)^{-1}$$

$$\times \mathbf{E}\,\mathrm{Tr}\,S_k R]^{-1} + 0(1) = \varepsilon_n,$$

where

$$\lim_{n\to\infty} \varepsilon_n = 0.$$

Besides, from (25.5.30), we obtain

$$E\delta_k^2 = \varepsilon^2\theta^{-2}\{E[(S_k b, \xi_k)^2|\xi_k]^2 - [E(S_k b, \xi_k)^2]^2\}$$
$$\times [1 + \varepsilon\theta^{-1}(n_1 + n_2 - 2)^{-1}E\operatorname{Tr} RS_k]^{-1}.$$
(25.5.33)

Hence, by virtue of Lemma (25.5.6), (25.5.29) holds. Lemma 25.5.7 is proved.

Lemma 25.5.8. *The distributions of the random variables*

$$[-2(S\nu, b) + (S\nu, \nu) - (n_1^{-1} + n_2^{-1})\operatorname{Tr} RS]\sqrt{n_1 + n_2 - 2}[2(n_1 + n_2 - 2)$$
$$\times (n_1^{-1} + n_2^{-1})^2 \operatorname{Tr} E(RS)^2 + 4E(SRSb, b)(n_1 + n_2 - 2)(n_1^{-1} + n_2^{-1})]^{-1/2}$$
(25.5.34)

converge to the standard normal law.

Proof. Lemma 25.5.8 obviously holds when the matrix $\widehat{R}$ is fixed. By taking into account the fact that for the expression entering the square root in formula (25.5.34) the law of large numbers is valid by Lemma 25.5.7, we get that Lemma 25.5.8 holds.

Lemma 25.5.9. *For any $0 < \tau < 1$,*

$$\operatorname{plim}_{m\to\infty}[\hat\theta_m - \theta_m][n_1 + n_2 - 2]^{1-\tau} = 0,$$
(25.5.35)
$$\overline{\lim_{m\to\infty}} E[\operatorname{Tr} S - E\operatorname{Tr} S]^2 < \infty.$$
(25.5.36)

Proof. The relation (25.5.36) follows from the fact that

$$E\gamma_k^2 \le c(n_1 + n_2 - 2)^{-2},$$

and (25.5.35) follows from formula (25.5.8). Lemma 25.5.9 is proved. By virtue of Lemma 25.5.9 in formula (25.5.23), the last four addends and the third one, being multiplied by $\sqrt{n_1 + n_2 - 2}$, tend in measure to zero. Therefore, from Lemmas 25.5.6, 25.5.8, and formula (25.5.33), the statement of the theorem follows.

§6 The G_{13}-Anderson–Fisher Statistics Estimate

By considering the observation classification in the case of two general populations that have normal distributions, the so-called discriminant function was used,

$$D(x) = (R^{-1}x, a_1 - a_2) - \frac{1}{2}(R^{-1}(a_1 - a_2), (a_1 + a_2)),$$

where R is a nonsingular matrix of order m; a_1 and a_2 are m-dimensional expectation vectors; x is an m-dimensional vector.

Let $x_1, \ldots, x_{n_1}, y_1, \ldots, y_{n_2}$ be observations over the m-dimensional random vectors ξ and η of the distributions $N(a_1, R)$, $N(a_2, R)$,

$$\hat{a}_1 = n_1^{-1} \sum_{i=1}^{n_1} x_i, \quad \hat{a}_2 = n_2^{-1} \sum_{i=1}^{n_2} y_i, \quad \hat{R} = (n_1 + n_2 - 2)^{-1}$$
$$\times \left\{ \sum_{k=1}^{n_1} (x_k - \hat{a}_1)(x_k - \hat{a}_1)' + \sum_{p=1}^{n_2} (y_p - \hat{a}_2)(y_p - \hat{a}_2)' \right\},$$
$$\hat{D}(x) = (\hat{R}^{-1}(\hat{a}_1 - \hat{a}_2), (x - \frac{1}{2}(\hat{a}_1 + \hat{a}_2)).$$

Lemma 25.6.1. *If R is a nonsingular matrix, then*

$$\hat{D}(x) \approx (n_1 + n_2 - 2)2^{-1}\{[(HH')_{11}^{-1} + (HH')_{22}^{-1}](c(x), b)$$
$$+ [(HH')_{11}^{-1} - (HH')_{22}^{-1}]\sqrt{(b,b)(c(x),c(x))}\},$$
$$b = R^{-1/2}[a_1 - a_2] + \nu_1 \sqrt{n_1^{-1} + n_2^{-1}},$$
$$c(x) = R^{-1/2}(x - \frac{1}{2}(a_1 + a_2)) + \frac{1}{2}\nu_2 \sqrt{n_1^{-1} + n_2^{-1}},$$

where ν_1 and ν_2 are independent m-dimensional random vectors, normally distributed by the law $N(0, I)$ and independent of the random matrix $H = (\nu_{ij})_{i=\overline{1,m}, j=\overline{1,n_1+n_2-2}}$ whose elements are independent and distributed according to the normal law $N(0, 1)$.

Proof. By using (25.3.2) for the function $\hat{D}(x)$, we have

$$\hat{D}(x) \approx \left(\left(\frac{HH'}{n_1 + n_2 - 2} \right)^{-1} b, c(x) \right), \qquad b = R^{-1/2}(\hat{a}_1 - \hat{a}_2),$$
$$c(x) = R^{-1/2}(x - \frac{1}{2}(\hat{a}_1 + \hat{a}_2)). \tag{25.6.1}$$

It is obvious that

$$((HH')^{-1}(n_1 + n_2 - 2)b, c(x)) \approx (n_1 + n_2 - 2) \operatorname{Tr}(HH')^{-1}K(x)/2,$$

where $K(x) = [bc'(x) + c(x)b']$. Since $HH' \approx THH'T$ for every orthogonal real matrix T of order m, from (25.6.1) we get

$$(n_1 + n_2 - 2)((HH')^{-1}b, c(x)) \approx (n_1 + n_2 - 2)[(HH')_{11}^{-1}$$
$$\times \lambda_1(x) + (HH')_{22}^{-1}\lambda_2(x)]/2, \tag{25.6.2}$$

where $\lambda_i(x)$, $i = \overline{1,2}$ are nonzero eigenvalues of the matrix $K(x)$.

We find them, considering that $c(x) = c$. It is obvious that

$$\det[Iz + K(x)] = \det[z + bc'][1 + ((Iz + bc')^{-1}c, b)]$$

$$= [z^n + z^{n-1}(b,c)]\left\{\frac{\partial}{\partial\varepsilon}\ln\det[Iz + bc' + \varepsilon cb']_{\varepsilon=0} + 1\right\}$$

$$= [z^n + z^{n-1}(b,c)]\left\{\frac{\partial}{\partial\varepsilon}\ln\det[Iz + \varepsilon cb']_{\varepsilon=0}\right.$$

$$\left. +\frac{\partial}{\partial\varepsilon}\ln[1 + ((Iz + \varepsilon cb')^{-1}b,c)]_{\varepsilon=0} + 1\right\}$$

$$= [z^n + z^{n-1}(b,c)]\left\{\frac{(c,b)}{z} - \frac{\frac{1}{z^2}(b,b)(c,c)}{1+z(b,c)} + 1\right\} = 0.$$

From this equation, we have

$$z^{n-2}\left[2z^2 + 2z(c,b) + \frac{1}{2}(b,c)^2 - (b,b)(c,c)\right] = 0.$$

Hence, we obtain

$$\lambda_{1,2}(x) = (c(x),b) \pm \sqrt{(b,b)(c(x),c(x))}.$$

Consequently, from (25.6.2) follows

$$\widehat{D}(x) \approx (n_1 + n_2 - 2)2^{-1}\{[(HH')_{11}^1 + (HH')_{22}^{-1}]$$
$$\times (c(x),b) + [(HH')_{11}^{-1}(HH')_{22}^{-1}\sqrt{(b,b)(c(x),c(x))}]\}.$$
$$(25.6.3)$$

Note, that

$$b \approx R^{-1/2}[a_1 - a_2] + \nu_1\sqrt{n_1^{-1} + n_2^{-1}},$$
$$c(x) \approx R^{-1/2}(x - \frac{1}{2}(a_1 + a_2)) + \nu_2\sqrt{n_1^{-1} + n_2^{-1}});$$
$$(25.6.4)$$

(25.6.2) and (25.6.3) imply the statement of Lemma 25.6.1.

Theorem 25.6.1. *Let $n_1 + n_2 - 2 > m$, and R_m be a nonsingular matrix,*

$$\lim_{m\to\infty}\frac{m}{n_1} = c_1, \quad \lim_{m\to\infty}\frac{m}{n_2} = c_2, \quad 0 \le c_1, c_2 < \infty, \quad c_1 + c_2 \ne c_1c_2,$$
$$(25.6.5)$$

$$\text{plim}_{m\to\infty}\{[(R^{-1}(a_1 - a_2),(a_1 - a_2)) + (R^{-1}c(\xi_i),c(\xi_i))]$$
$$\times [n_1^{-1} + n_2^{-1}] = 0, \quad i = \overline{1,2},$$
$$(25.6.6)$$

where ξ_i is random normally distributed $N(a_i, R)$ variable.

Then

$$\text{plim}_{n\to\infty}[\hat{D}(\xi_i)\frac{n_1+n_2-2-m}{n_1+n_2-2} - D(\xi_i)] = 0, \quad i = \overline{1,2}. \qquad (25.6.7)$$

Proof. By using the proof of the consistency for the estimate G_3 and condition (25.6.5), we get

$$\text{plim}_{m\to\infty}\left(\frac{HH'}{n_1+n_2-2}\right)^{-1}_{ii} = [1 - [c_1^{-1} + c_2^{-1}]^{-1}]^{-1}.$$

But the condition (25.6.6) implies (25.6.7). Theorem 25.6.1 is proved.

The estimate $G_{13} = \hat{D}(x)\frac{n_1+n_2-2-m}{n_1+n_2-2}$ will be referred to as the $G_{13}(x)$ estimate of the discriminant function $D(x)$.

Let us prove its asymptotic normality.

Theorem 25.6.2. *Let $n_1+n_2-2 > m$, and R nonsingular matrix and condition (25.6.5) hold. Then*

$$\lim_{m\to\infty} \mathbf{P}\{[G_{13}(\xi_i) - D(\xi_i)]\sqrt{n_1+n_2-m-2}d_n < x\} = \frac{1}{\sqrt{2\pi}}\int_{-\infty}^{x} e^{-y^2/2}dy,$$

where

$$d_n^{-2} = D^2(\xi_i) + \{(R^{-1}(a_1-a_2),(a_1-a_2)) + (R^{-1}(\xi_i - \frac{1}{2}(a_1-a_2)),$$

$$(\xi_i - \frac{1}{2}(a_1-a_2)) + (R^{-1}(a_1-a_2),(a_1-a_2))(R^{-1}(\xi_i - \frac{1}{2}(a_1-a_2)),$$

$$(\xi_i - \frac{1}{2}(a_1-a_2)))\}\frac{n_1+n_2-m-2}{(n_1^{-1}+n_2^{-1})^{-1}}.$$

The proof of Theorem 25.6.2 is similar to that of Theorem 25.4.1.

Similarly to that of §5, we consider the estimate G_{14} of the regularized discriminant function:

$$G_{14}(x) = ((I\varepsilon+\varepsilon\hat{\theta}_m^{-1}\hat{R})^{-1}\vec{x},(\hat{\vec{x}}_1-\hat{\vec{x}}_2))-2^{-1}((I\varepsilon+\varepsilon\hat{\theta}_m^{-1}\hat{R})^{-1}(\hat{\vec{x}}_1-\hat{\vec{x}}_2),(\hat{\vec{x}}_1+\hat{\vec{x}}_2)),$$

where θ_m is the nonnegative solution of the equation

$$1 - k_m + k_m\theta m^{-1}\text{Tr}(I\theta + \hat{R})^{-1} = \theta\varepsilon^{-1}, \quad \varepsilon > 0, \quad k_m = \frac{m}{n_1+n_2-2}.$$

As in the proof of Theorem 25.5.1, we prove that the estimate G_{14} is asymptotically normal.

§7 The G_{15}-Estimate of the Nonlinear Discriminant Function, Obtained by Observations over Random Vectors with Different Covariance Matrices

In the case of partitioning two objects based on the use of the normal distribution, the nonlinear discriminant function is (see §2):

$$2^{-1}[-(x-a_1)'R_1^{-1}(x-a_1) + (x-a_2)'R_2^{-1}(x-a_2) - \ln \det R_1 R_2^{-1}]. \quad (25.7.1)$$

As was mentioned in the previous chapters, the estimates of this function obtained by changing a_1, a_2, R_1, and R_2 with their empirical standard estimates are inadequate because of their slow convergence to the true ones. Therefore, instead of the two first terms in expression (25.7.1), it is appropriate to use G_{14}-estimates, and instead of the last one, to use the G_1-estimate:

$$G_1(R_1) = \ln \det \widehat{R}_1 - \ln \frac{(n-1)(n-2)\ldots(n-m)}{(n-1)^m}.$$

From Theorem 23.9.1 follows that, if instead of condition (23.9.2), we use

$$\lim_{n\to\infty} \ln \frac{n-m}{n} = 0,$$

then

$$\mathrm{plim}_{n\to\infty}[(G_1(R_1) - G_1(R_2)) - \ln \det R_1 R_2^{-1}] = 0.$$

Thus, the estimate G_{15} of expression (25.7.1), derived in the course of the compilation of G_{14} and G_1 and some conditions (that are strictly indicated within the correspondent theorems), will be consistent, where x is a normal random vector independent of $\widehat{R}_1$ and $\widehat{R}_2$.

CHAPTER 26

RANDOM DETERMINANTS IN
THE EXPERIMENT DESIGN

This chapter deals with problems of the experiment design under the G-condition:

$$\lim_{n \to \infty} \frac{m}{n} = c, \quad 0 < c < \infty.$$

Such a condition occurs when solving problems in which the number m of unknown parameters is large, and the number of experiments n has the same order.

Given the G-condition, the evaluation of every separate parameter yields under some standard conditions the value $c_1 n^{-1/2}$, where c_1 is some constant. In some cases, the total evaluation error is

$$mn^{-1/2}c_1.$$

In view of the above, it seems that it is impossible to obtain consistent estimates under the G-condition. However, for many problems it is necessary to evaluate not the parameters c_i, but some function of the estimates $\hat{c}_i$ of these parameters,

$$f_m(\hat{c}_1, \ldots, \hat{c}_m).$$

But it turns out that in many cases it is possible to find the limit of this function as $n \to \infty$,

$$\lim_{n \to \infty} [f_m(\hat{c}_1, \ldots, \hat{c}_m) - g(c_1, \ldots, c_m)] = 0.$$

The function g is known and is the solution of some equation. The function g differs from the true function f, but when these two functions are known, the expression to be found has the form

$$f\{g^{(-1)}\{f(\hat{c}_1, \ldots, \hat{c}_m)\}\},$$

where $g^{(-1)}$ is an inverse function.

Certainly, it is but a brief outline of applications of the G- analysis methods described in this chapter.

§1 The Resolvent Method in the Theory of Experiment Design

Let $\vec{x}$ be a vector of the system input parameters; y an output parameter connected with the vector $\vec{x}$ by some functional relation $f(\vec{x})$, the function f unknown. The problem is to find such values of parameters $\vec{x}$ which would allow to reach $\max f(x)$, where X is a certain range of parameter x. In accordance with the main postulates of the theory of the experiment design, the function $f(x)$ can be approximated by some linear function $\sum_{i=1}^{n} c_i x_i$, where c_i are unknown coefficients. This approximation may be sufficiently close only in that the domain D which includes the point of maximum to be found. Therefore, obtaining this domain D is the main problem in the theory of the experiment design. If an experiment is carried out in the points of this domain, then according to the method of least squares, we find for unknown constants c_i the following expression,

$$\vec{c} = \lim_{\alpha \downarrow 0}(I\alpha + X'X)^{-1}X'y, \tag{26.1.1}$$

where $X = (x_{ij})$, $i = \overline{1,n}$, $j = \overline{1,m}$, $\vec{x}_k = (x_{k1},\ldots,x_{kn})'$ is the Kth input vector, and y_k is the corresponding output value $\vec{x}_k \in D$.

If there is the inverse matrix $(X'X)^{-1}$, then instead of formula (26.1.1) we have the following one,

$$\vec{c} = (X'X)^{-1}X'y. \tag{26.1.2}$$

For some problems, it is not possible to carry out the experiment at the planned points. Instead of the matrix X in formulas (26.6.1) and (26.1.2), we consider the matrix $X + \Xi$, where $\Xi = (\xi_{ij})$ is some random matrix of noises with density p.

Consequently, in this case we have for the vector $\vec{c}$ the estimate

$$\hat{\vec{c}} = \lim_{\alpha \downarrow 0}((X + \Xi)'(X + \Xi) + I\alpha)^{-1}(X + \Xi)'\vec{y}.$$

Besides, the vector y, as a rule, is known with some error $\vec{\varepsilon}$.

Thus, the estimate of evaluation $\hat{c} - c$ is equal to

$$\hat{c} - c = \lim_{\alpha \downarrow 0}[((X + \Xi)'(X + \Xi) + I\alpha)^{-1}(X + \Xi)'\vec{\varepsilon}]. \tag{26.1.3}$$

From formula (26.1.3)], we obtain

$$\mathbf{E}\|\hat{c} - c\|^2 = \mathbf{E}\lim_{\alpha \downarrow 0}\mathrm{Tr}[(X + \Xi)'(X + \Xi) + I\alpha]^{-1}(X + \Xi)$$
$$\times \vec{\varepsilon}\vec{\varepsilon}'(X + \Xi)[(X + \Xi)'(X + \Xi) + I\alpha]^{-1}. \tag{26.1.4}$$

If noises (elements of random matrices) tend in probability to zero, the number of the parameters m is bounded (i.e., m has not trend to infinity when

n increases) then the influence of noises in formula (26.1.4) vanishes. However, at large dimensions of the matrix X, random noises Ξ can considerably distort the estimate. It is natural to assume that the number of parameters m and the number of experiments n are commensurable. As a measure of commensurablility, we can take the following G-condition:

$$\lim_{n \to \infty} \frac{m}{n} = c, \quad 0 < c < 1.$$

Define $X = A$.

In formula (26.1.4), we assume that $\alpha > 0$ is fixed, and components ε_i of the random vector $\vec{\varepsilon}$ are independent and do not depend on matrix Ξ, $\mathbf{E}\varepsilon_i = 0$, Var $\varepsilon_i = n^{-1}$. It is obvious that formula (26.1.4) is represented in the following form:

$$\mathbf{E}\|\hat{c}_\alpha - c\|^2 = n^{-1} \operatorname{Tr} \mathbf{E}(A + \Xi)'(A + \Xi)[(A + \Xi)'(A + \Xi) + I\alpha]^{-2}$$
$$+ \alpha^2 (R_\alpha^2 c, c),$$
$$R_\alpha = [(A + \Xi)'(A + \Xi) + I\alpha]^{-1}.$$

Theorem 26.1.1. *For every value of n, let elements of the matrix Ξ be independent and distributed according to the normal law with parameters $N(0, \sigma^2 n^{-1})$, $0 < \sigma^2 < \infty$, $\mu_n(x) \Rightarrow \mu(x)$, where*

$$\mu_n(x) = m^{-1} \sum_{k=1}^{m} F(x - \lambda_k),$$

$F(y) = 1$ at $y > 0$ and $F(y) = 0$ at $y \leq 0$, and λ_k are eigenvalues of matrix $A'A$. Then,

$$\lim_{n \to \infty} \mathbf{E} n^{-1} \operatorname{Tr}(A + \Xi)'(A + \Xi)[(A + \Xi)'(A + \Xi) + I\alpha]^{-2} = \alpha \frac{d}{d\alpha} m(\alpha) + m(\alpha),$$
$$(26.1.5)$$

where $m(\alpha)$ satisfies the equation

$$m(\alpha) = c \int_0^\infty [\alpha(1 + \sigma^2 m(\alpha)) + \sigma^2(1-c) + x(1 + \sigma^2 m(\alpha))^{-1}]^{-1} d\mu(x). \quad (26.1.6)$$

The solution of Eq. (26.1.6) exists and is unique in the class of analytic functions $m(\alpha) > 0$, $\alpha > 0$.

Proof. It is obvious that for any matrix X,

$$\operatorname{Tr} X'X(I\alpha + X'X)^{-2} = \alpha \frac{d}{d\alpha} \operatorname{Tr}(I\alpha + X'X)^{-1} + \operatorname{Tr}(I\alpha + X'X)^{-1}.$$

Matrix A can be represented in the form $A = U_1 \Lambda U_2$, where U_1 and U_2 are orthogonal matrices, and Λ is a diagonal matrix of the eigenvalues of matrix

$(A'A)^{1/2}$. Due to the fact that the distribution of matrix Ξ is invariant with respect to the orthogonal transformation, we obtain

$$\mathbf{E}\, \text{TR}(I\alpha + (A+\Xi)'(A+\Xi))^{-1} = \mathbf{E}\, \text{Tr}(I\alpha + (\tilde\Lambda + \Xi)'(\tilde\Lambda + \Xi))^{-1},$$
$$\tilde\Lambda' = (\Lambda, 0) - n \times m.$$

Let b_{ij} be elements of matrix $B = (\tilde\Lambda + \Xi)'(\tilde\Lambda + \Xi)$.

It is obvious that

$$\text{Tr}(I\alpha + B)^{-1} = \sum_{i=1}^{m_n} r_{ii},$$

where r_{ii} are diagonal elements of the matrix $R := (I\alpha + B)^{-1}$. The symbol ":=" means equality by definition. For the elements r_{ij}, the formula (see Chapter 9) holds:

$$r_{kk} = [\alpha + \sum_{l=1}^{n} b_{kl}^2 - \sum_{i\neq k, j\neq k} r_{ij}^k \sum_{l,p=1}^{n} b_{il} b_{kl} b_{jp} b_{kp}]^{-1} \qquad (26.1.7)$$

where r_{ij}^k are elements of the matrix $R_k := (I\alpha + (\Lambda + \Xi)_k (\Lambda + \Xi)_k')^{-1}$, and matrix $(\Lambda + \Xi)_k$ is obtained from matrix $\Lambda + \Xi$ by replacing elements of kth row by zeros. It is obvious that $r_{ik}^k = 0$, $i \neq k$. In this formula, we consider that the summation is taken over all $j,, i = \overline{1,n}$, except $i = k$, $j = k$.

Transform (26.1.7) into the following form:

$$r_{kk} = [\alpha + \sigma^2 + \lambda_k^2 - \sigma^2 n^{-1} \text{Tr}\, R_k (\Lambda + \Xi)_k' (\Lambda + \Xi)_k - \lambda_k^2 \sum_{i\neq k, j\neq k} r_{ij}^k b_{ik} b_{jk} + \varepsilon_{1n}]^{-1},$$

$$(26.1.8)$$

where

$$\varepsilon_{1n} = \sum_{l=1}^{n} b_{kl}^2 - \sigma^2 - \lambda_k^2 - \sum_{i\neq k, j\neq k} r_{ij}^k \sum_{l,p=1}^{n} b_{il} b_{ip} \xi_{kl} \xi_{kp}$$

$$+ \sigma^2 n^{-1} \text{Tr}\, R_k (\Lambda + \Xi)_k (\Lambda + \Xi)_k' - \lambda_k \sum_{p=1}^{n} \xi_{kp} \sum_{i,j\neq k} r_{ij}^k b_{ik} b_{jp}.$$

We introduce the notation $T_p^k = (b_{ip} b_{jp})$ as the square matrices of the nth order whose kth column and kth row have zero elements.

It is obvious that

$$\sum_{i\neq k, j\neq k} r_{ij}^k b_{ik} b_{jk} = \text{Tr}\, R_k T_k^k.$$

It is easy to see (see Chapter 9) that

$$\text{Tr}\, R_k T_k^k = (R_k^k \vec{b}_k, \vec{b}_k)[1 + (R_k^k \vec{b}_k, \vec{b}_k)]^{-1},$$

where

$$R_k^k = (I\alpha + \sum_{p \neq k} T_p^k)^{-1}, \quad \vec{b}_k = (b_{1k}, \dots, b_{nk}).$$

By taking this equality into consideration, we transform (26.1.8) to the following form:

$$r_{kk} = [\alpha + \alpha\sigma^2 n^{-1} \operatorname{Tr} R_k + \sigma^2 \frac{m}{n} + \lambda_k^2[1 + \sigma^2 n^{-1} \operatorname{Tr} R_k^k]^{-1} + \varepsilon_{2n}]^{-1}, \quad (26.1.9)$$

where

$$\varepsilon_{2n} = \varepsilon_{1n} + [1 + (R_k^k \vec{b}_k, \vec{b}_k)]^{-1} - [1 + \sigma^2 n^{-1} \operatorname{Tr} R_k^k]^{-1}.$$

Using (26.1.9), we obtain

$$n^{-1} \mathbf{E} \operatorname{Tr} R = n^{-1} \sum_{k=1}^{m} \mathbf{E}[\alpha(1 + \sigma^2 n^{-1} \operatorname{Tr} R_k) + \sigma^2(1 - c)$$
$$+ \lambda_k^2 (1 + \sigma^2 n^{-1} \operatorname{Tr} R_k^k)^{-1}]^{-1} + 0(1).$$

Therefore, for $m_n(\alpha) := n^{-1} \mathbf{E} \operatorname{Tr} R$, the equation

$$m_n(\alpha) = n^{-1} \sum_{k=1}^{m_n} [\alpha(1 + \sigma^2 m_n(\alpha)) + \sigma^2(1 - c) + \lambda_k^2(1 + \sigma^2 m_n(\alpha))^{-1}]^{-1} + 0(1)$$

$$(26.1.10)$$

holds. We introduce the functions

$$\mu_n(x) = n^{-1} \sum_{k=1}^{n} F(x - \lambda_k).$$

Then formula (26.1.10) takes the form

$$m_n(\alpha) = c \int_0^\infty [\alpha(1 + \sigma^2 m_n(\alpha)) + \sigma^2(1-c) + x(1 + \sigma^2 m_n(\alpha))^{-1}]^{-1} d\mu_n(x) + 0(1).$$

Choose a convergent subsequence of functions $m_n(\alpha)$. It is easy to do since $m_n(\alpha)$ is the Stieltjes transformation for some function of distribution. Then, passing to the limit as $n' \to \infty$, we have $m_{n'}(\alpha) \to m(\alpha)$, and $m(\alpha)$ satisfies the equation

$$m(\alpha) = c \int_0^\infty [\alpha(1 + \sigma^2 m(\alpha)) + \sigma^2(1 - c) + x(1 + \sigma^2 m(\alpha))^{-1}]^{-1} d\mu(x).$$

Due to the uniqueness of this solution of the equation, the whole sequence $m_n(\alpha)$ tends to the limit as $n \to \infty$. Theorem 26.1.1 is proved.

We generalize Theorem 26.1.1. Suppose that the random elements $\xi_{pl}^{(n)}$ and $\xi_{lp}^{(n)}$ of the matrix $H_n = (\xi_{pl}^{(n)})$ are dependent.

Again, without loss of generality we assume, that all matrices considered in this Theorem are square.

Theorem 26.1.2. *For any* $n = 1, 2, \ldots$, *let the random vectors* $(\xi_{lp}^{(n)}, \xi_{pl}^{(n)})$, $p \geq l$, $p, l = \overline{1, n}$ *be stochastically independent,*

$$\mathbf{E}|\xi_{pl}^{(n)}|^2 = n^{-1}, \qquad \mathbf{E}\xi_{pl}^{(n)}\xi_{lp}^{(n)} = \rho n^{-1}, \quad \rho > 0, \qquad (26.1.11)$$

let the Lindeberg condition hold, for any $\tau > 0$,

$$\lim_{n \to \infty} n^{-1} \sum_{p,l=1}^{n} \mathbf{E}|\xi_{pl}^{(n)}|^2 \chi(|\xi_{pl}^{(n)}| > \tau) = 0, \qquad (26.1.12)$$

let the matrices $A = (a_{ij})_{i,j=1}^{n}$ *be symmetric,* $\sup_n \sup_{i=\overline{1,n}} \sum_{j=1}^{n} a_{ij}^2 < \infty$, *and*

$$\nu_n(x) \Rightarrow \nu(x), \qquad (26.1.13)$$

where $\nu_n(x)$ *is the normalized spectral function of the matrix* A, *and* $\nu(x)$ *is a distribution function.*
   Then for $\theta > 0$,

$$\lim_{n \to \infty} \mathbf{E}n^{-1} \operatorname{Tr}(A + \Xi)'(A + \Xi)[(A + \Xi)'(A + \Xi) + I\theta]^{-2} = \frac{d}{d\theta}\theta m(\theta), \quad (26.1.14)$$

where $m(\theta)$ *satisfies the equation*

$$m(\theta) = \int \{\theta[1 + (1 - \rho)m(\theta) + y^2[1 + (1 - \rho)m(\theta)]^{-1}]^{-1}d\mu(y). \quad (26.1.15)$$

The Stieltjes transform $p(t) = \int (1 + ity)^{-1}d\mu(y)$ of the spectral function $\mu(y)$ satisfies the equation:

$$p(t) = \int [1 + ity + t^2 \rho p(t)]^{-1}d\nu(y).$$

Solutions of Eq. (26.1.15) exist and are unique in the class of analytic functions $m(\theta) > 0$, as $\theta > 0$, $p(t)$, $\operatorname{Re}p(t) \geq 0$.

Proof. Let the Lindeberg condition (26.1.12) hold. Consider the matrices $P_n = n^{-1/2}\rho^{1/2}L_n + (1 - \rho)^{1/2}\Xi n^{-1/2}$, where $L_n = (\nu_{ij})_{i,j=1}^{n}$ are real symmetric matrices whose entries ν_{ij}, $i \geq j$ are independent and distributed according to the normal law $N(0, 1)$, the matrix Ξ does not depend on the matrix A, and all its entries are independent and also distributed according to the normal law $N(0,1)$, the matrices P_n do not depend on the H_n.
   Let $Q = (I\theta + (A + P_n)(A + P_n)')^{-1}$ and show that

$$\lim_{n \to \infty} \mathbf{E}[n^{-1}\operatorname{Tr}(I\theta + BB')^{-1} - n^{-1}\operatorname{Tr}Q] = 0, \quad B = A + H_n. \quad (26.1.16)$$

For this, we introduce the matrices T_k whose entries in the first k columns and rows coincide with those of matrix $A + P_n$ and the rest of the rows and columns and vectors are equal to those of matrix B.

Consider the equality

$$
n^{-1} \operatorname{Tr}(I\theta + BB')^{-1} - n^{-1} \operatorname{Tr} Q
$$
$$
= n^{-1} \sum_{k=1}^{n} [\operatorname{Tr}(I\theta + T_{k-1}T'_{k-1})^{-1} - \operatorname{Tr}(I\theta + T_k T'_k)^{-1}]
$$
$$
= \sum_{k=1}^{n} n^{-1} [\operatorname{Tr}(I\theta + T_{k-1}T'_{k-1})^{-1} - \operatorname{Tr}(I\theta + \widetilde{T}_{k-1}\widetilde{T}'_{k-1})^{-1}
$$
$$
- \{\operatorname{Tr}(I\theta + T_k T'_k)^{-1} - \operatorname{Tr}(I\theta + \widetilde{T}_{k-1}\widetilde{T}'_{k-1})^{-1}\}],
\tag{26.1.17}
$$

where $\widetilde{T}_k$ is the matrix obtained from the matrix T_k by zeroes for the kth row. It follows from Chapter 17 that

$$
\operatorname{Tr}(I\theta + T_{k-1}T'_{k-1})^{-1} - \operatorname{Tr}(I\theta + \widetilde{T}_{k-1}\widetilde{T}'_{k-1})^{-1}
$$
$$
= (\frac{\partial}{\partial \theta}) \ln[\theta + (b_k, \bar{b}_k) - (R_k \widetilde{T}_{k-1} b_k, \widetilde{T}_{k-1} b_k)],
\tag{26.1.18}
$$

where $R_k = (I\theta + \widetilde{T}_{k-1}\widetilde{T}'_{k-1})^{-1}$,

$$
\operatorname{Tr}(I\theta + T_k T'_k)^{-1} - \operatorname{Tr}(I\theta + \widetilde{T}_{k-1}\widetilde{T}'_{k-1})^{-1}
$$
$$
= (\frac{\partial}{\partial \theta}) \ln[\theta + (\nu_k, \nu_k) - (R_k \widetilde{T}_{k-1}\nu_k, \widetilde{T}_{k-1}\nu_k)],
\tag{16.1.19}
$$

where ν_k and b_k are the kth row of the matrix T_k and B, respectively.

Since $(b_k, b_k) - (R_k \widetilde{T}_{k-1}b_k, \widetilde{T}_{k-1}b_k) = ((I\theta + \widetilde{T}'_{k-1}\widetilde{T}_{k-1})^{-1}b_k, b_k)$, it follows from Chapter 17 that for any $\varepsilon > 0$,

$$
\sup_{\theta > \varepsilon > 0} \{|(\frac{\partial}{\partial \theta}) \ln[\theta + (b_k, b_k) - (R_k \widetilde{T}_{k-1}b_k, \widetilde{T}_{k-1}b_k)]|
$$
$$
+ |(\frac{\partial}{\partial \theta}) \ln[\theta + (\nu_k, \nu_k) - (R_k \widetilde{T}_{k-1}\nu_k, \widetilde{T}_{k-1}\nu_k)]|\} < \infty.
\tag{26.1.20}
$$

Let us now consider the equality (26.1.18) more closely. Obviously,

$$
(R_k \widetilde{T}_{k-1}b_k, \widetilde{T}_{k-1}b_k) = \operatorname{Tr} R_k C,
\tag{26.1.21}
$$

where $C = \widetilde{T}_{k-1}b_k(\widetilde{T}_{k-1}b_k)' = \widetilde{T}_{k-1}b_k b'_k \widetilde{T}'_{k-1}$.

Let $\tilde{R}_k = (I\theta + \tilde{T}_{k-1}\tilde{T}'_{k-1})^{-1}$, where $\tilde{T}_{k-1}$ is the matrix obtained from the matrix T_{k-1} by deleting the kth column a_k and kth row b_k. Due to Chapter 17, (26.1.21) yields

$$\mathrm{Tr}\, R_k C - \mathrm{Tr}\, \tilde{R}_k C = -(\tilde{R}_k C \tilde{R}_k a_k, a_k)[1 + (\tilde{R}_k a_k, a_k)]^{-1}. \qquad (26.1.22)$$

Using (26.1.22), we find:

$$\mathrm{Tr}(I\theta + T_{k-1}T'_{k-1})^{-1} - \mathrm{Tr}(I\theta + \tilde{T}_{k-1}\tilde{T}'_{k-1})^{-1} = (\frac{\partial}{\partial\theta})\ln[\theta + (b_k, b_k)$$
$$+ (\tilde{R}_k C \tilde{R}_k a_k, a_k)[1 + (\tilde{R}_k a_k, a_k)]^{-1} - (\tilde{R}_k \tilde{T}_{k-1} b_k, \tilde{T}_{k-1} b_k)]. \qquad (26.1.23)$$

Note, that

$$b_k = (\xi_{1k} + a_{1k}, \ldots, \xi_{kk} + a_{kk} + \theta, \ldots, \xi_{nk} + a_{nk}),$$
$$\tilde{T}_{k-1} b_k = \tilde{T}_{k-1}\tilde{b}_k + (a_{kk} + \theta + \xi_{kk})a_k, \qquad (26.1.24)$$

where $\tilde{b}_k = b_k$ as $a_{kk} + \theta = 0$. (Index (n) in $\xi_{pl}^{(n)}$ will be omitted sometimes). Using (26.1.24), we have

$$(\tilde{R}_k \tilde{T}_{k-1} b_k, \tilde{T}_{k-1} b_k) = (\tilde{R}_k \tilde{T}_{k-1}\tilde{b}_k, \tilde{T}_{k-1}\tilde{b}_k) + (\tilde{R}_k \tilde{T}_{k-1}\tilde{b}_k, a_k)(\xi_{kk} + a_{kk})$$
$$+ (\tilde{R}_k a_k, \tilde{T}_{k-1}\tilde{b}_k)(\xi_{kk} + a_{kk}) + (a_{kk} + \xi_{kk})^2(\tilde{R}_k a_k, a_k). \qquad (26.1.25)$$

Substituting (26.1.25) to (26.1.23) and using (26.1.20) and (26.1.11), we obtain

$$\mathbf{E}n^{-1}\sum_{k=1}^{n}[\mathrm{Tr}(I\theta + T_{k-1}T'_{k-1})^{-1} - \mathrm{Tr}(I\theta + \tilde{T}_{k-1}\tilde{T}'_{k-1})^{-1}]$$
$$= n^{-1}\sum_{k=1}^{n}\mathbf{E}(\frac{\partial}{\partial\theta})\ln[\theta(1 + n^{-1})\,\mathrm{Tr}\,\tilde{R}_k$$
$$+ |\rho n^{-1}\,\mathrm{Tr}\,\tilde{R}_k\tilde{T}_{k-1} - \theta|^2(1 + n^{-1}\,\mathrm{Tr}\,\tilde{R}_k)^{-1}] + 0(1), \quad n \to \infty. \qquad (26.1.26)$$

Analogously, we prove that

$$\mathbf{E}n^{-1}\sum_{k=1}^{n}[\mathrm{Tr}(I\theta + T_{k-1}T'_{k-1})^{-1} - \mathrm{Tr}(I\theta + \tilde{T}_{k-1}\tilde{T}'_{k-1})^{-1}]$$

is equal to the same expression.

Therefore, because of (26.1.17), the equality (26.1.16) holds.

We can represent the matrix $K = A + n^{-1/2}\rho^{1/2}L$ in the form $H\Lambda H'$, where H is an orthogonal matrix of the eigenvectors, $\Lambda = (\lambda_i \delta_{ij})$ is a diagonal matrix of the eigenvalues. Since the distributions of the matrices $H\Xi H'$ and Ξ coincide,

$$\mathbf{E} n^{-1} \operatorname{Tr} Q = n^{-1} \mathbf{E} \operatorname{Tr}(I\theta + (\Lambda - (1-\rho)^{1/2}n^{-1/2}\Xi)(\Lambda - (1-\rho)^{1/2}n^{-1/2}\Xi)')^{-1}.$$

For this expression with fixed Λ, we can apply the proof of the circle law (Ch. 18).

By repeating almost literally the proof of the circle law and by using (26.1.16), we obtain

$$m_n(\theta) = n^{-1}\mathbf{E}\sum_{k=1}^{n}\{\theta[1 + (1-\rho)m_n(\theta)]$$
$$+ \lambda_k^2(1 + (1-\rho))m_n(\theta))^{-1}\}^{-1} + 0(1), \quad n \to \infty,$$

where $m_n(\theta) = n^{-1}\mathbf{E}\operatorname{Tr}(I\theta + BB')^{-1}$, $\theta > 0$.

This equation yields

$$m_n(\theta) = \int \{\theta[1 + (1-\rho)m_n(\theta)] + x^2(1 + (1-\rho)$$
$$\times m_n(\theta))^{-1}\}^{-1}d\mathbf{E}\mu_n(x) + 0(1), \quad n \to \infty,$$

(26.1.27)

where $\mu_n(x)$ is the normalized spectral function of the matrix K.

Let us choose a convergent subsequence $m_{n'}(\theta) \to m(\theta) n \to \infty$, $\theta > 0$. Since the functions $m_n(\theta)$ are Stieltjes transforms and analytic functions as $\theta > 0$, $m(\theta)$ will be analytic functions, too. Then, by using Eq. (9.5.3), we obtain (26.1.14) from (26.1.27). It is easy to verify that Eq. (26.1.15) has the unique solution in the class of analytical functions as $\theta > 0$. Therefore $\lim_{n\to\infty} m_n(\theta) = m(\theta)$, where $m(\theta)$ is the solution of the equation (26.1.15). Theorem 26.1.2 is proved.

§2 The G_{16}-Estimate of the Estimation Errors in the Theory of the Design of Experiments

Let us give the system

$$y = Ac + \xi,$$ (26.2.1)

where $A = (a_{ij})_{j=\overline{1,m_n}}$, $i = \overline{1,n}$ is a matrix of experiment design, which is unknown, and only the observation $A + \Xi$ over this matrix is known,

$\Xi = (\xi_{ij})_{j=\overline{1,m_n}}$, $\quad i = \overline{1,n}$ is a random matrix of noise,

$c = (c_i, i = \overline{1,m_n})'$ is a vector of unknown parameters,

$y = (y_i, i\overline{1,n})'$ is a vector of observations,

$\vec{\xi} = (\xi_i, i = \overline{1,n})'$ is a vector of observation errors,

which is stochastically independent of matrix Ξ.

Assume, that $\mathbf{E}\vec{\xi} = 0$, $\mathbf{E}\vec{\xi}\vec{\xi}' = n^{-1}I$, the random variables ξ_{ij} are independent and distributed concerning random variables are introduced for the sake of simplification of the vector $\vec{c}$ estimates. If the matrices $(A'A)$ are nonsingular, then as an estimate for vector $\vec{c}$ we can take the estimate obtained by the method of least squares.

$$\hat{c} = (A'A)^{-1}Ay. \qquad (26.2.2)$$

From (26.2.1) and (26.2.2) we obtain that estimation errors are

$$\mathbf{E}\|c - \hat{c}\|^2 = \mathrm{Tr}(A'A)^{-1}. \qquad (26.2.3)$$

In this section, we solve the following problem: how, with one observation $X = A + \Xi$, we can estimate expression (26.2.3) when the G-conditions hold: $\lim_{n\to\infty} m_n n^{-1} = c < 1$. For this purpose, instead of (26.2.3), we consider expression

$$\mathbf{E}\,\mathrm{Tr}[(A + \Xi)'(A + \Xi)]^{-1}.$$

Note, that for any real $\theta > 0$,

$$|n^{-1}\,\mathrm{Tr}[n^{-1}A'A]^{-1} - n^{-1}\,\mathrm{Tr}[I\theta + n^{-1}A'A]^{-1}| \leq \theta n^{-1}\,\mathrm{Tr}[n^{-1}A'A]^{-2}. \quad (26.2.4)$$

Using the proof of Theorem 26.1.1, we obtain that the function

$$m_n(\theta) = n^{-1}\mathbf{E}\,\mathrm{Tr}[I\theta + n^{-1}(A + \Xi)'(A + \Xi)]^{-1}$$

satisfies the equation

$$m_n(\theta) = c \int_0^\infty [\theta(1 + m_n(\theta)) + 1 - c + x(1 + m_n(\theta))^{-1}]^{-1}d\mu_n(x) + 0(1), \quad (26.2.5)$$

where $\mu_n(x)$ is a normalized spectral function of the matrix $A'An^{-1}$.

From equation (26.2.5), we obtain that if $\hat{\theta}$ is a solution of the equation

$$\hat{\theta}(1 + m_n(\hat{\theta}))^2 + (1 - c)(1 + m_n(\hat{\theta})) = \theta, \quad \theta > 0,$$

then Eq. (26.2.5) is

$$c^{-1}m_n(\hat{\theta})(1 + m_n(\hat{\theta}))^{-1} = \int_0^\infty (\theta + x)^{-1}d\mu_n(x) + 0(1).$$

Thus, as an estimate of the value $n^{-1}\,\mathrm{Tr}(I\theta + n^{-1}A'A)^{-1}$, we can choose the quantity

$$G_{16} = c^{-1}\hat{m}_n(\hat{\theta})(1 + \hat{m}_n(\hat{\theta}))^{-1},$$

where $\hat{m}_n(\theta) = n^{-1}\operatorname{Tr}[I\theta + X'X]^{-1}$, $X \approx A + \Xi$ is the observation over A, and $\hat{\theta}$ is a positive solution of the equation

$$\hat{\theta}(1 + \hat{m}_n(\hat{\theta}))^2 + (1 - c)(1 + \hat{m}_n(\hat{\theta})) = 0, \quad \theta > 0.$$

Suppose that a positive solution of the equation exists and is unique.
   As for any $\theta > 0$,

$$\operatorname{plim}_{n\to\infty}[\hat{m}_n(\theta) - m_n(\theta)] = 0,$$

(see Chapters 9–11), then it is easy to show, that

$$\operatorname{plim}_{n\to\infty}[c^{-1}\hat{m}_n(\hat{\theta})(1 + \hat{m}_n(\hat{\theta}))^{-1} - m^{-1}\operatorname{Tr}(I\theta + n^{-1}A'A)^{-1}] = 0.$$

But, if in addition,

$$\overline{\lim_{n\to\infty}} \, n\operatorname{Tr}(A'A)^{-2} < \infty,$$

then (26.2.4) implies that

$$\lim_{\theta\downarrow 0}\operatorname{plim}_{n\to\infty}[c^{-1}\hat{m}_n(\hat{\theta})(1 + \hat{m}_n(\hat{\theta}))^{-1} - n^{-1}\operatorname{Tr}(A'A)^{-1}] = 0.$$

CHAPTER 27

RANDOM DETERMINANTS IN PHYSICS

In this chapter we shall look at the application of random determinants in nuclear physics, theory of scattering, in some models of disordered crystals, and in statistical physics.

§1 The Wigner Hypothesis

Energy levels of heavy atomic nuclei under high energy are arranged rather close to one another, and it is practically impossible to find them even if the Hamiltonian of system is known. In this connection, Wigner proposed the statistical model of highly excited states of heavy atomic nuclei in which complex nuclei were regarded as some "black cavity," and particles, which made up a nucleus, interacted according to an unknown random law. The central problem is the choice of a mathematical model of such systems. As a model of an ensemble of complex nuclei, Wigner chose a Hermitian matrix of large dimensions, elements of which were independent random variables with zero means, identical variances, and limited moments. This model of the Hamiltonian of the system is the simplest one, and it is not surprising that the normalized spectral function of such a matrix has nothing in common with the distribution of levels in nuclei. The limit spectral function for normalized spectral functions of Hermitian random matrices described above was obtained by Wigner. Its density has the semicircle form; and therefore, the spectral function was called Wigner's semicircle law [167]. To reach coordination between theoretical and experimental results, one may now complicate the statistical model of the atomic nucleus by allowing the elements of the random matrix to have different distribution or being independent random variables [50]. But it turned out that the statistical model of the nucleus, proposed by Wigner, agreed with empirical densities of energy levels of nuclei, and was also rather suitable for the investigation of the theoretical law of distance (spacing) between neighbouring levels. Let us study the subject. In many papers, it was pointed out that the "repulsion" between levels of identical symmetry must lead to disappearance of levels with distances as small

627

as desired in spectra of atomic nuclei. Such behaviour of energy levels was also confirmed by experimental data. Wigner proposed the following hypothesis: In the sequence of great numbers of levels, on the average separated by distance D from one another and having identical values for all quantum numbers submitted to identification, such as moments of quantity of movements and even parity, the probability to find two levels at the distance between t and $t + \Delta t$ is equal to

$$Q(t)dt := (2D)^{-1}\pi t \exp(-\pi t^2 4^{-1} D^{-2})dt. \qquad (27.1.1)$$

As the mathematical model for checking this hypothesis, Wigner's model was chosen again. Mehtà and Gaudin [132] obtained the density $p(t)$ of the distance between two neighbouring eigenvalues of a random Hermitian matrix, elements of which were arranged on a diagonal and above it, were independent and distributed according to the standard normal law. This density differs from the value (27.1.1) but not too much, $\sup_t |Q(t) - p(t)| \le 0.0162$. Therefore, the Wigner hypothesis is suitable for practical purposes.

Below, we consider a formula for the average distribution function of distances between neighbouring eigenvalues of some random matrices. With the help of this formula and the limit theorems for Borel functions of random variables, the assertion specifying the results of Mehta and Gaudin was obtained.

Let $\lambda_1 \ge \lambda_2 \ge \cdots \ge \lambda_n$ be the eigenvalues of a symmetric random matrix Ξ and suppose the Ξ is such that the random variables λ_i have the density $p(x_1, \ldots, x_n)$, $x_1 > \cdots > x_n$, and the function p is symmetric, i.e., p is invariant under a simultaneous permutation of the variables. We consider the spectral function

$$\theta_n(x) = n^{-1} \sum_{i=1}^{n-1} \mathbf{E}F(x - (\lambda_i - \lambda_{i+1})),$$

where

$$F(x) = \begin{cases} 1, & x > 0 \\ 0, & x \le 0. \end{cases}$$

The function $\theta_n(x)$ is equal to the average distribution function of a random variable, which is equal to the distance between two neighbouring eigenvalues. Obviously owing to the fact that $p(y_1, \ldots, y_n)$ is symmetric,

$$\theta_n(x) = n^{-1} \sum_{i=1}^{n} \iint_{y_i > y_{i+1}, y_i - y_{i+1} < x} dy_i dy_{i+1} \iint_{y_1 > \cdots > y_i > y_{i+1} \cdots > y_n}$$

$$\times p(y_1, \ldots, y_n) \prod_{k \ne i, i+1}^{n} dy_k = n^{-1}[(n-2)!]^{-1} \sum_{i=1}^{n}$$

$$\times \iint_{y_i > y_{i+1}, y_i - y_{i+1} < x} dy_i dy_{i+1} \iint_{L_i} p(y_1, \ldots, y_n) \prod_{k \ne i, i+1}^{n} dy_k,$$

where L_i is the set of values of variables y_l, $l \neq i$, $i+1$, among which $i-1$ variables are greater than y_i and the rest are less than y_{i+1}. Then

$$\theta_n(x) = n^{-1}(n!)^{-1} \sum_{s \neq p} \iint_{y_s > y_p, y_s - y_p < x} dy_s dy_p$$

$$\times \int_{R \setminus]y_s, y_p[} \cdots \int_{R \setminus]y_s, y_p[} p(y_1, \ldots, y_n) \prod_{k \neq s, p} dy_k,$$

where R is the set of real numbers.

Differentiating the equation with respect to x, we have

$$(\frac{\partial}{\partial x})\theta_n(x) = n^{-1}(n!)^{-1} \sum_{i \neq j} \int dy_i \int_{R \setminus]y_j + x, y_j[} \cdots \int_{R \setminus]y_j + x, y_j[}$$

$$\times p(y_1, \ldots, y_{i-1}, y_j + x, y_{i+1}, \ldots, y_n) \prod_{k \neq i, j} dy_k. \qquad (27.1.2)$$

We write Eq. (27.1.2) in the following form,

$$(\frac{\partial}{\partial x})\theta_n(x) = n^{-1}(n!)^{-1} \sum_{i \neq j} \int_{-c}^{c} dy_i \int_{R \setminus]y_j + x, y_j[} \cdots \int_{R \setminus]y_j + x, y_j[}$$

$$\times p(y_1, \ldots, y_{i-1}, \ldots, y_n) \prod_{k \neq i, j} dy_k + \varepsilon_n(x),$$

where

$$\varepsilon_n(x) = (\frac{\partial}{\partial x})[n^{-1} E \sum_{i=1}^{n-1} F(x - (\lambda_i - \lambda_{i+1})) F(|\lambda_{i+1}| - c)], \quad c > 0. \quad (27.1.3)$$

In addition to $(\frac{\partial}{\partial x})\theta_n(x)$, we consider the spectral functions

$$(\frac{\partial}{\partial x})\kappa_n(x, c) = (\frac{\partial}{\partial x})\theta_n(x) - (\frac{\partial}{\partial x})\varepsilon_n(x) = (\frac{\partial}{\partial x})[n^{-1} E \sum_{i=1}^{n-1}$$

$$\times F(x - (\lambda_i - \lambda_{i+1})) F(c - |\lambda_{i+1}|)].$$

After obvious transformations, we obtain

$$(n!)^{-1} \sum_{i \neq j} \int_{R \setminus]y, v[} \int_{R \setminus]y, v[} p(y_1, \ldots, y_n)_{y_j = v, y_i = u} \prod_{k \neq j, i} dy_k$$

$$= (n!)^{-1} (\frac{\partial^2}{\partial u \partial v}) \int_{R \setminus]u, v[} \cdots \int_{R \setminus]u, v[} p(y_1, \ldots, y_n) \prod_{k=1}^{n} dy_k = (\frac{\partial^2}{\partial u \partial v}) p(u, v),$$

where $p(u, v)$ is the probability that all eigenvalues lie outside the interval $]u, v[$.

On the basis of this equation and Eq. (27.1.2), the formula (27.1.3) takes the following form,

$$\theta_n(x) = n^{-1} \int_0^x \int_{-c}^c (\frac{\partial^2}{\partial u \partial v}) p(u, v)_{u=y, v=z+y} \, dy \, dz + \int_0^x \varepsilon_n(z) \, dz. \quad (27.1.4)$$

A similar formula was derived by Gaudin and Mehta [132] "on the physical level of rigor."

Now we use the original method of investigation of the formula (27.1.4) proposed by Mehta and Gaudin.

Let $p(y_1, y_2, \ldots, y_n)$ be the density of the eigenvalues of a Gaussian Hermitian matrix:

$$c \exp(-\sum_{i=1}^n x_i^2) \prod_{i>j} (x_i - x_j)^2, \quad x_1 > \cdots > x_n,$$

$$c^{-1} = 2^{-n(n-1)/2} n! \pi^{n/2} \prod_{j=o}^{n-1} j!.$$

For this matrix,

$$p(u, v) = (n!)^{-1} \int \cdots \int_{x_j \in]u, v[} \det[\varphi_{k-1}(x_j)]_{k,j=\overline{1,n}}^2 \prod_{i=1}^n dx_i,$$

where

$$\varphi_j(x) = (2^j j! \sqrt{\pi})^{-1/2} e^{x^2/2} (-\frac{\partial}{\partial x})^j e^{-x^2}.$$

We use the following formula:

$$\det[\int_a^b \varphi_i(x)\varphi_j(x) \, dx] = (n!)^{-1} \int_a^b \cdots \int_a^b \overset{2}{\det}(\varphi_i(x_j)) \prod_{j=1}^n dx_j. \quad (27.1.5)$$

Then we obtain

$$p(u, v) = \det[\delta_{ij} - \int_u^v \varphi_i(x)\varphi_j(x) \, dx].$$

To study limit theorems for such determinants, we make use of their integral representations

$$p(u, v) = \left[\mathbf{E} \exp \left\{ \int_u^v \left(\sum_{i=1}^n \eta_i \varphi_i(x) \right)^2 dx \right\} \right]^{-2}, \quad (27.1.6)$$

where η_i, $i \in N$ are independent random variables distributed according to the normal law $N(0, 1/2)$.

Using the latter equation, we rewrite $\kappa_n(x, c)$ in the following form,

$$\mathbf{E}\kappa_n(x, c) = n^{-1} \int_0^x \int_{-c}^c dy[2p^{-3}(u, v) \left[\mathbf{E}\exp\left\{\int_u^v \left(\sum \eta_i \varphi_i(x)\right)^2 dx\right\}\right.$$

$$\times \left(\sum \eta_i \varphi_i(v)\right)^2 \left(\sum \eta_i \varphi_i(u)\right)^2 - 3p^{-1}(u, v)$$

$$\times \mathbf{E}\exp\left\{\int_u^v \left(\sum \eta_i \varphi_i(x)\right)^2 dx\right\} \left(\sum \eta_i \varphi_i(u)\right)^2$$

$$\left. \times \mathbf{E}\exp\left\{\int_u^v \left(\sum \eta_i \varphi_i(x)\right)^2 dx\right\} \left(\sum \eta_i \varphi_i(v)\right)^2\right]\bigg|_{v=y, u=z+y} dz.$$

Let

$$\exp\left\{\int_u^v \left(\sum \eta_i \varphi_i(x)\right)^2 dx\right\} = \xi(u, v).$$

Then

$$\mathbf{E}\kappa(z, c) = n^{-1} \int_0^z \int_{-c}^c dy \left\{2p^{-3}(y, x + y) \left[\mathbf{E}\xi(y, x + y) \left(\sum \eta_i \varphi_i(y)\right)^2\right.\right.$$

$$\times \left(\sum \eta_i \varphi_i(x + y)\right)^2 - 3p^{-1}(y, x + y)\mathbf{E}\xi(y, x + y) \left(\sum \eta_i \varphi_i(y)\right)^2$$

$$\left.\left. \times \mathbf{E}\xi(y, x + y) \left(\sum \eta_i \varphi_i(x + y)\right)^2\right]\right\} dx.$$

Substitute the values $zn^{-1/2}$, $cn^{-1/2}$ by z, c and replace the variables $y = un^{-1/2}$, $x = vn^{-1/2}$; we have as a result

$$\mathbf{E}\kappa(zn^{-1/2}, cn^{-1/2}) = n^{-1} \int_0^z \int_{-c}^c dy \left\{2p^{-3}(yn^{-1/2}, (x + y)n^{-1/2})\right.$$

$$\times \left[\mathbf{E}\xi(yn^{-1/2}, (x + y)n^{-1/2})n^{-1} \left(\sum \eta_i \varphi_i(yn^{-1/2})\right)^2\right.$$

$$\times \left(\sum \eta_i \varphi_i((x + y)n^{-1/2})\right)^2 - 3p^{-1}(yn^{-1/2}, (x + y)n^{-1/2})$$

$$\times \mathbf{E}\xi(yn^{-1/2}, (x + y)n^{-1/2}) \left(\sum \eta_i \varphi_i(yn^{-1/2})\right)^2 \mathbf{E}\xi(yn^{-1/2},$$

$$\left.\left. (x + y)n^{-1/2})n^{-1/2} \left(\sum \eta_i \varphi_i((x + y)n^{-1/2})\right)^2\right]\right\} \qquad (27.1.7)$$

(the variables u and v are denoted by y, x in the obtained interval). Find the limit $p(yn^{-1/2}, (x + y)n^{-1/2})$ as $n \to \infty$.

Obviously,

$$p(yn^{-1/2}, (x + y)n^{-1/2}) = \left[\mathbf{E}\exp\left\{\int_0^x \eta_n^2(z + y)dz\right\}\right]^{-2},$$

where

$$\eta_n(t) = \sum_{i=1}^{n} \eta_i \varphi_i(tn^{-1/2})n^{-1/4},$$

and $\eta_n(t)$ is a Gaussian continuous random function with probability 1 for which $\mathbf{E}\eta_n(t) = 0$, and the covariance function is

$$R_n(t,c) = n^{-1/2} \sum_{i=0}^{n-1} \varphi_i(tn^{-1/2})\varphi_i(sn^{-1/2}).$$

We prove the following lemma.

Lemma 27.1.1. $\eta_n(t) \Rightarrow \eta_n(t)$, $-\infty < t < \infty$ where $\eta(t)$ is the stationary Gaussian random function, $\mathbf{E}\eta(t) = 0$,

$$R(t,s) = [\pi^{-1} \sin(t-s)](t-s)^{-1}.$$

Proof. Let $p_0(x), \ldots, p_n(x)$ be n orthogonal polynomials and k_n be the coefficient of highest order of the polynomial $p_n(x)$. The Christoffel–Darboux formula states that

$$\sum_{i=0}^{n} p_i(x)p_i(y) = k_n k_{n+1}^{-1}(x-y)^{-1}[p_{n+1}(x)p_n(y) - p_n(x)p_{n+1}(y)].$$

By making use of this formula and of the definition of the Hermitian polynomials $H_n(x)$, we obtain

$$\sum_{i=0}^{n-1} \varphi_i(x)\varphi_i(y) = \sqrt{n/2}(x-y)^{-1}[\varphi_n(x)\varphi_{n-1}(y) - \varphi_n(y)\varphi_{n-1}(x)].$$
$$(27.1.8)$$

For the Hermitian polynomials $H_n(x)$, the following asymptotic formula

$$\Gamma(n/2+1)[\Gamma(n+1)]^{-1}\exp(-x^2/2)H_n(x)$$
$$= \cos((\sqrt{n}x - n\pi/2) + 6^{-1}x^3 n^{-1/2}\sin(\sqrt{n}x - n\pi/2) + 0(n^{-1}),$$

is valid.

Taking into consideration that

$$\varphi_i(x) = (2^j j!\sqrt{\pi})^{-1/2}\exp(x^2/2)(-d/dx)^j)\exp(-x^2),$$
$$\Gamma(n+1)(2^n n!\sqrt{\pi})^{-1}[\Gamma(n/2+1)]^{-1} = 2^{1/4}\pi^{-1/2}n^{-1/4} + 0(1),$$
$$(27.1.9)$$

we obtain

$$\sqrt{\pi}n^{1/4}2^{1/4}\varphi_j(xn^{-1/2}) = (-1)^{n/2}\cos x + 0(1) \qquad (27.1.10)$$

if n is an even number, and

$$\sqrt{\pi} n^{1/4} 2^{-1/4} \varphi_j(x n^{-1/2}) = (-1)^{(n-1)/2} \sin x + 0(1) \qquad (27.1.11)$$

if n is an odd number.

By proving Eq. (27.1.9), we make use of the Stirling formula

$$\ln \Gamma(n) = \ln \sqrt{2\pi} + (n - n^{-1}) \ln n - n + \theta(12n)^{-1}, 0 < \theta < 1.$$

Substituting values (27.1.10) and (27.1.11) in formula (27.1.8), we obtain

$$n^{-1/2} \sum_{j=0}^{n-1} \varphi_i(x n^{-1/2}) \varphi_j(y n^{-1/2})$$

$$= 2^{-1/2} \{[\sqrt{\pi} n^{1/4} 2^{-1/4} \varphi_n(x n^{-1/2})]$$
$$\times [\sqrt{\pi} n^{1/4} 2^{-1/4} \varphi_{n-1}(y n^{-1/2})] - [\sqrt{\pi} n^{1/4} 2^{-1/4} \varphi_n(y n^{-1/2})$$
$$\times \sqrt{\pi} n^{1/4} 2^{-1/4} \varphi_{n-1}(x n^{-1/2})]\} (x n^{-1/2} - y n^{-1/2})^{-1} \sqrt{2} n^{-1/2} \pi^{-1}$$
$$= \pi^{-1}(x - y)^{-1} \sin(x - y) + 0(1).$$

Lemma (27.1.1) is proved.

On the basis of Lemma (27.1.1),

$$\lim_{n \to \infty} p(y n^{-1/2}, (x+y) n^{-1/2}) = \left[\mathbf{E} \exp \left\{ \int_0^x \eta^2(z+y) dz \right\} \right]^{-2}, \qquad (27.1.12)$$

$$\xi(y n^{-1/2}, (x+y) n^{-1/2}) \Rightarrow \exp \left\{ \int_0^x \eta^2(z+y) dz \right\}, \qquad (27.1.13)$$

$$n^{-1/2} (\sum_i \eta_i \varphi_i((x+y) n^{-1/2}))^2 \Rightarrow \eta^2(x+y). \qquad (27.1.14)$$

The following assertions proved by Gaudin and Mehta are based on these formulae.

Theorem 27.1.1. *If Ξ is a Gaussian Hermitian random matrix, then the limit of the probability that the eigenvalues λ_i of Ξ do not lie in the interval $]\alpha n^{-1/2}$, $\beta n^{-1/2}[$ is equal to*

$$[\mathbf{E} \exp \left\{ \int_\alpha^\beta \eta^2(z) dz \right\}]^{-2} = \prod_{i=1}^\infty (1 - \mu_i), \qquad (27.1.15)$$

where μ_i are the eigenvalues of the integral equation

$$\pi^{-1} \int_\alpha^\beta (x - y)^{-1} \sin(x - y) t(y) dy = \lambda t(x), x \in]\alpha, \beta[. \qquad (27.1.16)$$

Proof. We can expand the process $\eta(z)$ into the series

$$\eta(z) = \sum_{l=0}^{\infty} \eta_l \sqrt{\mu_l} \varphi_l(z),$$

where $\varphi_l(z)$ and μ_l are the eigenfunctions and eigenvalues of the integral equation (27.1.16).

This implies the conclusion of Theorem (27.1.1).

Theorem 27.1.2. *If Ξ is a Gaussian Hermitian matrix, then*

$$\lim_{n \to \infty} n E \kappa_n(x n^{-1/2}, c n^{-1/2}) = 2c \int_0^x (d^2/dy^2) q(y) dy, \qquad (27.1.17)$$

where $q(z) = \prod_{i=1}^{n \to \infty}(1 - \mu_i(z))$ and $\mu_i(z)$ are the eigenvalues of the integral equation

$$\pi^{-1} \int_0^z (x - y)^{-1} \sin(x - y) f(y) dy = \lambda f(x), x \in [0, z]. \qquad (27.1.18)$$

Proof. By using formulas (27.1.12)–(27.1.14), we transform Eq. (27.1.7) to the following form,

$$n E \kappa_n(x n^{-1/2}, c n^{-1/2}) = \int_0^z \int_{-c}^c dy dx \{2\rho^{-3}(y, x + y)$$

$$\times [E\xi(y, x + y)\eta^2(y)\eta^2(x + y) - 3\rho^{-1}(y, x + y)$$

$$\times E\xi(y, x + y)\eta^2(y) E\xi(y, x + y)\eta^2(x + y)]\} + 0(1),$$

$$(27.1.19)$$

where

$$\rho(y, x + y) = \left[E\left\{ \int_0^x \eta^2(z + y) dz \right\} \right],$$

$$\xi(y, x + y) = \exp\left\{ \int_0^x \eta^2(z + y) dz \right\}.$$

Obviously, the expression on the right-hand side of Eq. (27.1.19) is equal to

$$\int_0^z dx \int_{-c}^c dy \left(\frac{\partial^2}{\partial u \partial v} \right) \rho(u, v)_{v=y, u=x+y}^{-2} + 0(1).$$

Since $\eta(t)$ is the stationary Gaussian process, $\rho(u, v) = \rho(u - v)$. Hence, on the basis of Theorem (27.1.1), we obtain equation (27.1.7).

To obtain the limit distribution of distances between two eigenvalues, we did not use the entire spectrum of the matrix Ξ but only some part of it. At first, we "stretched" the spectrum, that is, we multiplied the differences by $\lambda_{i+1} - \lambda_i$ to $\sqrt{n}$, and then confined ourselves only to those λ_i that were small in value.

Theorem (27.1.2) may be applied to the Wigner hypothesis in the following way. Suppose that in some domain (e.g., $]-\varepsilon n^{-1/2}, \varepsilon n^{-1/2}[$, $\varepsilon > 0$ is an arbitrary constant number), it is necessary to measure the energy level of the atomic nucleus (in our mathematical models, it means that we measure an eigenvalue), and then to measure its distance to the energy level nearest to it (on the right). In this case, we regard that the interval $]-\varepsilon n^{-1/2}, \varepsilon n^{-1/2}[$ includes only one energy level. This distance, obviously , will be some random variable, and in Wigner's model, its distribution will be equal to

$$\Delta_n(x) := \sum_{i=1}^{n} P\{\lambda_i - \lambda_{i+1} < x, |\lambda_i| < \varepsilon n^{-1/2}, |\lambda_k| > \varepsilon n^{-1/2}, k \neq i\}$$

$$\times (\sum_{i=1}^{n} P\{|\lambda_i| < \varepsilon n^{-1/2}, |\lambda_k| > \varepsilon n^{-1/2}, k \neq i\})^{-1}.$$

Making use of formula (27.1.17), we obtain

$$\lim_{n \to \infty} \Delta_n(xn^{-1/2}) = \left[\varepsilon \int_{\varepsilon}^{x} (d^2/dy^2)q(y)dy + \varepsilon \int_{0}^{\varepsilon} (d^2/dy^2)q(y)dy\right]$$

$$\times \left[\varepsilon \int_{0}^{\varepsilon} (d^2/dy^2)q(y)dy + \varepsilon \int_{\varepsilon}^{\infty} (d^2/dy^2)q(y)dy\right]^{-1}.$$

Passing to the limit as $\varepsilon \to 0$,

$$\lim_{\varepsilon \to 0} \lim_{n \to \infty} \Delta_n(xn^{-1/2}) = \int_{0}^{x} (d^2/dy^2)q(y)dy \left[\int_{0}^{\infty} (d^2/dy^2)q(y)dy\right]^{-1}.$$

§2 Some Properties of the Stochastic Scattering Matrix

Recently, the statistical theory of nuclear reactions was the focus of the attention of nuclear research. It was applied to the problem of formation and decay of a compound nucleus, to preequilibrium processes, and to deep inelastic collisions of heavy ions [1], [19]. This theory is based on a random matrix model for the nuclear Hamiltonian and ergodicity condition [19]. Mean values of the scattering matrix that model characteristics of reactions being observed can be calculated as an ensemble mean. However, our results were obtained with a different degree of accuracy, and therefore, the sphere of their application is not always defined. This paragraph gives proofs of limit theorems for a scattering matrix and shows functional equations defining their mean values.

According to [159], for data of the complete spin and parity, the symmetric and unitary matrix of scattering can be written in the form:

$$S = U\widetilde{S}U^T, \qquad \widetilde{S} = I - iR,$$
$$R = F(I\varepsilon - H_0 - V + iF^T F/2)^{-1}F^T,$$

where U is a unitary matrix of dimension $N \times N$; V and F are the real random matrices of dimensions $\Lambda \times \Lambda$ and $N \times \Lambda$ with the elements v_{ij} and γ_{ij}, respectively: I is the identity matrix of dimension $N \times N$; the matrices H_0 and V are nonrandom, and H_0, V, F do not depend on the energy ε. Here N is the number of open channels, and Λ is the number of so-called bound states "embedded" in the continuum.

In the framework of the shell-model approach to nuclear reactions, the values v_{ij} and γ_{ij} may be identified with those of the matrix elements of residual interactions between the states of the discrete spectrum and states of discrete and continuous spectra, respectively; and H_0 may be identified with the Hamiltonian of the mean field.

As in the above references, we shall consider the values v_{ij} and γ_{ij} as independent random values distributed according to the normal law with zero mean and second moments given as:

$$\mathbf{E}(v_{ij}v_{i'j'}) = (\delta_{ii'}\delta_{jj'} + \delta_{ij}\delta_{i'j'})\mathbf{E}v_{ij}^2, \quad \mathbf{E}v_{ij}^2 = \delta^2\Lambda^{-1},$$
$$\mathbf{E}(\gamma_{ij}\gamma_{i'j'}) = \delta_{ii'}\delta_{jj'}\mathbf{E}\gamma_{ij}^2, \quad \mathbf{E}\gamma_{ij}^2 = \sigma^2 N^{-1}. \tag{27.2.1}$$

It is possible to consider limit properties of the matrix $\widetilde{S}$. The problem is to find the limits of the expressions $\mathbf{E}\widetilde{S}_{ii}$ and $\mathbf{E}[(\widetilde{S} - \mathbf{E}\widetilde{S})_{ij}(\widetilde{S} - M\widetilde{S})_{pl}c_n]$ as $N \to \infty$, $\Lambda \to \infty$, where c_n is some sequence of normalizing values.

We shall prove limit theorems under the condition that

$$\lim_{\Lambda\to\infty, N\to\infty} \Lambda N^{-1} = c < 1, \qquad \nu_\Lambda(x) \Rightarrow \nu(x), \tag{27.2.2}$$

and $\nu(x)$ is the normalized spectral function of the matrix $I\varepsilon - h_0$.

Theorem 27.2.1. *Let conditions (27.2.1) and (27.2.2) hold. Then for any $k = \overline{1, N}$,*

$$\mathrm{plim}_{\Lambda\to\infty}[\widetilde{S}_{kk} - 1 + ic_K] = 0, \tag{27.2.3}$$

where

$$c_K = \sigma^2 mc(1 + i\sigma^2 mc/2)^{-1}, \quad m = \lim_{\theta\downarrow 0} m(\theta), \tag{27.2.4}$$

$$m(\theta) = \int_{-\infty}^{+\infty} \frac{d\mu(x)}{x + \dfrac{i/2\sigma^2}{1 + \frac{i\sigma^2}{2}cm(\theta)} + i\theta}, \tag{27.2.5}$$

$\mu(x)$ *is the distribution function whose Stieltjes transformation*

$$p(z) = \int_{-\infty}^{\infty} d\mu(x)(z-x)^{-1}, \quad \text{Im}\, z \neq 0$$

satisfies the equation

$$p(z) = \int_{-\infty}^{+\infty} d\nu(x)[z - x - \delta^2 p(z)]^{-1}. \tag{27.2.6}$$

Solutions of Eq. (27.2.5) and (27.2.6) exist and are unique in the class of analytic functions $\text{Im}\, p(z)\, \text{Im}\, z < 0$, $\text{Im}\, z \neq 0$, and $\text{Im}\, m(\theta) < 0$, $\theta > 0$.

Proof. Introduce the matrix $B = (\delta_{kp}\delta_{ik})$ of dimension $N \times N$, then

$$R_{kk} = \text{Tr}\, RB = -i\frac{\partial}{\partial\alpha} \ln\det[I\varepsilon - H_0 - V + \frac{i}{2}F^T(I + 2\alpha B)F], \quad \alpha \neq 0 \tag{27.2.7}$$

where

$$F^T(I + 2\alpha B)F = \sum_{p=1}^{N} \vec{\gamma}_p' \vec{\gamma}_p + 2\alpha \vec{\gamma}_k' \vec{\gamma}_k,$$

and $\vec{\gamma}_p$ is the vector row of the matrix F.

By using the identity,

$$\ln\det(A + \vec{x}^T\vec{x}) - \ln\det A = \ln(1 + (A^{-1}\vec{x}, \vec{x})),$$

where A is a positive-definite matrix of nth order, $\vec{x}$ is the n-dimensional vector; from (27.2.7) we obtain

$$R_{kk} = (Q_k\vec{\gamma}_k, \vec{\gamma}_k)[1 + i/2(Q_k\vec{\gamma}_k, \vec{\gamma}_k)]^{-1}, \tag{27.2.8}$$

$$Q_k = [\varepsilon I - H_0 - V + \frac{i}{2}\sum_{p\neq k}\vec{\gamma}_p'\vec{\gamma}_p]^{-1}.$$

We need the following auxiliary assertions.

Lemma 27.2.1. *If condition (27.2.2) holds,*

$$\varliminf_{N\to\infty} N^{-1}\mathbf{E}\,\text{Tr}\, Q_k Q_k^* < \infty. \tag{27.2.9}$$

Proof. It is obvious that

$$N^{-1}\mathbf{E}\,\text{Tr}\, Q_k Q_k^* \leq N^{-1}\mathbf{E}\,\text{Tr}\left[\sum_{p\neq k, p=1}^{L}\gamma_p\gamma_p'\right]^{-2}.$$

It follows from Chapter 23 that the diagonal elements b_{ii} of the matrix $B = (b_{ij}) = \left[\sum_{p\neq k,p=1}^{L}\gamma_p\gamma_p'\right]^{-1}$ are distributed as the random values

$$\left(\sum_{S=N-2}^{L}y_s^2 N^{-1}\right)^{-1},$$

and the nondiagonal elements b_{ij}, $i \neq j$ are distributed as the random values

$$N^{-1/2}y_{N-1}\left[\sum_{S=N-2}^{L}x_s^2 N^{-1}\right]^{-1/2}\left[\sum_{S=N-2}^{L}y_s^2 N^{-1}\right]^{-1},$$

where the random values x_s, y_s, $s = 1, 2, \ldots$ are independent and distributed according to the normal law $N(0,1)$. Taking these relations and condition (27.2.2) into consideration, we obtain

$$\varlimsup_{N\to\infty}\mathbf{E}\,\mathrm{Tr}\,Q_k Q_k^* \leq \varlimsup_{N\to\infty}N^{-1}\left[N\mathbf{E}\left(\sum_{S=N-2}^{L}y_s^2 N^{-1}\right)^{-2}\right.$$

$$\left.+N(N-1)\mathbf{E}y_{N-2}^2 N^{-1}\left[\sum_{S=N-2}^{L}x_s^2 N^{-1}\right]^{-1}\left[\sum_{S=N-2}^{L}y_s^2 N^{-1}\right]^2\right] < \infty.$$

Lemma 27.2.1 is proved.

By using Lemma 27.2.1, we obtain

$$\mathrm{plim}_{N\to\infty}[(Q_k\vec{\gamma}_k, \vec{\gamma}_k) - \frac{\sigma^2}{N}\mathrm{Tr}\,Q_k] = 0. \qquad (27.2.10)$$

Lemma 27.2.2. *If condition (27.2.2) holds,*

$$\lim_{\theta\downarrow 0}\varlimsup_{N\to\infty}|N^{-1}\mathrm{Tr}\,Q_k - N^{-1}\mathrm{Tr}\,T_k(\theta)| = 0, \qquad (27.2.11)$$

where $T_k^{(\theta)} = [I\varepsilon - H_0 - V + \frac{i}{2}\sum_{p\neq k}\gamma_p'\gamma_p + iI\theta]^{-1}$, $\theta > 0$.

Proof. It is easy to see that

$$|\mathrm{Tr}\,Q_k - \mathrm{Tr}\,T_k(\theta)| \leq \theta\,\mathrm{Tr}\,B^2.$$

Therefore, the assertion of Lemma 27.2.2 follows from Lemma 27.2.1.

From Chapter 9, if conditions (27.2.1) and (27.2.2) hold, we obtain that for $\theta > 0$,

$$\mathrm{plim}_{N\to\infty}\Lambda^{-1}[\mathrm{Tr}\,T_k(\theta) - \mathbf{E}\,\mathrm{Tr}\,T_k(\theta)] = 0, \qquad (27.2.12)$$

$$\mathrm{plim}_{N\to\infty}\Lambda^{-1}[\mathbf{E}\,\mathrm{Tr}\,T_k(\theta) - \mathbf{E}\,\mathrm{Tr}\,T(\theta)] = 0,$$

where $T(\theta) = [I\varepsilon - H_0 - V + \frac{i}{2}F'F + i\theta I]^{-1}$.

By using (27.2.8) and (27.2.12), we have

$$\lim_{\theta \downarrow 0} \text{plim}_{\Lambda \to \infty} [R_{kk} - \sigma^2 m_\Lambda(\theta) c (1 + \frac{i}{2}\sigma^2 c m_\Lambda(\theta))^{-1}] = 0, \qquad (27.2.13)$$

where

$$m_\Lambda(\theta) = \Lambda^{-1} \mathbf{E} \, \text{Tr} \, T(\theta).$$

Since the distributions of the vectors γ_k are invariant in relation to the orthogonal transformations, we consider that in expression (27.2.13) the matrix $I\varepsilon - H_0 - V$ is replaced by the diagonal matrix $X = (\delta_{ij} x_i)$ with its eigenvalues x_i.

By using the formula

$$\frac{\det A_k}{\det A} = [a_{kk} - (A^{-1}a_k, b_k)],$$

where $A = (a_{ij})_{i,j=1}^n$ is a square nonsingular matrix of nth order, the matrix A_k has been obtained from matrix A by deleting the kth row and the kth column, a_k and b_k are the vector row and the vector column of the matrix A_k, respectively, we get

$$T_{kk}(\theta) = [x_k + i\theta + \frac{i}{2}(p_k, p_k) - \left(\frac{i}{2}\right)^2 (F_k T^k(\theta)F_k' p_k, p_k)]^{-1}, \qquad (27.2.14)$$

where p_k is the kth vector column of the matrix F, F_k is the matrix obtained from matrix F by deleting the kth column, $T^k(\theta) = (X_k + i\theta I + \frac{i}{2}F_k'F_k)^{-1}$, and X_k is the matrix obtained from matrix X by deleting the kth column and the kth vector. It is obvious that

$$\text{plim}_{\Lambda \to \infty} [(p_k, p_k) - \sigma^2] = 0,$$

$$\text{plim}_{\Lambda \to \infty} [(F_k T^k(\theta)F_k' p_k, p_k)^{-1} - \frac{\sigma^2}{N} \, \text{Tr} \, F_k' F_k T^k(\theta)] = 0.$$

By using these expressions, we obtain from (27.2.14),

$$T_{kk}(\theta) = [x_k + \frac{i}{2}\sigma^2 + i\theta - \left(\frac{i}{2}\right)^2 \frac{\sigma^2}{N} \, \text{Tr} \, F_k' F_k T^k(\theta)]^{-1} + \varepsilon_N, \qquad (27.2.15)$$

where $\text{plim}_{N \to \infty} \varepsilon_N = 0$. From (27.2.15) and by using (27.2.12), we find

$$T_{kk}(\theta) = [x_k + \frac{i}{2}\sigma^2 - \frac{i}{2}\sigma^2 c + i\theta + \frac{i\sigma^2}{2} N^{-1} \sum_{k=1}^{\Lambda} (x_k + i\theta) T_{kk}(\theta)]^{-1} + \varepsilon_N. \qquad (27.2.16)$$

Let

$$a = N^{-1} \sum_{k=1}^{\Lambda} (x_k + i\theta) T_{kk}(\theta).$$

For a, from Eq. (27.2.16), we have the equation

$$a = N^{-1} \sum_{k=1}^{\Lambda} (x_k + i\theta)[x_k + i\theta + \frac{i}{2}\sigma^2 - \frac{i\sigma^2}{2}c + \frac{i\sigma^2}{2}a]^{-1} + \varepsilon_N$$

$$= c - \sigma^2 c[1 - c + a]m_\Lambda(\theta)\frac{i}{2} + \varepsilon_n.$$

From this equation, we obtain

$$a = \frac{c - \{\sigma^2 c(1 - c)\}m_\Lambda(\theta)\frac{i}{2}}{1 + \frac{i\sigma^2}{2}cm_\Lambda(\theta)} + \varepsilon_N.$$

But then, by using (27.2.16), we find

$$m_\Lambda(\theta) = \Lambda^{-1}\mathbf{E}\sum_{k=1}^{\Lambda} T_{kk}(\theta) = \Lambda^{-1}\sum_{k=1}^{\Lambda} \mathbf{E}[x_k + \frac{i}{2}\sigma^2 - \sigma^2 c + i\theta + \frac{i}{2}\sigma^2 a]^{-1}$$

$$+ 0(1) = \Lambda^{-1}\sum_{k=1}^{\Lambda} \mathbf{E}[x_k + i\theta + \frac{i}{2}\sigma^2[1 + \frac{i}{2}\sigma^2 cm]^{-1}]^{-1} + 0(1).$$

From this equation and by using (27.2.2), we get Eq. (27.2.5).

We can obtain Eq. (27.2.6) just as in Chapter 9 [see the derivation of Eq. (9.5.3)].

Let us now prove the theorem for the nondiagonal elements $\widetilde{S}_{pl}$, $p \neq l$.

Theorem 27.2.2. *Let the conditions of Theorem (27.2.1) hold. Then*

$$(\widetilde{S}_{pl} - \mathbf{E}\widetilde{S}_{pl})\sqrt{N} \sim i(\beta_1 + i\beta_2)\alpha, \tag{27.2.17}$$

where $\alpha = [1 + \frac{i}{2}\frac{\sigma^2}{N}\mathbf{E}\operatorname{Tr} Q]^{-2}$, $Q = [I\varepsilon - H_0 - V + \frac{i}{2}F'F]^{-1}$, *and* β_1 *and* β_2 *are the random values distributed according to the general normal law with zero mean values and a covariance matrix*

$$\begin{bmatrix} \frac{\sigma^2}{N}\mathbf{E}\operatorname{Tr}(\operatorname{Re} Q)^2 & \frac{\sigma^2}{N}\operatorname{Tr}(\operatorname{Re} Q \operatorname{Im} Q) \\ \sigma^2 N^{-1}\mathbf{E}\operatorname{Tr}(\operatorname{Re} Q \operatorname{Im} Q) & \sigma^2 N^{-1}\mathbf{E}\operatorname{Tr}(\operatorname{Im} Q)^2 \end{bmatrix}$$

Proof. For an arbitrary square matrix A of Λth order, the following formula (see Lemma 2.5.1) holds,

$$\operatorname{Tr} QA - \operatorname{Tr} Q_k A = -\frac{i}{2}(Q_k A Q_k \gamma_k, \gamma_k)[1 + \frac{i}{2}(Q_k \gamma_k, \gamma_k)]^{-1}. \tag{27.2.18}$$

By using this formula, we obtain

$$
\begin{aligned}
R_{lp} &= \operatorname{Tr} Q\gamma'_p\gamma_l = \operatorname{Tr} Q\gamma'_p\gamma_l - \operatorname{Tr} Q_p\gamma'_p\gamma_l + \operatorname{Tr} Q_p\gamma'_p\gamma_l \\
&= \frac{-\frac{i}{2}(Q_p\gamma'_p\gamma_l Q_p\gamma_p, \gamma_p)}{1 + \frac{i}{2}(Q_p\gamma_p, \gamma_p)} + \operatorname{Tr} Q_p\gamma'_p\gamma_l = \frac{(Q_p\gamma_p, \gamma_l)}{1 + \frac{i}{2}(Q_p\gamma_p, \gamma_p)}.
\end{aligned}
\tag{27.2.19}
$$

Again, by using formula (27.2.18),

$$
(Q_p\gamma_p, \gamma_l) = \operatorname{Tr} Q_p\gamma'_p\gamma_l - \operatorname{Tr} Q_{pl}\gamma'_p\gamma_l + \operatorname{Tr} Q_{pl}\gamma'_p\gamma_l = \frac{(Q_{pl}\gamma_p, \gamma_l)}{1 + \frac{i}{2}(Q_{pl}\gamma_p, \gamma_l)},
\tag{27.2.20}
$$

where

$$
Q_{pl} = \left[I\epsilon - H_0 - V + \frac{i}{2}\sum_{k \neq p, l} \gamma'_k\gamma_k \right]^{-1},
$$

$$
\begin{aligned}
(Q_p\gamma_p, \gamma_p) &= \operatorname{Tr} Q_p\gamma'_p\gamma_p - \operatorname{Tr} Q_{pl}\gamma'_p\gamma_p + \operatorname{Tr} Q_{pl}\gamma'_p\gamma_p \\
&= \frac{-\frac{i}{2}(Q_{pl}\gamma_p, \gamma_l)^2}{1 + \frac{i}{2}(Q_{pl}\gamma_l, \gamma_l)} + (Q_{pl}\gamma_p, \gamma_p).
\end{aligned}
\tag{27.2.21}
$$

Substituting (27.2.21) and (27.2.20) by (27.2.19), we have

$$
R_{lp} = \frac{(Q_{pl}\gamma_p, \gamma_l)[1 + \frac{i}{2}(Q_{pl}\gamma_l, \gamma_l)]^{-1}}{1 + \frac{i}{2}(Q_{pl}\gamma_p, \gamma_p) + \frac{1}{4}(Q_{pl}\gamma_p, \gamma_l)^2[1 + \frac{i}{2}(Q_{pl}\gamma_l, \gamma_l)]^{-1}}.
\tag{27.2.22}
$$

By using the expressions (27.2.12) in which we must take $\theta = 0$, as well as

$$
\mathbf{E}|(Q_{pl}\gamma_p, \gamma_l)|^2 \leq \sigma^4 N^{-2} \operatorname{Tr} Q_{pl}Q^*_{pl} \to 0 \qquad (N \to \infty),
$$

we obtain from formula (27.2.22):

$$
\operatorname{plim}_{N \to \infty}\{\sqrt{N}R_{lp} - \sqrt{N}(Q_{pl}\gamma_p, \gamma_l)[1 + \frac{i}{2}\sigma^{-2}N^{-1}\operatorname{Tr} Q_{pl}]^{-2}\} = 0.
\tag{27.2.23}
$$

It is easy to prove that the random value $\sqrt{N}(Q_{pl}\gamma_p, \gamma_l)$ for large values of L is distributed approximately like the random value $\beta_1 + i\beta_2$ (see Chapter 5). For this we must consider the common characteristic function of the random values $\sqrt{N}\operatorname{Re}(Q_{pl}\gamma_p, \gamma_l)$, $\sqrt{N}\operatorname{Im}(Q_{pl}\gamma_p, \gamma_l)$. Therefore, by using (27.2.3) and (27.2.12), we obtain (27.2.17). Theorem 27.2.2 is proved.

§3 Application of Random Determinants in Some Mathematical Models of Solid-State Physics

A great number of results connected with the study of random determinants have been obtained when solving some mathematical models in solid-state physics [122], [123]. In particular, Dyson considers the following problem: Let an n-chain of elastically connected particles be given with masses $m_1, m_2, \ldots, m_n$ being independent uniformly distributed random values. Longitudinal oscillations of such a chain with fastened ends may be described by a system of equations,

$$x_{k+1} - 2x_k + x_{k-1} = m_k \ddot{x}_k, \quad x_0 = x_{n+1} = 0,$$

in which the constant of elastic interaction of neighbouring particles is equal to 1, and x_k are displacements of particles from the equilibrium state.

The natural frequencies λ_{kn}, $k = \overline{1, n}$ of this oscillating system can be found from the solvability condition of the following system of linear equations,

$$u_{k+1} - 2u_k + u_{k-1} = -\lambda_{kn} m_k u_k, \quad u_0 = u_{n+1} = 0,$$

i.e., they can be obtained as eigenvalues of the Jacobian matrix

$$C_n = \left(-\frac{2\delta_{ij}}{m_i} + \frac{\delta_{ij-1}}{\sqrt{m_i m_{i-1}}} + \frac{\delta_{ij+1}}{\sqrt{m_i m_{i+1}}} \right).$$

Dyson suggested a study of the limit behaviour of the normalized spectral functions $\mu_n(x) = n^{-1} \sum F(x - \lambda_{kn})$, $F(x) = 1$ at $x > 0$, $F(x) = 0$ at $x \leq 0$ of the matrix C_n, with the help of the so-called logarithmic transformation,

$$\int_0^\infty \ln(z + x) d\mu_n(x) = n^{-1} \ln \det(Iz + C_n), \quad z = t + is.$$

For the random values $n^{-1} \ln \det(Iz + C_n)$, it is easy to prove the limit theorems of the type of the law of large numbers as well as to find the limit of their mathematical expectations, because the well-known recurrent relations will hold for the determinants of Jacobi matrices (see Chapter 12).

Dyson's theorem for the limit behaviour of the values $n^{-1} \ln \det(Iz + \Xi_n)$ is generalized in Chapter 12: Let the random values ξ_i, $i = 1, 2, \ldots$ of the matrices $\Xi_n = ((2 + \xi_i)d_{ij} - \delta_{ij+1} - \delta_{ij-1})$ be independent, nonnegative, uniformly distributed, and for some $\delta > 0$, $E|\ln \xi_i|^{1+\delta} < \infty$. Then,

$$\text{plim}_{n \to \infty} n^{-1} \ln \det \Xi_n = \int_1^\infty \ln x \, dF(x),$$

where the distribution function $F(x)$ satisfies Dyson's integral equation,

$$F(x) = \iint_{2+y-z^{-1}<x,z\geq 1} dF(z)d\mathbf{P}\{\xi_1 < y\}.$$

Using this equation, it is possible to find the limit spectral function by the recursion formula for "logarithmic transformation." We must mention here that in some cases, we can manage without this transformation if we use the Sturm oscillation theorem (Chapter 12). Analogous equations for determinants of nonsymmetric random Jacobi matrices were obtained in Chapter 17.

CHAPTER 28

RANDOM DETERMINANTS IN NUMERICAL ANALYSIS

One of the main problems of calculus is the problem of finding the solutions
of systems of linear algebraic equations

$$Ax = b, \qquad (28.0.1)$$

where $A = (a_{ij})$, $i = \overline{1,n}$, $j = \overline{1,m}$ is a rectangular matrix, $b' = (b_1, \ldots, b_n)$
is a known vector, and $x' = (x_1, \ldots, x_m)$ is an unknown solution.

This delusively simple problem is very complicated for the following reasons:

1) When solving practical problems, the entries of the matrix A are the
values of the quantities, obtained as a result of experiments for some real
systems. In many cases, the following hypothesis holds: These values are the
realizations of some independent random variables whose means are equal to
the corresponding entries of the matrix A. Thus, instead of the matrix, we
have some observations x_i, $i = 1, 2, \ldots$, such that $\mathbf{E}x_i = A$.

The standard approach to the solution of the problem is to choose as an
approximate solution of the system (28.0.1) the estimation $y_s = (zz')^{-1}z'b$,
where $z = s^{-1}\sum_{i=1}^{s} x_s$, with s being a number of independent observations
on the matrix A. If m and n do not depend on s, $\det AA' \neq 0$, then the
estimation y_s is consistent i.e., $\text{plim}_{s \to \infty} y_s = (AA')^{-1}A'b$. Even if the matrix
A is well-conditioned with "moderate" values of m and n, we need a great
number of observations. Note that many modern technical and economic
systems are so complicated that even carrying out several observations on
such a system can prove to be beyond the powers of those who had created it,
because of its unknown behaviour during the experiment. It appears that y_s
is not the best estimation in the sense of the minimum of its variance. There
exist considerably better estimations of the solution $(AA')^{-1}A'b$, obtained by
using only one observation X on the matrix A when the G-condition holds:

$$\varlimsup_{n \to \infty} mn^{-1} < c, \quad 0 < c < \infty.$$

This section is devoted to finding such estimations.

644

2) Since the matrix A is unknown but only an observation X is known on it, the system of equations $Xy = b$ is perhaps ill-conditioned. The solutions of such systems will be chosen in the regularized form

$$y_\alpha = (I\alpha_n + XX')^{-1}X'b,$$

where $\alpha_n > 0$ is a parameter of regularization and I is the identity matrix. The necessity to choose a regularized solution arises because under some conditions for components of the vector x, the so-called arctangent law holds (see Chapter 15).

From the arctangent law, it follows that limiting distributions of the variables X_k have no means, and for such a variable it is impossible to use the law of large numbers for the estimation of the real solution. Such "abnormal" behaviour of the errors of the solution is observed also for random entries with nonzero mathematical expectations and different variances. But if we take instead of y the regularized y_α, then we can use theorems of the type of the law of large numbers.

3) The solution of the system (28.0.1) is calculated by computer. But all the calculations by computer are carried out with some rounding errors. These errors do not influence the solutions when the system is of small dimension, but when the values m and n are large, they must be taken into consideration. Thus, even if the matrix A is known exactly, for large m and n, we obtain such solutions as if the entries of the matrix A were known with some errors.

To find a consistent estimation of the solution of the system (28.0.1) under the condition that the number of observations on the matrix A is bounded, it is necessary that the solution x be chosen in regularized form and that the G-condition holds. In general, this problem is not yet solved. However, in some cases, worked out by the author by means of G-analysis (see Chapter 23), the problem can be solved using only one observation x on the matrix A.

G-analysis has been worked out by the author when solving problems of multidimensional statistical analysis of observations of large dimensions. At present, classes of G_8-estimations of many quantities have been found, which are used in this analysis (see Chapter 23–26). The series of estimations of the class of G_8-estimations contains the estimations of solutions of the system (28.0.1).

§1 Consistent Estimations of the Solutions of Systems of Linear Algebraic Equations, Obtained during Observations of Independent Random Coefficients with Identical Variances

Let

$$x_\alpha = [I\alpha n + AA']^{-1}A'b$$

be a regularized solution of the system of equations $Ax = b$. The G_8-estimation of the solution x_α is

$$G_8 = [I\theta(\alpha)n + X'X]^{-1}X'b, \qquad (28.1.1)$$

where $\theta(\alpha)$ is a positive solution of the equation

$$\theta(\alpha)[1 + \sigma^2 a(\theta(\alpha))]^2 + \sigma^2(1 - mn^{-1})(1 + \sigma^2 a(\theta(\alpha)) = \alpha, \quad \alpha > 0,$$
$$a(y) = n^{-1} \operatorname{Tr}[Iy + n^{-1}X'X]^{-1}, \quad y > 0, \tag{28.1.2}$$

and X is a random matrix, $\mathbf{E}X = A$, $\operatorname{Var} x_{ij} = \sigma^2$.

Theorem 28.1.1. *Let the entries x_{ij} of the matrix X be independent, and the vectors $c \in R^m$ and b satisfy the conditions*

$$\sup_n \sup_{k=\overline{1,n}} [\|b\| + \|c\| + \|a_k\|] < \infty, \quad a_k = (a_{ki}, i = \overline{1,m}),$$
$$\tag{28.1.3}$$

$$\varlimsup_{n\to\infty} \sup_{k=1\div n} n^{-1}\lambda_k(A'A) < \infty, \tag{28.1.4}$$

$$\varlimsup_{n\to\infty} mn^{-1} \le 1, \tag{28.1.5}$$

$$\sup_n \sup_{i,j} \mathbf{E}|x_{ij} - a_{ij}|^4 < c < \infty, \tag{28.1.6}$$

where λ_k and φ_k are the eigenvalues and normalized eigenvectors of the matrix $A'A$, respectively; the solution $\theta(\alpha)$ of Eq. (28.1.2) is positive for $\alpha > 0$, beginning with some $n \ge n_0$.
     Then

$$\operatorname{plim}_{n\to\infty}[(G_8, c) - (x_\alpha, c)] = 0. \tag{28.1.7}$$

Proof. Let

$$f(z, \gamma) = n^{-1} \operatorname{Tr}[I\gamma + B'(z)B(z)]^{-1},$$

where

$$B(z) = n^{-1/2}X + zbc'n^{1/2},$$

$\gamma > 0, z$ are real parameters. By differentiating the function f by z at the point $z = 0$, we have

$$\frac{\partial}{\partial z}f(z, \gamma)_{z=0} = -n^{-1/2}\operatorname{Tr}[(bc')'B(0) + B'(0)bc']$$

$$\times [I\gamma + B'(0)B(0)]^{-2} = 2\frac{\partial}{\partial\gamma}(\hat{x}_\gamma, c),$$

where

$$\hat{x}_\gamma = (I\gamma n + X'X)^{-1}X'b.$$

By integrating this equation by γ, we obtain

$$(G_8, c) = -\frac{1}{2}\int_{\theta(\alpha)}^\infty \frac{\partial}{\partial z}f(z, y)_{z=0}dy. \tag{28.1.8}$$

Lemma 28.1.1. *When the conditions of Theorem (28.1.1) hold,*

$$f(z, y) = a_m(z, y) + \delta_n(z, y),$$

$$a_m(z, y) = \mathbf{E}n^{-1} \sum_{k=1}^{m} [y(1 + \sigma^2 a_m(z, y))$$

$$+ \sigma^2(1 - \frac{m}{n})\lambda_k(z)(1 + \sigma^2 a_m(z, y))^{-1} + \varepsilon_{kn}(z, y)],$$

$$(28.1.9)$$

where $\lambda_k(z)$ are singular eigenvalues of the matrix

$$K(z) = n^{-1/2}A + zbc'n^{1/2},$$

$$\delta_n(z, y) = n^{-1} \sum_{k=1}^{m} \left\{ \mathbf{E} \frac{([\frac{\partial}{\partial y}R_k^k(z, y)]p_k, p_k)}{1 + (R_k^k(z, y)p_k, p_k)} - \mathbf{E} \frac{([\frac{\partial}{\partial y}R_k^k(z, y)]q_k, q_k)}{1 + (R_k^k(z, y)q_k, q_k)} \right.$$

$$\left. + \mathbf{E}_{k-1} \frac{([\frac{\partial}{\partial y}T_k(z, y)]p_k, p_k)}{1 + (T_k(z, y)p_k, p_k)} - \mathbf{E}_k \frac{([\frac{\partial}{\partial y}T_k(z, y)]p_k, p_k)}{1 + (T_k(z, y)p_k, p_k)} \right\},$$

$$R_k^k(z, y) = (Iy + \sum_{s=1}^{k-1} q_s q_s' + \sum_{s=k+1}^{m} p_s p_s')^{-1},$$

$$T(z, y) = (Iy + \sum_{\substack{s \neq k \\ s=1}}^{m} p_s p_s')^{-1},$$

$$(28.1.10)$$

q_s is the sth vector column of the matrix $Q'(z)$,

$$Q(z) = n^{-1/2}(A + \Xi) + zbc'n^{1/2},$$

p_s is the sth vector column of the matrix $B'(z)$, $\Xi = (\xi_{ij})$, $i = \overline{1, n}$, $j = \overline{1, m}$ is a random matrix whose entries ξ_{ij} are independent of one another, do not depend on matrix X, and are distributed by the normal law $N(0, \sigma^2)$; and $\mathbf{E}_k$ is a conditional mathematical expectation under the fixed minimal σ-algebra with respect to which the random vectors q_s, p_s, $s = (k+1) \div m$ are measurable.

$$\varepsilon_{kn}(z, y) = (s_k, s_k) - \sigma^2 - \lambda_k(z) - (L_k C_k s_k, C_k s_k) + n^{-1} \operatorname{Tr} L_k C_k C_k'$$

$$+ yn^{-1}\sigma^2 \operatorname{Tr}(L_k - \mathbf{E}L) + \lambda_k(z)\{[1 + (L_k^k \tilde{s}_k, \tilde{s}_k)]^{-1}$$

$$- [1 + n^{-1}\sigma^2 \mathbf{E} \operatorname{Tr} L]^{-1}\}, \quad C = (\Lambda + Y),$$

$\Lambda = (\lambda_k^{1/2}(z)\delta_{kp})_{k,p=1}^{m}$ is a diagonal matrix of the eigenvalues of the matrix $[K'(z)K(z)]^{1/2}$, $K(z) = U_1\Lambda U_2$, U_1 and U_2 are orthogonal matrices of dimension $n \times m$ and $m \times m$, respectively, matrix C_k is obtained from matrix $\Lambda + Y$ by replacing the entries of the kth row by zeros,

$$L_k = (l_{ij}^k) = (Iy + C_k'C_k)^{-1}, \qquad L_k^k = (Iy + \sum_{p \neq k} T_p^k)^{-1},$$

$$T_p^k = (C_{pi}C_{pj})_{i,j=1}^{m}, \qquad L = (Iy + C'C)^{-1}, \qquad Y = U_1'\Xi U_2 n^{-1/2};$$

the entries of the kth column and kth row of the matrix T_p^k are zeros, and s_k is the sth vector row of the matrix C,

$$\tilde{s}_k = (s_{k1}, \ldots, s_{kk-1}, 0, s_{kk+1}, \ldots, s_{km}).$$

Proof.   Obviously,

$$f(z, y) = a_m(z, y) + \delta_m(z, y), \qquad (28.1.11)$$

where

$$a_m(z, y) = \mathbf{E} n^{-1} \operatorname{Tr}[Iy + Q(z)Q'(z)]^{-1} = \mathbf{E} n^{-1} \operatorname{Tr}[Iy + CC']^{-1},$$
$$\delta_m(z, y) = n^{-1} \operatorname{Tr}[Iy + B'(z)B(z)]^{-1} - \mathbf{E} n^{-1} \operatorname{Tr}[Iy + Q(z)Q'(z)]^{-1}$$
$$= \beta_1 + \beta_2, \qquad (28.1.12)$$
$$\beta_1 = n^{-1} \operatorname{Tr}[Iy + B'(z)B(z)]^{-1} - \mathbf{E} n^{-1} \operatorname{Tr}[Iy + B'(z)B(z)]^{-1},$$
$$\beta_2 = -\mathbf{E} n^{-1} \operatorname{Tr}[Iy + Q(z)Q'(z)]^{-1} + \mathbf{E} n^{-1} \operatorname{Tr}[Iy + B'(z)B(z)].$$

By using formula

$$\operatorname{Tr}(Iy + A + xx')^{-1} - \operatorname{Tr}(Iy + A)^{-1}$$
$$= \frac{\partial}{\partial y} \ln \det[Iy + A + x'x] - \frac{\partial}{\partial y} \ln \det[Iy + A]$$
$$= \frac{\partial}{\partial y} \ln[1 + x'(Iy + A)^{-1}x], \qquad (28.1.13)$$

where A is a nonnegative-definite matrix of order m, x is a vector of the dimension m, $y > 0$, we obtain

$$\beta_1 = n^{-1} \sum_{k=1}^{m} \{\mathbf{E}_{k-1} \operatorname{Tr}[Iy + B'(z)B(z)]^{-1} - \mathbf{E}_k \operatorname{Tr}[Iy + B'(z)B(z)]^{-1}\}$$
$$= n^{-1} \sum_{k=1}^{m} \left\{ \mathbf{E}_{k-1} \frac{([\frac{\partial}{\partial y} T_k(z, y)]q_k, q_k)}{1 + (T_k(z, y)q_k, q_k)} - \mathbf{E}_k \frac{([\frac{\partial}{\partial y} T_k(z, y)]q_k, q_k)}{1 + (T_k(z, y)q_k, q_k)} \right\},$$
$$\beta_2 = n^{-1} \sum_{k=1}^{n} \mathbf{E}\{\operatorname{Tr} R_{k-1} - \operatorname{Tr} R_k\}$$
$$= n^{-1} \sum_{k=1}^{m} \mathbf{E} \left\{ \frac{([\frac{\partial}{\partial y} R_k^k(z, y)]p_k, p_k)}{1 + (R_k^k(z, y)p_k, p_k)} - \frac{([\frac{\partial}{\partial y} R_k^k(z, y)]q_k, q_k)}{1 + (R_k^k(z, y)q_k, q_k)} \right\},$$

where $R_k = (Iy + \sum_{s=1}^{k} q_s q_s' + \sum_{s=k+1}^{m} p_s p_s')^{-1}$.

From these two equalities, formula (28.1.10) follows. We only have to determine the fact that formula (28.1.9) holds. Obviously, by (28.1.12),

$$a_m(z, y) = n^{-1} \sum_{k=1}^{m} \mathbf{E} l_{kk}. \qquad (28.1.14)$$

For the variables l_{kk}, the following formula holds (Chapter 2, §5),

$$l_{kk} = [y + (s_k, s_k) - (L_k C_k s_k, C_k s_k)]^{-1}.$$

We transform it into the following form,

$$l_{kk} = [y + \sigma^2 + \lambda_k(z) - n^{-1}\sigma^2 \operatorname{Tr} L_k C_k' C_k - \lambda_k(z) \operatorname{Tr} L_k T_k^k + \varepsilon_{1n}]^{-1}, \quad (28.1.15)$$

where

$$\varepsilon_{1n} = (s_k, s_k) - \sigma^2 - \lambda_k(z) - (L_k C_k s_k, C_k s_k) + \sigma^2 n^{-1} \operatorname{Tr} L_k C_k C_k'.$$

It is easy to verify that

$$\operatorname{Tr} L_k T_k^k = (L_k^k \tilde{s}_k, \tilde{s}_k)[1 + (L_k^k \tilde{s}_k, \tilde{s}_k)]^{-1}.$$

Taking this equality into account, we transform (28.1.15) to the form

$$l_{kk} = [y + \sigma^2(1 - mn^{-1}) + yn^{-1}\sigma^2 \operatorname{Tr} L_k + \lambda_k(z)[1 + n^{-1}\sigma^2 \operatorname{Tr} L_k^k]^{-1} + \varepsilon_{2n}]^{-1},$$

where

$$\varepsilon_{2n} = \varepsilon_{1n} + \lambda_k(z)\{[1 + (L_k^k \tilde{s}_k, \tilde{s}_k)]^{-1} - [1 + n^{-1}\sigma^2 \operatorname{Tr} L_k^k]^{-1}\}.$$

From this formula, we obtain

$$l_{kk} = [y(1 + n^{-1}\sigma^2 \mathbf{E} \operatorname{Tr} L) + \sigma^2(1 - mn^{-1}) + \lambda_k(z)[1 + n^{-1}\sigma^2 \mathbf{E} \operatorname{Tr} L]^{-1} + \varepsilon_{kn}]^{-1},$$

where

$$\varepsilon_{kn} = \varepsilon_{2n} + yn^{-1}\sigma^2 \operatorname{Tr}(L_k - \mathbf{E}L) + \lambda_k(z)\{[1 + n^{-1}\sigma^2$$
$$\times \operatorname{Tr} L_k^k]^{-1} - [1 + n^{-1}\mathbf{E}\sigma^2 \operatorname{Tr} L]^{-1}\}.$$

By using this formula and by letting $a_m(z, y) = \mathbf{E}n^{-1} \operatorname{Tr} L$, we deduce formula (28.1.9). Lemma 28.1.1 is proved.

Lemma 28.1.2. *When the conditions of Theorem 28.1.1 hold for any $y > 0$,*

$$\operatorname{plim}_{n\to\infty}[(\frac{\partial}{\partial z})\delta_n(z, y)]_{z=0} = 0, \quad (28.1.16)$$

$$\operatorname{plim}_{n\to\infty} n^{-1} \sum_{k=1}^{m}[|(\frac{\partial}{\partial z})\varepsilon_{kn}(z, y)|_{z=0} + |\varepsilon_{kn}(0, y)|] = 0. \quad (28.1.17)$$

Proof. By using formula (28.1.10), we find

$$\frac{\partial}{\partial z}\delta_n(z, y)|_{z=0} = n^{-1} \sum_{k=1}^{m}[\{\mathbf{E}\varphi(\xi_k n^{-1/2}, \Gamma_k, \tilde{\Gamma}_k) - \mathbf{E}\varphi(\eta_k n^{-1/2}, \Gamma_k, \tilde{\Gamma}_k)\}$$
$$+ \mathbf{E}_{k-1}\varphi(\eta_k n^{-1/2}, B_k, \tilde{B}_k) - \mathbf{E}_k\varphi(\eta_k n^{-1/2}, B_k, \tilde{B}_k)], \quad (28.1.18)$$

where ξ_k and η_k are the kth vector columns of the matrices $A + \Xi$ and X, respectively,

$$\varphi(x, \Gamma_k, \tilde{\Gamma}_k) = 2([Iy + \Gamma_k]^{-3}[\tilde{\Gamma}_k(bc')'_k + (bc')_k \Gamma'_k]n^{1/2}x, x)$$
$$\times [1 + ((Iy + \Gamma_k)^{-1}x, x)]^{-1} - 2c_k([Iy + \Gamma_k]^{-2}b, x)n^{1/2}$$
$$\times [1 + ((Iy + \Gamma_k)^{-1}x, x]^{-1} - ([Iy + \Gamma_k]^{-2}x, x)[1 + ([Iy + \Gamma_k]^{-1}x, x)]^{-2}$$
$$\times \{-([Iy + \Gamma_k]^{-2}[\tilde{\Gamma}_k(bc')'_k + (bc')_k \tilde{\Gamma}'_k]n^{1/2}x, x)$$
$$+ 2c_k([Iy + \Gamma_k]^{-1}b, x)n^{1/2},$$

$$\Gamma_k = n^{-1}[\sum_{s=1}^{k-1} \xi_s \xi'_s + \sum_{s=k+1}^{m} \eta_s \eta'_s], \quad x \in R^m, \quad (bc')_k = (b_i c_j)_{j \neq k}^{i=1 \div n, j=1 \div m},$$

$$\tilde{\Gamma}_k = n^{-1}\{\xi_1, \ldots, \xi_{k-1}, \eta_{k+1}, \ldots, \eta_m\},$$

$$B_k = n^{-1} \sum_{s \neq k, s=1}^{m} \eta_s \eta'_s, \quad \tilde{B}_k = n^{-1}\{\eta_i, i \neq k, i = 1 \div m\}.$$

Note that

$$[Iy + \Gamma_k]^{-3}\tilde{\Gamma}_k = [Iy + \Gamma_k]^{-3}\Gamma_k H,$$

where H is some orthogonal matrix.

Therefore, since the matrix Γ_k is nonnegative-definite, it can be represented in the form $U_1 \Lambda U_2$, where U_1 and U_2 are orthogonal matrices, and Λ is a diagonal matrix with nonnegative entries on its diagonal, which are less than some constant.

Taking this note into account, it is easy to obtain that

$$|\mathbf{E}\varphi(\xi_k n^{-1/2}, \tilde{\Gamma}_k, \Gamma_k) - \mathbf{E}\varphi(\eta_k n^{-1/2}, \Gamma_k \tilde{\Gamma}_k)| \leq \sum_{x = \xi_k, \eta_k}$$
$$\times \mathbf{E}\{[|(U_1 \Lambda U_2(bc')'_k x, x)n^{-1/2}| + |2c_k([Iy + \Gamma_k]^{-2}b, x)|]$$
$$\times [1 + ((Iy + \Gamma_k)^{-1}x, x)n^{-1}]^{-1} - [1 + n^{-1}\mathbf{E}\operatorname{Tr}(Iy + \Gamma_k)\xi_k \xi'_k]^{-1}|$$
$$+ [|(U_1 \tilde{\Lambda} U_2(bc')'_k x, x)n^{-1/2}| + |2c_k([Iy + \Gamma_k]^{-1}b, x|]$$
$$\times |([Iy + \Gamma_k]^{-2}x, x)n^{-1}[1 + ([Iy + \Gamma_k]^{-1}x, x)n^{-1}]^{-2}$$
$$- n^{-1}\mathbf{E}\operatorname{Tr}[Iy + \Gamma_k]^{-2}\xi_k \xi'_k[1 + n^{-1}\mathbf{E}\operatorname{Tr}[Iy + \Gamma_k]^{-1}\xi_k \xi'_k]^{-2}|\},$$

$$(28.1.19)$$

where $\tilde{\Lambda}$ is a diagonal matrix with bounded diagonal entries.

By using the Schwartz inequality, the inequalities

$$\mathbf{E}[[1 + ((Iy + \Gamma_k)^{-1}\xi_k, \xi_k)n^{-1}]^{-1} - [1 + n^{-1}\mathbf{E}\operatorname{Tr}(Iy + \Gamma_k)^{-1}\xi_k \xi'_k]] \leq \tilde{c}_1 n^{-1},$$
$$\mathbf{E}[([Iy + \Gamma_k]^{-2}\xi_k, \xi_k)n^{-1}[1 + ((Iy + \Gamma_k)^{-1}\xi_k \xi_k)n^{-1}]^{-2} - n^{-1}$$
$$\times \mathbf{E}\operatorname{Tr}[Iy + \Gamma_k]^{-2}\xi_k \xi'_k[1 + n^{-1}\mathbf{E}\operatorname{Tr}[Iy + \Gamma_k]^{-1}\xi_k \xi'_k]^{-2}\}^2, \quad \leq \tilde{c}_2 n^{-1},$$

inequality (28.1.19), and the fact that $\mathbf{E}|x_{ij}|^4 < \tilde{c} < \infty$,

$$|a_{ij}| < c < \infty, \quad j = 1 \div m,$$

we obtain that

$$|\mathbf{E}\varphi(\xi_k n^{-1/2}, \Gamma_k \tilde{\Gamma}_k) - \mathbf{E}\varphi(\eta_k n^{-1/2}, \Gamma_k, \tilde{\Gamma}_k)|$$
$$\leq \tilde{c}[n^{-1}\{(b,b)(a_k,a_k)^2(c,c)n^{-1} + (b,b)(c,c)(a_k,a_k)n^{-1}$$
$$+ (b,b)(c,c)n^{-1} + |c_k|^2(b,a_k)^2 + |c_k|^2(b,b)\}]^{1/2}.$$

From this inequality and by condition (28.1.3), it follows that

$$\lim_{n \to \infty} n^{-1} \sum_{k=1}^{m} [\{\mathbf{E}\varphi(\xi_k n^{-1/2}, \Gamma_k, \tilde{\Gamma}_k) - \mathbf{E}\varphi(\eta_k n^{-1/2}, \Gamma_k \tilde{\Gamma}_k)] = 0. \quad (28.1.20)$$

Since the random variables

$$\gamma_k = \mathbf{E}_{k-1}\varphi(\eta_k n^{-1/2}, B_k, \tilde{B}_k) - \mathbf{E}_k \varphi(\eta_k n^{-1/2}, B_k, \tilde{B}_k)$$

are noncorrelated, analogously, we get

$$\mathbf{E}\left[n^{-1} \sum_{k=1}^{m} \gamma_k\right]^2 = n^{-2} \sum_{k=1}^{m} \mathbf{E}\gamma_k^2 \leq \tilde{c}n^{-1} \max_k [\mathbf{E}(b, \xi_k)^2(c, \xi_k)^2$$
$$+ c_k \mathbf{E}(b, \xi_k)^2] \to 0(n \to \infty).$$

Thus, using the limit and (28.1.20), we make sure that (28.1.16) holds. Let us prove (28.1.17). Here we often use theorems of the type of law of large numbers. To simplify the calculations, we show how by means of these theorems it is possible to prove that

$$\lim_{n \to \infty} \mathbf{E}n^{-1} \sum_{k=1}^{m} |(s_k, s_k) - \sigma^2 - \lambda_k(z)|_{z=0} = 0.$$

In order to do this, we represent the expression of the left side of the relation in the following form,

$$\lim_{n \to \infty} n^{-1} \sum_{k=1}^{m} \left\{ \mathbf{E}|\sum_{j=1}^{n} \frac{y_{kj}^2 - \sigma^2}{n}| + 2\mathbf{E}|y_{kk}\sqrt{\lambda_k(0)}|n^{-1/2} \right\}$$

$$\leq \lim_{n \to \infty} n^{-1} \sum_{k=1}^{m} \left\{ \sqrt{\sum_{j=1}^{n} \frac{\mathbf{E}(y_{kj}^2 - \sigma^2)^2}{n^2}} + 2cn^{-1/2} \right\} = 0.$$

Here we make use of the fact that entries of the matrix Ξ are independent and distributed by the standard normal law; therefore, the entries y_{ij} of the matrix Y are also independent and distributed by the standard normal law.

Analogously, we prove that

$$\lim_{n \to \infty} \mathbf{E}|\varepsilon_{km}(0, y)| = 0. \tag{28.1.21}$$

After calculations of derivatives, we obtain

$$\frac{\partial}{\partial z}\varepsilon_{km}(z, y)_{z=0} = 2y_{kk}n^{-1/2}\sqrt{\lambda_k(z)_{z=0}} - \frac{\partial}{\partial z}\lambda_k(z)_{z=0} + (\tilde{L}_k^2[\left(\frac{\partial}{\partial z}\Lambda_k\right)Y_k$$

$$+ Y_k'\left(\frac{\partial}{\partial z}\Lambda_k\right)]\tilde{C}_k\tilde{s}_k, \tilde{C}_k\tilde{s}_k)_{z=0} - 2(\tilde{L}_k\left(\frac{\partial}{\partial z}\Lambda_k\right)_{z=0}\tilde{s}_k, \tilde{C}_k\tilde{s}_k)$$

$$- 2\sum_{i=1}^{m}(\tilde{L}_k\tilde{C}_k)_{km}\frac{\partial}{\partial z}\lambda_k(0)\tilde{s}_{ik} - yn^{-1}\sigma^2 \operatorname{Tr}[\tilde{L}_k^2\{\left(\frac{\partial}{\partial z}\Lambda_k\right)Y_k$$

$$+ Y_k'\left(\frac{\partial}{\partial z}\Lambda_k\right)\}_{z=0} - \mathbf{E}\tilde{L}^2\{\left(\frac{\partial}{\partial z}\Lambda\right)Y + Y\left(\frac{\partial}{\partial z}\Lambda\right)\}_{z=0} + \frac{\partial}{\partial z}\lambda_k(0)$$

$$\times \{[1 + (\tilde{L}_k^k\tilde{s}_k, \tilde{s}_k)]^{-1} - [1 + n^{-1}\sigma^2\mathbf{E}\operatorname{Tr}\tilde{L}]^{-1}\} + \lambda_k(0)$$

$$\times \{[1 + (L_k^k\tilde{s}_k, \tilde{s}_k)]^{-2}((L_k^k)^2\frac{\partial}{\partial z}\sum_{p \neq k}T_p^k\tilde{s}_k, \tilde{s}_k)_{z=0} - [1 + n^{-1}\sigma^2\mathbf{E}\operatorname{Tr}\tilde{L}]^2$$

$$\times \mathbf{E}\operatorname{Tr}\tilde{L}^2\{\left(\frac{\partial}{\partial z}\Lambda\right)Y + Y'\left(\frac{\partial}{\partial z}\Lambda\right)\}_{z=0}, \tag{28.1.22}$$

where Λ_k and Y_k are obtained from the matrices Λ and Y by deleting the kth column, the sign $\sim$ (a tilde) above the matrices and the vectors means that $z = 0$.

It is easy to check that

$$\left(\frac{\partial}{\partial z}\right)\lambda_k(0) = (B\varphi_k, \varphi_k), \tag{28.1.23}$$

where

$$B = (bc')'A + A'bc',$$

$$\left(\frac{\partial}{\partial z}\right)\sqrt{\lambda_k(0)} = \frac{1}{\sqrt{\lambda_k(0)}}(b, A\varphi_k)(c, \varphi_k). \tag{28.1.24}$$

By representing the matrix A in the form

$$A + A(A'A)^{-1/2}(A'A)^{1/2},$$

we transform formulas (28.1.23) and (28.1.24) to

$$\left(\frac{\partial}{\partial z}\right)\sqrt{\lambda_k(0)} = (b, H\varphi_k)(c, \varphi_k), \tag{28.1.25}$$

where $H = A(A'A)^{-1/2}$,

$$\left(\frac{\partial}{\partial z}\right)\lambda_k(0) = 2\lambda_k(0)(b, H\varphi_k)(c, \varphi_k). \qquad (28.1.26)$$

By using conditions (28.1.3)–(28.1.5), the limit (28.1.21) and formulas (28.1.22), (28.1.25), and (28.1.26) as in the proof of (28.1.16), we obtain that (28.1.17) holds. Lemma 28.1.2 is proved.

Lemma 28.1.3. *If the conditions of Theorem 28.1.1 hold and as $n \to \infty$, any measurable solution $\hat\theta(\alpha)$ of Eq. (28.1.2) approaches in probability the solution $\theta(\alpha)$ of this equation in which the function $a(y)$ is replaced by $Ea(y)$. The solution $\theta(\alpha)$ of Eq. (28.1.2) exists.*

Proof. From Lemma 28.1.1, it follows that the function $a(y) = a_m(0, y)$ satisfies the equation

$$a(y)[1 + \sigma^2 a(y)]^{-1} = En^{-1}\sum_{k=1}^{m}[y(1 + \sigma^2 a(y))^2 + \sigma^2(1 - mn^{-1})$$

$$\times (1 + \sigma^2 a(y)) + \lambda_k(0) + \varepsilon_{km}(y)(1 + \sigma^2 a(y))]^{-1},$$

where $\varepsilon_{km}(y) := \varepsilon_{km}(y, 0)$. Differentiating both sides of this equation by y, we obtain

$$-[1 + \sigma^2 a(y)]^{-2}a'(y) = En^{-1}\sum_{k=1}^{m}b_k^{-2}(y)\left(\frac{\partial}{\partial y}\right)\beta(y) + E\varepsilon_n, \qquad (28.1.27)$$

where

$$b_k(y) = \beta(y) + \varepsilon_{km}(y)(1 + \sigma^2 a(y)) + \lambda_k(0),$$
$$\beta(y) = y(1 + \sigma^2 a(y))^2 + \sigma^2(1 - mn^{-1})(1 + \sigma^2 a(y)),$$
$$\varepsilon_n = n^{-1}\sum_{k=1}^{m}b_k^{-2}(y)\left(\frac{\partial}{\partial y}\right)[\varepsilon_{km}(y)(1 + \sigma^2 a(y))].$$

By using (28.1.16) and (28.1.17), we obtain that

$$\lim_{n\to\infty} E|\varepsilon_n| = 0.$$

Therefore, from Eq. (28.1.27), it follows that for every $y > 0$, beginning with some $n \geq n(y)$,

$$\left(\frac{\partial}{\partial y}\right)\beta(y) > 0, \qquad (28.1.28)$$

the function $\beta(y)$ is equal to zero if y satisfies the equation

$$[y(1 + \sigma^2 a(y)) + \sigma^2(1 - mn^{-1})](1 + \sigma^2 a(y)) = 0.$$

The derivative of the function on the left-hand side of this equality is

$$1 + \sigma^2 a(y) + y\sigma^2 a'(y) = 1 + \sigma^2 \int (y+x)^{-1} d\mathbf{E}\mu(x) - \sigma^2 \int y(y+x)^{-2} d\mathbf{E}\mu(x),$$

where $\mu(x)$ is the normalized spectral function of the matrix $X'Xn^{-1}$.

Obviously, this expression is nonnegative for all $y > 0$.

Therefore, the solution of equation

$$y(1 + \sigma^2 a(y)) = -\sigma^2 (1 - mn^{-1})$$

exists, is unique, and consequently, by the inequality (28.1.28) as $n \geq n(y)$, the solution $\theta(\alpha)$ of Eq. (28.1.2) exists in which the function $a(y)$ is substituted by $\mathbf{E}a(y)$ and this solution is unique. After simple transforms, we make sure that

$$\text{plim}_{n\to\infty} [\hat{\theta}(\alpha) - \theta(\alpha)] = 0.$$

Lemma 28.1.3 is proved.

The derivative by z at the point $z = 0$ of the expression (28.1.9) is

$$q(y) = -n^{-1} \sum_{k=1}^{m} [y(1 + \sigma^2 a(y)) + \sigma^2(1 - mn^{-1}) + \varepsilon_{kn}(0)$$

$$+ \lambda_k(0)[1 + \sigma^2 a(y)]^{-1}]^{-2} \{ y\sigma^2 q(y) + \left(\frac{\partial}{\partial z} \right) \lambda_k(0)$$

$$\times [1 + \sigma^2 a(y)]^{-1} - \lambda_k(0)[1 + \sigma^2 a(y)]^{-2} \sigma^2 q(y) + \left(\frac{\partial}{\partial z} \right) \varepsilon_{kn}(0) \}. \tag{28.1.29}$$

Since

$$\left(\frac{\partial}{\partial z} \right) \lambda_k(0) = (B\varphi_k, \varphi_k), \qquad B = (bc')'A + A'bc',$$

the conditions (28.1.3)–(28.1.5) hold; after simple calculations we find

$$q(y) = \frac{-\frac{1}{n} \sum_{k=1}^{m} \frac{(B\varphi_k, \varphi_k)}{1 + \sigma^2 a(y)} [y(1+\sigma^2 a(y)) + \sigma^2(1-mn^{-1}) + \lambda_k(1+\sigma^2 a(y))^{-1}]^{-2}}{1 + \frac{1}{n} \sum_{k=1}^{m} \frac{y\sigma^2 - \lambda_k \sigma^2 (1+\sigma^2 a(y))^{-2}}{[y(1+\sigma^2 a(y)) + \sigma^2(1-mn^{-1}) + \lambda_k(1+\sigma^2 a(y))^{-1}]^{2}}}. \tag{28.1.30}$$

By using Lemma 28.1.1 and Eq. (28.1.2), we obtain

$$a(y) = n^{-1} \sum_{k=1}^{m} [y(1 + \sigma^2 a(y)) + \sigma^2(1 - mn^{-1}) + \lambda_k(1 + \sigma^2 a(y))^{-1}]^{-1},$$

$$\theta(\alpha)[1 + \sigma^2 a(\theta(\alpha))]^2 + \sigma^2(1 - mn^{-1})(1 + \sigma^2 a(\theta(\alpha)) = \alpha.$$

By making use of Lemma 28.1.3, we find from these two equations that equation

$$a(\theta(y))[1 + \sigma^2 a(\theta(y))]^{-1} = n^{-1} \sum_{k=1}^{m} (y + \lambda_k)^{-1}$$

holds. Then Eq. (28.1.30) implies

$$q(\theta(y)) = -n^{-1} \sum_{k=1}^{m} \frac{(B\varphi_k, \varphi_k)}{(y + \lambda_k)^2} [1 + \sigma^2 a(\theta(y))][(1 + \sigma^2 a(\theta(y)))^{-1}$$

$$- (\theta + y(1 + \sigma^2 a(\theta(y))^{-1}) \sigma^2 \frac{\partial}{\partial y} a(\theta(y))]^{-1}. \qquad (28.1.31)$$

But

$$\theta(1 + \sigma^2 a(\theta))^2 + \sigma^2 (1 - mn^{-1})(1 + \sigma^2 a(\theta)) = y.$$

Consequently,

$$(1 + \sigma^2 a(\theta)) \left(\frac{\partial}{\partial y}\right) \theta = (1 + \sigma^2 a(\theta))^{-1} - (\theta + y(1 + \sigma^2 a(\theta))^{-1}) \sigma^2 \frac{\partial}{\partial y} a(\theta).$$

Therefore, from (28.1.31), we obtain

$$q(\theta(y)) = n^{-1} \sum_{k=1}^{m} (B\varphi_k; \varphi_k)(y + \lambda_k)^{-2} \left(\frac{\partial}{\partial y} \theta(y)\right)^{-1}$$

$$= 2[\frac{\partial}{\partial y}(Iy + n^{-1}A'A)^{-1} n^{-1} A'b, c)] \left[\frac{\partial}{\partial y}\theta(y)\right]^{-1}.$$

Consequently,

$$(G_8, \vec{c}) = \int_{\theta(\alpha)}^{\infty} q_m(y) dy = \int_{\alpha}^{\infty} q_m(\theta(z)) \theta'(z) dz + \varepsilon_m = (\hat{x}_\alpha, c) + \varepsilon_n,$$

where $\text{plim}_{n\to\infty} \varepsilon_n = 0$.
Theorem 28.1.1 is proved.

§2 Consistent Estimations of the Solutions of a System of Linear Algebraic Equations with a Symmetric Matrix of Coefficients

In this paragraph, we suppose that the matrix A in the system of equations (28.1.1) is symmetric. For such systems, finding the G_8-estimations of solutions is considerably simplified, since for the regularized solutions, the formula

$$X_\alpha = [I\alpha n + A^2]^{-1} Ab\sqrt{n} = -\operatorname{Im}[I\sqrt{\alpha} + n^{-1/2}iA]^{-1}b, \quad \alpha > 0,$$

is valid.

We can apply the results of Chapters 7–11 for such expressions. Let us explain the idea of finding the G-estimations in the following example. Let X be a random matrix represented in the form $X = A + \Xi$, where A is a real symmetric matrix of order n, Ξ is a random symmetric matrix whose entries ξ_{ij} on the diagonal and above it are independent and distributed by the normal laws $N(0, (1 + \delta_{ij})/2)$.

Then for any real vector c of order n,

$$(x_\alpha, c) \approx -\operatorname{Im}([I\alpha^{1/2} + in^{-1/2}\Lambda + in^{-1/2}\Xi]^{-1}\tilde{b}, \tilde{c}),$$

where $\tilde{b} = H'b'$, $\tilde{c} = H'c$, and H is an orthogonal matrix, whose columns are equal to the eigenvectors of the matrix A, $\Lambda = \{\lambda_i \delta_{ij}\}_{i,j=1}^n$, and λ_i are eigenvalues of the matrix A.

If $\sup_n[(b,b) + (c,c)] < c_1 < \infty$, $\sup_n \sup_{i=\overline{1,n}} |n^{-1/2}\lambda_i| < c_2 < \infty$, then as in Chapter 9, we can prove that

$$\operatorname{plim}_{n\to\infty}[(x_\alpha, c) - \mathbf{E}(x_\alpha, c)] = 0,$$

and

$$\mathbf{E}(x_\alpha, c) = -\operatorname{Im}\sum_{k=1}^n \mathbf{E}r_{kk}\tilde{b}_k\tilde{c}_k,$$

where r_{kk} is the kth diagonal entry of the matrix $[I\alpha^{1/2}+in^{-1/2}\Lambda+in^{-1/2}\Xi]^{-1}$.

For the quantities r_{kk}, Corollary 9.5.1 is valid. Therefore,

$$\mathbf{E}(x_\alpha, c) = -\operatorname{Im}\sum_{k=1}^n [\alpha^{1/2} + in^{-1/2}\lambda_i + m_n(\alpha)]^{-1}\tilde{b}_k\tilde{c}_k,$$

where the function $m_n(\alpha)$ satisfies the Pastur equation [see (9.5.3)]

$$m_n(\alpha) = \frac{1}{n}\sum_{k=1}^n [\alpha^{1/2} + in^{-1/2}\lambda_i + m_n(\alpha)]^{-1} + 0(1). \tag{28.2.1}$$

By using this equation, it is easy to obtain the G-estimation of solution x_α. As the G_8-estimation, we take the expression

$$G_8 = -\operatorname{Im}\sum_{k=1}^m [I\hat{\alpha}^{1/2} + n^{-1/2}iX]^{-1}b,$$

where X is an observation on the matrix A distributed as well as the matrix $A + \Xi$, and $\hat{\alpha}$ is a measurable solution of equation $\hat{\alpha}^{1/2} + m_n(\hat{\alpha}) = \sqrt{\alpha}$, $\alpha > 0$, where $m_n(y) = n^{-1}\operatorname{Tr}(I\sqrt{y} + n^{-1/2}iX)^{-1}$.

As in §1, we prove that this estimation is consistent.

REFERENCES

1. Agassi, D., Weidenmüller, H. A. and Mautzouranis, C. (1975). "The Statistical Theory of Nuclear Reactions for Strongly Overlapping Resonances as a Theory of Transport Phenomena," *Phys. Rev. C.* **22**, N3, 145–179.
2. Anderson, G. A. (1970). "An Asymptotic Expansion for the Noncentral Wishart Distribution," *Ann. Math. Stat.* **41**, 5, 1700–1707.
3. Anderson, T. W. (1958). *An Introduction to Multivariate Statistical Analysis.* John Wiley & Sons, New York, London.
4. Arnold, L. (1960). "On the Asymptotic Distribution of the Eigenvalues of Random Matrices," *J. Math. Anal. and Appl.* **20**, 262–268.
5. Arnold, L. (1971). "On Wigner's Semicircle Law for the Eigenvalues of Random Matrices," *Z. Wahrschein. Theorie verw. Geb.* **19**, 191–198.
6. Aivazyan, S. A., Yenukov, I. S., and Meshalkin, L. D. (1985). *Applied Statistics. Reference edition M., Finances and Statistics* (Russian).
7. Bagai, O. P. (1965). "The Distribution of the Generalized Variance," *Ann. Math. Stat.* **36**, 1, 120–130.
8. Barut, A. O. (1967). *The Theory of the Scattering Matrix.* McMillan Co., New York.
9. Baxter, R. (1982). *Exactly Solved Models in Statistical Mechanics.* Academic Press.
10. Beckenbach, E. F. and Bellman, R. (1961). *Inequalities.* Springer-Verlag.
11. Bellman, R. (1955). "A Note on the Mean Value of Random Determinants," *Quart. Appl. Math.* **13**, 3, 322–324.
12. Bellman, R. (1960). *Introduction to Matrix Analysis.* McGraw-Hill, New York.
13. Bennet, B. M. (1955). "On the Cumulants of the Logarithmic Generalized Variance and Variance Ratio," *Skand. actuarietidskr.* **38**, 1–2, 17–21.
14. Berezin, F. A. (1973). "Some Remarks on the Wigner Distribution," *Theoret. and Math. Phys.* **17**, 1163–1175. English transl. from the Soviet journal *Teoret. Mat. Fiz.* **17** (1973), 305–318.
15. Berlin, T. H. and Kac, M. (1952). "The Spherical Model of a Ferromagnet," *Phys. Rev.* **86**.
16. Bernshtein, S. N. (1964). "On the Dependencies among Random Variables," *Collected Works, IV.* Nauka, Moscow (Russian).
17. Bharucha-Reid, A. T. (1970). *Probabilistic Method in Applied Mathematics, Vol. 2.* New York, Academic Press.

18. Bohigas, O. and Giannoni, M. J. (1975). "Level Density Fluctuations and Random Matrix Theory," *Ann. Phys.* (N.Y.) **89**, 393.

19. Brody, T. A., Flores, J., French, J. B., Mello, P. A., Pandey, A., and Wong, S. (1981). "Random-matrix Physics: Spectrum and Strength Fluctuations," *Rev. Mod. Phys.* **53**, 385.

20. Brown, B. M. and Eagleson, G. K. (1971). "Martingale Convergence to Infinitely Divisible Laws with Finite Variances," *Trans. Amer. Math. Soc.* **162**, 449–453.

21. Burkholder, D. L. (1973). "Distribution Function Inequalities for Martingales," *Ann. Probab.* **1**, N1, 19–42.

22. Butkovskii, A. G. (1977). *Structural Theory of Distributed Systems.* Nauka, Moscow. (Russian).

23. Butkovskii, A. G. and Samoilenko, Yu. I. (1979) "Control of Quantum Plants, I." *Avtomat. Telemeks.* N4 5–25.

24. Chahdra, P. C. (1965). "Distribution of the Determinant of the Sum of Products Matrix in the Non-central Linear Case," *Math. Nachr.* **28**, N3–4, 169–179.

25. Cramer, H. (1946). *Mathematical Methods in Statistics.* Princeton Univ. Press, Princeton, New Jersey.

26. Deemer, W. L. and Olkin, I. (1951). "The Jacobians of Certain Matrix Transformations Useful in Multivariate Analysis," *Biometrika* **38**, 345–367.

27. Dharmadhikari, S., Fabian, V., and Jogdeo, K. (1968) "Bounds on the Moments of Martingales," *Ann. Math. Statist.* **39**, N5, 1717–1723.

28. Dobrushin, R. L. (1956). "Central Limit Theorem for Nonstationary Markov Chains. II.," *Theor. Probab. Appl.* **1** English. transl. from the Soviet journal *Teor. Veroyatnst i Primenen.*

29. Doob, J. L. (1953). *Stochastic Processes.* John Wiley, New York.

30. Dyson, F. J. (1953). "The Dynamics of Disordered Linear Chain," *Phys. Rev.* **92**, N6, 1331–1338.

31. Dyson, F. J. (1962–63). "Statistical Theory of the Energy Levels of Complex Systems. I–V," *J. Mathematical Phys.* **3** (1962), 140–156, 157–165, 166–175; **3** (1963), 701-712, 713–719 (Parts IV and V Coauthored with Madan Lab Mehta).

32. Dyson, F. J. (1962). "A Brownian-motion Model for the Eigenvalues of a Random Matrix," *J. Math. Phys.* **3**, N6, 31–60.

33. Erdesh, P. (1963). "Some Unsolved Problems," *Mathematics* **7**, N4, 109–143 (Russian).

34. Feller, W. (1950). *Probability Theory and its Applications.* John Wiley and Sons, New York.

35. Fihtengolc, G. M. (1962). *A Course of Differential and Integral Calculus.* Vol. II, 5th ed. Fizmatigiz, Moscow; German transl., VEB Deutscher Verlag, Berlin, 1966.

36. Fortet, R. (1951). "Random Determinants," *J. Research Nat. Bur. Standards* **47**, 465–470.

37. Gaenssler, P., Strobel, J., and Stute, W. (1978). "On Central Limit Theorems for Martingale Triangular Arrays," *Acta Math. Acad. Sci. Hungar.* **31**, 205–216.

38. Gantmacher, F. R. (1959). *Applications of the Theory of Matrices*. Chelsea New York.

39. Gantmacher, F. R. (1959). *Matrix Theory*. Vol. 1 Chelsea, New York. (transl. from Russian).

40. Gihman, I. I., and Skorohod, A. V. (1965). *Introduction to the Theory of Random Processes*. Nauka, Moscow, English transl., Saunders, Philadelphia, Pa., 1969.

41. Girko, V. L. (1975). "Limit Theorems for a Random Determinant," *Theor. Probability and Math. Statist.* **6**, 39–46. English transl. from the Soviet journal *Theor. Verojatnost i Mat. Statist. Vyp.* **6** (1972) 41–48.

42. Girko, V. L. (1973). "Limit Theorems for Determinants of Dominant Random Matrices," *Calculative and Applied Mathematics* **19**, 130–136 (Russian).

43. Girko, V. L. (1973). "Limit Theorems for Solution of Systems of Linear Random Equations and the Eigenvalues and Determinant of Random Matrices," *Soviet Math. Dokl.* **14** 1508–1511. English transl. from the Soviet journal *Dok. Akad. Nauk SSSR* **212**, 1039–1042.

44. Girko, V. L. (1974). "Limit Theorems of General Form for the Spectral Functions of Random Matrices," *Dopovidi Akad. Nauk Ukrain RSR Ser. A.* **10**, 874–876 (Russian).

45. Girko, V. L. (1974). "Inequalities for a Random Determinant and a Random Permanent," *Theor. Probability Math. Statist.* N4, 42–50. English transl. from the Soviet Journal *Teor. Verojatnost. i Mat. Statist.* **4**, 1971.

46. Girko, V. L. (1974). "On the Distribution of Solutions of Systems of Linear Equations with Random Coefficients," *Theor. Probability and Mat. Statist.* N2, 41–44. English transl. from the Soviet journal *Teor. Verojatnost. i Mat. Statist.* **2**.

47. Girko, V. L. (1974). "Limit Theorems for Eigenvalues of Random Matrices," *Soviet Math. Dokl.* **15**, N2, 636–639. English transl. from the Soviet journal *Dokl. Akad. Nauk SSSR*, **215**.

48. Girko, V. L. (1975). "Limit Theorems for a Random Determinant. III.," *Theor. Probability and Math. Statist.* N8, 25–30. English transl. from the Soviet journal *Teor. Veroyatnost. i Mat. Statist.* **8**.

49. Girko, V. L. (1975). "Limit Theorems for a Random Determinant. I.," *Theory Probability and Math. Statist.* N5, 25–31. English transl. from the Soviet journal *Teor. Veroyatnost. i Mat. Statist.* **5**.

50. Girko, V. L. (1975). "Random Matrices," *Vishcha Shkola* (Izdat. Kiev. Univ.), Kiev (Russian).

51. Girko, V. L. (1975). "Limit Theorems for the Permanent of a Random Matrix," *Theor. Probability Math. Statist.* N3, 28–32. English transl. from the Soviet journal *Teor. Veroyatnost. i Mat. Statist.* **3**.

52. Girko, V. L. (1975). "Refinement of Some Theorems for a Random Determinant and Permanent," *Theor. Probability Math. Statist.* N7, English transl. from the Soviet journal *Teor. Veroyatnost. i Mat. Statist.* **7**.

53. Girko, V. L. (1976). "Limit Theorems for Random Quadratic Forms. I," *Theor. Probability Math. Statist.* N9. English transl. from the Soviet journal *Teor. Veroyatnost. i Mat. Statist.* **9**.

54. Girko, V. L. (1976). "Limit Theorems for Random Quadratic Forms. II," *Theor. Probability Math. Statist.* N10. English transl. from the Soviet journal *Teor. Veroyatnost. i Mat. Statist.* **10**.

55. Girko, V. L. (1976). "Random Jacobi Matrices. I," *Theor. Probability Math. Statist.* **12**, 23–33. English transl. from the Soviet journal *Teor. Veroyatnost. i Mat. Statist.* **12**, 25–35.

56. Girko, V. L. (1976). "A Limit Theorem for Products of Random Matrices," *Theor. Probability Appl.* **21**. English transl. from the Soviet journal *Teor. Veroyatnost. i Primenen.* **21**, 201–202.

57. Girko, V. L. "The Eigenvalues of Random Matrices. I," *Theor. Probability Math. Statist.* N1, 8–14. English transl. from the Soviet journal *Teor. Veroyatnost. i Mat. Statist.* N11, 10–16.

58. Girko, V. L. (1977). "Limit Theorems of General Form for the Normalized Spectral Functions of Symmetric Random Matrices," in *Limit Theorems for Random Processes, Akad. Nauk Ukrain. SSR*, Kiev, 50–70 (Russian).

59. Girko, V. L. (1977). "Limit Theorems of General Type for Spectral Functions of Random Matrices," *Theor. Probability Appl.* **22**. English transl. from the Soviet journal *Teor. Veroyatnost. i Primenen.* **22**, 160–163.

60. Girko, V. L. (1977). "Limit Theorems of General Form for the Spectral Functions of Random Matrices," *Theory Probab. Appl.* **22**, 156–160. English transl. from the Soviet journal *Teor. Veroyatnost. i Primenen.* **22**, 160–164

61. Girko, V. L. and Vinogradskaya, A. V. (1979). "Spectral Control of Linear Operators in Hilbert Space," *Vychisl. Prikl. Mat.* N38, 111-114.

62. Girko, V. L. (1983). "Limit Theorems for Sums of Random Variables, Connected in a Markov Chain. I," *Theor. Probability Math. Statist.* N26. English transl. from the Soviet journal *Teor. Veroyatnost. i Mat. Statist.* **22**.

63. Girko, V. L. (1979). "The Distribution of the Eigenvalues and Eigenvectors of Hermitian Random Matrices," *Ukr. Math. J.* **31**, 533–537. English transl. from the Soviet journal *Ukr. Mat. Zh.* **31**.

64. Girko, V. L. (1979). "The Logarithm Law," *Dokl. Akad. Nauk Ukrain. SSR, Ser. A.* **4**, 241-242 (Russian).

65. Girko, V. L. (1979). "The Central Limit Theorem for Random Determinants," *Theory Prob. Appl.* **24**, 729–740.

66. Girko, V. L. (1979). "Necessary and Sufficient Conditions of Limit Theorems for Borel Functions of Independent Random Variables," *Reports of the Ukrainian Academy of Sciences, Series Mathematics, Mechanics* **N10** (Russian).

67. Girko, V. L. and Smirnova V. V. (1980). "Limit Theorems for Stochastic Leont'ev Systems," *Reports of the Ukrainian Academy of Sciences, Series Mathematics, Mechanics* **N12** (Russian).

68. Girko, V. L. (1980), "The Wigner Conjecture," *Vychisl. i Prikladn. Mat.* **41**, 71–79 (Russian).

69. Girko, V. L. (1980). "On the Uniqueness of the Solution of the Canonical Spectral Equation," *Ukrain. Math. J.* **N6** 546–548. English transl. from the Soviet journal *Ukrain Mat. Zh.* **N6**, 802–805.

70. Girko, V. L. (1980). "Theory of Random Determinants," *Vishcha Shkola* (Izdat. Kiev. Univ.), Kiev (Russian).

71. Girko, V. L. (1980). "The Central Limit Theorem for Random Determinants," *Theory of Probability and its Applications* **24**, N4. English transl. from the Soviet journal *Teor. Veroyatnost i Primenen* **24**, N4, 728–740.

72. Girko, V. L. (1981). "On Normalized Spectral Functions of Random Matrices," *Theor. Probability Math. Statist.* **22**, 31–34. English transl. from the Soviet journal *Teor. Veroyatnost. i Mat. Statist.* **22**.

73. Girko, V. L. (1980). "Arctangent Law," *Reports of the Ukrainian Academy of Sciences, Series Mathematics, Mechanics* **4** (Russian), 7–9.

74. Girko, V. L. (1980). "Stochastic Spectral Equation," *Lecture Notes in Control and Information Sciences. Stochastic Differential Systems. Filtering and Control. Proceedings of the FP-WG 7/1, Working Conference,* Vilnius, Lithuania, USSR, Aug 28–Sept 2 1978. Springer-Verlag, Berlin, Heidelberg, New York.

75. Girko, V. L. (1980). "The Stochastic Ljapunov Problem," *Theor. Probability Math. Statist.* **20**, English transl. from the Soviet journal *Teor. Veroyatnost. i Mat. Statist.* **20**, 42–44.

76. Girko, V. L. and Vinogradskaya, A. V. (1981). "Spectrum Control of Hilbert–Schmidt Operators," *Vychist. Pricl. Mat.* **45**, 89–92 (Russian).

77. Girko, V. L. (1981). "Distribution of Eigenvalues and Eigenvectors of Unitary Random Matrices," *Theor. Probability Math. Statist.* **25**, 13–16. English transl. from the Soviet journal *Teor. Veroyatnost. i Mat. Statist.* **25**.

78. Girko, V. L. (1981). "The Polar Decomposition of Random Matrices," *Theor. Probability Math. Statist.* **23**, 21–31. English transl. from the Soviet journal *Teor. Veroyatnost. i Mat. Statist.* **23**, 20–30.

79. Girko, V. L. and Vasil'ev, V. V. (1982). "Limit Theorems for Determinants of Random Jacobi Matrices," *Theor. Probability Math. Statist.* **24**, English transl. from the Soviet journal *Teor. Veroyatnost. i Mat. Statist.* **24**, 16–27.

80. Girko, V. L. (1982). "V-transforms," *Dokl. Akad. Nauk Ukr. SSR, Ser. A.* **N3**, 5–6 (Russian).

81. Girko, V. L. and Vasil'ev, V. V. (1983). "The Central Limit Theorem for Normalized Spectral Functions of Random Jacobi Matrices," *Theor. Probability Math. Statist.* **29** (Russian).

82. Girko, V. L. and Smirnova, V. V. (1983). "A Method of Integral Representations of Solution of the Linear Stochastic Programming Problems," *Kibernetika* **6**, 122–124 (Russian).

83. Girko, V. L. and Smirnova, B. B. (1983). *Asymptotic Methods of Solving Some Problems of Linear Stochastic Programming and Models of Expenditure—Output Type.* Institute of Cybernetics Ukrainian Academy of Sciences, Preprint 83–28, 22 (Russian).

84. Girko, V. L. and Litvin, I. N. (1983). "Calman Stochastic Condition," *Calculative and Applied Mathematics* **49**, 135–138 (Russian).

85. Girko, V. L. (1983). "Limit Theorems for Functions of Random Variables," *Vishcha Shkola* (Izdat. Kiev. Univ.), Kiev (Russian).

86. Girko, V. L. (1984). "On the Circle Law," *Theor. Probability Math. Statist.* **28**, 15–23. English transl. of the Soviet journal *Teor. Veroyatnost. i Mat. Statist.* **28**.

87. Girko, V. L. (1983). "Spectral Theory of Nonself-adjoint Random Matrices. Probability Theory and Mathematical Statistics," *Proc. USSR-Japan Symp. Tbilis, Lect. Notes Math.* **1021**, 153–156.

88. Girko, V. L. and Onsha Y. M. (1984). "Resolvent Method of Solving Problems in the Theory of Planning the Experiments," *Calculative and Applied Mathematics* **52**, 129–132 (Russian).

89. Girko, V. L. and Litvin I. N. (1984). "The Integral Representation of Hyperdeterminants and its Application to the Research of Stability of Stochastic Systems, Control Dynamic Systems with Continuous-Discrete Parameters." Kiev, *Naukova Dumka*, 97–102 (Russian).

90. Girko, V. L., Kokobinadze, T. S., and Chaika O. G. (1984). "Distribution of the Eigenvalues of Gaussian Random Matrices," *Ukrainian Mathematical Journal* **36**, N1, 9–12. English transl. from the Soviet journal *Ukrainskii Matematicheskii Zhurnal* **36**, N1.

91. Girko, V. L. (1985). "Spectral Theory of Random Matrices," *Successes of Mathematical Sciences* **40**, N1 (241), 67–106 (Russian).

92. Girko, V. L. (1985). "Spectral Theory of Random Matrices." *Problems of Non-Linear and Turbulent Processes in Physics*, part 2. Kiev, Naukova Dumka, 35–37 (Russian).

93. Girko, V. L. (1985). "Distribution of Eigenvalues and Eigenvectors of Orthogonal Random Matrices," *Ukrainian Mathematical Journal* **37** N5, 568–575 (Russian).

94. Girko, V. L. and Vasil'ev, V. V. (1985). "Limit Theorems for the Normalized Spectral Functions of Nonself-Adjoint Random Jacobian Matrices," *Teor. Verojatnost. i Primenen.* **30**, N1, 3–9 (Russian).

95. Girko, V. L. (1985). "The Elliptic Law," *Teor. Verojatnost. i Primenen.* **30**, 4, 640–651 (Russian).

96. Girko, V. L. and Krak, Y. V. (1986). "Asymptotical Normality of Estimations of States of Adaptive Models of Robots—Manipulators," *Calculative and Applied Mathematics* **60**, 89–94 (Russian).

97. Girko, V. L. (1986). "G-analysis of Observations of Enormous Dimensionality," *Calculative and Applied Mathematics* **60**, 115–121 (Russian).

98. Girko, V. L. (1986). "G₂-estimations of Spectral Functions of Covariance Matrices," *Theor. Probability Math. Statist.* **35**, 28–31 (Russian).

99. Girko, V. L. (1986). "Random Determinants, The Results of Science and Technology," *Theor. Probability Math. Statist. Theoretical Cybernetics.* **24**. Moscow, 3–57 (Russian).

100. Gnedenko, B. V. (1967). *Course in the Theory of Probability.* 4th ed. Nauka, Moscow, 1965; English Transl., Chelsea, New York.

101. Gnedenko, B. V. and Kolmogorov, A. N. (1949). *Limit Distributions for Sums of Independent Random Variables.* GITTL, Moscow; English transl., Addison-Wesley, New York 1954, rev. ed. 1968.

102. Goodman, N. R. (1963). "Distribution of the Determinant of a Complex Wishart Distributed Matrix," *Ann. Math. Statistics* **34**, N1, 178–180.

103. Gokhberg, I. C. and Krein, M. G. (1972). *Introduction to the Theory of Linear Nonselfadjoint Operations.* Nauka, Moscow (Russian).

104. Gradshteyn, J. S. and Ryzhik, J. M. (1965). *Tables of Integrals, Series, and Products.* Academic Press, New York.

105. Grenander, U. (1963). *Probabilities on Algebraic Structures.* John Wiley, New York, London.

106. Grenander, U. and Silverstein, J. W. (1977). "Spectral Analysis of Networks with Topologies," *SIAM J. Appl. Math.* **32**, 499–519.

107. Grenander, U. and Szego, G. (1958). *Toeplitz Forms and their Applications.* Univ. of California Press, Berkeley, California.

108. Gunson, J. (1962). "Proof of a Conjecture by Dyson in the Statistical Theory of Energy Levels," *J. Math. Phys.* **3**, 752.

109. Hammersley, I. M. (1957). "Zeros of a Random Polynomial," *Third Berkeley Symposium.*

110. Hincin, A. Ja. (1961). *Continued Fractions.* 3rd ed. Fizmatgiz, Moscow; English transl., Noordhoff, 1963, and Univ. of Chicago Press, Chicago, Ill.

111. Hoefding, W. (1948). "A Class of Statistics with Asymptotically Normal Distribution," *Ann. Math. Statist.* **19**, N3, 489–494.

112. James, A. T. (1955). "The Non-Central Wishart Distribution," *Proc. Roy. Soc. Ser. A,* **229**, N8, 364–368.

113. Jonsson, D. (1982). "Some Limit Theorems for the Eigenvalues of a Sample Covariance Matrix," *J. Multivariate Anal.* **12**, 1–38.

114. Judickii, M. I. (1973). "Optimal Design of Regression Experiments in the Presence of Random Errors in the Levels of Factors, Questions of

Statistics and of the Control of Random Processes," *Izdanie Inst. Mat. Akad. Nauk. Ukrain. SSR.* Kiev, 251–271 (Russian).

115. Kac, M. (1957). *Probability and Related Topics in Physical Sciences.* Interscience Publishers, London, New York.

116. Kato, T. (1966). *Perturbation Theory for Linear Operators.* Springer-Verlag.

117. Klopotowski, A. (1977). "Limit Theorems for Sums of Dependent Random Vectors," in *R Dissertation Mathematical.* Warszawa, Panstwowe Wydawnictwo Naukowe, 5–58.

118. Koivo, A. J. and Guo, T. H. (1983). "Adaptive Linear Controller for Robotic Manipulators," *IEEE. Trans. Automat. Contr.* **28**, N2, 162–171.

119. Komlos, J. (1968). "On the Determinant of Random Matrices," *Studia Sci. Math. Hung.* **3**, N4, 387–399.

120. Landau, L. D. and Smorodinskii, Ya. (1958). *Lectures on Nuclear Theory.* Consultants Bureau, New York; English Transl. of the Soviet book *Lektsii po Teorii Atomnogo Yadra.* Costekhizdat, Moscow, Leningrad.

121. Lifshits, I. M., Gredeskul, S. A. and Pastur, L. A. (1982). *Introduction to the Theory of Disordered Systems.* Nauka, Moscow, 358 (Russian).

122. Lifshits, I. M. (1965). "On the Structure of the Energy Spectrum and Quantum State of Disordered Condenser Systems," *Soviet Physics Uspekhi* **7**, 549–573. English transl. from the Soviet journal *Uspekhi Fiz. Nauk.* **83**, 617–655.

123. Loeve, M. (1963). *Probability Theory.* 3rd ed. Van Nostrand., Princeton, New Jersey.

124. Mahaux, C. and Weidenmüller, H. A. (1969). *Shell-Model Approach to Nuclear Reactions.* Amsterdam. North-Holland.

125. Mahaux, C. and Weidenmüller, H. A. (1979). "Recent Developments in Compound-Nucleus Theory," *Ann. Rev. Nucl. Part. Sci.* **29**, N1, 1–31.

126. Mandelbrot, B. (1961). "Stable Paretian Random Function and the Multiplicative Variation of Income," *Econometrica* **29**, N4, 517–543.

127. Maradudin, A. A., Montroll, E. W., and Weiss, G. W. (1963). *Theory of Lattice Dynamics in the Harmonic Approximation.* Academic Press, New York, London.

128. Marchenko, V. A. and Pastur, L. A. (1967). "Distribution of the Eigenvalues in Certain Sets of Random Matrices," *Math. USSR-Sb.* **1**, 457–483. English transl. from the Soviet journal *Mat. Sb.* **72** (1968), 507–536.

129. Marcus, M. and Minc, H. (1964). *A Survey of Matrix Theory and Matrix Inequalities.* Allyn and Bacon, Boston.

130. McKean, H. P. (1969). *Stochastic Integrals.* Academic Press, New York, London.

131. Mehta, M. L. and Rosenzweig, N. (1968). "Distribution Laws for the Roots of a Random Antisymmetric Hermitian Matrix," *Nuclear Physics* **A 109**, 2.

132. Mehta, M. L. (1967). *Random Matrices and the Statistical Theory of Energy Levels.* Academic Press, New York, London.

133. Molchanov, S. A. (1978). "The Structure of the Eigenfunctions of One-dimensional Unordered Structures," *Math. USSR - Izv.* 12, 69–101. English transl. from the Soviet journal *Izv. Akad. Nauk SSSR. Ser. Mat.* 42.

134. Mott, N. F. and Davis, E. A. (1971). *Electron Processes in Non-Crystalline Materials.* Clarendon Press, Oxford.

135. Murnaghan, F. D. (1938). *The Theory of Group Representations.* Johns Hopkins Press, Baltimore, Maryland. Reprint Dover, New York, 1963.

136. Mushelishvili, N. I. (1966). *Some Principal Problems of Mathematical Theory of Elasticity.* Moscow, Nauka (Russian).

137. Nicholson, W. L. (1958). "On the Distribution of 2×2 Random Normal Determinants," *Ann. Math. Statist.* 29, 2, 575–580.

138. Nyguist, H., Rice, S. and Riordan, J. (1954). "The Distribution of Random Determinants," *Quart. Appl. Math.* 12, N2, 97–104.

139. Olkin, I. (1952). "Note on the Jacobians of Certain Matrix Transformations Useful in Multivariate Analysis," *Biometrika* 40, 43–46.

140. Pastur, L. A. (1972). "On the Spectrum of Random Matrices," *Theoret. and Math. Phys.* 10, 67–74. English transl. from the Soviet journal *Teoret. Mat. Fiz.* 10, 102–112.

141. Pastur, L. A. (1973). "The Spectra of Random Self-Adjoint Operators," *Russian Math. Surveys* 28, 1, 1–67. English transl. from the Soviet journal *Uspekhi Mat. Nauk* 28, 1, 4–63.

142. Petrov, V. V. (1975). *Sums of Independent Random Variables.* Springer-Verlag, New York, Berlin, Heidelberg.

143. Porter, C. E. (1965). *Statistical Theories of Spectra: Fluctuations.* Acad. Press, New York, London.

144. Prekopa, A. (1967). "On Random Determinants. I," *Studia Sci. Math. Hung.* 2, N1–2, 125–132.

145. Prokhorov, Yu. V. and Rozanov, Yu. A. (1969). *Probability Theory.* Springer-Verlag, Berlin, New York.

146. Reznikova, A. Y. (1980). "The Central Limit Theorem for the Spectrum of Random Jacobi Matrices," *Theory Probability and its Application* 25, 3, 513–522 (Russian).

147. Rotar, V. I. (1979). "Limit Theorems for Polylinear Forms," *Journal of Multivariate Analysis* 9, 511–530.

148. Rove-Beketov, J. S. (1960). "On the Limit Distribution of the Eigenfrequencies of a Disordered Chain," *Zap. Mat. Otdel Fiz. - Mat. Fak. i Harcov. Mat. Obsc.* 4, 26, 143–153 (Russian).

149. Ryazanov, B. V. 1978 "Spectra of Haussian Cyclic Matrices," *Theory Probability and Its Application* 23, 3, 564–579.

150. Schmidt, H. (1957). "Disordered One-dimensional Crystals," *Physical Review* 105, 2, 425–441.

151. Selberg, A. A. (1944). "Bemerkinger om et Multipled Integral," *Norsk Matematik Tidsskrift* 26, 71–78.
152. Sendler, W. (1977). "On the Moments of a Certain Class of Random Permanents," *Trans. 7th Prague Conf. Inform. Theory, Statist. Decision Funct.*, **vol. A.**
153. Shubin, N. Yu. (1975). "Statistical Methods in the Theory of Nucleus," *Soviet J. Particles Nuclei* 5, 413–433. English transl. from the Soviet journal *Fizika Elementarnykh Chastits i Atomnogo Yadra, Atomizdat*, Moscow, 5, 1023–1074.
154. Skorohod, A. V. (1965). *Studies in Stochastic Processes.* Addison-Wesley, Reading, Mass.
155. Skorohod, A. V. and Slobodenjuk, N. P. (1970). *Limit Theorems for Random Walks.* Naukova Dumka, Kiev (Russian).
156. Sokolov, N. P. (1972). *Introduction to the Theory of Multivariate Matrices.* Naukova Dumka, Kiev (Russian).
157. Szego, G. (1959). *Orthogonal Polynomials.* American Mathematical Society, New York.
158. Takemura, A. (1984). "An Orthogonally Invariant Minimax Estimator of the Covariance Matrix of a Multivariate Normal Population," *Tsukuba J. Math.* 8, N2, 367–376.
159. Verbaarschot, J. J. M., Weidenmüller, H. A. and Zirnbauer, M. R. (1984). "Grassman Integration and the Theory of Compound-Nuclear Reactions," *Phys. Lett.* B 149, N4–5, 263–266.
160. Vinogradskaya, A. V. and Girko, V. L. (1983). "Spectrum Control in Systems Described by Linear Equations in Hilbert Spaces," *Automation and Remote Control* 44 N5, Part 1, May.
161. Vorontsov, M. A. (1970). "Design of Optimal Control for a Class of Distributed-parameter Systems under Random Disturbances," *Tekh. Kibern.* N5, 200–205.
162. Wachter, K. W. (1978). "The Strong Limit of Random Spectra for Sample Matrices of Independent Elements," *Ann. Prob.*, 6 1–18.
163. Wachter, K. W. (1980). "The Limiting Empirical Measure of Multiple Discriminant Ratios," *Ann. Statist.*, 8 937–957.
164. Weyl, A. (1940). *L'integration dans les Groupes Topologiques et ses Applications.* Paris, Publications de l'Institut de Mathematiques de Clermont - Ferrand.
165. Weyl, H. (1946). *The Classical Groups.* Princeton University Press, Princeton, New Jersey.
166. Wigner, E. P. (1967). "Random Matrices in Physics," *SIAM Rev.* 9, N1, 1–23.
167. Wigner, E. P. (1968). "On the Distribution of the Roots of Certain Symmetric Matrices," *Ann. Math.* 67, 2, 325–327.
168. Wilson, K. (1962). "Proof of a Conjecture by Dyson," *J. Math. Phys.* 3, 1040.

169. Yin, Y. Q., Bai, Z. D., and Krishnaiah, P. R. (1983) "Limiting Behavior of the Eigenvalues of a Multivariate F Matrix," *J. Multivariate Anal.* **13**, 500–516.

170. Yin, Y. Q. and Krishnaiah, P. R. (1983). "A Limit Theorem for the Eigenvalues of the Product of Two Random Matrices," *J. Multivariate Anal.* **13**, 489–507.

171. Yin, Y. Q. and Krishnaiah, P. R. (1985). "Limit Theorem of the Eigenvalues of the Sample Covariance Matrix when the Underlying Distribution is Isotropic," *Teor. Verojatnost. i prim.* **30**, 4, 810–816.

172. Zubov, V. J. (1975). *Lectures on Control Theory.* Nauka, Moscow (Russian).

173. Zurbenko, I. G. (1968). "Certain Moments of Random Determinants," *Teor. Verojatnost. i Primenen.* **13**, 720–725; English transl. in *Theor. Probab. Appl.* **13**.

INDEX

668